Control System Fundamentals

The Electrical Engineering Handbook Series

Series Editor
Richard C. Dorf
University of California, Davis

Titles Included in the Series

The Control Handbook

Second Edition

Edited by

William S. Levine

University of Maryland

College Park, MD, USA

Control System Fundamentals

Control System Applications

Control System Advanced Methods

Control System Fundamentals

Edited by

William S. Levine

University of Maryland

College Park, MD, USA

CRC Press is an imprint of the
Taylor & Francis Group, an **informa** business

MATLAB® and Simulink® are trademarks of The MathWorks, Inc. and are used with permission. The MathWorks does not warrant the accuracy of the text or exercises in this book. This book's use or discussion of MATLAB® and Simulink® software or related products does not constitute endorsement or sponsorship by The MathWorks of a particular pedagogical approach or particular use of the MATLAB® and Simulink® software.

CRC Press
Taylor & Francis Group
6000 Broken Sound Parkway NW, Suite 300
Boca Raton, FL 33487-2742

© 2011 by Taylor and Francis Group, LLC
CRC Press is an imprint of Taylor & Francis Group, an Informa business

No claim to original U.S. Government works

Printed in the United States of America on acid-free paper
10 9 8 7 6 5 4 3 2 1

International Standard Book Number: 978-1-4200-7362-1 (Hardback)

Library of Congress Cataloging-in-Publication Data

Control system fundamentals / editor, William S. Levine. -- 2nd ed.
    p. cm. -- (The electrical engineering handbook series)
   Includes bibliographical references and index.
   ISBN 978-1-4200-7362-1
   1. Automatic control. 2. Control theory. I. Levine, W. S. II. Title. III. Series.

TJ213.L419 2011
629.8--dc22
                          2010026365

Visit the Taylor & Francis Web site at
http://www.taylorandfrancis.com

and the CRC Press Web site at
http://www.crcpress.com

Contents

SECTION I   Mathematical Foundations

SECTION II   Models for Dynamical Systems

SECTION III   Analysis and Design Methods for Continuous-Time Systems

SECTION IV   Digital Control

SECTION V   Analysis and Design Methods for Nonlinear Systems

Preface to the Second Edition

As you may know, the first edition of *The Control Handbook* was very well received. Many copies were sold and a gratifying number of people took the time to tell me that they found it useful. To the publisher, these are all reasons to do a second edition. To the editor of the first edition, these same facts are a modest disincentive. The risk that a second edition will not be as good as the first one is real and worrisome. I have tried very hard to insure that the second edition is at least as good as the first one was. I hope you agree that I have succeeded.

I have made two major changes in the second edition. The first is that all the *Applications* chapters are new. It is simply a fact of life in engineering that once a problem is solved, people are no longer as interested in it as they were when it was unsolved. I have tried to find especially inspiring and exciting applications for this second edition.

Secondly, it has become clear to me that organizing the *Applications* book by academic discipline is no longer sensible. Most control applications are interdisciplinary. For example, an automotive control system that involves sensors to convert mechanical signals into electrical ones, actuators that convert electrical signals into mechanical ones, several computers and a communication network to link sensors and actuators to the computers does not belong solely to any specific academic area. You will notice that the applications are now organized broadly by application areas, such as automotive and aerospace.

One aspect of this new organization has created a minor and, I think, amusing problem. Several wonderful applications did not fit into my new taxonomy. I originally grouped them under the title Miscellaneous. Several authors objected to the slightly pejorative nature of the term "miscellaneous." I agreed with them and, after some thinking, consulting with literate friends and with some of the library resources, I have renamed that section "Special Applications." Regardless of the name, they are all interesting and important and I hope you will read those articles as well as the ones that did fit my organizational scheme.

There has also been considerable progress in the areas covered in the *Advanced Methods* book. This is reflected in the roughly two dozen articles in this second edition that are completely new. Some of these are in two new sections, "Analysis and Design of Hybrid Systems" and "Networks and Networked Controls."

There have even been a few changes in the *Fundamentals*. Primarily, there is greater emphasis on sampling and discretization. This is because most control systems are now implemented digitally.

I have enjoyed editing this second edition and learned a great deal while I was doing it. I hope that you will enjoy reading it and learn a great deal from doing so.

William S. Levine

MATLAB® and Simulink® are registered trademarks of The MathWorks, Inc. For product information, please contact:

The MathWorks, Inc.
3 Apple Hill Drive
Natick, MA, 01760-2098 USA
Tel: 508-647-7000
Fax: 508-647-7001
E-mail: info@mathworks.com
Web: www.mathworks.com

Acknowledgments

The people who were most crucial to the second edition were the authors of the articles. It took a great deal of work to write each of these articles and I doubt that I will ever be able to repay the authors for their efforts. I do thank them very much.

The members of the advisory/editorial board for the second edition were a very great help in choosing topics and finding authors. I thank them all. Two of them were especially helpful. Davor Hrovat took responsibility for the automotive applications and Richard Braatz was crucial in selecting the applications to industrial process control.

It is a great pleasure to be able to provide some recognition and to thank the people who helped bring this second edition of *The Control Handbook* into being. Nora Konopka, publisher of engineering and environmental sciences for Taylor & Francis/CRC Press, began encouraging me to create a second edition quite some time ago. Although it was not easy, she finally convinced me. Jessica Vakili and Kari Budyk, the project coordinators, were an enormous help in keeping track of potential authors as well as those who had committed to write an article. Syed Mohamad Shajahan, senior project executive at Techset, very capably handled all phases of production, while Richard Tressider, project editor for Taylor & Francis/CRC Press, provided direction, oversight, and quality control. Without all of them and their assistants, the second edition would probably never have appeared and, if it had, it would have been far inferior to what it is.

Most importantly, I thank my wife Shirley Johannesen Levine for everything she has done for me over the many years we have been married. It would not be possible to enumerate all the ways in which she has contributed to each and everything I have done, not just editing this second edition.

William S. Levine

Editorial Board

Editor

William S. Levine received B.S., M.S., and Ph.D. degrees from the Massachusetts Institute of Technology. He then joined the faculty of the University of Maryland, College Park where he is currently a research professor in the Department of Electrical and Computer Engineering. Throughout his career he has specialized in the design and analysis of control systems and related problems in estimation, filtering, and system modeling. Motivated by the desire to understand a collection of interesting controller designs, he has done a great deal of research on mammalian control of movement in collaboration with several neurophysiologists.

He is co-author of *Using MATLAB to Analyze and Design Control Systems*, March 1992. Second Edition, March 1995. He is the coeditor of *The Handbook of Networked and Embedded Control Systems,* published by Birkhauser in 2005. He is the editor of a series on control engineering for Birkhauser. He has been president of the IEEE Control Systems Society and the American Control Council. He is presently the chairman of the SIAM special interest group in control theory and its applications.

He is a fellow of the IEEE, a distinguished member of the IEEE Control Systems Society, and a recipient of the IEEE Third Millennium Medal. He and his collaborators received the Schroers Award for outstanding rotorcraft research in 1998. He and another group of collaborators received the award for outstanding paper in the *IEEE Transactions on Automatic Control*, entitled "Discrete-Time Point Processes in Urban Traffic Queue Estimation."

Contributors

Juan C. Agüero
Centre for Complex Dynamic Systems
    and Control
The University of Newcastle
Callaghan, New South Wales, Australia

Anders Ahlén
Department of Technology
Uppsala University
Uppsala, Sweden

Karl J. Åström
Department of Automatic Control
Lund Institute of Technology
Lund, Sweden

Derek P. Atherton
School of Engineering
The University of Sussex
Brighton, United Kingdom

David M. Auslander
Department of Mechanical Engineering
University of California, Berkeley
Berkeley, California

Robert H. Bishop
College of Engineering
The University of Texas at Austin
Austin, Texas

Richard D. Braatz
Department of Chemical Engineering
University of Illinois at Urbana–Champaign
Urbana, Illinois

Charles M. Close
Department of Electrical, Computer, and
    Systems Engineering
Rensselaer Polytechnic Institute
Troy, New York

John J. D'Azzo
Department of Electrical and
    Computer Engineering
Air Force Institute of Technology
Wright-Patterson Air Force Base, Ohio

Bradley W. Dickinson
Department of Electrical Engineering
Princeton University
Princeton, New Jersey

Richard C. Dorf
College of Engineering
University of California, Davis
Davis, California

A. Feuer
Electrical Engineering Department
Technion–Israel Institute of Technology
Haifa, Israel

Dean K. Frederick
Department of Electrical, Computer,
    and Systems Engineering
Rensselaer Polytechnic Institute
Troy, New York

James T. Gillis
The Aerospace Corporation
Los Angeles, California

Graham C. Goodwin
Centre for Complex Dynamic Systems
    and Control
The University of Newcastle
Callaghan, New South Wales, Australia

Stefan F. Graebe
PROFACTOR GmbH
Steyr, Austria

C. W. Gray
The Aerospace Corporation
El Segundo, California

Tore Hägglund
Department of Automatic Control
Lund Institute of Technology
Lund, Sweden

Richard Hill
Mechanical Engineering Department
University of Detroit Mercy
Detroit, Michigan

Constantine H. Houpis
Department of Electrical and
    Computer Engineering
Air Force Institute of Technology
Wright-Patterson Air Force Base, Ohio

Jason C. Jones
SunPower Corporation
Richmond, California

Edward W. Kamen
School of Electrical and Computer
    Engineering
Georgia Institute of Technology
Atlanta, Georgia

Masako Kishida
Department of Chemical Engineering
University of Illinois at Urbana–Champaign
Urbana, Illinois

B. P. Lathi
Department of Electrical and
    Electronic Engineering
California State University
Sacramento, California

William S. Levine
Department of Electrical Engineering
University of Maryland
College Park, Maryland

Mohamed Mansour
Automatic Control Laboratory
Swiss Federal Institute of Technology
Zurich, Switzerland

R. H. Middleton
The Hamilton Institute
National University of Ireland, Maynooth
Maynooth, Ireland

Norman S. Nise
Electrical and Computer Engineering Department
California State Polytechnic University
Pomona, California

Katsuhiko Ogata
Department of Mechanical Engineering
University of Minnesota
Minneapolis, Minnesota

Gustaf Olsson
Department of Industrial Electrical
    Engineering and Automation
Lund University
Lund, Sweden

Z. J. Palmor
Faculty of Mechanical Engineering
Technion–Israel Institute of Technology
Haifa, Israel

John R. Ridgely
Department of Mechanical Engineering
California Polytechnic State University
San Luis Obispo, California

Charles E. Rohrs
Rohrs Consulting
Newton, Massachusetts

Mario E. Salgado
Department of Electronic Engineering
Federico Santa María Technical University
Valparaíso, Chile

Michael Santina
The Boeing Company
Seal Beach, California

Jeff S. Shamma
Department of Aerospace Engineering and
    Engineering Mechanics
The University of Texas at Austin
Austin, Texas

Raymond T. Stefani
Electrical Engineering Department
California State University
Long Beach, California

Allen R. Stubberud
Department of Electrical Engineering and
  Computer Science
University of California, Irvine
Irvine, California

Peter Stubberud
Department of Electrical and
  Computer Engineering
The University of Nevada, Las Vegas
Las Vegas, Nevada

Harry L. Trentelman
Research Institute of Mathematics and
  Computer Science
University of Groningen
Groningen, The Netherlands

William A. Wolovich
School of Engineering
Brown University
Providence, Rhode Island

Jiann-Shiou Yang
Department of Electrical and Computer
  Engineering
University of Minnesota
Duluth, Minnesota

Juan I. Yuz
Department of Electronic Engineering
Federico Santa María Technical University
Valparaíso, Chile

I

Mathematical Foundations

1

Ordinary Linear Differential and Difference Equations

B.P. Lathi
California State University

1.1 Differential Equations

A function containing variables and their derivatives is called a *differential expression*, and an equation involving differential expressions is called a *differential equation*. A differential equation is an *ordinary* differential equation if it contains only one independent variable; it is a *partial* differential equation if it contains more than one independent variable. We shall deal here only with ordinary differential equations.

In the mathematical texts, the independent variable is generally *x*, which can be anything such as time, distance, velocity, pressure, and so on. In most of the applications in control systems, the independent variable is time. For this reason we shall use here independent variable *t* for time, although it can stand for any other variable as well.

The following equation

$$\left(\frac{d^2 y}{dt^2}\right)^4 + 3\frac{dy}{dt} + 5y^2(t) = \sin t$$

is an ordinary differential equation of second *order* because the highest derivative is of second order. An *n*th-order differential equation is *linear* if it is of the form

$$a_n(t)\frac{d^n y}{dt^n} + a_{n-1}(t)\frac{d^{n-1} y}{dt^{n-1}} + \cdots + a_1(t)\frac{dy}{dt} + a_0(t)y(t) = r(t) \qquad (1.1)$$

where the coefficients $a_i(t)$ are not functions of $y(t)$. If these coefficients (a_i) are constants, the equation is linear with *constant coefficients*. Many engineering (as well as nonengineering) systems can be modeled by these equations. Systems modeled by these equations are known as *linear time-invariant* (LTI) systems.

In this chapter we shall deal exclusively with linear differential equations with constant coefficients. Certain other forms of differential equations are dealt with elsewhere in this volume.

1.1.1 Role of Auxiliary Conditions in Solution of Differential Equations

We now show that a differential equation does not, in general, have a unique solution unless some additional constraints (or conditions) on the solution are known. This fact should not come as a surprise. A function $y(t)$ has a unique derivative dy/dt, but for a given derivative dy/dt, there are infinite possible functions $y(t)$. If we are given dy/dt, it is impossible to determine $y(t)$ uniquely unless an additional piece of information about $y(t)$ is given. For example, the solution of a differential equation

$$\frac{dy}{dt} = 2 \tag{1.2}$$

obtained by integrating both sides of the equation is

$$y(t) = 2t + c \tag{1.3}$$

for any value of c. Equation 1.2 specifies a function whose slope is 2 for all t. Any straight line with a slope of 2 satisfies this equation. Clearly the solution is not unique, but if we place an additional constraint on the solution $y(t)$, then we specify a unique solution. For example, suppose we require that $y(0) = 5$; then out of all the possible solutions available, only one function has a slope of 2 and an intercept with the vertical axis at 5. By setting $t = 0$ in Equation 1.3 and substituting $y(0) = 5$ in the same equation, we obtain $y(0) = 5 = c$ and

$$y(t) = 2t + 5$$

which is the unique solution satisfying both Equation 1.2 and the constraint $y(0) = 5$.

In conclusion, differentiation is an irreversible operation during which certain information is lost. To reverse this operation, one piece of information about $y(t)$ must be provided to restore the original $y(t)$. Using a similar argument, we can show that, given d^2y/dt^2, we can determine $y(t)$ uniquely only if two additional pieces of information (constraints) about $y(t)$ are given. In general, to determine $y(t)$ uniquely from its nth derivative, we need n additional pieces of information (constraints) about $y(t)$. These constraints are also called *auxiliary conditions*. When these conditions are given at $t = 0$, they are called *initial conditions*.

We discuss here two systematic procedures for solving linear differential equations of the form in Equation 1.1. The first method is the *classical method*, which is relatively simple, but restricted to a certain class of inputs. The second method (the convolution method) is general and is applicable to all types of inputs. A third method (Laplace transform) is discussed elsewhere in this volume. Both the methods discussed here are classified as *time-domain* methods because with these methods we are able to solve the above equation directly, using t as the independent variable. The method of Laplace transform (also known as the *frequency-domain* method), on the other hand, requires transformation of variable t into a frequency variable s.

In engineering applications, the form of linear differential equation that occurs most commonly is given by

$$\frac{d^n y}{dt^n} + a_{n-1}\frac{d^{n-1} y}{dt^{n-1}} + \cdots + a_1\frac{dy}{dt} + a_0 y(t) = b_m\frac{d^m f}{dt^m} + b_{m-1}\frac{d^{m-1} f}{dt^{m-1}} + \cdots + b_1\frac{df}{dt} + b_0 f(t) \tag{1.4a}$$

where all the coefficients a_i and b_i are constants. Using operational notation D to represent d/dt, this equation can be expressed as

$$(D^n + a_{n-1}D^{n-1} + \cdots + a_1 D + a_0)y(t) = (b_m D^m + b_{m-1}D^{m-1} + \cdots + b_1 D + b_0)f(t) \tag{1.4b}$$

or

$$Q(D)y(t) = P(D)f(t) \qquad (1.4c)$$

where the polynomials $Q(D)$ and $P(D)$, respectively, are

$$Q(D) = D^n + a_{n-1}D^{n-1} + \cdots + a_1 D + a_0$$
$$P(D) = b_m D^m + b_{m-1}D^{m-1} + \cdots + b_1 D + b_0$$

Observe that this equation is of the form of Equation 1.1, where $r(t)$ is in the form of a linear combination of $f(t)$ and its derivatives. In this equation, $y(t)$ represents an output variable, and $f(t)$ represents an input variable of an LTI system. Theoretically, the powers m and n in the above equations can take on any value. Practical noise considerations, however, require [1] $m \leq n$.

1.1.2 Classical Solution

When $f(t) \equiv 0$, Equation 1.4a is known as the *homogeneous* (or complementary) equation. We shall first solve the homogeneous equation. Let the solution of the homogeneous equation be $y_c(t)$, that is,

$$Q(D)y_c(t) = 0$$

or

$$(D^n + a_{n-1}D^{n-1} + \cdots + a_1 D + a_0)y_c(t) = 0$$

We first show that if $y_p(t)$ is the solution of Equation 1.4a, then $y_c(t) + y_p(t)$ is also its solution. This follows from the fact that

$$Q(D)y_c(t) = 0$$

If $y_p(t)$ is the solution of Equation 1.4a, then

$$Q(D)y_p(t) = P(D)f(t)$$

Addition of these two equations yields

$$Q(D)[y_c(t) + y_p(t)] = P(D)f(t)$$

Thus, $y_c(t) + y_p(t)$ satisfies Equation 1.4a and therefore is the general solution of Equation 1.4a. We call $y_c(t)$ the *complementary* solution and $y_p(t)$ the *particular* solution. In system analysis parlance, these components are called the *natural* response and *the forced* response, respectively.

1.1.2.1 Complementary Solution (The Natural Response)

The complementary solution $y_c(t)$ is the solution of

$$Q(D)y_c(t) = 0 \qquad (1.5a)$$

or

$$(D^n + a_{n-1}D^{n-1} + \cdots + a_1 D + a_0)y_c(t) = 0 \qquad (1.5b)$$

A solution to this equation can be found in a systematic and formal way. However, we will take a short cut by using heuristic reasoning. Equation 1.5b shows that a linear combination of $y_c(t)$ and its n successive derivatives is zero, not at some values of t, but for all t. This is possible if and only if $y_c(t)$ and all its n

successive derivatives are of the same form. Otherwise their sum can never add to zero for all values of t. We know that only an exponential function $e^{\lambda t}$ has this property. So let us assume that

$$y_c(t) = ce^{\lambda t}$$

is a solution of Equation 1.5b. Now

$$Dy_c(t) = \frac{dy_c}{dt} = c\lambda e^{\lambda t}$$

$$D^2 y_c(t) = \frac{d^2 y_c}{dt^2} = c\lambda^2 e^{\lambda t}$$

$$\cdots \cdots \cdots \cdots$$

$$D^n y_c(t) = \frac{d^n y_c}{dt^n} = c\lambda^n e^{\lambda t}$$

Substituting these results in Equation 1.5b, we obtain

$$c(\lambda^n + a_{n-1}\lambda^{n-1} + \cdots + a_1\lambda + a_0)e^{\lambda t} = 0$$

For a nontrivial solution of this equation,

$$\lambda^n + a_{n-1}\lambda^{n-1} + \cdots + a_1\lambda + a_0 = 0 \tag{1.6a}$$

This result means that $ce^{\lambda t}$ is indeed a solution of Equation 1.5, provided that λ satisfies Equation 1.6a. Note that the polynomial in Equation 1.6a is identical to the polynomial $Q(D)$ in Equation 1.5b, with λ replacing D. Therefore, Equation 1.6a can be expressed as

$$Q(\lambda) = 0 \tag{1.6b}$$

When $Q(\lambda)$ is expressed in factorized form, Equation 1.6b can be represented as

$$Q(\lambda) = (\lambda - \lambda_1)(\lambda - \lambda_2) \cdots (\lambda - \lambda_n) = 0 \tag{1.6c}$$

Clearly λ has n solutions: $\lambda_1, \lambda_2, \ldots, \lambda_n$. Consequently, Equation 1.5 has n possible solutions: $c_1 e^{\lambda_1 t}, c_2 e^{\lambda_2 t}, \ldots, c_n e^{\lambda_n t}$, with $c_1, c_2, \ldots, c_n$ as arbitrary constants. We can readily show that a general solution is given by the sum of these n solutions,* so that

$$y_c(t) = c_1 e^{\lambda_1 t} + c_2 e^{\lambda_2 t} + \cdots + c_n e^{\lambda_n t} \tag{1.7}$$

where $c_1, c_2, \ldots, c_n$ are arbitrary constants determined by n constraints (the auxiliary conditions) on the solution.

---

* To prove this fact, assume that $y_1(t), y_2(t), \ldots, y_n(t)$ are all solutions of Equation 1.5. Then

$$Q(D)y_1(t) = 0$$
$$Q(D)y_2(t) = 0$$
$$\cdots \cdots \cdots \cdots$$
$$Q(D)y_n(t) = 0$$

Multiplying these equations by $c_1, c_2, \ldots, c_n$, respectively, and adding them together yields

$$Q(D)[c_1 y_1(t) + c_2 y_2(t) + \cdots + c_n y_n(t)] = 0$$

This result shows that $c_1 y_1(t) + c_2 y_2(t) + \cdots + c_n y_n(t)$ is also a solution of the homogeneous Equation 1.5.

The polynomial $Q(\lambda)$ is known as the *characteristic polynomial*. The equation

$$Q(\lambda) = 0 \tag{1.8}$$

is called the *characteristic* or *auxiliary* equation. From Equation 1.6c, it is clear that $\lambda_1, \lambda_2, \ldots, \lambda_n$ are the roots of the characteristic equation; consequently, they are called the *characteristic roots*. The terms *characteristic values*, *eigenvalues*, and *natural frequencies* are also used for characteristic roots.* The exponentials $e^{\lambda_i t} (i = 1, 2, \ldots, n)$ in the complementary solution are the *characteristic modes* (also known as *modes* or *natural modes*). There is a characteristic mode for each characteristic root, and the *complementary solution is a linear combination of the characteristic modes.*

Repeated Roots

The solution of Equation 1.5 as given in Equation 1.7 assumes that the n characteristic roots $\lambda_1, \lambda_2, \ldots, \lambda_n$ are distinct. If there are repeated roots (the same root occurring more than once), the form of the solution is modified slightly. By direct substitution we can show that the solution of the equation

$$(D - \lambda)^2 y_c(t) = 0$$

is given by

$$y_c(t) = (c_1 + c_2 t)e^{\lambda t}$$

In this case, the root λ repeats twice. Observe that the characteristic modes in this case are $e^{\lambda t}$ and $te^{\lambda t}$. Continuing this pattern, we can show that for the differential equation

$$(D - \lambda)^r y_c(t) = 0 \tag{1.9}$$

the characteristic modes are $e^{\lambda t}, te^{\lambda t}, t^2 e^{\lambda t}, \ldots, t^{r-1} e^{\lambda t}$, and the solution is

$$y_c(t) = (c_1 + c_2 t + \cdots + c_r t^{r-1})e^{\lambda t} \tag{1.10}$$

Consequently, for a characteristic polynomial

$$Q(\lambda) = (\lambda - \lambda_1)^r (\lambda - \lambda_{r+1}) \cdots (\lambda - \lambda_n)$$

the characteristic modes are $e^{\lambda_1 t}, te^{\lambda_1 t}, \ldots, t^{r-1} e^{\lambda t}, e^{\lambda_{r+1} t}, \ldots, e^{\lambda_n t}$. and the complementary solution is

$$y_c(t) = (c_1 + c_2 t + \cdots + c_r t^{r-1})e^{\lambda_1 t} + c_{r+1} e^{\lambda_{r+1} t} + \cdots + c_n e^{\lambda_n t}$$

1.1.2.2 Particular Solution (The Forced Response): Method of Undetermined Coefficients

The particular solution $y_p(t)$ is the solution of

$$Q(D)y_p(t) = P(D)f(t) \tag{1.11}$$

It is a relatively simple task to determine $y_p(t)$ when the input $f(t)$ is such that it yields only a finite number of independent derivatives. Inputs having the form $e^{\zeta t}$ or t^r fall into this category. For example, $e^{\zeta t}$ has only one independent derivative; the repeated differentiation of $e^{\zeta t}$ yields the same form, that is, $e^{\zeta t}$. Similarly, the repeated differentiation of t^r yields only r independent derivatives. The particular solution to such an input can be expressed as a linear combination of the input and its independent derivatives. Consider, for example, the input $f(t) = at^2 + bt + c$. The successive derivatives of this input are $2at + b$ and $2a$. In this case, the input has only two independent derivatives. Therefore the particular solution can

---

* The term *eigenvalue* is German for characteristic value.

TABLE 1.1

Input $f(t)$	Forced Response
1. $e^{\zeta t} \zeta \neq \lambda_i \ (i = 1, 2, \cdots, n)$	$\beta e^{\zeta t}$
2. $e^{\zeta t} \quad \zeta = \lambda_i$	$\beta t e^{\zeta t}$
3. $k \quad$ (a constant)	$\beta \quad$ (a constant)
4. $\cos(\omega t + \theta)$	$\beta \cos(\omega t + \phi)$
5. $(t^r + \alpha_{r-1} t^{r-1} + \cdots + \alpha_1 t + \alpha_0) e^{\zeta t}$	$(\beta_r t^r + \beta_{r-1} t^{r-1} + \cdots + \beta_1 t + \beta_0) e^{\zeta t}$

be assumed to be a linear combination of $f(t)$ and its two derivatives. The suitable form for $y_p(t)$ in this case is therefore

$$y_p(t) = \beta_2 t^2 + \beta_1 t + \beta_0$$

The undetermined coefficients β_0, β_1, and β_2 are determined by substituting this expression for $y_p(t)$ in Equation 1.11 and then equating coefficients of similar terms on both sides of the resulting expression.

Although this method can be used only for inputs with a finite number of derivatives, this class of inputs includes a wide variety of the most commonly encountered signals in practice. Table 1.1 shows a variety of such inputs and the form of the particular solution corresponding to each input. We shall demonstrate this procedure with an example.

Note: By definition, $y_p(t)$ cannot have any characteristic mode terms. If any term $p(t)$ shown in the right-hand column for the particular solution is also a characteristic mode, the correct form of the forced response must be modified to $t^i p(t)$, where i is the smallest possible integer that can be used and still can prevent $t^i p(t)$ from having a characteristic mode term. For example, when the input is $e^{\zeta t}$, the forced response (right-hand column) has the form $\beta e^{\zeta t}$. But if $e^{\zeta t}$ happens to be a characteristic mode, the correct form of the particular solution is $\beta t e^{\zeta t}$ (see Pair 2). If $t e^{\zeta t}$ also happens to be a characteristic mode, the correct form of the particular solution is $\beta t^2 e^{\zeta t}$, and so on.

Example 1.1:

Solve the differential equation

$$(D^2 + 3D + 2)y(t) = Df(t) \tag{1.12}$$

if the input

$$f(t) = t^2 + 5t + 3$$

and the initial conditions are $y(0^+) = 2$ and $\dot{y}(0^+) = 3$.
   The characteristic polynomial is

$$\lambda^2 + 3\lambda + 2 = (\lambda + 1)(\lambda + 2)$$

Therefore the characteristic modes are e^{-t} and e^{-2t}. The complementary solution is a linear combination of these modes, so that

$$y_c(t) = c_1 e^{-t} + c_2 e^{-2t} \quad t \geq 0$$

Here the arbitrary constants c_1 and c_2 must be determined from the given initial conditions.

The particular solution to the input $t^2 + 5t + 3$ is found from Table 1.1 (Pair 5 with $\zeta = 0$) to be

$$y_p(t) = \beta_2 t^2 + \beta_1 t + \beta_0$$

Moreover, $y_p(t)$ satisfies Equation 1.11, that is,

$$(D^2 + 3D + 2)y_p(t) = Df(t) \tag{1.13}$$

Now

$$Dy_p(t) = \frac{d}{dt}(\beta_2 t^2 + \beta_1 t + \beta_0) = 2\beta_2 t + \beta_1$$

$$D^2 y_p(t) = \frac{d^2}{dt^2}(\beta_2 t^2 + \beta_1 t + \beta_0) = 2\beta_2$$

and

$$Df(t) = \frac{d}{dt}[t^2 + 5t + 3] = 2t + 5$$

Substituting these results in Equation 1.13 yields

$$2\beta_2 + 3(2\beta_2 t + \beta_1) + 2(\beta_2 t^2 + \beta_1 t + \beta_0) = 2t + 5$$

or

$$2\beta_2 t^2 + (2\beta_1 + 6\beta_2)t + (2\beta_0 + 3\beta_1 + 2\beta_2) = 2t + 5$$

Equating coefficients of similar powers on both sides of this expression yields

$$2\beta_2 = 0$$
$$2\beta_1 + 6\beta_2 = 2$$
$$2\beta_0 + 3\beta_1 + 2\beta_2 = 5$$

Solving these three equations for their unknowns, we obtain $\beta_0 = 1$, $\beta_1 = 1$, and $\beta_2 = 0$. Therefore,

$$y_p(t) = t + 1 \quad t > 0$$

The total solution $y(t)$ is the sum of the complementary and particular solutions. Therefore,

$$y(t) = y_c(t) + y_p(t)$$
$$= c_1 e^{-t} + c_2 e^{-2t} + t + 1 \quad t > 0$$

so that

$$\dot{y}(t) = -c_1 e^{-t} - 2c_2 e^{-2t} + 1$$

Setting $t = 0$ and substituting the given initial conditions $y(0) = 2$ and $\dot{y}(0) = 3$ in these equations, we have

$$2 = c_1 + c_2 + 1$$
$$3 = -c_1 - 2c_2 + 1$$

The solution to these two simultaneous equations is $c_1 = 4$ and $c_2 = -3$. Therefore,

$$y(t) = 4e^{-t} - 3e^{-2t} + t + 1 \quad t \geq 0$$

1.1.2.3 The Exponential Input $e^{\zeta t}$

The exponential signal is the most important signal in the study of LTI systems. Interestingly, the particular solution for an exponential input signal turns out to be very simple. From Table 1.1 we see that the particular solution for the input $e^{\zeta t}$ has the form $\beta e^{\zeta t}$. We now show that $\beta = Q(\zeta)/P(\zeta)$*. To determine the constant β, we substitute $y_p(t) = \beta e^{\zeta t}$ in Equation 1.11, which gives us

$$Q(D)[\beta e^{\zeta t}] = P(D)e^{\zeta t} \tag{1.14a}$$

Now observe that

$$De^{\zeta t} = \frac{d}{dt}\left(e^{\zeta t}\right) = \zeta e^{\zeta t}$$

$$D^2 e^{\zeta t} = \frac{d^2}{dt^2}\left(e^{\zeta t}\right) = \zeta^2 e^{\zeta t}$$

$$\cdots\cdots\cdots\cdots$$

$$D^r e^{\zeta t} = \zeta^r e^{\zeta t}$$

Consequently,

$$Q(D)e^{\zeta t} = Q(\zeta)e^{\zeta t} \quad \text{and} \quad P(D)e^{\zeta t} = P(\zeta)e^{\zeta t}$$

Therefore, Equation 1.14a becomes

$$\beta Q(\zeta)e^{\zeta t} = P(\zeta)e^{\zeta t} \tag{1.14b}$$

and

$$\beta = \frac{P(\zeta)}{Q(\zeta)}$$

Thus, for the input $f(t) = e^{\zeta t}$, the particular solution is given by

$$y_p(t) = H(\zeta)e^{\zeta t} \quad t > 0 \tag{1.15a}$$

where

$$H(\zeta) = \frac{P(\zeta)}{Q(\zeta)} \tag{1.15b}$$

This is an interesting and significant result. It states that for an exponential input $e^{\zeta t}$, the particular solution $y_p(t)$ is the same exponential multiplied by $H(\zeta) = P(\zeta)/Q(\zeta)$. The total solution $y(t)$ to an exponential input $e^{\zeta t}$ is then given by

$$y(t) = \sum_{j=1}^{n} c_j e^{\lambda_j t} + H(\zeta)e^{\zeta t}$$

where the arbitrary constants $c_1, c_2, \ldots, c_n$ are determined from auxiliary conditions.

Recall that the exponential signal includes a large variety of signals, such as a constant ($\zeta = 0$), a sinusoid ($\zeta = \pm j\omega$), and an exponentially growing or decaying sinusoid ($\zeta = \sigma \pm j\omega$). Let us consider the forced response for some of these cases.

1.1.2.4 The Constant Input f(t) = C

Because $C = Ce^{0t}$, the constant input is a special case of the exponential input $Ce^{\zeta t}$ with $\zeta = 0$. The particular solution to this input is then given by

$$y_p(t) = CH(\zeta)e^{\zeta t} \quad \text{with} \quad \zeta = 0$$
$$= CH(0) \tag{1.16}$$

---

* This is true only if ζ is not a characteristic root.

1.1.2.5 The Complex Exponential Input $e^{j\omega t}$

Here $\zeta = j\omega$, and

$$y_p(t) = H(j\omega)e^{j\omega t} \tag{1.17}$$

1.1.2.6 The Sinusoidal Input $f(t) = \cos \omega_0 t$

We know that the particular solution for the input $e^{\pm j\omega t}$ is $H(\pm j\omega)e^{\pm j\omega t}$. Since $\cos \omega t = (e^{j\omega t} + e^{-j\omega t})/2$, the particular solution to $\cos \omega t$ is

$$y_p(t) = \frac{1}{2}[H(j\omega)e^{j\omega t} + H(-j\omega)e^{-j\omega t}]$$

Because the two terms on the right-hand side are conjugates,

$$y_p(t) = \text{Re}[H(j\omega)e^{j\omega t}]$$

But

$$H(j\omega) = |H(j\omega)|e^{j\angle H(j\omega)}$$

so that

$$\begin{aligned} y_p(t) &= \text{Re}\,\{|H(j\omega)|e^{j[\omega t + \angle H(j\omega)]}\} \\ &= |H(j\omega)|\cos\,[\omega t + \angle H(j\omega)] \end{aligned} \tag{1.18}$$

This result can be generalized for the input $f(t) = \cos(\omega t + \theta)$. The particular solution in this case is

$$y_p(t) = |H(j\omega)|\cos[\omega t + \theta + \angle H(j\omega)] \tag{1.19}$$

Example 1.2:

Solve Equation 1.12 for the following inputs:

(a) $10e^{-3t}$ (b) 5 (c) e^{-2t} (d) $10\cos(3t + 30°)$.

The initial conditions are $y(0^+) = 2, \dot{y}(0^+) = 3$.
The complementary solution for this case is already found in Example 1.1 as

$$y_c(t) = c_1 e^{-t} + c_2 e^{-2t} \quad t \geq 0$$

For the exponential input $f(t) = e^{\zeta t}$, the particular solution, as found in Equation 1.15 is $H(\zeta)e^{\zeta t}$, where

$$H(\zeta) = \frac{P(\zeta)}{Q(\zeta)} = \frac{\zeta}{\zeta^2 + 3\zeta + 2}$$

(a) For input $f(t) = 10e^{-3t}, \zeta = -3$, and

$$\begin{aligned} y_p(t) &= 10H(-3)e^{-3t} \\ &= 10\left[\frac{-3}{(-3)^2 + 3(-3) + 2}\right]e^{-3t} \\ &= -15e^{-3t} \quad t > 0 \end{aligned}$$

The total solution (the sum of the complementary and particular solutions) is

$$y(t) = c_1 e^{-t} + c_2 e^{-2t} - 15e^{-3t} \quad t \geq 0$$

and

$$\dot{y}(t) = -c_1 e^{-t} - 2c_2 e^{-2t} + 45e^{-3t} \quad t \geq 0$$

The initial conditions are $y(0^+) = 2$ and $\dot{y}(0^+) = 3$. Setting $t = 0$ in the above equations and substituting the initial conditions yields

$$c_1 + c_2 - 15 = 2 \quad \text{and} \quad -c_1 - 2c_2 + 45 = 3$$

Solution of these equations yields $c_1 = -8$ and $c_2 = 25$. Therefore,

$$y(t) = -8e^{-t} + 25e^{-2t} - 15e^{-3t} \quad t \geq 0$$

(b) For input $f(t) = 5 = 5e^{0t}$, $\zeta = 0$, and

$$y_p(t) = 5H(0) = 0 \quad t > 0$$

The complete solution is $y(t) = y_c(t) + y_p(t) = c_1 e^{-t} + c_2 e^{-2t}$. We then substitute the initial conditions to determine c_1 and c_2 as explained in Part a.

(c) Here $\zeta = -2$, which is also a characteristic root. Hence (see Pair 2, Table 1.1, or the comment at the bottom of the table),

$$y_p(t) = \beta t e^{-2t}$$

To find β, we substitute $y_p(t)$ in Equation 1.11, giving us

$$(D^2 + 3D + 2)y_p(t) = Df(t)$$

or

$$(D^2 + 3D + 2)[\beta t e^{-2t}] = De^{-2t}$$

But

$$D\left[\beta t e^{-2t}\right] = \beta(1 - 2t)e^{-2t}$$

$$D^2\left[\beta t e^{-2t}\right] = 4\beta(t - 1)e^{-2t}$$

$$De^{-2t} = -2e^{-2t}$$

Consequently,

$$\beta(4t - 4 + 3 - 6t + 2t)e^{-2t} = -2e^{-2t}$$

or

$$-\beta e^{-2t} = -2e^{-2t}$$

This means that $\beta = 2$, so that

$$y_p(t) = 2t e^{-2t}$$

The complete solution is $y(t) = y_c(t) + y_p(t) = c_1 e^{-t} + c_2 e^{-2t} + 2t e^{-2t}$. We then substitute the initial conditions to determine c_1 and c_2 as explained in Part a.

(d) For the input $f(t) = 10\cos(3t + 30°)$, the particular solution [see Equation 1.19] is

$$y_p(t) = 10|H(j3)| \cos[3t + 30° + \angle H(j3)]$$

where

$$H(j3) = \frac{P(j3)}{Q(j3)} = \frac{j3}{(j3)^2 + 3(j3) + 2}$$

$$= \frac{j3}{-7 + j9} = \frac{27 - j21}{130} = 0.263e^{-j37.9°}$$

Therefore,

$$|H(j3)| = 0.263, \qquad \angle H(j3) = -37.9°$$

and

$$y_p(t) = 10(0.263)\cos(3t + 30° - 37.9°)$$

$$= 2.63\cos(3t - 7.9°)$$

The complete solution is $y(t) = y_c(t) + y_p(t) = c_1 e^{-t} + c_2 e^{-2t} + 2.63\cos(3t - 7.9°)$. We then substitute the initial conditions to determine c_1 and c_2 as explained in Part a.

1.1.3 Method of Convolution

In this method, the input $f(t)$ is expressed as a sum of impulses. The solution is then obtained as a sum of the solutions to all the impulse components. The method exploits the superposition property of the linear differential equations. From the sampling (or sifting) property of the impulse function, we have

$$f(t) = \int_0^t f(x)\delta(t-x)\,dx \quad t \geq 0 \tag{1.20}$$

The right-hand side expresses $f(t)$ as a sum (integral) of impulse components. Let the solution of Equation 1.4 be $y(t) = h(t)$, when $f(t) = \delta(t)$ and all the initial conditions are zero. Then use of the linearity property yields the solution of Equation 1.4 to input $f(t)$ as

$$y(t) = \int_0^t f(x)h(t-x)\,dx \tag{1.21}$$

For this solution to be general, we must add a complementary solution. Thus, the general solution is given by

$$y(t) = \sum_{j=1}^n c_j e^{\lambda_j t} + \int_0^t f(x)h(t-x)\,dx \tag{1.22}$$

where the lower limit 0 is understood to be 0^- in order to ensure that impulses, if any, in the input $f(t)$ at the origin are accounted for side of Equation 1.22 is well known in the literature as the *convolution integral*. The function $h(t)$ appearing in the integral is the solution of Equation 1.4 for the impulsive input $[f(t) = \delta(t)]$. It can be shown that [1]

$$h(t) = P(D)[y_o(t)u(t)] \tag{1.23}$$

where $y_o(t)$ is a linear combination of the characteristic modes subject to initial conditions

$$y_o^{(n-1)}(0) = 1$$
$$y_o(0) = y_o^{(1)}(0) = \cdots = y_o^{(n-2)}(0) = 0 \tag{1.24}$$

The function $u(t)$ appearing on the right-hand side of Equation 1.23 represents the unit step function, which is unity for $t \geq 0$ and is 0 for $t < 0$.

The right-hand side of Equation 1.23 is a linear combination of the derivatives of $y_o(t)u(t)$. Evaluating these derivatives is clumsy and inconvenient because of the presence of $u(t)$. The derivatives will generate an impulse and its derivatives at the origin [recall that $\frac{d}{dt}u(t) = \delta(t)$]. Fortunately when $m \leq n$ in Equation 1.4, the solution simplifies to

$$h(t) = b_n \delta(t) + [P(D)y_o(t)]u(t) \tag{1.25}$$

Example 1.3:

Solve Example 1.1.2.6, Part a using method of convolution.

We first determine $h(t)$. The characteristic modes for this case, as found in Example 1.1.2.2, are e^{-t} and e^{-2t}. Since $y_o(t)$ is a linear combination of the characteristic modes

$$y_o(t) = K_1 e^{-t} + K_2 e^{-2t} \quad t \geq 0$$

Therefore,

$$\dot{y}_o(t) = -K_1 e^{-t} - 2K_2 e^{-2t} \quad t \geq 0$$

The initial conditions according to Equation 1.24 are $\dot{y}_o(0) = 1$ and $y_o(0) = 0$. Setting $t = 0$ in the above equations and using the initial conditions, we obtain

$$K_1 + K_2 = 0 \qquad \text{and} \qquad -K_1 - 2K_2 = 1$$

Solution of these equations yields $K_1 = 1$ and $K_2 = -1$. Therefore,

$$y_o(t) = e^{-t} - e^{-2t}$$

Also, in this case, the polynomial $P(D) = D$ is of first order, and $b_2 = 0$. Therefore, from Equation 1.25

$$h(t) = [P(D)y_o(t)]u(t) = [Dy_o(t)]u(t)$$
$$= \left[\frac{d}{dt}(e^{-t} - e^{-2t})\right]u(t)$$
$$= (-e^{-t} + 2e^{-2t})u(t)$$

and

$$\int_0^t f(x)h(t-x)\,dx = \int_0^t 10e^{-3x}\left[-e^{-(t-x)} + 2e^{-2(t-x)}\right]dx$$
$$= -5e^{-t} + 20e^{-2t} - 15e^{-3t}$$

The total solution is obtained by adding the complementary solution $y_c(t) = c_1 e^{-t} + c_2 e^{-2t}$ to this component. Therefore,

$$y(t) = c_1 e^{-t} + c_2 e^{-2t} - 5e^{-t} + 20e^{-2t} - 15e^{-3t}$$

Setting the conditions $y(0^+) = 2$ and $y(0^+) = 3$ in this equation (and its derivative), we obtain $c_1 = -3$, $c_2 = 5$ so that

$$y(t) = -8e^{-t} + 25e^{-2t} - 15e^{-3t} \quad t \geq 0$$

which is identical to the solution found by the classical method.

1.1.3.1 Assessment of the Convolution Method

The convolution method is more laborious compared to the classical method. However, in system analysis, its advantages outweigh the extra work. The classical method has a serious drawback because it yields the total response, which cannot be separated into components arising from the internal conditions and the external input. In the study of systems it is important to be able to express the system response to an input $f(t)$ as an explicit function of $f(t)$. This is not possible in the classical method. Moreover, the classical method is restricted to a certain class of inputs; it cannot be applied to any input.*

If we must solve a particular linear differential equation or find a response of a particular LTI system, the classical method may be the best. In the theoretical study of linear systems, however, it is practically useless. General discussion of differential equations can be found in numerous texts on the subject [2].

---

* Another minor problem is that because the classical method yields total response, the auxiliary conditions must be on the total response, which exists only for $t \geq 0^+$. In practice we are most likely to know the conditions at $t = 0^-$ (before the input is applied). Therefore, we need to derive a new set of auxiliary conditions at $t = 0^+$ from the known conditions at $t = 0^-$. The convolution method can handle both kinds of initial conditions. If the conditions are given at $t = 0^-$, we apply these conditions only to $y_c(t)$ because by its definition the convolution integral is 0 at $t = 0^-$.

1.2 Difference Equations

The development of difference equations is parallel to that of differential equations. We consider here only linear difference equations with constant coefficients. An nth-order difference equation can be expressed in two different forms; the first form uses delay terms such as $y[k-1]$, $y[k-2]$, $f[k-1]$, $f[k-2]$, ..., and so on, and the alternative form uses advance terms such as $y[k+1]$, $y[k+2]$, ..., and so on. Both forms are useful. We start here with a general nth-order difference equation, using advance operator form

$$y[k+n] + a_{n-1}y[k+n-1] + \cdots + a_1 y[k+1] + a_0 y[k]$$
$$= b_m f[k+m] + b_{m-1} f[k+m-1] + \cdots + b_1 f[k+1] + b_0 f[k] \qquad (1.26)$$

1.2.1 Causality Condition

The left-hand side of Equation 1.26 consists of values of $y[k]$ at instants $k+n$, $k+n-1$, $k+n-2$, and so on. The right-hand side of Equation 1.26 consists of the input at instants $k+m$, $k+m-1$, $k+m-2$, and so on. For a causal equation, the solution cannot depend on future input values. This shows that when the equation is in the advance operator form of the Equation 1.26, causality requires $m \leq n$. For a general causal case, $m = n$, and Equation 1.26 becomes

$$y[k+n] + a_{n-1}y[k+n-1] + \cdots + a_1 y[k+1] + a_0 y[k]$$
$$= b_n f[k+n] + b_{n-1} f[k+n-1] + \cdots + b_1 f[k+1] + b_0 f[k] \qquad (1.27a)$$

where some of the coefficients on both sides can be zero. However, the coefficient of $y[k+n]$ is normalized to unity. Equation 1.27a is valid for all values of k. Therefore, the equation is still valid if we replace k by $k-n$ throughout the equation. This yields the alternative form (the delay operator form) of Equation 1.27a

$$y[k] + a_{n-1}y[k-1] + \cdots + a_1 y[k-n+1] + a_0 y[k-n]$$
$$= b_n f[k] + b_{n-1} f[k-1] + \cdots + b_1 f[k-n+1] + b_0 f[k-n] \qquad (1.27b)$$

We designate the form of Equation 1.27a the *advance operator form*, and the form of Equation 1.27b the *delay operator form*.

1.2.2 Initial Conditions and Iterative Solution

Equation 1.27b can be expressed as

$$y[k] = -a_{n-1}y[k-1] - a_{n-2}y[k-2] - \cdots - a_0 y[k-n]$$
$$+ b_n f[k] + b_{n-1} f[k-1] + \cdots + b_0 f[k-n] \qquad (1.27c)$$

This equation shows that $y[k]$, the solution at the kth instant, is computed from $2n+1$ pieces of information. These are the past n values of $y[k] : y[k-1], y[k-2], \ldots, y[k-n]$ and the present and past n values of the input: $f[k], f[k-1], f[k-2], \ldots, f[k-n]$. If the input $f[k]$ is known for $k = 0, 1, 2, \ldots$, then the values of $y[k]$ for $k = 0, 1, 2, \ldots$ can be computed from the $2n$ initial conditions $y[-1], y[-2], \ldots, y[-n]$ and $f[-1], f[-2], \ldots, f[-n]$. If the input is causal, that is, if $f[k] = 0$ for $k < 0$, then $f[-1] = f[-2] = \ldots = f[-n] = 0$, and we need only n initial conditions $y[-1], y[-2], \ldots, y[-n]$. This allows us to compute iteratively or recursively the values $y[0], y[1], y[2], y[3], \ldots$, and so on.* For

---

* For this reason, Equation 1.27 is called a *recursive difference equation*. However, in Equation 1.27, if $a_0 = a_1 = a_2 = \cdots = a_{n-1} = 0$, then it follows from Equation 1.27c that determination of the present value of $y[k]$ does not require the past values $y[k-1], y[k-2], \ldots$, and so on. For this reason, when $a_i = 0$, $(i = 0, 1, \ldots, n-1)$, the difference Equation 1.27 is *nonrecursive*. This classification is important in designing and realizing digital filters. In this discussion, however, this classification is not important. The analysis techniques developed here apply to general recursive and nonrecursive equations. Observe that a nonrecursive equation is a special case of recursive equation with $a_0 = a_1 = \ldots = a_{n-1} = 0$.

instance, to find $y[0]$ we set $k = 0$ in Equation 1.27c. The left-hand side is $y[0]$, and the right-hand side contains terms $y[-1], y[-2], \ldots, y[-n]$, and the inputs $f[0], f[-1], f[-2], \ldots, f[-n]$. Therefore, to begin with, we must know the n initial conditions $y[-1], y[-2], \ldots, y[-n]$. Knowing these conditions and the input $f[k]$, we can iteratively find the response $y[0], y[1], y[2], \ldots$, and so on. The following example demonstrates this procedure. This method basically reflects the manner in which a computer would solve a difference equation, given the input and initial conditions.

Example 1.4:

Solve iteratively

$$y[k] - 0.5y[k-1] = f[k] \tag{1.28a}$$

with initial condition $y[-1] = 16$ and the input $f[k] = k^2$ (starting at $k = 0$). This equation can be expressed as

$$y[k] = 0.5y[k-1] + f[k] \tag{1.28b}$$

If we set $k = 0$ in this equation, we obtain

$$\begin{aligned} y[0] &= 0.5y[-1] + f[0] \\ &= 0.5(16) + 0 = 8 \end{aligned}$$

Now, setting $k = 1$ in Equation 1.28b and using the value $y[0] = 8$ (computed in the first step) and $f[1] = (1)^2 = 1$, we obtain

$$y[1] = 0.5(8) + (1)^2 = 5$$

Next, setting $k = 2$ in Equation 1.28b and using the value $y[1] = 5$ (computed in the previous step) and $f[2] = (2)^2$, we obtain

$$y[2] = 0.5(5) + (2)^2 = 6.5$$

Continuing in this way iteratively, we obtain

$$\begin{aligned} y[3] &= 0.5(6.5) + (3)^2 = 12.25 \\ y[4] &= 0.5(12.25) + (4)^2 = 22.125 \end{aligned}$$

. . . .

This iterative solution procedure is available only for difference equations; it cannot be applied to differential equations. Despite the many uses of this method, a closed-form solution of a difference equation is far more useful in the study of system behavior and its dependence on the input and the various system parameters. For this reason, we shall develop a systematic procedure to obtain a closed-form solution of Equation 1.27.

1.2.2.1 Operational Notation

In difference equations it is convenient to use operational notation similar to that used in differential equations for the sake of compactness and convenience. For differential equations, we use the operator D to denote the operation of differentiation. For difference equations, we use the operator E to denote the operation for advancing the sequence by one time interval. Thus,

$$Ef[k] \equiv f[k+1]$$
$$E^2 f[k] \equiv f[k+2]$$
$$\cdots\cdots\cdots\cdots\cdots$$
$$E^n f[k] \equiv f[k+n]$$

(1.29)

A general *n*th-order difference Equation 1.27a can be expressed as

$$(E^n + a_{n-1} E^{n-1} + \cdots + a_1 E + a_0) y[k] = (b_n E^n + b_{n-1} E^{n-1} + \cdots + b_1 E + b_0) f[k]$$

(1.30a)

or

$$Q[E] y[k] = P[E] f[k]$$

(1.30b)

where $Q[E]$ and $P[E]$ are *n*th-order polynomial operators, respectively,

$$Q[E] = E^n + a_{n-1} E^{n-1} + \cdots + a_1 E + a_0$$

(1.31a)

$$P[E] = b_n E^n + b_{n-1} E^{n-1} + \cdots + b_1 E + b_0$$

(1.31b)

1.2.3 Classical Solution

Following the discussion of differential equations, we can show that if $y_p[k]$ is a solution of Equation 1.27 or Equation 1.30, that is,

$$Q[E] y_p[k] = P[E] f[k]$$

(1.32)

then $y_p[k] + y_c[k]$ is also a solution of Equation 1.30, where $y_c[k]$ is a solution of the homogeneous equation

$$Q[E] y_c[k] = 0$$

(1.33)

As before, we call $y_p[k]$ the particular solution and $y_c[k]$ the complementary solution.

1.2.3.1 Complementary Solution (The Natural Response)

By definition

$$Q[E] y_c[k] = 0$$

(1.33a)

or

$$(E^n + a_{n-1} E^{n-1} + \cdots + a_1 E + a_0) y_c[k] = 0$$

(1.33b)

or

$$y_c[k+n] + a_{n-1} y_c[k+n-1] + \cdots + a_1 y_c[k+1] + a_0 y_c[k] = 0$$

(1.33c)

We can solve this equation systematically, but even a cursory examination of this equation points to its solution. This equation states that a linear combination of $y_c[k]$ and delayed $y_c[k]$ is zero *not for some values of k, but for all k*. This is possible *if and only if* $y_c[k]$ and delayed $y_c[k]$ have the same form. Only an exponential function γ^k has this property as seen from the equation

$$\gamma^{k-m} = \gamma^{-m}\gamma^k$$

This shows that the delayed γ^k is a constant times γ^k. Therefore, the solution of Equation 1.33 must be of the form

$$y_c[k] = c\gamma^k \tag{1.34}$$

To determine c and γ, we substitute this solution in Equation 1.33. From Equation 1.34, we have

$$Ey_c[k] = y_c[k+1] = c\gamma^{k+1} = (c\gamma)\gamma^k$$
$$E^2 y_c[k] = y_c[k+2] = c\gamma^{k+2} = (c\gamma^2)\gamma^k$$
$$\dots\dots\dots\dots\dots\dots\dots\dots\dots \tag{1.35}$$
$$E^n y_c[k] = y_c[k+n] = c\gamma^{k+n} = (c\gamma^n)\gamma^k$$

Substitution of this in Equation 1.33 yields

$$c(\gamma^n + a_{n-1}\gamma^{n-1} + \cdots + a_1\gamma + a_0)\gamma^k = 0 \tag{1.36}$$

For a nontrivial solution of this equation

$$(\gamma^n + a_{n-1}\gamma^{n-1} + \cdots + a_1\gamma + a_0) = 0 \tag{1.37a}$$

or

$$Q[\gamma] = 0 \tag{1.37b}$$

Our solution $c\gamma^k$ [Equation 1.34] is correct, provided that γ satisfies Equation 1.37a. Now, $Q[\gamma]$ is an nth-order polynomial and can be expressed in the factorized form (assuming all distinct roots):

$$(\gamma - \gamma_1)(\gamma - \gamma_2)\cdots(\gamma - \gamma_n) = 0 \tag{1.37c}$$

Clearly γ has n solutions $\gamma_1, \gamma_2, \cdots, \gamma_n$ and, therefore, Equation 1.33 also has n solutions $c_1\gamma_1^k, c_2\gamma_2^k, \ldots, c_n\gamma_n^k$. In such a case we have shown that the general solution is a linear combination of the n solutions. Thus,

$$y_c[k] = c_1\gamma_1^k + c_2\gamma_2^k + \cdots + c_n\gamma_n^k \tag{1.38}$$

where $\gamma_1, \gamma_2, \ldots, \gamma_n$ are the roots of Equation 1.37a and $c_1, c_2, \ldots, c_n$ are arbitrary constants determined from n auxiliary conditions. The polynomial $Q[\gamma]$ is called the *characteristic polynomial*, and

$$Q[\gamma] = 0 \tag{1.39}$$

is the *characteristic equation*. Moreover, $\gamma_1, \gamma_2, \cdots, \gamma_n$, the roots of the characteristic equation, are called *characteristic roots* or *characteristic values* (also *eigenvalues*). The exponentials $\gamma_i^k (i = 1, 2, \ldots, n)$ are the *characteristic* modes or *natural* modes. A characteristic mode corresponds to each characteristic root, and the complementary solution is a linear combination of the characteristic modes of the system.

Repeated Roots

For repeated roots, the form of characteristic modes is modified. It can be shown by direct substitution that if a root γ repeats r times (root of multiplicity r), the characteristic modes corresponding to this root are $\gamma^k, k\gamma^k, k^2\gamma^k, \ldots, k^{r-1}\gamma^k$. Thus, if the characteristic equation is

$$Q[\gamma] = (\gamma - \gamma_1)^r (\gamma - \gamma_{r+1})(\gamma - \gamma_{r+2})\cdots(\gamma - \gamma_n) \tag{1.40}$$

the complementary solution is

$$y_c[k] = (c_1 + c_2 k + c_3 k^2 + \cdots + c_r k^{r-1})\gamma_1^k$$
$$+ c_{r+1}\gamma_{r+1}^k + c_{r+2}\gamma_{r+2}^k + \cdots + c_n\gamma_n^k \tag{1.41}$$

TABLE 1.2

Input $f[k]$	Forced Response $y_p[k]$
1. r^k $r \neq \gamma_i$ $(i = 1, 2, \ldots, n)$	βr^k
2. r^k $r = \gamma_i$	$\beta k r^k$
3. $\cos(\Omega k + \theta)$	$\beta \cos(\Omega k + \phi)$
4. $\left(\sum\limits_{i=0}^{m} \alpha_i k^i\right) r^k$	$\left(\sum\limits_{i=0}^{m} \beta_i k^i\right) r^k$

1.2.3.2 Particular Solution

The particular solution $y_p[k]$ is the solution of

$$Q[E]y_p[k] = P[E]f[k] \tag{1.42}$$

We shall find the particular solution using the method of undetermined coefficients, the same method used for differential equations. Table 1.2 lists the inputs and the corresponding forms of solution with undetermined coefficients. These coefficients can be determined by substituting $y_p[k]$ in Equation 1.42 and equating the coefficients of similar terms.

Note: By definition, $y_p[k]$ cannot have any characteristic mode terms. If any term $p[k]$ shown in the right-hand column for the particular solution should also be a characteristic mode, the correct form of the particular solution must be modified to $k^i p[k]$, where i is the smallest integer that will prevent $k^i p[k]$ from having a characteristic mode term. For example, when the input is r^k, the particular solution in the right-hand column is of the form cr^k. But if r^k happens to be a natural mode, the correct form of the particular solution is $\beta k r^k$ (see Pair 2).

Example 1.5:

Solve

$$(E^2 - 5E + 6)y[k] = (E - 5)f[k] \tag{1.43}$$

if the input $f[k] = (3k + 5)u[k]$ and the auxiliary conditions are $y[0] = 4, y[1] = 13$.

The characteristic equation is

$$\gamma^2 - 5\gamma + 6 = (\gamma - 2)(\gamma - 3) = 0$$

Therefore, the complementary solution is

$$y_c[k] = c_1(2)^k + c_2(3)^k$$

To find the form of $y_p[k]$ we use Table 1.2, Pair 4 with $r = 1, m = 1$. This yields

$$y_p[k] = \beta_1 k + \beta_0$$

Therefore,

$$y_p[k + 1] = \beta_1(k + 1) + \beta_0 = \beta_1 k + \beta_1 + \beta_0$$
$$y_p[k + 2] = \beta_1(k + 2) + \beta_0 = \beta_1 k + 2\beta_1 + \beta_0$$

Also,

$$f[k] = 3k + 5$$

and

$$f[k + 1] = 3(k + 1) + 5 = 3k + 8$$

Substitution of the above results in Equation 1.43 yields

$$\beta_1 k + 2\beta_1 + \beta_0 - 5(\beta_1 k + \beta_1 + \beta_0) + 6(\beta_1 k + \beta_0) = 3k + 8 - 5(3k + 5)$$

or

$$2\beta_1 k - 3\beta_1 + 2\beta_0 = -12k - 17$$

Comparison of similar terms on two sides yields

$$\left. \begin{array}{rcl} 2\beta_1 & = & -12 \\ -3\beta_1 + 2\beta_0 & = & -17 \end{array} \right\} \implies \begin{array}{rcl} \beta_1 & = & -6 \\ \beta_2 & = & -\dfrac{35}{2} \end{array}$$

This means

$$y_p[k] = -6k - \frac{35}{2}$$

The total response is

$$y[k] = y_c[k] + y_p[k]$$
$$= c_1(2)^k + c_2(3)^k - 6k - \frac{35}{2} \quad k \geq 0 \tag{1.44}$$

To determine arbitrary constants c_1 and c_2 we set $k = 0$ and 1 and substitute the auxiliary conditions $y[0] = 4, y[1] = 13$ to obtain

$$\left. \begin{array}{l} 4 = c_1 + c_2 - \dfrac{35}{2} \\ 13 = 2c_1 + 3c_2 - \dfrac{47}{2} \end{array} \right\} \implies \begin{array}{l} c_1 = 28 \\ c_2 = \dfrac{-13}{2} \end{array}$$

Therefore,

$$y_c[k] = 28(2)^k - \frac{13}{2}(3)^k \tag{1.45}$$

and

$$y[k] = \underbrace{28(2)^k - \frac{13}{2}(3)^k}_{y_c[k]} \underbrace{- 6k - \frac{35}{2}}_{y_p[k]} \tag{1.46}$$

1.2.4 A Comment on Auxiliary Conditions

This method requires auxiliary conditions $y[0], y[1], \ldots, y[n-1]$ because the total solution is valid only for $k \geq 0$. But if we are given the initial conditions $y[-1], y[-2], \ldots, y[-n]$, we can derive the conditions $y[0], y[1], \ldots, y[n-1]$ using the iterative procedure discussed earlier.

1.2.4.1 Exponential Input

As in the case of differential equations, we can show that for the equation

$$Q[E]y[k] = P[E]f[k] \tag{1.47}$$

the particular solution for the exponential input $f[k] = r^k$ is given by

$$y_p[k] = H[r]r^k \quad r \neq \gamma_i \tag{1.48}$$

where

$$H[r] = \frac{P[r]}{Q[r]} \tag{1.49}$$

The proof follows from the fact that if the input $f[k] = r^k$, then from Table 1.2 (Pair 4), $y_p[k] = \beta r^k$. Therefore,

$$E^i f[k] = f[k+i] = r^{k+i} = r^i r^k \quad \text{and} \quad P[E]f[k] = P[r]r^k$$
$$E^j y_p[k] = \beta r^{k+j} = \beta r^j r^k \quad \text{and} \quad Q[E]y[k] = \beta Q[r]r^k$$

so that Equation 1.47 reduces to

$$\beta Q[r]r^k = P[r]r^k$$

which yields $\beta = P[r]/Q[r] = H[r]$.

This result is valid only if r is not a characteristic root. If r is a characteristic root, the particular solution is $\beta k r^k$ where β is determined by substituting $y_p[k]$ in Equation 1.47 and equating coefficients of similar terms on the two sides. Observe that the exponential r^k includes a wide variety of signals such as a constant C, a sinusoid $\cos(\Omega k + \theta)$, and an exponentially growing or decaying sinusoid $|\gamma|^k \cos(\Omega k + \theta)$.

1.2.4.2 A Constant Input $f(k) = C$

This is a special case of exponential Cr^k with $r = 1$. Therefore, from Equation 1.48 we have

$$y_p[k] = C\frac{P[1]}{Q[1]}(1)^k = CH[1] \tag{1.50}$$

1.2.4.3 A Sinusoidal Input

The input $e^{j\Omega k}$ is an exponential r^k with $r = e^{j\Omega}$. Hence,

$$y_p[k] = H[e^{j\Omega}]e^{j\Omega k} = \frac{P[e^{j\Omega}]}{Q[e^{j\Omega}]}e^{j\Omega k}$$

Similarly for the input $e^{-j\Omega k}$

$$y_p[k] = H[e^{-j\Omega}]e^{-j\Omega k}$$

Consequently, if the input

$$f[k] = \cos \Omega k = \frac{1}{2}(e^{j\Omega k} + e^{-j\Omega k})$$
$$y_p[k] = \frac{1}{2}\left\{H[e^{j\Omega}]e^{j\Omega k} + H[e^{-j\Omega}]e^{-j\Omega k}\right\}$$

Since the two terms on the right-hand side are conjugates

$$y_p[k] = \text{Re}\left\{H[e^{j\Omega}]e^{j\Omega k}\right\}$$

If

$$H[e^{j\Omega}] = |H[e^{j\Omega}]|e^{j\angle H[e^{j\Omega}]}$$

then

$$y_p[k] = \text{Re}\left\{|H[e^{j\Omega}]|e^{j(\Omega k + \angle H[e^{j\Omega}])}\right\}$$
$$= |H[e^{j\Omega}]| \cos(\Omega k + \angle H[e^{j\Omega}]) \tag{1.51}$$

Using a similar argument, we can show that for the input

$$f[k] = \cos(\Omega k + \theta)$$
$$y_p[k] = |H[e^{j\Omega}]| \cos(\Omega k + \theta + \angle H[e^{j\Omega}]) \tag{1.52}$$

Example 1.6:

Solve

$$(E^2 - 3E + 2)y[k] = (E + 2)f[k]$$

for $f[k] = (3)^k u[k]$ and the auxiliary conditions $y[0] = 2$, $y[1] = 1$.

In this case

$$H[r] = \frac{P[r]}{Q[r]} = \frac{r+2}{r^2 - 3r + 2}$$

and the particular solution to input $(3)^k u[k]$ is $H[3](3)^k$; that is,

$$y_p[k] = \frac{3+2}{(3)^2 - 3(3) + 2}(3)^k = \frac{5}{2}(3)^k$$

The characteristic polynomial is $(\gamma^2 - 3\gamma + 2) = (\gamma - 1)(\gamma - 2)$. The characteristic roots are 1 and 2. Hence, the complementary solution is $y_c[k] = c_1 + c_2(2)^k$ and the total solution is

$$y[k] = c_1(1)^k + c_2(2)^k + \frac{5}{2}(3)^k$$

Setting $k = 0$ and 1 in this equation and substituting auxiliary conditions yields

$$2 = c_1 + c_2 + \frac{5}{2} \quad \text{and} \quad 1 = c_1 + 2c_2 + \frac{15}{2}$$

Solution of these two simultaneous equations yields $c_1 = 5.5$, $c_2 = -5$. Therefore,

$$y[k] = 5.5 - 6(2)^k + \frac{5}{2}(3)^k \quad k \geq 0$$

1.2.5 Method of Convolution

In this method, the input $f[k]$ is expressed as a sum of impulses. The solution is then obtained as a sum of the solutions to all the impulse components. The method exploits the superposition property of the linear difference equations. A discrete-time unit impulse function $\delta[k]$ is defined as

$$\delta[k] = \begin{cases} 1 & k = 0 \\ 0 & k \neq 0 \end{cases} \tag{1.53}$$

Hence, an arbitrary signal $f[k]$ can be expressed in terms of impulse and delayed impulse functions as

$$f[k] = f[0]\delta[k] + f[1]\delta[k-1] + f[2]\delta[k-2] + \cdots + f[k]\delta[0] + \cdots \quad k \geq 0 \tag{1.54}$$

The right-hand side expresses $f[k]$ as a sum of impulse components. If $h[k]$ is the solution of Equation 1.30 to the impulse input $f[k] = \delta[k]$, then the solution to input $\delta[k-m]$ is $h[k-m]$. This follows from the fact that because of constant coefficients, Equation 1.30 has time-invariance property. Also, because Equation 1.30 is linear, its solution is the sum of the solutions to each of the impulse components of $f[k]$ on the right-hand side of Equation 1.54. Therefore,

$$y[k] = f[0]h[k] + f[1]h[k-1] + f[2]h[k-2] + \cdots + f[k]h[0] + f[k+1]h[-1] + \cdots$$

All practical systems with time as the independent variable are causal, that is, $h[k] = 0$ for $k < 0$. Hence, all the terms on the right-hand side beyond $f[k]h[0]$ are zero. Thus,

$$y[k] = f[0]h[k] + f[1]h[k-1] + f[2]h[k-2] + \cdots + f[k]h[0]$$
$$= \sum_{m=0}^{k} f[m]h[k-m] \tag{1.55}$$

The general solution is obtained by adding a complementary solution to the above solution. Therefore, the general solution is given by

$$y[k] = \sum_{j=1}^{n} c_j \gamma_j^k + \sum_{m=0}^{k} f[m]h[k-m] \tag{1.56}$$

The last sum on the right-hand side is known as the *convolution sum* of $f[k]$ and $h[k]$.

The function $h[k]$ appearing in Equation 1.30 is the solution of Equation 1.30 for the impulsive input ($f[k] = \delta[k]$) when all initial conditions are zero, that is, $h[-1] = h[-2] = \cdots = h[-n] = 0$. It can be shown that [2] $h[k]$ contains an impulse and a linear combination of characteristic modes as

$$h[k] = \frac{b_0}{a_0} \delta[k] + A_1 \gamma_1^k + A_2 \gamma_2^k + \cdots + A_n \gamma_n^k \tag{1.57}$$

where the unknown constants A_i are determined from n values of $h[k]$ obtained by solving the equation $Q[E]h[k] = P[E]\delta[k]$ iteratively.

Example 1.7:

Solve Example 1.5 using convolution method. In other words solve

$$(E^2 - 3E + 2)y[k] = (E + 2)f[k]$$

for $f[k] = (3)^k u[k]$ and the auxiliary conditions $y[0] = 2$, $y[1] = 1$.

The unit impulse solution $h[k]$ is given by Equation 1.57. In this case, $a_0 = 2$ and $b_0 = 2$. Therefore,

$$h[k] = \delta[k] + A_1(1)^k + A_2(2)^k \tag{1.58}$$

To determine the two unknown constants A_1 and A_2 in Equation 1.58, we need two values of $h[k]$, for instance $h[0]$ and $h[1]$. These can be determined iteratively by observing that $h[k]$ is the solution of $(E^2 - 3E + 2)h[k] = (E + 2)\delta[k]$, that is,

$$h[k+2] - 3h[k+1] + 2h[k] = \delta[k+1] + 2\delta[k] \tag{1.59}$$

subject to initial conditions $h[-1] = h[-2] = 0$. We now determine $h[0]$ and $h[1]$ iteratively from Equation 1.59. Setting $k = -2$ in this equation yields

$$h[0] - 3(0) + 2(0) = 0 + 0 \implies h[0] = 0$$

Next, setting $k = -1$ in Equation 1.59 and using $h[0] = 0$, we obtain

$$h[1] - 3(0) + 2(0) = 1 + 0 \implies h[1] = 1$$

Setting $k = 0$ and 1 in Equation 1.58 and substituting $h[0] = 0$, $h[1] = 1$ yields

$$0 = 1 + A_1 + A_2 \quad \text{and} \quad 1 = A_1 + 2A_2$$

Solution of these two equations yields $A_1 = -3$ and $A_2 = 2$. Therefore,

$$h[k] = \delta[k] - 3 + 2(2)^k$$

and from Equation 1.56

$$y[k] = c_1 + c_2(2)^k + \sum_{m=0}^{k} (3)^m \left[\delta[k-m] - 3 + 2(2)^{k-m} \right]$$

$$= c_1 + c_2(2)^k + 1.5 - 4(2)^k + 2.5(3)^k$$

The sums in the above expression are found by using the geometric progression sum formula

$$\sum_{m=0}^{k} r^m = \frac{r^{k+1} - 1}{r - 1} \quad r \neq 1$$

Setting $k = 0$ and 1 and substituting the given auxiliary conditions $y[0] = 2$, $y[1] = 1$, we obtain

$$2 = c_1 + c_2 + 1.5 - 4 + 2.5 \quad \text{and} \quad 1 = c_1 + 2c_2 + 1.5 - 8 + 7.5$$

Solution of these equations yields $c_1 = 4$ and $c_2 = -2$. Therefore,

$$y[k] = 5.5 - 6(2)^k + 2.5(3)^k$$

which confirms the result obtained by the classical method.

1.2.5.1 Assessment of the Classical Method

The earlier remarks concerning the classical method for solving differential equations also apply to difference equations. General discussion of difference equations can be found in texts on the subject [3].

References

1. Birkhoff, G. and Rota, G.C., *Ordinary Differential Equations*, 3rd ed., John Wiley & Sons, New York, 1978.
2. Lathi, B.P., *Linear Systems and Signals*, Berkeley-Cambridge Press, Carmichael, CA, 1992.
3. Goldberg, S., *Introduction to Difference Equations*, John Wiley & Sons, New York, 1958.

2

The Fourier, Laplace, and z-Transforms

Edward W. Kamen
Georgia Institute of Technology

2.1 Introduction

The study of signals and systems can be carried out in terms of either a time-domain or a transform-domain formulation. Both approaches are often used together in order to maximize our ability to deal with a particular problem arising in applications. This is very much the case in controls engineering where both time-domain and transform-domain techniques are extensively used in analysis and design. The transform-domain approach to signals and systems is based on the transformation of functions using the Fourier, Laplace, and z-transforms. The fundamental aspects of these transforms are presented in this section along with some discussion on the application of these constructs.

The development in this chapter begins with the Fourier transform (FT), which can be viewed as a generalization of the Fourier series representation of a periodic function. The FT and Fourier series are named after Jean Baptiste Joseph Fourier (1768–1830), who first proposed in a 1807 paper that a series of sinusoidal harmonics could be used to represent the temperature distribution in a body. In 1822 Fourier wrote a book on his work, which was translated into English many years later (see [1]). It was also during the first part of the 1800s that Fourier was successful in constructing a frequency-domain representation for aperiodic (nonperiodic) functions. This resulted in the FT, which provides a representation of a function $f(t)$ of a real variable t in terms of the frequency components comprising the function. Much later (in the 1900s), an FT theory was developed for functions $f(k)$ of an integer variable k. This resulted in the discrete-time Fourier transform (DTFT) and the N-point discrete Fourier transform (N-point DFT), both of which are briefly considered in this section.

Also during the early part of the 1800s, Pierre Simon Laplace (1749–1827) carried out his work on the generalization of the FT, which resulted in the transform that now bears his name. The Laplace transform can be viewed as the FT with the addition of a real exponential factor to the integrand of the integral operation. This results in a transform that is a function of a complex variable $s = \sigma + j\omega$. Although the modification to the FT may not seem to be very major, in fact the Laplace transform is an extremely powerful tool in many application areas (such as controls) where the utility of the FT is somewhat limited. In this section, a brief presentation is given on the one-sided Laplace transform with much of the focus on rational transforms.

The discrete-time counterpart to the Laplace transform is the z-transform which was developed primarily during the 1950s (e.g., see [2–4]). The one-sided z-transform is considered, along with the connection to the DTFT.

Applications and examples involving the Fourier, Laplace, and z-transforms are given in the second part of this section. There the presentation centers on the relationship between the pole locations of a rational transform and the frequency spectrum of the transformed function; the numerical computation of the FT; and the application of the Laplace and z-transforms to solving differential and difference equations. The application of the transforms to systems and controls is pursued in other chapters in this handbook.

2.2 Fundamentals of the Fourier, Laplace, and z-Transforms

Let $f(t)$ be a real-valued function of the real-valued variable t; that is, for any real number t, $f(t)$ is a real number. The function $f(t)$ can be viewed as a signal that is a function of the continuous-time variable t (in units of seconds) and where t takes values from $-\infty$ to ∞. The FT $F(\omega)$ of $f(t)$ is defined by

$$F(\omega) = \int_{-\infty}^{\infty} f(t)e^{-j\omega t}\,dt, \quad -\infty < \omega < \infty \tag{2.1}$$

where ω is the frequency variable in radians per second (rad/s), $j = \sqrt{-1}$ and $e^{-j\omega t}$ is the complex exponential given by Euler's formula

$$e^{-j\omega t} = \cos(\omega t) - j\sin(\omega t) \tag{2.2}$$

Inserting Equation 2.2 into Equation 2.1 results in the following expression for the FT:

$$F(\omega) = R(\omega) + jI(\omega) \tag{2.3}$$

where $R(\omega)$ and $I(\omega)$ are the real and imaginary parts, respectively, of $F(\omega)$ given by

$$R(\omega) = \int_{-\infty}^{\infty} f(t)\cos(\omega t)\,dt$$
$$I(\omega) = -\int_{-\infty}^{\infty} f(t)\sin(\omega t)\,dt \tag{2.4}$$

From Equation 2.3, it is seen that in general the FT $F(\omega)$ is a complex-valued function of the frequency variable ω. For any value of ω, $F(\omega)$ has a magnitude $|F(\omega)|$ and an angle $\angle F(\omega)$ given by

$$\angle F(\omega) = \begin{cases} \tan^{-1}\left(\dfrac{I(\omega)}{R(\omega)}\right), & R(\omega) \geq 0 \\[3mm] \pi + \tan^{-1}\left(\dfrac{I(\omega)}{R(\omega)}\right), & R(\omega) < 0 \end{cases} \tag{2.5}$$

where again $R(\omega)$ and $I(\omega)$ are the real and imaginary parts defined by Equation 2.4. The function $|F(\omega)|$ represents the magnitude of the frequency components comprising $f(t)$, and thus the plot of $|F(\omega)|$ versus

ω is called the *magnitude spectrum* of $f(t)$. The function $\angle F(\omega)$ represents the phase of the frequency components comprising $f(t)$, and thus the plot of $\angle F(\omega)$ versus ω is called the *phase spectrum* of $f(t)$. Note that $F(\omega)$ can be expressed in the polar form

$$F(\omega) = |F(\omega)| \exp[j\angle F(\omega)] \tag{2.6}$$

whereas the rectangular form of $F(\omega)$ is given by Equation 2.3.

The function (or signal) $f(t)$ is said to have an FT in the ordinary sense if the integral in Equation 2.1 exists for all real values of ω. Sufficient conditions that ensure the existence of the integral are that $f(t)$ have only a finite number of discontinuities, maxima, and minima over any finite interval of time and that $f(t)$ be absolutely integrable. The latter condition means that

$$\int_{-\infty}^{\infty} |f(t)| \, dt < \infty \tag{2.7}$$

There are a number of functions $f(t)$ of interest for which the integral in Equation 2.1 does not exist; for example, this is the case for the constant function $f(t) = c$ for $-\infty < t < \infty$, where c is a nonzero real number. Since the integral in Equation 2.1 obviously does not exist in this case, the constant function does not have an FT in the ordinary sense, but it does have an FT in the generalized sense, given by

$$F(\omega) = 2\pi c\delta(\omega) \tag{2.8}$$

where $\delta(\omega)$ is the impulse function. If Equation 2.8 is inserted into the inverse FT given by Equation 2.11, the result is the constant function $f(t) = c$ for all t. This observation justifies taking Equation 2.8 as the definition of the (generalized) FT of the constant function.

The FT defined by Equation 2.1 can be viewed as an operator that maps a time function $f(t)$ into a frequency function $F(\omega)$. This operation is often given by

$$F(\omega) = \Im[f(t)] \tag{2.9}$$

where $\Im$ denotes the FT operator. From Equation 2.9, it is clear that $f(t)$ can be recomputed from $F(\omega)$ by applying the inverse FT operator denoted by $\Im^{-1}$; that is,

$$f(t) = \Im^{-1}[F(\omega)] \tag{2.10}$$

The inverse operation is given by

$$f(t) = \frac{1}{2\pi} \int_{-\infty}^{\infty} F(\omega)e^{j\omega t} \, d\omega \tag{2.11}$$

The FT satisfies a number of properties that are very useful in applications. These properties are listed in Table 2.1 in terms of functions $f(t)$ and $g(t)$ whose transforms are $F(\omega)$ and $G(\omega)$, respectively. Appearing in this table is the convolution $f(t) * g(t)$ of $f(t)$ and $g(t)$, defined by

$$f(t) * g(t) = \int_{-\infty}^{\infty} f(\tau)g(t - \tau) \, d\tau \tag{2.12}$$

Also, in Table 2.1 is the convolution $F(\omega) * G(\omega)$ given by

$$F(\omega) * G(\omega) = \int_{-\infty}^{\infty} F(\lambda)G(\omega - \lambda) \, d\lambda \tag{2.13}$$

From the properties in Table 2.1 and the generalized transform given by Equation 2.8, it is possible to determine the FT of many common functions. A list of FTs of some common functions is given in Table 2.2.

TABLE 2.1   Properties of FT

Property	Transform/Property
Linearity	$\Im[af(t) + bg(t)] = aF(\omega) + bG(\omega)$ for any scalars a, b
Right or left shift in t	$\Im[f(t - t_0)] = F(\omega)\exp(-j\omega t_0)$ for any t_0
Time scaling	$\Im[f(at)] = (1/a)F(\omega/a)$ for any real number $a > 0$
Time reversal	$\Im[f(-t)] = F(-\omega) = \overline{F(\omega)} = $ complex conjugate of $F(\omega)$
Multiplication by a power of t	$\Im[t^n f(t)] = j^n \dfrac{d^n}{d\omega^n} F(\omega), n = 1, 2, \ldots$
Multiplication by $\exp(j\omega_0 t)$	$\Im[f(t)\exp(j\omega_0 t)] = F(\omega - \omega_0)$ for any real number ω_0
Multiplication by $\sin(\omega_0 t)$	$\Im[f(t)\sin(\omega_0 t)] = (j/2)[F(\omega + \omega_0) - F(\omega - \omega_0)]$
Multiplication by $\cos(\omega_0 t)$	$\Im[f(t)\cos(\omega_0 t)] = (1/2)[F(\omega + \omega_0) + F(\omega - \omega_0)]$
Differentiation in the time domain	$\Im\left[\dfrac{d^n}{dt^n} f(t)\right] = (j\omega)^n F(\omega), n = 1, 2, \ldots$
Multiplication in the time domain	$\Im[f(t)g(t)] = \dfrac{1}{2\pi}[F(\omega) * G(\omega)]$
Convolution in the time domain	$\Im[f(t) * g(t)] = F(\omega)G(\omega)$
Duality	$\Im[F(t)] = 2\pi f(-\omega)$
Parseval's theorem	$\displaystyle\int_{-\infty}^{\infty} f(t)g(t)\,dt = \dfrac{1}{2\pi}\int_{-\infty}^{\infty} F(-\omega)G(\omega)\,d\omega$
Special case of Parseval's theorem	$\displaystyle\int_{-\infty}^{\infty} f^2(t)\,dt = \dfrac{1}{2\pi}\int_{-\infty}^{\infty} |F(\omega)|^2\,d\omega$

2.2.1 Laplace Transform

Given the real-valued function $f(t)$, the *two-sided* (or *bilateral*) Laplace transform $F(s)$ of $f(t)$ is defined by

$$F(s) = \int_{-\infty}^{\infty} f(t)e^{-st}\,dt \tag{2.14}$$

TABLE 2.2   Common FTs

$f(t)$	$F(\omega)$
$\delta(t) = $ unit impulse	1
$c, -\infty < t < \infty$	$2\pi c\delta(\omega)$
$f(t) = \begin{cases} 1, & -T/2 \leq t \leq T/2 \\ 0, & \text{all other } t \end{cases}$	$\dfrac{2}{\omega}\sin\left(\dfrac{T\omega}{2}\right)$
$\dfrac{\sin(at)}{t}$	$\begin{cases} \pi, & -a < \omega < a \\ 0, & \text{all other } \omega \end{cases}$
$e^{-b|t|}$, any $b > 0$	$\dfrac{2b}{\omega^2 + b^2}$
e^{-bt^2} any $b > 0$	$\sqrt{\dfrac{\pi}{b}}\, e^{-\omega^2/4b}$
$e^{j\omega_0 t}$	$2\pi\delta(\omega - \omega_0)$
$\cos(\omega_0 t + \theta)$	$\pi[e^{-j\theta}\delta(\omega + \omega_0) + e^{j\theta}\delta(\omega - \omega_0)]$
$\sin(\omega_0 t + \theta)$	$j\pi[e^{-j\theta}\delta(\omega + \omega_0) - e^{j\theta}\delta(\omega - \omega_0)]$

where s is a complex variable. The *one-sided* (or *unilateral*) Laplace transform of $f(t)$ is defined by

$$F(s) = \int_0^\infty f(t)e^{-st}\,dt \tag{2.15}$$

Note that if $f(t) = 0$ for all $t < 0$, the one-sided and two-sided Laplace transforms of $f(t)$ are identical. In controls engineering, the one-sided Laplace transform is primarily used, and thus our presentation focuses on only the one-sided Laplace transform, which is referred to as the Laplace transform.

Given a function $f(t)$, the set of all complex numbers s such that the integral in Equation 2.15 exists is called the *region of convergence* of the Laplace transform of $f(t)$. For example, if $f(t)$ is the unit-step function $u(t)$ given by $u(t) = 1$ for $t \geq 0$ and $u(t) = 0$ for $t < 0$, the integral in Equation 2.15 exists for any $s = \sigma + j\omega$ with real part $\sigma > 0$. Hence, the region of convergence is the set of all complex numbers s with positive real part, and, for any such s, the transform of the unit-step function $u(t)$ is equal to $1/s$.

Given a function $f(t)$, if the region of convergence of the Laplace transform $F(s)$ includes all complex numbers $s = j\omega$ for ω ranging from $-\infty$ to ∞, then $F(j\omega) = F(s)|_{s=j\omega}$ is well defined (i.e., exists) and is given by

$$F(j\omega) = \int_0^\infty f(t)e^{-j\omega t}\,dt \tag{2.16}$$

Then if $f(t) = 0$ for $t < 0$, the right-hand side of Equation 2.16 is equal to the FT $F(\omega)$ of $f(t)$ (see Equation 2.1). Hence, the FT of $f(t)$ is given by

$$F(\omega) = F(s)|_{s=j\omega} \tag{2.17}$$

(Note that we are denoting $F(j\omega)$ by $F(\omega)$.) This fundamental result shows that the FT of $f(t)$ can be computed directly from the Laplace transform $F(s)$ if $f(t) = 0$ for $t < 0$ and the region of convergence includes the imaginary axis of the complex plane (all complex numbers equal to $j\omega$).

The Laplace transform defined by Equation 2.15 can be viewed as an operator, denoted by $F(s) = L[f(t)]$ that maps a time function $f(t)$ into the function $F(s)$ of the complex variable s. The inverse Laplace transform operator is often denoted by L^{-1}, and is given by

$$f(t) = L^{-1}[F(s)] = \frac{1}{2\pi j}\int_{c-j\infty}^{c+j\infty} X(s)e^{st}\,ds \tag{2.18}$$

The integral in Equation 2.18 is evaluated along the path $s = c + j\omega$ in the complex plane from $c - j\infty$ to $c + j\infty$, where c is any real number for which the path $c + j\omega$ lies in the region of convergence of the transform $F(s)$. It is often possible to determine $f(t)$ without having to use Equation 2.18; for example, this is the case when $F(s)$ is a rational function of s. The computation of the Laplace transform or the inverse transform is often facilitated by using the properties of the Laplace transform, which are listed in Table 2.3. In this table, $f(t)$ and $g(t)$ are two functions with Laplace transforms $F(s)$ and $G(s)$, respectively, and $u(t)$ is the unit-step function defined by $u(t) = 1$ for $t \geq 0$ and $u(t) = 0$ for $t < 0$. Using the properties in Table 2.3, it is possible to determine the Laplace transform of many common functions without having to use Equation 2.15. A list of common Laplace transforms is given in Table 2.4.

2.2.2 Rational Laplace Transforms

The Laplace transform $F(s)$ of a function $f(t)$ is said to be a *rational function* of s if it can be written as a ratio of polynomials in s; that is,

$$F(s) = \frac{N(s)}{D(s)}, \tag{2.19}$$

where $N(s)$ and $D(s)$ are polynomials in the complex variable s given by

$$N(s) = b_m s^m + b_{m-1} s^{m-1} + \cdots + b_1 s + b_0 \tag{2.20}$$

$$D(s) = s^n + a_{n-1} s^{n-1} + \cdots + a_1 s + a_0 \tag{2.21}$$

TABLE 2.3 Properties of the (One-Sided) Laplace Transform

Property	Transform/Property
Linearity	$L[af(t) + bg(t)] = aF(s) + bG(s)$ for any scalars a, b
Right shift in t	$L[f(t - t_0)u(t - t_0)] = F(s)\exp(-st_0)$ for any $t_0 > 0$
Time scaling	$L[f(at)] = (1/a)F(s/a)$ for any real number $a > 0$
Multiplication by a power of t	$L[t^n f(t)] = (-1)^n \dfrac{d^n}{ds^n} F(s), n = 1, 2, \ldots$
Multiplication by $e^{\alpha t}$	$L[f(t)e^{\alpha t}] = F(s - \alpha)$ for any real or complex number α
Multiplication by $\sin(\omega_0 t)$	$L[f(t)\sin(\omega_0 t)] = (j/2)[F(s + j\omega_0) - F(s - j\omega_0)]$
Multiplication by $\cos(\omega_0 t)$	$L[f(t)\cos(\omega_0 t)] = (1/2)[F(s + j\omega_0) + F(s - j\omega_0)]$
Differentiation in the time domain	$L\left[\dfrac{d}{dt}f(t)\right] = sF(s) - f(0)$
Second derivative	$L\left[\dfrac{d^2}{dt^2}f(t)\right] = s^2 F(s) - sf(0) - \dfrac{d}{dt}f(0)$
nth derivative	$L\left[\dfrac{d^n}{dt^n}f(t)\right] = s^n F(s) - s^{n-1}f(0) - s^{n-2}\dfrac{d}{dt}f(0) - \cdots - \dfrac{d^{n-1}}{dt^{n-1}}f(0)$
Integration	$L\left[\displaystyle\int_0^t f(\tau)d\tau\right] = \dfrac{1}{s}F(s)$
Convolution in the time domain	$L[f(t) * g(t)] = F(s)G(s)$
Initial-value theorem	$f(0) = \lim\limits_{s \to \infty} sF(s)$
Final-value theorem	If $f(t)$ has a finite limit $f(\infty)$ as $t \to \infty$, then $f(\infty) = \lim\limits_{s \to 0} sF(s)$

In Equations 2.20 and 2.21, m and n are positive integers and the coefficients $b_m, b_{m-1}, \ldots, b_1, b_0$ and $a_{n-1}, \ldots, a_1, a_0$ are real numbers. In Equation 2.19, it is assumed that $N(s)$ and $D(s)$ do not have any common factors. If there are common factors, they should always be cancelled. Also note that the polynomial $D(s)$ is monic; that is, the coefficient of s^n is equal to 1. A rational function $F(s)$ can always be written with a monic denominator polynomial $D(s)$. The integer n, which is the degree of $D(s)$, is called the order of the rational function $F(s)$. It is assumed that $n \geq m$, in which case $F(s)$ is said to be a *proper rational function*. If $n > m$, $F(s)$ is said to be *strictly proper*.

Given a rational transform $F(s) = N(s)/D(s)$ with $N(s)$ and $D(s)$ defined by Equations 2.20 and 2.21, let $z_1, z_2, \ldots, z_m$ denote the roots of the polynomial $N(s)$, and let $p_1, p_2, \ldots, p_n$ denote the roots of $D(s)$; that is, $N(z_i) = 0$ for $i = 1, 2, \ldots, m$ and $D(p_i) = 0$ for $i = 1, 2, \ldots, n$. In general, z_i and p_i may be real or complex numbers, but if any are complex, they must appear in complex conjugate pairs. The numbers $z_1, z_2, \ldots, z_m$ are called the *zeros* of the rational function $F(s)$ since $F(s) = 0$ when $s = z_i$ for $i = 1, 2, \ldots, m$; and the numbers $p_1, p_2, \ldots, p_n$ are called the *poles* of $F(s)$ since the magnitude $|F(s)|$ becomes infinite as s approaches p_i for $i = 1, 2, \ldots, n$.

If $F(s)$ is strictly proper ($n > m$) and the poles $p_1, p_2, \ldots, p_n$ of $F(s)$ are distinct (nonrepeated), then $F(s)$ has the partial fraction expansion

$$F(s) = \frac{c_1}{s - p_1} + \frac{c_2}{s - p_2} + \cdots + \frac{c_n}{s - p_n}, \tag{2.22}$$

where the c_i are the *residues* given by

$$c_i = [(s - p_i)F(s)]_{s=p_i}, \quad i = 1, 2, \ldots, n \tag{2.23}$$

For a given value of i, the residue c_i is real if and only if the corresponding pole p_i is real, and c_i is complex if and only if p_i is complex.

TABLE 2.4   Common Laplace Transforms

$f(t)$	Laplace Transform $F(s)$
$u(t) =$ unit-step function	$\dfrac{1}{s}$
$u(t) - u(t - T)$ for any $T > 0$	$\dfrac{1 - e^{-Ts}}{s}$
$\delta(t) =$ unit impulse	1
$\delta(t - t_0)$ for any $t_0 > 0$	$e^{-t_0 s}$
$t^n, t \geq 0$	$\dfrac{n!}{s^{n+1}}, n = 1, 2, \ldots$
e^{-at}	$\dfrac{1}{s + a}$
$t^n e^{-at}$	$\dfrac{n!}{(s + a)^{n+1}}, n = 1, 2, \ldots$
$\cos(\omega t)$	$\dfrac{s}{s^2 + \omega^2}$
$\sin(\omega t)$	$\dfrac{\omega}{s^2 + \omega^2}$
$\cos^2 \omega t$	$\dfrac{s^2 + 2\omega^2}{s(s^2 + 4\omega^2)}$
$\sin^2 \omega t$	$\dfrac{2\omega^2}{s(s^2 + 4\omega^2)}$
$\sinh(at)$	$\dfrac{a}{s^2 - a^2}$
$\cosh(at)$	$\dfrac{s}{s^2 - a^2}$
$e^{-at} \cos(\omega t)$	$\dfrac{s + a}{(s + a)^2 + \omega^2}$
$e^{-at} \sin(\omega t)$	$\dfrac{\omega}{(s + a)^2 + \omega^2}$
$t \cos(\omega t)$	$\dfrac{s^2 - \omega^2}{(s^2 + \omega^2)^2}$
$t \sin(\omega t)$	$\dfrac{2\omega s}{(s^2 + \omega^2)^2}$
$te^{-at} \cos(\omega t)$	$\dfrac{(s + a)^2 - \omega^2}{\left[(s + a)^2 + \omega^2\right]^2}$
$te^{-at} \sin(\omega t)$	$\dfrac{2\omega(s + a)}{\left[(s + a)^2 + \omega^2\right]^2}$

From Equation 2.22 we see that the inverse Laplace transform $f(t)$ is given by the following sum of exponential functions:

$$f(t) = c_1 e^{p_1 t} + c_2 e^{p_2 t} + \cdots + c_n e^{p_n t} \tag{2.24}$$

If all the poles $p_1, p_2, \ldots, p_n$ of $F(s)$ are real numbers, then $f(t)$ is a sum of real exponentials given by Equation 2.24. If $F(s)$ has a pair of complex poles $p = \sigma \pm j\omega$, then $f(t)$ contains the term

$$ce^{(\sigma + j\omega)t} + \bar{c}e^{(\sigma - j\omega)t} \tag{2.25}$$

where $\bar{c}$ is the complex conjugate of c. Then writing c in the polar form $c = |c|e^{j\theta}$, we have

$$
\begin{aligned}
ce^{(\sigma+j\omega)t} + \bar{c}e^{(\sigma-j\omega)t} &= |c|e^{j\theta}e^{(\sigma+j\omega)t} + |\bar{c}|e^{-j\theta}e^{(\sigma-j\omega)t} \\
&= |c|e^{\sigma t}\left[e^{j(\omega t+\theta)} + e^{-j(\omega t+\theta)} \right]
\end{aligned}
\tag{2.26}
$$

Finally, using Euler's formula, Equation 2.26 can be written in the form

$$
ce^{(\sigma+j\omega)t} + \bar{c}e^{(\sigma-j\omega)t} = 2|c|e^{\sigma t}\cos(\omega t + \theta)
\tag{2.27}
$$

From Equation 2.27 it is seen that if $F(s)$ has a pair of complex poles, then $f(t)$ contains a sinusoidal term with an exponential amplitude factor $e^{\sigma t}$. Note that if $\sigma = 0$ (so that the poles are purely imaginary), Equation 2.27 is purely sinusoidal.

If one of the poles (say p_1) is repeated r times and the other $n - r$ poles are distinct, $F(s)$ has the partial fraction expansion

$$
F(s) = \frac{c_1}{s - p_1} + \frac{c_2}{(s - p_1)^2} + \cdots + \frac{c_r}{(s - p_1)^r} + \frac{c_{r+1}}{s - p_{r+1}} + \cdots + \frac{c_n}{s - p_n}
\tag{2.28}
$$

In Equation 2.28, the residues $c_{r+1}, c_{r+2}, \ldots, c_n$ are calculated as in the distinct-pole case; that is,

$$
c_i = [(s - p_i)F(s)]_{s=p_i}, \quad i = r+1, r+2, \ldots, n
\tag{2.29}
$$

and the residues $c_1, c_2, \ldots, c_r$ are given by

$$
c_i = \frac{1}{(r - i)!} \left\{ \frac{d^{r-i}}{ds^{r-i}} \left[(s - p_1)^r F(s) \right] \right\}_{s=p_1}
\tag{2.30}
$$

Then, taking the inverse transform of Equation 2.28 yields

$$
f(t) = c_1 e^{p_1 t} + c_2 t e^{p_1 t} + \cdots + \frac{c_r}{(r-1)!} t^{r-1} e^{p_1 t} + c_{r+1} e^{p_{r+1} t} + \cdots + c_n e^{p_n t}
\tag{2.31}
$$

The above results reveal that the analytical form of the function $f(t)$ depends directly on the poles of $F(s)$. In particular, if $F(s)$ has a nonrepeated real pole p, then $f(t)$ contains a real exponential term of the form ce^{pt} for some real constant c. If a real pole p is repeated r times, then $f(t)$ contains terms of the form $c_1 e^{pt}, c_2 t e^{pt}, \ldots, c_r t^{r-1} e^{pt}$ for some real constants $c_1, c_2, \ldots, c_r$. If $F(s)$ has a nonrepeated complex pair $\sigma \pm j\omega$ of poles, then $f(t)$ contains a term of the form $ce^{\sigma t} \cos(\omega t + \theta)$ for some real constants c and θ. If the complex pair $\sigma \pm j\omega$ is repeated r times, $f(t)$ contains terms of the form $c_1 e^{\sigma t} \cos(\omega t + \theta_1), c_2 t e^{\sigma t} \cos(\omega t + \theta_2), \ldots, c_r t^{r-1} e^{\sigma t} \cos(\omega t + \theta_r)$ for some real constants $c_1, c_2, \ldots, c_r$ and $\theta_1, \theta_2, \ldots, \theta_r$. These results are summarized in Table 2.5.

TABLE 2.5 Relationship between the Poles of $F(s)$ and the Form of $f(t)$

Pole Locations of $F(s)$	Corresponding Terms in $f(t)$
Nonrepeated real pole at $s = p$	ce^{pt}
Real pole at $s = p$ repeated r times	$\sum_{i=1}^{r} c_i t^{i-1} e^{pt}$
Nonrepeated complex pair at $s = \sigma \pm j\omega$	$ce^{\sigma t} \cos(\omega t + \theta)$
Complex pair at $s = \sigma \pm j\omega$ repeated r times	$\sum_{i=1}^{r} c_i t^{i-1} e^{\sigma t} \cos(\omega t + \theta_i)$

If $F(s)$ is proper, but not strictly proper (so that $n = m$ in Equations 2.20 and 2.21), then using long division $F(s)$ can be written in the form

$$F(s) = b_n + \frac{R(s)}{D(s)} \qquad (2.32)$$

where the degree of $R(s)$ is strictly less than n. Then $R(s)/D(s)$ can be expanded via partial fractions as was done in the case when $F(s)$ is strictly proper. Note that for $F(s)$ given by Equation 2.32, the inverse Laplace transform $f(t)$ contains the impulse $b_n \delta(t)$. Hence, having $n = m$ in $F(s)$ results in an impulsive term in the inverse transform.

From the relationship between the poles of $F(s)$ and the analytical form of $f(t)$, it follows that $f(t)$ converges to zero as $t \to \infty$ if and only if all the poles $p_1, p_2, \ldots, p_n$ of $F(s)$ have real parts that are strictly less than zero; that is, $Re(p_i) < 0$ for $i = 1, 2, \ldots, n$. This condition is equivalent to requiring that all the poles be located in the *open left half-plane (OLHP)*, which is the region in the complex plane to the left of the imaginary axis.

It also follows from the relationship between the poles of $F(s)$ and the form of $f(t)$ that $f(t)$ has a finite limit $f(\infty)$ as $t \to \infty$ if and only if all the poles of $F(s)$ have real parts that are less than zero, except that $F(s)$ may have a nonrepeated pole at $s = 0$. In mathematical terms, the conditions for the existence of a finite limit $f(\infty)$ are

$$Re(p_i) < 0 \quad \text{for all poles } p_i \neq 0 \qquad (2.33)$$
$$\text{If } p_i = 0 \quad \text{is a pole of } F(s), \text{then } p_i \text{ is nonrepeated} \qquad (2.34)$$

If the conditions in Equations 2.33 and 2.34 are satisfied, the limiting value $f(\infty)$ is given by

$$f(\infty) = [sF(s)]_{s=0} \qquad (2.35)$$

The relationship in Equation 2.35 is a restatement of the final-value theorem (given in Table 2.3) in the case when $F(s)$ is rational and the poles of $F(s)$ satisfy the conditions in Equations 2.33 and 2.34.

2.2.3 Irrational Transforms

The Laplace transform $F(s)$ of a function $f(t)$ is said to be an *irrational function* of s if it is not rational; that is, $F(s)$ cannot be expressed as a ratio of polynomials in s. For example, $F(s) = e^{-t_0 s}/s$ is irrational since the exponential function $e^{-t_0 s}$ cannot be expressed as a ratio of polynomials in s. In this case, the inverse transform $f(t)$ is equal to $u(t - t_0)$, where $u(t)$ is the unit-step function.

Given any function $f(t)$ with transform $F(s)$ and given any real number $t_0 > 0$, the transform of the time-shifted (or time-delayed) function $f(t - t_0)u(t - t_0)$ is equal to $F(s)e^{-t_0 s}$. Time-delayed signals arise in systems with time delays, and thus irrational transforms appear in the study of systems with time delays. Also, any function $f(t)$ that is of finite duration in time has a transform $F(s)$ that is irrational. For instance, suppose that

$$f(t) = \gamma(t)[u(t - t_0) - u(t - t_1)], \quad 0 \le t_0 < t_1 \qquad (2.36)$$

so that $f(t) = \gamma(t)$ for $t_0 \le t < t_1$, and $f(t) = 0$ for all other t. Then $f(t)$ can be written in the form

$$f(t) = \gamma_0(t - t_0)u(t - t_0) - \gamma_1(t - t_1)u(t - t_1) \qquad (2.37)$$

where $\gamma_0(t) = \gamma(t + t_0)$ and $\gamma_1(t) = \gamma(t + t_1)$. Taking the Laplace transform of Equation 2.37 yields

$$F(s) = \Gamma_0(s)e^{-t_0 s} - \Gamma_1(s)e^{-t_1 s} \qquad (2.38)$$

where $\Gamma_0(s)$ and $\Gamma_1(s)$ are the transforms of $\gamma_0(t)$ and $\gamma_1(t)$, respectively. Note that by Equation 2.38, the transform $F(s)$ is an irrational function of s.

To illustrate the above constructions, suppose that

$$f(t) = e^{-at}[u(t-1) - u(t-2)] \tag{2.39}$$

Writing $f(t)$ in the form of Equation 2.37 gives

$$f(t) = e^{-a}e^{-a(t-1)}u(t-1) - e^{-2a}e^{-a(t-2)}u(t-2) \tag{2.40}$$

Then, transforming Equation 2.40 yields

$$F(s) = [e^{-(s+a)} - e^{-2(s+a)}]\frac{1}{s+a} \tag{2.41}$$

Clearly, $F(s)$ is an irrational function of s.

2.2.4 Discrete-Time FT

Let $f(k)$ be a real-valued function of the integer-valued variable k. The function $f(k)$ can be viewed as a discrete-time signal; in particular, $f(k)$ may be a sampled version of a continuous-time signal $f(t)$. More precisely, $f(k)$ may be equal to the sample values $f(kT)$ of a signal $f(t)$ with t evaluated at the sample times $t = kT$, where T is the sampling interval. In mathematical terms, the sampled signal is given by

$$f(k) = f(t)|_{t=kT} = f(kT), \quad k = 0, \pm1, \pm2, \ldots \tag{2.42}$$

Note that we are denoting $f(kT)$ by $f(k)$. The FT of a function $f(k)$ of an integer variable k is defined by

$$F(\Omega) = \sum_{k=-\infty}^{\infty} f(k)e^{-j\Omega k}, \quad -\infty < \Omega < \infty \tag{2.43}$$

where Ω is interpreted as the real frequency variable. The transform $F(\Omega)$ is called the DTFT since it can be viewed as the discrete-time counterpart of the FT defined above. The DTFT is directly analogous to the FT, so that all the properties of the FT discussed above carry over to the DTFT. In particular, as is the case for the FT, the DTFT $F(\Omega)$ is in general a complex-valued function of the frequency variable Ω, and thus $F(\Omega)$ must be specified in terms of a magnitude function $|F(\Omega)|$ and an angle function $\angle F(\Omega)$. The magnitude function $|F(\Omega)|$ (respectively, the angle function $\angle F(\Omega)$) displays the magnitude (respectively, the phase) of the frequency components comprising $f(k)$. All of the properties of the FT listed in Table 2.1 have a counterpart for the DTFT, but this will not be pursued here.

In contrast to the FT, the DTFT $F(\Omega)$ is always a periodic function of the frequency variable Ω with period 2π; that is,

$$F(\Omega + 2\pi) = F(\Omega) \quad \text{for} -\infty < \Omega < \infty \tag{2.44}$$

As a result of the periodicity property in Equation 2.44, it is necessary to specify $F(\Omega)$ over a 2π interval only, such as 0 to 2π or $-\pi$ to π. Given $F(\Omega)$ over any 2π interval, $f(k)$ can be recomputed using the inverse DTFT. In particular, if $F(\Omega)$ is specified over the interval $-\pi < \Omega < \pi$, $f(k)$ can be computed from the relationship

$$f(k) = \frac{1}{2\pi}\int_{-\pi}^{\pi} F(\Omega)e^{jk\Omega}\, d\Omega \tag{2.45}$$

In practice, the DTFT $F(\Omega)$ is usually computed only for a discrete set of values of the frequency variable Ω. This is accomplished by using the N-point discrete Fourier transform (N-point DFT) of $f(k)$

given by

$$F_n = \sum_{k=0}^{N-1} f(k)e^{-j2\pi kn/N}, \quad n = 0, 1, \ldots, N-1 \tag{2.46}$$

where N is a positive integer. If $f(k) = 0$ for $k < 0$ and $k \geq N$, comparing Equations 2.46 and 2.43 reveals that

$$F_n = F\left(\frac{2\pi n}{N}\right), \quad n = 0, 1, \ldots, N-1 \tag{2.47}$$

Hence, the DFT F_n is equal to the values of the DTFT $F(\Omega)$ with Ω evaluated at the discrete points $\Omega = 2\pi n/N$ for $n = 0, 1, 2, \ldots, N-1$.

The computation of the DFT F_n given by Equation 2.46 can be carried out using a fast algorithm called the Fast Fourier transform (FFT). The inverse FFT can be used to compute $f(k)$ from F_n. A development of the FFT is beyond the scope of this section (see "Further Reading").

2.2.5 z-Transform

Given the function $f(k)$, the *two-sided* (or *bilateral*) z-transform $F(z)$ of $f(k)$ is defined by

$$F(z) = \sum_{k=-\infty}^{\infty} f(k)z^{-k} \tag{2.48}$$

where z is a complex variable. The *one-sided* (or *unilateral*) z-transform of $f(k)$ is defined by

$$F(z) = \sum_{k=0}^{\infty} f(k)z^{-k} \tag{2.49}$$

Note that if $f(k) = 0$ for $k = -1, -2, \ldots$, the one-sided and two-sided z-transforms of $f(k)$ are the same. As is the case with the Laplace transform, in controls engineering the one-sided version is the most useful, and thus the development given below is restricted to the one-sided z-transform, which is referred to as the z-transform.

Given $f(k)$, the set of all complex numbers z such that the summation in Equation 2.49 exists is called the region of convergence of the z-transform of $f(k)$. If the region of convergence of the z-transform includes all complex numbers $z = e^{j\Omega}$ for Ω ranging from $-\infty$ to ∞, then $F(e^{j\Omega}) = F(z)|_{z=e^{j\Omega}}$ is well defined (i.e., exists) and is given by

$$F(e^{j\Omega}) = \sum_{k=0}^{\infty} f(k)(e^{j\Omega})^{-k} \tag{2.50}$$

But $(e^{j\Omega})^{-k} = e^{-j\Omega k}$, and thus Equation 2.50 can be rewritten as

$$F(e^{j\Omega}) = \sum_{k=0}^{\infty} f(k)e^{-j\Omega k} \tag{2.51}$$

Then if $f(k) = 0$ for all $k < 0$, the right-hand side of Equation 2.51 is equal to the DTFT $F(\Omega)$ of $f(k)$ (see Equation 2.43). Therefore, the DTFT of $f(k)$ is given by

$$F(\Omega) = F(z)|_{z=e^{j\Omega}} \tag{2.52}$$

This result shows that the DTFT of $f(k)$ can be computed directly from the z-transform $F(z)$ if $f(k) = 0$ for all $k < 0$ and the region of convergence includes all complex numbers $z = e^{j\Omega}$ with $-\infty < \Omega < \infty$. Note that since $|e^{j\Omega}| = 1$ for any value of Ω and $\angle e^{j\Omega} = \Omega$, the set of complex numbers $z = e^{j\Omega}$ comprises the unit circle of the complex plane. Hence, the DTFT of $f(k)$ is equal to the values of the z-transform on

TABLE 2.6   Properties of the (One-Sided) z-Transform

Property	Transform/Property
Linearity	$Z[af(k) + bg(k)] = aF(z) + bG(z)$ for any scalars a, b
Right shift of $f(k)u(k)$	$Z[f(k-q)u(k-q)] = z^{-q}F(z)$ for any integer $q \geq 1$
Right shift of $f(k)$	$Z[f(k-1)] = z^{-1}F(z) + f(-1)$
	$Z[f(k-2)] = z^{-2}F(z) + f(-2) + z^{-1}f(-1)$
	$\vdots$
	$Z[f(k-q)] = z^{-q}F(z) + f(-q) + z^{-1}f(-q+1) + \cdots + z^{-q+1}f(-1)$
Left shift in time	$Z[f(k+1)] = zF(z) - f(0)z$
	$Z[f(k+2)] = z^2F(z) - f(0)z^2 - f(1)z$
	$\vdots$
	$Z[f(k+q)] = z^qF(z) - f(0)z^q - f(1)z^{q-1} - \cdots - f(q-1)z$
Multiplication by k	$Z[kf(k)] = -z\dfrac{d}{dz}F(z)$
Multiplication by k^2	$Z[k^2f(k)] = z\dfrac{d}{dz}F(z) + z^2\dfrac{d^2}{dz^2}F(z)$
Multiplication by a^k	$Z[a^kf(k)] = F\left(\dfrac{z}{a}\right)$
Multiplication by $\cos(\Omega k)$	$Z[\cos(\Omega k)f(k)] = \dfrac{1}{2}[F(e^{j\Omega}z) + F(e^{-j\Omega}z)]$
Multiplication by $\sin(\Omega k)$	$Z[\sin(\Omega k)f(k)] = \dfrac{j}{2}[F(e^{j\Omega}z) - F(e^{-j\Omega}z)]$
Summation	$\left[\displaystyle\sum_{i=0}^{k} f(i)\right] = \dfrac{z}{z-1}F(z)$
Convolution	$Z[f(k) * g(k)] = F(z)G(z)$
Initial-value theorem	$f(0) = \displaystyle\lim_{z\to\infty} F(z)$
Final-value theorem	If $f(k)$ has a finite limit $f(\infty)$ as $k \to \infty$, then $f(\infty) = \displaystyle\lim_{z\to1}(z-1)F(z)$

the unit circle of the complex plane, assuming that the region of convergence of $F(z)$ includes the unit circle.

The z-transform defined by Equation 2.49 can be viewed as an operator, denoted by $F(z) = Z[f(k)]$, that maps a discrete-time function $f(k)$ into the function $F(z)$ of the complex variable z. The inverse z-transform operation is denoted by $f(k) = Z^{-1}[F(z)]$. As discussed below, when $F(z)$ is a rational function of z, the inverse transform can be computed using long division or by carrying out a partial fraction expansion of $F(z)$. The computation of the z-transform or the inverse z-transform is often facilitated by using the properties of the z-transform given in Table 2.6. In this table, $f(k)$ and $g(k)$ are two functions with z-transforms $F(z)$ and $G(z)$, respectively, and $u(k)$ is the unit-step function defined by $u(k) = 1$ for $k \geq 0$ and $u(k) = 0$ for $k < 0$. A list of common z-transforms is given in Table 2.7. In Table 2.7, the function $\delta(k)$ is the unit pulse defined by $\delta(0) = 1$, $\delta(k) = 0$ for $k \neq 0$.

2.2.6 Rational z-Transforms

As is the case for the Laplace transform, the z-transform $F(z)$ is often a rational function of z; that is, $F(z)$ is given by

$$F(z) = \frac{N(z)}{D(z)} \tag{2.53}$$

TABLE 2.7 Common z-Transform Pairs

$f(k)$	z-Transform $F(z)$
$u(k) =$ unit-step function	$\dfrac{z}{z-1}$
$u(k) - u(k-N), N = 1, 2, \ldots$	$\dfrac{z^N - 1}{z^{N-1}(z-1)}, N = 1, 2, \ldots$
$\delta(k) =$ unit pulse	1
$\delta(k-q), q = 1, 2, \ldots$	$\dfrac{1}{z^q}, q = 1, 2, \ldots$
a^k, a real or complex	$\dfrac{z}{z-a}$
k	$\dfrac{z}{(z-1)^2}$
$k+1$	$\dfrac{z^2}{(z-1)^2}$
k^2	$\dfrac{z(z+1)}{(z-1)^3}$
ka^k	$\dfrac{az}{(z-a)^2}$
$k^2 a^k$	$\dfrac{az(z+a)}{(z-a)^3}$
$k(k+1)a^k$	$\dfrac{2az^2}{(z-a)^3}$
$\cos(\Omega k)$	$\dfrac{z^2 - (\cos \Omega)z}{z^2 - (2\cos \Omega)z + 1}$
$\sin(\Omega k)$	$\dfrac{(\sin \Omega)z}{z^2 - (2\cos \Omega)z + 1}$
$a^k \cos(\Omega k)$	$\dfrac{z^2 - (a\cos \Omega)z}{z^2 - (2a\cos \Omega)z + a^2}$
$a^k \sin(\Omega k)$	$\dfrac{(a\sin \Omega)z}{z^2 - (2a\cos \Omega)z + a^2}$

where $N(z)$ and $D(z)$ are polynomials in the complex variable z given by

$$N(z) = b_m z^m + b_{m-1} z^{m-1} + \cdots + b_1 z + b_0 \tag{2.54}$$

$$D(z) = z^n + a_{n-1} z^{n-1} + \cdots + a_1 z + a_0 \tag{2.55}$$

It is assumed that the order n of $F(z)$ is greater than or equal to m, and thus $F(z)$ is proper. The poles and zeros of $F(z)$ are defined in the same way as given above for rational Laplace transforms.

When the transform $F(z)$ is in the rational form in Equation 2.53, the inverse z-transform $f(k)$ can be computed by expanding $F(z)$ into a power series in z^{-1} by dividing $D(z)$ into $N(z)$ using long division. The values of the function $f(k)$ are then read off from the coefficients of the power series expansion. The first few steps of the process are carried out below:

$$
z^n + a_{n-1}z^{n-1} + \cdots + a_1 z + a_0 \overline{\big)} \begin{array}{l} b_m z^{m-n} + (b_{m-1} - a_{n-1}b_m)z^{m-n-1} + \cdots \\ \hline b_m z^m + b_{m-1}z^{m-1} \qquad\qquad\qquad\qquad\quad +\cdots \\ b_m z^m + a_{n-1}b_m z^{m-1} \qquad\qquad\qquad\qquad +\cdots \\ \hline (b_{m-1} - a_{n-1}b_m)z^{m-1} \qquad\qquad\qquad +\cdots \\ (b_{m-1} - a_{n-1}b_m)z^{m-1} \qquad\qquad\qquad +\cdots \\ \hline \vdots \end{array}
$$

Since the value of $f(k)$ is equal to the coefficient of z^{-k} in the power series expansion of $F(z)$, it follows from the above division process that $f(n-m) = b_m, f(n-m+1) = b_{m-1} - a_{n-1}b_m$, and so on.

To express the inverse z-transform $f(k)$ in closed form, it is necessary to expand $F(z)$ via partial fractions. It turns out that the form of the inverse z-transform $f(k)$ is simplified if $F(z)/z = N(z)/D(z)z$ is expanded by partial fractions. Note that $F(z)/z$ is strictly proper since $F(z)$ is assumed to be proper.

Letting $p_1, p_2, \ldots, p_n$ denote the poles of $F(z)$, if the p_i are distinct and are nonzero, then $F(z)/z$ has the partial fraction expansion

$$
\frac{F(z)}{z} = \frac{c_0}{z} + \frac{c_1}{z - p_1} + \frac{c_2}{z - p_2} + \cdots + \frac{c_n}{z - p_n} \tag{2.56}
$$

where the residues are given by

$$
c_0 = F(0) \tag{2.57}
$$

$$
c_i = \left[(z - p_i)\frac{F(z)}{z} \right]_{z=p_i}, \quad i = 1, 2, \ldots, n \tag{2.58}
$$

Then multiplying both sides of Equation 2.56 by z gives

$$
F(z) = c_0 + \frac{c_1 z}{z - p_1} + \frac{c_2 z}{z - p_2} + \cdots + \frac{c_n z}{z - p_n} \tag{2.59}
$$

and taking the inverse z-transform gives

$$
f(k) = c_0\delta(k) + c_1 p_1^k + c_2 p_2^k + \cdots + c_n p_n^k \tag{2.60}
$$

If the poles $p_1, p_2, \ldots, p_n$ of $F(z)$ are real numbers, then from Equation 2.60 it is seen that $f(k)$ is the sum of geometric functions of the form cp^k, plus a pulse function $c_0\delta(k)$ if $c_0 \neq 0$. If $F(z)$ has a pair of complex poles given in polar form by $\sigma e^{\pm j\Omega}$, then it can be shown that $f(k)$ contains a sinusoidal term of the form

$$
c\sigma^k \cos(\Omega k + \theta) \tag{2.61}
$$

for some constants c and θ.

If $F(z)$ has a real pole p that is repeated r times, then $f(k)$ contains the terms $c_1 p^k, c_2 k p^k, \ldots, c_r k^{r-1}p^k$; and if $F(z)$ has a pair of complex poles given by $\sigma e^{\pm j\Omega}$ that is repeated r times, then $f(k)$ contains terms of the form $c_1\sigma^k \cos(\Omega k + \theta_1), c_2 k\sigma^k \cos(\Omega k + \theta_2), \ldots, c_r k^{r-1}\sigma^k \cos(\Omega k + \theta_r)$ for some constants $c_1, c_2, \ldots, c_r$ and $\theta_1, \theta_2, \ldots, \theta_r$. These results are summarized in Table 2.8.

From the relationship between the poles of $F(z)$ and the analytical form of $f(k)$, it follows that $f(k)$ converges to zero as $k \to \infty$ if and only if all the poles $p_1, p_2, \ldots, p_n$ of $F(z)$ have magnitudes that are strictly less than one; that is, $|p_i| < 1$ for $i = 1, 2, \ldots, n$. This is equivalent to requiring that all the poles be located inside the *open unit disk* of the complex plane, which is the region of the complex plane consisting of all complex numbers whose magnitude is strictly less than one.

It also follows from the relationship between pole locations and the form of the function that $f(k)$ has a finite limit $f(\infty)$ as $k \to \infty$ if and only if all the poles of $F(z)$ have magnitudes that are less than one,

TABLE 2.8 Relationship between the Poles of $F(z)$ and the Form of $f(k)$

Pole Locations of $F(z)$	Corresponding Terms in $f(k)$
Nonrepeated real pole at $z = p$	cp^k
Real pole at $z = p$ repeated r times	$\sum\limits_{i=1}^{r} c_i k^{i-1} p^k$
Nonrepeated complex pair at $z = \sigma e^{\pm j\Omega}$	$c\sigma^k \cos(\Omega k + \theta)$
Complex pair at $z = \sigma e^{\pm j\Omega}$ repeated r times	$\sum\limits_{i=1}^{r} c_i k^{i-1} \sigma^k \cos(\Omega k + \theta_i)$

except that $F(z)$ may have a nonrepeated pole at $z = 1$. In mathematical terms, the conditions for the existence of a finite limit $f(\infty)$ are:

$$|p_i| < 1 \quad \text{for all poles } p_i \neq 1 \tag{2.62}$$

$$\text{If } p_i = 1 \quad \text{is a pole of } F(z), \text{ then } p_i \text{ is nonrepeated} \tag{2.63}$$

If the conditions in Equations 2.62 and 2.63 are satisfied, the limiting value $f(\infty)$ is given by

$$f(\infty) = [(z - 1)F(z)]_{z=1} \tag{2.64}$$

The relationship in Equation 2.64 is a restatement of the final-value theorem (given in Table 2.6) in the case when $F(z)$ is rational and the poles of $F(z)$ satisfy the conditions in Equations 2.62 and 2.63.

2.3 Applications and Examples

Given a real-valued signal $f(t)$ of the continuous-time variable t, the FT $F(\omega)$ reveals the frequency spectrum of $f(t)$; in particular, the plot of $|F(\omega)|$ versus ω is the magnitude spectrum of $f(t)$, and the plot of $\angle F(\omega)$ versus ω is the phase spectrum of $f(t)$. The magnitude function $|F(\omega)|$ is sometimes given in decibels (dB) defined by

$$|F(\omega)|_{dB} = 20\log_{10}|F(\omega)| \tag{2.65}$$

Given a signal $f(t)$ with FT $F(\omega)$, if there exists a positive number B such that $|F(\omega)|$ is zero (or approximately zero) for all $\omega > B$, the signal $f(t)$ is said to be *band limited* or to have a finite bandwidth; that is, the frequencies comprising $f(t)$ are limited (for the most part) to a finite range from 0 to B rad/s. The *3-dB bandwidth* of such a signal is the smallest positive value B_{3dB} such that

$$|F(\omega)| \leq .707F_{\max} \quad \text{for all } \omega > B_{3dB} \tag{2.66}$$

where $F_{\max}$ is the maximum value of $|F(\omega)|$. The inequality in Equation 2.66 is equivalent to requiring that the magnitude $|F(\omega)|_{dB}$ in decibels be down from its peak value by 3dB or more. For example, suppose that $f(t)$ is the T-second rectangular pulse defined by

$$f(t) = \begin{cases} 1, & -T/2 \leq t \leq T/2 \\ 0, & \text{all other } t \end{cases}$$

From Table 2.2, the FT is

$$F(\omega) = \frac{2}{\omega} \sin\left(\frac{T\omega}{2}\right) \tag{2.67}$$

Note that by l'Hôpital's rule, $F(0) = T$. A plot of $|F(\omega)|$ vs. ω is given in Figure 2.1.

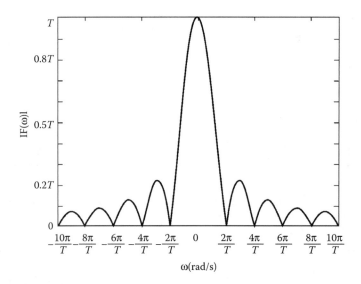

FIGURE 2.1 Magnitude spectrum of the T-second rectangular pulse.

From Figure 2.1, it is seen that most of the frequency content of the rectangular pulse is contained in the main lobe, which runs from $-2\pi/T$ to $2\pi/T$ rad/s. Also, the plot shows that there is no finite positive number B such that $|F(\omega)|$ is zero for all $\omega > B$. However, $|F(\omega)|$ is converging to zero as $\omega \to \infty$, and thus this signal can still be viewed as being bandlimited. Since the maximum value of $|F(\omega)|$ is $F_{\max} = T$, the 3-dB bandwidth of the T-second rectangular pulse is the smallest positive number $B_{3\mathrm{dB}}$ for which

$$\left| \frac{2}{\omega} \sin\left(\frac{T\omega}{2} \right) \right| \le .707T \quad \text{for all } \omega > B_{3\mathrm{dB}} \tag{2.68}$$

From Figure 2.1 it is clear that if the duration T of the rectangular pulse is decreased, the magnitude spectrum spreads out, and thus the 3-dB bandwidth increases. Hence, a shorter duration pulse has a wider 3-dB bandwidth. This result is true in general; that is, signals with shorter time durations have wider bandwidths than signals with longer time durations.

2.3.1 Spectrum of a Signal Having a Rational Laplace Transform

Now suppose that the signal $f(t)$ is zero for all $t < 0$, and that the Laplace transform $F(s)$ of $f(t)$ is rational in s; that is $F(s) = N(s)/D(s)$, where $N(s)$ and $D(s)$ are polynomials in s given by Equations 2.20 and 2.21. It was noted in the previous section that if the region of convergence of $F(s)$ includes the imaginary axis ($j\omega$-axis) of the complex plane, then the FT $F(\omega)$ is equal to the Laplace transform $F(s)$ with $s = j\omega$. When $F(s)$ is rational, it turns out that the region of convergence includes the $j\omega$-axis if and only if all the poles of $F(s)$ lie in the OLHP; thus, in this case, the FT is given by

$$F(\omega) = F(s)|_{s=j\omega} \tag{2.69}$$

For example, if $f(t) = ce^{-at}$ for $t \ge 0$ with $a > 0$, then $F(s) = c/(s+a)$ which has a single pole at $s = -a$. Since the point $-a$ lies in the OLHP, the FT of the exponential function ce^{-at} is

$$F(\omega) = \frac{c}{j\omega + a} \tag{2.70}$$

It follows from Equation 2.70 that the 3-dB bandwidth is equal to a. Hence, the farther over in the OLHP the pole $-a$ (i.e., the larger a), the larger the bandwidth of the signal. Since the rate of decay to zero of

ce^{-at} increases as a is increased, this result again confirms the property that shorter duration time signals have wider bandwidths. In the case when $c = a = 1$, a plot of the magnitude spectrum $|F(\omega)|$ is shown in Figure 2.2. For any real values of a and c, the magnitude spectrum rolls off to zero at the rate of 20 dB/decade where a decade is a factor of 10 in frequency.

As another example, consider the signal $f(t)$ whose Laplace transform is

$$F(s) = \frac{c}{s^2 + 2\zeta\omega_n s + \omega_n^2} \tag{2.71}$$

where ω_n is assumed to be strictly positive ($\omega_n > 0$). In this case, $F(s)$ has two poles p_1 and p_2 given by

$$\begin{aligned} p_1 &= -\zeta\omega_n + \omega_n\sqrt{\zeta^2 - 1} \\ p_2 &= -\zeta\omega_n - \omega_n\sqrt{\zeta^2 - 1} \end{aligned} \tag{2.72}$$

When $\zeta > 1$, both poles are real, nonrepeated, and lie in the OLHP (assuming that $\omega_n > 0$). As $\zeta \to \infty$, the pole p_1 moves along the negative real axis to the origin of the complex plane and the pole p_2 goes to $-\infty$ along the negative axis of the complex plane. For $\zeta > 1$, $F(s)$ can be expanded by partial fractions as follows:

$$F(s) = \frac{c}{(s - p_1)(s - p_2)} = \frac{c}{p_1 - p_2}\left[\frac{1}{s - p_1} - \frac{1}{s - p_2}\right] \tag{2.73}$$

Taking the inverse Laplace transform gives

$$f(t) = \frac{c}{p_1 - p_2}[e^{p_1 t} - e^{p_2 t}] \tag{2.74}$$

and thus $f(t)$ is a sum of two decaying real exponentials. Since both poles lie in the OLHP, the FT $F(\omega)$ is given by

$$F(\omega) = \frac{c}{\omega_n^2 - \omega^2 + j(2\zeta\omega_n\omega)} \tag{2.75}$$

For the case when $c = \omega_n^2 = 100$ and $\zeta = 2$, the plot of the magnitude spectrum $|F(\omega)|$ is given in Figure 2.3. In this case, the spectral content of the signal $f(t)$ rolls off to zero at the rate of 40dB/decade, starting with the peak magnitude of 1 at $\omega = 0$.

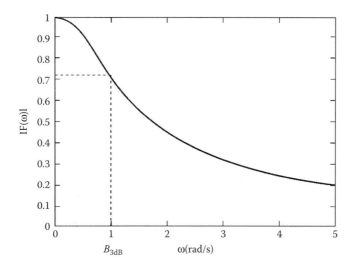

FIGURE 2.2 Magnitude spectrum of the exponential function e^{-t}.

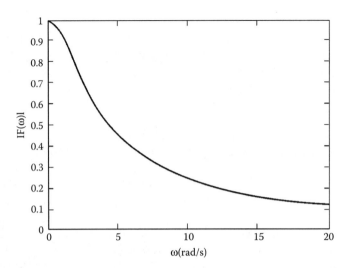

FIGURE 2.3 Magnitude spectrum of the signal with transform $F(s) = 100/(s^2 + 40s + 100)$.

When $\zeta = 1$, the poles p_1 and p_2 of $F(s)$ are both equal to $-\omega_n$, and $F(s)$ becomes

$$F(s) = \frac{c}{(s + \omega_n)^2} \tag{2.76}$$

Taking the inverse transform gives

$$f(t) = cte^{-\omega_n t} \tag{2.77}$$

Since ω_n is assumed to be strictly positive, when $\zeta = 1$ both the poles are in the OLHP; in this case, the FT is

$$F(\omega) = \frac{c}{(j\omega + \omega_n)^2} \tag{2.78}$$

As ζ varies from 1 to -1, the poles of $F(s)$ trace out a circle in the complex plane with radius ω_n. The loci of pole locations is shown in Figure 2.4. Note that the poles begin at $-\omega_n$ when $\zeta = 1$, then split apart and approach the $j\omega$-axis at $\pm j\omega_n$ as $\zeta \to 0$ and then move to ω_n as $\zeta \to -1$. For $-1 < \zeta < 1$, the inverse transform of $F(s)$ can be determined by first completing the square in the denominator of $F(s)$:

$$F(s) = \frac{c}{(s + \zeta\omega_n)^2 + \omega_d^2} \tag{2.79}$$

where

$$\omega_d = \omega_n \sqrt{1 - \zeta^2} > 0 \tag{2.80}$$

Note that ω_d is equal to the imaginary part of the pole p_1 given by Equation 2.72. Using Table 2.4, we have that the inverse transform of $F(s)$ is

$$f(t) = \frac{c}{\omega_d} e^{-\zeta\omega_n t} \sin \omega_d t \tag{2.81}$$

From Equation 2.81, it is seen that $f(t)$ now contains a sinusoidal factor. When $0 < \zeta < 1$, the poles lie in the OLHP, and the signal is a decaying sinusoid. In this case, the FT is

$$F(\omega) = \frac{c}{(j\omega + \zeta\omega_n)^2 + \omega_d^2} \tag{2.82}$$

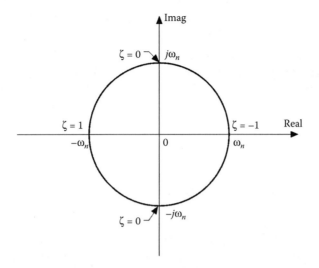

FIGURE 2.4 Loci of poles of $F(s)$ as ζ varies from 1 to -1.

The magnitude spectrum $|F(\omega)|$ is plotted in Figure 2.5 for the values $c = \omega_n^2 = 100$ and $\zeta = 0.3, 0.5, 0.7$. Note that for $\zeta = 0.5$ and 0.3, a peak appears in $|F(\omega)|$. This corresponds to the sinusoidal oscillation resulting from the $\sin(\omega_d t)$ factor in $f(t)$. Also note that as ζ is decreased from 0.5 to 0.3, the peak increases in magnitude, which signifies a longer duration oscillation in the signal $f(t)$. This result is expected since the poles are approaching the $j\omega$-axis of the complex plane as $\zeta \to 0$. As $\zeta \to 0$, the peak in $|F(\omega)|$ approaches ∞, so that $|F(\omega)|$ does not exist (in the ordinary sense) in the limit as $\zeta \to 0$. When $\zeta \to 0$, the signal $f(t)$ is purely oscillatory and does not have an FT in the ordinary sense. In addition, when $\zeta < 0$ there is no FT (in any sense) since there is a pole of $F(s)$ in the open right half-plane (ORHP).

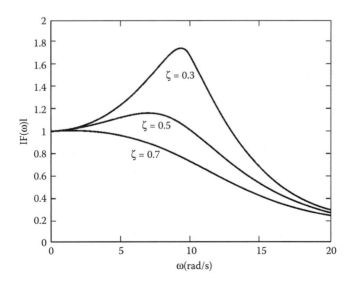

FIGURE 2.5 Magnitude spectrum of the signal with transform $F(s) = 100/(s^2 + 20\zeta s + 100)$ and with $\zeta = 0.7, 0.5, 0.3$.

The above results lead to the following generalized properties of the magnitude spectrum $|F(\omega)|$ of a signal $f(t)$ whose Laplace transform $F(s)$ is rational with all poles in the OLHP:

- If the poles of $F(s)$ are real, the magnitude spectrum $|F(\omega)|$ simply rolls off to zero as $\omega \to \infty$, starting with a peak value at $\omega = 0$ of $F(0)$.
- If $F(s)$ has a pair of complex conjugate poles at $s = \sigma \pm j\omega_d$ with the ratio σ/ω_d sufficiently small and $F(s)$ has no zeros located near the poles $\sigma \pm j\omega_d$ in the complex plane, then $|F(\omega)|$ will have a peak located approximately at $\omega = \omega_d$.

2.3.2 Numerical Computation of the FT

In many applications, the signal $f(t)$ cannot be given in function form; rather, all have a set of sample values $f(k) = f(kT)$, where k ranges over a subset of integers and T is the sampling interval. Without loss of generality, we can assume that k starts with $k = 0$. Also, since all signals arising in practice are of finite duration in time, we can assume that $f(k)$ is zero for all $k \geq N$ for some positive integer N. The problem is then to determine the FT of $f(t)$ using the sample values $f(k) = f(kT)$ for $k = 0, 1, 2, \ldots, N - 1$.

One could also carry out a discrete-time analysis by taking the N-point DFT F_n of the sampled signal $f(k)$. In the previous section, it was shown that F_n is equal to $F(\frac{2\pi n}{N})$ for $n = 0, 1, 2, \ldots, N - 1$, where $F(\frac{2\pi n}{N})$ is the DTFT $F(\Omega)$ of $f(k)$ with the frequency variable Ω evaluated at $2\pi n/N$. Hence the discrete-time counterpart of the frequency spectrum can be determined from F_n. For details on this, see "Further Reading."

Again letting $F(\omega)$ denote the FT of $f(t)$, we can carry out a numerical computation of the FT as follows. First, since $f(t)$ is zero for $t < 0$ and $t \geq NT$, from the definition of the FT in Equation 2.1 we have

$$F(\omega) = \int_0^{NT} f(t)e^{-j\omega t}\, dt \tag{2.83}$$

Assuming that $f(t)$ is approximately constant over each T-second interval $[(k-1)T, kT]$, we obtain the following approximation to Equation 2.83:

$$F(\omega) = \sum_{k=0}^{N-1} \left[\int_{kT}^{kT+T} e^{-j\omega t}\, dt \right] f(k) \tag{2.84}$$

Then carrying out the integration in the right-hand side of Equation 2.84 gives

$$F(\omega) = \frac{1 - e^{-j\omega T}}{j\omega} \sum_{k=0}^{N-1} e^{-j\omega kT} f(k) \tag{2.85}$$

Finally, setting $\omega = 2\pi n/NT$ in Equation 2.85 yields

$$F\left(\frac{2\pi n}{NT}\right) = \frac{1 - e^{-j2\pi n/N}}{j2\pi n/NT} F_n \tag{2.86}$$

where F_n is the DFT of $f(k)$ given by Equation 2.46.

It should be stressed that the relationship in Equation 2.86 is only an approximation; that is, the right-hand side of Equation 2.86 is an approximation to $F(2\pi n/NT)$. In general, the approximation is more accurate the larger the N is and/or the smaller the T is. For a good result, it is also necessary that $f(t)$ be suitably small for $t < 0$ and $t \geq NT$.

As an example, let $f(t)$ be the 2-second pulse given by $f(t) = 1$ for $0 \leq t \leq 2$ and $f(t) = 0$ for all other t. Using the time shift property in Table 2.1 and the FT of a rectangular pulse given in Table 2.2, we have that the FT of $f(t)$ is

$$F(\omega) = \frac{2}{\omega} \sin(\omega)e^{-j\omega} \tag{2.87}$$

A MATLAB program (adapted from [5]) for computing the exact magnitude spectrum $|F(\omega)|$ and the approximation based on Equation 2.86 is given in Figure 2.6. The program was run for the case when

$N = 128$ and $T = 0.1$, with the plots shown in Figure 2.7. Note that the approximate values are fairly close, at least for frequencies over the span of the main lobe. A better approximation can be achieved by increasing N and/or decreasing T. The reader is invited to try this using the program in Figure 2.6.

```
N = input('Input N:');
T = input('Input T:');
t = 0:T:2;
f=[ones(1,length(t)) zeros(1,N-length(t))];
Fn=fft(f);
gam=2*pi/N/T;
n=0:10/gam;
Fapp=(1-exp(-j*n*gam*T))/j/n/gam*Fn;
w=0:.05:10;
Fexact=2*sin(w)./w;
plot(n*gam,abs(Fapp(1:length(n))),'og',w,abs(Fexact),'b')
```

FIGURE 2.6   MATLAB program for computing the FT of the 2-second pulse.

2.3.3 Solution of Differential Equations

One of the major applications of the Laplace transform is in solving linear differential equations. To pursue this, we begin by considering the first-order linear constant-coefficient differential equation given by

$$\dot{f}(t) + af(t) = w(t), \quad t \geq 0 \tag{2.88}$$

where $\dot{f}(t)$ is the derivative of $f(t)$ and $w(t)$ is an arbitrary real-valued function of t. To solve Equation 2.88, we apply the Laplace transform to both sides of the equation. Using linearity and the derivative

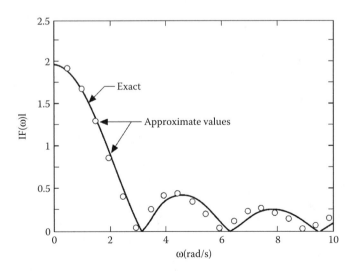

FIGURE 2.7   Exact and approximate magnitude spectra of the 2-second pulse.

properties of the Laplace transform given in Table 2.3, we have

$$sF(s) - f(0) + aF(s) = W(s) \tag{2.89}$$

where $F(s)$ is the transform of $f(t)$, $W(s)$ is the transform of $w(t)$, and $f(0)$ is the initial condition. Then solving Equation 2.89 for $F(s)$ gives

$$F(s) = \frac{1}{s+a}[f(0) + W(s)] \tag{2.90}$$

Taking the inverse Laplace transform of $F(s)$ then yields the solution $f(t)$. For example, if $w(t)$ is the unit-step function $u(t)$ and $a \neq 0$, then $W(s) = 1/s$ and $F(s)$ becomes

$$F(s) = \frac{1}{s+a}\left[f(0) + \frac{1}{s}\right] = \frac{f(0)s + 1}{(s+a)s} = \frac{f(0) - \frac{1}{a}}{s+a} + \frac{\frac{1}{a}}{s} \tag{2.91}$$

Taking the inverse transform gives

$$f(t) = \left[f(0) - \frac{1}{a}\right]e^{-at} + \frac{1}{a}, \quad t \geq 0 \tag{2.92}$$

Now consider the second-order differential equation

$$\ddot{f}(t) + a_1\dot{f}(t) + a_0 f(t) = w(t) \tag{2.93}$$

Again using the derivative property of the Laplace transform, taking the transform of both sides of Equation 2.93 we obtain

$$s^2 F(s) - sf(0) - \dot{f}(0) + a_1[sF(s) - f(0)] + a_0 F(s) = W(s) \tag{2.94}$$

where $f(0)$ and $\dot{f}(0)$ are the initial conditions. Solving Equation 2.94 for $F(s)$ yields

$$F(s) = \frac{1}{s^2 + a_1 s + a_0}[f(0)s + \dot{f}(0) + a_1 f(0) + W(s)] \tag{2.95}$$

For example, if $a_0 = 2$, $a_1 = 3$, and $w(t) = u(t)$, then

$$F(s) = \frac{1}{(s+1)(s+2)}\left[f(0)s + \dot{f}(0) + 3f(0) + \frac{1}{s}\right]$$

$$F(s) = \frac{f(0)s^2 + \left[\dot{f}(0) + 3f(0)\right]s + 1}{(s+1)(s+2)s}$$

$$F(s) = \frac{2f(0) + \dot{f}(0) - 1}{s+1} + \frac{-f(0) - \dot{f}(0) + 0.5}{s+2} + \frac{0.5}{s} \tag{2.96}$$

Inverse transforming Equation 2.96 gives

$$f(t) = [2f(0) + \dot{f}(0) - 1]e^{-t} + [-f(0) - \dot{f}(0) + 0.5]e^{-2t} + 0.5, \quad t \geq 0 \tag{2.97}$$

For the general case, consider the nth-order linear constant-coefficient differential equation

$$f^{(n)}(t) + \sum_{i=0}^{n-1} a_i f^{(i)}(t) = w(t) \tag{2.98}$$

where $f^{(i)}(t)$ is the ith derivative of $f(t)$. Given $w(t)$ and the initial conditions $f(0), \dot{f}(0), \ldots, f^{(n-1)}(0)$, the solution $f(t)$ to Equation 2.98 is unique. The solution can be determined by taking the transform of both sides of Equation 2.98 and solving for $F(s)$. This yields

$$F(s) = \frac{1}{D(s)}[N(s) + W(s)] \tag{2.99}$$

where $D(s)$ is the polynomial

$$D(s) = s^n + a_{n-1}s^{n-1} + \cdots + a_1 s + a_0 \tag{2.100}$$

and $N(s)$ is a polynomial in s of the form

$$N(s) = b_{n-1}s^{n-1} + b_{n-2}s^{n-2} + \cdots + b_1 s + b_0 \tag{2.101}$$

The coefficients $b_0, b_1, \ldots, b_{n-1}$ of $N(s)$ depend on the values of the n initial conditions $f(0), \dot{f}(0), \ldots, f^{(n-1)}(0)$. The relationship between the b_i and the initial conditions is given by the matrix equation

$$b = Px \tag{2.102}$$

where b and x are the column vectors

$$b = \begin{bmatrix} b_0 \\ b_1 \\ \vdots \\ b_{n-2} \\ b_{n-1} \end{bmatrix} \qquad x = \begin{bmatrix} f(0) \\ \dot{f}(0) \\ \vdots \\ f^{(n-2)}(0) \\ f^{(n-1)}(0) \end{bmatrix} \tag{2.103}$$

and P is the n-by-n matrix given by

$$P = \begin{bmatrix} a_1 & a_2 & \cdots & a_{n-2} & a_{n-1} & 1 \\ a_2 & a_3 & \cdots & a_{n-1} & 1 & 0 \\ \vdots & \vdots & & \vdots & \vdots & \vdots \\ a_{n-1} & 1 & \cdots & 0 & 0 & 0 \\ 1 & 0 & \cdots & 0 & 0 & 0 \end{bmatrix} \tag{2.104}$$

The matrix P given by Equation 2.104 is invertible for any values of the a_i, and thus there is a one-to-one and onto correspondence between the set of initial conditions and the coefficients of the polynomial $N(s)$ in Equation 2.101. In particular, this implies that for any given vector b of coefficients of $N(s)$, there is a vector x of initial conditions that results in the polynomial $N(s)$ with the given coefficients. From Equation 2.102, it is seen that $x = P^{-1}b$, where P^{-1} is the inverse of P.

Once $N(s)$ is computed using Equation 2.102, the solution $f(t)$ to Equation 2.98 can then be determined by inverse transforming Equation 2.99. If $W(s)$ is a rational function of s, then the right-hand side of Equation 2.99 is rational in s and thus, in this case, $f(t)$ can be computed via a partial fraction expansion.

An interesting consequence of the above constructions is the following characterization of a real-valued function $f(t)$ whose Laplace transform $F(s)$ is rational. Suppose that

$$F(s) = \frac{N(s)}{D(s)} \tag{2.105}$$

where $D(s)$ and $N(s)$ are given by Equations 2.100 and 2.101, respectively. Then comparing Equations 2.105 and 2.99 shows that $f(t)$ is the solution to the nth-order homogeneous equation

$$f^{(n)}(t) + \sum_{i=0}^{n-1} a_i f^{(i)}(t) = 0 \tag{2.106}$$

with the initial conditions given by $x = P^{-1}b$, where x and b are defined by Equation 2.103. Hence, any function $f(t)$ having a rational Laplace transform is the solution to a homogeneous differential equation. This result is of fundamental importance in the theory of systems and controls.

2.3.4 Solution of Difference Equations

The discrete-time counterpart to the solution of differential equations using the Laplace transform is the solution of difference equations using the z-transform. We begin by considering the first-order linear constant-coefficient difference equation

$$f(k+1) + af(k) = w(k), \quad k \geq 0 \tag{2.107}$$

where $w(k)$ is an arbitrary real-valued function of the integer variable k. Taking the z-transform of Equation 2.107 using the linearity and left shift properties given in Table 2.6 yields

$$zF(z) - f(0)z + aF(z) = W(z) \tag{2.108}$$

where $F(z)$ is the z-transform of $f(k)$ and $f(0)$ is the initial condition. Then solving Equation 2.108 for $F(z)$ gives

$$F(z) = \frac{1}{z+a}[f(0)z + W(z)] \tag{2.109}$$

For example, if $w(k)$ is the unit-step function $u(k)$ and $a \neq 1$, then $W(z) = z/(z-1)$ and $F(z)$ becomes

$$F(z) = \frac{1}{z+a}\left[f(0)z + \frac{z}{z-1}\right]$$

$$F(z) = \frac{f(0)z(z-1) + z}{(z+a)(z-1)} \tag{2.110}$$

Then

$$\frac{F(z)}{z} = \frac{f(0)(z-1) + 1}{(z+a)(z-1)} \tag{2.111}$$

and expanding by partial fractions gives

$$\frac{F(z)}{z} = \frac{f(0) - \frac{1}{1+a}}{z+a} + \frac{\frac{1}{1+a}}{z-1} \tag{2.112}$$

Thus

$$F(z) = \frac{\left[f(0) - \frac{1}{1+a}\right]z}{z+a} + \frac{\frac{1}{1+a}z}{z-1} \tag{2.113}$$

and taking the inverse z-transform gives

$$f(k) = \left[f(0) - \frac{1}{1+a}\right](-a)^k + \frac{1}{1+a}, \quad k \geq 0 \tag{2.114}$$

For the general case, consider the nth-order linear constant-coefficient difference equation

$$f(k+n) + \sum_{i=0}^{n-1} a_i f(k+i) = w(k) \tag{2.115}$$

The initial conditions for Equation 2.115 may be taken to be the n values $f(0), f(1), \ldots, f(n-1)$. Another choice is to take the initial values to be $f(-1), f(-2), \ldots, f(-n)$. We prefer the latter choice since the initial values are given for negative values of the time index k. In this case, the use of the z-transform to solve Equation 2.115 requires that the equation be time shifted. This is accomplished by replacing k by $k - n$ in Equation 2.115, which yields

$$f(k) + \sum_{i=1}^{n} a_{n-i} f(k-i) = w(k-n) \tag{2.116}$$

Then using the right-shift property of the z-transform and transforming Equation 2.116 yields

$$F(z) = \frac{1}{D(z^{-1})} \left[N(z^{-1}) + z^{-n} W(z) + \sum_{i=1}^{n} w(-i) z^{-n+i} \right] \tag{2.117}$$

where $D(z^{-1})$ and $N(z^{-1})$ are polynomials in z^{-1} given by

$$D(z^{-1}) = 1 + a_{n-1} z^{-1} + \cdots + a_1 z^{-n+1} + a_0 z^{-n} \tag{2.118}$$
$$N(z^{-1}) = b_{n-1} + b_{n-2} z^{-1} + \cdots + b_1 z^{-n+2} + b_0 z^{-n+1} \tag{2.119}$$

The coefficients b_i of $N(z^{-1})$ are related to the initial values by the matrix equation

$$b = Q\phi \tag{2.120}$$

where b and ϕ are the column vectors

$$b = \begin{bmatrix} b_0 \\ b_1 \\ \vdots \\ b_{n-2} \\ b_{n-1} \end{bmatrix} \qquad \phi = \begin{bmatrix} f(-1) \\ f(-2) \\ \vdots \\ f(-n+1) \\ f(-n) \end{bmatrix} \tag{2.121}$$

and Q is the n-by-n matrix given by

$$Q = \begin{bmatrix} a_0 & 0 & \cdots & 0 & 0 & 0 \\ a_1 & a_0 & \cdots & 0 & 0 & 0 \\ \vdots & \vdots & & \vdots & \vdots & \vdots \\ a_{n-2} & a_{n-3} & \cdots & a_1 & a_0 & 0 \\ a_{n-1} & a_{n-2} & \cdots & a_2 & a_1 & a_0 \end{bmatrix} \tag{2.122}$$

The matrix Q given by Equation 2.122 is invertible for any values of the a_i as long as $a_0 \neq 0$, and thus for any given vector b of coefficients of the polynomial $N(z^{-1})$, there is a vector ϕ of initial conditions that results in the polynomial $N(z^{-1})$ with the given coefficients. Clearly, if $a_0 \neq 0$, then $\phi = Q^{-1}b$, where Q^{-1} is the inverse of Q.

Once $N(z^{-1})$ is computed using Equation 2.120, the solution $f(k)$ to Equation 2.115 or Equation 2.116 can then be determined by inverse transforming Equation 2.117. If $W(z)$ is a rational function of z, then the right-hand side of Equation 2.117 is rational in z, and in this case, $f(k)$ can be computed via a partial fraction expansion.

Finally, it is worth noting (in analogy with the Laplace transform) that any function $f(k)$ having a rational z-transform $F(z)$ is the solution to a homogeneous difference equation of the form

$$f(k+n) + \sum_{i=0}^{n-1} a_i f(k+i) = 0 \tag{2.123}$$

where the initial conditions are determined using Equations 2.120 through 2.122.

2.3.5 Defining Terms

3-dB bandwidth: For a bandlimited signal, this is the smallest value B_{3dB} for which the magnitude spectrum $|F(\omega)|$ is down by 3 dB or more from the peak magnitude for all $\omega > B_{3dB}$.

Bandlimited signal: A signal $f(t)$ whose FT $F(\omega)$ is zero (or approximately zero) for all $\omega > B$, where B is a finite positive number.

Irrational function: A function $F(s)$ of a complex variable s that cannot be expressed as a ratio of polynomials in s.

Magnitude spectrum: The magnitude $|F(\omega)|$ of the FT of a function $f(t)$.

One-sided (or unilateral) transform: A transform that operates on a function $f(t)$ defined for $t \geq 0$.

Open left half-plane (OLHP): The set of all complex numbers having negative real part.

Open unit disk: The set of all complex numbers whose magnitude is less than 1.

Phase spectrum: The angle $\angle F(\omega)$ of the FT of a function $f(t)$.

Poles of a rational function $N(s)/D(s)$**:** The values of s for which $D(s) = 0$, assuming that $N(s)$ and $D(s)$ have no common factors.

Proper rational function: A rational function $N(s)/D(s)$ where the degree of $N(s)$ is less than or equal to the degree of $D(s)$.

Rational function: A ratio of two polynomials $N(s)/D(s)$ where s is a complex variable.

Region of convergence: The set of all complex numbers for which a transform exists (i.e., is well defined) in the ordinary sense.

Residues: The values of the numerator constants in a partial fraction expansion of a rational function.

Strictly proper rational function: A rational function $N(s)/D(s)$ where the degree of $N(s)$ is strictly less than the degree of $D(s)$.

Two-sided (or bilateral) transform: A transform that operates on a function $f(t)$ defined for $-\infty < t < \infty$.

Zeros of a rational function $N(s)/D(s)$**:** The values of s for which $N(s) = 0$, assuming that $N(s)$ and $D(s)$ have no common factors.

References

1. Fourier, J.B.J., *The Analytical Theory of Heat*, Cambridge, (trans. A. Freeman) 1878.
2. Barker, R.H., The pulse transfer function and its applications to sampling servo systems, *Proc. IEEE*, 99, Part IV, 302–317, 1952.
3. Jury, E.I., Analysis and synthesis of sampled-data control systems, *Communications and Electronics*, 1954, 1–15.
4. Ragazzini, J.R. and Zadeh, L. A., The analysis of sampled-data systems, *Trans. AIEE*, 71, Part II:225–232, 1952.
5. Kamen, E.W. and Heck, B.S., *Fundamentals of Signals and Systems using the Web and MATLAB*, 3rd ed., Prentice Hall, Upper Saddle River, NJ, 2007.

Further Reading

Rigorous developments of the FT can be found in:

6. Papoulis, A., *The Fourier Integral and Its Applications*, McGraw-Hill, New York, 1962.
7. Bracewell, R. N., *The Fourier Transform and Its Applications*, 3rd ed., McGraw-Hill, New York, 1999.

For an in-depth treatment of the FFT and its applications, see:

8. Brighman, E. O., *The Fast Fourier Transform and Its Applications*, Prentice-Hall, Englewood Cliffs, NJ, 1988.
9. Rao, K. R., Kim, D. N., and Hwang, J. J., *Fast Fourier Transform*, Springer, New York, 2009.

An extensive development of the Laplace transform is given in:

10. Schiff, J. L., *The Laplace Transform*, Springer, New York, 1999.

For a thorough development of the z-transform, see:

11. Jury, E. I., *Theory and Application of the z-Transform Method*, Wiley, New York, 1964.

Treatments of the Fourier, Laplace and z-transforms can be found in textbooks on signals and systems, such as [5] given in the references above. Additional textbooks on signals and systems are listed below:

12. Haykin, S. and Veen, B. V., *Signals and Systems*, 2nd ed.,Wiley, New York, 2002.
13. Lathi, B. P., *Linear Systems and Signals*, 2nd ed., Oxford Press, Oxford, UK, 2005.
14. Lee, E. A. and Varaiya, P., *Structure and Interpretation of Signals and Systems*, Addison-Wesley, Reading, MA, 2003.
15. Oppenheim, A. V. and Willsky, A. V. with Nawab, S. H., *Signals and Systems*, 2nd ed., Prentice Hall, Upper Saddle River, NJ, 1997.
16. Phillips, C.L., Parr, J., and Riskin, E., *Signals, Systems, and Transforms*, 4th ed., Prentice Hall, Upper Saddle River, NJ, 2007.
17. Roberts, M. J., *Fundamentals of Signals and Systems*, McGraw-Hill, Boston, MA, 2008.
18. Ziemer, R. E., Tranter, W. H., and Fannin, D. R., *Signals and Systems: Continuous and Discrete*, 4th ed., Prentice-Hall, Upper Saddle River, NJ, 1998.

3

Matrices and Linear Algebra

Bradley W. Dickinson
Princeton University

3.1 Introduction

Matrices and linear algebra are indispensible tools for analysis and computation in problems involving systems and control. This chapter presents an overview of these subjects that highlights the main concepts and results.

3.2 Matrices

To introduce the notion of a *matrix*, we start with some notation that will be used for describing matrices and for presenting the rules for manipulating matrices in algebraic expressions.

Let $\mathcal{R}$ be a *ring*, a set of quantities together with definitions for addition and multiplication operations. Standard notations for some examples of rings arising frequently in control systems applications include $\mathbb{R}$ (the real numbers), $\mathbb{C}$ (the complex numbers), $\mathbb{R}[s]$ (the set of polynomials in the variable s having real coefficients), and $\mathbb{R}(s)$ (the set of rational functions, i.e., ratios of polynomials). Each of these rings has distinguished elements 0 and 1, the identity elements for addition and multiplication, respectively. Rings for which addition and multiplication are commutative operations and for which multiplicative inverses of all nonzero quantities exist are known as *fields*; in the examples given, the real numbers, the complex numbers, and the rational functions are fields.

A *matrix* (more descriptively an $\mathcal{R}$-matrix, e.g., a complex matrix or a real matrix) is a rectangular array of *matrix elements* that belong to the ring $\mathcal{R}$. When $\mathbf{A}$ is a matrix with m rows and n columns, denoted $\mathbf{A} \in \mathcal{R}^{m \times n}$, $\mathbf{A}$ is said to be "an m by n (written $m \times n$) matrix," and its matrix elements are indexed with a double subscript, the first indicating the row position and the second indicating the column position. The notation used is

$$\mathbf{A} = \begin{bmatrix} a_{11} & a_{12} & \cdot & \cdot & \cdot & a_{1n} \\ a_{21} & a_{22} & \cdot & \cdot & \cdot & a_{2n} \\ \cdot & \cdot & & & & \cdot \\ \cdot & \cdot & & & & \cdot \\ \cdot & \cdot & & & & \cdot \\ a_{m1} & a_{m2} & \cdot & \cdot & \cdot & a_{mn} \end{bmatrix} \quad \text{and} \quad (\mathbf{A})_{ij} = a_{ij} \tag{3.1}$$

Three special "shapes" of matrices commonly arise and are given descriptive names. $\mathcal{R}^{m \times 1}$ is the set of column matrices, also known as *column m-vectors*, *m-vectors*, or when no ambiguity results simply as *vectors*. Similarly, $\mathcal{R}^{1 \times n}$ is the set of *row vectors*. Finally, $\mathcal{R}^{n \times n}$ is the set of *square matrices* of size n.

3.3 Matrix Algebra

Since matrix elements belong to a ring $\mathcal{R}$, they may be combined in algebraic expressions involving addition and multiplication operations. This provides the means for defining algebraic operations for matrices. The usual notion of equality is adopted: two $m \times n$ matrices are equal if and only if they have the same elements.

3.3.1 Scalar Multiplication

The product of a matrix $\mathbf{A} \in \mathcal{R}^{m \times n}$ and a *scalar* $z \in \mathcal{R}$ may always be formed. The resulting matrix, also in $\mathcal{R}^{m \times n}$, is obtained by elementwise multiplication:

$$(z\mathbf{A})_{ij} = za_{ij} = (\mathbf{A}z)_{ij} \tag{3.2}$$

3.3.2 Matrix Addition

Two matrices, both in $\mathcal{R}^{m \times n}$, say $\mathbf{A}$ and $\mathbf{B}$, may be added to produce a third matrix $\mathbf{C} \in \mathcal{R}^{m \times n}$, $\mathbf{A} + \mathbf{B} = \mathbf{C}$, where $\mathbf{C}$ is the matrix of elementwise sums

$$(\mathbf{C})_{ij} = c_{ij} = a_{ij} + b_{ij} \tag{3.3}$$

3.3.3 Matrix Multiplication

Two matrices, say $\mathbf{A}$ and $\mathbf{B}$, may be multiplied with $\mathbf{A}$ as the left factor and $\mathbf{B}$ as the right factor if and only if their sizes are compatible: if $\mathbf{A} \in \mathcal{R}^{m_A \times n_A}$ and $\mathbf{B} \in \mathcal{R}^{m_B \times n_B}$, then it is required that $n_A = m_B$. That is, the number of columns of the left factor must equal the number of rows of the right factor. When this is the case, the product matrix $\mathbf{C} = \mathbf{AB}$ is $m_A \times n_B$, that is, the product has the same number of rows as the left factor and the same number of columns as the right factor. With simpler notation, if $\mathbf{A} \in \mathcal{R}^{m \times n}$ and $\mathbf{B} \in \mathcal{R}^{n \times p}$, then the product matrix $\mathbf{C} \in \mathcal{R}^{m \times p}$ is given by

$$(\mathbf{C})_{ij} = c_{ij} = \sum_{k=1}^{n} a_{ik} b_{kj} \tag{3.4}$$

Using the interpretation of the rows and the columns of a matrix as matrices themselves, several important observations follow from the defining equation for the elements of a matrix product.

1. The columns of the product matrix $\mathbf{C}$ are obtained by multiplying the matrix $\mathbf{A}$ times the corresponding columns of $\mathbf{B}$.
2. The rows of the product matrix $\mathbf{C}$ are obtained by multiplying the corresponding rows of $\mathbf{A}$ times the matrix $\mathbf{B}$.
3. The (i, j)th element of the product matrix, $(\mathbf{C})_{ij}$, is the product of the ith row of $\mathbf{A}$ times the jth column of $\mathbf{B}$.
4. The product matrix $\mathbf{C}$ may be expressed as the sum of products of the kth column of $\mathbf{A}$ times the kth row of $\mathbf{B}$.

Unlike matrix addition, matrix multiplication is generally not commutative. If the definition of matrix multiplication allows for both of the products $\mathbf{AB}$ and $\mathbf{BA}$ to be formed, the two products are square matrices, but they are not necessarily equal nor even the same size.

The addition and multiplication operations for matrices obey familiar rules of associativity and distributivity: (a) $(\mathbf{A} + \mathbf{B}) + \mathbf{C} = \mathbf{A} + (\mathbf{B} + \mathbf{C})$; (b) $(\mathbf{AB})\mathbf{C} = \mathbf{A}(\mathbf{BC})$; (c) $(\mathbf{A} + \mathbf{B})\mathbf{C} = \mathbf{AC} + \mathbf{BC}$; and (d) $\mathbf{A}(\mathbf{B} + \mathbf{C}) = \mathbf{AB} + \mathbf{AC}$.

3.3.4 The Zero Matrix and the Identity Matrix

The *zero matrix*, denoted $\mathbf{0}$, is any matrix whose elements are all zero:

$$(\mathbf{0})_{ij} = 0 \tag{3.5}$$

Usually the number of rows and columns of $\mathbf{0}$ will be understood from context; $\mathbf{0}_{m \times n}$ will specifically denote the $m \times n$ zero matrix. Clearly, $\mathbf{0}$ is the additive identity element for matrix addition: $\mathbf{0} + \mathbf{A} = \mathbf{A} = \mathbf{A} + \mathbf{0}$; indeed, $\mathcal{R}^{m \times n}$ with the operation of matrix addition is a group. For matrix multiplication, if $\mathbf{A}$ is $m \times n$ then $\mathbf{0}_{m \times m}\mathbf{A} = \mathbf{0}_{m \times n} = \mathbf{A}\mathbf{0}_{n \times n}$.

The *identity matrix*, denoted $\mathbf{I}$, is a square matrix whose only nonzero elements are the ones along its main diagonal:

$$(\mathbf{I})_{ij} = \begin{cases} 1 & \text{for } i = j \\ 0 & \text{for } i \neq j \end{cases} \tag{3.6}$$

Again, the dimensions of $\mathbf{I}$ are usually obtained from context; the $n \times n$ identity matrix is specifically denoted $\mathbf{I}_n$. The identity matrix serves as an identity element for matrix multiplication: $\mathbf{I}_m\mathbf{A} = \mathbf{A} = \mathbf{A}\mathbf{I}_n$ for any $\mathbf{A} \in \mathcal{R}^{m \times n}$. This has an important implication for square matrices: $\mathcal{R}^{n \times n}$, with the operations of matrix addition and matrix multiplication, is a (generally noncommutative) ring.

3.3.5 Matrix Inverse

Closely related to matrix multiplication is the notion of *matrix inverse*. If $\mathbf{A}$ and $\mathbf{X}$ are square matrices of the same size and they satisfy $\mathbf{AX} = \mathbf{I} = \mathbf{XA}$, then $\mathbf{X}$ is called the matrix inverse of $\mathbf{A}$, and is denoted by $\mathbf{A}^{-1}$. The inverse matrix satisfies

$$\mathbf{AA}^{-1} = \mathbf{I} = \mathbf{A}^{-1}\mathbf{A} \tag{3.7}$$

For a square matrix $\mathbf{A}$, if $\mathbf{A}^{-1}$ exists, it is unique and $(\mathbf{A}^{-1})^{-1} = \mathbf{A}$. If $\mathbf{A}$ has a matrix inverse, $\mathbf{A}$ is said to be *invertible*; the terms *nonsingular* and *regular* are also used as synonyms for invertible. If $\mathbf{A}$ has no matrix inverse, it is said to be *noninvertible* or *singular*.

The invertible matrices in $\mathcal{R}^{n\times n}$, along with the operation of matrix multiplication, form a group, the *general linear group*, denoted by $GL(\mathcal{R}, n)$; $\mathbf{I}_n$ is the identity element of the group.

If $\mathbf{A}$ and $\mathbf{B}$ are square, invertible matrices of the same size, then their products are invertible also and

$$(\mathbf{AB})^{-1} = \mathbf{B}^{-1}\mathbf{A}^{-1}; \; (\mathbf{BA})^{-1} = \mathbf{A}^{-1}\mathbf{B}^{-1} \tag{3.8}$$

This extends to products of more than two factors, giving the *product rule for matrix inverses*: The inverse of a product of square matrices is the product of their inverses taken in reverse order, provided the inverses of all of the factors exist.

3.3.5.1 Some Useful Matrix Inversion Identities

1. If the $n \times n$ matrices $\mathbf{A}$, $\mathbf{B}$, and $\mathbf{A} + \mathbf{B}$ are all invertible, then

$$(\mathbf{A}^{-1} + \mathbf{B}^{-1})^{-1} = \mathbf{A}(\mathbf{A} + \mathbf{B})^{-1}\mathbf{B} \tag{3.9}$$

2. Assuming that the matrices have suitable dimensions and that the indicated inverses all exist:

$$(\mathbf{A} + \mathbf{BCD})^{-1} = \mathbf{A}^{-1} - \mathbf{A}^{-1}\mathbf{B}(\mathbf{C}^{-1} + \mathbf{DA}^{-1}\mathbf{B})^{-1}\mathbf{DA}^{-1} \tag{3.10}$$

This simplifies in the important special case when $\mathbf{C} = 1$, $\mathbf{B}$ is a column vector and $\mathbf{D}$ is a row vector.

Determining whether a square matrix is invertible and, if so, finding its matrix inverse, are important for a variety of applications. The determinant turns out to be a means of characterizing invertibility.

3.3.6 The Determinant

The *determinant* of a square matrix $\mathbf{A} \in \mathcal{R}^{n\times n}$, denoted $\det \mathbf{A}$, is a scalar function taking the form of a sum of signed products of n matrix elements. While an explicit formula for $\det \mathbf{A}$ can be given [7], it is common to define the determinant inductively as follows. For $\mathbf{A} = [a_{11}] \in \mathcal{R}^{1\times 1}$, $\det \mathbf{A} = a_{11}$. For $\mathbf{A} \in \mathcal{R}^{n\times n}$, with $n > 1$,

$$\det \mathbf{A} = \sum_{k=1}^{n}(-1)^{i+k}a_{ik}\Delta_{ik}(\mathbf{A}) \;\; \text{or} \;\; \det \mathbf{A} = \sum_{k=1}^{n}(-1)^{i+k}a_{ki}\Delta_{ki}(\mathbf{A}) \tag{3.11}$$

These are the *Laplace expansions* for the determinant corresponding to the ith row and ith column of $\mathbf{A}$ respectively. In these formulas, the quantity $\Delta_{ik}(\mathbf{A})$ is the determinant of the $(n-1) \times (n-1)$ square matrix obtained by deleting the ith row and kth column of $\mathbf{A}$, and similarly for $\Delta_{ki}(\mathbf{A})$.

The quantities $\Delta_{ik}(\mathbf{A})$ and $\Delta_{ki}(\mathbf{A})$ are examples of $(n-1) \times (n-1)$ *minors* of $\mathbf{A}$; for any k, $1 \leq k \leq n-1$, an $(n-k) \times (n-k)$ minor of $\mathbf{A}$ is the determinant of an $(n-k) \times (n-k)$ square matrix obtained by deleting some set of k rows and k columns of $\mathbf{A}$.

For any n, $\det \mathbf{I}_n = 1$. For $\mathbf{A} \in \mathcal{R}^{2\times 2}$, the Laplace expansions lead to the well-known formula: $\det \mathbf{A} = a_{11}a_{22} - a_{12}a_{21}$.

3.3.6.1 Properties of the Determinant

Many properties of determinants can be verified directly from the Laplace expansion formulas. For example, consider the *elementary row and column operations*: replacing any row of a matrix by its sum with another row does not change the value of the determinant, and, likewise, replacing any column of a matrix by its sum with another column does not change the value of the determinant; replacing a row (or a column) of a matrix with a nonzero multiple of itself changes the determinant by the same factor; interchanging two rows (or columns) of a matrix changes only the sign of the determinant (i.e., the determinant is multiplied by -1).

If $\mathbf{A} \in \mathcal{R}^{n \times n}$ and $z \in \mathcal{R}$, then $\det(z\mathbf{A}) = z^n \det \mathbf{A}$. If $\mathbf{A}$ and $\mathbf{B}$ are matrices for which both products $\mathbf{AB}$ and $\mathbf{BA}$ are defined, then $\det(\mathbf{AB}) = \det(\mathbf{BA})$. If in addition, both matrices are square then

$$\det(\mathbf{AB}) = \det(\mathbf{BA}) = \det \mathbf{A} \, \det \mathbf{B} = \det \mathbf{B} \, \det \mathbf{A} \tag{3.12}$$

This is the *product rule for determinants*.

3.3.7 Determinants and Matrix Inverses

3.3.7.1 Characterization of Invertibility

The determinant of an invertible matrix and the determinant of its inverse are reciprocals. If $\mathbf{A}$ is invertible, then

$$\det(\mathbf{A}^{-1}) = 1/\det \mathbf{A} \tag{3.13}$$

This result indicates that invertibility of matrices is related to existence of multiplicative inverses in the underlying ring $\mathcal{R}$. In ring-theoretic terminology, the *units* of $\mathcal{R}$ are those ring elements having multiplicative inverses. When $\mathcal{R}$ is a field, all nonzero elements are units, but for $\mathcal{R} = \mathbb{R}[s]$ (or $\mathbb{C}[s]$), the ring of polynomials with real (or complex) coefficients, only the nonzero constants (i.e., the nonzero polynomials of degree 0) are units.

Determinants provide a characterization of invertibility as follows:

The matrix $\mathbf{A} \in \mathcal{R}^{n \times n}$ is invertible if and only if $\det \mathbf{A}$ is a unit in $\mathcal{R}$.

When $\mathcal{R}$ is a field, all nonzero ring elements are units and the criterion for invertibility takes a simpler form:

When $\mathcal{R}$ is a field, the matrix $\mathbf{A} \in \mathcal{R}^{n \times n}$ is invertible if and only if $\det \mathbf{A} \neq 0$.

3.3.8 Cramer's Rule and PLU Factorization

Cramer's rule provides a general formula for the elements of $\mathbf{A}^{-1}$ in terms of a ratio of determinants:

$$(\mathbf{A}^{-1})_{ij} = (-1)^{i+j} \Delta_{ji}(\mathbf{A}) / \det \mathbf{A} \tag{3.14}$$

where $\Delta_{ji}(\mathbf{A})$ is the $(n-1) \times (n-1)$ minor of $\mathbf{A}$ in which the jth row and ith column of $\mathbf{A}$ are deleted.

If $\mathbf{A}$ is a 1×1 matrix over $\mathcal{R}$, then it is invertible if and only if it is a unit; when $\mathbf{A}$ is invertible, $\mathbf{A}^{-1} = 1/\mathbf{A}$. (For instance, the 1×1 matrix s over the ring of polynomials, $\mathbb{R}[s]$, is not invertible; however, as a matrix over $\mathbb{R}(s)$, the field of rational functions, it is invertible with inverse $1/s$.)

If $\mathbf{A} \in \mathcal{R}^{2 \times 2}$, then $\mathbf{A}$ is invertible if and only if $\det \mathbf{A} = \Delta = a_{11}a_{22} - a_{21}a_{12}$ is a unit. When $\mathbf{A}$ is invertible,

$$\mathbf{A} = \begin{bmatrix} a_{11} & a_{12} \\ a_{21} & a_{22} \end{bmatrix} \quad \text{and} \quad \mathbf{A}^{-1} = \begin{bmatrix} a_{22}/\Delta & -a_{12}/\Delta \\ -a_{21}/\Delta & a_{11}/\Delta \end{bmatrix} \tag{3.15}$$

A 2×2 polynomial matrix has a polynomial matrix inverse just in case Δ equals a nonzero constant.

Cramer's rule is almost never used for computations because of its computational complexity and numerical sensitivity. When a matrix of real or complex numbers needs to be inverted, certain matrix factorization methods are employed; such factorizations also provide the best methods for numerical computation of determinants.

Inversion of upper and lower triangular matrices is done by a simple process of back-substitution; the inverses have the same triangular form. This may be exploited in combination with the product rule for inverses (and for determinants) since any invertible matrix $\mathbf{A} \in \mathbb{R}^{n \times n}$ ($\mathbb{R}$ can be replaced by another field $\mathcal{F}$) can be factored into the form

$$\mathbf{A} = \mathbf{PLU} \tag{3.16}$$

where the factors on the right side are, respectively, a permutation matrix, a lower triangular matrix, and an upper triangular matrix. The computation of this *PLU factorization* is equivalent to the process of Gaussian elimination with pivoting [6]. The resulting expression for the matrix inverse (usually kept in its factored form) is

$$\mathbf{A}^{-1} = \mathbf{U}^{-1}\mathbf{L}^{-1}\mathbf{P}^{-1} \tag{3.17}$$

whereas $\det \mathbf{A} = \det \mathbf{P} \det \mathbf{L} \det \mathbf{U}$. ($\det \mathbf{P} = \pm 1$, since $\mathbf{P}$ is a permutation matrix.)

3.3.9 Matrix Transposition

Another operation on matrices that is useful in a number of applications is *matrix transposition*. If $\mathbf{A}$ is an $m \times n$ matrix with $(\mathbf{A})_{ij} = a_{ij}$, the *transpose* of $\mathbf{A}$, denoted $\mathbf{A}^{\mathrm{T}}$, is the $n \times m$ matrix given by

$$(\mathbf{A}^{\mathrm{T}})_{ij} = a_{ji} \tag{3.18}$$

Thus, the transpose of a matrix is formed by interchanging its rows and columns.

If a square matrix $\mathbf{A}$ satisfies $\mathbf{A}^{\mathrm{T}} = \mathbf{A}$, it is called a *symmetric matrix*. If a square matrix $\mathbf{A}$ satisfies $\mathbf{A}^{\mathrm{T}} = -\mathbf{A}$, it is called a *skew-symmetric matrix*.

For matrices whose elements may possibly be complex numbers, a generalization of transposition is often more appropriate. The *Hermitian transpose* of matrix $\mathbf{A}$, denoted $\mathbf{A}^{\mathrm{H}}$, is formed by interchanging rows and columns and replacing each element by its complex conjugate:

$$(\mathbf{A}^{\mathrm{H}})_{ij} = a_{ji}^* \tag{3.19}$$

The matrix $\mathbf{A}$ is *Hermitian symmetric* if $\mathbf{A}^{\mathrm{H}} = \mathbf{A}$.

3.3.9.1 Properties of Transposition

Several relationships between transposition and other matrix operations are noteworthy. For any matrix, $(\mathbf{A}^{\mathrm{T}})^{\mathrm{T}} = \mathbf{A}$; for $\mathbf{A} \in \mathcal{R}^{m \times n}$ and $z \in \mathcal{R}$, $(z\mathbf{A})^{\mathrm{T}} = z\mathbf{A}^{\mathrm{T}}$. With respect to algebraic operations, $(\mathbf{A} + \mathbf{B})^{\mathrm{T}} = \mathbf{A}^{\mathrm{T}} + \mathbf{B}^{\mathrm{T}}$ and $(\mathbf{AB})^{\mathrm{T}} = \mathbf{B}^{\mathrm{T}}\mathbf{A}^{\mathrm{T}}$. (The products $\mathbf{AA}^{\mathrm{T}}$ and $\mathbf{A}^{\mathrm{T}}\mathbf{A}$ are always defined.) With respect to determinants and matrix inversion, if $\mathbf{A}$ is a square matrix, $\det(\mathbf{A}^{\mathrm{T}}) = \det \mathbf{A}$, and if $\mathbf{A}$ is an invertible matrix, $\mathbf{A}^{\mathrm{T}}$ is also invertible, with $(\mathbf{A}^{\mathrm{T}})^{-1} = (\mathbf{A}^{-1})^{\mathrm{T}}$. A similar list of properties holds for Hermitian transposition.

3.3.9.2 Orthogonal and Unitary Matrices

Even for 2×2 matrices, transposition appears to be a much simpler operation than inversion. Indeed, the class of matrices for which $\mathbf{A}^{\mathrm{T}} = \mathbf{A}^{-1}$ is quite remarkable. A real matrix whose transpose is also its inverse is known as an *orthogonal matrix*. (This terminology is in common use, although it would be preferable to use *real unitary matrix* as will become apparent later.) The set of $n \times n$ orthogonal matrices, along with the operation of matrix multiplication, is a group; it is a subgroup of the group of invertible matrices, $GL(\mathbb{R}, n)$. For complex matrices, when $\mathbf{A}$ satisfies $\mathbf{A}^{\mathrm{H}} = \mathbf{A}^{-1}$, it is called a *unitary matrix*; the unitary matrices form a subgroup of $GL(\mathbb{C}, n)$.

3.3.10 Block Matrices

It is sometimes convenient to partition the rows and columns of a matrix so that the matrix elements are grouped into submatrices. For example, a matrix $A \in \mathcal{R}^{m \times n}$ may be partitioned into n columns (submatrices in $\mathcal{R}^{m \times 1}$) or into m rows (submatrices in $\mathcal{R}^{1 \times n}$). More generally

$$A = \begin{bmatrix} A_{11} & \cdots & A_{1q} \\ \vdots & & \vdots \\ A_{p1} & \cdots & A_{pq} \end{bmatrix} \tag{3.20}$$

where all submatrices in each block row have the same number of rows and all submatrices in each block column have the same number of columns; that is, submatrix A_{ij} is $m_i \times n_j$, with $m_1 + \cdots + m_p = m$ and $n_1 + \cdots + n_q = n$. Such a matrix A is said to be a $p \times q$ *block matrix*, and it is denoted by $A = (A_{ij})$ for simplicity.

Matrix addition can be carried out blockwise for $p \times q$ block matrices with *conformable* partitions, where the corresponding submatrices have the same number of rows and columns. Matrix multiplication can also be carried out blockwise provided the left factor's column partition is *compatible* with the right factor's row partition: it is required that if $A = (A_{ij})$ is a $p_A \times q_A$ block matrix with block column i having n_i columns, and $B = (B_{ij})$ is a $p_B \times q_B$ block matrix with block row j having m_j rows, then when $q_A = p_B$ and, in addition, $n_i = m_i$ for each i, the product matrix $C = AB$ is a $p_A \times q_B$ block matrix $C = (C_{ij})$, where block C_{ij} is given by

$$C_{ij} = \sum_{k=1}^{r} A_{ik} B_{kj} \tag{3.21}$$

where $r = q_A = p_B$.

For square matrices written as $p \times p$ block matrices having square "diagonal blocks" A_{ii}, the determinant has a blockwise representation. For a square 2×2 block matrix,

$$\det A = \det \begin{bmatrix} A_{11} & A_{12} \\ A_{21} & A_{22} \end{bmatrix} = \det A_{11} \det(A_{22} - A_{21} A_{11}^{-1} A_{12}) \tag{3.22}$$

provided $\det A_{11} \neq 0$. If this block matrix is invertible, its inverse may be expressed as a conformable block matrix:

$$A^{-1} = \begin{bmatrix} A_{11} & A_{12} \\ A_{21} & A_{22} \end{bmatrix}^{-1} = \begin{bmatrix} S_{11} & S_{12} \\ S_{21} & S_{22} \end{bmatrix} \tag{3.23}$$

and assuming A_{11} is invertible, the blocks of the inverse matrix are: $S_{11} = A_{11}^{-1} + A_{11}^{-1} A_{12} \Phi^{-1} A_{21} A_{11}^{-1}$; $S_{21} = -\Phi^{-1} A_{21} A_{11}^{-1}$; $S_{12} = -A_{11}^{-1} A_{12} \Phi^{-1}$; $S_{22} = \Phi^{-1} = (A_{22} - A_{21} A_{11}^{-1} A_{12})^{-1}$.

3.3.11 Matrix Polynomials and the Cayley–Hamilton Theorem

If $A \in \mathcal{R}^{n \times n}$, define $A^0 = I_n$, and A^r equal to the product of r factors of A, for integer $r \geq 1$. When A is invertible, A^{-1} has already been introduced as the notation for the inverse matrix. Nonnegative powers of A^{-1} provide the means for defining $A^{-r} = (A^{-1})^r$.

For any polynomial, $p(s) = p_0 s^k + p_1 s^{k-1} + \cdots + p_{k-1} s + p_k$, with coefficients $p_i \in \mathcal{R}$, the *matrix polynomial* $p(A)$ is defined as $p(A) = p_0 A^k + p_1 A^{k-1} + \cdots + p_{k-1} A + p_k I$. When the ring of scalars, $\mathcal{R}$, is a field (and in some more general cases), $n \times n$ matrices obey certain polynomial equations of the form $p(A) = 0$; such a polynomial $p(s)$ is an *annihilating polynomial* of A. The monic annihilating polynomial of least degree is called the *minimal polynomial* of A; the minimal polynomial is the (monic) greatest common divisor of all annihilating polynomials. The degree of the minimal polynomial of an $n \times n$ matrix is never larger than n because of the following remarkable result.

3.3.11.1 Cayley–Hamilton Theorem

Let $\mathbf{A} \in \mathcal{R}^{n \times n}$, where $\mathcal{R}$ is a field. Let $\chi(s)$ be the nth degree monic polynomial defined by

$$\chi(s) = \det(s\mathbf{I} - \mathbf{A}) \qquad (3.24)$$

Then $\chi(\mathbf{A}) = \mathbf{0}$.

The polynomial $\chi(s) = \det(s\mathbf{I} - \mathbf{A})$ is called the *characteristic polynomial* of $\mathbf{A}$.

3.3.12 Equivalence for Polynomial Matrices

Multiplication by $\mathbf{A}^{-1}$ transforms an invertible matrix $\mathbf{A}$ to a simple form: $\mathbf{A}\mathbf{A}^{-1} = \mathbf{I}\mathbf{A}\mathbf{A}^{-1} = \mathbf{I}$. For $\mathbf{A} \in \mathcal{R}^{n \times n}$ with $\det \mathbf{A} \neq 0$ but $\det \mathbf{A}$ not equal to a unit in $\mathcal{R}$, transformations of the form $\mathbf{A} \mapsto \mathcal{P}\mathbf{A}\mathcal{Q}$, where $\mathcal{P}, \mathcal{Q} \in \mathcal{R}^{n \times n}$ are invertible matrices, produce $\det \mathbf{A} \mapsto \det \mathcal{P} \det \mathbf{A} \det \mathcal{Q}$, that is, the determinant is multiplied by the invertible element $\det \mathcal{P} \det \mathcal{Q} \in \mathcal{R}$. Thus, invertible matrices $\mathcal{P}$ and $\mathcal{Q}$ can be sought to bring the product $\mathcal{P}\mathbf{A}\mathcal{Q}$ to some simplified form even when $\mathbf{A}$ is not invertible; $\mathcal{P}\mathbf{A}\mathcal{Q}$ and $\mathbf{A}$ are said to be related by $\mathcal{R}$-*equivalence*.

For equivalence of polynomial matrices (see [5] for details), where $\mathcal{R} = \mathbb{R}[s]$ (or $\mathbb{C}[s]$), let $\mathcal{P}(s)$ and $\mathcal{Q}(s)$ be invertible $n \times n$ polynomial matrices. Such matrices are called *unimodular*; they have constant, nonzero determinants. Let $\mathcal{A}(s)$ be an $n \times n$ polynomial matrix with nonzero determinant. Then for the equivalent matrix $\bar{\mathcal{A}}(s) = \mathcal{P}(s)\mathcal{A}(s)\mathcal{Q}(s)$, $\det \bar{\mathcal{A}}(s)$ differs from $\det \mathcal{A}(s)$ only by a constant factor; with no loss of generality $\bar{\mathcal{A}}(s)$ may be assumed to be scaled so that $\det \bar{\mathcal{A}}(s)$ is a *monic* polynomial, that is, so that the coefficient of the highest power of s in $\det \bar{\mathcal{A}}(s)$ is 1.

In forming $\bar{\mathcal{A}}(s)$, the multiplication of $\mathcal{A}(s)$ on the left by unimodular $\mathcal{P}(s)$ corresponds to performing a sequence of elementary row operations on $\mathcal{A}(s)$, and the multiplication of $\mathcal{A}(s)$ on the right by unimodular $\mathcal{Q}(s)$ corresponds to performing a sequence of elementary column operations on $\mathcal{A}(s)$. By suitable choice of $\mathcal{P}(s)$ and $\mathcal{Q}(s)$, $\mathcal{A}(s)$ may be brought to the *Smith canonical form*, $\mathcal{A}_S(s)$, a diagonal polynomial matrix whose diagonal elements are monic polynomials $\{\phi_i(s) : 1 \leq i \leq n\}$ satisfying the following divisibility conditions: $\phi_k(s)$ is a factor of $\phi_{k-1}(s)$, for $1 < k \leq n$.

The polynomials in the Smith canonical form $\mathcal{A}_S(s)$ are the *invariant polynomials* of $\mathcal{A}(s)$, and they may be obtained from $\mathcal{A}(s)$ as follows. Let $\epsilon_0(s) = 1$, and for $1 \leq i \leq n$, let $\epsilon_i(s)$ be the monic greatest common divisor of all nonzero $i \times i$ minors of $\mathcal{A}(s)$. Then the invariant polynomials are given by $\phi_i(s) = \epsilon_i(s)/\epsilon_{i-1}(s)$. It follows that $\det \mathcal{A}(s)$ is a constant multiple of the polynomial $\epsilon_n(s) = \phi_1(s)\phi_2(s) \cdots \phi_n(s)$.

Example

As an example of the Smith canonical form, consider

$$\mathcal{A}(s) = \begin{bmatrix} s(s+1)(s^2+s+1) & s^2(s+1)^2 \\ s(s+1)^2/3 & s(s+1)^2/3 \end{bmatrix}$$

The 1×1 minors are the matrix elements: $\Delta_{11} = a_{22}(s)$, $\Delta_{12} = a_{21}(s)$, $\Delta_{21} = a_{12}(s)$, and $\Delta_{22} = a_{11}(s)$. The sole 2×2 minor is $\det \mathcal{A}(s)$. So the invariant polynomials are found from $\epsilon_1(s) = s(s+1)$ and $\epsilon_2(s) = s^2(s+1)^3$, giving $\phi_1(s) = s(s+1)$ and $\phi_2(s) = s(s+1)^2$. The corresponding Smith canonical form is indeed equivalent to $\mathcal{A}(s)$:

$$\mathcal{A}_S(s) = \begin{bmatrix} s(s+1) & 0 \\ 0 & s(s+1)^2 \end{bmatrix} = \begin{bmatrix} 1 & -s \\ 0 & 3 \end{bmatrix} \mathcal{A}(s) \begin{bmatrix} 1 & 0 \\ -1 & 1 \end{bmatrix}$$

3.4 Vector Spaces

3.4.1 Definitions

A *vector space* consists of an ordered tuple $(\mathcal{V}, \mathcal{F}, +, \cdot)$ having the following list of attributes:

1. $\mathcal{V}$ is a set of elements called vectors, containing a distinguished vector $\mathbf{0}$, the zero vector.
2. $\mathcal{F}$ is a field of scalars; most commonly $\mathcal{F} = \mathbb{R}$ or $\mathbb{C}$, the real or complex numbers.
3. The $+$ operation is a vector addition operation defined on $\mathcal{V}$. For all $\mathbf{v}_1, \mathbf{v}_2, \mathbf{v}_3 \in \mathcal{V}$, the following properties must hold: (a) $\mathbf{v}_1 + \mathbf{v}_2 = \mathbf{v}_2 + \mathbf{v}_1$, (b) $\mathbf{v}_1 + \mathbf{0} = \mathbf{v}_1$, and (c) $(\mathbf{v}_1 + \mathbf{v}_2) + \mathbf{v}_3 = \mathbf{v}_1 + (\mathbf{v}_2 + \mathbf{v}_3)$.
4. The $\cdot$ operation is a scalar multiplication of vectors (and usually the $\cdot$ is not written explicitly). For all $\mathbf{v}_1, \mathbf{v}_2 \in \mathcal{V}$ and $\alpha_1, \alpha_2 \in \mathcal{F}$, the following properties must hold: (a) $0\mathbf{v}_1 = \mathbf{0}$, (b) $1\mathbf{v}_1 = \mathbf{v}_1$, (c) $\alpha_1(\mathbf{v}_1 + \mathbf{v}_2) = \alpha_1\mathbf{v}_1 + \alpha_1\mathbf{v}_2$, (d) $(\alpha_1 + \alpha_2)\mathbf{v}_1 = \alpha_1\mathbf{v}_1 + \alpha_2\mathbf{v}_1$, and (e) $\alpha_1(\alpha_2\mathbf{v}_1) = (\alpha_1\alpha_2)\mathbf{v}_1$.

These conditions formalize the idea that a vector space is a set of elements closed under the operation of taking linear combinations.

3.4.2 Examples and Fundamental Properties

The conventional notation for the vector space $\mathcal{V}$ consisting of (column) n-vectors of elements of $\mathcal{F}$ is $\mathcal{F}^n$; thus, $\mathbb{C}^n$ and $\mathbb{R}^n$ denote the spaces of complex and real n-vectors, respectively. To show that the theory is widely applicable, some other examples of vector spaces will be mentioned. Still others will arise later on.

1. The set of $m \times n$ matrices over a field $\mathcal{F}$, with the usual rules for scalar multiplication and matrix addition, forms a vector space, denoted $\mathcal{F}^{m \times n}$.
2. The set of polynomial functions of a complex variable $\mathcal{P} = \{p(s) : p(s) = p_0 s^k + p_1 s^{k-1} + \cdots + p_{k-1}s + p_k\}$ is a vector space over $\mathbb{C}$ because addition of two polynomials produces another polynomial, as does multiplication of a polynomial by a complex number.
3. The set $\mathcal{C}[0, T]$ of real-valued continuous functions defined on the closed interval $0 \le t \le T$ is a real vector space because the sum of two continuous functions is another continuous function and scalar multiplication also preserves continuity.

A common form of vector space is the *direct product space*. If $\mathcal{V}_1$ and $\mathcal{V}_2$ are vector spaces over a common field $\mathcal{F}$, the direct product, $\mathcal{V}_1 \times \mathcal{V}_2$, is the vector space whose elements are ordered pairs $(\mathbf{v}_1, \mathbf{v}_2)$, where $\mathbf{v}_1 \in \mathcal{V}_1$ and $\mathbf{v}_2 \in \mathcal{V}_2$, and where the vector space operations are defined elementwise. The extension to n-fold direct products is straightforward; there is an obvious correspondence between the n-fold direct product $\mathcal{F} \times \mathcal{F} \times \cdots \mathcal{F}$ and $\mathcal{F}^n$.

A number of important concepts from the theory of vector spaces will now be introduced.

3.4.2.1 Subspaces

If $\mathcal{V}$ is a vector space and $\mathcal{W}$ is a subset of vectors from $\mathcal{V}$, $\mathcal{W}$ is called a *subspace* of $\mathcal{V}$ if $\mathbf{0} \in \mathcal{W}$ and if $\alpha_1 \mathbf{v}_1 + \alpha_2 \mathbf{v}_2 \in \mathcal{W}$ for all $\mathbf{v}_1$ and $\mathbf{v}_2 \in \mathcal{W}$ and all α_1 and $\alpha_2 \in \mathcal{F}$. Notice that this means that $\mathcal{W}$ is a vector space itself.

The geometric intuition of subspaces is that they consist of "planes" (often called "hyperplanes" in spaces of high dimension) passing through the origin $\mathbf{0}$. The set $\mathcal{W} = \{\mathbf{0}\}$ is always a subspace, and $\mathcal{V}$ is a subspace of itself. If $\mathbf{v}$ is a nonzero vector in a vector space $\mathcal{V}$, then the set $\{\alpha\mathbf{v} : \alpha \in \mathcal{F}\}$ is a subspace of $\mathcal{V}$. For two subspaces $\mathcal{W}_1$ and $\mathcal{W}_2$, the intersection $\mathcal{W}_1 \cap \mathcal{W}_2$ is a subspace, and the sum $\mathcal{W}_1 + \mathcal{W}_2 = \{\mathbf{w}_1 + \mathbf{w}_2 : \mathbf{w}_1 \in \mathcal{W}_1 \text{ and } \mathbf{w}_2 \in \mathcal{W}_2\}$ is a subspace. When $\mathcal{W}_1 \cap \mathcal{W}_2 = \{\mathbf{0}\}$, the sum is said to be a *direct sum* and is denoted by $\mathcal{W}_1 \oplus \mathcal{W}_2$. When $\mathcal{V} = \mathcal{W}_1 \oplus \mathcal{W}_2$, the subspaces are said to be *complementary*.

3.4.2.2 Quotient Spaces

Let $\mathcal{V}$ be a vector space and let $\mathcal{W}$ be a subspace of $\mathcal{V}$. Associated with $\mathcal{W}$ is an *equivalence relation* on $\mathcal{V}$ defined by the condition that vectors $\mathbf{v}_1$ and $\mathbf{v}_2$ are equivalent just in case $(\mathbf{v}_1 - \mathbf{v}_2) \in \mathcal{W}$. The subset consisting of a vector $\mathbf{v}$ and all equivalent vectors in $\mathcal{V}$ is the *equivalence class* containing $\mathbf{v}$; the equivalence classes form a partition of $\mathcal{V}$, with every vector belonging to exactly one equivalence class. The set of equivalence classes inherits from $\mathcal{V}$ and $\mathcal{W}$, the structure of a vector space known as the *quotient space*, denoted by $\mathcal{V}/\mathcal{W}$. Scalar multiplication and vector addition in $\mathcal{V}$ extend to the equivalence classes comprising $\mathcal{V}/\mathcal{W}$ in the natural way, and $\mathcal{W}$ is the equivalence class serving as the zero element of $\mathcal{V}/\mathcal{W}$.

3.4.2.3 Linear Independence

A set of vectors $\{\mathbf{v}_1, \mathbf{v}_2, \ldots, \mathbf{v}_k\}$ is called *linearly independent* when the equation

$$\sum_{i=1}^{k} \alpha_i \mathbf{v}_i = \mathbf{0} \tag{3.25}$$

is satisfied only by the trivial choice of the scalars: $\alpha_1 = \alpha_2 = \cdots = \alpha_k = 0$. No nontrivial linear combination of linearly independent vectors equals the zero vector. A set of vectors that is not linearly independent is called *linearly dependent*. Any set containing $\mathbf{0}$ is linearly dependent.

3.4.2.4 Spanning Set

For any subset of vectors $\{\mathbf{v}_1, \mathbf{v}_2, \ldots, \mathbf{v}_k\}$ from $\mathcal{V}$, the *span* of the subset is the subspace of elements $\{\mathbf{v} = \sum_{i=1}^{k} \alpha_i \mathbf{v}_i : \alpha_i \in \mathcal{F}, 1 \leq i \leq k\}$, denoted sp $\{\mathbf{v}_1, \mathbf{v}_2, \ldots, \mathbf{v}_k\}$. If $\mathbf{v} \in \text{sp}\{\mathbf{v}_1, \mathbf{v}_2, \ldots, \mathbf{v}_k\}$ for every vector $\mathbf{v} \in \mathcal{V}$, the subset is called a *spanning set* for $\mathcal{V}$, and the vectors of the set *span* $\mathcal{V}$.

3.4.2.5 Basis

A *basis* for a vector space $\mathcal{V}$ is any spanning set for $\mathcal{V}$ consisting of linearly independent vectors.

3.4.2.6 Dimension

If a vector space has a spanning set with finitely many vectors, then the number of vectors in every basis is the same and this number is the *dimension* of the vector space. (By convention, the vector space consisting of $\mathbf{0}$ alone has no basis, and has dimension zero.) A vector space having no spanning set with finitely many vectors is called infinite dimensional.

All subspaces of a finite-dimensional vector space $\mathcal{V}$ are finite dimensional; if $\mathcal{V}$ has dimension n and $\mathcal{W}$ is a subspace of $\mathcal{V}$, then dim $\mathcal{W} \leq$ dim $\mathcal{V}$, and dim $\mathcal{V}/\mathcal{W} =$ dim $\mathcal{V} -$ dim $\mathcal{W}$.

3.4.2.7 Coordinates

If $\mathcal{S}$ is a linearly independent set of vectors, and $\mathbf{v} \in \text{sp} \, \mathcal{S}$, then there is a *unique* way of expressing $\mathbf{v}$ as a linear combination of the vectors in $\mathcal{S}$. Thus, given a basis, every vector in a vector space has a unique representation as a linear combination of the vectors in the basis. If $\mathcal{V}$ is a vector space of dimension n with basis $\mathcal{B} = \{\mathbf{b}_1, \mathbf{b}_2, \ldots, \mathbf{b}_n\}$, there is a natural correspondence between $\mathcal{V}$ and the vector space $\mathcal{F}^n$ defined by

$$\mathbf{v} \mapsto \mathbf{v}_{\mathcal{B}} = \begin{bmatrix} \alpha_1 \\ \alpha_2 \\ \vdots \\ \alpha_n \end{bmatrix} \tag{3.26}$$

where the elements of the *coordinate vector* $\mathbf{v}_B$ give the representation of $\mathbf{v} \in \mathcal{V}$ with respect to the basis B:

$$\mathbf{v} = \sum_{i=1}^{n} \alpha_i \mathbf{b}_i \tag{3.27}$$

In particular, the basis vectors of $\mathcal{V}$ correspond to the *standard basis vectors* of $\mathcal{F}^n$: $\mathbf{b}_i \mapsto \mathbf{e}_i$, where $\mathbf{e}_i \in \mathcal{F}^n$ is the vector whose ith element is 1 and all of its other elements are 0; the ith element of the standard basis of $\mathcal{F}^n$, $\mathbf{e}_i$, is called the ith *unit vector* or ith *principal axis vector*.

3.4.3 Linear Functions

Let $\mathcal{X}$ and $\mathcal{Y}$ be two vector spaces over a common field, $\mathcal{F}$. A *linear function* (sometimes called a linear transformation, linear operator, or linear mapping), denoted $f : \mathcal{X} \to \mathcal{Y}$, assigns to each $\mathbf{x} \in \mathcal{X}$ an element $\mathbf{y} \in \mathcal{Y}$ so as to make

$$f(\alpha_1 \mathbf{x}_1 + \alpha_2 \mathbf{x}_2) = \alpha_1 f(\mathbf{x}_1) + \alpha_2 f(\mathbf{x}_2) \tag{3.28}$$

for every $\mathbf{x}_1$ and $\mathbf{x}_2 \in \mathcal{X}$ (the domain of the function) and for every choice of scalars α_1 and $\alpha_2 \in \mathcal{F}$.

For any $\alpha \in \mathcal{F}$, the function $f : \mathbf{x} \mapsto \alpha \mathbf{x}$ is a linear functions. For $\alpha = 0$, it is the *zero function*; for $\alpha = 1$ it is the *identity function*.

When its domain, $\mathcal{X}$, is finite dimensional, a linear function is uniquely determined by its values on any basis for $\mathcal{X}$.

Composition of linear functions preserves linearity. If $h : \mathcal{X} \to \mathcal{Z}$ is defined as $h(\mathbf{x}) = g(f(\mathbf{x}))$, the composition of two other linear functions, $f : \mathcal{X} \to \mathcal{Y}$ and $g : \mathcal{Y} \to \mathcal{Z}$, then h is a linear function.

If f is a linear function, it necessarily maps subspaces of $\mathcal{X}$ to subspaces of $\mathcal{Y}$. There are two subspaces of particular importance associated with a linear function $f : \mathcal{X} \to \mathcal{Y}$. The *nullspace* or *kernel* of f is the subspace

$$\ker f = \{\mathbf{x} \in \mathcal{X} : f(\mathbf{x}) = \mathbf{0} \in \mathcal{Y}\} \tag{3.29}$$

The *range* or *image* of f is the subspace

$$\operatorname{im} f = \{\mathbf{y} \in \mathcal{Y} : \mathbf{y} = f(\mathbf{x}) \text{ for some } \mathbf{x} \in \mathcal{X}\} \tag{3.30}$$

If $\operatorname{im} f = \mathcal{Y}$, f is *surjective* or *onto*; if $\ker f = \{\mathbf{0}\}$, f is *injective* or *1 to 1*. When f is both surjective and injective, f is *invertible*, and there is an inverse function, $f^{-1} : \mathcal{Y} \to \mathcal{X}$ so that the compositions satisfy $f(f^{-1}(\mathbf{y})) = \mathbf{y}$ and $f^{-1}(f(\mathbf{x})) = \mathbf{x}$ for all $\mathbf{y} \in \mathcal{Y}$ and all $\mathbf{x} \in \mathcal{X}$. The inverse function f^{-1} is a linear function, when f is a linear function. When the linear function $f : \mathcal{X} \to \mathcal{Y}$ is invertible, the vector spaces $\mathcal{X}$ and $\mathcal{Y}$ are said to be *isomorphic*.

If $\mathcal{W}$ is a subspace of $\mathcal{X}$, the function $f|_{\mathcal{W}} : \mathcal{W} \to \mathcal{Y}$, called the *restriction of* f *to* $\mathcal{W}$ and defined by $f|_{\mathcal{W}}(\mathbf{w}) = f(\mathbf{w})$, is a linear function from $\mathcal{W}$ to $\mathcal{Y}$. For example, $f|_{\ker f}$ is the zero function on $\mathcal{W} = \ker f$.

When f is a linear function from a vector space $\mathcal{X}$ to itself, $f : \mathcal{X} \to \mathcal{X}$, the composite function $f(f(\mathbf{x}))$ is denoted succinctly as $f^2(\mathbf{x})$; similarly, for $n > 1$, $f^{n+1}(\mathbf{x}) = f(f^n(\mathbf{x})) = f^n(f(\mathbf{x}))$. Both $\ker f$ and $\operatorname{im} f$ are subspaces of $\mathcal{X}$. The quotient space $\mathcal{X}/\ker f$ is the vector space of equivalence classes of vectors in $\mathcal{X}$, defined by the condition that vectors $\mathbf{x}_1$ and $\mathbf{x}_2$ belong to the same equivalence class just in case $(\mathbf{x}_1 - \mathbf{x}_2) \in \ker f$.

For finite-dimensional $\mathcal{X}$, with $f : \mathcal{X} \to \mathcal{X}$, the quotient space $\mathcal{X}/\ker f$ and the subspace $\operatorname{im} f$ are isomorphic; the linear function $g : \operatorname{im} f \to \mathcal{X}/\ker f$ mapping $\mathbf{x}$ to its equivalence class in $\mathcal{X}/\ker f$ for every $\mathbf{x} \in \operatorname{im} f$ is easily shown to be invertible. It follows that $\dim \ker f + \dim \operatorname{im} f = \dim \mathcal{X}$.

A linear function $f : \mathcal{X} \to \mathcal{X}$ is called a *projection* if $f(f(\mathbf{x})) = f(\mathbf{x})$ for all $\mathbf{x}$; more precisely, f is a projection onto $\operatorname{im} f$. For a projection f, the subspaces $\ker f$ and $\operatorname{im} f$ are complementary; the linear function defined as $\mathbf{x} \mapsto \mathbf{x} - f(\mathbf{x})$ is called the complementary projection, and it is a projection onto $\ker f$.

For $f : \mathcal{X} \to \mathcal{X}$, any subspace mapped into itself is said to be f-*invariant*. For a subspace $\mathcal{W} \subseteq \mathcal{X}$, $\mathcal{W}$ is f-invariant if $f(\mathbf{w}) \in \mathcal{W}$ for all $\mathbf{w} \in \mathcal{W}$. $\ker f$ is f-invariant, and for any $\mathbf{x} \in \mathcal{X}$, the subspace spanned by the vectors $\{\mathbf{x}, f(\mathbf{x}), f(f(\mathbf{x})), \ldots\}$ is f-invariant. When $\mathcal{W}$ is f-invariant, $f|_{\mathcal{W}}$ is a linear function from $\mathcal{W}$ to $\mathcal{W}$.

Let $\mathcal{L}(\mathcal{X}, \mathcal{Y})$ denote the set of all linear functions from $\mathcal{X}$ to $\mathcal{Y}$; $\mathcal{L}(\mathcal{X}, \mathcal{Y})$ is a vector space over $\mathcal{F}$. If $\mathcal{X}$ has dimension n and $\mathcal{Y}$ has dimension m, then $\mathcal{L}(\mathcal{X}, \mathcal{Y})$ has dimension mn. If $\{\mathbf{x}_1, \ldots, \mathbf{x}_n\}$ is a basis for $\mathcal{X}$ and $\{\mathbf{y}_1, \ldots, \mathbf{y}_m\}$ is a basis for $\mathcal{Y}$, then a basis for $\mathcal{L}(\mathcal{X}, \mathcal{Y})$ is the set of linear functions $\{f_{ij}(\mathbf{x}) : 1 \leq i \leq n, \ 1 \leq j \leq m\}$, where basis function $f_{ij}(\mathbf{x})$ takes the value $\mathbf{y}_j$ for basis vector $\mathbf{x}_i$ and the value $\mathbf{0}$ for all other basis vectors $\mathbf{x}_k$, $k \neq i$.

For the special case of $\mathcal{F}$-valued linear functions, $\mathcal{L}(\mathcal{X}, \mathcal{F})$ is known as the space of *linear functionals* on $\mathcal{X}$ or the *dual space* of $\mathcal{X}$. If $\mathcal{X}$ has dimension n, $\mathcal{L}(\mathcal{X}, \mathcal{F})$ also has dimension n. The basis of functions described above is called the *dual basis*: with basis $\{\mathbf{x}_1, \ldots, \mathbf{x}_n\}$ for $\mathcal{X}$, the dual basis functions of $\mathcal{L}(\mathcal{X}, \mathcal{F})$ are the linear functionals $\{f_i(\mathbf{x}) : 1 \leq i \leq n\}$, where $f_i(\mathbf{x})$ takes the value 1 for basis vector $\mathbf{x}_i$ and the value 0 for all other basis vectors $\mathbf{x}_k$, $k \neq i$.

3.4.3.1 Matrix Representations

Let $\mathbf{A}$ be an $m \times n$ matrix over $\mathcal{F}$. Then the function $f : \mathcal{F}^n \to \mathcal{F}^m$, defined in terms of matrix multiplication, $\mathbf{x} \mapsto \mathbf{A}\mathbf{x}$ is a linear function. Indeed, every linear function $f : \mathcal{F}^n \to \mathcal{F}^m$ takes this form for a unique matrix $\mathbf{A}_f \in \mathcal{F}^{m \times n}$. Specifically, for $1 \leq i \leq n$, the ith column of $\mathbf{A}_f$ is defined to be the vector $f(\mathbf{e}_i) \in \mathcal{F}^m$, where $\mathbf{e}_i$ is the ith unit vector in $\mathcal{F}^n$. With this definition, for any $\mathbf{x} \in \mathcal{F}^n, f(\mathbf{x}) = \mathbf{A}_f \mathbf{x}$; for the case when $f(\mathbf{x}) = \mathbf{A}\mathbf{x}$, $\mathbf{A}_f = \mathbf{A}$, as expected.

This same idea can be extended using bases and coordinate vectors to provide a matrix representation for any linear function. Let f be a linear function mapping $\mathcal{X}$ to $\mathcal{Y}$, let $\mathcal{B}_X$ be a basis for $\mathcal{X}$, and let $\mathcal{B}_Y$ be a basis for $\mathcal{Y}$. Suppose $\mathcal{X}$ has dimension n and $\mathcal{Y}$ has dimension m. Then there is a unique matrix $\mathbf{A}_f \in \mathcal{F}^{m \times n}$ giving $\mathbf{A}_f \mathbf{x}_{\mathcal{B}_X} = \mathbf{y}_{\mathcal{B}_Y}$ if and only if $f(\mathbf{x}) = \mathbf{y}$, where $\mathbf{x}_{\mathcal{B}_X} \in \mathcal{F}^n$ is the coordinate vector of $\mathbf{x}$ with respect to the basis $\mathcal{B}_X$ and $\mathbf{y}_{\mathcal{B}_Y} \in \mathcal{F}^m$ is the coordinate vector of $\mathbf{y}$ with respect to the basis $\mathcal{B}_Y$. Thus the ith column of $\mathbf{A}_f$ is the coordinate vector (with respect to the basis $\mathcal{B}_Y$) of the vector $f(\mathbf{b}_i) \in \mathcal{Y}$, where $\mathbf{b}_i \in \mathcal{X}$ is the ith vector of the basis $\mathcal{B}_X$ for $\mathcal{X}$. $\mathbf{A}_f$ is called the *matrix representation of f with respect to the bases $\mathcal{B}_X$ and $\mathcal{B}_Y$*. When $\mathcal{X} = \mathcal{Y}$ and $\mathcal{B}_X = \mathcal{B}_Y$, $\mathbf{A}_f$ is simply called the *matrix representation of f with respect to the basis $\mathcal{B}_X$*.

If $h : \mathcal{X} \to \mathcal{Z}$ is defined as $h(\mathbf{x}) = g(f(\mathbf{x}))$, the composition of two other linear functions, $f : \mathcal{X} \to \mathcal{Y}$ and $g : \mathcal{Y} \to \mathcal{Z}$, and if the three vector spaces are finite dimensional and bases are chosen, the corresponding relationship between the matrix representations of the linear functions is given by $\mathbf{A}_h = \mathbf{A}_g \mathbf{A}_f$, and so composition of linear functions corresponds to matrix multiplication. For the case of $\mathcal{Z} = \mathcal{X}$ and $h(\mathbf{x}) = f^n(\mathbf{x})$ for some $n > 1$, the corresponding matrix representation is $\mathbf{A}_{f^n} = \mathbf{A}_f^n$.

Certain matrix representations arise from the relationship between two bases of a single vector space. If $\mathcal{B} = \{\mathbf{b}_1, \mathbf{b}_2, \ldots, \mathbf{b}_n\}$ and $\widehat{\mathcal{B}} = \{\widehat{\mathbf{b}}_1, \widehat{\mathbf{b}}_2, \ldots, \widehat{\mathbf{b}}_n\}$ are two bases for a vector space $\mathcal{X}$, let $t : \mathcal{X} \to \mathcal{X}$ be the identity function on $\mathcal{X}$, so that $t(\mathbf{x}) = \mathbf{x}$. Its matrix representation using basis $\mathcal{B}$ for $t(\mathbf{x})$ and $\widehat{\mathcal{B}}$ for $\mathbf{x}$ is the $n \times n$ matrix $\mathbf{T}$ whose ith column is the coordinate vector of $\widehat{\mathbf{b}}_i$ with respect to the basis $\mathcal{B}$. Thus the coordinate vectors satisfy the equation

$$\mathbf{T}\mathbf{x}_{\widehat{\mathcal{B}}} = \mathbf{x}_{\mathcal{B}} \tag{3.31}$$

$\mathbf{T}$ is an invertible matrix because the identity function is invertible: $t^{-1}(\mathbf{x}) = t(\mathbf{x})$. Also, the matrix $\mathbf{T}^{-1}$ is the matrix representation of the identity function on $\mathcal{X}$ with respect to the basis $\widehat{\mathcal{B}}$ for $t^{-1}(\mathbf{x})$ and basis $\mathcal{B}$ for $\mathbf{x}$.

Let $\mathcal{X} = \mathcal{F}^n$, and let $\mathcal{B}$ be the standard basis (i.e., the basis of standard unit vectors). Then every $\mathbf{x}$ is its own coordinate vector: $\mathbf{x} = \mathbf{x}_{\mathcal{B}}$. Let $\mathbf{T}$ be an invertible $n \times n$ matrix with columns $\mathbf{t}_1, \ldots, \mathbf{t}_n$. These n vectors in $\mathcal{F}^n$ are their own $\mathcal{B}$ coordinate vectors, so $\mathbf{T}$ is the matrix representation of the identity function on $\mathcal{F}^n$ corresponding to the standard basis $\mathcal{B}$ and basis $\widehat{\mathcal{B}} = \{\mathbf{t}_1, \ldots, \mathbf{t}_n\}$. Clearly, $\widehat{\mathcal{B}}$ and $\mathcal{B}$ coordinate vectors are related by the equation $\mathbf{T}\mathbf{x}_{\widehat{\mathcal{B}}} = \mathbf{x}_{\mathcal{B}}$; in other words, matrix $\mathbf{T}$ is the transformation of coordinates from the basis for $\mathcal{F}^n$ given by its columns to the standard basis.

Now consider a linear function $f : \mathcal{X} \to \mathcal{X}$ along with bases $\mathcal{B}$ and $\widehat{\mathcal{B}}$ for $\mathcal{X}$. Let $\mathbf{A}_f$ denote the matrix representation of f with respect to basis $\mathcal{B}$, and let $\widehat{\mathbf{A}}_f$ denote the matrix representation of f with respect to basis $\widehat{\mathcal{B}}$. The matrix representation $\mathbf{T}$ introduced above may be used to express the relationship between the two matrix representations of f:

$$\mathbf{T}^{-1}\mathbf{A}_f\mathbf{T} = \widehat{\mathbf{A}}_f \qquad (3.32)$$

or equivalently

$$\mathbf{A}_f = \mathbf{T}\widehat{\mathbf{A}}_f\mathbf{T}^{-1} \qquad (3.33)$$

Two $n \times n$ matrices $\mathbf{A}$ and $\widehat{\mathbf{A}}$ related by the equation $\mathbf{T}^{-1}\mathbf{A}\mathbf{T} = \widehat{\mathbf{A}}$, for some invertible matrix $\mathbf{T}$, are said to be *similar matrices*, with $\mathbf{T}$ being called a *similarity transformation*. Keeping in mind the reasoning that led to the similarity equation, we have the interpretation of $\mathbf{T}$ as the matrix formed from column vectors selected as basis vectors, so that the product $\mathbf{A}\mathbf{T}$, viewed columnwise, gives the vectors obtained by applying the linear function $f(\mathbf{x}) = \mathbf{A}\mathbf{x}$ to these basis vectors. Then, the multiplication by $\mathbf{T}^{-1}$ produces the matrix, $\mathbf{T}^{-1}\mathbf{A}\mathbf{T} = \widehat{\mathbf{A}}$, whose columns are the coordinate vectors with respect to this basis.

Similarity is an equivalence relation on $\mathcal{F}^{n \times n}$, and a complete characterization of the similarity equivalence classes is of major importance.

3.4.3.2 Linear Functionals

The matrix representation of linear functionals has some noteworthy aspects. Let $\mathcal{B}_X = \{\mathbf{x}_1, \ldots, \mathbf{x}_n\}$ be a basis for $\mathcal{X}$ and let the dual basis of $\mathcal{L}(\mathcal{X}, \mathcal{F})$ be $\{h_1(\mathbf{x}), \ldots, h_n(\mathbf{x})\}$. Let the basis of $\mathcal{F}$ as a 1-dimensional vector space over itself be $\{1\}$. From the definition of the dual basis functions, the matrix representation of $h(\mathbf{x}) \in \mathcal{L}(\mathcal{X}, \mathcal{F})$, where $h(\mathbf{x}) = \alpha_1 h_1(\mathbf{x}) + \cdots + \alpha_n h_n(\mathbf{x})$ is given by $\mathbf{A}_h = [\alpha_1 \alpha_2 \cdots \alpha_n] \in \mathcal{F}^{1 \times n}$. Notice that $\mathbf{A}_h^T$ is the vector of coordinates of $h(\mathbf{x})$ with respect to the dual basis.

Specializing to $\mathcal{X} = \mathcal{F}^n$, there is a connection with matrix inversion. As in our discussion of similarity, let $\mathbf{T}$ be the matrix formed from column vectors selected as basis vectors. The ith basis vector may be expressed as $\mathbf{T}\mathbf{e}_i$, where $\mathbf{e}_i$ is the ith unit vector. Since $\mathbf{T}^{-1}\mathbf{T} = \mathbf{I}$, it follows that the linear functional determined by the ith row of $\mathbf{T}^{-1}$, $h_i(\mathbf{x}) = \mathbf{e}_i^T\mathbf{T}^{-1}\mathbf{x}$, is precisely the ith dual basis functional.

Matrix transposition also has its origins in this property of linear functions as shown by a consideration of dual spaces. Let f be a linear function mapping $\mathcal{X}$ to $\mathcal{Y}$, let $\mathcal{B}_X$ be a basis for $\mathcal{X}$, and let $\mathcal{B}_Y$ be a basis for $\mathcal{Y}$. Suppose $\mathcal{X}$ has dimension n and $\mathcal{Y}$ has dimension m. The matrix representation of f is $\mathbf{A}_f \in \mathcal{F}^{m \times n}$. Now consider the dual spaces $\mathcal{L}(\mathcal{X}, \mathcal{F})$ and $\mathcal{L}(\mathcal{Y}, \mathcal{F})$; each dual space has its associated dual basis. Take any linear functional $g \in \mathcal{L}(\mathcal{Y}, \mathcal{F})$ and consider its composition with f, $g(f(\mathbf{x}))$. For the matrix representations, $\mathbf{A}_h = \mathbf{A}_g\mathbf{A}_f$. Using properties of matrix transposition, $\mathbf{A}_h^T = (\mathbf{A}_g\mathbf{A}_f)^T = \mathbf{A}_f^T\mathbf{A}_g^T$. As noted above, $\mathbf{A}_h^T$ and $\mathbf{A}_g^T$ are the coordinate vectors of $h(\mathbf{x})$ and $g(\mathbf{y})$, respectively. So $\mathbf{A}_f^T$ is the matrix representation of a linear function, the *adjoint function* of $f(\mathbf{x})$. $f^T : \mathcal{L}(\mathcal{Y}, \mathcal{F}) \to \mathcal{L}(\mathcal{X}, \mathcal{F})$, given by $f^T(g) : g(\mathbf{y}) \mapsto g(f(\mathbf{x}))$, so that $f^T(g)(\mathbf{x}) = g(f(\mathbf{x}))$.

3.4.3.3 Lie Algebras of Linear Functions

For two linear functions $f_1(\mathbf{x})$ and $f_2(\mathbf{x})$ in $\mathcal{L}(\mathcal{X}, \mathcal{X})$, the composite functions $f_1(f_2(\mathbf{x}))$ and $f_2(f_1(\mathbf{x}))$ are also linear functions. If $\mathcal{X}$ has dimension n and basis $\mathcal{B}$, the matrix representations of f_1 and f_2 will be denoted by $\mathbf{A}_1$ and $\mathbf{A}_2$, respectively; then $\mathbf{A}_1\mathbf{A}_2$ and $\mathbf{A}_2\mathbf{A}_1$ are the corresponding matrix representations of the composite functions. We conclude that square matrices *commute*, that is, $\mathbf{A}_1\mathbf{A}_2 = \mathbf{A}_2\mathbf{A}_1$ if and only if the associated composite functions are the same. To simplify notation, denote $\mathbf{A}_1\mathbf{A}_2 - \mathbf{A}_2\mathbf{A}_1 = [\mathbf{A}_1, \mathbf{A}_2]$, the *Lie product* of the matrices $\mathbf{A}_1$ and $\mathbf{A}_2$. $[\mathbf{A}_1, \mathbf{A}_2]$ is the matrix representation of the linear function $f_1(f_2(\mathbf{x})) - f_2(f_1(\mathbf{x}))$, the *Lie bracket* of the linear functions, denoted by $[f_1, f_2](\mathbf{x})$. Thus the matrices $\mathbf{A}_1$ and $\mathbf{A}_2$ commute if and only if the corresponding Lie bracket is the zero function.

In $\mathcal{L}(\mathbb{R}^n, \mathbb{R}^n)$, the real vector space of all linear functions mapping $\mathbb{R}^n$ to itself, a subspace $\mathcal{A}$ for which the Lie bracket maps $\mathcal{A} \times \mathcal{A}$ into $\mathcal{A}$ is called a *Lie algebra* (over $\mathbb{R}$). In the simplest cases, the Lie bracket of every pair of functions in $\mathcal{A}$ is the zero function, but more general subspaces may also be Lie algebras. The dimension of $\mathcal{A}$ is its dimension as a subspace of linear functions.

For every point $\widehat{\mathbf{x}} \in \mathbb{R}^n$, a real Lie algebra $\mathcal{A}$ associates a subspace of $\mathbb{R}^n$ given by $\widehat{\mathbf{x}} \mapsto \{f(\widehat{\mathbf{x}}) \in \mathbb{R}^n : f(\mathbf{x}) \in \mathcal{A}\}$. Under certain conditions, this subspace is the space of tangent vectors at the point $\widehat{\mathbf{x}}$ to a manifold ("surface") in $\mathbb{R}^n$ described implicitly as the solution space of an associated nonlinear equation.

Example

For $\mathbf{x} \in \mathbb{R}^n$ and $\mathbf{S} \in \mathbb{R}^{n \times n}$, $\mathcal{A}_{\text{skew}} = \{\mathbf{Sx} : \mathbf{S}^T = -\mathbf{S}\}$ is the $n(n-1)/2$-dimensional Lie algebra of skew symmetric linear functions. Any two linearly independent functions in $\mathcal{A}_{\text{skew}}$ have a nonzero Lie bracket. For $n = 3$ the tangent vectors at each point $\widehat{\mathbf{x}}$ form a subspace of $\mathbb{R}^3$ having dimension 2, the tangent plane at point $\widehat{\mathbf{x}}$ to the solution manifold of $\mathbf{F}(\mathbf{x}) = x_1^2 + x_2^2 + x_3^2 - (\widehat{x}_1^2 + \widehat{x}_2^2 + \widehat{x}_3^2) = 0$. This manifold is recognized as the surface of the sphere whose squared radius is $\widehat{x}_1^2 + \widehat{x}_2^2 + \widehat{x}_3^2$.

3.4.4 Norms and Inner Products

Vectors in $\mathbb{R}^2$ and $\mathbb{R}^3$ are often viewed as points in 2- and 3-dimensional Euclidean space, respectively. The resulting geometric intuition may be extended to other vector spaces by developing more general notions of length and angle.

3.4.4.1 Vector Norms

The notion of vector norm is introduced to play the role of length. For a vector space $\mathcal{V}$ over the real or complex numbers, the notation $\|\mathbf{v}\|$ is used to denote the *norm* of vector $\mathbf{v}$; $\|\mathbf{v}\| \in \mathbb{R}$. To qualify as a norm, three properties must hold.

N1. For all $\mathbf{v} \in \mathcal{V}$, $\|\mathbf{v}\| \geq 0$ with equality holding only for $\mathbf{v} = \mathbf{0}$

N2. For all $\mathbf{v} \in \mathcal{V}$ and all $\alpha \in \mathcal{F}$, $\|\alpha\mathbf{v}\| = |\alpha|\|\mathbf{v}\|$

N3. (*Triangle inequality*) For all $\mathbf{v}_1$ and $\mathbf{v}_2 \in \mathcal{V}$, $\|\mathbf{v}_1 + \mathbf{v}_2\| \leq \|\mathbf{v}_1\| + \|\mathbf{v}_2\|$

In N2, $|\alpha|$ denotes the absolute value when the field of scalars $\mathcal{F} = \mathbb{R}$, and it denotes the modulus (or magnitude) when $\mathcal{F} = \mathbb{C}$.

Examples

The *Euclidean norm* on $\mathbb{R}^n$ is given by

$$\|\mathbf{v}\| = (\mathbf{v}^T\mathbf{v})^{1/2} = \left(\sum_{i=1}^{n} v_i^2\right)^{1/2} \tag{3.34}$$

It corresponds to Euclidean length for vectors in $\mathbb{R}^2$ and $\mathbb{R}^3$. Other norms for $\mathbb{R}^n$ are the *uniform norm*, which will be denoted by $\|\mathbf{v}\|_\infty$, with

$$\|\mathbf{v}\|_\infty = \max\{|v_i| : 1 \leq i \leq n\} \tag{3.35}$$

and the family of p-norms, defined for real numbers $1 \leq p < \infty$ by

$$\|\mathbf{v}\|_p = \left(\sum_{i=1}^{n} |v_i|^p\right)^{1/p} \tag{3.36}$$

The Euclidean norm is the p-norm for $p = 2$.

Various norms turn out to be appropriate for applications involving vectors in other vector spaces; as an example, a suitable norm for $C[0, T]$, the space of real-valued continuous functions on the interval $0 \leq t \leq T$ is the uniform norm:

$$\|c(t)\|_\infty = \max\{|c(t)| : 0 \leq t \leq T\} \tag{3.37}$$

The name and notation are the same as used for the uniform norm on $\mathbb{R}^n$ since the analogy is apparent. A notion of p-norm for vector spaces of functions can also be established. The 2-norm is the natural generalization of Euclidean norm:

$$\|c(t)\|_2 = \left(\int_0^T |c(t)|^2 \, dt \right)^{1/2} \tag{3.38}$$

As a final example, the *Frobenius norm* of a matrix $\mathbf{A} \in \mathbb{R}^{m \times n}$, denoted $\|\mathbf{A}\|_F$, is the Euclidean norm of the nm-vector consisting of all of the elements of $\mathbf{A}$:

$$\|\mathbf{A}\|_F = \left(\sum_{i=1}^m \sum_{j=1}^n a_{ij}^2 \right)^{1/2} \tag{3.39}$$

3.4.4.2 Norms of Linear Functions

When matrices are viewed as representations of linear functions, it is more appropriate to employ a different kind of norm, one that arises from the role of a linear function as a mapping between vector spaces. For example, consider the linear function $f : \mathbb{R}^n \to \mathbb{R}^n$ given by $f(\mathbf{v}) = \mathbf{Av}$. When $\mathbb{R}^n$ is equipped with the Euclidean norm, the *induced Euclidean norm* of $\mathbf{A}$, is defined by

$$\|\mathbf{A}\| = \max\{\|\mathbf{Av}\| : \|\mathbf{v}\| = 1\} \tag{3.40}$$

This is easily generalized. For any linear function $f : \mathcal{X} \to \mathcal{Y}$ and norms $\|\cdot\|_\mathcal{X}$ and $\|\cdot\|_\mathcal{Y}$, the induced norm of f takes the form:

$$\|f\| = \max\{\|f(\mathbf{x})\|_\mathcal{Y} : \|\mathbf{x}\|_\mathcal{X} = 1\} \tag{3.41}$$

The subscripts are commonly suppressed when the choice of norms is readily determined from context. The induced matrix norm $\|\mathbf{A}\|$ for $\mathbf{A} \in \mathcal{F}^{n \times m}$ is simply the induced norm of the linear function $\mathbf{Av}$.

A consequence of the definition of induced norm is the inequality

$$\|\mathbf{Av}\| \leq \|\mathbf{A}\| \, \|\mathbf{v}\| \tag{3.42}$$

which holds for all vectors $\mathbf{v}$. This inequality also implies the following inequality for the induced norm of a matrix product:

$$\|\mathbf{AB}\| \leq \|\mathbf{A}\| \, \|\mathbf{B}\| \tag{3.43}$$

Examples

Explicit expressions for three of the most important induced matrix norms can be determined. Suppose $\mathbf{A} \in \mathbb{R}^{n \times n}$. For the induced Euclidean norm,

$$\|\mathbf{A}\| = \sigma_1(\mathbf{A}) \tag{3.44}$$

the largest *singular value* of the matrix. (Singular values are discussed later. Lacking an explicit formula for $\|\mathbf{A}\|$, in some cases it suffices to have the easily evaluated bounds $\|\mathbf{A}\| \leq \|\mathbf{A}\|_F \leq \sqrt{n}\|\mathbf{A}\|$.)

For the induced uniform norm,

$$\|\mathbf{A}\|_\infty = \max_{1 \le i \le n} \sum_{j=1}^{n} |a_{ij}| \tag{3.45}$$

which is the largest of the absolute row-sums of **A**. For the induced 1-norm,

$$\|\mathbf{A}\|_1 = \max_{1 \le j \le n} \sum_{i=1}^{n} |a_{ij}| \tag{3.46}$$

which is the largest of the absolute column-sums of **A**.

3.4.4.3 Metric Spaces

It is sometimes useful to have a measure of the distance between two vectors $\mathbf{v}_1$ and $\mathbf{v}_2$ without imposing the limitation of property N2 of norms. As an alternative, a distance function or *metric* is defined on a set $\mathcal{S}$ of vectors as a real-valued function of vectors, $d(\mathbf{v}_1, \mathbf{v}_2) : \mathcal{S} \times \mathcal{S} \to \mathbb{R}$, satisfying three properties:

M1. For all $\mathbf{v}_1$ and $\mathbf{v}_2 \in \mathcal{S}$, $d(\mathbf{v}_1, \mathbf{v}_2) = d(\mathbf{v}_2, \mathbf{v}_1)$
M2. For all $\mathbf{v}_1$ and $\mathbf{v}_2 \in \mathcal{S}$, $d(\mathbf{v}_1, \mathbf{v}_2) \ge 0$ with equality holding only for $\mathbf{v}_1 = \mathbf{v}_2$
M3. (*triangle inequality*) For all $\mathbf{v}_1$, $\mathbf{v}_2$, and $\mathbf{v}_3 \in \mathcal{S}$, $d(\mathbf{v}_1, \mathbf{v}_2) \le d(\mathbf{v}_1, \mathbf{v}_3) + d(\mathbf{v}_3, \mathbf{v}_2)$

A *metric space* is a vector space with associated metric. Since the definition of a metric makes no use of vector space operations (scalar multiplication and vector addition), applications involving less structured subsets $\mathcal{S}$ are also possible. For example, if $\mathcal{S}$ is the unit circle in $\mathbb{R}^2$, that is, the points having Euclidean norm equal to 1, then the (shortest) arc length between two points in $\mathcal{S}$ is a metric.

3.4.4.4 Inner Products and Orthogonality

For two vectors in $\mathcal{F}^n$, the dot product (also called the *scalar product*) is the function defined by $\mathbf{v}_1 \cdot \mathbf{v}_2 = \mathbf{v}_1^{\mathrm{T}} \mathbf{v}_2$.

For nonzero vectors $\mathbf{v}_1$ and $\mathbf{v}_2$ in $\mathbb{R}^2$ or $\mathbb{R}^3$, the Euclidean geometric notion of angle is easily expressed in terms of the dot product. With $\mathbf{v}_1 \cdot \mathbf{v}_2 = \mathbf{v}_1^{\mathrm{T}} \mathbf{v}_2$, the angle between $\mathbf{v}_1$ and $\mathbf{v}_2$, θ, satisfies

$$\cos(\theta) = \frac{\mathbf{v}_1^{\mathrm{T}} \mathbf{v}_2}{((\mathbf{v}_1^{\mathrm{T}} \mathbf{v}_1)(\mathbf{v}_2^{\mathrm{T}} \mathbf{v}_2))^{1/2}} = \frac{\mathbf{v}_1^{\mathrm{T}} \mathbf{v}_2}{\|\mathbf{v}_1\| \, \|\mathbf{v}_2\|} \tag{3.47}$$

where the Euclidean norm is used in the second expression.

The notion of *inner product* of vectors is used to obtain a generalization of the dot product and thereby to provide a geometric interpretation for angles between vectors in other vector spaces. If $\mathcal{X}$ is a vector space over $\mathbb{R}$, the mapping from the direct product space (ordered pairs of vectors) $\mathcal{X} \times \mathcal{X}$ to $\mathbb{R}$ defined by $(\mathbf{x}_1, \mathbf{x}_2) \mapsto \langle \mathbf{x}_1, \mathbf{x}_2 \rangle$ is an inner product if the following properties are satisfied:

I1. For all $\mathbf{x} \in \mathcal{X}$, $\langle \mathbf{x}, \mathbf{x} \rangle \ge 0$ with equality holding only for $\mathbf{x} = \mathbf{0}$
I2. For all $\mathbf{x}_1$ and $\mathbf{x}_2 \in \mathcal{X}$, $\langle \mathbf{x}_1, \mathbf{x}_2 \rangle = \langle \mathbf{x}_2, \mathbf{x}_1 \rangle$
I3. For all $\mathbf{x}_1$ and $\mathbf{x}_2 \in \mathcal{X}$ and $\alpha \in \mathbb{R}$, $\langle \alpha \mathbf{x}_1, \mathbf{x}_2 \rangle = \alpha \langle \mathbf{x}_1, \mathbf{x}_2 \rangle$
I4. For all $\mathbf{x}_1$, $\mathbf{x}_2$, and $\mathbf{x}_3 \in \mathcal{X}$, $\langle \mathbf{x}_1 + \mathbf{x}_2, \mathbf{x}_3 \rangle = \langle \mathbf{x}_1, \mathbf{x}_3 \rangle + \langle \mathbf{x}_2, \mathbf{x}_3 \rangle$

Inner products for complex vector spaces are complex valued, and they satisfy similar properties (but involving complex conjugation).

The definition $\|\mathbf{x}\| = (\langle \mathbf{x}, \mathbf{x} \rangle)^{1/2}$ provides $\mathcal{X}$ with a *compatible* norm. Furthermore, the *Schwarz inequality*

$$|\langle \mathbf{x}_1, \mathbf{x}_2 \rangle| \leq \|\mathbf{x}_1\| \, \|\mathbf{x}_2\| \tag{3.48}$$

allows for the interpretation of the angle θ between vectors $\mathbf{x}_1$ and $\mathbf{x}_2$ through

$$\cos(\theta) = \frac{\langle \mathbf{x}_1, \mathbf{x}_2 \rangle}{\|\mathbf{x}_1\| \, \|\mathbf{x}_2\|} \tag{3.49}$$

For the vector space $\mathbb{R}^n$, the definition $\langle \mathbf{v}_1, \mathbf{v}_2 \rangle = \mathbf{v}_1^T \mathbf{v}_2$, coincides with the dot product. Furthermore, the Euclidean norm on $\mathbb{R}^n$ is *compatible* with this inner product.

With the notion of angle now defined in terms of inner product, two vectors are said to be *orthogonal* if their inner product is zero.

For example, in the vector space $C[0, T]$ of real-valued continuous functions on the interval $0 \leq t \leq T$, the inner product of two functions $c_1(t)$ and $c_2(t)$ is defined by

$$\langle c_1(t), c_2(t) \rangle = \int_0^T c_1(t) c_2(t) \, dt \tag{3.50}$$

and the Euclidean norm on $C[0, T]$ defined earlier is compatible with this inner product.

While an inner product provides a norm, there are many cases of vector spaces having norms that are not compatible with any definition of an inner product. For example, on $\mathbb{R}^n$ the uniform norm does not correspond to any inner product.

3.4.4.5 Inner Product Spaces

Let $\mathcal{V}$ be an *inner product space*, a vector space with an inner product and compatible norm. A set of mutually orthogonal vectors is known as an orthogonal set, and a basis consisting of mutually orthogonal vectors is known as an *orthogonal basis*. An orthogonal basis consisting of vectors whose norms are all one (i.e. consisting of vectors having unit length) is called an *orthonormal basis*.

Given an orthonormal basis, any vector is easily expressed as a linear combination of the orthonormal basis vectors. If $\{\mathbf{w}_1, \ldots, \mathbf{w}_k\}$ is an orthonormal basis, the vector $\mathbf{v}$ is given by

$$\mathbf{v} = \sum_{i=1}^k \alpha_i \mathbf{w}_i \tag{3.51}$$

where $\alpha_i = \langle \mathbf{v}, \mathbf{w}_i \rangle$. Also, as a generalization of the Pythagorean theorem,

$$\|\mathbf{v}\|^2 = \sum_{i=1}^k \alpha_i^2 \tag{3.52}$$

3.4.4.6 Orthogonal Projections

Suppose $\mathcal{V}$ is an inner product space and let $\mathcal{W}$ be a finite-dimensional subspace of $\mathcal{V}$. The subspace $\mathcal{W}^\perp = \{\mathbf{v} : \langle \mathbf{v}, \mathbf{w} \rangle = 0 \text{ for all } \mathbf{w} \in \mathcal{W}\}$ is called the *orthogonal complement of $\mathcal{W}$ in $\mathcal{V}$*. Since every $\mathbf{v} \in \mathcal{V}$ can be written uniquely in the form $\mathbf{v} = \mathbf{w} + \mathbf{w}^\perp$, where $\mathbf{w} = \hat{\mathbf{v}} \in \mathcal{W}$ and $\mathbf{w}^\perp = (\mathbf{v} - \hat{\mathbf{v}}) \in \mathcal{W}^\perp$, $\mathcal{W} \cap \mathcal{W}^\perp = \{\mathbf{0}\}$, and $\mathcal{V} = \mathcal{W} \oplus \mathcal{W}^\perp$ is an orthogonal direct sum decomposition of $\mathcal{V}$. The function $p_{\mathcal{W}} : \mathcal{V} \to \mathcal{V}$ defined in terms of the unique decomposition, $\mathbf{v} \mapsto \mathbf{w} = \hat{\mathbf{v}}$ is a linear function. Furthermore, $p_{\mathcal{W}}(p_{\mathcal{W}}(\mathbf{v})) = p_{\mathcal{W}}(\mathbf{v})$, so the function is a projection; it is called the *orthogonal projection of $\mathcal{V}$ onto $\mathcal{W}$*. The orthogonal direct sum decomposition of $\mathcal{V}$ may be written as $\mathcal{V} = \text{im } p_{\mathcal{W}} \oplus \ker p_{\mathcal{W}}$. The *complementary orthogonal projection* of $p_{\mathcal{W}}$ is the orthogonal projection of $\mathcal{V}$ onto $\mathcal{W}^\perp$, denoted as $p_{\mathcal{W}^\perp}$; its kernel and image provide another orthogonal direct sum representation of $\mathcal{V}$, whose terms correspond to the image and kernel of $p_{\mathcal{W}}$, respectively.

3.4.4.7 Linear Least-Squares Approximation

Suppose $\mathcal{V}$ is an inner product space and let $\mathcal{W}$ be a finite-dimensional subspace of $\mathcal{V}$. (Other assumptions about the subspace $\mathcal{W}$ can deal with some more general situations [10].) Every vector $\mathbf{v} \in \mathcal{V}$ has a unique *linear least-squares approximation* in $\mathcal{W}$, that is, a unique vector $\widehat{\mathbf{v}} \in \mathcal{W}$ satisfying $\|\mathbf{v} - \widehat{\mathbf{v}}\| \leq \|\mathbf{v} - \mathbf{w}\|$ for any choice of $\mathbf{w} \in \mathcal{W}$. Indeed, $\widehat{\mathbf{v}} = p_{\mathcal{W}}(\mathbf{v})$, the orthogonal projection of $\mathbf{v}$ onto $\mathcal{W}$. The solution is characterized by the *projection theorem*:

$$(\mathbf{v} - \widehat{\mathbf{v}}) \in \mathcal{W}^{\perp} \tag{3.53}$$

which expresses the condition that the error in the linear least-squares approximation of $\mathbf{v}$ is orthogonal to every vector in $\mathcal{W}$. More concretely, the linear least-squares approximation is found as the solution to the *normal equations*

$$\langle \mathbf{v} - \widehat{\mathbf{v}}, \mathbf{w}_i \rangle = 0 \quad \text{for all } \mathbf{w}_i \in \mathcal{W} \tag{3.54}$$

(For finite-dimensional $\mathcal{W}$, it suffices to check the finite number of equations obtained by selecting $\{\mathbf{w}_i\}$ to be a basis for $\mathcal{W}$, and the equations take their simplest form for an orthonormal basis of $\mathcal{W}$.)

3.4.4.8 Gram–Schmidt Orthogonalization and QR Factorization

There is a constructive procedure for obtaining an orthonormal basis starting from an arbitrary basis, the *Gram–Schmidt procedure*. Starting with a basis of k vectors $\{\mathbf{v}_1, \mathbf{v}_2, \ldots, \mathbf{v}_k\}$, the orthonormal basis $\{\mathbf{w}_1, \mathbf{w}_2, \ldots, \mathbf{w}_k\}$ is constructed sequentially according to the following steps:

1. $\mathbf{w}_1 = \mathbf{z}_1 / \|\mathbf{z}_1\|$, where $\mathbf{z}_1 = \mathbf{v}_1$
2. For $2 \leq i \leq k$, $\mathbf{w}_i = \mathbf{z}_i / \|\mathbf{z}_i\|$, where $\mathbf{z}_i = \mathbf{v}_i - \sum_{j=1}^{i-1} \langle \mathbf{v}_i, \mathbf{w}_j \rangle \mathbf{w}_j$

A geometric interpretion of the Gram–Schmidt procedure is enlightening. With subspaces $\mathcal{W}_0 = \{\mathbf{0}\}$ and $\mathcal{W}_i = \mathrm{sp}\{\mathbf{v}_1, \ldots, \mathbf{v}_i\}$ for $i \geq 1$, the vectors $\mathbf{z}_i$ are the orthogonal projections of $\mathbf{v}_i$ onto $\mathcal{W}_{i-1}^{\perp}$, and the orthonormal basis vectors are obtained by scaling these projections to vectors of unit length.

For k vectors in $\mathbb{R}^m$, $k \leq m$, take the vectors as columns of $\mathbf{V} \in \mathbb{R}^{m \times k}$. Then the Gram–Schmidt procedure produces the matrix factorization $\mathbf{V} = \mathbf{W}\mathbf{U}$, where $\mathbf{W} \in \mathbb{R}^{m \times k}$ and $\mathbf{U} \in \mathbb{R}^{k \times k}$ is an upper triangular matrix. The columns of the matrix $\mathbf{W}$ are orthonormal so that $\mathbf{W}^{\mathrm{T}}\mathbf{W} = \mathbf{I}_k$.

The factorization of $\mathbf{V}$ into a product of a matrix with orthonormal columns times an upper triangular matrix, $\mathbf{W}\mathbf{U}$, is traditionally known as the QR factorization [6]. It is rarely computed column-by-column because better numerical accuracy can be achieved by taking a different approach. For simplicity, assume that $\mathbf{V}$ is $m \times m$. If any sequence of orthogonal matrices $\mathbf{W}_1, \mathbf{W}_2, \ldots, \mathbf{W}_j$ can be found so that $\mathbf{V}$ is transformed to an upper triangular matrix,

$$\mathbf{W}_j \cdots \mathbf{W}_2 \mathbf{W}_1 \mathbf{V} = \mathbf{U} \tag{3.55}$$

then multiplying both sides of this equation by $\mathbf{W} = \mathbf{W}_1^{\mathrm{T}} \mathbf{W}_2^{\mathrm{T}} \cdots \mathbf{W}_j^{\mathrm{T}}$ produces the QR factorization.

A commonly applied computational procedure for QR factorization involves a certain sequence of $j = m$ symmetric orthogonal matrices known as Householder transformations [8], matrices of the form $\mathbf{W}(\mathbf{y}) = \mathbf{I} - 2\mathbf{y}\mathbf{y}^{\mathrm{T}} / \|\mathbf{y}\|^2$. The matrix $\mathbf{W}_i$ is chosen to be the Householder transformation that produces all subdiagonal elements of the ith column of $\mathbf{W}_i \cdots \mathbf{W}_1 \mathbf{V}$ equal to zero without changing any of the zeros that are subdiagonal elements of the first $i - 1$ columns of $\mathbf{W}_{i-1} \cdots \mathbf{W}_1 \mathbf{V}$.

3.4.4.9 Orthogonal Transformations, Orthogonal Matrices, and Orthogonal Projection Matrices

A linear function $f : \mathcal{V} \to \mathcal{V}$, where $\mathcal{V}$ is an inner product space over $\mathbb{R}$, is an *orthogonal transformation* if it maps an orthonormal basis to an orthonormal basis. If $\mathcal{V}$ is n dimensional, the matrix representation of f with respect to an orthonormal basis is an *orthogonal matrix*, an $n \times n$ matrix $\mathbf{O}$ satisfying $\mathbf{O}^{-1} = \mathbf{O}^{\mathrm{T}}$. The columns of an orthogonal matrix form an orthonormal basis for $\mathbb{R}^n$ (with the usual inner product,

$\langle \mathbf{x}_1, \mathbf{x}_2 \rangle = \mathbf{x}_1^T \mathbf{x}_2$). For an orthogonal transformation, $\|f(\mathbf{v})\| = \|\mathbf{v}\|$, and for an orthogonal matrix $\|\mathbf{Ox}\| = \|\mathbf{x}\|$, for all $\mathbf{x}$. Any orthogonal matrix has induced Euclidean norm equal to 1.

A matrix formulation is also useful in studying orthogonal projections in the inner product space $\mathbb{R}^n$. A matrix $\mathbf{A} \in \mathbb{R}^{n \times n}$ is called a projection matrix when $\mathbf{A}^2 = \mathbf{A}$, which corresponds to $\mathbf{A}$ being the matrix representation of a projection f on an n-dimensional real vector space with respect to some basis. The matrix representation of the complementary projection is $\mathbf{I} - \mathbf{A}$. In other words, $\mathbf{Ax}$ is a projection defined on (the coordinate space) $\mathbb{R}^n$. For the subspace $\mathcal{W}$ of $\mathbb{R}^n$ having an orthonormal basis $\{\mathbf{w}_1, \ldots, \mathbf{w}_k\}$, an orthogonal projection onto $\mathcal{W}$ is given by $\mathbf{WW}^T = \mathbf{A}$, where $\mathbf{W} \in \mathbb{R}^{n \times k}$ is the matrix whose columns are the orthonormal basis vectors. Notice that this orthogonal projection matrix is symmetric. In fact, the choice of orthonormal basis does not matter; uniqueness of the orthogonal projection onto any subspace can be shown directly from the defining equations $\mathbf{A}^2 = \mathbf{A}$ and $\mathbf{A}^T = \mathbf{A}$.

3.5 Linear Equations

For a linear function $f : \mathcal{X} \to \mathcal{Y}$, it is frequently of interest to find a vector $\mathbf{x} \in \mathcal{X}$ whose image under f is some given vector $\mathbf{y} \in \mathcal{Y}$, that is, to *solve* the equation $f(\mathbf{x}) = \mathbf{y}$ for $\mathbf{x}$. By resorting to the matrix representation of f if necessary, there is no loss of generality in assuming that the problem is posed in the framework of matrices and vectors, a framework that is suited to numerical computation as well as to theoretical analysis using matrix algebra. With $\mathbf{A} \in \mathcal{F}^{m \times n}$, $\mathbf{x} \in \mathcal{F}^n$, and $\mathbf{y} \in \mathcal{F}^m$. the equation

$$\mathbf{Ax} = \mathbf{y} \tag{3.56}$$

specifies m *linear equations* in n unknowns (the elements of $\mathbf{x}$).

3.5.1 Existence and Uniqueness of Solutions

From the definition of matrix multiplication, the left-hand side of Equation 3.56 is a linear combination of the n columns of $\mathbf{A}$, the unknown coefficients of the linear combination being the elements of the vector $\mathbf{x}$. Thus for a given $\mathbf{y} \in \mathcal{F}^m$, there will be a solution if and only if $\mathbf{y} \in \operatorname{im} \mathbf{Ax}$; since this subspace, the image or range of the linear function $\mathbf{Ax}$, is spanned by the columns of $\mathbf{A}$, it is conventionally called the *range* of $\mathbf{A}$, denoted by $\mathrm{R}(\mathbf{A})$.

In order that a solution $\mathbf{x}$ can be found for every possible choice of $\mathbf{y}$, it is necessary and sufficient that $\mathrm{R}(\mathbf{A})$ has dimension m, or equivalently that there are m linearly independent columns among the n columns of $\mathbf{A}$. When $\mathbf{A}$ has fewer than m linearly independent columns, the linear equations will be inconsistent for some choices of $\mathbf{y}$, and for such $\mathbf{A}$ the terminology *overdetermined linear equations* is commonly used.

Solutions to linear equations are not necessarily unique; the terminology *underdetermined linear equations* is used in the case when the columns of $\mathbf{A}$ are not linearly independent, since uniqueness holds if and only if the columns of $\mathbf{A}$ are linearly independent vectors. In the case $n > m$, uniqueness never holds, whereas in the case $n = m$, the uniqueness condition coincides with the existence condition: the matrix $\mathbf{A}$ must be invertible, and this is equivalent to the condition that the determinant of $\mathbf{A}$ be nonzero.

If a linear equation has two distinct solutions, $\mathbf{x}_1$ and $\mathbf{x}_2$, then the difference vector $\mathbf{x} = \mathbf{x}_1 - \mathbf{x}_2$ is a *nontrivial* (nonzero) solution to the related homogeneous equation

$$\mathbf{Ax} = \mathbf{0} \tag{3.57}$$

This equation shows that a nontrivial solution to the homogeneous equation may be found if and only if the columns of $\mathbf{A}$ are not linearly independent. For $\mathbf{A} \in \mathcal{F}^{n \times n}$ this is equivalent to the condition $\det \mathbf{A} = 0$.

The set of all solutions to the homogeneous equation forms a subspace of $\mathcal{F}^n$, $\ker \mathbf{Ax}$, the kernel or nullspace of the linear function $\mathbf{Ax}$. It is conventionally called the *nullspace* of $\mathbf{A}$, denoted $\mathrm{N}(\mathbf{A})$. By considering the quotient space $\mathcal{F}^n/N(A)$. We find the following fundamental result relating the subspaces associated with A: $\dim R(A) + \dim N(A) = n$. Indeed, $N(A)$ provides a means of describing the entire set of solutions to underdetermined linear equations. If $Ax = y$ has a solution, say x_0, the set of all solutions can be expressed as

$$S_x = \{x : x = x_0 + \sum_{i=1}^{k} \alpha_i x_i\} \tag{3.58}$$

where the α_i are arbitrary scalars and $\{x_1, \ldots, x_k\}$ is a basis for $N(A)$. When $N(A) = \{0\}$ consistent linear equations have a unique solution.

The dimension of $R(A)$ (a subspace of $\mathcal{F}^m$) is called the *rank* of A. Thus the rank of A is the number of its linearly independent columns. Similarly, the dimension of $N(A)$ (a subspace of $\mathcal{F}^n$) is called the *nullity* of A. As already noted, for $m \times n$ matrices, these quantities are related by the equation $\text{rank}(A) + \text{nullity}(A) = n$.

Remarkably, the rank of a matrix and the rank of its transpose are the same; in other words, for any $m \times n$ matrix A, the number of linearly independent rows equals the number of linearly independent columns. This follows from an examination of the duality between subspaces associated with A and those associated with A^T. In $\mathcal{F}^n$, the subspace $(R(A^T))^\perp$ is the same as the subspace $N(A)$. Similarly, in $\mathcal{F}^m$, the subspace $(R(A))^\perp$ is the same as the subspace $N(A^T)$. Since $\dim R(A^T) + \dim(R(A^T))^\perp = n$ and since $\dim R(A) + \dim N(A) = n$, that is, $\text{rank}(A) + \text{nullity}(A) = n$, it follows that $\dim R(A^T) = \dim R(A)$, that is, $\text{rank}(A^T) = \text{rank}(A)$

When $\mathcal{F}^m$ is an inner product space, a useful characterization of solvability of linear equations arises. In this case, $(R(A))^\perp = \{0\}$ when $R(A) = \mathcal{F}^m$ so a solution x to $Ax = y$ can be found for every possible choice of y if and only if $N(A^T) = \{0\}$.

3.5.2 Solving Linear Equations

For $A \in \mathcal{F}^{n \times n}$, when A is invertible the solution of $Ax = y$ can be written explicitly as a linear function of y, $y = A^{-1}x$; this relation also shows that the ith column of A^{-1} is the solution to the linear equation $Ax = e_i$, where e_i is the ith unit vector.

For numerical computation of the solution of $Ax = y$, the PLU factorization of A (i.e., Gaussian elimination) may be used. Given the factorization $A = PLU$, the solution x may be obtained from solving two triangular sets of linear equations: solve $PLz = y$ (which is triangular after a reordering of the equations), and then solve $Ux = z$.

For the case of linear equations over the field of real numbers, additional results may be developed by employing geometric concepts. Let $A \in \mathbb{R}^{m \times n}$, $x \in \mathbb{R}^n$, and $y \in \mathbb{R}^m$. The usual inner products will be used for $\mathbb{R}^n$ and $\mathbb{R}^m$: $\langle x_1, x_2 \rangle = x_1^T x_2$ and $\langle y_1, y_2 \rangle = y_1^T y_2$. In this framework, the matrix $A^T A \in \mathbb{R}^{n \times n}$ is called the *Gram matrix* associated with A; the Gram matrix is symmetric and its (i,j)th element is the inner product of the ith and jth columns of A. The Gram matrix is invertible if and only if A has linearly independent columns, and so to test for uniqueness of solutions to consistent linear equations, it suffices to verify that $\det A^T A$ is nonzero. In this case, premultiplying both sides of the linear equation $Ax = y$ by A^T produces the equation (the *normal equations* for the components of x), $A^T Ax = A^T y$, which has the solution $x = (A^T A)^{-1} A^T y$. An alternative approach with better inherent numerical accuracy is to use the QR factorization $A = WU$, premultiplying both sides of the linear equation by W^T to give an easily solved triangular system of linear equations, $Ux = W^T y$.

3.5.2.1 Numerical Conditioning of Linear Equations

Geometric methods are useful in sensitivity analysis for linear equations. For $A \in \mathbb{R}^{n \times n}$, consider the linear equations $Ax = y$, and suppose that the vector y is perturbed to become $y + \Delta y$. Then $A(x + \Delta x) = y + \Delta y$, where $\Delta x = A^{-1} \Delta y$. Using norms to quantify the relative change in x arising from the relative change in **y**, leads to the inequality

$$\frac{\|\Delta\mathbf{x}\|}{\|\mathbf{x}\|} \leq \kappa\left(\mathbf{A}\right)\frac{\|\Delta\mathbf{y}\|}{\|\mathbf{y}\|} \tag{3.59}$$

where $\kappa(\mathbf{A})$ denotes the *condition number* of **A**, defined as

$$\kappa(\mathbf{A}) = \|\mathbf{A}\|\|\mathbf{A}^{-1}\| \tag{3.60}$$

Since $\kappa(\mathbf{A}) = \|\mathbf{A}\|\|\mathbf{A}^{-1}\| \geq \|\mathbf{A}\mathbf{A}^{-1}\| = 1$, when $\kappa(\mathbf{A}) \approx 1$, the matrix **A** is well-conditioned, but when $\kappa(\mathbf{A}) \gg 1$, the matrix **A** is ill-conditioned. The condition number of **A** also serves as the multiplier scaling relative errors in **A**, measured by the induced norm, to relative errors in **x** [7].

3.5.3 Approximate Solutions and the Pseudoinverse

A geometric approach provides a means of avoiding issues of existence and uniqueness of solutions by replacing a consideration of the linear equation $\mathbf{A}\mathbf{x} = \mathbf{y}$ with the following more general problem: among the vectors $\widehat{\mathbf{x}}$ that minimize the Euclidean norm of the error vector $\mathbf{A}\widehat{\mathbf{x}} - \mathbf{y}$, find that vector **x** of smallest Euclidean norm. The resulting vector **x** provides an approximate solution to inconsistent linear equations and, alternatively, a solution in the case of consistent linear equations.

A unique solution to the general problem always exists and takes the form

$$\mathbf{x} = \mathbf{A}^{\dagger}\mathbf{y} \tag{3.61}$$

where the matrix $\mathbf{A}^{\dagger}$ is called the *pseudoinverse* of **A** [7] because it coincides with $\mathbf{A}^{-1}$, when **A** is square and nonsingular. $\mathbf{A}^{\dagger} \in \mathbb{R}^{n \times m}$, and it is the unique solution to the following set of matrix equations:

$$\mathbf{A}\mathbf{A}^{\dagger}\mathbf{A} = \mathbf{A} \tag{3.62a}$$

$$\mathbf{A}^{\dagger}\mathbf{A}\mathbf{A}^{\dagger} = \mathbf{A}^{\dagger} \tag{3.62b}$$

$$(\mathbf{A}\mathbf{A}^{\dagger}) \text{ is symmetric} \tag{3.62c}$$

$$(\mathbf{A}^{\dagger}\mathbf{A}) \text{ is symmetric} \tag{3.62d}$$

From these equations it is easily seen that $(\mathbf{A}^{\dagger})^{\dagger} = \mathbf{A}$ and that $(\mathbf{A}^{T})^{\dagger} = (\mathbf{A}^{\dagger})^{T}$.

The geometric interpretation of these equations is particularly significant. From Equation 3.62a, $(\mathbf{A}\mathbf{A}^{\dagger})^{2} = \mathbf{A}\mathbf{A}^{\dagger}$, so it is a projection. By Equation 3.62c, $\mathbf{A}\mathbf{A}^{\dagger}$ is an orthogonal projection; Equation 3.62a further implies that it is the orthogonal projection onto R(**A**). By similar reasoning, using Equation 3.62d, $\mathbf{A}^{\dagger}\mathbf{A}$ is the orthogonal projection onto R($\mathbf{A}^{T}$).

The complemetary orthogonal projection of $\mathbf{A}\mathbf{A}^{\dagger}$ is $\mathbf{I} - \mathbf{A}\mathbf{A}^{\dagger}$, and the orthogonal complement of R(**A**) in $\mathbb{R}^{m}$ is N($\mathbf{A}^{T}$).

The roles of **A** and $\mathbf{A}^{\dagger}$ are interchanged in going from Equation 3.62a to Equation 3.62b. It follows that $\mathbf{A}^{\dagger}\mathbf{A}$ is the orthogonal projection onto R($\mathbf{A}^{\dagger}$) and so R($\mathbf{A}^{\dagger}$) = R($\mathbf{A}^{T}$). [Similarly, $\mathbf{A}\mathbf{A}^{\dagger}$ is the orthogonal projection onto R($(\mathbf{A}^{\dagger})^{T}$), etc.]

Closed-form expressions for $\mathbf{A}^{\dagger}$ are available in some cases. If **A** has linearly independent columns, then the Gram matrix $\mathbf{A}^{T}\mathbf{A}$ is invertible and $\mathbf{A}^{\dagger} = (\mathbf{A}^{T}\mathbf{A})^{-1}\mathbf{A}^{T}$; for this case, $\mathbf{A}^{\dagger}$ may also be expressed in terms of the QR factorization of $\mathbf{A} = \mathbf{W}\mathbf{U}$: $\mathbf{A}^{\dagger} = \mathbf{U}^{-1}\mathbf{W}^{T}$. If $\mathbf{A}^{T}$ has linearly independent columns, then the Gram matrix associated with $\mathbf{A}^{T}$, $\mathbf{A}\mathbf{A}^{T}$, is invertible and $\mathbf{A}^{\dagger} = \mathbf{A}^{T}(\mathbf{A}\mathbf{A}^{T})^{-1}$. When neither **A** nor $\mathbf{A}^{T}$ has linearly independent columns, no simple expression for $\mathbf{A}^{\dagger}$ is available. In a later section, a matrix factorization of **A** known as the *singular value decomposition* (SVD) will be used to provide an expression for the pseudoinverse.

3.5.3.1 Sparse Solutions of Underdetermined Linear Equations

For consistent underdetermined linear equations, the solution $\mathbf{x} = \mathbf{A}^{\dagger}\mathbf{y}$ produces the solution of smallest Euclidean norm. However, in some applications, a solution of a different character is desired.

Any inherent nonuniqueness of solutions can be resolved by selecting a solution with *sparseness*, meaning that a solution vector with many zero components is preferred. In this formulation, the number of unknowns (elements of **x**) required to express the known quantities (elements of **y**) using the structural relationships inherent in **A** can be minimized.

To solve the combinatorial optimization problem of maximizing the number of zero components over the vectors in the solution set S_x in Equation 3.58 appears to be an intractable problem in general. However, it turns out that finding the solution vector having minimum 1 norm produces sparse solutions in many situations. Furthermore, the approach of minimizing $\|x\|_1$ also extends to solving for sparse approximate solutions when inconsistencies, as suitably measured by the Euclidean norm, are small [3], much like the problem formulation leading to the pseudoinverse solution. However, the computation of sparse solutions relies on computational techniques for convex optimization problems, with no "closed-form" solution like Equation 3.61.

3.6 Eigenvalues and Eigenvectors

Scalar multiplication, $v \mapsto \alpha v$, is the simplest kind of linear function that maps a vector space into itself. The zero function, $v \mapsto 0$, and the identity function, $v \mapsto v$, are two special cases. For a general linear function, $f : \mathcal{V} \to \mathcal{V}$, it is natural to investigate whether or not there are vectors, and hence subspaces of vectors, on which f is equivalent to scalar multiplication.

3.6.1 Definitions and Fundamental Properties

If $\mathcal{W}_\lambda$ is a nonzero subspace such that $f(w) = \lambda w$ for all $w \in \mathcal{W}_\lambda$, then it is called an *eigenspace* of f corresponding to *eigenvalue* λ. The nonzero vectors in $\mathcal{W}_\lambda$ are called *eigenvectors* of f corresponding to eigenvalue λ.

To study eigenvalues, eigenvectors, and eigenspaces it is customary to use matrix representations and coordinate spaces. In this framework, the equation determining an eigenvector and its corresponding eigenvalue takes the form

$$\mathbf{A}\mathbf{u} = \lambda \mathbf{u} \text{ for } \mathbf{u} \neq 0 \tag{3.63}$$

where $\mathbf{A} \in \mathcal{F}^{n \times n}$, $\mathbf{u} \in \mathcal{F}^n$, and $\lambda \in \mathcal{F}$. Equivalently,

$$(\lambda \mathbf{I} - \mathbf{A})\mathbf{u} = 0 \tag{3.64}$$

A nontrivial solution of this homogeneous linear equation will exist if and only if $\det(\lambda \mathbf{I} - \mathbf{A}) = 0$. This equation is called the *characteristic equation* of the matrix **A**, since it involves the monic nth degree characteristic polynomial of **A**,

$$\det(\lambda \mathbf{I} - \mathbf{A}) = \chi(\lambda) = \lambda^n + \chi_1 \lambda^{n-1} + \cdots + \chi_{n-1}\lambda + \chi_n \tag{3.65}$$

Eigenvalues are zeros of the characteristic polynomial, that is, roots of the characteristic equation.

Depending on the field $\mathcal{F}$, roots of the characteristic equation may or may not exist; that is, $(\lambda \mathbf{I} - \mathbf{A})$ may be invertible for all $\lambda \in \mathcal{F}$. For a characteristic polynomial such as $\lambda^2 + 1$, there are no real zeros even though the polynomial has real coefficients; on the other hand, this polynomial has two complex zeros. Indeed, by the Fundamental Theorem of Algebra, every nth degree polynomial with complex coefficients has n complex zeros, implying that

$$\det(\lambda \mathbf{I} - \mathbf{A}) = (\lambda - \lambda_1)(\lambda - \lambda_2) \cdots (\lambda - \lambda_n) \tag{3.66}$$

for some set of complex numbers $\lambda_1, \ldots, \lambda_n$, not necessarily distinct. Thus, for finding eigenvalues and eigenvectors of $\mathbf{A} \in \mathbb{R}^{n \times n}$ it is sometimes convenient to regard **A** as an element of $\mathbb{C}^{n \times n}$.

The eigenvalues and eigenvectors of real matrices are constrained by conjugacy conditions. If $\mathbf{A}$ is real and λ is an eigenvalue with nonzero imaginary part, then λ^*, the complex conjugate of λ, is also an eigenvalue of $\mathbf{A}$. (The characteristic polynomial of a real matrix has real coefficients and its complex zeros occur in conjugate pairs.) If $\mathbf{u}$ is an eigenvector of the real matrix $\mathbf{A}$ corresponding to eigenvalue λ having nonzero imaginary part, then $\mathbf{u}^*$ (component wise conjugation) is an eigenvector of $\mathbf{A}$ corresponding to eigenvalue λ^*.

Some classes of real matrices have real eigenvalues. Since the diagonal elements of any upper triangular matrix are its eigenvalues, every real upper triangular matrix has real eigenvalues. The same is true of lower triangular matrices and diagonal matrices. More surprisingly, any *normal matrix*, a matrix $\mathbf{A} \in \mathbb{C}^{n \times n}$ with $\mathbf{A}^H \mathbf{A} = \mathbf{A} \mathbf{A}^H$, has real eigenvalues. A matrix $\mathbf{A} \in \mathbb{R}^{n \times n}$ is normal when $\mathbf{A}^T \mathbf{A} = \mathbf{A} \mathbf{A}^T$, and thus any real symmetric matrix, that is, $\mathbf{Q} \in \mathbb{R}^{n \times n}$ with $\mathbf{Q}^T = \mathbf{Q}$, has real eigenvalues.

3.6.2 Eigenvector Bases and Diagonalization

When λ is an eigenvalue of $\mathbf{A} \in \mathbb{C}^{n \times n}$, the subspace of $\mathbb{C}^n$ given by $\mathcal{W}_\lambda = \mathrm{N}(\lambda \mathbf{I} - \mathbf{A})$ is the associated *maximal eigenspace*; it has dimension greater than zero. If λ_1 and λ_2 are two eigenvalues of $\mathbf{A}$ with $\lambda_1 \neq \lambda_2$, corresponding eigenvectors $\mathbf{u}(\lambda_1) \in \mathcal{W}_{\lambda_1}$ and $\mathbf{u}(\lambda_2) \in \mathcal{W}_{\lambda_2}$ are linearly independent. This leads to a sufficient condition for existence of a basis of $\mathbb{C}^n$ consisting of eigenvectors of $\mathbf{A}$. If $\mathbf{A}$ has n distinct eigenvalues, the set of n corresponding eigenvectors forms a basis for $\mathbb{C}^n$.

More generally, if $\mathbf{A}$ has $r \leq n$ distinct eigenvalues, $\{\lambda_1, \ldots, \lambda_r\}$, with associated maximal eigenspaces $\mathcal{W}_1, \mathcal{W}_2, \ldots, \mathcal{W}_r$ having dimensions $d_1, d_2, \ldots, d_r$ equal to the algebraic multiplicities of the eigenvalues (as zeros of the characteristic polynomial), respectively, then $d_1 + \cdots + d_r = n$ and $\mathbb{C}^n$ has a basis consisting of eigenvectors of $\mathbf{A}$. This case always holds for real symmetric matrices.

Let $\mathbf{A} \in \mathbb{C}^{n \times n}$ be a matrix whose eigenvectors $\{\mathbf{u}_1, \mathbf{u}_2, \ldots, \mathbf{u}_n\}$ form a basis $\mathcal{B}$ for $\mathbb{C}^n$; let $\{\lambda_1, \lambda_2, \ldots, \lambda_n\}$ be the corresponding eigenvalues. Let $\mathbf{T}$ be the invertible $n \times n$ matrix whose ith column is $\mathbf{u}_i$. Then $\mathbf{AT} = \mathbf{T\Lambda}$, where $\mathbf{\Lambda}$ is a diagonal matrix formed from the eigenvalues:

$$
\mathbf{\Lambda} =
\begin{bmatrix}
\lambda_1 & 0 & \cdot & \cdot & \cdot & 0 \\
0 & \lambda_2 & & & & \cdot \\
\cdot & & \cdot & & & \cdot \\
\cdot & & & \cdot & & \cdot \\
\cdot & & & & \cdot & 0 \\
0 & \cdot & \cdot & \cdot & 0 & \lambda_n
\end{bmatrix}
\tag{3.67}
$$

Solving for $\mathbf{\Lambda}$ gives

$$
\mathbf{T}^{-1} \mathbf{A} \mathbf{T} = \mathbf{\Lambda}
\tag{3.68}
$$

Thus $\mathbf{A}$ is similar to the diagonal matrix of its eigenvalues, $\mathbf{\Lambda}$, and $\mathbf{T\Lambda T}^{-1} = \mathbf{A}$. Also, $\mathbf{\Lambda}$ is the matrix representation of the linear function $f(\mathbf{x}) = \mathbf{Ax}$ with respect to the eigenvector basis $\mathcal{B}$ of $\mathbb{C}^n$.

For any matrix whose eigenvectors form a basis of $\mathbb{C}^n$, the similarity equation $\mathbf{A} = \mathbf{T\Lambda T}^{-1}$ may be rewritten using the definition of matrix multiplication, giving

$$
\mathbf{A} = \mathbf{T\Lambda T}^{-1} = \sum_{i=1}^{n} \lambda_i \mathbf{u}_i \mathbf{v}_i
\tag{3.69}
$$

where $\mathbf{v}_i$ is the ith row of $\mathbf{T}^{-1}$. This is called the *spectral representation* of $\mathbf{A}$. The row vector $\mathbf{v}_i$ is called a *left eigenvector* of $\mathbf{A}$ since it satisfies $\mathbf{v}_i \mathbf{A} = \lambda_i \mathbf{v}_i$.

3.6.3 More Properties of Eigenvalues and Eigenvectors

From the factored form of the characteristic polynomial it follows that $\det \mathbf{A} = \lambda_1 \lambda_2 \cdots \lambda_n$, the product of the eigenvalues. Thus $\mathbf{A}$ will be invertible if and only if it has no zero eigenvalue. If $\lambda = 0$ is an eigenvalue of $\mathbf{A}$, $\mathrm{N}(\mathbf{A})$ is the associated maximal eigenspace.

Let λ be an eigenvalue of $\mathbf{A}$ with corresponding eigenvector $\mathbf{u}$. For integer $k \geq 0$, λ^k is an eigenvalue of $\mathbf{A}^k$ with corresponding eigenvector $\mathbf{u}$; more generally, for any polynomial $\alpha(s) = \alpha_0 s^d + \cdots + \alpha_{d-1} s + \alpha_d$, $\alpha(\lambda)$ is an eigenvalue of $\alpha(\mathbf{A})$ with corresponding eigenvector $\mathbf{u}$. If $\mathbf{A}$ is invertible, $1/\lambda$ is an eigenvalue of $\mathbf{A}^{-1}$ with corresponding eigenvector $\mathbf{u}$. The eigenvalues of $\mathbf{A}$ are the same as the eigenvalues of $\mathbf{A}^{\mathrm{T}}$. Every eigenvalue of an orthogonal matrix (or a unitary matrix) has unit magnitude.

For two square matrices of the same size, $\mathbf{A}_1$ and $\mathbf{A}_2$, the eigenvalues of $\mathbf{A}_1 \mathbf{A}_2$ are the same as the eigenvalues of $\mathbf{A}_2 \mathbf{A}_1$. However, the Lie product, $[\mathbf{A}_1, \mathbf{A}_2] = \mathbf{A}_1 \mathbf{A}_2 - \mathbf{A}_2 \mathbf{A}_1$ is zero only when the matrices *commute*. If $\mathbf{A}_1$ and $\mathbf{A}_2$ are $n \times n$ and have a common set of n linearly independent eigenvectors, then they commute. Conversely, if $\mathbf{A}_1$ and $\mathbf{A}_2$ commute and if $\mathbf{A}_1$ has distinct eigenvalues, then $\mathbf{A}_1$ has n linearly independent eigenvectors that are also eigenvectors of $\mathbf{A}_2$.

The *spectral radius* of a square matrix (over $\mathbb{C}$ or $\mathbb{R}$) is defined as $\rho(\mathbf{A}) = \lim_{k \to \infty} \|\mathbf{A}^k\|^{1/k}$. Any induced matrix norm can be used in the definition; the same limit value always exists. The spectral radius equals the largest eigenvalue magnitude; $\rho(\mathbf{A}) = \max\{|\lambda| : \det(\lambda \mathbf{I} - \mathbf{A}) = 0\}$. All eigenvalues of $\mathbf{A}$ lie within the disc $\{s \in \mathbb{C} : |s| \leq \rho(\mathbf{A})\}$ in the complex plane. If $\mathbf{A}$ is invertible, all eigenvalues of $\mathbf{A}$ lie within the annulus $\{s \in \mathbb{C} : \rho(\mathbf{A}^{-1}) \leq |s| \leq \rho(\mathbf{A})\}$ in the complex plane.

For an upper or lower triangular matrix, including a diagonal matrix, the diagonal elements are its eigenvalues. In general, the diagonal elements of a real or complex matrix $\mathbf{A}$ define the centers of circular *Gershgorin discs* in the complex plane, $G_i(a_{i,i}) = \{s \in \mathbb{C} : |s - a_{i,i}| \leq \Sigma_{k \neq i} |a_{i,k}|\}$, and every eigenvalue lies in at least one Gershgorin disc.

Asymptotically stable linear time-invariant sytems in discrete time and in continuous time correspond, respectively, to two classes of matrices characterized by their eigenvalue locations. A real or complex matrix $\mathbf{A}$ is said to be a *Schur matrix* if its spectral radius is less than one; $\mathbf{A}$ is said to be a *Hurwitz matrix* if its eigenvalues all have negative real parts.

3.6.4 Symmetric Matrices

For a real symmetric matrix $\mathbf{Q}$, all eigenvalues are real and the corresponding eigenvectors may be chosen with real components. For this case, if λ_1 and λ_2 are two eigenvalues with $\lambda_1 \neq \lambda_2$, the corresponding real eigenvectors $\mathbf{u}(\lambda_1) \in \mathcal{W}_{\lambda_1}$ and $\mathbf{u}(\lambda_2) \in \mathcal{W}_{\lambda_2}$ are not only linearly independent, they are also orthogonal, $\langle \mathbf{u}(\lambda_1), \mathbf{u}(\lambda_2) \rangle = 0$. Further, each maximal eigenspace has dimension equal to the algebraic multiplicity of the associated eigenvalue as a zero of the characteristic polynomial, and each maximal eigenspace has an orthogonal basis of eigenvectors. Thus, for any real symmetric matrix $\mathbf{Q}$, there is an orthogonal basis for $\mathbb{R}^n$ consisting of eigenvectors; by scaling the lengths of the basis vectors to one, an orthonormal basis of eigenvectors is obtained. Thus $\mathbf{\Lambda} = \mathbf{O}^{\mathrm{T}} \mathbf{Q} \mathbf{O}$, where $\mathbf{O}$ is an orthogonal matrix. (This may be generalized. If $\mathbf{A}$ is a complex Hermitian matrix, that is, $\mathbf{A}^{\mathrm{H}} = \mathbf{A}$, where $\mathbf{A}^{\mathrm{H}}$, denotes the combination of conjugation and transposition: $\mathbf{A}^{\mathrm{H}} = (\mathbf{A}^*)^{\mathrm{T}}$. Then $\mathbf{A}$ has real eigenvalues and there is a basis of $\mathbb{C}^n$ comprised of normalized eigenvectors so that $\mathbf{\Lambda} = \mathbf{U}^{\mathrm{H}} \mathbf{A} \mathbf{U}$, where $\mathbf{U}$ is a unitary matrix.)

For symmetric $\mathbf{Q}$, when the eigenvectors are chosen to be orthonormal, the spectral representation simplifies to

$$\mathbf{Q} = \mathbf{O} \mathbf{\Lambda} \mathbf{O}^{\mathrm{T}} = \sum_{i=1}^{n} \lambda_i \mathbf{u}_i \mathbf{u}_i^{\mathrm{T}} \tag{3.70}$$

A real symmetric matrix $\mathbf{Q}$ is said to be *positive definite* if the inequality $\mathbf{x}^{\mathrm{T}} \mathbf{Q} \mathbf{x} > 0$ holds for all $\mathbf{x} \in \mathbb{R}^n$ with $\mathbf{x} \neq 0$; $\mathbf{Q}$ is said to be *nonnegative definite* if $\mathbf{x}^{\mathrm{T}} \mathbf{Q} \mathbf{x} \geq 0$ for all $\mathbf{x} \in \mathbb{R}^n$. When all eigenvalues of $\mathbf{Q}$ are positive (nonnegative), $\mathbf{Q}$ is positive definite (nonnegative definite). When $\mathbf{Q}$ is nonnegative definite, let $\mathbf{V} = \mathbf{\Lambda}^{1/2} \mathbf{O}$, where $\mathbf{\Lambda}^{1/2}$ denotes the diagonal matrix of nonnegative square roots of the eigenvalues; then $\mathbf{Q}$ may be written in factored form as $\mathbf{Q} = \mathbf{V}^{\mathrm{T}} \mathbf{V}$. This shows that if $\mathbf{Q}$ is positive definite, it is the Gram matrix of a set of linearly independent vectors, the columns of $\mathbf{V}$. If the QR factorization of $\mathbf{V}$ is $\mathbf{V} = \mathbf{W} \mathbf{U}$, then the *Cholesky factorization* of $\mathbf{Q}$ is $\mathbf{Q} = \mathbf{U}^{\mathrm{T}} \mathbf{U}$ [6], which is also seen to be the (symmetric) PLU factorization of $\mathbf{Q}$ (i.e., $\mathbf{P} = \mathbf{I}$ and $\mathbf{L} = \mathbf{U}^{\mathrm{T}}$).

Conversely, for any matrix $\mathbf{H} \in \mathbb{R}^{m \times n}$, the symmetric matrices $\mathbf{Q}_{n \times n} = \mathbf{H}^{\mathsf{T}}\mathbf{H}$ (the Gram matrix of the columns of $\mathbf{H}$) and $\mathbf{Q}_{m \times m} = \mathbf{H}\mathbf{H}^{\mathsf{T}}$ (the Gram matrix of the columns of $\mathbf{H}^{\mathsf{T}}$) are both nonnegative definite since $\mathbf{x}^{\mathsf{T}}\mathbf{Q}_{m \times m}\mathbf{x} = \langle \mathbf{H}^{\mathsf{T}}\mathbf{x}, \mathbf{H}^{\mathsf{T}}\mathbf{x} \rangle \geq 0$ for all $\mathbf{x} \in \mathbb{R}^m$ and $\mathbf{x}^{\mathsf{T}}\mathbf{Q}_{n \times n}\mathbf{x} = \langle \mathbf{H}\mathbf{x}, \mathbf{H}\mathbf{x} \rangle \geq 0$ for all $\mathbf{x} \in \mathbb{R}^n$.

3.7 The Jordan Form and Similarity of Matrices

If the matrix $\mathbf{A}$ does not have a set of n linearly independent eigenvectors, it is not similar to a diagonal matrix, but eigenvalues and eigenvectors still play a role in providing various useful representations; for at least one of its eigenvalues λ, the dimension of $\mathcal{W}_\lambda$ is smaller than the algebraic multiplicity of λ as a zero of the characteristic polynomial. For ease of notation let $\mathbf{A}_\lambda = (\lambda \mathbf{I} - \mathbf{A})$ so that $\mathcal{W}_\lambda = \mathsf{N}(\mathbf{A}_\lambda)$. Then for $k \geq 1$, $\mathsf{N}(\mathbf{A}_\lambda^k) \subseteq \mathsf{N}(\mathbf{A}_\lambda^{k+1})$. Let $\mathrm{I}(\mathbf{A}_\lambda)$ be the *index* of $\mathbf{A}_\lambda$, the smallest positive integer k such that $\mathsf{N}(\mathbf{A}_\lambda^k) = \mathsf{N}(\mathbf{A}_\lambda^{k+1})$. Then the subspace $\mathcal{W}_{\lambda,\mathrm{I}} = \mathsf{N}(\mathbf{A}_\lambda^{\mathrm{I}})$ has dimension equal to the algebraic multiplicity of λ as a zero of the characteristic polynomial; when $\mathrm{I}(\mathbf{A}_\lambda) = 1$, $\mathcal{W}_{\lambda,\mathrm{I}} = \mathcal{W}_\lambda$, the associated maximal eigenspace.

For each eigenvalue λ, $\mathcal{W}_{\lambda,\mathrm{I}}$ is an $\mathbf{A}$-invariant subspace (i.e., f-invariant for $f(\mathbf{x}) = \mathbf{A}\mathbf{x}$). For eigenvalues $\lambda_1 \neq \lambda_2$, the corresponding "generalized eigenspaces" are independent, that is, for nonzero vectors $\mathbf{v}_1 \in \mathcal{W}_{\lambda_1, \mathrm{I}_1}$ and $\mathbf{v}_2 \in \mathcal{W}_{\lambda_2, \mathrm{I}_2}$, $\mathbf{v}_1$ and $\mathbf{v}_2$ are linearly independent. The vectors obtained by choosing bases of all of the generalized eigenspaces, $\{\mathcal{W}_{\lambda_i, \mathrm{I}_i}\}$, may be collected together to form a basis of $\mathbb{C}^n$ consisting of eigenvectors and "generalized eigenvectors," and the general form of the matrix representation of the linear function $\mathbf{A}\mathbf{x}$ with respect to such a basis is a block diagonal matrix called the *Jordan form* [5]. Using the basis vectors as the columns of $\mathbf{T}$, the Jordan form of $\mathbf{A}$ is obtained by a similarity transformation,

$$\mathbf{T}^{-1}\mathbf{A}\mathbf{T} = \begin{bmatrix} \mathbf{M}(\lambda_1, d_1) & 0 & \cdot & \cdot & \cdot & 0 \\ 0 & \mathbf{M}(\lambda_2, d_2) & & & & \cdot \\ \cdot & & \cdot & & & \cdot \\ \cdot & & & \cdot & & \cdot \\ \cdot & & & & \cdot & 0 \\ 0 & & \cdot & & 0 & \mathbf{M}(\lambda_r, d_r) \end{bmatrix} \tag{3.71}$$

where block $\mathbf{M}(\lambda_i, d_i)$ has dimension $d_i \times d_i$ and d_i is the algebraic multiplicity of eigenvalue λ_i, assuming there are r distinct eigenvalues.

The matrix $\mathbf{M}(\lambda_i, d_i)$ is the matrix representation of $(\mathbf{A}\mathbf{x})|_{\mathcal{W}_{\lambda_i, \mathrm{I}_i}}$ with respect to the basis chosen for $\mathcal{W}_{\lambda_i, \mathrm{I}_i}$. If there are e_i linearly independent eigenvalues corresponding to eigenvalue λ_i then the basis vectors can be chosen so that $\mathbf{M}(\lambda_i, d_i)$ takes the block diagonal form

$$\mathbf{M}(\lambda_i, d_i) = \begin{bmatrix} \mathbf{J}_1(\lambda_i) & 0 & \cdot & \cdot & \cdot & 0 \\ 0 & \mathbf{J}_2(\lambda_i) & & & & \cdot \\ \cdot & & \cdot & & & \cdot \\ \cdot & & & \cdot & & \cdot \\ \cdot & & & & \cdot & 0 \\ 0 & & \cdot & & 0 & \mathbf{J}_{e_i}(\lambda_i) \end{bmatrix} \tag{3.72}$$

where the *Jordan blocks* $\mathbf{J}_k(\lambda_i)$ take one of two forms: if the kth block has size $\delta_{i,k} = 1$, then $\mathbf{J}_k(\lambda_i) = [\lambda_i]$; if the kth block has size $\delta_{i,k} > 1$, $\mathbf{J}_k(\lambda_i)$ is given elementwise as

$$(\mathbf{J}_k(\lambda_i))_{p,q} = \begin{cases} \lambda_i, & \text{for } p = q \\ 1, & \text{for } p = q+1 \\ 0, & \text{otherwise} \end{cases} \tag{3.73}$$

The sizes of the Jordan blocks can be expressed in terms of the dimensions of the subspaces $\mathcal{W}_{\lambda_i, j}$ for $1 \leq j \leq \mathrm{I}_i$. The largest block size is equal to $\mathrm{I}_i = \mathrm{I}(\mathbf{A}_{\lambda_i})$. Of course $\sum_{k=1}^{e_i} \delta_{i,k} = d_i$. If every Jordan block has size 1, $e_i = d_i$ and the Jordan form is a diagonal matrix.

3.7.1 Invariant Factors and the Rational Canonical Form

Eigenvalues are invariant under similarity transformation. For any invertible matrix $\mathbf{T}$, when $\mathbf{T}^{-1}\mathbf{A}_1\mathbf{T} = \mathbf{A}_2$, $\mathbf{A}_1$ and $\mathbf{A}_2$ have the same eigenvalues. Let λ be an eigenvalue of $\mathbf{A}_1$ and of $\mathbf{A}_2$. If $\mathbf{u}$ is a corresponding eigenvector of $\mathbf{A}_1$, then $\mathbf{T}^{-1}\mathbf{u}$ is a corresponding eigenvector of $\mathbf{A}_2$. Likewise, $\mathbf{T}^{-1}$ transforms generalized eigenvectors of $\mathbf{A}_1$ to generalized eigenvectors of $\mathbf{A}_2$. Given a convention for ordering the Jordan blocks, the Jordan form is a canonical form for matrices under the equivalence relation of similarity, and two matrices are similar if and only if they have the same Jordan form.

The use of eigenvectors and generalized eigenvectors, and related methods involving invariant subspaces, is one approach to studying similarity of matrices. Polynomial matrix methods offer a second approach, based on the transformation of the polynomial matrix $(s\mathbf{I} - \mathbf{A})$ corresponding to similarity: $(s\mathbf{I} - \mathbf{A}) \mapsto \mathbf{T}^{-1}(s\mathbf{I} - \mathbf{A})\mathbf{T}$. This is a special case of the general equivalence transformation for polynomial matrices $\mathcal{A}(s) \mapsto \mathcal{P}(s)\mathcal{A}(s)\mathcal{Q}(s)$, whereby unimodular matrices $\mathcal{P}(s)$ and $\mathcal{Q}(s)$ transform $\mathcal{A}(s)$; two polynomial matrices are equivalent if and only if they have the same invariant polynomials [11].

Two real or complex matrices $\mathbf{A}_1$ and $\mathbf{A}_2$ are similar if and only if the polynomial matrices $(s\mathbf{I} - \mathbf{A}_1)$ and $(s\mathbf{I} - \mathbf{A}_2)$ have the same invariant polynomials [11]. The invariant polynomials of $(s\mathbf{I} - \mathbf{A})$ are also known as the *invariant factors* of $\mathbf{A}$ since they are factors of the characteristic polynomial $\det(s\mathbf{I} - \mathbf{A})$: $\det(s\mathbf{I} - \mathbf{A}) = \phi_1(s)\phi_2(s)\ldots\phi_n(s)$. Also, $\phi_{i+1}(s)$ is a factor of $\phi_i(s)$, and $\phi_1(s)$ is the *minimal polynomial* of $\mathbf{A}$, the monic polynomial $p(s)$ of least degree such that $p(\mathbf{A}) = 0$.

If $\mathbf{A}$ has q nontrivial invariant factors, then by similarity transformation it can be brought to the *rational canonical form* by a suitable choice of $\mathbf{T}$ [5]:

$$\mathbf{T}^{-1}\mathbf{A}\mathbf{T} = \begin{bmatrix} \mathbf{C}(\phi_1(s)) & 0 & \cdot & \cdot & \cdot & 0 \\ 0 & \mathbf{C}(\phi_2(s)) & & & & \cdot \\ \cdot & & \cdot & & & \cdot \\ \cdot & & & \cdot & & 0 \\ \cdot & & & & \cdot & \\ 0 & & \cdot & \cdot & 0 & \mathbf{C}(\phi_q(s)) \end{bmatrix} \tag{3.74}$$

where the ith diagonal block of this matrix, $\mathbf{C}(\phi_i(s))$, is a *companion matrix* associated with ith invariant factor: for a polynomial $\pi(s) = s^m + \pi_1 s^{m-1} + \cdots + \pi_{m-1}s + \pi_m$,

$$\mathbf{C}(\pi(s)) = \begin{bmatrix} 0 & 0 & \cdot & \cdot & \cdot & 0 & -\pi_m \\ 1 & 0 & \cdot & \cdot & \cdot & 0 & -\pi_{m-1} \\ 0 & 1 & & & & \cdot & \cdot \\ \cdot & & \cdot & & & & \cdot \\ \cdot & & & \cdot & & & \cdot \\ \cdot & & & & 1 & 0 & -\pi_2 \\ 0 & \cdot & \cdot & \cdot & 0 & 1 & -\pi_1 \end{bmatrix} \tag{3.75}$$

The characteristic polynomial of $\mathbf{C}(\pi(s))$ is $\pi(s)$.

3.8 Singular Value Decomposition

Another approach may be taken to the transformation of a matrix to a diagonal form. Using two orthonormal bases obtained from eigenvectors of the symmetric nonnegative definite matrices $\mathbf{A}^T\mathbf{A}$ and $\mathbf{A}\mathbf{A}^T$ provides a representation known as the *singular value decomposition* (SVD) of $\mathbf{A}$ [6]:

$$\mathbf{A} = \mathbf{V}\mathbf{\Sigma}\mathbf{U}^T \tag{3.76}$$

In this equation, $\mathbf{V}$ is an orthogonal matrix of eigenvectors of the product $\mathbf{A}\mathbf{A}^T$ and $\mathbf{U}$ is an orthogonal matrix of eigenvectors of the product $\mathbf{A}^T\mathbf{A}$. $\mathbf{\Sigma}$ is a diagonal matrix of nonnegative quantities

$\sigma_1 \geq \sigma_2 \geq \cdots \geq \sigma_n \geq 0$ known as the *singular values* of $\mathbf{A}$. The singular values are obtained from the eigenvalues of the nonnegative definite matrix $\mathbf{A}\mathbf{A}^T$ (or those of $\mathbf{A}^T\mathbf{A}$) by taking positive square roots and reordering if necessary.

While not a similarity invariant, singular values are invariant under left and right orthogonal transformations. For any orthogonal matrices $\mathbf{O}_1$ and $\mathbf{O}_2$, when $\mathbf{O}_1\mathbf{A}_1\mathbf{O}_2 = \mathbf{A}_2$, $\mathbf{A}_1$ and $\mathbf{A}_2$ have the same singular values. The SVD holds for rectangular matrices in exactly the same form as for square matrices: $\mathbf{A} = \mathbf{V}\boldsymbol{\Sigma}\mathbf{U}^T$, where $\mathbf{V}$ is an orthogonal matrix of eigenvectors of the product $\mathbf{A}\mathbf{A}^T$ and $\mathbf{U}$ is an orthogonal matrix of eigenvectors of the product $\mathbf{A}^T\mathbf{A}$. For $\mathbf{A} \in \mathbb{R}^{m \times n}$, the matrix $\boldsymbol{\Sigma} \in \mathbb{R}^{m \times n}$ contains the singular values (the positive square roots of the common eigenvalues of $\mathbf{A}\mathbf{A}^T$ and $\mathbf{A}^T\mathbf{A}$) as diagonal elements of a square submatrix of size $\min(m, n)$ located in its upper left corner, with all other elements of $\boldsymbol{\Sigma}$ being zero.

The SVD is useful in a number of applications. If $\text{rank}(\mathbf{A}) = k$ (i.e., $\mathbf{A}$ has k linearly independent columns or rows), then k is the number of its nonzero singular values. When $\boldsymbol{\Sigma}$ must be determined by numerical techniques and is therefore subject to computational inaccuracies, judging the size of elements of $\boldsymbol{\Sigma}$ in comparison to the "machine accuracy" of the computer being used provides a sound basis for computing rank.

The SVD can be used to generate "low rank" approximations to the matrix $\mathbf{A}$. If $\text{rank}(\mathbf{A}) = k$, then for $\kappa < k$ the best rank κ approximation, that is, the one minimizing the induced Euclidean norm $\|\mathbf{A} - \widehat{\mathbf{A}}\|$ over rank κ matrices $\widehat{\mathbf{A}}$, is obtained by setting the smallest $k - \kappa$ nonzero elements of $\boldsymbol{\Sigma}$ to zero and multiplying by $\mathbf{V}$ and $\mathbf{U}^T$ as in the defining equation for the SVD.

Finally, the SVD also provides a way of computing the pseudoinverse:

$$\mathbf{A}^\dagger = \mathbf{U}\mathbf{D}^\dagger\mathbf{V}^T \tag{3.77}$$

where $\mathbf{D}^\dagger$, the pseudoinverse of $\boldsymbol{\Sigma}$, is obtained from $\boldsymbol{\Sigma}^T$ by inverting its nonzero elements.

The largest singular value of $\mathbf{A} \in \mathbb{R}^{n \times n}$ equals $\|\mathbf{A}\|$, its induced Euclidean norm. When $\mathbf{A}$ is invertible, its smallest singular value gives the distance of $\mathbf{A}$ from the set of singular matrices, and the ratio of its largest and smallest singular values is equal to its condition number, $\kappa(\mathbf{A})$.

The singular values of a real symmetric matrix $\mathbf{Q}$ are equal to the absolute values of its eigenvalues and, in particular, the spectral radius equals its induced Euclidean norm, $\rho(\mathbf{Q}) = \|\mathbf{Q}\|$.

3.9 Matrices and Multivariable Functions

It has already been noted that the most general linear function $f : \mathbb{R}^n \to \mathbb{R}^m$ can be expressed in terms of matrix multiplication as $\mathbf{y} = \mathbf{A}\mathbf{x}$, for $\mathbf{y} \in \mathbb{R}^m$, $\mathbf{x} \in \mathbb{R}^n$, and where $\mathbf{A} \in \mathbb{R}^{m \times n}$ is the matrix representation of $f(\mathbf{x})$ with respect to the standard bases of $\mathbb{R}^n$ and $\mathbb{R}^m$. An important generalization involves linear matrix equations; the case involving square matrices will be described below. Matrices also find application involving a variety of more general *nonlinear* functions, and some important examples will be described.

3.9.1 Quadratic Forms

Homogeneous quadratic functionals arise in many important applications. A general setting would involve a (real) inner product space $\mathcal{X}$, with inner product $\langle \cdot, \cdot \rangle$, along with a linear function $f : \mathcal{X} \to \mathcal{X}$. The associated homogeneous quadratic functional is $q : \mathcal{X} \to \mathbb{R}$ defined as $q(\mathbf{x}) = \langle \mathbf{x}, f(\mathbf{x}) \rangle$.

In the case of $\mathcal{X} = \mathbb{R}^n$, the quadratic functional is often called a *quadratic form* and can be expressed in terms of matrix multiplication as

$$q(\mathbf{x}) = \mathbf{x}^T\mathbf{Q}\mathbf{x} \tag{3.78}$$

where $\mathbf{Q} \in \mathbb{R}^{n \times n}$ is the matrix representation of $f(\mathbf{x})$ with respect to the standard basis of $\mathbb{R}^n$. Without loss of generality, it may be assumed that $\mathbf{Q}$ is symmetric, since the symmetric part of $\mathbf{Q}$, $(\mathbf{Q} + \mathbf{Q}^T)/2$ yields the same quadratic form.

Since every symmetric matrix has real eigenvalues and can be diagonalized by an orthogonal similarity transformation, let $\mathbf{O}^T\mathbf{Q}\mathbf{O} = \mathbf{\Lambda}$. Taking $\mathbf{y} = \mathbf{O}^T\mathbf{x}$ gives $Q(\mathbf{x}) = \mathbf{x}^T\mathbf{Q}\mathbf{x} = \mathbf{y}^T\mathbf{\Lambda}\mathbf{y} = \sum_{i=1}^n \lambda_i y_i^2$. Thus, the quadratic form may be expressed as a weighted sum of squares of certain linear functionals of $\mathbf{x}$. The quadratic form $q(\mathbf{x})$ is positive (for all nonzero $\mathbf{x}$), or equivalently the matrix $\mathbf{Q}$ is *positive definite*, when all of the eigenvalues of $\mathbf{Q}$ are positive.

Another characterization of positive definiteness is given in terms of determinants. A set of principal minors of $\mathbf{Q}$ consists of the determinants of a nested set of n submatrices of $\mathbf{Q}$ formed as follows. Let $(\pi_1, \pi_2, \ldots, \pi_n)$ be a permutation of $(1, 2, \ldots, n)$. Let $\Delta_0 = \det \mathbf{Q}$, and for $1 \leq i < n$ let Δ_i be the $(n - i) \times (n - i)$ minor given by the determinant of the submatrix of $\mathbf{Q}$ obtained by deleting rows and columns $\pi_1, \ldots, \pi_i$. $\mathbf{Q}$ is positive definite if and only if any set of principal minors has all positive elements.

A final characterization of positive definiteness is expressed in terms of Gram matrices. $\mathbf{Q}$ is positive definite if and only if it can be written as the Gram matrix of a set of n linearly independent vectors; taking such a set of vectors to be columns of a matrix $\mathbf{H}$, $\mathbf{Q} = \mathbf{H}^T\mathbf{H}$ and thus $Q(\mathbf{x}) = (\mathbf{H}\mathbf{x})^T(\mathbf{H}\mathbf{x})$, which expresses $Q(\mathbf{x})$ as $\|\mathbf{H}\mathbf{x}\|^2$, the squared Euclidean norm of the vector $\mathbf{H}\mathbf{x}$.

Starting with the quadratic form $\mathbf{x}^T\mathbf{Q}\mathbf{x}$, an invertible linear change of variables $\mathbf{x} = \mathbf{C}^T\mathbf{y}$ produces another quadratic form, $\mathbf{y}^T\mathbf{C}\mathbf{Q}\mathbf{C}^T\mathbf{y}$. The corresponding *congruence* transformation of symmetric matrices, whereby $\mathbf{Q} \mapsto \mathbf{C}\mathbf{Q}\mathbf{C}^T$, does not necessarily preserve eigenvalues; however, the signs of eigenvalues are preserved. The number of positive, negative, and zero eigenvalues of $\mathbf{Q}$ characterizes its equivalence class under congruence transformations.

3.9.2 Matrix-Valued Functions

The algebra of matrices provides a direct means for defining a polynomial function of a square matrix. More generally, functions can be defined explicitly, like polynomial functions, or implicitly as the solution to some matrix equation(s).

3.9.2.1 Matrix Functions

Let $p : \mathbb{C} \to \mathbb{C}$, be a polynomial function, with $p(s) = p_0 s^m + p_1 s^{m-1} + \cdots + p_{m-1} s + p_m$, $p_i \in \mathbb{C}$, $0 \leq i \leq m$. When $\mathbf{A} \in \mathbb{C}^{n \times n}$, $p(\mathbf{A}) = p_0 \mathbf{A}^m + p_1 \mathbf{A}^{m-1} + \cdots + p_{m-1}\mathbf{A} + p_m\mathbf{I}$.

For a function $w : \mathbb{C} \to \mathbb{C}$ given by a power series $w(s) = \sum_0^\infty w_i s^i$, convergent in some region $\{s \in \mathbb{C} : |s| < R\}$, the corresponding matrix function $w(\mathbf{A}) = \sum_0^\infty w_i \mathbf{A}^i$ is defined for matrices $\mathbf{A}$ with $\rho(\mathbf{A}) < R$, where $\rho(\mathbf{A})$ denotes the spectral radius of $\mathbf{A}$. Under this condition, the sequence of partial sums, $\mathbf{S}_k = \sum_0^k w_i \mathbf{A}^i$, converges to a limiting matrix $\mathbf{S}_\infty$, meaning that $\lim_{k \to \infty} \|\mathbf{S}_\infty - \mathbf{S}_n\| = 0$.

Similarity transformations can simplify the evaluation of matrix functions defined by power series. Suppose $\mathbf{A}$ is similar to a diagonal matrix of its eigenvalues, $\mathbf{\Lambda} = \mathbf{T}^{-1}\mathbf{A}\mathbf{T}$. Then $\mathbf{A}^i = \mathbf{T}\mathbf{\Lambda}^i\mathbf{T}^{-1}$ and for a function $w(\mathbf{A})$ defined as a power series, $w(\mathbf{A}) = \mathbf{T}w(\mathbf{\Lambda})\mathbf{T}^{-1}$. Since $w(\mathbf{\Lambda})$ is the diagonal matrix of values $w(\lambda_i)$, $w(\mathbf{A})$ is determined by (a) the values of $w(s)$ on the eigenvalues of $\mathbf{A}$ and (b) the similarity transformation $\mathbf{T}$ whose columns are linearly independent eigenvectors of $\mathbf{A}$. When $\mathbf{A}$ is not similar to a diagonal matrix, $w(\mathbf{A})$ may be still be evaluated using a similarity transformation to Jordan form.

When the eigenvalues of $\mathbf{A}$ are distinct, $w(\mathbf{A})$ can be obtained by finding an interpolating polynomial. Denote the characteristic polynomial by $\chi(s) = \det(s\mathbf{I} - \mathbf{A})$; its zeros are the eigenvalues, $\{\lambda_i, 1 \leq i \leq n\}$. Define polynomials $\xi_k(s) = \chi(s)/(s - \lambda_k)$, $1 \leq k \leq n$. Then each $\xi_k(s)$ has degree $n - 1$ and zeros $\{\lambda_j : j \neq k\}$; let $\xi_k = \xi_k(\lambda_k) \neq 0$. Then the polynomial $L(s) = \sum_{k=1}^n (w(\lambda_k)/\xi_k)\xi_k(s)$ is the unique polynomial of degree $< n$ interpolating the function values; $L(\lambda_i) = w(\lambda_i)$, $1 \leq i \leq n$. Thus, $w(\mathbf{A}) = L(\mathbf{A})$, since $\mathbf{T}^{-1}f(\mathbf{A})\mathbf{T} = w(\mathbf{\Lambda}) = L(\mathbf{\Lambda}) = \mathbf{T}^{-1}L(\mathbf{A})\mathbf{T}$, where $\mathbf{\Lambda} = \mathbf{T}^{-1}\mathbf{A}\mathbf{T}$.

This shows that functions of a matrix $\mathbf{A}$ are not as general as might seem at first. Indeed, it is the Cayley–Hamilton Theorem that implies every polynomial or power-series expression $w(\mathbf{A})$ may be expressed as a polynomial function of degree less than n; writing $w(s) = \chi(s)q(s) + r_w(s)$ for a (unique) polynomial $r_w(s)$ having degree less than n, the degree of the characteristic polynomial $\chi(s)$, it follows that $w(\mathbf{A}) = r_w(\mathbf{A})$, since $\chi(\mathbf{A}) = \mathbf{0}$.

However, by using parametric functions $w(s)$, a variety of useful matrix-valued functions of a real or complex variable may be constructed. The polynomial dependence of the functions on $\mathbf{A}$ involves, for these cases, parametric coefficients that depend on the characteristic polynomial coefficients of $\mathbf{A}$ and the power series coefficients of the function $w(s)$.

For example, consider the exponential function $w(s) = \exp(st)$, with parameter $t \in \mathbb{R}$, defined by the power series

$$\exp(st) = \sum_{i=0}^{\infty} \frac{t^i}{i!} s^i \tag{3.79}$$

For all $t \in \mathbb{R}$, this series converges for all $s \in \mathbb{C}$ so that the matrix exponential function $\exp(\mathbf{A}t)$ is given by

$$\exp(\mathbf{A}t) = \sum_{i=0}^{\infty} \frac{t^i}{i!} \mathbf{A}^i \tag{3.80}$$

for all $\mathbf{A} \in \mathbb{C}^{n \times n}$. With further effort, this expression may be reduced to a polynomial function of $\mathbf{A}$ having degree $n - 1$, whose coefficients are functions of t.

Alternatively, for any (fixed) $\mathbf{A} \in \mathbb{C}^{n \times n}$, the power series in t defines $\exp(\mathbf{A}t)$ as a matrix-valued function of t, $\exp(\mathbf{A}t) : \mathbb{R} \to \mathbb{C}^{n \times n}$. Analogs of the usual properties of the exponential function include $\exp(\mathbf{0}) = \mathbf{I}$; $\exp(\mathbf{A}(t + \tau)) = \exp(\mathbf{A}t)\exp(\mathbf{A}\tau)$; $(\exp(\mathbf{A}t))^{-1} = \exp(-\mathbf{A}t)$. However, $\exp((\mathbf{A}_1 + \mathbf{A}_2)t) \neq \exp(\mathbf{A}_1 t)\exp(\mathbf{A}_2 t)$, unless $\mathbf{A}_1$ and $\mathbf{A}_2$ commute; indeed, $\exp(-\mathbf{A}_2 t)\exp((\mathbf{A}_1 + \mathbf{A}_2)t)\exp(-\mathbf{A}_1 t) = \mathbf{I} + [\mathbf{A}_1, \mathbf{A}_2]t^2/2 + \cdots$, where $[\mathbf{A}_1, \mathbf{A}_2] = \mathbf{A}_1 \mathbf{A}_2 - \mathbf{A}_2 \mathbf{A}_1$, the Lie product of $\mathbf{A}_1$ and $\mathbf{A}_2$.

Another important parametric matrix function is obtained from the power series for $w(s) = (\lambda - s)^{-1}$,

$$(\lambda - s)^{-1} = \sum_{i=0}^{\infty} \lambda^{-(i+1)} s^i \tag{3.81}$$

which, given $\lambda \in \mathbb{C}$, converges for those $s \in \mathbb{C}$ satisfying $|s| < |\lambda|$. The resulting matrix function is

$$(\lambda \mathbf{I} - \mathbf{A})^{-1} = \sum_{i=0}^{\infty} \lambda^{-(i+1)} \mathbf{A}^i \tag{3.82}$$

which is defined for $\mathbf{A} \in \mathbb{C}^{n \times n}$ satisfying $\rho(\mathbf{A}) < |\lambda|$.

On the other hand, for any fixed $\mathbf{A} \in \mathbb{C}^{n \times n}$, the power series in λ defines $(\lambda \mathbf{I} - \mathbf{A})^{-1}$ as a matrix-valued function of λ known as the *resolvent matrix* of $\mathbf{A}$, $(\lambda \mathbf{I} - \mathbf{A})^{-1} : \mathcal{D} \to \mathbb{C}^{n \times n}$ with the domain $\mathcal{D} = \{\lambda \in \mathbb{C} : |\lambda| > \rho(\mathbf{A})\}$. Additional properties of the resolvent matrix arise because $(\lambda \mathbf{I} - \mathbf{A})$ is a matrix of rational functions, that is, $(\lambda \mathbf{I} - \mathbf{A}) \in \mathcal{F}^{n \times n}$, where $\mathcal{F} = \mathbb{C}(\lambda)$, the field of rational functions of the complex variable λ with complex coefficients. Over $\mathbb{C}(\lambda)$, $(\lambda \mathbf{I} - \mathbf{A})$ is invertible because its determinant is the characteristic polynomial of $\mathbf{A}$, $\chi(\lambda)$, and is therefore a nonzero rational function. By Cramer's rule, the form of $(\lambda \mathbf{I} - \mathbf{A})^{-1}$ is

$$(\lambda \mathbf{I} - \mathbf{A})^{-1} = \frac{\mathbf{\Psi}(\lambda)}{\det(\lambda \mathbf{I} - \mathbf{A})} \tag{3.83}$$

where $\mathbf{\Psi}(\lambda)$ is a matrix whose elements are polynomials having degree $< n$. Multiplying both sides of this equation by $(\det(\lambda \mathbf{I} - \mathbf{A}))(\lambda \mathbf{I} - \mathbf{A})$ and equating coefficients of powers of λ leads to an explicit form for $\mathbf{\Psi}(\lambda)$:

$$\mathbf{\Psi}(\lambda) = \mathbf{I}\lambda^{n-1} + (\mathbf{A} + \chi_1 \mathbf{I})\lambda^{n-2} + (\mathbf{A}^2 + \chi_1 \mathbf{A} + \chi_2 \mathbf{I})\lambda^{n-3} + \cdots$$
$$+ (\mathbf{A}^{n-1} + \chi_1 \mathbf{A}^{n-2} + \chi_2 \mathbf{A}^{n-3} + \cdots + \chi_{n-1}\mathbf{I}) \tag{3.84}$$

Expressing $(\lambda \mathbf{I} - \mathbf{A})^{-1}$ as a matrix of rational functions thus provides a means of defining it as a matrix-valued function for all $\lambda \in \mathbb{C}$ except for the zeros of $\det(\lambda \mathbf{I} - \mathbf{A})$, that is, except for the eigenvalues of $\mathbf{A}$. Indeed, the explicit form of the resolvent is a polynomial of degree $n - 1$ in $\mathbf{A}$ where the functions of λ giving the polynomial coefficients are rational functions of λ defined for all $\lambda \in \mathbb{C}$ except for the zeros of $\det(\lambda \mathbf{I} - \mathbf{A})$.

3.9.2.2 Matrix Functions from Solution of Matrix Equations

Matrix functions are not always given by explicit formulae; they can be defined implicitly as solutions to algebraic equations. An important result for the study of matrix equations is the *Contraction Mapping Theorem* [10]. With $\| \cdot \|$ a norm on $\mathbb{R}^{n \times n}$, suppose $g : \mathbb{R}^{n \times n} \to \mathbb{R}^{n \times n}$, and suppose that $\mathcal{S}$ is a closed g-invariant subset (not necessarily a subspace) so that when $\mathbf{X} \in \mathcal{S}$ then $g(\mathbf{X}) \in \mathcal{S}$. If $\|g(\mathbf{X})\| < \gamma \|\mathbf{X}\|$ for some γ with $0 < \gamma < 1$, then g is called a contraction mapping on $\mathcal{S}$ with contraction constant γ. If g is a contraction mapping on $\mathcal{S}$, then a solution to the *fixed-point equation* $\mathbf{X} = g(\mathbf{X})$ exists and is unique in $\mathcal{S}$. The solution may be found by the method of successive approximation: for an arbitrary $\mathbf{X}_0 \in \mathcal{S}$ let

$$\mathbf{X}_i = g(\mathbf{X}_{i-1}) \qquad \text{for} \quad i > 0 \tag{3.85}$$

Then $\mathbf{X}_\infty = \lim_{i \to \infty} \mathbf{X}_i$ exists and satisfies $\mathbf{X}_\infty = g(\mathbf{X}_\infty)$.

3.9.2.3 Linear Matrix Equations

For solving a linear equation of the form $f(\mathbf{X}) = \mathbf{Y}$, where f is a linear function and $\mathbf{X}$ and $\mathbf{Y} \in \mathcal{F}^{n \times n}$, the selection of a basis leads to a matrix representation $\mathbf{A}_f \in \mathcal{F}^{n^2 \times n^2}$, and hence to a corresponding linear equation involving coordinate vectors: $\mathbf{A}_f \mathbf{x} = \mathbf{y}$, with $\mathbf{x}$ and $\mathbf{y} \in \mathcal{F}^{n^2}$.

For the linear function having the form $f(\mathbf{X}) = \mathbf{A}_1 \mathbf{X} \mathbf{A}_2$, with $\mathbf{A}_1$ and $\mathbf{A}_2 \in \mathcal{F}^{n \times n}$, an equivalent linear equation for coordinate vectors can be expressed concisely using Kronecker products. The *Kronecker product*, or *tensor product*, for vectors and matrices provides a useful, systematic means of expressing products of matrix elements [7,9], particularly useful for transforming linear matrix equations to the form $\mathbf{y} = \mathbf{A} \mathbf{x}$. For any two matrices $\mathbf{A} \in \mathcal{R}^{m_A \times n_A}$ and $\mathbf{B} \in \mathcal{R}^{m_B \times n_B}$, the Kronecker product, denoted $\mathbf{A} \otimes \mathbf{B}$, is the $m_A m_B \times n_A n_B$ matrix written in block-matrix form as

$$\mathbf{A} \otimes \mathbf{B} = \begin{bmatrix} a_{11}\mathbf{B} & \cdots & a_{1n_A}\mathbf{B} \\ \vdots & & \vdots \\ a_{m_A 1}\mathbf{B} & \cdots & a_{m_A n_A}\mathbf{B} \end{bmatrix} \tag{3.86}$$

and with this definition, it satisfies (K1) associativity: $(\mathbf{A} \otimes \mathbf{B}) \otimes \mathbf{C} = \mathbf{A} \otimes (\mathbf{B} \otimes \mathbf{C})$, so that $\mathbf{A} \otimes \mathbf{B} \otimes \mathbf{C}$ is unambiguous; (K2) $(\mathbf{A} + \mathbf{B}) \otimes (\mathbf{C} + \mathbf{D}) = (\mathbf{A} \otimes \mathbf{C}) + (\mathbf{A} \otimes \mathbf{D}) + (\mathbf{B} \otimes \mathbf{C}) + (\mathbf{B} \otimes \mathbf{D})$; (K3) $(\mathbf{AB}) \otimes (\mathbf{CD}) = (\mathbf{A} \otimes \mathbf{C})(\mathbf{B} \otimes \mathbf{D})$. (It is assumed that the numbers of rows and columns of the various matrices allow the matrix additions and matrix multiplications to be carried out.)

Some further properties of the Kronecker product of square matrices include: (K4) if $\mathbf{A}$ and $\mathbf{B}$ are invertible, then $(\mathbf{A} \otimes \mathbf{B})^{-1} = \mathbf{A}^{-1} \otimes \mathbf{B}^{-1}$; (K5) $\mathbf{A} \otimes \mathbf{B}$ has eigenvalues given by the distinct products of eigenvalues of $\mathbf{A}$ and $\mathbf{B}$: $\lambda_i(\mathbf{A})\lambda_j(\mathbf{B})$; (K6) $(\mathbf{I} \otimes \mathbf{A} + \mathbf{B} \otimes \mathbf{I})$ has eigenvalues given by the distinct sums of eigenvalues of $\mathbf{A}$ and $\mathbf{B}$: $\lambda_i(\mathbf{A}) + \lambda_j(\mathbf{B})$.

To use the Kronecker product formulation for solving $f(\mathbf{X}) = \mathbf{A}_1 \mathbf{X} \mathbf{A}_2 = \mathbf{Y}$, first form $\mathbf{x}^{\mathrm{T}}$ by concatenating the rows of $\mathbf{X}$; similarly for $\mathbf{y}^{\mathrm{T}}$. Then the matrix equation $\mathbf{A}_1 \mathbf{X} \mathbf{A}_2 = \mathbf{Y}$ is transformed to the equivalent form $(\mathbf{A}_1 \otimes \mathbf{A}_2^{\mathrm{T}}) \mathbf{x} = \mathbf{y}$.

When the linear function $f(\mathbf{X})$ takes a more complicated form, which may always be expressed as a sum of terms,

$$f(\mathbf{X}) = \sum_i \mathbf{A}_{1,i} \mathbf{X} \mathbf{A}_{2,i} \tag{3.87}$$

the Kronecker product approach may provide important insight. For example, the linear matrix equation $\mathbf{A}_1 \mathbf{X} - \mathbf{X} \mathbf{A}_2 = \mathbf{Y}$ becomes $(\mathbf{A}_1 \otimes \mathbf{I}_n - \mathbf{I}_n \otimes \mathbf{A}_2^{\mathrm{T}}) \mathbf{x} = \mathbf{y}$. To characterize invertibility of the resulting $n^2 \times n^2$ matrix, it is most convenient to use the condition that it has no zero eigenvalue. From property (K6), its eigenvalues are given by the differences of the eigenvalues of $\mathbf{A}_1$ and those of $\mathbf{A}_2$, $\lambda_i(\mathbf{A}_1) - \lambda_j(\mathbf{A}_2)$; thus there will be no zero eigenvalue unless some eigenvalue of $\mathbf{A}_1$ is also an eigenvalue of $\mathbf{A}_2$.

As a second example, the linear matrix equation $\mathbf{X} - \mathbf{A}_1 \mathbf{X} \mathbf{A}_2 = \mathbf{Y}$ becomes $(\mathbf{I} - \mathbf{A}_1 \otimes \mathbf{A}_2^{\mathrm{T}})\,\mathbf{x} = \mathbf{y}$, and from property (K5), the resulting $n^2 \times n^2$ matrix is invertible unless some eigenvalue of $\mathbf{A}_1$ is the multiplicative inverse of some eigenvalue of $\mathbf{A}_2$. Interestingly, under suitable conditions the contraction mapping approach may be applied to find an expression for the solution in this example. Rewriting the equation to be solved as $\mathbf{X} = g(\mathbf{X}) = \mathbf{A}_1 \mathbf{X} \mathbf{A}_2 + \mathbf{Y}$ and applying successive approximation with initial trial solution $\mathbf{X}_0 = \mathbf{Y}$, leads to the solution

$$\mathbf{X} = \sum_{i=0}^{\infty} \mathbf{A}_1^{i} \mathbf{Y} \mathbf{A}_2^{i} \tag{3.88}$$

provided that every eigenvalue of $\mathbf{A}_1$ and of $\mathbf{A}_2$ has magnitude less than unity; that is, the matrices must be Schur matrices.

There is a characterization of real Schur matrices involving the linear matrix equation:

$$\mathbf{A}^{\mathrm{T}} \mathbf{X} \mathbf{A} - \mathbf{X} = -\mathbf{Y} \tag{3.89}$$

$\mathbf{A}$ is a real Schur matrix if and only if (*i*) Equation 3.89 has a unique solution $\mathbf{X}$ for every choice of $\mathbf{Y}$ and (*ii*) whenever $\mathbf{Y}$ is a real (symmetric) positive definite matrix, the solution $\mathbf{X}$ is a real (symmetric) positive definite matrix.

Similarly, there is a characterization of real Hurwitz matrices involving another linear matrix equation:

$$\mathbf{A}^{\mathrm{T}} \mathbf{X} + \mathbf{X} \mathbf{A} = -\mathbf{Y} \tag{3.90}$$

$\mathbf{A}$ is a real Hurwitz matrix if and only if (i) Equation 3.90 has a unique solution $\mathbf{X}$ for every choice of $\mathbf{Y}$ and (ii) whenever $\mathbf{Y}$ is a real (symmetric) positive definite matrix, the solution $\mathbf{X}$ is a real (symmetric) positive definite matrix.

Bibliography

1. Bellman, R. 1995. *Introduction to Matrix Analysis.* Society for Industrial and Applied Mathematics, Philadelphia, PA.
2. Bernstein, D.S. 2005. *Matrix Mathematics: Theory, Facts, and Formulas with Application to Linear Systems Theory.* Princeton University Press, Princeton, NJ.
3. Candès, E., Romberg, J., and Tao, T. 2006. Stable signal recovery from incomplete and inaccurate measurements. *Communications on Pure and Applied Mathematics*, vol. 59, no. 8, pp. 1207–1223.
4. Fuhrmann, P.A. 1996. *A Polynomial Approach to Linear Algebra.* Springer-Verlag, New York, NY.
5. Gantmacher, F.R. 1959. *The Theory of Matrices*, 2 vols. Chelsea Publishing Co., New York, NY. (American Mathematical Society reprint, 2000.)
6. Golub, G.H. and Van Loan, C.F. 1996. *Matrix Computations*, 3rd ed. The Johns Hopkins University Press, Baltimore, MD.
7. Halmos, P.R. 1987. *Finite-Dimensional Vector Spaces.* Springer-Verlag, New York, NY.
8. Householder, A.S. 2006. *The Theory of Matrices in Numerical Analysis.* Dover Publications, New York, NY.
9. Laub, A.J. 2005. *Matrix Analysis for Scientists and Engineers.* Society for Industrial and Applied Mathematics, Philadelphia, PA.
10. Luenberger, D.G. 1969. *Optimization by Vector Space Methods.* John Wiley and Sons, New York, NY.
11. MacDuffee, C.C. 1946. *The Theory of Matrices.* Chelsea Publishing Co., New York, NY. (Dover Phoenix Edition reprint, 2004.)
12. Meyer, C.D. 2000. *Matrix Analysis and Applied Linear Algebra.* Society for Industrial and Applied Mathematics, Philadelphia, PA.
13. Strang, G. 2006. *Linear Algebra and Its Applications*, 4th ed. Brooks/Cole, Belmont, CA.

Further Reading

The bibliography, a mix of "classics" (current editions and reprints) and more current titles, comprises a set of personal favorites. Citations in the text have been used sparingly, mostly to indicate books where more detailed discussions about a topic can be found.

Most books on linear systems at the introductory graduate level cover basic material on linear algebra and matrices. Roger Brockett's *Finite Dimensional Linear Systems* (John Wiley, 1970), David Delchamps' *State Space and Input-Output Linear Systems* (Springer-Verlag, 1988), Thomas Kailath's *Linear Systems* (Prentice-Hall, 1980), W.J. (Jack) Rugh's *Linear Systems* (Prentice-Hall, 1995), and Eduardo Sontag's *Mathematical Control Theory: Deterministic Finite Dimensional Systems* (Springer-Verlag, 1998) are recommended.

For coverage of current research on matrices and linear algebra the following journals are recommended: *Linear Algebra and Its Applications*, published by Elsevier Science, Inc., New York, NY, *SIAM Journal on Matrix Analysis and Applications*, published by the Society for Industrial and Applied Mathematics, Philadelphia, PA, and *Electronic Journal of Linear Algebra (ELA)*, www.math.technion.ac.il/iic/ela, published online by ILAS, the International Linear Algebra Society.

A wealth of information is accessible in electronic form online. MathWorld (www.mathworld.wolfram.com), PlanetMath (planetmath.org), and Wikipedia (www.wikipedia.org) are valuable sources for finding specific topical information. Other online resources of interest are the following:

www.matrixcookbook.com
The Matrix Cookbook, compiled by Kaare Brandt Petersen and Michael Syskind Pedersen, 2008.

www.mathworks.com
Home page of The MathWorks, producers of the MATLAB software package.

An online resource with examples and some additional material is available. See
www.princeton.edu/~bradley/MatricesandLinearAlgebra.pdf

4

Complex Variables

C.W. Gray
The Aerospace Corporation

4.1 Complex Numbers

From elementary algebra, the reader should be familiar with the *imaginary* number i where

$$i^2 = -1 \tag{4.1}$$

Historically, in engineering mathematics, the square root of -1 is often denoted by j to avoid notational confusion with current i.

Every new number system in the history of mathematics created cognitive problems which were often not resolved for centuries. Even the terms for the *irrational number* $\sqrt{2}$, *transcendental number* π, and the *imaginary number* $i = \sqrt{-1}$ bear witness to the conceptual difficulties. Each system was encountered in the solution or completeness of a classical problem. Solutions to the quadratic and cubic polynomial equations were presented by Cardan in 1545, who apparently regarded the complex numbers as fictitious but used them formally. Remarkably, A. Girald (1590–1633) conjectured that any polynomial of the nth degree would have n roots in the complex numbers. This conjecture which is known as the *fundamental theorem of algebra* become famous and withstood false proofs from d'Alembert (1746) and Euler (1749). In fact, the dissertation of Gauss (1799) contains five different proofs of the conjecture, two of which are flawed [2].

4.1.1 The Algebra of Complex Numbers

A complex number is formed from a pair of real numbers (x, y) where the complex number is

$$z = x + iy \tag{4.2}$$

and where

$$x = \Re(z), y = \Im(z) \tag{4.3}$$

are called the *real* and *imaginary* parts of z. A complex number z which has no real part $\Re(z) = 0$ is called *purely imaginary*. Two complex numbers are said to be *equal* if both their real and imaginary parts are equal. Assuming the ordinary rules of arithmetic, one derives the rules for addition and multiplication of complex numbers as follows:

$$(x + iy) + (u + iv) = (x + u) + i(y + v) \tag{4.4}$$

$$(x + iy) \cdot (u + iv) = (xu - yv) + i(xv + yu) \tag{4.5}$$

Complex numbers form a *field* satisfying the *commutative, associative,* and *distributive laws.* The real numbers 0 and 1 are the additive and multiplicative *identities.* Assuming these laws, the rule for division is easily derived:

$$\frac{x + iy}{u + iv} = \frac{x + iy}{u + iv} \cdot \frac{u - iv}{u - iv} = \frac{(xu + yv) + i(yu - xv)}{u^2 + v^2} \tag{4.6}$$

4.1.2 Conjugation and Modulus

The formula for complex multiplication employs the fact that $i^2 = -1$. The transformation,

$$z = x + iy \rightarrow \bar{z} = x - iy \tag{4.7}$$

is called *complex conjugation* and has the fundamental properties associated with an isomorphism

$$\overline{a + b} = \bar{a} + \bar{b} \tag{4.8}$$

$$\overline{ab} = \bar{a} \cdot \bar{b} \tag{4.9}$$

The formulas

$$\Re(z) = \frac{z + \bar{z}}{2} \quad \text{and} \quad \Im(z) = \frac{z - \bar{z}}{2i} \tag{4.10}$$

express the real and imaginary parts of z terms of conjugation. Consider the polynomial equation

$$a_n z^n + a_{n-1} z^{n-1} + \cdots + a_1 z + a_0 = 0$$

Taking the complex conjugate of both sides,

$$\bar{a}_n \bar{z}^n + \bar{a}_{n-1} \bar{z}^{n-1} + \cdots + \bar{a}_1 \bar{z} + \bar{a}_0 = 0$$

If the coefficients $a_i = \bar{a}_i$ are real, ξ and $\bar{\xi}$ are roots of the same equation, and hence the nonreal roots of a polynomial with real coefficients occur in conjugate pairs. The product $z\bar{z} = x^2 + y^2$ is always positive if $z \neq 0$. The *modulus* or *absolute value* is defined as

$$|z| = \sqrt{z\bar{z}} = \sqrt{x^2 + y^2} \tag{4.11}$$

Properties of conjugation can be employed to obtain

$$|ab| = |a| \cdot |b| \quad \text{and} \quad \left|\frac{a}{b}\right| = \frac{|a|}{|b|} \tag{4.12}$$

Formulas for the sum and difference follow from expansion:

$$|a + b|^2 = (a + b) \cdot (\bar{a} + \bar{b}) = a\bar{a} + (a\bar{b} + b\bar{a}) + b\bar{b}$$

$$|a + b|^2 = |a|^2 + |b|^2 + 2\Re(a\bar{b}) \tag{4.13}$$

The fact

$$\Re(a\bar{b}) \leq |ab|$$

can be combined with Equation 4.13

$$|a+b|^2 \leq (|a|+|b|)^2$$

to yield the *triangle inequality*

$$|a+b| \leq |a|+|b| \tag{4.14}$$

Cauchy's inequality is true for complex numbers:

$$\left| \sum_{i=1}^{n} a_i b_i \right|^2 \leq \left(\sum_{i=1}^{n} |a_i|^2 \right) \cdot \left(\sum_{i=1}^{n} |b_i|^2 \right) \tag{4.15}$$

4.1.3 Geometric Representation

A complex number $z = x + iy$ can be represented as a pair (x, y) on the *complex plane*. The x-axis is called the *real axis* and the y-axis the *imaginary axis*. The addition of complex numbers can be viewed as vector addition in the plane. The *modulus* or absolute value $|z|$ is interpreted as the *length* of the vector. The product of two complex numbers can be evaluated geometrically if we introduce polar coordinates:

$$x = r\cos\theta \tag{4.16}$$

and

$$y = r\sin\theta \tag{4.17}$$

Hence, $z = r(\cos\theta + i\sin\theta)$. This trigonometric form has the property that $r = |z|$ is the modulus and θ is called the *argument*,

$$\theta = \arg z \tag{4.18}$$

Consider two complex numbers z_1, z_2 where, for $k = 1, 2$,

$$z_k = r_k(\cos\theta_k + i\sin\theta_k)$$

The product is easily computed as

$$z_1 z_2 = r_1 r_2 [(\cos\theta_1\cos\theta_2 - \sin\theta_1\sin\theta_2) + i(\sin\theta_1\cos\theta_2 + \cos\theta_1\sin\theta_2)]$$

The standard addition formulas yield

$$z = z_1 z_2 = r_1 r_2 [\cos(\theta_1 + \theta_2) + i\sin(\theta_1 + \theta_2)] = r(\cos\theta + i\sin\theta) \tag{4.19}$$

The geometric interpretation of the product $z = z_1 z_2$ of complex numbers can be reduced to the dilation or stretch/contraction given by the product $r = r_1 r_2$ and the sum of the rotations

$$\arg(z_1 z_2) = \arg z_1 + \arg z_2 \tag{4.20}$$

The argument of the product is equal to the sum of the arguments. The argument of 0 is not defined and the polar angle θ in Equations 4.16 and 4.17 is only defined to a multiple of 2π.

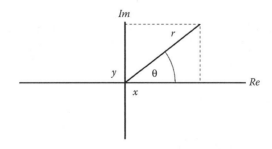

FIGURE 4.1 Polar representation.

The trigonometric form for the division $z = z_1/z_2$ can be derived by noting that the modulus is $r = r_1/r_2$ and

$$\arg \frac{z_1}{z_2} = \arg z_1 - \arg z_2 \tag{4.21}$$

From the preceding discussion, we can derive the powers of $z = r(\cos\theta + i\sin\theta)$ given by

$$z^n = r^n(\cos n\theta + i\sin n\theta) \tag{4.22}$$

For a complex number on the unit circle $r = 1$, we obtain *de Moivre's formula* (1730):

$$(\cos\theta + i\sin\theta)^n = \cos n\theta + i\sin n\theta \tag{4.23}$$

The above formulas can be applied to find the roots of the equation $z^n = a$, where

$$a = r(\cos\theta + i\sin\theta)$$

and

$$z = \rho(\cos\phi + i\sin\phi)$$

Then Equation 4.22 yields

$$\rho^n(\cos n\phi + i\sin n\phi) = r(\cos\theta + i\sin\theta)$$

or

$$\rho = \sqrt[n]{r} \tag{4.24}$$

and

$$\phi = \frac{\theta}{n} + k \cdot \frac{2\pi}{n} \tag{4.25}$$

for $k = 0, 1, \ldots, n-1$. We have found n roots to the equation $z^n = a$. If $a = 1$, then all of the roots lie on the unit circle and we can define the *primitive nth root of unity* ξ:

$$\xi = \cos\frac{2\pi}{n} + i\sin\frac{2\pi}{n} \tag{4.26}$$

The roots of the equation $z^n = 1$ are easily expressed as $1, \xi, \xi^2, \ldots, \xi^{n-1}$.

4.2 Complex Functions

Let $\Omega \subseteq C$ be a subset of the complex plane. A rule of correspondence which associates each element $z = x + iy \in \Omega$ with a unique $w = f(z) = u(x,y) + iv(x,y)$ is called a *single-valued* complex function. Functions like $f(z) = \sqrt{z}$ are called *multiple-valued* and can be considered as a collection of single-valued functions. Definitions of the concepts of limit and continuity are analogous to those encountered in the functions of a real variable. The modulus function is employed as the *metric*.

Definition 4.1: Open Region

A subset $\Omega \subseteq C$ of the complex plane is called an open region or domain if, for every $z_0 \in \Omega$, there exists a $\delta > 0$ exists so that the circular disk $|z - z_0| < \delta$, centered at z_0, is contained in Ω.

Definition 4.2: Limit

$$\lim_{z \to z_0} f(z) = w_0 \tag{4.27}$$

if, for every $\epsilon > 0$, a $\delta > 0$ exists so that $|f(z) - w_0| < \epsilon$ for all z satisfying $0 < |z - z_0| < \delta$

Definition 4.3: Continuity

The function $f(z)$ is continuous at the point z_0 if

$$\lim_{z \to z_0} f(z) = f(z_0) \tag{4.28}$$

The function is said to be continuous in a region Ω if it is continuous at each point $z_0 \in \Omega$.

Definition 4.4: Derivative

If $f(z)$ is a single-valued complex function in some region Ω of the complex plane, the derivative of $f(z)$ is

$$f'(z) = \lim_{\Delta z \to 0} \frac{f(z + \Delta z) - f(z)}{\Delta z} \tag{4.29}$$

The function is called differentiable provided that the limit exists and is the same regardless of the manner in which the complex number $\Delta z \to 0$.

A point where $f(z)$ is not differentiable is called a *singularity*. As in the theory of real-valued functions, the sums, differences, products, and quotients (provided the divisor is not equal to zero) of continuous or differentiable complex functions are continuous or differentiable. It is important to note that the function $f(z) = \bar{z}$ is an example of a function which is continuous but nowhere differentiable.

$$\lim_{\Delta z \to 0} \frac{f(z + \Delta z) - f(z)}{\Delta z} = \pm 1 \tag{4.30}$$

depending upon whether the limit is approached through purely real or imaginary sequences Δz.

4.2.1 Cauchy–Riemann Equations

A function $f(z)$ is said to be *analytic* in a region Ω if it is differentiable and $f'(z)$ is continuous at every point $z \in \Omega$. Analytic functions are also called *regular* or *holomorphic*. A region Ω, for which a complex-valued function is analytic, is called *a region of analyticity*. The previous example showed that the function $\bar{z}$ is not an analytic function. The requirement that a function be analytic is extremely strong. Consider an analytic function $w = f(z) = u(x, y) + iv(x, y)$. The derivative $f'(z)$ can be found by computing the limit through real variations $\Delta z = \Delta x \to 0$:

$$f'(z) = \frac{\partial f}{\partial x} = \frac{\partial u}{\partial x} + i \frac{\partial v}{\partial x} \tag{4.31}$$

or through purely imaginary variations $\Delta z = i\Delta y$:

$$f'(z) = \frac{1}{i} \frac{\partial f}{\partial y} = -i \frac{\partial f}{\partial y} = \frac{\partial v}{\partial y} - i \frac{\partial u}{\partial y} \tag{4.32}$$

Since the function is differentiable,

$$\frac{\partial f}{\partial x} = -i\frac{\partial f}{\partial y} \tag{4.33}$$

Equating expressions (Equations 4.31 and 4.32), one obtains the *Cauchy–Riemann* differential equations

$$\frac{\partial u}{\partial x} = \frac{\partial v}{\partial y}, \quad \frac{\partial u}{\partial y} = -\frac{\partial v}{\partial x} \tag{4.34}$$

Conversely, if the partial derivatives in Equation 4.34 are continuous and u, v satisfy the Cauchy–Riemann equations, then the function $f(z) = u(x, y) + iv(x, y)$ is analytic. If the second derivatives of u and v relative to x, y exist and are continuous, then, by differentiation and use of Equation 4.34,

$$\frac{\partial^2 u}{\partial x^2} + \frac{\partial^2 u}{\partial y^2} = 0, \quad \frac{\partial^2 v}{\partial x^2} + \frac{\partial^2 v}{\partial y^2} = 0 \tag{4.35}$$

The real part u and the imaginary part v satisfy *Laplace's equation* in two dimensions. Functions satisfying Laplace's equation are called *harmonic functions*.

4.2.2 Polynomials

The constant functions and the function $f(z) = z$ are analytic functions. Since the product and the sum of analytic functions are analytic, it follows that any polynomial,

$$p(z) = a_n z^n + a_{n-1} z^{n-1} + \cdots + a_1 z + a_0 \tag{4.36}$$

with complex coefficients a_i is also an analytic function on the entire complex plane. If $a_n \neq 0$, the polynomial $p(z)$ is said to be of degree n. If $a_n = 1$, then $p(z)$ is called a *monic* polynomial.

Theorem 4.1: Fundamental Theorem of Algebra

Every polynomial equation $p(z) = 0$ of degree n has exactly n complex roots ξ_i, $i = 1, \ldots, n$. The polynomial $p(z)$ can be uniquely factored as

$$p(z) = \prod_{i=1}^{n} (z - \xi_i) \tag{4.37}$$

The roots ξ are not necessarily distinct. Roots of $p(z)$ are commonly called *zeros*. If the root ξ_i appears k times in the factorization, it is called a *zero of order k*.

4.2.2.1 Bernoulli's Method

The following numerical method, attributed to Bernoulii, can be employed to find the dominant (largest in modulus) root of a polynomial. The method can be employed as a quick numerical method to check if a discrete-time system is stable (all roots of the characteristic polynomial lie in the unit circle). If there are several roots of the same modulus, then the method is modified and shifts are employed.

Given $a \in C^n$, define the monic polynomial:

$$p_a(z) = z^n + a_{n-1}z^{n-1} + \cdots + a_0$$

Let $\{x_k\}$ be a nonzero solution to the difference equation

$$x_k = -a_{n-1}x_{k-1} - \cdots - a_0 x_{k-n}$$

If $p_a(z)$ has a single largest dominant root r, then in general

$$r = \lim_{k \to \infty} \frac{x_{k+1}}{x_k}$$

If a complex conjugate pair of roots r_1, r_2 is dominant and the coefficients are real $a \in R^{n+1}$, then $r_1, r_2 = r(\cos \theta \pm \sin \theta)$, where

$$r^2 = \lim_{k \to \infty} \frac{x_k^2 - x_{k+1}x_{k-1}}{x_{k-1}^2 - x_k x_{k-2}}$$

and

$$2r \cos \theta = \lim_{k \to \infty} \frac{x_{k+1}x_{k-1} - x_{k-1}x_k}{x_k x_{k-2} - x_{k-1}^2}$$

We sketch the proof of Bernoulli's method for a single real dominant root. The typical response of the difference equation to a set of initial conditions can be written as

$$x_k = c_1 r_1^k + c_2 r_2^k + \cdots + c_n r_n^k$$

where $r_1, \ldots r_n$ are roots of the characteristic equation $p_a(x)$ with $|r_1| > |r_2| \geq \cdots \geq |r_n|$. If the initial conditions are selected properly, $c_1 \neq 0$ and

$$\frac{x_{k+1}}{x_k} = r_1 \frac{1 + (c_2/c_1)(r_2/r_1)^{k+1} + \cdots (c_n/c_1)(r_n/r_1)^{k+1}}{1 + (c_2/c_1)(r_2/r_1)^k + \cdots (c_n/c_1)(r_n/r_1)^k}$$

If r_1 is dominant, then $|r_j/r_1| < 1$ and the fractional expression tends toward 1. Hence x_{k+1}/x_k tends toward r_1. The proof of the complex dominant root formula is a slight generalization.

4.2.2.2 Genji's Formula

The following polynomial root perturbation formula can be employed with the root locus method to adjust or *tweak* the gains of a closed-loop system.

Let $a \in C^{n+1}$ be a vector. Define the polynomial $p_a(z) = a_n z^n + a_{n-1}z^{n-1} + \cdots + a_0$. If $r \in C$ is a root $p_a(r) = 0$, then the following formula relates a perturbation of the root dr to a perturbation of the coefficients $da \in C^{n+1}$:

$$dr = -\frac{p_{da}(r)}{p_a'(r)} \tag{4.38}$$

The formula follows from taking the total differential of the expression $p_a(r) = 0$,

$$[da_n r^n + da_{n-1}r^{n-1} + \cdots + da_0] + [na_n r^{n-1} dr + (n-1)a_{n-1}r^{n-2} dr + \cdots + a_1 dr]$$

Hence,

$$p_{da}(r) + p_a'(r)dr = 0$$

4.2.2.3 Lagrange's Interpolation Formula

Suppose that $z_0, z_1, \ldots, z_n$ are $n + 1$ distinct complex numbers. Given w_i, where $0 \leq i \leq n$, we wish to find the polynomial $p(z)$ of degree n so that $p(z_i) = w_i$. The polynomial $p(z)$ can be employed as a method of interpolation. For $0 \leq i \leq n$, define

$$p_i(z) = \prod_{j \neq i} \left(\frac{z - z_j}{z_i - z_j} \right) \tag{4.39}$$

Clearly, $p_i(z_i) = 1$, and $p_i(z_j) = 0$, for $i \neq j$. Hence the interpolating polynomial can be found by

$$p(z) = \sum_{i=0}^{n} w_i p_i(z) \tag{4.40}$$

4.2.3 Zeros and Poles

The notion of repeated root can be generalized:

Definition 4.5: Zeros

An analytic function $f(z)$ has a zero at $z = a$ of order $k > 0$ if the following limit exists and is nonzero:

$$\lim_{z \to a} \frac{f(z)}{(z - a)^k} \neq 0 \tag{4.41}$$

A *singular point* of a function $f(z)$ is a value of z at which $f(z)$ fails to be analytic. If $f(z)$ is analytic in a region Ω, except at an *interior point* $z = a$, the point $z = a$ is called an *isolated singularity*. For example,

$$f(z) = \frac{1}{z - a}$$

The concept of a *pole* is analogous to that of a zero.

Definition 4.6: Poles

A function $f(z)$ with an isolated singularity at $z = a$ has a pole of order $k > 0$ if the following limit exists and is nonzero:

$$\lim_{z \to a} f(z)(z - a)^k \neq 0 \tag{4.42}$$

A pole of order 1 is called a simple pole.

Clearly, if $f(z)$ has a pole of order k at $z = a$, then

$$f(z) = \frac{g(z)}{(z - a)^k} \tag{4.43}$$

and, if $f(z)$ is analytic and has a zero of order k, then

$$f(z) = (z - a)^k g(z) \tag{4.44}$$

where $g(z)$ is analytic in a region including $z = a$ and $g(a) \neq 0$.

The function

$$f(z) = \frac{\sin z}{z}$$

is not defined at $z = 0$, but could be extended to an analytic function which takes the value 1 at $z = 0$. If a function can be extended to be analytic at a point $z = a$, then $f(z)$ is said to have a *removable singularity*. Hence, if a function $f(z)$ has a pole of order k at $z = a$, then the function $f(z)(z-a)^k$ has a removable singularity at $z = a$. A singularity at a point $z = a$, which is neither removable nor a pole of finite order k, is called *an essential singularity*.

4.2.4 Rational Functions

A *rational* function $H(z)$ is a quotient of two polynomials $N(z)$ and $D(z)$.

$$H(z) = \frac{N(z)}{D(z)} = \frac{b_m z^m + b_{m-1} z^{m-1} + \cdots b_1 z + b_0}{z^n + a_{n-1} z^{n-1} + \cdots + a_1 z + a_0} \tag{4.45}$$

We shall assume in the discussion that the quotient is in *reduced form* and there are no common factors and hence no common zeros. A rational function is called *proper* if $m \leq n$. If the degrees satisfy $m < n$, then $H(z)$ is called *strictly proper*. In control engineering, rational functions most commonly occur as the transfer functions of linear systems. Rational functions $H(z)$ of a variable z denote the transfer functions of *discrete* systems and transfer functions $H(s)$ of the variable s are employed for *continuous* systems. Strictly proper functions have the property that

$$\lim_{z \to \infty} H(z) = 0$$

and hence *roll off* the power at high frequencies. Roots of the numerator and denominator are the zeros and poles of the corresponding rational function, respectively.

4.2.4.1 Partial Fraction Expansion

Consider a rational function $H(s) = N(s)/D(s)$ where the denominator $D(s)$ is a polynomial with distinct zeros $\xi_1, \xi_2, \ldots, \xi_n$. $H(s)$ can be expressed in a partial fraction expansion as

$$\frac{N(s)}{D(s)} = \frac{A_1}{s - \xi_1} + \frac{A_2}{s - \xi_2} + \cdots + \frac{A_n}{s - \xi_n} \tag{4.46}$$

Multiplying both sides of the equation by $s - \xi_i$ and letting $s \to \xi_i$,

$$A_i = \lim_{s \to \xi_i} (s - \xi_i) H(s) \tag{4.47}$$

Applying L'Hospital's rule,

$$A_i = \lim_{s \to \xi_i} N(s) \frac{(s - \xi_i)}{D(s)} = N(\xi_i) \lim_{s \to \xi_i} \frac{1}{D'(s)} = \frac{N(\xi_i)}{D'(\xi_i)}$$

Thus

$$\frac{N(s)}{D(s)} = \frac{N(\xi_1)}{D'(\xi_1)} \cdot \frac{1}{s - \xi_1} + \frac{N(\xi_2)}{D'(\xi_2)} \cdot \frac{1}{s - \xi_2} + \cdots + \frac{N(\xi_n)}{D'(\xi_n)} \cdot \frac{1}{s - \xi_n} \tag{4.48}$$

This formula is commonly called *Heaviside's expansion formula*, and it can be employed for computing the inverse Laplace transform of rational functions when the roots of $D(s)$ are distinct.

In general, any strictly proper rational function $H(s)$ can be written as a sum of the strictly proper rational functions

$$\frac{A_{\xi,r}}{(s-\xi)^r} \tag{4.49}$$

where ξ is a zero of $D(s)$ of order k, where $r \leq k$. If ξ is a repeated zero of $D(s)$ of order k, the coefficient $A_{\xi,r}$ corresponding to the power $r \leq k$ can be found,

$$A_{\xi,r} = \lim_{s \to \xi_i} \frac{1}{(k-r)!} \frac{d^{k-r}}{ds^{k-r}} [(s-\xi_i)^k H(s)] \tag{4.50}$$

4.2.4.2 Lucas' Formula

The *Nyquist stability criterion* or the *principle of the argument* relies upon a generalization of Lucas's formula. The derivative of a factored polynomial of the form,

$$P(s) = a_n(s-\xi_1)(s-\xi_2)\cdots(s-\xi_n)$$

yields Lucas' formula,

$$\frac{P'(s)}{P(s)} = \frac{1}{s-\xi_1} + \frac{1}{s-\xi_2} + \cdots + \frac{1}{s-\xi_n} \tag{4.51}$$

Let $z = a$ be a zero of order k of the function $f(z)$. Application of Equation 4.44

$$f(z) = (z-a)^k g(z)$$
$$f'(z) = k(z-a)^{k-1} g(z) + (z-a)^k g'(z)$$

gives

$$\frac{f'(z)}{f(z)} = \frac{k}{z-a} + \frac{g'(z)}{g(z)} \tag{4.52}$$

where $g(a) \neq 0$ and $g(z)$ is analytic at $z = a$. For a pole at $z = a$ of order k of a function $f(z)$, Equation 4.43 yields

$$f(z) = (z-a)^{-k} g(z)$$
$$f'(z) = -k(z-a)^{-k-1} g(z) + (z-a)^{-k} g'(z)$$

This gives

$$\frac{f'(z)}{f(z)} = -\frac{k}{z-a} + \frac{g'(z)}{g(z)} \tag{4.53}$$

where $g(a) \neq 0$ and $g(z)$ is analytic around $z = a$. Inductive use of the above expressions results in a generalization of Lucas' formula (Equation 4.51). For a rational function with zeros α_j and poles ξ_i,

$$H(s) = \frac{N(s)}{D(s)} = \frac{\prod_{j=1}^{m}(s-\alpha_j)}{\prod_{i=1}^{n}(s-\xi_i)}$$
$$= \frac{(s-\alpha_1)(s-\alpha_2)\cdots(s-\alpha_m)}{(s-\xi_1)(s-\xi_2)\cdots(s-\xi_n)} \tag{4.54}$$

and

$$\frac{H'(s)}{H(s)} = \sum_{j=1}^{m} \frac{1}{s-\alpha_j} - \sum_{i=1}^{n} \frac{1}{s-\xi_i} \tag{4.55}$$

Rational functions can be generalized:

Definition 4.7: Meromorphic function

A function $f(z)$, which is analytic in an open region Ω and whose every singularity is an isolated pole, is said to be meromorphic.

The transfer function of every continuous time-invariant linear system is meromorphic in the complex plane. Systems or block diagrams which employ the Laplace transform for a delay e^{-sT} result in meromorphic transfer functions. If the meromorphic function $f(z)$ has a finite number of zeros α_j and poles ξ_i in a region Ω, then Equations 4.55, 4.52, and 4.53 yield a generalized Lucas formula,

$$\frac{f'(z)}{f(z)} = \sum_{j=1}^{m} \frac{1}{z - \alpha_j} - \sum_{i=1}^{n} \frac{1}{z - \xi_i} + \frac{g'(z)}{g(z)} \tag{4.56}$$

where $g(z) \neq 0$ is analytic in Ω.

4.2.5 Power Series Expansions

A *power series* is of the form

$$f(z) = a_0 + a_1 z + a_2 z^2 + \cdots + a_n z^n + \cdots = \sum_{n=0}^{\infty} a_n z^n \tag{4.57}$$

In general, a power series can be expanded around a point $z = z_0$,

$$f(z) = \sum_{n=0}^{\infty} a_n (z - z_0)^n \tag{4.58}$$

Series expansions do not always converge for all values of z. For example, the *geometric series*

$$\frac{1}{1 - z} = 1 + z + z^2 + \cdots + z^n + \cdots \tag{4.59}$$

converges when $|z| < 1$. Every power series has a *radius of convergence* ρ. In particular, Equation 4.58 converges for all $|z - z_0| < \rho$ where, by *Hadamard's formula*,

$$\frac{1}{\rho} = \lim_{n \to \infty} \sup \sqrt[n]{|a_n|} \tag{4.60}$$

Historically, two different approaches have been taken to set forth the fundamental theorems in the theory of analytic functions of a single complex variable. Cauchy's approach (1825) defines an analytic function as in Subsection 4.2.1 employing the Cauchy-Riemann equations 4.34 and Green's theorem in the plane to derive the famous integral formulas (Section 4.3.1). The existence of a power series expansion follows directly from the integral formulas.

Theorem 4.2: Taylor's Series

Let $f(z)$ be an analytic function on a circular region Ω centered at $z = z_0$. For all points in the circle,

$$f(z) = \sum_{n=0}^{\infty} a_n (z - z_0)^n \quad where \quad a_n = \frac{f^{(n)}(z_0)}{n!} \tag{4.61}$$

This expansion agrees with the form for the Taylor series expansion of a function of a real variable. Most texts base their exposition of the theory of a complex variable on Cauchy's approach incorporating a slight weakening of the definition due to Goursat. Weierstrauss' approach defines a function as analytic at a point z_0, if there is a convergent power series expansion (Equation 4.58). If one accepts the relevant theorems concerning the ability to move integrals and derivatives through power series expansions, the Cauchy integral formulas are easily demonstrated.

4.2.5.1 The Exponential Function

The exponential function is defined by the power series

$$e^z = 1 + \frac{z}{1!} + \frac{z^2}{2!} + \cdots + \frac{z^n}{n!} + \cdots \tag{4.62}$$

which converges for all complex values of z. Familiar identities of the form $e^{a+b} = e^a \cdot e^b$ are true by virtue of the formal power series expansion. Euler's formula (1749)

$$e^{i\theta} = \cos\theta + i\sin\theta \tag{4.63}$$

is easily derived from substitution in Equation 4.62,

$$e^{i\theta} = \left(1 - \frac{\theta^2}{2!} + \cdots\right) + i\left(\frac{\theta}{1!} - \frac{\theta^3}{3!} + \cdots\right)$$

Thus, the polar form $z = r(\cos\theta + i\sin\theta)$ can be compactly expressed as $z = re^{i\theta}$. De Moivre's formula (Equation 4.23) states the obvious relationship

$$(e^{i\theta})^n = e^{in\theta}$$

The unit circle $|z| = 1$ can be parameterized as $z = e^{i\theta}$, where $0 \le \theta < 2\pi$. Substituting in a power series expansion of the form (Equation 4.57) yields a Fourier series expansion,

$$f(z) = \sum_{n=0}^{\infty} a_n z^n = \sum_{n=0}^{\infty} a_n e^{in\theta}$$

Curves of the form,

$$\gamma(t) = r_1 e^{i\omega_1 t} + r_2 e^{i\omega_2 t} + \cdots + r_m e^{i\omega_m t}$$

are epicycles and examples of almost periodic functions. The ancient approach of employing epicycles to describe the motions of the planets can be viewed as an exercise in Fourier approximation. If $z = x + iy$, then $e^z = e^x(\cos y + i\sin y)$.

The multiple-valued *logarithm* function $\ln z$ is defined as the inverse of the exponential e^z. Hence, if $z = re^{i\theta}$ is in polar form and n is an integer,

$$\ln z = \ln r + i(\theta + 2\pi n) \tag{4.64}$$

The imaginary part of the logarithm is the same as the argument function. The addition theorem of the exponential implies

$$\ln(z_1 z_2) = \ln z_1 + \ln z_2 \tag{4.65}$$

which makes sense only if both sides of the equation represent the same infinite set of complex numbers.

4.2.5.2 Trigonometric Functions

The trigonometric functions are defined as

$$\cos z = \frac{e^{iz} + e^{-iz}}{2} \quad \text{and} \quad \sin z = \frac{e^{iz} - e^{-iz}}{2i} \tag{4.66}$$

Note that the sine and cosine functions are periodic,

$$\sin(z + 2\pi n) = \sin z \quad \text{and} \quad \cos(z + 2\pi n) = \cos z$$

The expressions for the trigonometric functions (4.66) can be employed to deduce the addition formulas

$$\cos(z_1 + z_2) = \cos z_1 \cos z_2 - \sin z_1 \sin z_2 \tag{4.67}$$
$$\sin(z_1 + z_2) = \cos z_1 \sin z_2 + \sin z_1 \cos z_2 \tag{4.68}$$

and the modulation formula

$$\cos \omega_1 t \cos \omega_2 t = \frac{1}{2}[\cos(\omega_1 + \omega_2)t + \cos(\omega_1 - \omega_2)t] \tag{4.69}$$

If signals of frequencies ω_1, ω_2 are modulated with each other, they produce energy at the sum $\omega_1 + \omega_2$ and difference frequencies $\omega_1 - \omega_2$.

4.3 Complex Integrals

If $f(z) = u(x, y) + iv(x, y)$ is defined and continuous in a region Ω, we define the *integral of* $f(z)$ along some curve $\gamma \subseteq \Omega$ by

$$\int_\gamma f(z)\, dz = \int_\gamma (u + iv)(dx + i\, dy)$$
$$= \int_\gamma u\, dx - v\, dy + i \int_\gamma v\, dx + u\, dy \tag{4.70}$$

These expressions depend upon line integrals for real-valued functions. If the curve $\gamma(t)$ is a piecewise differentiable arc $\gamma(t)$ for $a \leq t \leq b$, then Equation 4.70 is equivalent to

$$\int_\gamma f(z)\, dz = \int_a^b f(\gamma(t))\gamma'(t)\, dt \tag{4.71}$$

The most important property of the line integral (4.71) is its invariance with a change of parameter. Hence, if two curves start and end at the same points and trace out the same curve γ, the value of the integrals (Equation 4.71) will be the same. Distinctions are made in terms of the direction of travel,

$$\int_{-\gamma} f(z)\, dz = - \int_\gamma f(z)\, dz$$

A curve or arc $\gamma(t)$ is said to be *closed* if the endpoints coincide $\gamma(a) = \gamma(b)$. A closed curve is called *simple* if it does not intersect itself.

All points to the left of a curve as it is traversed are said to be *enclosed* by it. A counterclockwise (CCW) traverse around a contour is said to be *positive*. A closed curve $\gamma(t)$ is said to make n *positive encirclements* of the origin $z = 0$ if vector $\gamma(t)$ rotates in a CCW direction and completes n rotations. A *negative encirclement* is obtained if the path is traversed in a clockwise (CW) directions.

The notions of enclosement or encirclement have different conventions in the mathematical literature and in engineering expositions of classical control theory. Most texts in classical control state that a point is enclosed or encircled by a contour if it lies to the right of the curve as it is traversed, and CW contours and rotations are called positive.

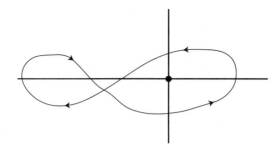

FIGURE 4.2   A single positive encirclement of the origin.

4.3.1 Integral Theorems

4.3.1.1 Cauchy's Theorem

Suppose $f(z)$ is analytic in a region Ω bounded by a simple closed curve γ. *Cauchy's theorem* states that

$$\int_\gamma f(z)dz = \oint_\gamma f(z)dz = 0 \qquad (4.72)$$

This equation is equivalent to saying that $\int_{z_1}^{z_2} f(z)dz$ is unique and is *independent of the path joining z_1 and z_2.*

Let $\gamma(t)$ for $0 \le t \le 1$ be a closed curve which does not pass through the origin $z = 0$. Consider the line integral for an integer n,

$$\int_\gamma z^n dz \quad \text{for } n \ne -1$$

By Cauchy's theorem, this integral is zero if $n \ge 0$. By computation,

$$\int_\gamma z^n dz = \left.\frac{z^{n+1}}{n+1}\right|_\gamma = \frac{\gamma(1)^{n+1}}{n+1} - \frac{\gamma(0)^{n+1}}{n+1}$$

Because the curve is closed, $\gamma(0) = \gamma(1)$, and

$$\int_\gamma z^n dz = 0 \quad \text{for } n \ne -1 \qquad (4.73)$$

This argument can be generalized: for any closed curve $\gamma(t)$ not passing through the point $z = a$,

$$\int_\gamma (z - a)^n dz = 0 \quad \text{for } n \ne -1 \qquad (4.74)$$

Let $f(z)$ be a power series expansion of the form

$$f(z) = \sum_{n=0}^{\infty} a_n(z - a)^n$$

and let $\gamma(t)$ lie within the radius of convergence. Applying Equation 4.74 and moving the integration through the expansion

$$\int_\gamma f(z)dz = \sum_{n=0}^{\infty} \int_\gamma a_n(z - a)^n dz = 0 \qquad (4.75)$$

gives a version of Cauchy's theorem.

4.3.1.2 Cauchy's Integral Formulas

Consider the closed curve $\gamma(t) = a + e^{it}$ for $0 \le t \le 2\pi k$. The curve lies on a unit circle centered at $z = a$ and completes k CCW positive encirclements of $z = a$. Consider the line integral

$$\int_\gamma \frac{1}{z-a} dz \tag{4.76}$$

By computation,

$$\int_\gamma \frac{1}{z-a} dz = \ln(z-a)|_\gamma$$

is a multivalued function. To obtain the integral, one must consider the expression

$$\ln e^{it} - \ln e^0 = it$$

for $0 \le t \le 2\pi k$. Thus

$$\int_\gamma \frac{1}{z-a} dz = 2\pi ki \tag{4.77}$$

The equation can be generalized as

$$n(\gamma, a) = \frac{1}{2\pi i} \int_\gamma \frac{1}{z-a} dz \tag{4.78}$$

where $n(\gamma, a)$ is called *the winding number of γ around $z = a$*. The integral counts the number of CCW encirclements of the point $z = a$.

If $f(z)$ is analytic within and on a region Ω bounded by a simple closed curve γ and $a \in \Omega$ is a point interior to γ, then

$$f(a) = \frac{1}{2\pi i} \oint_\gamma \frac{f(z)}{z-a} dz \tag{4.79}$$

where γ is traversed in the CCW direction. Higher-order derivatives $f^{(r)}(a)$ can be expressed as

$$f^{(r)}(a) = \frac{r!}{2\pi i} \oint_\gamma \frac{f(z)}{(z-a)^{r+1}} dz \tag{4.80}$$

Equations 4.79 and 4.80 are known as the *Cauchy integral formulas*. The formulas imply that, if the analytic function $f(z)$ is known on a simple closed curve γ, then its value (and, its higher derivatives) in the interior of γ are preordained by the behavior of the function along γ. This quite remarkable fact is contrary to any intuition that one might infer from real-valued functions.

If $f(z)$ has a Taylor power series expansion around the point $z = a$,

$$f(z) = \sum_{n=0}^\infty a_n (z-a)^n, \quad \text{where } a_n = \frac{f^{(n)}(a)}{n!} \tag{4.81}$$

and the closed curve γ is contained in the radius of convergence, then

$$\frac{1}{2\pi i} \int_\gamma \frac{f(z)}{(z-a)} dz = \frac{1}{2\pi i} \sum_{n=0}^\infty \int_\gamma a_n (z-a)^{n-1} dz$$

The terms corresponding to $n > 0$ are zero by Cauchy's theorem (Equation 4.74), and the use of Equation 4.78 yields a version of the Cauchy integral formula (Equation 4.79):

$$n(\gamma, a) f(a) = n(\gamma, a) a_0 = \frac{1}{2\pi i} \int_\gamma \frac{f(z)}{z-a} dz \tag{4.82}$$

Formal division of the power series expansion (Equation 4.81) by the term $(z - a)^{r+1}$ yields the higher derivative formulas

$$n(\gamma, a) f^{(r)}(a) = \frac{1}{2\pi i} \int_\gamma \frac{f(z)}{(z-a)^{r+1}} dz \tag{4.83}$$

4.3.2 The Argument Principle

Every rational transfer function $H(s)$ is meromorphic. The transfer function of a time-invariant linear system is meromorphic, even when it employs delays of the form e^{-sT}. Let $f(z)$ be a meromorphic function in a region Ω which contains a finite number of zeros α_j and poles ξ_k. The generalized Lucas formula (Equation 4.56) gives

$$\frac{f'(z)}{f(z)} = \sum_{j=1}^m \frac{1}{z - \alpha_j} - \sum_{k=1}^n \frac{1}{z - \xi_k} + \frac{g'(z)}{g(z)} \tag{4.84}$$

where $g(z) \neq 0$ is analytic in Ω. Since $g'(z)/g(z)$ is analytic in Ω, one can apply Cauchy's theorem (Equation 4.72) to deduce

$$\frac{1}{2\pi i} \int_\gamma \frac{g'(z)}{g(z)} dz = 0$$

By Equations 4.78 and 4.84

$$\frac{1}{2\pi i} \int_\gamma \frac{f'(z)}{f(z)} dz = \sum_j^m n(\gamma, \alpha_j) - \sum_{k=1}^n n(\gamma, \xi_k) \tag{4.85}$$

The function $w = f(z)$ maps γ onto a closed curve $\Gamma(t) = f(\gamma(t))$ and, by a change in variables,

$$\frac{1}{2\pi i} \int_\Gamma \frac{1}{w} dw = \frac{1}{2\pi i} \int_\gamma \frac{f'(z)}{f(z)} dz \tag{4.86}$$

hence

$$n(\Gamma, 0) = \sum_j n(\gamma, \alpha_j) - \sum_k n(\gamma, \xi_k) \tag{4.87}$$

The left-hand side of Equation 4.85 can be viewed as the number of CCW encirclements of the origin $n(\Gamma, 0)$. If γ is a simple closed curve, then Equation 4.85 computes the difference $m - n$ between the number of zeros and number of poles. Equation 4.87 is known as the *principle of the argument*.

For example, consider the simple closed curve γ of Figure 4.3. If a function $f(z)$ has three poles and a single zero enclosed by the curve γ, then the argument principle states that the curve $\Gamma = f(\gamma)$ must make two negative encirclements of the origin.

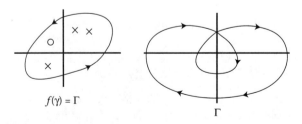

$f(\gamma) = \Gamma$

Γ

FIGURE 4.3 The number of encirclements of the origin by Γ is equal to the difference between the number encirclements of the zeros and poles.

The argument principle (Equation 4.85) can be generalized. If $g(z)$ is analytic in a region Ω and $f(z)$ is meromorphic in Ω with a finite number of zeros and poles, then, for any closed curve γ,

$$\frac{1}{2\pi i} \int_\gamma g(z)\frac{f'(z)}{f(z)} dz = \sum_{j=1}^m n(\gamma, \alpha_j)g(\alpha_j) - \sum_{k=1}^n n(\gamma, \xi_k)g(\xi_k) \tag{4.88}$$

The case $g(z) = z$ is of interest. Suppose that $f(z)$ is analytic in a circular region Ω of radius $r > 0$ around a. Ω is bounded by the simple closed curve $\gamma(t) = a + re^{it}$ for $0 \leq t \leq 2\pi$. Suppose the function $f(z)$ has an inverse in Ω, then $f(z) - w$ has only a single zero in Ω, and Equation 4.88 yields the *inversion formula*

$$f^{-1}(w) = \frac{1}{2\pi i} \oint_\gamma \frac{zf'(z)}{f(z) - w} dz \tag{4.89}$$

4.3.2.1 Other Important Theorems

The following theorems are employed in H^∞ control theory.

1. *Liouville's Theorem*: If $f(z)$ is analytic and $|f(z)| < M$ is bounded in the entire complex plane, then $f(z)$ must be a constant.
2. *Cauchy's Estimate*: Suppose the analytic function $f(z)$ is bounded, $|f(z)| < M$, on and inside a circular region of radius r centered at $z = a$, then the kth derivative satisfies

$$|f^{(k)}(a)| \leq \frac{Mk!}{r^k} \tag{4.90}$$

3. *Maximum Modulus Theorem*: If $f(z)$ is a nonconstant analytic function inside and on a simple closed curve γ, then the maximum value of $|f(z)|$ occurs on γ and is not achieved on the interior.
4. *Minimum Modulus Theorem*: If $f(z)$ is a nonzero analytic function inside and on a simple closed curve γ, then the minimum value of $|f(z)|$ occurs on γ.
5. *Rouche's Theorem*: If $f(z), g(z)$ are analytic on a simple closed curve γ, then $f(z)$ and the sum $f(z) + g(z)$ have the same number of zeros inside γ.
6. *Gauss' Mean Value Theorem*: If $f(z)$ is analytic inside and on the circle of radius r centered at $z = a$, then $f(a)$ is the average value of $f(z)$ along the circle,

$$f(a) = \frac{1}{2\pi} \int_0^{2\pi} f(a + re^{i\theta})d\theta \tag{4.91}$$

4.3.3 The Residue Theorem

Let $f(z)$ be analytic in a region Ω except at a pole at $z = a \in \Omega$ of order k. By Equation 4.43,

$$g(z) = (z - a)^k f(z)$$

has a removable singularity at $z = a$ and can be viewed as analytic over Ω. Thus, $g(z)$ may be expanded in a Taylor series about $z = a$. Dividing by $(z - a)^k$ yields the *Laurent expansion*

$$f(z) = \frac{a_{-k}}{(z - a)^k} + \cdots + \frac{a_{-1}}{z - a} + a_0 + a_1(z - a) + a_2(z - a)^2 + \cdots \tag{4.92}$$

In general, a series of the form

$$\sum_{r=-\infty}^{\infty} a_r(z - a)^r$$

is called a *Laurent series*. Because power series expansions have a radius of convergence, a Laurent series can be viewed as the expansion of two analytic functions $h_1(z), h_2(z)$ where

$$H(z) = h_1\left(\frac{1}{z-a}\right) + h_2(z-a) = \sum_{r=-\infty}^{\infty} a_r(z-a)^r$$

The series converges for values of z which lie in an annular region $\rho_1 < |z-a| < \rho_2$ where ρ_1 can be zero and ρ_2 could be infinite. The *principal part* $h_1(1/(z-a))$ corresponds to the coefficients a_r, where $r < 0$ and the *analytic part* $h_2(z)$ corresponds to the coefficients a_r, where $r \geq 0$. If the principal part has infinitely many nonzero terms $a_r \neq 0$, then $z = a$ is said to be an, *essential singularity* of the function $H(z)$. The coefficient a_{-1} is called the *residue of H(z) at the point z = a*.

If $f(z)$ is analytic within and on a simple closed curve γ except for an isolated singularity at $z = a$, then it has a Laurent series expansion around $z = a$ where, by Equation 4.74 and the Cauchy integral formula (4.79),

$$f(z) = \sum_{r=-\infty}^{\infty} a_r(z-a)^r, \quad \text{where } a_{r-1} = \frac{1}{2\pi i} \oint_\gamma \frac{f(z)}{(z-a)^r} dz \tag{4.93}$$

The *residue* is defined as

$$\text{Res}(f, a) = a_{-1} = \frac{1}{2\pi i} \oint_\gamma f(z)\, dz \tag{4.94}$$

and, for an arbitrary curve γ where $z = a$ is the only singularity enclosed by γ,

$$n(\gamma, a)\text{Res}(f, a) = \frac{1}{2\pi i} \int_\gamma f(z)\, dz$$

At a simple pole of $f(z)$ at $z = a$,

$$\text{Res}(f, a) = \lim_{z \to a}(z-a)f(z) \tag{4.95}$$

and, at a pole of order k,

$$\text{Res}(f, a) = \lim_{z \to a} \frac{1}{(k-1)!} \frac{d^{k-1}}{dz^{k-1}}[(z-a)^k f(z)] \tag{4.96}$$

For a simple pole, Equation 4.95 is identical to Equation 4.47 and, for a pole of order k, Equation 4.96 is identical to Equation 4.50 with $r = 1$.

The *residue theorem* states that, if $f(z)$ is analytic within and on a region Ω defined by a simple closed curve γ except at a finite number of isolated singularities $\xi_1, \ldots, \xi_k$, then,

$$\oint_\gamma f(z)\, dz = 2\pi i[\text{Res}(f, \xi_1) + \cdots + \text{Res}(f, \xi_k)] \tag{4.97}$$

Cauchy's theorem (Equation 4.72) and the integral theorems can be viewed as special cases of the residue theorem.

The residue theorem can be employed to find the values of various integrals. For example, consider the integral

$$\int_{-\infty}^{\infty} \frac{1}{1+z^4}\, dz$$

The poles of the function $f(z) = 1/(1+z^4)$ occur at the points

$$e^{i\pi/4}, \quad e^{i3\pi/4}, \quad e^{i5\pi/4}, \quad e^{i7\pi/4}$$

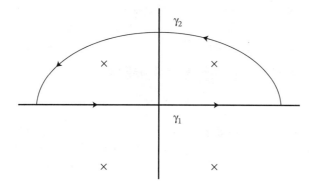

FIGURE 4.4 Residue theorem example.

Two of the poles lie in the upper half-plane and two lie in the lower half-plane. Employing Equation 4.95, one can compute the residues of the poles in the upper half-plane

$$\text{Res}(f, e^{i\pi/4}) = \frac{e^{-i3\pi/4}}{4}$$

and

$$\text{Res}(f, e^{i3\pi/4}) = \frac{e^{-i\pi/4}}{4}$$

and the sum of the residues in the upper half-plane is $-i\sqrt{2}/4$. Consider the contour integral of Figure 4.4. The curve γ consists of two curves γ_1 and γ_2, and hence,

$$\oint_\gamma f(z)dz = \int_{\gamma_1} f(z)dz + \int_{\gamma_2} f(z)dz$$

By the residue theorem (4.97),

$$\oint_\gamma \frac{1}{1+z^4}dz = 2\pi i(-i\sqrt{2}/4) = \frac{\pi\sqrt{2}}{2}$$

One can show that the limit of the line integral

$$\int_{\gamma_2} f(z)dz \to 0$$

as the radius of the semicircle approaches infinity and the curve γ_1 approaches the interval $(-\infty, \infty)$. Thus

$$\int_{-\infty}^{\infty} \frac{1}{1+z^4}dz = \frac{\pi\sqrt{2}}{2}$$

4.4 Conformal Mappings

Every analytic function $w = f(z)$ can be viewed as a mapping from the z plane to the w plane. Suppose $\gamma(t)$ is a differentiable curve passing through a point z_0 at time $t = 0$. The curve $\Gamma = f(\gamma)$ is a curve passing through the point $w_0 = f(z_0)$. An application of the chain rule gives

$$\Gamma'(0) = f'(z_0)\gamma'(0) \tag{4.98}$$

Taking the argument and assuming $f'(z_0) \neq 0$ and $\gamma'(0) \neq 0$, then

$$\arg \Gamma'(0) = \arg f'(z_0) + \arg \gamma'(0)$$

Hence the angle between the directed tangents of γ and Γ at the point z_0 is the angle $\arg f'(z_0)$. Thus, if $f(z_0) \neq 0$, two curves γ_1, γ_2 which intersect at angle are mapped by $f(z)$ to two curves which intersect at the same angle. A mapping with this property is called *conformal*, and hence, if $f(z)$ is analytic and $f'(z) \neq 0$ in a region Ω, then $f(z)$ is conformal on Ω.

Equation 4.98 has an additional geometric interpretation. The quantity $|f'(z_0)|^2$ can be viewed as the *area dilation factor* at z_0. Infinitesimal area elements $dxdy$ around z_0 are expanded (or contracted) by a factor of $|f'(z_0)|^2$.

4.4.1 Bilinear or Linear Fractional Transformations

A transformation of the form,

$$w = T(z) = \frac{az + b}{cz + d} \tag{4.99}$$

where $ad - bc \neq 0$ is called a *linear fractional or bilinear transformation*. This important class of transformations occurs in control theory in developing Padé delay approximations, transformations between continuous s-domain and discrete z-domain realizations, and lead-lag compensators. There are four *fundamental transformations*:

1. Translation. $w = z + b$.
2. Rotation. $w = az$ where $a = e^{i\theta}$.
3. Dilation: $w = az$ where $a = r$ is real. If $a < 1$, the mapping contracts; if $a > 1$, its expands.
4. Inversion: $w = 1/z$.

Every fractional transformation can be decomposed into a combination of translations, rotations, dilations, and inversions. In fact, every linear fractional transformation of the form (4.99) can be associated with a 2×2 complex matrix A_T,

$$A_T = \begin{bmatrix} a & b \\ c & d \end{bmatrix} \quad \text{where } \det A_T = ad - bc \neq 0 \tag{4.100}$$

By direct substitution, one can show that, if T, S are two bilinear transformations, then the composition $T \cdot S = T(S(z))$ is bilinear and

$$A_{T \cdot S} = A_T A_S \tag{4.101}$$

holds for the corresponding 2×2 matrix multiplication. The fundamental transformations correspond to the matrices

$$\begin{bmatrix} 1 & b \\ 0 & 1 \end{bmatrix}, \begin{bmatrix} e^{i\theta} & 0 \\ 0 & 1 \end{bmatrix}, \begin{bmatrix} r & 0 \\ 0 & 1 \end{bmatrix}, \quad \text{and} \quad \begin{bmatrix} 0 & 1 \\ 1 & 0 \end{bmatrix}$$

For a fractional transformation $w = T(z)$, if γ is a curve which describes a circle or a line in the z plane, then $\Gamma = T(\gamma)$ is a circle or a line in the w plane. This follows from the fact that it is valid for the fundamental transformations and, hence, for any composition.

Any scalar multiple αA, where $\alpha \neq 0$, corresponds to the same linear fractional transformation as A.

Hence, one could assume that the matrix A_T is *unimodular* or $\det A_T = ad - bc = 1$; if $\alpha = \sqrt{\det A_T}$, then the linear transformation given by (a, b, c, d) is identical to one given by $(a, b, c, d)/\alpha$.

Every linear fractional transformation T (Equation 4.99) has an inverse which is a linear fractional transformation

$$z = T^{-1}(w) = \frac{dw - b}{-cw + a} \tag{4.102}$$

If A is unimodular, then A^{-1} is unimodular and

$$A^{-1} = \begin{bmatrix} d & -b \\ -c & a \end{bmatrix}$$

For example, for a sample time Δt, the *Tustin or bilinear transformation* is the same as the Padé approximation,

$$z^{-1} = e^{-s\Delta t} \approx \frac{1 - s\Delta t/2}{1 + s\Delta t/2}$$

or

$$z = T(s) = \frac{1 + s\Delta t/2}{1 - s\Delta t/2} \tag{4.103}$$

and Equation 4.102 yields

$$s = T^{-1}(z) = \frac{z - 1}{z\Delta t/2 + \Delta t/2} = \frac{2}{\Delta t} \cdot \frac{z - 1}{z + 1} \tag{4.104}$$

The Tustin transformation conformally maps the left half-plane of the s-domain onto the unit disk in the z-domain. Thus if one designs a stable system with a transfer function $H(s)$ and discretizes the system by the Tustin transformation, one obtains a stable z-domain system with transfer function $G(z) = H[T^{-1}(z)]$.

4.4.2 Applications to Potential Theory

The real and the imaginary parts of an analytic function $f(z)$ satisfy *Laplace's equation*,

$$\nabla^2 \Phi = \frac{\partial^2 \Phi}{\partial x^2} + \frac{\partial^2 \Phi}{\partial y^2} = 0 \tag{4.105}$$

Solutions to Laplace's equation are called *harmonic*. Laplace's equation occurs in electromagnetics and the velocity potential of stationary fluid flow. An equation of the form

$$\nabla^2 \Phi = f(x, y) \tag{4.106}$$

is called *Poisson's equation* commonly occurring in problems solving for the potential derived from Gauss' law of electrostatics. Let Ω be a region bounded by a simple closed curve γ. Two types of *boundary-value problem* are commonly associated with Laplace's equation:

1. *Dirichlet's Problem*: Determine a solution to Laplace's equation subject to a set of prescribed values along the boundary γ.
2. *Neumann's Problem*: Determine a solution to Laplace's equation so that the derivative normal to the curve $\partial \Phi / \partial n$ takes prescribed values along γ.

Conformal mapping can be employed to find a solution of Poisson's or Laplace's equation. In general, one attempts to find an analytic or meromorphic function $w = f(z)$ which maps the region Ω to the interior of the unit circle or the upper half-plane. The mapped boundary-valued problem is then solved on the w plane for the unit circle or upper half-plane and is then transformed via $f^{-1}(w)$ to solve the problem on Ω.

Let $f(z) = u(x, y) + iv(x, y)$ be analytic on a region Ω. Both $u(x, y)$ and $u(x, y)$ and $v(x, y)$ satisfy Equation 4.105. The function $v(x, y)$ is called a conjugate harmonic function to $u(x, y)$. Since the mapping $f(z)$ is conformal, the curves

$$u(x, y) = a, v(x, y) = b$$

for a fixed a, b are orthogonal. The first curve $u(x, y) = a$ is often called the *equipotential line* and the curve $v(x, y) = b$ is called the *streamline* of the flow.

References

1. Ahlfors, L.V., *Complex Analysis*, McGraw-Hill, New York, 1966.
2. Bell, E. T., *The Development of Mathematics*, Dover, New York, NY, 1940.
3. Cartan, H., *Elementary Theory of Analytic Functions of One or Several Complex Variables*, Hermann, Addison-Wesley, 1963.
4. Churchill, R. V., *Introduction to Complex Variables and Applications*, McGraw-Hill, 2nd ed., 1962.
5. Knopp, K., *Theory of Functions*, Dover, New York, NY, 1945.
6. Marsden, J. and Hoffman, M., *Complex Analysis*, W. H. Freeman, 2nd ed., 1987.

II

Models for
Dynamical
Systems

5

Standard Mathematical Models

William S. Levine
University of Maryland

James T. Gillis
The Aerospace Corporation

Graham C. Goodwin
The University of Newcastle

Juan C. Agüero
The University of Newcastle

Juan I. Yuz
Federico Santa María Technical University

Harry L. Trentelman
University of Groningen

Richard Hill
University of Detroit Mercy

5.1 Input–Output Models

William S. Levine

5.1.1 Introduction

A fundamental problem in science and engineering is to predict the effect a particular action will have on a physical system. This problem can be posed more precisely as follows. What will be the response, $y(t)$, for all times t in the interval $t_0 \leq t < t_f$, of a specified system to an arbitrary input $u(t)$ over the same time interval ($t_0 \leq t < t_f$)? Because the question involves the future behavior of the system, its answer requires some sort of model of the system.

Engineers use many different kinds of models to predict the results of applying inputs to physical systems. One extreme example is the 15-acre scale model, scaled at one human stride to the mile, of the drainage area of the Mississippi river that the U.S. Corps of Engineers uses to predict the effect of flood control actions [1]. Such models, although interesting, are not very general. It is more useful, in a chapter such as this one, to concentrate on classes of models that can be used for a wide variety of problems.

The input–output models form just such a class. The fundamental idea behind input–output models is to try to model only the relation between inputs and outputs. No attempt is made to describe the "internal" behavior of the system. For example, electronic amplifiers are often described only by input–output models. The many internal voltages and currents are ignored by the model. Because of this concentration on the external behavior, many different physical systems can, and do, have the same input–output models. This is a particular advantage in design. Given that a particular input–output behavior is required and specified, the designer can choose the most advantageous physical implementation.

This chapter is restricted to input–output models. Section 5.2 deals with state-space models. A complete discussion of all types of input–output models would be virtually impossible. Instead, this chapter concentrates on an exemplary subclass of input–output models, those that are linear invariant and time invariant (LTI). Although no real system is either linear invariant or time invariant, many real systems are well approximated by LTI models within the time duration and range of inputs over which they are used. Even when LTI models are somewhat inaccurate, they have so many advantages over more accurate models that they are often still used, albeit cautiously. These advantages will be apparent from the ensuing discussion.

LTI ordinary differential and difference equation (ODE) models will be introduced in Section 5.1.2. The same acronym is used for both differential and difference equations because they are very similar, have analogous properties, and it will be clear from the context which is meant. ODE LTI models are very often obtained from the physics of a given system. For example, ODE models for electrical circuits and many mechanical systems can be deduced directly from the physics. The section concludes with a brief introduction to nonlinear and time-varying ODE input–output models.

Section 5.1.3 deals with continuous-time and discrete-time impulse response models. These are slightly more general than the ODE models. Such models are primarily used for LTI systems. An introduction to impulse response models for time-varying linear systems concludes the section.

Section 5.1.4 describes transfer function models of LTI systems. Transfer functions are very important in classical control theory and practice. They have the advantage of being directly measurable. That is, given a physical system that is approximately LTI, its transfer function can be determined experimentally.

The chapter concludes with Section 5.1.5, in which the equivalence among the different descriptions of LTI models is discussed.

5.1.2 Ordinary Differential and Difference Equation Models

Consider a simple example of a system, such as an electronic amplifier. Such a system normally has an input terminal pair and an output terminal pair, as illustrated in Figure 5.1. Also, there is often a line cord that must be connected to a wall outlet to provide power for the amplifier. Typically, the amplifier

FIGURE 5.1   A representation of an electronic amplifier.

is designed to have very high impedance at the input and very low impedance at the output. Because of this, the input voltage is not affected by the amplifier and the output voltage is determined only by the amplifier. Thus, in normal operation, the amplifier can be regarded as a system with input $u(t) = v_{in}(t)$ and output $y(t) = v_0(t)$. The relationship between $y(t)$ and $u(t)$ is designed to be approximately $y(t) = au(t)$, where a is some real constant. Notice that the power supply is ignored, along with the currents, in this simplified model of the amplifier. Furthermore, the facts that the amplifier saturates and that the gain, a, generally depends on the input frequency have also been ignored. This illustrates a fundamental aspect of modeling. Those features of the real system that are deemed unimportant should be left out of the model. This requires considerable judgement. The best models include exactly those details that are essential and no others. This is context dependent. Much of the modeling art is in deciding which details to include in the model.

It is useful to generalize this simple electronic example to a large class of single-input single-output (SISO) models. Consider the structure depicted in Figure 5.2, where $u(t)$ and $y(t)$ are both scalars. In many physical situations, the relation between $u(t)$ and $y(t)$ is a function of the derivatives of both functions. For example, consider the RC circuit of Figure 5.3, where $u(t) = v(t)$ and $y(t) = i(t)$. It is well known that a mathematical model for this circuit is [2]

$$v(t) = Ri(t) + \frac{1}{C} \int_{-\infty}^{t} i(\tau) \, d\tau \tag{5.1}$$

Differentiating both sides once with respect to t, replacing $v(t)$ by $u(t)$, $i(t)$ by $y(t)$ and dividing by R gives

$$\frac{1}{R} \dot{u}(t) = \dot{y}(t) + \frac{1}{RC} y(t) \tag{5.2}$$

This example illustrates two important points. First, Equations 5.1 and 5.2 are approximations to reality. Real RC circuits behave linearly only if the input voltage is not too large. Real capacitors include some leakage (large resistor in parallel with the capacitance) and real resistors include some small inductance. The conventional model is a good model in the context of inputs, $v(t)$, that are not too large (so the nonlinearity can be ignored) and, not too high frequency (so the inductance can be ignored). The leakage current can be ignored whenever the capacitor is not expected to hold its charge for a long time.

Second, Equation 5.1 implicitly contains an assumption that the input and output are defined for past times, τ, $-\infty < \tau < t$. Otherwise, the integral in Equation 5.1 would be meaningless. This has apparently disappeared in Equation 5.2. However, in order to use Equation 5.2 to predict the response of the system to a given input, one would also need to know an "initial condition," such as $y(t_0)$ for some specific t_0. In

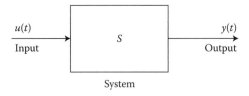

FIGURE 5.2   A standard input–output representation of a continuous-time system.

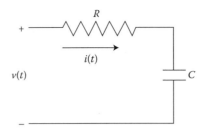

FIGURE 5.3 A series RC circuit.

the context of input–output models it is preferable not to have to specify separately the initial conditions and the input. The separate specification of initial conditions can be avoided, for systems that are known to be stable, by assuming the system is "initially at rest"—that is, by assuming the input is zero prior to some time t_0 (which may be $-\infty$) and that the initial conditions are all zero up to t_0 ($y(t)$ and all its derivatives are zero prior to t_0). If the response to nonzero initial conditions is important, as it is in many control systems, nonzero initial conditions can be specified. Given a complete set of initial conditions, the input prior to t_0 is irrelevant.

The RC circuit is an example of a stable system. If the input is zero for a time duration longer than 5RC, the charge on the capacitor and the current, $y(t)$, will decay virtually to zero. The choice of 5RC, is somewhat arbitrary. The time constant of the transient response of the RC circuit is RC and 5 time constants are commonly used as the time at which the response is approximately zero. If the input subsequently changes from zero, say, at time t_0, the RC circuit can be modeled by a system that is "initially at rest" even though it may have had a nonzero input at some earlier time.

A simple generalization of the RC circuit example provides a large, and very useful class of input–output models for systems. This is the class of models of the form

$$\frac{d^n y(t)}{dt^n} + a_{n-1}\frac{d^{n-1}y(t)}{dt^{n-1}} + \cdots + a_0 y(t) = b_m \frac{d^m u(t)}{dt^m} + \cdots + b_0 u(t) \tag{5.3}$$

where the $a_i, i = 0, 1, 2, \ldots, n-1$, and the $b_j, j = 0, 1, 2, \ldots, m$ are real numbers.

The reader is very likely to have seen such models before because they are common in many branches of engineering and physics. Both the previous examples are special cases of Equation 5.3. Equations of this form are also studied extensively in mathematics.

Several features of Equation 5.3 are important. Both sides of Equation 5.3 could be multiplied by any nonzero real number without changing the relation between $y(t)$ and $u(t)$. In order to eliminate this ambiguity, the coefficient of $dy^n(t)/dt^n$ is always made to be one by convention.

Models of the form of Equation 5.3 are known as linear systems for the following reason. Assume, in addition to Equation 5.3, that the system is at rest prior to some time, t_0. Assume also that input $u_i(t)$, $t_0 \le t < t_f$, produces the response $y_i(t)$, $t_0 \le t < t_f$, for $i = 1, 2$. That is,

$$\frac{d^n y_i(t)}{dt^n} + a_{n-1}\frac{d^{n-1}y_i(t)}{dt^{n-1}} t \cdots + a_0 y_i(t) = b_m \frac{d^m u_i(t)}{dt^m} + \cdots + b_0 u_i(t) \quad \text{for } i = 1, 2 \tag{5.4}$$

Then, if α and β are arbitrary constants (physically, α and β must be real, but mathematically they can be complex), then the input

$$u_s(t) = \alpha u_1(t) + \beta u_2(t) \tag{5.5}$$

to the system (Equation 5.3) produces the response

$$y_s(t) = \alpha y_1(t) + \beta y_2(t) \tag{5.6}$$

A proof is elementary. Substitute Equations 5.5 and 5.6 into Equation 5.3 and rearrange the terms [3].

The mathematical definition of linearity requires that the superposition property described by Equations 5.5 and 5.6 be valid. More precisely, a system is linear if and only if the system's response, $y_s(t)$, to any linear combination of inputs $(u_s(t) = \alpha u_1(t) + \beta u_2(t))$ is the same linear combination of the responses to the inputs taken one at a time $(y_s(t) = \alpha y_1(t) + \beta y_2(t))$.

If nonzero initial conditions are included in Equation 5.3, as is often the case in control, the input $u_k(t), k = 1, 2$, will produce output

$$y_k(t) = y_{ic}(t) + y_{u_k}(t) \quad k = 1, 2 \tag{5.7}$$

where $y_{ic}(t)$ denotes the response to initial conditions with $u(t) = 0$, and $y_{u_k}(t)$ denotes the response to $u_k(t)$ with zero initial conditions. When the input $u_s(t)$ in Equation 5.5 is applied to Equation 5.3 and the initial conditions are not zero, the resulting output is

$$y_s(t) = y_{ic}(t) + \alpha y_{u_1}(t) + \beta y_{u_2}(t) \tag{5.8}$$

When $y_s(t)$ is computed by means of Equations 5.6 and 5.7, the result is

$$y_s(t) = (\alpha + \beta)y_{ic}(t) + \alpha y_{u_1}(t) + \beta y_{u_2}(t) \tag{5.9}$$

The fact that Equation 5.8, the correct result, and Equation 5.9 are different proves that nonzero initial conditions invalidate the strict mathematical linearity of Equation 5.3. However, systems having the form of Equation 5.3 are generally known as linear systems, even when they have nonzero initial conditions.

Models of the form of Equation 5.3 are known as time-invariant systems for the following reason. Assume, in addition to Equation 5.3, that the system is at rest prior to the time, t_0, at which the input is applied. Assume also that the input $u(t)$, $t_0 \leq t < \infty$, produces the output $y(t), t_0 \leq t < \infty$. Then, applying the same input shifted by any amount T produces the same output shifted by an amount T. More precisely, letting (remember that $u(t) = 0$ for $t < t_0$ as part of the "at rest" assumption)

$$u_d(t) = u(t - T), \quad t_0 + T < t < \infty \tag{5.10}$$

Then, the response, $y_d(t)$, to the input $u_d(t)$ is

$$y_d(t) = y(t - T), \quad t_0 + T < t < \infty \tag{5.11}$$

A proof that systems described by Equation 5.3 are time-invariant is simple; substitute $u(t - T)$ into Equation 5.3 and use the uniqueness of the solution to linear ODEs to show that the resulting response must be $y(t - T)$.

Of course, many physical systems are neither linear nor time invariant. A simple example of a nonlinear system can be obtained by replacing the resistor in the RC circuit example by a nonlinear resistance, a diode for instance. Denoting the resistor current–voltage relationship by $v(t) = f(i(t))$, where $f(\cdot)$ is some differentiable function, Equation 5.1 becomes

$$v(t) = f(i(t)) + \frac{1}{C} \int_{-\infty}^{t} i(\tau)\, d\tau \tag{5.12}$$

Differentiating both sides with respect to t, replacing $v(t)$ by $u(t)$, and $i(t)$ by $y(t)$ gives

$$\dot{u}(t) = \frac{df}{dy}(y(t))\dot{y}(t) + \frac{1}{C}y(t) \tag{5.13}$$

The system in Equation 5.12 is not linear because an input of the form (Equation 5.5) would not produce an output of the form (Equation 5.6).

One could also allow the coefficients in Equation 5.3 (a_is and b_js) to depend on time. The result would still be linear but would no longer be time invariant. There will be a brief discussion of such systems in the following section.

Before introducing the impulse response, the class of discrete-time models will be introduced. Consider the ODE

$$y(k+n) + a_{n-1}y(k+n-1) + \cdots + a_0 y(k) = b_m u(k+m) + \cdots + b_0 u(k) \tag{5.14}$$

where the $a_i, i = 0, 1, 2, \ldots, n-1$ and $b_j, j = 0, 1, 2, \ldots, m$ are real numbers and $k = k_0, k_0 + 1, k_0 + 2, \ldots, k_f$ are integers. Such models commonly arise from either sampling a continuous-time physical system or as digital simulations of physical systems. The properties of Equation 5.14 are similar to those of Equation 5.3. The leading coefficient (the coefficient of $y(k+n)$) is conventionally taken to be one. The computation of $y(k)$ given $u(k), k = k_0, k_0 + 1, \ldots, k_f$ also requires n initial conditions. The need for initial conditions can be addressed, for stable systems, by assuming the system is "initially at rest" prior to the instant, k_0, at which an input is first applied. This initially "at rest" assumption means that (1) $u(k) = 0$ for $k < k_0$ and (2) that the initial conditions ($y(0), y(-1), \ldots, y(-n+1)$ for example) are all zero. When the system is initially at rest, analogous arguments to those given for Equation 5.3 show that Equation 5.14 is linear and time invariant. Generally, even when the system is not initially at rest, systems of the form of Equation 5.14 are known as LTI discrete-time systems.

5.1.3 Impulse Response

The starting point for discussion of the impulse response is not the system but the signal, specifically the input $u(t)$ or $u(k)$. Generally, inputs and outputs, as functions of time, are called signals. The discrete-time case, $u(k)$, will be described first because it is mathematically much simpler. The first step is to raise an important question that has been heretofore ignored. What is the collection of possible inputs to a system? This collection will be a set of signals. In Section 5.1.2, it was assumed that any $u(k)$, such that $u(k) = 0$, for $k < k_0$, could be an input. In reality this is not so. For example, it would be physically impossible to create the following signal:

$$u(k) = \begin{cases} 0, & k \le 0 \\ k^2, & k \ge 0 \end{cases} \tag{5.15}$$

Some energy is required to produce any physical signal; the energy needed for the signal in Equation 5.15 would be infinite.

There was a second assumption about the collection of signals described in Section 5.1.2. Equation 5.5 and the definition of linearity assume that $u_s(t)$ ($u_s(k)$ gives the discrete-time version), defined only as a linear combination of possible inputs, is also a possible input. Mathematically, this amounts to assuming that the collection of possible input signals forms a vector space. For engineering purposes the requirement is the following.

If $u_i(k), i = 1, 2, 3, \ldots$ belong to a collection of possible input signals and $\alpha_i, i = 1, 2, 3, \ldots$ are real numbers, then

$$u_t(k) = \sum_{i=1}^{\infty} \alpha_i u_i(k) \tag{5.16}$$

also belongs to the collection of possible input signals.

Equation 5.16 provides the first key to an economical description of LTI discrete-time systems. The second key is the signal known as the unit impulse or the unit pulse,

$$\delta(k) = \begin{cases} 1, & k = 0 \\ 0 & \text{all other integers} \end{cases} \tag{5.17}$$

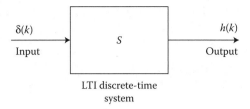

FIGURE 5.4 An input–output representation of a discrete-time LTI system showing a unit pulse as input and the discrete-time impulse response as output.

Using $\delta(k)$ and Equation 5.16, any signal

$$u(k) = u_k \ (u_k \text{ a real number}) \ -\infty < k < \infty \tag{5.18}$$

can be rewritten as a sum of unit pulses

$$u(k) = \sum_{i=-\infty}^{\infty} u_i \delta(k - i) \tag{5.19}$$

This initially seems to be a ridiculous thing to do. Equation 5.19 is just a complicated way to write Equation 5.18. However, suppose you are given an LTI discrete-time system, S, and that the response of this system to an input $\delta(k)$ is $h(k)$, as illustrated in Figure 5.4. Because the system is time invariant, its response to an input $\delta(k - i)$ is just $h(k - i)$. Because the u_i in Equation 5.19 are constants, like α and β in Equations 5.5 and 5.6, because the system is linear, and because Equations 5.5 and 5.6 can be extended to infinite sums by induction, the following argument is valid. Denote the action of S on $u(k), -\infty < k < \infty$, by

$$y(k) = S(u(k))$$

Then,

$$y(k) = S(\sum_{i=-\infty}^{\infty} u_i \delta(k - i))$$

$$= \sum_{i=-\infty}^{\infty} u_i S(\delta(k - i))$$

$$y(k) = \sum_{i=-\infty}^{\infty} u_i h(k - i) \quad -\infty < k < \infty \tag{5.20}$$

Equation 5.20 demonstrates that the response of the LTI discrete-time system, S, to any possible input, $u(k), -\infty < k < \infty$, can be computed from *one* output signal, $h(k), -\infty < k < \infty$, known as the impulse (or unit pulse) response of the system. Thus, the impulse response, $h(k), -\infty < k < \infty$, is an input–output model of S.

The main uses of impulse response models are theoretical. This is because using Equation 5.20 involves a very large amount of computation and there are several better ways to compute $y(k), -\infty < k < \infty$, when the impulse response and the input are given. One example of the use of the impulse response is in the determination of causality.

A system is said to be causal if and only if the output at any time $k, y(k)$, depends only on the input at times up to and including time k, that is, on the set of $u(\ell)$ for $-\infty < \ell \leq k$. Real systems must be causal. However, it is easy to construct mathematical models that are not causal.

It is evident from Equation 5.20 that an impulse response, $h(k)$, is causal if and only if

$$h(k) = 0 \quad \text{for all } k < 0 \tag{5.21}$$

For causal impulse responses, Equation 5.20 becomes

$$y(k) = \sum_{i=-\infty}^{k} u_i h(k - i) \tag{5.22}$$

because $h(k - i) = 0$ by Equation 5.21 for $i > k$.

The development of the continuous-time impulse response as an input–output model for LTI continuous-time systems is analogous to that for discrete time. The ideas are actually easy to understand, especially after seeing the discrete-time version. However, the underlying mathematics are very technical. Thus, proofs will be omitted.

A natural beginning is with the second key to an economical description of LTI systems, the definition of the unit impulse. For reasons that will be explained below, the continuous-time unit impulse is usually defined by the way it operates on smooth functions, rather than as an explicit function such as Equation 5.17.

Definition 5.1:

Let $f(t)$ be any function that is continuous on the interval $-\epsilon < t < \epsilon$ for every $0 < \epsilon < \epsilon_m$ and some ϵ_m. Then the unit impulse (also known as the Dirac delta function) $\delta(t)$ satisfies

$$f(0) = \int_{-\infty}^{\infty} f(\tau)\delta(\tau)\, d\tau \tag{5.23}$$

To see why Equation 5.23 is used and not something like Equation 5.17, try to construct a $\delta(t)$ that would satisfy Equation 5.23. The required function must be zero everywhere but at $t = 0$ and its integral over any interval that includes $t = 0$ must be one.

Given a signal, $u(t)$ $-\infty < t < \infty$, the analog to Equation 5.19 then becomes, using Equation 5.23,

$$u(t) = \int_{-\infty}^{\infty} u(\tau)\delta(t - \tau)\, d\tau \tag{5.24}$$

Equation 5.24 can then be used analogously to Equation 5.19 to derive the continuous-time equivalent of Equation 5.20. That is, suppose you are given an LTI continuous-time system, S, and that the response of this system to a unit impulse applied at $t = 0$, $\delta(t)$, is $h(t)$ for all t, $-\infty < t < \infty$. In other words,

$$h(t) = S(\delta(t)) \quad -\infty < t < \infty$$

To compute the response of S, $y(t)(-\infty < t < \infty)$, to an input $u(t)(-\infty < t < \infty)$, proceed as follows.

$$y(t) = S(u(t))$$
$$= S(\int_{-\infty}^{\infty} u(\tau)\delta(t - \tau)\, d\tau)$$

Because $S(\cdot)$ acts on signals (functions of time, t), because $u(\tau)$ acts as a constant (not a function of t), and because integration commutes with the action of linear systems (think of integration as the limit of a sequence of sums),

$$y(t) = \int_{-\infty}^{\infty} u(\tau) \mathcal{S}(\delta(t-\tau)) \, d\tau$$

$$y(t) = \int_{-\infty}^{\infty} u(\tau) h(t-\tau) \, d\tau \quad -\infty < t < \infty \tag{5.25}$$

As in the discrete-time case, the primary use of Equation 5.25 is theoretical. There are better ways to compute $y(t)$ than by direct computation of the integral in Equation 5.25. Specifically, the Laplace or Fourier transform provides an efficient means to compute $y(t)$ from knowledge of $h(t)$ and $u(t)$, $-\infty < t < \infty$. The transforms also provide a good vehicle for the discussion of physical signals that can be used as approximations to the unit impulse. These transforms will be discussed in the next section.

Two apparently different classes of input–output models have been introduced for both continuous-time and discrete-time LTI systems. A natural question is whether the ODE models are equivalent to the impulse response models. It is easy to see that, in continuous time, there are impulse responses for which there are not equivalent ODEs. The simplest such example is a pure delay. That is

$$y(t) = u(t - t_d) \quad -\infty < t < \infty \tag{5.26}$$

where $t_d > 0$ is a fixed time delay.

The impulse response for a pure delay is

$$h(t) = \delta(t - t_d) \quad -\infty < t < \infty \tag{5.27}$$

but there is no ODE that exactly matches Equation 5.26 or 5.27. Note that there are real systems for which a pure delay is a good model. Electronic signals travel at a finite velocity. Thus, long transmission paths correspond to pure delays.

The converse is different. Every ODE has a corresponding impulse response. It is easy to demonstrate this in discrete time. Simply let $\delta(k)$ be the input in Equation 5.14 with $n = 1$ and $m = 0$. Assuming Equation 5.14 is initially at rest results in a recursive calculation for $h(k)$. For example, let

$$y(k+1) + ay(k) = bu(k) \quad -\infty < k < \infty$$

Replacing $u(k)$ by $\delta(k)$ gives an input that is zero prior to $k = 0$. Assuming the system is initially at rest makes $y(k) = 0$ for $k < 0$. Then

$$y(0) + ay(-1) = bu(-1)$$

gives

$$y(0) = h(0) = 0, \quad y(1) + ay(0) = bu(0) = b$$

gives

$$y(1) = h(1) = b, \quad y(2) + ay(1) = bu(1) = 0$$

gives

$$y(2) = h(2) = -ab,$$

and so on.

A similar recursion results in the general case when $\delta(k)$ is input to an arbitrary ODE.

The result in the continuous-time case is the same—every ODE has a corresponding impulse response—but the mathematics is more complicated unless one uses transforms.

The response of a system that is linear but time varying to a unit impulse depends on the time at which the impulse is applied. Thus, the impulse response of a linear time-varying system must be denoted in the continuous-time case by $h(t, \tau)$ where τ is the time at which the impulse is applied and t is the time at which the impulse response is recorded. Because the system is linear, the argument that produced Equation 5.25 also applies in the time-varying case. The result is

$$y(t) = \int_{-\infty}^{\infty} u(\tau)h(t, \tau)\, d\tau \quad -\infty < t < \infty \tag{5.28}$$

The analogous result holds in discrete time.

$$y(k) = \sum_{i=-\infty}^{\infty} u(i)h(k, i) \quad -\infty < k < \infty \tag{5.29}$$

See *Control System Advanced Methods*, Chapter 3, for more information about linear time-varying systems.

There are forms of impulse response models that are useful in the study of nonlinear systems. See *Control System Advanced Methods*, Chapter 40.

5.1.4 Transfer Functions

The other common form for LTI input–output models is the transfer function. The transfer function, as an input–output model, played a very important role in communications and in the development of feedback control theory in the 1930s. Transfer functions are still very important and useful. One reason is that for asymptotically stable, continuous-time LTI systems, the transfer function can be measured easily. To see this, suppose that such a system is given. Assume that this system has an impulse response, $h(t)$, and that it is causal ($h(t) = 0$ for $t < 0$). Suppose that this system is excited with the input

$$u(t) = \cos \omega t \quad -\infty < t < \infty \tag{5.30}$$

Equation 5.30 is a mathematical idealization of a situation where the input cosinusoid started long enough in the past that all the transients have decayed to zero. The corresponding output is

$$y(t) = \int_{-\infty}^{\infty} h(t - \tau) \cos \omega \tau \, d\tau \tag{5.31}$$

$$= \int_{-\infty}^{\infty} h(t - \tau) \left(\frac{e^{j\omega t} + e^{-j\omega \tau}}{2} \right) d\tau$$

$$= \frac{1}{2} \int_{-\infty}^{\infty} h(t - \tau)e^{j\omega \tau} d\tau + \frac{1}{2} \int_{-\infty}^{\infty} h(t - \tau)e^{-j\omega \tau} d\tau$$

$$= \frac{1}{2} \int_{-\infty}^{\infty} h(\sigma)e^{j\omega(t - \sigma)} d\sigma + \frac{1}{2} \int_{-\infty}^{\infty} h(\sigma)e^{-j\omega(t - \sigma)} d\sigma$$

$$y(t) = \left(\int_{-\infty}^{\infty} h(\sigma)e^{-j\omega \sigma} d\sigma \right) \frac{e^{j\omega t}}{2} + \left(\frac{1}{2} \int_{-\infty}^{\infty} h(\sigma)e^{j\omega \sigma} d\sigma \right) \frac{e^{-j\omega t}}{2} \tag{5.32}$$

Define for all real ω, $-\infty < \omega < \infty$,

$$H(j\omega) = \int_{-\infty}^{\infty} h(\sigma)e^{-j\omega \sigma} \, d\sigma \tag{5.33}$$

Notice that $H(j\omega)$ is a complex number for every real ω and that the complex conjugate of $H(j\omega)$, denoted $H^*(j\omega)$, is $H(-j\omega)$. Then, Equation 5.32 becomes

$$y(t) = \frac{H(j\omega)e^{j\omega t} + H^*(j\omega)e^{-j\omega t}}{2} \quad \text{or,} \quad y(t) = |H(j\omega)| \cos(\omega t + \angle H(j\omega)) \tag{5.34}$$

where

$$|H(j\omega)| = \text{magnitude of } H(j\omega)$$
$$\measuredangle H(j\omega) = \text{angle of } H(j\omega).$$

Of course, $H(j\omega)$, for all real ω, $-\infty < \omega < \infty$, is the transfer function of the given LTI system. It should be noted that some authors call $H(j\omega)$ the frequency response of the system and reserve the term "transfer function" for $H(s)$ (to be defined shortly). Both $H(j\omega)$ and $H(s)$ will be called transfer functions in this article. Equation 5.34 shows how to measure the transfer function; evaluate Equation 5.34 experimentally for every value of ω. Because it is impossible to measure $H(j\omega)$ for every ω, $-\infty < \omega < \infty$, what is actually done is to measure $H(j\omega)$ for a finite collection of ω's and interpolate.

Suppose the object of the experiment was to measure the impulse response. Recognize that Equation 5.33 defines $H(j\omega)$ to be the Fourier transform of $h(t)$. The inverse of the Fourier transform is given by

$$x(t) \triangleq \int_{-\infty}^{\infty} X(j\omega)e^{j\omega t}\frac{d\omega}{2\pi} \tag{5.35}$$

where $X(j\omega)$ is a function of ω, $-\infty < \omega < \infty$.

Applying Equation 5.35 to Equation 5.33 gives

$$h(t) = \int_{-\infty}^{\infty} H(j\omega)e^{j\omega t}\frac{d\omega}{2\pi} \tag{5.36}$$

Equation 5.36 provides a good way to determine $h(t)$ for asymptotically stable continuous-time LTI systems. Measure $H(j\omega)$ and then compute $h(t)$ from Equation 5.36. Of course, it is not possible to measure $H(j\omega)$ for all ω, $-\infty < \omega < \infty$. It is possible to measure $H(j\omega)$ for enough values of ω to compute a good approximation to $h(t)$. In many applications, control design using Bode, Nichols, or Nyquist plots for example, knowing $H(j\omega)$ is sufficient.

Having just seen that the transfer function can be measured when the system is asymptotically stable, it is natural to ask what can be done when the system is unstable. The integral in Equation 5.33 blows up; because of this the Fourier transform of $h(t)$ does not exist. However, the Laplace transform of $h(t)$ is defined for unstable as well as stable systems and is given by

$$H(s) = \int_{-\infty}^{\infty} h(t)e^{-st}\,dt \tag{5.37}$$

for all complex s such that the integral in Equation 5.37 is finite.

Transfer functions have several important and useful properties. For example, it is easy to prove that the transfer function for a continuous-time LTI system satisfies

$$Y(s) = H(s)U(s) \tag{5.38}$$

where $Y(s)$ is the Laplace transform of the output $y(t)$, $-\infty < t < \infty$, and $U(s)$ is the Laplace transform of the input $u(t)$, $-\infty < t < \infty$.

To prove Equation 5.38, take the Laplace transform of both sides of Equation 5.25 to obtain

$$Y(s) = \int_{-\infty}^{\infty} \left(\int_{-\infty}^{\infty} u(\tau)h(t-\tau)d\tau \right) e^{-st}\,dt$$
$$= \int_{-\infty}^{\infty} \int_{-\infty}^{\infty} u(\tau)h(t-\tau)e^{st}\,dt\,d\tau$$

Make the change of variables $\sigma = t - \tau$

$$= \int_{-\infty}^{\infty} \int_{-\infty}^{\infty} u(\tau)h(\sigma)e^{-s(\sigma+\tau)}d\sigma\,d\tau$$

$$Y(s) = \int_{-\infty}^{\infty} h(\sigma)e^{-s\sigma}d\sigma \int_{-\infty}^{\infty} u(\tau)e^{-st}d\tau = H(s)U(s) \tag{5.39}$$

The Laplace transform provides an easy means to demonstrate the relationship between transfer functions, $H(s)$ or $H(j\omega)$, and ODE models of LTI continuous-time systems. Take the Laplace transform of both sides of Equation 5.3, assuming that the system is initially at rest. The result is

$$(s^n + a_{n-1}s^{n-1} + \cdots + a_0)Y(s) = (b_m s^m + \cdots + b_0)U(s) \tag{5.40}$$

where the fact that the Laplace transform of $\dot{y}(t)$ is $sY(s)$ has been used repeatedly. Dividing through gives

$$Y(s) = \frac{b_m s^m + b_{m-1}s^{m-1} + \cdots + b_0}{s^n + a_{n-1}s^{n-1} + \cdots + a_0}U(s) \tag{5.41}$$

Equation 5.41 shows that, for a continuous-time ODE of the form of Equation 5.3,

$$H(s) = \frac{b_m s^m + b_{m-1}s^{m-1} + \cdots + b_0}{s^n + a_{n-1}s^{n-1} + \cdots + a_0} \tag{5.42}$$

The discrete-time case is very similar. The discrete-time analog of the Fourier transform is the discrete Fourier transform. The discrete-time analog of the Laplace transform is the Z-transform. The results and their derivations parallel those for continuous-time systems. Refer any textbook on signals and systems such as [3] or [4] for details.

5.1.5 Conclusions

Although this chapter has treated the ODE, impulse response, and transfer function descriptions of LTI systems separately, it should be apparent that they are equivalent descriptions for a large class of systems. A demonstration that the impulse response and transfer function descriptions are more general than the ODE descriptions has already been given; there is no continuous-time ODE corresponding to $H(s) = e^{-sT}$. However, all three descriptions are equivalent whenever $H(s)$ can be written as a rational function, that is, as a ratio of polynomials in s.

There is a result known as Runge's theorem [5, p. 258] that proves that any $H(s)$ that is analytic in a region of the s-plane can be approximated to uniform accuracy, in that region, by a rational function. A family of such approximations is known as the Padé approximants [6]. The basic idea is to expand the given analytic function in a Taylor series (this is always possible) and then choose the coefficients of the rational function so as to match as many terms of the series as possible. For example,

$$e^{-sT} = 1 - Ts + \frac{T^2}{2!}s^2 + \cdots + \frac{T^2}{n!}s^n + \cdots \approx \frac{b_m s^m + b_{m-1}s^{m-1} + \ldots + b_0}{s^m + a_{m-1}s^{m-1} + \cdots + a_0}$$

The $2m - 1$ coefficients $(b_0, b_1, \ldots, b_m, a_0, \ldots, a_{m-1})$ can then be selected to match the first $2m - 1$ coefficients of the Taylor series. The result is known as the Padé approximation to a pure delay of duration T in the control literature [7, p. 332]. The approximation improves with increasing m.

There are still examples for which approximation by ODEs is problematic. An example is the flow of heat in a long solid rod. Letting x denote the displacement along the rod and assuming that the input is applied at $x = 0$, that the initial temperature of the rod is zero, and that the output is the temperature of the rod at point x, then the transfer function observed at x is [8, pp. 182–184] and [9, pp. 145–150]

$$H(s, x) = e^{-x\sqrt{s/a}} \tag{5.43}$$

where a is the thermal diffusivity.

Even for a fixed x, this is an example of a transfer function to which Runge's theorem does not apply in any region of the complex plane that includes the origin. The reason is that $H(s, x)$ is not differentiable at $s = 0$ for any $x > 0$ and is therefore not analytic in any region containing the origin. In many applications it is nonetheless adequate to approximate this transfer function by a simple low-pass filter,

$$H(s) = \frac{b_0}{s + a_0} \tag{5.44}$$

This example emphasizes the difficulty of making general statements about modeling accuracy. Deciding whether a given model is adequate for some purpose requires a great deal of expertise about the physical system and the intended use of the model. The decision whether to use an input–output model is somewhat easier. Input–output models are appropriate whenever the internal operation of the physical system is irrelevant to the problem of interest. This is true in many systems problems, including many control problems.

References

1. McPhee, J., *The Control of Nature*, The Noonday Press, Farrar Strauss, and Giroux, 1989, 50.
2. Bose, A.G. and Stevens, K.N., *Introductory Network Theory*, Harper & Row, New York, NY, 1965.
3. Oppenheim, A.V. and Willsky, A.S., with Nawab, S.H., *Signals and Systems*, 2nd ed., Prentice-Hall, Englewood Cliffs, NJ, 1997.
4. Lathi, B.P., *Linear Systems and Signals*, Berkeley-Cambridge Press, Carmichael, CA, 1992.
5. Rudin, W., *Real and Complex Analysis*, McGraw-Hill, New York, NY, 1966.
6. Baker, Jr., G.A., *Essentials of Padé Approximants*, Academic Press, New York, NY, 1975.
7. Franklin, G.F., Powell, J.D., and Emami-Naeni, A., *Feedback Control of Dynamic Systems*, 4th ed., Prentice-Hall, Englewood Cliffs, NJ, 2002.
8. Aseltine, J.A., *Transform Method in Linear Systems Analysis*, McGraw-Hill, New York, NY, 1958.
9. Yosida, K., *Operational Calculus: A Theory of Hyperfunctions*, Springer-Verlag, New York, NY, 1984.

Further Reading

There are literally hundreds of textbooks on the general topic of signals and systems. Most have useful sections on input–output descriptions of LTI systems. References [3] and [4] are good examples. Reference [3] has a particularly good and well-organized bibliography.

A particularly good book for those interested in the mathematical technicalities of Fourier transforms and the impulse response is

10. Lighthill, M.J., *Introduction to Fourier Analysis and Generalized Functions*, Cambridge Monographs on Mechanics and Applied Mathematics, London, England, 1958.

Those interested in nonlinear input–output models of systems should read *Control System Advanced Methods*, Chapter 40. Those interested in linear time-varying systems should read *Control System Advanced Methods*, Chapter 3. Chapter 57 in *Control System Advanced Methods*, is a good starting point for those interested in the experimental determination of LTI input–output models.

5.2 State Space

James T. Gillis

5.2.1 Introduction

This chapter introduces the state-space methods used in control systems; it is an approach deeply rooted in the techniques of differential equations, linear algebra, and physics.

Webster's Ninth College Dictionary [1] defines a "state" as a "mode or condition of being" (1a), and "a condition or state of physical being of something" (2a). By a *state-space approach* one means a description of a system in which the "state" gives a complete description of the system at a given time; it implies that there are orderly rules for the transition from one state to another. For example, if the system is a particle governed by Newton's Law: $F = ma$, then the state could be the position and velocity of the particle or the position and momentum of the particle. These are both state descriptions of such a system. Thus, state-space descriptions of a system are not unique.

5.2.2 States

5.2.2.1 Basic Explanation

In this section the concepts of state and state space are introduced in an intuitive manner, then formally. Several examples of increasing complexity are discussed. A method for conversion of an ordinary differential equation (ODE) state-space model to a transfer function model is discussed along with several conversions of a rational transfer function to a state-space model.

The key concept is the *state* of a system, which is a set of variables which, along with the current time, summarizes the current configuration of a system. While some texts require it, there are good reasons for not requiring the variables to be a minimal set.* It is often desirable to work with a minimal set of variables, e.g., to improve numerical properties or to minimize the number of components used to build a system.

Given the state of a system at a given time, the prior history is of no additional help in determining the future behavior of the system. The state summarizes all the past behavior for the purposes of determining future behavior. The *state space* is the set of allowable values. The state space defines the topological, algebraic, and geometric properties associated with the evolution of the system over time. The state-space description carries an internal model of the system dynamics. For example, one familiar equation is Newton's equation: $F(t) = ma(t) = m\ddot{x}(t)$. The Hamiltonian formulation for this problem, yields a set of coupled set of first-order equations for position ($q = x$) and momentum ($p = m\dot{x}$):

$$\frac{d}{dt}\begin{bmatrix} q \\ p \end{bmatrix} = \begin{bmatrix} \dfrac{p}{m} \\ F(t) \end{bmatrix} \tag{5.45}$$

$$\begin{bmatrix} q(0) & p(0) \end{bmatrix}^T = \begin{bmatrix} x(0) & m\dot{x}(0) \end{bmatrix}^T. \tag{5.46}$$

---

* A typical example is the evolution of a direction cosine matrix. This is a three by three matrix that gives the orientation of one coordinate system with respect to another. Such matrices have two restrictions on them; they have determinant 1, and their transpose is their inverse. This is also called SO(3), the special orthogonal group of order 3. A smaller set of variables is pitch, roll, yaw ($\mathbb{R}^3$); however, this description is only good for small angles as $\mathbb{R}^3$ is commutative and rotations are not. When the relationship is a simple rotation with angular momentum $\omega = [\omega_x, \omega_y, \omega_z]$, the dynamics can be described with a state space in $\mathbb{R}^9 \sim \mathbb{R}^{3\times 3}$, as

$$\frac{d}{dt}A = -\begin{bmatrix} 0 & -\omega_z & \omega_y \\ \omega_z & 0 & -\omega_x \\ -\omega_y & \omega_x & 0 \end{bmatrix} A$$

which is not a minimal representation, but is simple to deal with. There is no worry about large angles in this representation.

This is indeed a state-space description of Newton's equations. The state at a given time is the vector $[q(t) \quad p(t)]^T$. The state space is $\mathbb{R}^2$. Examples of topological properties of the state space are: it is continuous, it has dimension two, etc. Of course, $\mathbb{R}^2$ enjoys many algebraic and geometric properties also. One could also integrate Newton's equations twice and get

$$x(t) = x(0) + t\dot{x}(0) + \frac{1}{m} \int_0^t \int_0^s F(\tau) \, d\tau ds. \tag{5.47}$$

While very useful, this is an example of an "input–output" model of the system. The single variable "$x(t)$" is not sufficient to characterize the future behavior; clearly, one needs $\dot{x}(t)$. Many facts about Equation 5.47 can be deduced by examining the solution; however, methods such as phase portraits (plot of q -vs- p) are frequently helpful in elucidating information about the behavior of the system without solving it. See [2, Chapter 2, Section 1].

Often, the structure of the state space can be guessed by the structure of the initial conditions for the problem. This is because the initial conditions summarize the behavior of the system up to the time that they are given. In most mechanical systems, the state space is twice the number of the *degrees of freedom*; this assumes that the dynamics are second order. The degrees of freedom are positions,"x_i," and the additional variables needed to make up a state description are the velocities, "$\dot{x}_i$."

State-space descriptions can be quite complicated, as is the case when two connected bodies separate (or collide), in which case the initial dimension of the state space would double (or half). Such problems occur in the analysis of the motions of launch vehicles, such as the space shuttle which uses, and ejects, solid rocket motors as well as a large liquid tank on its ascent into orbit. In such a case two different models are often created, and an attempt is made to capture and reconcile the forces of separation. This approach is not feasible for some systems, such as gimballed arms in robotics, which experience gimbal lock or the dropping or catching of gripped objects. Such problems are, usually, intrinsically difficult.

5.2.2.2 Reduction to First Order

In the case of Newton's laws, the state-space description arose out of a reduction to a system of first-order differential equations. This technique is quite general. Given a higher-order differential equation:*

$$y^{(n)} = f(t, u(t), y, \dot{y}, \ddot{y}, \ldots, y^{(n-1)}) \tag{5.48}$$

with initial conditions:

$$y(t_0) = y_0, \ \dot{y}(t_0) = y_1, \ldots, y^{(n-1)}(t_0) = y_{n-1}. \tag{5.49}$$

Consider the vector $x \in \mathbb{R}^n$ with $x_1 = y(t)$, $x_2 = \dot{y}(t), \ldots, x_n = y^{(n-1)}(t)$. Then Equation 5.48 can be written as:

$$\frac{d}{dt}x = \begin{bmatrix} x_2 \\ x_3 \\ \vdots \\ x_n \\ f(t, u(t), x_1, x_2, x_3, \ldots, x_n) \end{bmatrix} \tag{5.50}$$

with initial conditions:

$$x(t_0) = [y_0, \ y_1, \ldots, y_{n-1}]^T. \tag{5.51}$$

---

* In this section superscripts in parenthesis represent derivatives, as do over–dots, e.g., $\ddot{y} = y^{(2)} = d^2/dt^2 y$.

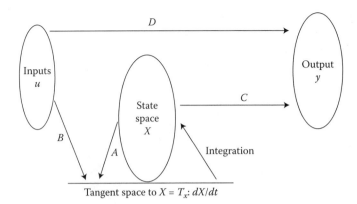

FIGURE 5.5 A pictorial view of the state-space model for a system with dynamics: $\dot{x} = A(x) + B(u)$ and measurement equation $y = C(x) + D(u)$. Note that the tangent space for $\mathbb{R}^n$ is identified with $\mathbb{R}^n$ itself, but this is not true for other objects. For example, SO(3) is a differentiable manifold, and its tangent space is composed of skew symmetric 3×3 matrices noted earlier.

Here x is the state and the state space is $\mathbb{R}^n$. This procedure is known as *reduction to first order*. It is the center of discussion for most of what follows.

Another case that arises in electric circuits and elsewhere is the integro-differential equation, such as that associated with Figure 5.6:

$$e(t) = RI(t) + L\frac{d}{dt}I(t) + \frac{1}{C}\int_0^t I(\tau)\,d\tau. \tag{5.52}$$

Letting $x_1(t) = \int_0^t I(\tau)d\tau$, $x_2(t) = I(t)$, so that $\dot{x}_1 = x_2$. Then the state-space formulation is:

$$\frac{d}{dt}x = \begin{bmatrix} 0 & 1 \\ -\frac{1}{LC} & -\frac{R}{L} \end{bmatrix} x + \begin{bmatrix} 0 \\ \frac{1}{L} \end{bmatrix} e(t) \tag{5.53}$$

$$I(t) = \begin{bmatrix} 0 & 1 \end{bmatrix} x. \tag{5.54}$$

This is a typical linear system with an output map.

Reduction to first order can also handle f (Equation 5.48) which depends on integrals of the inputs, $\int_0^t u(\tau)\,d\tau$, by setting $x_{n+1} = \int_0^t u(\tau)\,d\tau$ so that $\dot{x}_{n+1}(t) = u(t)$, etc. However, unless $\dot{u}$ can be considered as the input, it cannot be handled in the reduction to first order. The sole exception is if f is linear, in

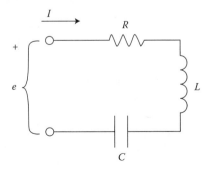

FIGURE 5.6 An RLC circuit, with impressed voltage $e(t)$ (input function), yields an equation for the current, $I(t)$.

which case the equation can be integrated until no derivatives of the input appear in it. This procedure will fail if there are more derivatives of the input than the original variable since there will not be enough initial conditions to be compatible with the integrations! That is, differential equations of the form $\sum_0^N \frac{d^i}{dt^i} y_i(t) = \sum_0^M \frac{d^i}{dt^i} u_i(t)$, discussed in Section 5.2.2.5, must have $N \geq M$ in order to develop a state-space model for the system.

Equivalent reduction to first order works for higher-order difference equations (see Section 5.2.2.3).

Several other forms are commonly encountered. The first is the delay, e.g., $\dot{x}(t) = ax(t) + bx(t - T)$ for some fixed T. This problem is discussed in Section 5.2.2.8. Another form, the transfer function, is discussed in Section 5.2.2.5.

5.2.2.3 ARMA

The standard ARMA (auto regressive moving average) model is given by:

$$y_t = \sum_{l=1}^{n} a_l y_{t-l} + \sum_{l=0}^{n} b_l u_{t-l}. \tag{5.55}$$

The "$a_j y_{t-j}$" portion of this is the auto regressive part and the "$b_k u_{t-k}$" portion is the moving average part. This can always be written as a state space model,

$$x_{t+1} = Ax_t + Bu_t \tag{5.56}$$

$$y_t = Cx_t + b_0 u_t, \tag{5.57}$$

with $x \in \mathbb{R}^n$, as follows.

Using the "Z" transform, where $y_{t-1} \to z^{-1} Y(z)$, Equation 5.55 transforms to

$$[1 - a_1 z^{-1} \cdots - a_n z^{-n}]Y(z) = [b_0 + b_1 z^{-1} \cdots + b_n z^{-n}]U(z).$$

$$\frac{Y(z)}{U(z)} = \frac{b_0 + b_1 z^{-1} \cdots + b_n z^{-n}}{1 - a_1 z^{-1} \cdots - a_n z^{-n}}$$

$$= b_0 + \frac{c_1 z^{-1} \cdots + c_{n-1} z^{-(n-1)}}{1 - a_1 z^{-1} \cdots - a_n z^{-n}}$$

The last step is by division (multiply top and bottom by z^n, do the division, and multiply top and bottom by z^{-n}). This is the transfer function representation; then, apply the process directly analogous to the one explained in Section 5.2.2.5 for continuous systems to get the state-space model. The resulting state-space model is

$$A = \begin{bmatrix} 0 & 1 & 0 & 0 & \cdots & 0 & 0 \\ 0 & 0 & 1 & 0 & \cdots & 0 & 0 \\ 0 & 0 & 0 & 1 & \cdots & 0 & 0 \\ \vdots & & & \vdots & & & \vdots \\ 0 & 0 & 0 & 0 & \cdots & 0 & 1 \\ a_n & a_{n-1} & & & \cdots & a_2 & a_1 \end{bmatrix}$$

$$B = [0\ 0\ \cdots\ 0\ 1]^T$$

$$C = [c_1\ c_2\ \cdots\ c_{n-1}]$$

The solution operator for this equation is given by the analog of the variation of constants formula:

$$x_t = A^t x_0 + \sum_{l=0}^{t-1} A^{t-1-l} Bu_l \tag{5.58}$$

5.2.2.4 Ordinary Differential Equation

An ordinary differential equation, such as: $\ddot{y} + 2\omega\xi\dot{y} + \omega^2 y = u(t)$, with $y(0) = y_0$; $\dot{y}(0) = \dot{y}_0$, is actually the prototype for this model. Here $0 < \xi < 1$ and $\omega > 0$. Consider $x = \begin{bmatrix} y & \dot{y} \end{bmatrix}^T$, then the equation could be written

$$\dot{x} = \begin{bmatrix} 0 & 1 \\ -\omega^2 & -2\omega\xi \end{bmatrix} x + \begin{bmatrix} 0 \\ 1 \end{bmatrix} u(t)$$

$$y(t) = \begin{bmatrix} 1 & 0 \end{bmatrix} x$$

$$x|_{t=0} = x_0 = \begin{bmatrix} y_0 & \dot{y}_0 \end{bmatrix}^T.$$

The state transition function is given by the variation of constants formula and the matrix exponential:*

$$x(t) = \exp\left\{\begin{bmatrix} 0 & 1 \\ -\omega^2 & -2\omega\xi \end{bmatrix} t\right\} x_0 + \int_0^t \exp\left\{\begin{bmatrix} 0 & 1 \\ -\omega^2 & -2\omega\xi \end{bmatrix} (t-\tau)\right\} \begin{bmatrix} 0 \\ 1 \end{bmatrix} u(\tau)\, d\tau \qquad (5.59)$$

$$= \Gamma(0, t, x_0, u) \qquad (5.60)$$

$$w(t) = y(t) = \begin{bmatrix} 1 & 0 \end{bmatrix} x(t).$$

It is easy to confirm that the variation of constants formula meets the criterion to be a state transition function. The system is linear and time invariant.

This is a specific case of the general first-order, linear time-invariant vector system which is written

$$\dot{x}(t) = Ax(t) + Bu(t) \qquad (5.61)$$

$$y(t) = Cx(t) + Du(t) \qquad (5.62)$$

where $x(t) \in \mathbb{R}^k$, $u(t) \in \mathbb{R}^p$, $y(t) \in \mathbb{R}^m$, and the quadruple $[A, B, C, D]$ are compatible constant matrices.

5.2.2.5 Transfer Functions and State-Space Models

This section is an examination of linear time-invariant differential equations (of finite order). All such systems can be reduced to first order, as already explained. Given a state-space description with $x(t) \in \mathbb{R}^k$, $u(t) \in \mathbb{R}^p$, $y(t) \in \mathbb{R}^m$ and

$$\dot{x} = Ax + Bu \quad \text{with initial conditions } x(t_0) = x_0 \qquad (5.63)$$

$$y = Cx + Du \qquad (5.64)$$

where $[A, B, C, D]$ are constant matrices of appropriate dimensions, the transfer function is given by: $G(s) = C(sI - A)^{-1}B + D$. This is arrived at by taking the Laplace transform of the equation and substituting—using initial conditions of zero. While this expression is simple, it can be deceiving as numerically stable computation of $(sI - A)^{-1}$ may be difficult. For small systems, one often computes $(sI - A)^{-1}$ by the usual methods (cofactor expansion, etc.). Other methods are given in [3]. One consequence of this expression is that state-space models are proper ($\lim_{s\to\infty} G(s) = D$).

Given a transfer function $G(s)$, which is a rational function, how does one develop a state-space realization of the system? First the single–input–single–output (SISO) case is treated, then extend it to

---

* Since it is useful, let $C(t) = \cos\omega\sqrt{1-\xi^2}t$ and $S(t) = \sin\omega\sqrt{1-\xi^2}t$, then

$$\exp\left\{\begin{bmatrix} 0 & 1 \\ -\omega^2 & -2\omega\xi \end{bmatrix} t\right\} = \exp\{-\xi\omega t\} \begin{bmatrix} C(t) + \dfrac{\xi}{\sqrt{1-\xi^2}}S(t) & \dfrac{1}{\omega\sqrt{1-\xi^2}}S(t) \\ -\dfrac{\omega}{\sqrt{1-\xi^2}}S(t) & C(t) - \dfrac{\xi}{\sqrt{1-\xi^2}}S(t) \end{bmatrix}$$

the multi–input–multi–output (MIMO) case (where $G(s)$ is a matrix of rational functions dimension m (outputs) by p (inputs)). The idea is to seek a quadruple $[A, B, C, D]$, representing the state equations as in Equations 5.63 and 5.64, so that $G(s) = C(sI - A)^{-1}B + D$, where A maps $\mathbb{R}^k \to \mathbb{R}^k$; C maps $\mathbb{R}^k \to \mathbb{R}^m$; B maps $\mathbb{R}^p \to \mathbb{R}^k$; and D maps $\mathbb{R}^p \to \mathbb{R}^m$. Here the state space is $\mathbb{R}^k$.

Let

$$G(s) = n(s)/d(s) + e(s) \tag{5.65}$$

where n, d, and e are polynomials with $deg(n) < deg(d)$, and this can be constructed by the Euclidean algorithm [4]. The coefficient convention is $d(s) = d_0 + d_1 s + \cdots + s^k$ for $deg(d) = k$. Notice that the leading order term of the denominator is normalized to one, and n is a polynomial of degree $k - 1$. The transfer function is said to be *proper* if $e(s)$ is constant and *strictly proper* if $e(s)$ is zero. If a transfer function is strictly proper, then the input does not directly appear in the output $D \equiv 0$; if it is proper then D is a constant. If the transfer function is not proper, then derivatives of the input appear in the output.

Given the transfer function 5.65, by taking the inverse Laplace transform of $\hat{y}(s) = G(s)\hat{u}(s)$, and substituting, one has the dynamics.*

$$d\left(\frac{d}{dt}\right)y(t) = \left(n\left(\frac{d}{dt}\right) + e_0 d(\frac{d}{dt})\right)u(t) \tag{5.66}$$

It is possible to work with this expression directly; however, the introduction of an auxiliary variable leads to two standard state-space representations for transfer functions. Introduce the variable $z(t)$ as

$$d\left(\frac{d}{dt}\right)z(t) = u(t) \tag{5.67}$$

$$y(t) = n\left(\frac{d}{dt}\right)z + e_0 u \tag{5.68}$$

or

$$d\left(\frac{d}{dt}\right)z(t) = n\left(\frac{d}{dt}\right)u \tag{5.69}$$

$$y(t) = z(t) + e_0 u. \tag{5.70}$$

Both of these expressions can be seen to be equivalent to Equation 5.66, by substitution.

Equations 5.67 and 5.68 can be reduced to first order by writing

$$\frac{d^k}{dt^k}(z) + \frac{d^{k-1}}{dt^{k-1}}(d_{k-1}z) + \cdots + d_0 z = u.$$

Then let $x_1 = z, x_2 = \dot{z}$, so that $\frac{d}{dt}x_1 = x_2$, $\frac{d}{dt}x_2 = x_3$, etc. This can be written in first-order matrix form as

$$\dot{x} = \begin{bmatrix} 0 & 1 & 0 & \cdots & 0 & 0 \\ 0 & 0 & 1 & \cdots & 0 & 0 \\ 0 & 0 & 0 & \cdots & 0 & 0 \\ \vdots & \vdots & \vdots & \vdots & \vdots & \vdots \\ 0 & 0 & 0 & \cdots & 1 & 0 \\ 0 & 0 & 0 & \cdots & 0 & 1 \\ -d_0 & -d_1 & -d_2 & \cdots & -d_{k-2} & -d_{k-1} \end{bmatrix} x + \begin{bmatrix} 0 \\ 0 \\ 0 \\ \vdots \\ 0 \\ 0 \\ 1 \end{bmatrix} u(t)$$

$$y(t) = [n_0 \; n_1 \; \cdots \; n_{k-2} \; n_{k-1}] \, x + e_0 \, u(t).$$

This is known as the *controllable canonical form.*

---

* Here the initial conditions are taken as zero when taking the inverse Laplace transform to get the dynamics; they must be incorporated as the initial conditions of the state-space model. As a notational matter, by $d(\frac{d}{dt})y(t)$ one means $d_0 y(t) + d_1 \dot{y}(t) + \cdots + d_{k-1} y^{(k-1)} + y^{(k)}$. Thus, $d(\frac{d}{dt})$ is a polynomial in the operator $\frac{d}{dt}$ and the same is true for $n(\frac{d}{dt})$.

The second pair of equations can be reduced to first order by writing Equation 5.69 as

$$0 = d_0 z - n_0 u + \frac{d}{dt}\left(d_1 z - n_1 u + \frac{d}{dt}\left(d_2 z - n_2 u \frac{d}{dt}\left(\cdots d_{k-2} z - n_{k-2} u \right. \right. \right.$$
$$\left. \left. \left. + \frac{d}{dt}(d_{k-1} z - n_{k-1} u + \frac{d}{dt}(z)) \cdots \right) \right) \right) \tag{5.71}$$

Let $x_1 =$ everything past the first $\frac{d}{dt}$, then the complete equation is $\dot{x}_1 = -d_0 z + n_0 u$. Let $x_2 =$ everything past the second $\frac{d}{dt}$, then $x_1 = d_1 z - n_1 u + \dot{x}_2$, which is the same as: $\dot{x}_2 = x_1 - d_1 z + n_1 u$. Proceeding in this fashion, $\dot{x}_l = x_{l-1} - d_{l-1} z + n_{l-1} u$; however, the interior of the last $\frac{d}{dt}$ is the variable z so that $x_k = z$! Substituting and writing in matrix form

$$\dot{x} = \begin{bmatrix} 0 & 0 & 0 & \cdots & 0 & 0 & -d_0 \\ 1 & 0 & 0 & \cdots & 0 & 0 & -d_1 \\ 0 & 1 & 0 & \cdots & 0 & 0 & -d_2 \\ \vdots & \vdots & \vdots & \vdots & \vdots & & \\ 0 & 0 & 0 & \cdots & 1 & 0 & -d_{k-2} \\ 0 & 0 & 0 & \cdots & 0 & 1 & -d_{k-1} \end{bmatrix} x + \begin{bmatrix} n_0 \\ n_1 \\ n_2 \\ \vdots \\ n_{k-1} \end{bmatrix} u(t)$$
$$y(t) = [0\ 0\ \cdots\ 0\ 1]\, x + e_0\, u(t) \tag{5.72}$$

This is known as *observable canonical form*. This form is usually reduced from a block diagram making the choice of variables more obvious.

Notice that Controllable Form is equal to Observable Form transposed,

$$[A, B, C, D] = [A^T, C^T, B^T, D^T]$$

which is the beginning of duality for controllability and observability. The A matrix in these two examples is in companion form, the negative of the coefficients of the characteristic equation on the outside edge, with ones off the diagonal. There are two other companion forms, and these are also related to controllability and observability.

If $G(s)$ is strictly improper, that is, $deg(e) \geq 1$, there is no state-space model; however, there is at least one trick used to circumvent this. As an example, one might deal with this by modeling $e(s) = s^2$ with $s^2/s^2 + as + b$, moving the poles outside the range of the system. Since $s^2/s^2 + as + b$ is proper, it has a state-space model; then adjoin this model to the original, feeding its output into the output original system. One effect is the bandwidth (dynamic range) of the system is greatly increased. This can result in "stiff" ODEs which have numerical problems associated with them. Obviously, this technique could be applied to more complex transfer functions $e(s)$. However, improper systems should be closely examined as such a system has an increasingly larger response to higher frequency inputs!

5.2.2.6 Miscellanea

Given a state-space representation of the form $\dot{x} = Ax + Bu$; $y = Cx + Du$, one could introduce a change of variables $z = Px$, with P^{-1} existing. Then the system can be written, by substitution: $\dot{z} = PAP^{-1}x + PBu$; $y = CP^{-1}z + Du$. Hence $Px \rightarrow z$ induces $[A, B, C, D] \rightarrow [PAP^{-1}, PB, CP^{-1}, D]$. Note that such a transformation does not change the characteristic equation; and hence the eigenvalues associated with A, since

$$det(sI - A) = det(P)det(P^{-1})det(sI - A)$$
$$= det(P)det(sI - A)det(P^{-1})$$
$$= det(P(sI - A)P^{-1})$$
$$= det(sPP^{-1} - PAP^{-1})$$
$$= det(sI - PAP^{-1}).$$

A little more algebra shows that the transfer function associated with the two systems is the same. The transformation $A \to P^{-1}AP$ is known as a *similarity transformation*.

Finding a state-space representation of smaller dimension has two guises: the first is the search for a minimal realization, touched on in Example 5.4. The second is the problem of model reduction.

Usual operations on transfer functions involve combining them, usually adding or cascading them. In both of these cases the state-space model can be determined from the state-space models of the component transfer functions. The total state-space size is usually the sum of the dimensions of the component state-space models. Here x_i is taken as the state (subvector) corresponding to transfer function G_i, rather than a single component of the state vector. The exception to this rule about component size has to do with repeated roots and will be illustrated below. The size of the state space is the degree of the denominator polynomial of the combined system—note that cancellation can occur between the numerator and the denominator of the combined system.

When two transfer functions are cascaded (the output of the first is taken as input to the second), the resulting transfer function is the product of the two transfer functions: $H(s) = G_2(s)G_1(s)$. To build the corresponding state-space model, one could simply generate a state-space model for $H(s)$; however, if state-space models exist for G_1 and G_2, there are sound reasons for proceeding to work in the state-space domain. This is because higher-order transfer functions are more likely to lead to poor numerical conditioning (see the discussion at the end of Section 5.2.2.8). Given $G_1 \sim [A_1, B_1, C_1, D_1]$ and $G_2 \sim [A_2, B_2, C_2, D_2]$, then simple algebra shows that

$$H = G_2 G_1 \sim \tag{5.73}$$

$$\dot{x} = \begin{bmatrix} A_1 & 0 \\ B_2 C_1 & A_2 \end{bmatrix} x(t) + \begin{bmatrix} B_1 \\ B_2 D_1 \end{bmatrix} u(t) \tag{5.74}$$

$$y(t) = [D_2 C_1 \ C_2] x(t) + D_2 D_1 u(t). \tag{5.75}$$

The state space used in this description is $\begin{bmatrix} x_1 & x_2 \end{bmatrix}^T$, where $x_1 \sim G_1$ and $x_2 \sim G_2$.

When two transfer functions are added, $H(s) = G_1(s) + G_2(s)$, and state-space models are available the equivalent construction yields:

$$H = G_1 + G_2 \sim \tag{5.76}$$

$$\dot{x} = \begin{bmatrix} A_1 & 0 \\ 0 & A_2 \end{bmatrix} x(t) + \begin{bmatrix} B_1 \\ B_2 \end{bmatrix} u(t) \tag{5.77}$$

$$y(t) = \begin{bmatrix} C_1 & C_2 \end{bmatrix} x(t) + [D_1 + D_2] u(t). \tag{5.78}$$

If state feedback is used, that is $u = Fx(t) + v(t)$, then,

$$[A, B, C, D] \to [A + BF, B, C + DF, D]$$

is the state-space system acting on $v(t)$. If output feedback is used, that is, $u = Fy(t) + v(t)$, then

$$[A, B, C, D] \to [A + BFC, B, C + DFC, D]$$

is the state-space system acting on $v(t)$. Both of these are accomplished by simple substitution. Clearly, there is a long list of such expressions, each one equivalent to a transfer function manipulation.

To illustrate these ideas, consider the system

$$G(s) = \frac{2(s^2+4)(s-2)}{(s+3)(s+2)^2} = \frac{2s^3 - 4s^2 + 8s - 16}{s^3 + 7s^2 + 16s + 12} \tag{5.79}$$

$$= 2 + \frac{-18s^2 - 24s - 40}{s^3 + 7s^2 + 16s + 12} \tag{5.80}$$

$$= 2 - \frac{130}{s+3} + \frac{112}{s+2} - \frac{64}{(s+2)^2} \tag{5.81}$$

This can be realized as a state-space model using four techniques. The last expression is the partial fractions expansion of $G(s)$ (see Chapter 4).

Example 5.1:

Controllable canonical form; $G(s) \sim [A_c, B_c, C_c, D_c]$, where

$$A_c = \begin{bmatrix} 0 & 1 & 0 \\ 0 & 0 & 1 \\ -12 & -16 & -7 \end{bmatrix}, \quad B_c = \begin{bmatrix} 0 \\ 0 \\ 1 \end{bmatrix}, \quad C_c^T = \begin{bmatrix} -40 \\ -24 \\ -18 \end{bmatrix}, \quad D_c = [2].$$

Example 5.2:

Observable canonical form; $G(s) \sim [A_o, B_o, C_o, D_o]$, where

$$A_o = \begin{bmatrix} 0 & 0 & -12 \\ 1 & 0 & -16 \\ 0 & 1 & -7 \end{bmatrix}, \quad B_o = \begin{bmatrix} -40 \\ -24 \\ -18 \end{bmatrix}, \quad C_o^T = \begin{bmatrix} 0 \\ 0 \\ 1 \end{bmatrix}, \quad D_o = [2].$$

Note that $[A_c, B_c, C_c, D_c] = [A_o^T, C_o^T, B_o^T, D^T]$, verifying the Controllable–Observable duality in this case.

Example 5.3:

As a product, $G = G_2 G_1$, $G_1(s) = \frac{s-2}{s+3} = 1 + \frac{-5}{s+3}$, $G_2(s) = 2\frac{s^2+4}{(s+2)^2} = 2 + \frac{-8s}{(s+2)^2}$. $G_1(s) \sim [A_1, B_1, C_1, D_1]$
$= [-3, 1, -5, 1]$;
    $G_2(s) \sim [A_2, B_2, C_2, D_2]$, where

$$A_2 = \begin{bmatrix} 0 & 1 \\ -4 & -4 \end{bmatrix}, \quad B_2 = \begin{bmatrix} 0 \\ 1 \end{bmatrix}, \quad C_2^T = \begin{bmatrix} 0 \\ 8 \end{bmatrix}, \quad D_2 = [2].$$

The combined system is

$$A_{21} = \begin{bmatrix} -3 & 0 & 0 \\ 0 & 0 & 1 \\ -5 & -4 & -4 \end{bmatrix}, \quad B_{21} = \begin{bmatrix} 1 \\ 0 \\ 1 \end{bmatrix}, \quad C_{21}^T = \begin{bmatrix} -10 \\ 0 \\ 8 \end{bmatrix}, \quad D_{21} = [2]$$

Example 5.4:

As a sum, $G = G_1 + G_2 + G_3$, where the summands are derived by the partial fractions method (see Chapter 8). This case is used to illustrate the *Jordan* representation of the state space. If one were to proceed hastily, using the addition method, then the resulting systems would have $1 + 1 + 2 = 4$ states. The other realizations have three states. Why? Since the term $s + 2$ appears in the last two transfer functions, it is possible to "reuse" it. The straightforward combination is not wrong, it simply fails to be minimal. The transfer function for $\frac{1}{(s+a)^2}$ is given by the product rule, Equations 5.74 and 5.75: Let $G^2(s) = G(s)G(s)$, where $G(s) = \frac{1}{s+a} \sim [-a, 1, 1, 0]$, so that $G^2 \sim [A, B, C, D]$ with:

$$A = \begin{bmatrix} -a & 0 \\ 1 & -a \end{bmatrix} \quad B = \begin{bmatrix} 1 \\ 0 \end{bmatrix} \quad C^T = \begin{bmatrix} 0 \\ 1 \end{bmatrix} \quad D = [0]$$

The third power, $G^3 = GG^2$ $[A, B, C, D]$, with

$$A = \begin{bmatrix} -a & 0 & 0 \\ 1 & -a & 0 \\ 0 & 1 & -a \end{bmatrix}, \quad B = \begin{bmatrix} 1 \\ 0 \\ 0 \end{bmatrix}, \quad C^T = \begin{bmatrix} 0 \\ 0 \\ 1 \end{bmatrix}, \quad D = [0]$$

This can be continued, and the A matrix will have $-a$ on the diagonal and 1 on the subdiagonal; $B = [1\ 0\ \cdots\ 0]^T$, $C = [0\ 0\ \cdots\ 0\ 1]$, $D = 0$. This is very close to the standard matrix form known as Jordan form. For G^2 a change of variables $x' = \begin{bmatrix} 0 & 1 \\ 1 & 0 \end{bmatrix} x$ takes the system to

$$A = \begin{bmatrix} -a & 1 \\ 0 & -a \end{bmatrix} \quad B = \begin{bmatrix} 0 \\ 1 \end{bmatrix} \quad C^T = \begin{bmatrix} 1 \\ 0 \end{bmatrix} \quad D = [0] \tag{5.82}$$

So that this is the system in Jordan form (this is just the exchange of variables $x' = [x_2\ x_1]^T$). Notice that if $C = \begin{bmatrix} 0 & 1 \end{bmatrix}$ in system 5.82, the transfer function is $\frac{1}{s+a}$! Thus if $G(s) = \frac{\alpha_1}{s+a} + \frac{\alpha_2}{(s+a)^2}$, use system 5.82 with $C = \begin{bmatrix} \alpha_2 & \alpha_1 \end{bmatrix}$.

In general, consider $G^k \sim [A, B, C, D]$, obtained by the product rules Equations 5.74 and 5.75, apply the conversion $x' = Px$, with $P = antidiagonal(1)$ (note that $P = P^{-1}$). The resulting system is

$$\dot{x} = \begin{bmatrix} -a & 1 & 0 & \cdots & 0 & 0 & 0 \\ 0 & -a & 1 & \cdots & 0 & 0 & 0 \\ 0 & 0 & -a & \cdots & 0 & 0 & 0 \\ \vdots & & \vdots & \vdots & \vdots & \vdots \\ 0 & 0 & 0 & \cdots & 0 & -a & 1 \\ 0 & 0 & 0 & \cdots & 0 & 0 & -a \end{bmatrix} x + \begin{bmatrix} 0 \\ 0 \\ 0 \\ \vdots \\ 0 \\ 1 \end{bmatrix} u(t)$$

$$y(t) = [1\ 0\ \cdots\ 0\ 0]\, x$$

And if the desired transfer function is $G(s) = \sum_{l=1}^{k} \frac{\alpha_l}{(s+a)^l}$, then $C = [\alpha_k\ \alpha_{k-2}\ \cdots\ \alpha_1]$.

Returning to the example, Equation 5.81, $G_1(s) = 2 - \frac{130}{s+3} \sim [-3, 1, -130, 2]$, where:

The last two transfer functions are in Jordan form: $G_2(s) + G_3(s) = \frac{112}{s+2} - \frac{64}{(s+2)^2} \sim [A_2, B_2, C_2, D_2]$, where

$G(s) \sim [A, B, C, D]$, where

$$A = \begin{bmatrix} -2 & 1 \\ 0 & -2 \end{bmatrix} \quad B = \begin{bmatrix} 0 \\ 1 \end{bmatrix} \quad C^T = \begin{bmatrix} -64 \\ 112 \end{bmatrix} \quad D = [0]$$

Since G_1 is already in Jordan form (trivially), the combined system is $G_{321}(s) \sim [A_{321}, B_{321}, C_{321}, D_{321}]$, where

$$
A_{321} = \begin{bmatrix} -2 & 1 & 0 \\ 0 & -2 & 1 \\ 0 & 0 & -2 \end{bmatrix}, \quad B_{321} = \begin{bmatrix} 1 \\ 0 \\ 1 \end{bmatrix}, \quad C_{321}^T = \begin{bmatrix} -64 \\ 112 \\ -130 \end{bmatrix}, \quad D_{321} = [2]
$$

Here A is also in Jordan form; this representation is very nice as it displays the eigenvalues and their multiplicities.

5.2.2.7 MIMO Transfer Functions to State Space

One of the historical advantages of the state-space methods was the ability to deal with multi-input-multi-output systems. The frequency-domain methods have matured to deal with this case also, and the following section deals with how to realize a state-space model from a transfer function $G(s)$, which is a matrix with entries that are rational functions.

The methods discussed here are straightforward generalizations of Equations 5.69 and 5.70 and Equations 5.67 and 5.68. A variety of other methods are discussed in [3] and [5], where numerical considerations are considered. The reader is referred to the references for more detailed information, including the formal proofs of the methods being presented.

A useful approach is to reduce the multi-input–multi-output system to a series of single-input–multi-output systems and then combine the results. This means treating the columns of $G(s)$ one at a time. If $[A_i, B_i, C_i, D_i]$ are the state-space descriptions associated with the ith column of $G(s)$, denoted $G_i(s)$, then the state-space description of $G(s)$ is

$$
\left[diag(A_1, A_2, \ldots, A_n), [B_1, B_2, \ldots, B_n], [C_1 C_2 \cdots C_n], [D_1 D_2 \cdots D_n] \right].
$$

where

$$
diag(A_1, A_2) = \begin{bmatrix} A_1 & 0 \\ 0 & A_2 \end{bmatrix}
$$

etc. The input to this system is the vector $[u_1 \ u_2 \cdots u_n]^T$. With this in mind, consider the development of a single-input–multi-output transfer function to a state-space model.

Given $G(s)$, a column vector, first subtract off the vector $E = G(s)|_{s=\infty}$ from G leaving strictly proper rational functions as entries, then find the least common denominator of all of the entries, $d(s) = s^k + \sum_0^{k-1} d_l s^l$, and factor it out. This leaves a vector of polynomials $n_i(s)$ as entries of the vector. Thus $G(s)$ has been decomposed as

$$
G(s) = \begin{bmatrix} g_1(s) \\ g_2(s) \\ \cdots \\ g_q(s) \end{bmatrix} = \begin{bmatrix} e_1 \\ e_2 \\ \cdots \\ e_q \end{bmatrix} + \frac{1}{d(s)} \begin{bmatrix} n_1(s) \\ n_2(s) \\ \cdots \\ n_q(s) \end{bmatrix}
$$

Writing $n_j(s) = \sum_{l=0}^{k-1} v_{j,l} s^l$, then the state-space realization is

$$
\frac{d}{dt} x = \begin{bmatrix} 0 & 1 & 0 & \cdots & 0 & 0 \\ 0 & 0 & 1 & \cdots & 0 & 0 \\ 0 & 0 & 0 & \cdots & 0 & 0 \\ \vdots & \vdots & \vdots & \vdots & \vdots & \vdots \\ 0 & 0 & 0 & \cdots & 1 & 0 \\ 0 & 0 & 0 & \cdots & 0 & 1 \\ -d_0 & -d_1 & -d_2 & \cdots & -d_{k-2} & -d_{k-1} \end{bmatrix} x + \begin{bmatrix} 0 \\ 0 \\ 0 \\ \vdots \\ 0 \\ 0 \\ 1 \end{bmatrix} u(t)
$$

$$\begin{bmatrix} y_1 \\ y_2 \\ \cdots \\ y_q \end{bmatrix} = \begin{bmatrix} v_{1,k-1} & v_{1,k-2} & \cdots & v_{1,0} \\ v_{2,k-1} & v_{2,k-2} & \cdots & v_{2,0} \\ \cdots & \cdots & \cdots & \cdots \\ v_{q,k-1} & v_{q,k-2} & \cdots & v_{q,0} \end{bmatrix} x + \begin{bmatrix} e_1 \\ e_2 \\ \cdots \\ e_q \end{bmatrix} u(t)$$

This a controllable realization of the transfer function. In order to get an observable realization, treat the transfer function by rows and proceed in a similar manner (this is multi-input–single output approach, see [3]). The duality of the two realizations still holds.

Additional realizations are possible, perhaps the most important being Jordan form. This is handled like the single-input-single-output case, using partial fractions with column coefficients rather than scalars. That is, expanding the entries of $G(s)$, a single-input–multi-output (column) transfer function, as partial fractions with (constant) vector coefficients and then using Equations 5.77 and 5.78 for dealing with the addition of transfer functions. Multiple inputs are handled one at a time and stacked appropriately.

Example 5.5:

Let

$$G(s) = \begin{bmatrix} \dfrac{s+3}{s^2+2s+2} \\ \dfrac{s^2+4}{(s+1)^2} \end{bmatrix} = \begin{bmatrix} \dfrac{2}{s+1} - \dfrac{1}{s+2} \\ 1 - \dfrac{2}{s+1} + \dfrac{5}{(s+1)^2} \end{bmatrix}$$

$$= \begin{bmatrix} 0 \\ 1 \end{bmatrix} + \frac{1}{s+1}\begin{bmatrix} 2 \\ -2 \end{bmatrix} - \frac{1}{s+2}\begin{bmatrix} 1 \\ 0 \end{bmatrix} + \frac{1}{(s+1)^2}\begin{bmatrix} 0 \\ 5 \end{bmatrix}$$

This results in the two systems

$$\dot{x_1} = \begin{bmatrix} -1 & 1 \\ 0 & -1 \end{bmatrix} x_1 + \begin{bmatrix} 0 \\ 1 \end{bmatrix} u(t)$$

$$y_1(t) = \begin{bmatrix} -1 & 0 \\ 0 & 5 \end{bmatrix} x_1 + \begin{bmatrix} 0 \\ 1 \end{bmatrix} u$$

$$\dot{x_2} = -2x_2 + u$$

$$y_2(t) = \begin{bmatrix} 2 \\ -2 \end{bmatrix} x_2$$

which combine to form

$$\frac{d}{dt}\begin{bmatrix} x_1 \\ x_2 \end{bmatrix} = \begin{bmatrix} -1 & 1 & 0 \\ 0 & -1 & 0 \\ 0 & 0 & -2 \end{bmatrix}\begin{bmatrix} x_1 \\ x_2 \end{bmatrix} + \begin{bmatrix} 0 \\ 1 \\ 1 \end{bmatrix} u(t)$$

$$y(t) = \begin{bmatrix} -1 & 0 & 2 \\ 0 & 5 & -2 \end{bmatrix}\begin{bmatrix} x_1 \\ x_2 \end{bmatrix} + \begin{bmatrix} 0 \\ 1 \end{bmatrix} u(t)$$

5.2.2.8 Padè Approximation of Delay

The most common nonrational transfer function model encountered is that resulting from a delay in the system:[*] $e^{-sT}\hat{f}(s) = L\{f(t-T)\}$. The function e^{-sT} is clearly transcendental, and this changes the

---

[*] $L\{\cdot\}$ is used to denote the Laplace transform of the argument with respect to the variable t.

fundamental character of the system if treated rigorously. However, it is fortunate that there is an often useful approximation of the transfer function which is rational and hence results in the addition of finite states to the differential equation.

The general idea is to find a rational function approximation to a given function, in this case, e^{-sT}, and replace the function with the rational approximation. Then use the methods of Section 5.2.2.5 to construct a state-space description. Specifically, given $f(s) = \sum_0^\infty a_n s^n$ find $P_M^N(s)$,

$$P_M^N(s) = \frac{\sum_0^N A_n s^n}{1 + \sum_1^M B_n s^n} \tag{5.83}$$

so that the coefficients of the first $N + M + 1$ terms of the two Taylor series match.* It is usual to look at the diagonal approximate $P_N^N(s)$. As will be shown, the diagonal Padé approximation of e^{-sT} shares with e^{-sT} the fact that it has unit magnitude along the imaginary axis. Thus, the Bode magnitude plot (log magnitude of the transfer function vs. log frequency) is the constant zero. However, the phase approximation deviates dramatically, asymptotically. Put another way: the approximation is perfect in magnitude; all of the errors is in phase. The difference in the phase (or angle) part of the Bode plot (which is phase vs. log frequency) is displayed in Figure 5.7, which compares the pure delay, e^{-sT}, with several Padé approximations.

The coefficients of Padé approximation to the delay can be in calculated in the following way which takes advantage of the properties of the exponential function. Here it is easier to allow the B_0 termto be

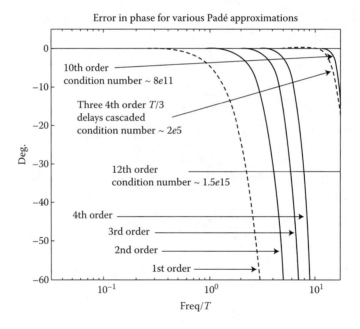

FIGURE 5.7 Comparison of e^{-sT} and various order Padé approximations.

---

* There is an extensive discussion of Padé approximations in [6]. This includes the use of Taylor expansions at two points to generate Padé approximates. While the discussion is not aimed at control systems, it is wide ranging and otherwise complete. Let: $f(s) \sim \sum_0^\infty a_n(s - s_0)^n$ as $s \to s_0$ and $f(s) \sim \sum_0^\infty b_n(s - s_1)^n$ as $s \to s_1$ then choose $P_M^N(s)$, as before a rational function, so that the first J terms of the Taylor series match the s_0 representation and so the first K terms of the s_1 representation likewise agree, where $J + K = N + M + 1$. The systems of equations used to solve for A_n and B_n are given in [6] in both the simple and more complex circumstance. In the case of matching a single Taylor series, the relationship between Padé approximations and continued fractions is exploited for a recursive method of computing the Padé coefficients. A simpler trick will suffice for the case at hand.

something other than unity; it will work out to be one:

$$e^{-sT} = \frac{e^{-s\frac{T}{2}}}{e^{s\frac{T}{2}}} \tag{5.84}$$

Therefore:

$$\frac{e^{-s\frac{T}{2}}}{e^{s\frac{T}{2}}} = \frac{\sum_0^N A_n s^n}{\sum_0^M B_n s^n} + O(M+N+1) \tag{5.85}$$

Hence:

$$e^{-s\frac{T}{2}} \times \sum_0^M B_n s^n = e^{s\frac{T}{2}} \times \left(\sum_0^N A_n s^n + O(M+N+1) \right). \tag{5.86}$$

Equality will hold if $\sum_0^M B_n s^n$ is the first $M+1$ terms in the Taylor expansion of $e^{s\frac{T}{2}}$ and $\sum_0^N A_n s^n$ is the first $N+1$ terms in the Taylor expansion of $e^{-s\frac{T}{2}}$. This yields the expression usually found in texts:

$$P_M^N(s) = \frac{\sum_0^N (-1)^n \frac{\left(\frac{sT}{2}\right)^n}{n!}}{\sum_0^M \frac{\left(\frac{sT}{2}\right)^n}{n!}}. \tag{5.87}$$

Notice that $P_N^N(s)$ is of the form $p(-s)/p(s)$ so that it is an *all pass filter*; that is, it has magnitude one for all $j\omega$ on the imaginary axis. To see this, let $s = j\omega$ and multiply the numerator and denominator by their conjugates. Then note that $(-j\omega)^n = (-1)^n (j\omega)^n$ and that $(-1)^{2n} = 1$. Hence the diagonal approximation has magnitude one for all $s = j\omega$ on the imaginary axis, just as e^{-sT} does. This is one of the reasons that this approximation is so useful for delays.

The approximation in Equation 5.87 has a zeroth-order term of 1 and the highest-order term contains $\frac{1}{n!}$; thus, numerical stability of the approximation is a concern, especially for high order (meaning over several decades of frequency). It has already been noted that $e^{-sT} = e^{-s\frac{T}{2}} e^{-s\frac{T}{2}}$, so that an approximation is possible as cascaded lower-order approximations. In practice, it is best not to symmetrically divide the delay, thus avoiding repeated roots in the system. For example,

$$e^{-sT} = e^{-s\frac{1.1T}{3}} e^{-s\frac{T}{3}} e^{-s\frac{.99T}{3}} \tag{5.88}$$

and approximate each of the delays using a fourth-order approximation. This gives a twelve state model, using cascaded state-space systems (see Equations 5.74 and 5.75). This model is compared in Figure 5.7, where it is seen to be not as good as an order twelve Padé approximation; however, the condition number* of the standard twelfth order Padé approximation is on the order of 10^{14}. Breaking the delay into three unequal parts resulting in no repeated roots and a condition number for the A matrix of about 10^4, quite a bit more manageable numerically. The resulting twelve state model is about as accurate as the tenth order approximation, which has a condition of about 10^{10}.

A nice property of the Padé approximation of a delay is that the poles are all in the left half-plane; however, the zeros are all in the right half-plane which is difficult (Problem 8.61 [6]). This does reflect the fact that delay is often destabilizing in feedback loops.

There are other common nonrational transfer functions, such as those that arise from partial differential equations (PDEs) and from the study of turbulence (e.g., the von Karman model for turbulence, which involves fractional powers of s). In such cases, use the off-diagonal approximates to effect an appropriate roll-off at high frequency (for the exponential function, this forces the choice of the diagonal approximate).

---

* The condition number is the ratio of the largest to the smallest magnitude of the eigenvalue; see [7] for more on condition numbers.

5.2.2.8.1 Remarks on Padé Approximations

The Padé approximation is one of the most common approximation methods; however, in control systems, it has several drawbacks that should be mentioned. It results in a finite-dimensional model, which may hide fundamental behavior of a PDE. While accurately curve fitting the transfer function in the s-domain, it is not an approximation that has the same pole–zero structure as the original transfer function. Therefore, techniques that rely on pole–zero cancellation will not result in effective control. In fact, the Padé approximation of a stable transfer function is not always stable!

A simple way to try to detect the situation where pole–zero cancellation is being relied upon is to use different models for design and for validation of the control system. The use of several models provides a simple, often effective method for checking the robustness of a control system design.

5.2.3 Linearization

5.2.3.1 Preliminaries

This section is a brief introduction to linearization of nonlinear systems. The point of the section is to introduce enough information to allow linearization, when possible, in a way that permits for stabilization of a system by feedback. The proofs of the theorems are beyond the scope of this section, since they involve finding generalized energy functions, known as Lyapunov* functions. A good reference for this information is [8, Chapter 5].

Let $x = [x_1 x_2 \cdots x_n]^T \in \mathbb{R}^n$ then the Jacobian of a function $\mathbf{f}(\mathbf{x}, t) = [f_1 f_2 \cdots f_m]^T$ is the $m \times n$ matrix of partials:

$$\left[\frac{\partial \mathbf{f}}{\partial \mathbf{x}} \right] = \begin{bmatrix} \dfrac{\partial f_1}{\partial x_1} & \dfrac{\partial f_1}{\partial x_2} & \cdots & \dfrac{\partial f_1}{\partial x_n} \\ \dfrac{\partial f_2}{\partial x_1} & \dfrac{\partial f_2}{\partial x_2} & \cdots & \dfrac{\partial f_2}{\partial x_n} \\ \vdots & \vdots & \vdots & \vdots \\ \dfrac{\partial f_m}{\partial x_1} & \dfrac{\partial f_m}{\partial x_2} & \cdots & \dfrac{\partial f_m}{\partial x_n} \end{bmatrix} \tag{5.89}$$

$$= \left[\frac{\partial \mathbf{f}}{\partial \mathbf{x}} \right]. \tag{5.90}$$

So that the Taylor expansion of $\mathbf{f}(\mathbf{x}, t)$ about $\mathbf{x}_0$ is

$$\mathbf{f}(\mathbf{x}, t) = \mathbf{f}(\mathbf{x}_0, t) + \left[\frac{\partial f}{\partial x} \right]\Big|_{\mathbf{x}=\mathbf{x}_0} \mathbf{x} + O(|\mathbf{x} - \mathbf{x}_0|^2).$$

The stability of a system of differential equations is discussed at length in Chapter 1. A brief review is provided here for ease of reference. A differential equation, $\frac{d}{dt}\mathbf{x} = \mathbf{f}(\mathbf{x}, t)$, is said to have an *equilibrium point* at $\mathbf{x}_0$ if $\mathbf{f}(\mathbf{x}_0, t) = \mathbf{0}$ for all t, thus $\frac{d}{dt}\mathbf{x}\Big|_{\mathbf{x}_0} = \mathbf{0}$ and the differential equation would have solution $\mathbf{x}(t) \equiv \mathbf{x}_0$.

A differential equation, $\frac{d}{dt}\mathbf{x} = \mathbf{f}(\mathbf{x}, t)$, with an equilibrium point at the origin is said to be *stable* if for every $\epsilon > 0$ and $t > 0$ there is a $\delta(\epsilon, t_0)$ so that $|\mathbf{x}_0| < \delta(\epsilon, t_0)$ implies that $|\mathbf{x}(t)| < \epsilon$ for all t. If δ depends only on ϵ, then the system is said to be *uniformly stable*. The equation is said to be *exponentially stable* if there is a ball $B_r = |\mathbf{x}| \leq r$ so that for all $\mathbf{x}_0 \in B_r$ the solution obeys the following bound for some

---

* The methods used here were developed in Lyapunov's monograph of 1892. There are several spellings of his name, benign variations on the transliteration of a Russian name. The other common spelling is Liapunov, and this is used in the translations of V. I. Arnold's text [2] (Arnold's name is also variously transliterated!).

$a, b > 0$: $|\mathbf{x}_0| < a|\mathbf{x}_0| \exp\{-bt\}$ for all t. This is a strong condition. For autonomous linear systems, all of these forms collapse to exponential stability, for which all of the eigenvalues having negative real parts is necessary and sufficient. However, as Examples 5.1 and 5.2 show, this simple condition is not universal. For time-varying systems, checking exponential stability can be quite difficult.

It is true that if the system is "slowly varying," things do work out. That is, if $\mathbf{x}_0$ is an equilibrium point and $\left[\dfrac{\partial \mathbf{f}}{\partial \mathbf{x}}\right]$ has eigenvalues with negative real parts bounded away from the imaginary axis and additional technical conditions hold, then the differential equation is exponentially stable. See Theorem 15, Chapter 5, Section 8 in [8]. Note that Example 5.2 is *not* slowly varying. The physical example of a pendulum in which the length is varied (even slightly) close to twice the period (which is a model of a child on a swing) shows that rapidly varying dynamics can take an exponentially stable system (a pendulum with damping) and change it into an unstable system. See Section 25 in [2] where there is an analysis of Mathieu's equation: $\ddot{q} = -\omega_0(1 + \epsilon \cos(t))q$, $\epsilon \ll 1$, which is unstable for $\omega_0 = \frac{k}{2}$, $k = 1, 2, \ldots$. This instability continues in the presence of a little damping. This phenomenon is known as *parametric resonance*.

5.2.3.2 Lyapunov's Linearization Method

Consider $\frac{d}{dt}\mathbf{x} = \mathbf{f}(\mathbf{x}, t)$, with an equilibrium point at $\mathbf{x}_0$ and $\mathbf{f}$ continuously differentiable in both arguments. Then

$$\frac{d}{dt}\mathbf{x} = \mathbf{f}(\mathbf{x}, t) \tag{5.91}$$

$$= \left\{\left[\frac{\partial \mathbf{f}}{\partial \mathbf{x}}\right]\Bigg|_{\mathbf{x}=\mathbf{x}_0}\right\}\mathbf{x} + \mathbf{f}_1(\mathbf{x}, t) \tag{5.92}$$

$$= \mathbf{A}(t)\mathbf{x} + \mathbf{f}_1(\mathbf{x}, t). \tag{5.93}$$

Here $\frac{d}{dt}\mathbf{x} = \mathbf{A}(t)\mathbf{x}$ is the candidate for the linearization of the system. The additional condition which must hold is that the approximation be uniform in t, that is

$$\lim_{|\mathbf{x}| \to \mathbf{x}_0} \sup_{t \geq 0} \frac{|\mathbf{f}_1(\mathbf{x}, t)|}{|\mathbf{x}|} = \mathbf{0}. \tag{5.94}$$

Note that if $\mathbf{f}_1(\mathbf{x}, t) = \mathbf{f}_1(\mathbf{x})$, then the definition of the Jacobian guarantees that condition 5.94 holds. (see Example 5.1 for an example where uniform convergence fails.)

Under these conditions, the system $\frac{d}{dt}\mathbf{x} = \mathbf{A}(t)\mathbf{x}$ represents a linearization of system about $\mathbf{x}_0$.

If it is desired to hold the system at $\mathbf{x}_0$ which is not an equilibrium point, then the system must be modified to subtract off $\mathbf{f}(\mathbf{x}_0, t)$. That is, $\tilde{\mathbf{f}}(\mathbf{x}, t) = \mathbf{f}(\mathbf{x}, t) - \mathbf{f}(\mathbf{x}_0, t)$. Thus, $\tilde{\mathbf{f}}$ has an equilibrium point at $\mathbf{x}_0$, so the theorems apply.

Theorem 5.1:

Consider $\frac{d}{dt}\mathbf{x} = \mathbf{f}(\mathbf{x}, t)$, with an equilibrium point at the origin, and

1. $\mathbf{f}(\mathbf{x}, t) - \mathbf{A}(t)\mathbf{x}$ converges uniformly to zero (i.e., that 5.94 holds).
2. $\mathbf{A}(t)$ is bounded.
3. The system $\frac{d}{dt}\mathbf{x} = \mathbf{A}(t)\mathbf{x}$ is exponentially stable.

Then $\mathbf{x}_0$ is an exponentially stable equilibrium of $\frac{d}{dt}\mathbf{x} = \mathbf{f}(\mathbf{x}, t)$.

If the system does not depend explicitly on time (i.e., $\mathbf{f}(\mathbf{x}, t) = \mathbf{f}(\mathbf{x})$), it is said to be *autonomous*. For autonomous systems, conditions 1 and 2 are automatically true, and exponential stability of 3 is simply the condition that all of the eigenvalues of $\mathbf{A}$ have negative real parts.

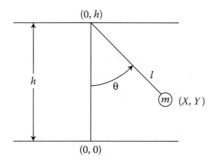

FIGURE 5.8 Simple pendulum.

Theorem 5.2:

Suppose $\frac{d}{dt}\mathbf{x} = \mathbf{f}(\mathbf{x}, t)$ has an equilibrium point at $\mathbf{x}_0$ and $\mathbf{f}(\mathbf{x}, t)$ is continuously differentiable, further that the Jacobian $\mathbf{A}(t) = \mathbf{A}$ is constant, and that the uniform approximation condition Equation 5.94 holds. Then, the equilibrium is unstable if $\mathbf{A}$ has at least one eigenvalue with positive real part.

5.2.3.3　Failures of Linearization

Example 5.6:

This example shows what happens in the case that there is no uniform approximation (i.e., that condition 5.94 fails).

$$\frac{dx}{dt} = -x + tx^2 \tag{5.95}$$

The right-hand side of this equation is clearly differentiable. The candidate for linearization is the first term of Equation 5.95, *it is not a linearization*—this system has no linearization. It is clear as t grows large the quadratic term will dominate the linear part. Thus, the system will grow rather than having zero as a stable equilibrium, as would be predicted by Theorem 5.1, if the uniform approximation condition were not necessary.

Example 5.7:

This example has fixed eigenvalues with negative real parts; nonetheless it is unstable.

$$\frac{dx}{dt}\begin{bmatrix} x_1(t) \\ x_2(t) \end{bmatrix} = \begin{bmatrix} -1 & 0 \\ e^{at} & -2 \end{bmatrix}\begin{bmatrix} x_1(t) \\ x_2(t) \end{bmatrix},$$

$$\mathbf{x} = \begin{bmatrix} 1 \\ 1 \end{bmatrix} \tag{5.96}$$

The solution to this equation is $x_1(t) = exp\{-t\}$ and $\dot{x}_2(t) = -2x_2(t) + exp\{(a - 1)t\}$. So that $x_2(t) = exp\{-2t\} + \int_0^t exp\{-2(t - s)\} exp\{(a - 1)s\}ds$, or $x_2(t) = exp\{-2t\} + exp\{-2t\}\int_0^t exp\{(a + 1)s\}ds$. Clearly, for $a > 1$, the system is unstable. This example is simple. Other slightly more complex examples show that systems with bounded $A(t)$ can have fixed eigenvalues and still result in an unstable system (see Example 90, Chapter 5, Section 4 in [8]).

5.2.3.4 Example of Linearization

Consider a simple pendulum and allow the length of the pendulum to vary (see Figure 5.8). Let (X, Y) be the position of the pendulum at time t. Then, $X = l \sin \theta$, $Y = h - l \cos \theta$. Further $\dot{X} = \dot{l} \sin \theta + l\dot{\theta} \cos \theta$, $\dot{Y} = -\dot{l} \cos \theta + l\dot{\theta} \sin \theta$. The kinetic energy (K.E.) is given by $(1/2)m(\dot{X}^2 + \dot{Y}^2)$, or where m is the mass at the end of the pendulum (and this is the only mass). The potential energy (P.E.) is: $mgY = mg(h - l \cos \theta)$, g being the gravitational constant (about 9.8 m/s^2 at sea level). So, both g and l are nonnegative.

The Lagrangian is given by $L(\theta, l, \dot{\theta}, \dot{l}) = L(\mathbf{x}, \dot{\mathbf{x}}) = K.E. - P.E.$; and the Euler–Lagrange equations

$$0 = \frac{d}{dt}\left[\frac{\partial L}{\partial \dot{\mathbf{x}}}\right] - \left[\frac{\partial L}{\partial \mathbf{x}}\right] \tag{5.97}$$

give the motions of the system (See [2]).

Here

$$\left[\frac{\partial L}{\partial \dot{\mathbf{x}}}\right] = \begin{bmatrix} ml^2\dot{\theta} \\ m\dot{l} \end{bmatrix}, \tag{5.98}$$

so

$$\frac{d}{dt}\left[\frac{\partial L}{\partial \dot{\mathbf{x}}}\right] = \begin{bmatrix} 2ml\dot{l}\dot{\theta} + ml^2\ddot{\theta} \\ m\ddot{l} \end{bmatrix} \tag{5.99}$$

and

$$\left[\frac{\partial L}{\partial \mathbf{x}}\right] = \begin{bmatrix} -mgl \sin \theta \\ m\dot{\theta}^2 l + mg \cos \theta \end{bmatrix} \tag{5.100}$$

Combined, these yield

$$\mathbf{0} = \begin{bmatrix} 2ml\dot{l}\dot{\theta} + ml^2\ddot{\theta} \\ m\ddot{l} \end{bmatrix} - \begin{bmatrix} -mgl \sin \theta \\ m\dot{\theta}^2 l + mg \cos \theta \end{bmatrix} \tag{5.101}$$

After a bit of algebraic simplification

$$0 = \ddot{\theta} + 2\frac{\dot{l}}{l}\dot{\theta} + \frac{g}{l} \sin \theta \tag{5.102}$$

$$0 = \ddot{l} - \dot{\theta}^2 l - g \cos \theta \tag{5.103}$$

Remarks

If the motion for l is prescribed, then the first equation of the Lagrangian is correct, but l and $\dot{l}$ are not degrees of freedom in the Lagrangian (i.e., $L = L(\theta, \dot{\theta})$). This happens because the rederivation of the system yields the first equation of our system. Using the above derivation and setting $l, \dot{l}$ to their prescribed values *will not* result in the correct equations. Consider Case I below, if $l = l_0, \dot{l} = \ddot{l} = 0$ is substituted, then the second equation becomes $\dot{\theta}^2 l_0 + g \cos \theta = 0$. Taking $\frac{d}{dt}$ of this results in $\dot{\theta}(\ddot{\theta} + \frac{g}{2l_0} \sin \theta) = 0$—which is inconsistent with the first equation: $\ddot{\theta} + \frac{g}{l_0} \sin \theta = 0$.

Case I

Let $\dot{l} \equiv 0$, then $\ddot{l} \equiv 0$, and let l be a positive constant; equations of motion are (see the above remark)

$$0 = \ddot{\theta} + \frac{g}{l} \sin \theta \tag{5.104}$$

Writing Equation 5.104 in first-order form, by letting $\mathbf{x} = \begin{bmatrix} \theta & \dot{\theta} \end{bmatrix}^T$, then

$$\dot{\mathbf{x}} = \mathbf{f}(\mathbf{x}) = \begin{bmatrix} x_2 \\ \frac{g}{l} \sin x_1 \end{bmatrix} \tag{5.105}$$

It is easy to check that $\mathbf{f}(\mathbf{0}) = \mathbf{0}$, and that $\mathbf{f}$, otherwise meets the linearization criterion, so that a linearization exists; it is given by

$$\dot{\mathbf{x}} = \begin{bmatrix} 0 & 1 \\ -\frac{g}{l} & 0 \end{bmatrix} \mathbf{x} \tag{5.106}$$

Case II

Let $\mathbf{x} = [\theta, l, \dot{\theta}, \dot{l}]^T$, then Equations 5.102 and 5.103 are given by

$$\dot{\mathbf{x}} = \begin{bmatrix} x_3 \\ x_4 \\ -2\frac{x_4}{x_2}x_3 - \frac{g}{x_2} \sin x_1 \\ x_3^2 x_2 + g \cos x_1 \end{bmatrix} \tag{5.107}$$

Clearly, $\mathbf{0}$ is not an equilibrium point of this system. Physically, $l = 0$ doesn't make much sense anyway. Let us seek an equilibrium at $\mathbf{x}_0 = [0, l_0, 0, 0]$, which is a normal pendulum of length l_0, at rest. Note $\mathbf{f}(\mathbf{x}_0) = [0, 0, 0, g]^T$; hence open-loop feedback of the form $[0, 0, 0, -g]^T$ will give the equilibrium at the desired point. Let $\mathbf{F}(\mathbf{x}) = \mathbf{f}(\mathbf{x}) - [0, 0, 0, g]^T$, then $\mathbf{F}(\mathbf{x}_0) = \mathbf{0}$, and it can be verified that $\dot{\mathbf{x}} = \mathbf{F}(\mathbf{x})$ meets all the linearization criteria at $\mathbf{x}_0$. Thus, $-\mathbf{f}(\mathbf{x}_0)$ is the open-loop control that must be applied to the system. The linearized system is given by $\mathbf{z} = \mathbf{x} - \mathbf{x}_0$, $\dot{\mathbf{z}} = \left[\frac{\partial \mathbf{F}}{\partial \mathbf{x}} \right] |_{\mathbf{x}_0} \mathbf{z}$.

$$\left[\frac{\partial \mathbf{F}}{\partial \mathbf{x}} \right] = \begin{bmatrix} 0 & 0 & 1 & 0 \\ 0 & 0 & 0 & 1 \\ -\frac{g}{x_2}\cos x_1 & \frac{2x_3 x_4 + g \sin x_1}{x_2^2} & -2\frac{x_4}{x_2} & -2\frac{x_3}{x_2} \\ -g \sin x_1 & x_3^2 & 2x_3 x_2 & 0 \end{bmatrix} \tag{5.108}$$

$$\left[\frac{\partial \mathbf{F}}{\partial \mathbf{x}} \right] |_{\mathbf{x}_0} = \begin{bmatrix} 0 & 0 & 1 & 0 \\ 0 & 0 & 0 & 1 \\ -\frac{g}{l_0} & 0 & 0 & 0 \\ 0 & 0 & 0 & 0 \end{bmatrix} \tag{5.109}$$

To determine the eigenvalues, examine $det(sI - A)$; in this case expansion is easy around the last row and $det(sI - A) = s^2(s^2 + \frac{g}{l_0})$. Thus, the system is not exponentially stable.

References

1. *Webster's Ninth New Collegiate Dictionary*, First Digital Edition, Merriam-Webster and NeXT Comuter, Inc., 1988.
2. Arnold, V.I., *Mathematical Methods of Classical Mechanics*, Second Edition, Springer-Verlag, 1989.
3. Chen, C.T., *Linear Systems Theory and Design*, Holt, Rinehart and Winston, 1984.
4. Wolovich, W., *Automatic Control Systems*, Harcourt Brace, 1994.
5. Kailith, T., *Linear Systems Theory*, Prentice Hall, 1980.
6. Bender C. and Orzag, S., *Nonlinear Systems Analysis*, Second Edition, McGraw-Hill, 1978.
7. Press, W., et. al., *Numerical Recipes in C*, Cambridge, 1988.
8. Vidyasagar, M., *Nonlinear Systems Analysis*, Second ed., Prentice Hall, 1993.
9. Balakrishnan, A.V., *Applied Functional Analysis*, Second Edition, Springer-Verlag, 1981.
10. Dorf, R.C., *Modern Control Systems*, 3rd ed., Addison-Wesley, 1983.
11. Fattorini, H., *The Cauchy Problem*, Addison-Wesley, 1983.
12. Kalman, R.E., Falb, P.E., and Arbib, M.A., *Topics in Mathematical Systems Theory*, McGraw-Hill, 1969.
13. Reed, M. and Simon, B., *Functional Analysis*, Academic Press, 1980.

5.3 Models for Sampled-Data Systems

Graham C. Goodwin, Juan I. Yuz, and Juan C. Agüero

5.3.1 Introduction

Models for dynamical systems usually arise from the application of physical laws such as conservation of mass, momentum, and energy. These models typically take the form of linear or nonlinear differential equations, where the parameters involved can usually be interpreted in terms of physical properties of the system. In practice, however, these kinds of models are not appropriate to interact with digital devices. For example, digital controllers can only act on a real system at specific time instants. Similarly, information from signals of a given system can usually only be recorded (and stored) at specific instants. This constitutes an unavoidable paradigm: continuous-time systems interact with actuators and sensors that are accessible only at discrete-time instants. As a consequence, *sampling* of continuous-time systems is a key problem both for estimation and control purposes [1,2].

The sampling process for a continuous-time system is represented schematically in Figure 5.9. In this figure we see that there are three basic elements involved in the sampling process. All of these elements play a core role in determining an appropriate discrete-time input–output description:

- The *hold device*, used to generate the continuous-time input $u(t)$ of the system, based on a discrete-time sequence u_k defined at specific time instants t_k;

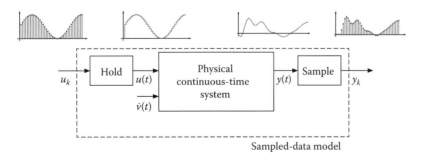

Sampled-data model

FIGURE 5.9 Sampling scheme of a continuous-time system.

- The *continuous-time system*, in general, can be defined by a set of linear or nonlinear differential equations, which generate the continuous-time output $y(t)$ from the continuous-time input $u(t)$, initial conditions, and (possibly) unmeasured disturbances; and
- The *sampling device*, which generates an output sequence of samples y_k from the continuous-time output $y(t)$, possibly including some form of *anti-aliasing* filter prior to instantaneous sampling.

For linear systems, it is possible to obtain *exact* sampled-data models from the sampling scheme shown in Figure 5.9. In particular, given a deterministic continuous-time system, it is possible to obtain a discrete-time model, which replicates the sequence of output samples. In the stochastic case, where the input of the system is assumed to be a *continuous-time white-noise* (CTWN) process, a sampled-data model can be obtained such that its output sequence has the same second-order properties as the continuous-time output at the sampling instants.

In this chapter we will review the above concepts. These ideas are central to modern control and measuring devices. We will focus on the linear SISO case for simplicity. We will briefly discuss extensions to the nonlinear case. In the linear case, we can use superposition to consider deterministic inputs and stochastic inputs separately. We begin in the next section with the deterministic case.

5.3.2 Sampled-Data Models for Linear Systems Having Deterministic Inputs

We begin our development by describing a general linear time-invariant continuous-time system by a transfer function

$$G(s) = \frac{F(s)}{E(s)} = \frac{f_m s^m + \cdots + f_0}{s^n + e_{n-1}s^{n-1} + \cdots + e_0}, \quad f_m \neq 0 \tag{5.110}$$

where s denotes the Laplace transform variable. For the moment we include the antialiasing filter in this description. The system is of order n, and has relative degree $r = n - m > 0$. Such a system can be equivalently expressed in state-space form as

$$\dot{x}(t) = Ax(t) + Bu(t) \tag{5.111}$$

$$y(t) = Cx(t) \tag{5.112}$$

Remark 5.1

For future use, it is also insightful to express model 5.111 in incremental form as

$$dx = Ax(t)dt + Bu(t)\, dt \tag{5.113}$$

An exact discrete-time representation of the system can be obtained under appropriate assumptions. A common assumption is that the continuous-time input signal is generated by a zero-order hold (ZOH), that is,

$$u(t) = u_k; t \in [k\Delta, k\Delta + \Delta) \tag{5.114}$$

and that the output is (instantaneously) sampled at uniform times

$$y_k = y(k\Delta) \tag{5.115}$$

where Δ is the sampling period. We then have the following standard result [2].

Lemma 5.1:

If the input of the continuous-time systems (Equations 5.111 and 5.112) is generated from the input sequence u_k using a ZOH, then a state-space representation of the resulting sampled-data model is given by

$$q\,x_k = x_{k+1} = A_q x_k + B_q u_k \tag{5.116}$$

$$y_k = C x_k \tag{5.117}$$

where the sampled output is $y_k = y(k\Delta)$, q denotes the forward shift operator, and the (discrete) system matrices are

$$A_q = e^{A\Delta}, \quad B_q = \int_0^\Delta e^{A\eta} B \, d\eta \tag{5.118}$$

A discrete-time transfer function representation of the sampled-data system can be readily obtained from Lemma 5.1 as

$$G_q(z) = C(zI_n - A_q)^{-1} B_q \tag{5.119}$$

$$G_q(z) = \frac{F_q(z)}{E_q(z)} = \frac{\bar{f}_{n-1} z^{n-1} + \cdots + \bar{f}_0}{z^n + \bar{e}_{n-1} z^{n-1} + \cdots + \bar{e}_0} \tag{5.120}$$

where z denotes the $\mathcal{Z}$-transform variable. Note that if $s = \lambda_\ell$ is a continuous-time pole (i.e., an eigenvalue of the matrix A), then $z = e^{\lambda_\ell \Delta}$ is a pole of the discrete-time transfer function (i.e., an eigenvalue of A_q).

The expression obtained in Equations 5.119 and 5.120 is equivalent to the pulse transfer function obtained directly from the continuous-time transfer function, as stated in the following lemma.

Lemma 5.2:

The sampled-data transfer function 5.119 can be obtained using the inverse Laplace transform of the continuous-time step response, computing its $\mathcal{Z}$-transform, and dividing it by the $\mathcal{Z}$-transform of a discrete-time step:

$$G_q(z) = (1 - z^{-1}) \mathcal{Z} \left\{ \mathcal{L}^{-1} \left\{ \frac{G(s)}{s} \right\}_{t=k\Delta} \right\} \tag{5.121}$$

$$= (1 - z^{-1}) \frac{1}{2\pi j} \int_{\gamma - j\infty}^{\gamma + j\infty} \frac{e^{s\Delta}}{z - e^{s\Delta}} \frac{G(s)}{s} \, ds \tag{5.122}$$

where Δ is the sampling period and $\gamma \in \mathbb{R}$ is such that all poles of $G(s)/s$ have real part less than γ. Furthermore, if the integration path in Equation 5.122 is closed by a semicircle to the right, we obtain

$$G_q(z) = (1 - z^{-1}) \sum_{\ell=-\infty}^{\infty} \frac{G((\log z + 2\pi j\ell)/\Delta)}{\log z + 2\pi j\ell} \tag{5.123}$$

Equation 5.123, when considered in the frequency-domain, substituting $z = e^{j\omega\Delta}$, illustrates the well-known aliasing effect: the frequency response of the sampled-data system is obtained by folding of the continuous-time frequency response, that is,

$$G_q(e^{j\omega\Delta}) = \frac{1}{\Delta} \sum_{\ell=-\infty}^{\infty} H_{\mathrm{ZOH}}\left(j\omega + j\frac{2\pi}{\Delta}\ell\right) G\left(j\omega + j\frac{2\pi}{\Delta}\ell\right) \tag{5.124}$$

where $H_{\mathrm{ZOH}}(s)$ is the Laplace transform of the ZOH impulse response, that is,

$$H_{\mathrm{ZOH}}(s) = \frac{1 - e^{-s\Delta}}{s} \tag{5.125}$$

Equation 5.124 can be also derived from Equation 5.119 using the state-space matrices in Equation 5.118 (see, e.g., [2, Lemma 4.6.1]).

Example 5.8:

Consider a system with transfer function

$$G(s) = \frac{6(s+5)}{(s+2)(s+3)(s+4)} \tag{5.126}$$

The exact sampled-data model for a sampling period $\Delta = 0.01$ is given by

$$G_q(z) = \frac{2.96 \times 10^{-4}(z+0.99)(z-0.95)}{(z-0.98)(z-0.97)(z-0.96)} \tag{5.127}$$

An important observation from the previous example is that the sampled-data system 5.120 will, in general, have relative degree 1. This means that there are *extra zeros* in the discrete-time model with no continuous-time counterpart. These are often called the *sampling zeros*. We will discuss these zeros in more detail in the next section.

5.3.3 Asymptotic Sampling Zeros

As we have seen in Section 5.3.2, the poles of a sampled-data model can be readily characterized in terms of the sampling period, Δ, and the continuous-time system poles. However, the relation between the discrete- and continuous-time zeros is much more involved.

In this section, we review results concerning the asymptotic behavior of the zeros in sampled-data models, as the sampling period goes to zero. These results follow the seminal work presented in [3], where the asymptotic location of the *intrinsic* and *sampling* zeros was first described, for the ZOH case, using shift operator models.

The next result characterizes the sampled-data model (and the sampling zeros) corresponding to a special model, namely, an rth-order integrator.

Remark 5.2

Throughout this chapter we will see that the sampled-data model for an rth-order integrator plays a very important role in obtaining asymptotic results. Indeed, as the sampling rate increases, a system of relative degree r, *behaves* as an rth-order integrator. This will be a recurrent and insightful interpretation for deterministic and stochastic systems.

Lemma 5.3: [3]

For sampling period Δ, the pulse transfer function corresponding to the rth-order integrator $G(s) = s^{-r}$, is given by

$$G_q(z) = \frac{\Delta^r}{r!} \frac{B_r(z)}{(z-1)^r} \tag{5.128}$$

where

$$B_r(z) = b_1^r z^{r-1} + b_2^r z^{r-2} + \cdots + b_r^r \tag{5.129}$$

$$b_k^r = \sum_{\ell=1}^{k} (-1)^{k-\ell} \ell^r \binom{r+1}{k-\ell} \tag{5.130}$$

Remark 5.3

The polynomials defined in Equations 5.129 and 5.130 correspond, in fact, to the Euler–Fröbenius polynomials (also called *reciprocal polynomials* [4]) and are known to satisfy several properties [5]:

1. Their coefficients can be computed recursively:

$$b_1^r = b_r^r = 1; \quad \forall r \geq 1 \tag{5.131}$$

$$b_k^r = k b_k^{r-1} + (r - k + 1) b_{k-1}^{r-1}; \quad k = 2, \ldots, r - 1 \tag{5.132}$$

2. Their roots always are negative real.
3. From the symmetry of the coefficients in Equation 5.130, that is, $b_k^r = b_{r+1-k}^r$, it follows that, if $B_r(z_0) = 0$, then $B_r(z_0^{-1}) = 0$.
4. They satisfy an interlacing property, namely, every root of the polynomial $B_{r+1}(z)$ lies between every two adjacent roots of $B_r(z)$, for $r \geq 2$.
5. The following recursive relation holds:

$$B_{r+1}(z) = z(1 - z)B_r{}'(z) + (rz + 1)B_r(z); \quad \forall r \geq 1 \tag{5.133}$$

where $B_r{}' = \dfrac{dB_r}{dz}$.

We list below the first of these polynomials:

$$B_1(z) = 1 \tag{5.134}$$

$$B_2(z) = z + 1 \tag{5.135}$$

$$B_3(z) = z^2 + 4z + 1 = (z + 2 + \sqrt{3})(z + 2 - \sqrt{3}) \tag{5.136}$$

$$B_4(z) = z^3 + 11z^2 + 11z + 1 = (z + 1)(z + 5 + 2\sqrt{6})(z + 5 - 2\sqrt{6}) \tag{5.137}$$

In the frequency-domain, a special case of interest is when we combine the infinite sum (Equation 5.123) with the result in Lemma 5.3, as presented in the following result.

Lemma 5.4: [3]

The following identity holds for $z = e^{j\omega\Delta}$:

$$\sum_{k=-\infty}^{\infty} \frac{1}{\left(\frac{\log z + j 2\pi k}{\Delta}\right)^r} = \frac{\Delta^r}{(r-1)!} \frac{z B_{r-1}(z)}{(z-1)^r} \tag{5.138}$$

where $B_{r-1}(z)$ is the Euler–Fröbenius polynomial.

We next extend Lemma 5.3 to a general transfer function when the sampling period Δ tends to zero.

Lemma 5.5: [3, Theorem 1]

Let $G(s)$ be a rational function:

$$G(s) = \frac{F(s)}{E(s)} = K \frac{(s - z_1)(s - z_2) \cdots (s - z_m)}{(s - p_1)(s - p_2) \cdots (s - p_n)} \tag{5.139}$$

and $G_q(z)$ the corresponding pulse transfer function. Assume that $m < n$, that is, $G(s)$ is strictly proper having relative degree $r \geq 1$. Then, as the sampling period Δ goes to 0, then m zeros of $G_q(z)$ go to 1 as $e^{z_i\Delta}$, *and the remaining* $r - 1$ *zeros of* $G_q(z)$ *go to the zeros of the polynomial* $B_r(z)$ *defined in Lemma 5.3, that is,*

$$G_q(z) \xrightarrow{\Delta \approx 0} K \frac{\Delta^r (z-1)^m B_r(z)}{(r)!(z-1)^n} \tag{5.140}$$

Example 5.9:

Consider again the system in Example 5.8 defined by the transfer function (5.126). By Lemma 5.5, if we use a ZOH to generate the input, then, as the sampling period Δ tends to zero, the associated sampled-data transfer function is given by

$$G_q(z) \xrightarrow[\text{(ZOH)}]{\Delta \approx 0} 6 \frac{\Delta^2(z+1)(z-1)}{2!(z-1)^3} \tag{5.141}$$

In fact, for $\Delta = 0.001$, the exact sampled data model is given by

$$G_q(z) = \frac{2.996 \times 10^{-6}(z + 0.999)(z - 0.995)}{(z - 0.998)(z - 0.997)(z - 0.996)} \tag{5.142}$$

Note that the resulting discrete-time model has two zeros, even though the continuous-time system has only one finite zero. Also, the discrete poles and one of the zeros now all appear very close to $z = 1$. Actually, the latter is a difficulty that we will address in the next section.

Remark 5.4

Obviously, a key question that arises from Lemma 5.5 is the properties of the asymptotic convergence result (Equation 5.141), that is, how the approximation error behaves as $\Delta \to 0$. We will address this key question in Section 5.3.6. Before doing that, in the next section we will take a small diversion to better understand the limiting process.

5.3.4 Incremental Models

A fundamental problem with the model (Equation 5.116) is that, as we take the limit as $\Delta \to 0$, we lose all information about the underlying continuous-time system. Indeed, it is easily seen from Equation 5.118 that the following result holds for **all** linear systems:

$$\lim_{\Delta \to 0} A_q = I \quad \text{and} \quad \lim_{\Delta \to 0} B_q = 0 \tag{5.143}$$

The origin of this difficulty is that Equation 5.116 describes the next value of x_k. However, it is clear that $x_{k+1} \to x_k$ as the sampling period $\Delta \to 0$. This difficulty is fundamental and intrinsic to shift operator model descriptions. However, the problem is readily bypassed if we, instead, express Equation 5.116 in *incremental* form. In order to do this, we subtract x_k from both sides of the equation and factor Δ out of the right-hand side. This leads to

$$dx_k = x_{k+1} - x_k = A_i x_k \Delta + B_i u_k \Delta \tag{5.144}$$

where

$$A_i = \frac{A_q - I}{\Delta} \quad \text{and} \quad B_i = \frac{B_q}{\Delta} \tag{5.145}$$

Remark 5.5

The above model can be seen to have the same structure as the incremental continuous-time model 5.113. Moreover, and importantly, if we now take the limit as the sampling period tends to zero, we obtain

$$A_i \xrightarrow{\Delta \to 0} A \quad \text{and} \quad B_i \xrightarrow{\Delta \to 0} B \tag{5.146}$$

where A and B are the corresponding continuous-time matrices. This is a pleasing by-product of the use of the incremental model 5.144. Indeed, we will see in the sequel that use of the incremental model is crucial in obtaining meaningful connections between continuous models and their discrete counterparts at fast sampling rates.

We see that we have achieved our objective of having a well-defined limit as the sampling rate is increased. Moreover, the limiting model takes us back to continuous time in a heuristically satisfactory fashion.

For future use, it will be convenient to define the discrete *delta operator* as

$$\delta x_k = \frac{dx_k}{\Delta} = \frac{x_{k+1} - x_k}{\Delta} = \frac{q-1}{\Delta} x_k \tag{5.147}$$

Also, we introduce γ as the complex variable associated with the δ-operator [1,2]:

$$\delta = \frac{q-1}{\Delta} \quad \Longleftrightarrow \quad \gamma = \frac{z-1}{\Delta} \tag{5.148}$$

Use of this operator makes the sampling period Δ explicit, and is crucial in showing how discrete-time results converge to their continuous-time counterpart when $\Delta \to 0$.

Example 5.10:

Consider the exact sampled-data (ESD) model (Equation 5.127) obtained in Example 5.8. We can reparameterize the system to express it in incremental form by using the change of variable $z = (1 + \Delta\gamma)$ where, for this particular example, the sampling period is $\Delta = 0.01$. This yields

$$G_\delta(\gamma) = G_q(1 + \Delta\gamma) = \frac{0.030(\gamma + 198.7)(\gamma + 4.88)}{(\gamma + 1.98)(\gamma + 2.96)(\gamma + 3.92)} \tag{5.149}$$

Remark 5.6

We wish to clarify a point of confusion that exists in some areas. The use of the *delta* operator is simply a way of reparameterizing discrete models via the transformation $q = \delta\Delta + 1$ or $\delta = (q-1)/\Delta$. This has the advantage of highlighting the link between discrete- and continuous-time domain and also achieving improved numerical properties [6]. Of course, any shift-domain model can be converted to delta form and vice versa. This is totally different to the use of Euler integration which, by chance, happens to have the property that continuous poles and zeros appear in the same location in the corresponding delta-domain discrete model.

5.3.5 Asymptotic Sampling Zeros for Incremental Models

We next use the incremental form (or equivalent δ-operator parameterization) to reexpress the results of Section 5.3.3. We begin with the following incremental counterpart to Lemma 5.3.

Lemma 5.6: [7]

Given a sampling period Δ, the exact sampled-data model corresponding to the rth-order integrator $G(s) = s^{-r}$, $r \geq 1$, when using a ZOH input, is given by

$$G_\delta(\gamma) = \frac{p_r(\Delta\gamma)}{\gamma^r} \tag{5.150}$$

where the polynomial $p_r(\Delta\gamma)$ is given by

$$p_r(\Delta\gamma) = \det \begin{bmatrix} 1 & \dfrac{\Delta}{2!} & \cdots & \dfrac{\Delta^{r-2}}{(r-1)!} & \dfrac{\Delta^{r-1}}{r!} \\ -\gamma & 1 & \cdots & \dfrac{\Delta^{r-3}}{(r-2)!} & \dfrac{\Delta^{r-2}}{(r-1)!} \\ \vdots & \ddots & \ddots & \vdots & \vdots \\ 0 & \cdots & -\gamma & 1 & \dfrac{\Delta}{2!} \\ 0 & \cdots & 0 & -\gamma & 1 \end{bmatrix} = \frac{B_r(1 + \Delta\gamma)}{r!} \tag{5.151}$$

where $B_r(\cdot)$ is the Euler–Fröbenius polynomial defined in Equation 5.129.

From 5.151 it is straightforward to see that

$$p_r(0) = 1 \iff B_r(1) = r! \tag{5.152}$$

A more interesting situation arises when we consider a general continuous-time transfer function. We saw in Lemma 5.5 that when the shift operator model was used, the m continuous zeros and n continuous poles all converged to $z = 1$ in the discrete model as $\Delta \to 0$. We see below that when incremental discrete models are used, then a much more pleasing result is obtained, that is, these *intrinsic* zeros and poles converge to their continuous locations.

We recall the sampled-data model given in Equation 5.144, that is, expressed in incremental (or δ-)form as

$$\frac{x_{k+1} - x_k}{\Delta} = \delta x_k = A_i x_k + B_i u_k \tag{5.153}$$

In transfer function form, we have that Equation 5.120 can be rewritten as

$$G_\delta(\gamma) = \frac{F_\delta(\gamma)}{E_\delta(\gamma)} = \frac{\tilde{f}_{n-1}\gamma^{n-1} + \cdots + \tilde{f}_0}{\gamma^n + \tilde{e}_{n-1}\gamma^{n-1} + \cdots + \tilde{e}_0} \tag{5.154}$$

Lemma 5.7:

In incremental form we have the following convergence result

$$F_\delta(\gamma) \xrightarrow{\Delta \to 0} F(\gamma) \tag{5.155}$$

$$E_\delta(\gamma) \xrightarrow{\Delta \to 0} E(\gamma) \tag{5.156}$$

where

$$F_\delta(\gamma) \xrightarrow{\Delta \approx 0} p_r(\Delta\gamma) F(\gamma) \tag{5.157}$$

and where $p_r(\Delta\gamma)$ is the equivalent to the Euler–Fröbenius polynomial in the γ-domain in Equation 5.151.

Example 5.11:

Consider again the sampled-data model in Example 5.10 expressed in the incremental form 5.149. Note that it can be rewritten as

$$G_\delta(\gamma) = \frac{5.882(1 + 0.005\gamma)(\gamma + 4.88)}{(\gamma + 1.98)(\gamma + 2.96)(\gamma + 3.92)} \qquad (5.158)$$

If we compare Equation 5.158 with the transfer function 5.126, we can notice that the sampled-data model expressed in incremental form recovers approximately the continuous-time poles, zeros and gain. Moreover, we can see that the extra zero due to sampling appears at the location predicted by Lemmas 5.6 and 5.7 if we notice that $p_2(\Delta\gamma) = 1 + \frac{\Delta}{2}\gamma$.

The convergence results in Lemmas 5.5 and 5.7 are fundamentally different. In particular, in Lemma 5.5, the intrinsic poles and zeros all converge to $z = 1$, irrespective of the underlying continuous system, whereas in Lemma 5.7, the intrinsic poles and zeros converge to their continuous locations, in particular, we see from Equations 5.156 and 5.157 in Lemma 5.7 that the discrete poles converge to their continuous counterparts as $\Delta \to 0$. Also, from Equation 5.157, we see that the discrete zeros split into m zeros which converge to their continuous counterparts plus $r - 1$ extra zeros arising from the sampling process.

5.3.6 Model Error Quantification

In this section we examine various aspects of the errors that exist between the limiting model described in Sections 5.3.3 and 5.3.5, and the true discrete model at different sampling rates. We will use the result in Section 5.3.5 in preference to the results in Section 5.3.3. We will consider both absolute and relative errors. Relative errors are of particular interest in applications. For example, it is typically more useful to know that a model has 1% accuracy as opposed to knowing that the absolute error is 0.1 which leaves open the question of whether the true value is 1 (corresponding to 10% error) or 0.01 (corresponding to a 1000% error).

Of course, in the linear case, one can always calculate the sampled-data model to any desired degree of accuracy by using the exact transformations given in Section 5.3.2. However, our goal here is to study the degree of model approximation required to achieve relative error convergence properties as the sampling period is reduced. This has the advantage of giving insights into different simple models. Of particular interest is the *sampling zeros* as previously described in Sections 5.3.3 and 5.3.5. The location of these zeros is highlighted in the approximate models, but their presence is blurred in the exact representation.

Our principal interest here will be in various simplified models and their associated relative error properties. The models that we will compare are the following.

1. *Exact sampled-data model* (ESD model): This is not an approximate model, but the exact model that is obtained for a linear deterministic system using the expressions in Section 5.3.2. This model is given in Equation 5.120, or, in state-space form, by Equations 5.116 and 5.117. We can write the model as

$$G_q^{\text{ESD}}(z) = \mathcal{Z}\left\{\frac{1 - e^{s\Delta}}{s}G(s)\right\} \qquad (5.159)$$

where $\mathcal{Z}\{H(s)\} = \sum_{k=0}^{\infty} h_k z^{-k}$ denotes the $\mathcal{Z}$-transform of h_k, the sampled impulse response of the transfer function $H(s)$. In the sequel, we consider $G(s)$ given by

$$G(s) = \frac{\prod_{i=1}^{m}(s - c_i)}{\prod_{i=1}^{n}(s - p_i)} \qquad (5.160)$$

Notice that, without loss of generality, we have not included a gain K in the transfer function $G(s)$. This choice will not affect the relative error analysis in the sequel, provided we adjust the gain of the approximated models such that d.c. gain matches the d.c. gain of the continuous-time system.

2. *Simple derivative replacement model* (SDR model): We obtain this model by simply replacing derivatives by divided differences. Note that this corresponds to the use of simple Euler integration or, equivalently, the use of a first-order Taylor expansion:

$$G_q^{\text{SDR}}(z) = \frac{\prod_{i=1}^{m}(\frac{z-1}{\Delta} - c_i)}{\prod_{i=1}^{n}(\frac{z-1}{\Delta} - p_i)}; \quad G_\delta^{\text{SDR}}(\gamma) = \frac{\prod_{i=1}^{m}(\gamma - c_i)}{\prod_{i=1}^{n}(\gamma - p_i)} \tag{5.161}$$

Note that this model does not include any sampling zeros. Also note that this model is not equivalent to a delta model; see Remark 5.6.

3. *Asymptotic sampling zeros model* (ASZ model): In this case, we use a discrete-time transfer function with sampling zeros located at their asymptotic location, and the intrinsic poles and zeros are placed corresponding to their location given by Euler integration:

$$G_q^{\text{ASZ}}(z) = \frac{B_r(z)\prod_{i=1}^{m}(\frac{z-1}{\Delta} - c_i)}{r!\prod_{i=1}^{n}(\frac{z-1}{\Delta} - p_i)}; \quad G_\delta^{\text{ASZ}}(\gamma) = \frac{p_r(\gamma\Delta)\prod_{i=1}^{m}(\gamma - c_i)}{\prod_{i=1}^{n}(\gamma - p_i)} \tag{5.162}$$

Note that by using the fact that $B_r(1) = r!$, we adjusted the d.c. gain of this model in order to match the continuous-time d.c. gain.

4. *Corrected sampling zero model* (CSZ model): In this case, we place the (shift operator domain) sampling zero near -1 (if one exists) at locations such that errors are of $\mathcal{O}(\Delta^2)$ while other sampling zeros and the intrinsic poles and zeros are located at the values given by Euler integration and the d.c. gain is matched to that of the continuous model. Note that there exists a sampling zero at $z = -1$ if and only if the relative degree r is even. In such cases, we use the sampled-data model

$$G_q^{\text{CSZ}}(z) = \frac{\tilde{B}_r(z)\prod_{i=1}^{m}(\frac{z-1}{\Delta} - c_i)}{r!\prod_{i=1}^{n}(\frac{z-1}{\Delta} - p_i)} \tag{5.163}$$

where

$$\tilde{B}_r(z) = B_r(z)\frac{z+1+\sigma_\Delta}{z+1} \cdot \frac{2}{2+\sigma_\Delta} \tag{5.164}$$

For r odd, then $\sigma_\Delta = 0$. For r even, then we choose σ_Δ as follows [8]:

$$\sigma_\Delta = \frac{\Delta}{r+1}\left\{\sum_{i=1}^{n}p_i - \sum_{i=1}^{n}z_i\right\} \tag{5.165}$$

where r is the relative degree and p_i, c_i denote the continuous poles and zeros. In particular, we have

$$(r=2) \quad \sigma_\Delta = \frac{\Delta}{3}\left(\sum_{i=1}^{m}c_i - \sum_{i=1}^{n}p_i\right) \tag{5.166}$$

$$(r=4) \quad \sigma_\Delta = \frac{\Delta}{5}\left(\sum_{i=1}^{m}c_i - \sum_{i=1}^{n}p_i\right) \tag{5.167}$$

To compare the relative errors between the various sampled-data models, we use two possible choices for the normalizing transfer function, namely, ESD or the approximate model. This leads to two $\mathcal{H}_\infty$ relative error functions

$$R_1^i(\Delta) = \left\| \frac{G_q^i(z) - G_q^{\text{ESD}}(z)}{G_q^{\text{ESD}}(z)} \right\|_\infty \tag{5.168}$$

$$R_2^i(\Delta) = \left\| \frac{G_q^i(z) - G_q^{\text{ESD}}(z)}{G_q^i(z)} \right\|_\infty \tag{5.169}$$

where the superscript i refers to the model types SDR, ASZ, and CSZ. The error function $R_2(\Delta)$ is closely related to control where relative errors of this type appear in robustness analysis [9].

The key result of this section is described in the following theorem.

Theorem 5.3: [10]

The relative error performance of the different discrete models is as follows:

	Relative error	r: odd	r: even
Normalizing	$R_1^{SDR}(\Delta)$	$\mathcal{O}(1)$	$\mathcal{O}(1/\Delta)$
by ESD	$R_1^{ASZ}(\Delta)$	$\mathcal{O}(\Delta)$	$\mathcal{O}(1)$
$G_q^{ESD}(z)$	$R_1^{CSZ}(\Delta)$	$\mathcal{O}(\Delta)$	$\mathcal{O}(\Delta)$
Normalizing	$R_2^{SDR}(\Delta)$	$\mathcal{O}(1)$	$\mathcal{O}(1)$
by	$R_2^{ASZ}(\Delta)$	$\mathcal{O}(\Delta)$	∞
$G_q^i(z)$	$R_2^{CSZ}(\Delta)$	$\mathcal{O}(\Delta)$	$\mathcal{O}(\Delta)$

Note that, for odd relative degree, the relative error is of order Δ for both the ASZ and CSZ models. However, for even relative degree, we need to use the CSZ model to obtain relative errors of order Δ. We illustrate these ideas by a simple example.

Example 5.12:

In this example we consider a third-order system with one finite zero. Such structure has also been considered in the previous examples but, in this case, we will express it as the more general transfer function:

$$G(s) = \frac{K(s - c_1)}{(s - p_1)(s - p_2)(s - p_3)} \tag{5.170}$$

Under the ZOH-input assumption, we discretize Equation 5.170 to obtain the exact sampled-data model (compare with Examples 5.8 and 5.9):

$$G_q^{ESD}(z) = K \frac{b_2(\Delta)z^2 + b_1(\Delta)z + b_0(\Delta)}{(z - e^{p_1\Delta})(z - e^{p_2\Delta})(z - e^{p_3\Delta})} \tag{5.171}$$

where the coefficients $b_\ell(\Delta)$ depend on the system coefficients and the sampling period. The exact transfer function 5.171 can also be expressed in the γ-domain, corresponding to the use of operator δ, as (compare with Examples 5.10 and 5.11)

$$G_\delta^{ESD}(\gamma) = K \frac{\beta_0(\Delta) + \beta_1(\Delta)\gamma + + \beta_2(\Delta)\gamma^2}{\left(\gamma - \dfrac{e^{p_1\Delta} - 1}{\Delta}\right)\left(\gamma - \dfrac{e^{p_2\Delta} - 1}{\Delta}\right)\left(\gamma - \dfrac{e^{p_3\Delta} - 1}{\Delta}\right)} \tag{5.172}$$

We also have that

$$G_q^{SDR}(z) = \frac{K\left(\dfrac{z-1}{\Delta} - c_1\right)}{\left(\dfrac{z-1}{\Delta} - p_1\right)\left(\dfrac{z-1}{\Delta} - p_2\right)\left(\dfrac{z-1}{\Delta} - p_3\right)} \tag{5.173}$$

This model is equivalent to replacing derivatives by the forward Euler operator in Equation 5.170. The equivalent δ-domain form is given by

$$G_\delta^{SDR}(\gamma) = \frac{K(\gamma - c_1)}{(\gamma - p_1)(\gamma - p_2)(\gamma - p_3)} \tag{5.174}$$

Also,

$$G_q^{ASZ}(z) = \frac{K(z+1)\left(\frac{z-1}{\Delta} - c_1\right)}{2\left(\frac{z-1}{\Delta} - p_1\right)\left(\frac{z-1}{\Delta} - p_2\right)\left(\frac{z-1}{\Delta} - p_3\right)} \qquad (5.175)$$

or, equivalently,

$$G_\delta^{ASZ}(\gamma) = \frac{K\left(\gamma - c_1\right)\left(1 + \gamma\frac{\Delta}{2}\right)}{\left(\gamma - p_1\right)\left(\gamma - p_2\right)\left(\gamma - p_3\right)} \qquad (5.176)$$

Finally, we have that

$$G_q^{CSZ}(z) = \frac{N^{CSZ}(z)}{D^{CSZ}(z)} \qquad (5.177)$$

where

$$N^{CSZ}(z) = K\left(z + 1 + \frac{\Delta(-c_1 + p_1 + p_2 + p_3)}{3}\right)\left(\frac{z-1}{\Delta} - c_1\right) \qquad (5.178)$$

$$D^{CSZ}(z) = \left(\frac{z-1}{\Delta} - p_1\right)\left(\frac{z-1}{\Delta} - p_2\right)\left(\frac{z-1}{\Delta} - p_3\right)\left(2 + \frac{\Delta(-c_1 + p_1 + p_2 + p_3)}{3}\right) \qquad (5.179)$$

The equivalent δ-domain form is in this case

$$G_\delta^{CSZ}(\gamma) = \frac{K\left(1 + \gamma\frac{\Delta}{2 + \frac{\Delta(-c_1 + p_1 + p_2 + p_3)}{3}}\right)(\gamma - c_1)}{\left(\gamma - p_1\right)\left(\gamma - p_2\right)\left(\gamma - p_3\right)} \qquad (5.180)$$

We compute the relative error between the ESD model 5.172 and the three *approximate* sampled-data models 5.174, 5.176, and 5.180 via $R_2^i(\Delta)$ (i.e., normalizing by $G_q^i(z)$). To be specific, we choose $K = 6, c_1 = -5, p_1 = -2, p_2 = -3$, and $p_3 = -4$, as in the previous examples.

The relative errors are shown in Figure 5.10, for three different sampling periods: $\Delta = 0.1, 0.01$, and 0.001. From this figure, we can clearly see the relative error of the Euler model 5.174 is of the order of 1, whereas for models that include the corrected sampling zero (models in 5.176 and 5.180) the relative error decreases as the sampling period decreases (a factor of 0.1 is equivalent to -20 dB).

A surprising observation from the above example is that the Euler model (i.e., SDR) gives the smallest relative errors up to a frequency which is about ten times the open-loop poles and zeros. Thus, provided one samples quickly but restricts the bandwidth to about 10 times the location of open-loop poles and zeros, then one can use simple Euler models with confidence. At higher frequencies, the relative error of Euler models converges to order 1 when the relative degree is even. On the other hand, the model using asymptotic sampling zeros gives good performance up to the vicinity of the folding frequency at which time the relative error diverges to ∞ for even relative degree. The model with corrected asymptotic sampling zero has relative errors that are of the order of Δ in all cases.

5.3.7 Stochastic Systems

We next consider the case of sampled stochastic linear systems described as

$$y(t) = H(\rho)\dot{v}(t) \qquad (5.181)$$

where $\dot{v}(t)$ is a CTWN input process. These models are sometimes simply called *noise models*. We will show how a sampled-data model can be obtained from Equation 5.181 that is *exact*, in the sense that

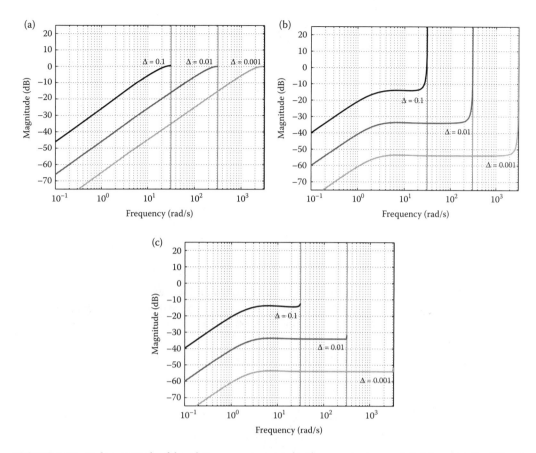

FIGURE 5.10 Bode magnitude of the relative errors associated with approximate sampled-data models for different sampling frequencies in Example 5.12. (a) SDR model; (b) ASZ model; and (c) CSZ model.

the second-order properties (i.e., spectrum) of its output sequence are the same as the second-order properties of the output of the continuous-time system at the sampling instants.

We first review the relationship between the spectrum of a continuous-time process and the associated discrete-time spectrum of the sequence of samples. We then briefly discuss the difficulties that may arise when dealing with a white-noise process in continuous time. Next we show how sampled-data models can be obtained for system 5.181. Finally, we characterize the asymptotic sampling zeros that appear in the sampled spectrum in a similar way as we did for the deterministic case in previous sections.

5.3.8 Spectrum of a Sampled Process

We consider a stationary continuous-time stochastic process $y(t)$, with zero mean and covariance function:

$$r_y(\tau) = E\{y(t + \tau)y(t)\} \tag{5.182}$$

The associated spectral density, or *spectrum*, of this process is given by the Fourier transform of the covariance function 5.182, that is,

$$\Phi_y(\omega) = \mathcal{F}\left\{r_y(\tau)\right\} = \int_{-\infty}^{\infty} r_y(\tau)e^{j\omega\tau}\,d\tau; \quad \omega \in \left(-\infty, \infty\right) \tag{5.183}$$

If we instantaneously sample the continuous-time signal, with sampling period Δ, we obtain the sequence $y_k = y(k\Delta)$. The covariance of this sequence, $r_y^d[\ell]$, is equal to the continuous-time signal covariance *at the sampling instants*:

$$r_y^d[\ell] = E\{y_{k+\ell}\,y_k\} = E\{y(k\Delta + \ell\Delta)\,y(k\Delta)\} = r_y(\ell\Delta) \tag{5.184}$$

The power spectral density of the sampled signal is given by the discrete-time fourier transform (DTFT) of the covariance function, namely:

$$\Phi_y^d(\omega) = \Delta \sum_{k=-\infty}^{\infty} r_y^d[k]e^{-j\omega k\Delta}; \quad \omega \in \left[\frac{-\pi}{\Delta}, \frac{-\pi}{\Delta}\right] \tag{5.185}$$

Remark 5.7

Note that we have used the DTFT as defined in [2], which includes the sampling period Δ as a scaling factor. We have included this factor because the DTFT defined in this fashion converges to the continuous-time Fourier transform as the sampling period Δ goes to zero.

Remark 5.8

The continuous- and discrete-time spectral densities, in Equations 5.183 and 5.185, respectively, are *real* functions of the frequency ω. However, to make the connections to the deterministic case apparent, we will sometimes express the continuous-time spectra in terms of the complex variable $s = j\omega$, that is,

$$\Phi_y(\omega) = \bar{\Phi}_y(j\omega) = \bar{\Phi}_y(s)\big|_{s=j\omega} \quad \text{(CT spectrum)} \tag{5.186}$$

Similarly, we express the discrete-time spectrum in terms of $z = e^{j\omega\Delta}$ or $\gamma = \gamma_\omega = \frac{e^{j\omega\Delta}-1}{\Delta}$, for shift and delta operator models, respectively, that is,

$$\Phi_y^d(\omega) = \Phi_y^q(e^{j\omega\Delta}) = \Phi_y^\delta(\gamma_\omega) \tag{5.187}$$

where

$$\Phi_y^q(e^{j\omega\Delta}) = \Phi_y^q(z)\big|_{z=e^{j\omega\Delta}} \quad (q\text{-domain DT spectrum}) \tag{5.188}$$

$$\Phi_y^\delta(\gamma_\omega) = \Phi_y^\delta(\gamma)\big|_{\gamma=\frac{e^{j\omega\Delta}-1}{\Delta}} \quad (\delta\text{-domain DT spectrum}) \tag{5.189}$$

The following lemma relates the spectrum of the sampled sequence to the spectrum of the original continuous-time process.

Lemma 5.8:

Let us consider a stochastic process $y(t)$, with spectrum given by Equation 5.183, together with its sequence of samples $y_k = y(k\Delta)$, with discrete-time spectrum given by Equation 5.185. Then the following relationship holds:

$$\Phi_y^d(\omega) = \sum_{\ell=-\infty}^{\infty} \Phi_y\left(\omega + \frac{2\pi}{\Delta}\ell\right) \tag{5.190}$$

Equation 5.190 reflects the well-known consequence of the sampling process: the aliasing effect. For deterministic systems, an analogous result was presented earlier in Equation 5.124. In the stochastic case considered here, the discrete-time spectrum is obtained by folding high-frequency components of the continuous-time spectrum back onto the range $(0, \frac{\pi}{\Delta})$.

5.3.9 Continuous-Time White Noise

The input $\dot{v}(t)$ to system 5.181 is modeled as zero mean *white-noise* process in continuous time. This means that it is a stochastic process that satisfies the following two conditions:

i. $E\{\dot{v}(t)\} = 0$, for all t and
ii. $\dot{v}(t)$ is independent of $\dot{v}(s)$, that is, $E\{\dot{v}(t)\dot{v}(s)\} = 0$, for all $t \neq s$.

Loosely, we can model a continuous stochastic process in state-space form as follows:

$$\frac{dx(t)}{dt} = Ax(t) + B\dot{v}(t) \tag{5.191}$$

$$y(t) = Cx(t) \tag{5.192}$$

where the system state vector is $x(t) \in \mathbb{R}^n$, the matrices are $A \in \mathbb{R}^{n \times n}$ and $B, C^T \in \mathbb{R}^n$, and the input $\dot{v}(t)$ is a CTWN process with (constant) spectral density σ_v^2. However, if we look for a stochastic process with continuous paths that satisfies the two conditions (i) and (ii), this happens to be equal to zero in the mean square sense, that is, $E\{\dot{v}(t)^2\} = 0$, for all t. This suggests that difficulties will arise since the process $\dot{v}(t)$ does not exist in a meaningful sense. However, we can circumvent these difficulties by expressing Equation 5.191 in incremental form as a stochastic differential equation:

$$dx = A\,x\,dt + B\,dv \tag{5.193}$$

Remark 5.9

Note that the connections between this model and the deterministic incremental model given in Remark 5.1.

The process $v(t)$ corresponds to a *Wiener* process and has the following properties:

i.   It has zero mean, that is, $E\{v(t)\} = 0$, for all t;
ii. Its increments are independent, that is, $E\{(v(t_1) - v(t_2))(v(s_1) - v(s_2))\} = 0$, for all $t_1 > t_2 > s_1 > s_2 \geq 0$; and
iii. For every s and t, $s \leq t$, the increments $v(t) - v(s)$ have a Gaussian distribution with zero mean and variance $E\{(v(t) - v(s))^2\} = \sigma_v^2|t - s|$.

This process is not differentiable anywhere. However, the CTWN, $\dot{v}(t)$, formally defined as the derivative of $v(t)$ is a useful heuristic device in the linear case. Note that the third condition above implies that CTWN will have infinite variance:

$$E\{dv\,dv\} = E\{(v(t + dt) - v(t))^2\} = \sigma_v^2\,dt \implies E\{\dot{v}^2\} = \infty \tag{5.194}$$

Remark 5.10

A CTWN process is a mathematical abstraction, but it can be approximated to any desired degree of accuracy by conventional stochastic processes with broadband spectra.

We now give two alternative interpretations of σ_v^2:

i.   Equation 5.194 suggests that one may consider σ_v^2 as the *incremental variance* of the Wiener process $v(t)$. Moreover, we can think of $\dot{v}(t)$ as a *generalized* process, introducing a Dirac delta function to define its covariance structure:

$$r_{\dot{v}}(t - s) = E\{\dot{v}(t)\,\dot{v}(s)\} = \sigma_v^2\delta(t - s) \tag{5.195}$$

ii. In the frequency-domain, $\sigma_{\dot{v}}^2$ can be thought of as the *power spectral density* of $\dot{v}(t)$ [2]. Indeed, from 5.195 we have that the spectral density satisfies

$$\Phi_{\dot{v}}(\omega) = \int_{-\infty}^{\infty} r_{\dot{v}}(\tau) e^{-j\omega\tau} \, d\tau = \sigma_{\dot{v}}^2 \quad \forall \omega \in (-\infty, \infty) \tag{5.196}$$

We see that the spectral density of $\dot{v}(t)$ is constant for all ω, which corresponds to the usual heuristic notion of white noise.

For simplicity of exposition, in the sequel, we will use models 5.191 and 5.192.

5.3.10 Stochastic Sampled-Data Models

For the moment we will assume that the process $y(t)$ in Equation 5.181 does not contain any unfiltered white-noise components. In practice, this can be guaranteed by the use of an antialiasing filter. As a consequence, we assume that $H(\rho)$ in Equation 5.181 is a strictly proper transfer function that can be represented in state-space form as in Equations 5.191 and 5.192. The following result gives the appropriate sampled-data model when considering instantaneous sampling of the output 5.192.

Lemma 5.9: [2]

Consider the stochastic system defined in state-space forms (Equations 5.191 and 5.192) where the input $\dot{v}(t)$ is a CTWN process with (constant) spectral density $\sigma_{\dot{v}}^2$. When the output $y(t)$ is instantaneously sampled, with sampling period Δ, an equivalent discrete-time model is given by

$$\delta x_k = A_i x_k + v_k \tag{5.197}$$

$$y_k = C x_k \tag{5.198}$$

where $A_i = (e^{A\Delta} - I_n)/\Delta$, and the sequence v_k is a discrete-time white-noise (DTWN) process, with zero mean and covariance structure given by

$$E\{v_k v_\ell^T\} = \Omega_i \frac{\delta_K[k-\ell]}{\Delta} \tag{5.199}$$

where:

$$\Omega_i = \frac{\sigma_{\dot{v}}^2}{\Delta} \int_0^{\Delta} e^{A\eta} BB^T e^{A^T \eta} \, d\eta \tag{5.200}$$

and $\delta_K(k)$ is the Kronecker delta given by

$$\delta_K[k] = \begin{cases} 1, & k = 0 \\ 0, & k \neq 0 \end{cases} \tag{5.201}$$

Remark 5.11

The matrix Ω_i is in fact the (constant) spectral density of the noise vector v_k, as can be seen by applying the discrete-time Fourier transform to Equation 5.199:

$$\mathcal{F}_d \left\{ \Omega_i \frac{\delta_K[k]}{\Delta} \right\} = \Delta \sum_{k=-\infty}^{\infty} \Omega_i \frac{\delta_K[k]}{\Delta} e^{-j\omega k \Delta} = \Omega_i; \quad \omega \in \left[-\frac{\pi}{\Delta}, \frac{\pi}{\Delta} \right] \tag{5.202}$$

Remark 5.12

Note that the previous result allows us to recover the continuous-time stochastic description (Equation 5.191), as the sampling period Δ goes to zero. In particular, the covariance (Equation 5.199) corresponds (in continuous-time) to the covariance of the vector process $B\dot{v}(t)$ in model 5.191, as it can be readily seen on noting that

$$\lim_{\Delta \to 0} \Omega_i = \sigma_v^2 BB^T = \Omega_c \tag{5.203}$$

$$\lim_{\Delta \to 0} \frac{1}{\Delta} \delta_K[k - \ell] = \delta(t_k - t_\ell) \tag{5.204}$$

Given the continuous-time system, Lemma 5.9 provides a sampled-data model expressed in terms of the δ-operator. A corresponding shift operator model can readily be obtained by rewriting Equation 5.197 as

$$q\, x_k = x_{k+1} = A_q\, x_k + \tilde{v}_k \tag{5.205}$$

where $\tilde{v}_k = v_k\,\Delta$ and, as before, $A_q = 1 + A_i\Delta$. Note that, for this model, the covariance structure of the noise sequence is given by

$$E\{\tilde{v}_k\, \tilde{v}_\ell^T\} = \Delta^2 E\{v_k\, v_\ell^T\} = \Delta\, \Omega_i\, \delta_K[k - \ell] = \Omega_q\, \delta_K[k - \ell] \tag{5.206}$$

where we have defined $\Omega_q = \Omega_i\,\Delta$.

Remark 5.13

The matrix Ω_q in Equation 5.206 can be computed by solving the discrete-time Lyapunov equation

$$\Omega_q = P - A_q P A_q^T \tag{5.207}$$

or, equivalently, in the δ-domain:

$$\Omega_i = A_i P + P A_i^T + \Delta A_i P A_i^T \tag{5.208}$$

where P satisfies the continuous-time Lyapunov equation $AP + PA^T + \Omega_c = 0$, for stable systems, or $AP + PA^T - \Omega_c = 0$, for anti stable systems. For Lemma 5.9 we have, in particular, $\Omega_c = \sigma_v^2 BB^T$.

The sampled-data models (Equations 5.197 and 5.198) are driven by a vector white-noise process v_k. The covariance of this process is determined by the matrix Ω_i in Equation 5.200, which will generically be *full rank*. However, we can gain additional insights by describing the sampled process $y_k = y(k\Delta)$ as the output of an equivalent sampled-data model driven by a *single* scalar noise source. This can be achieved by, first, obtaining the discrete-time spectrum of the sampled sequence y_k, and then performing *spectral factorization*.

The output spectrum of the sampled-data model is given in the following result.

Lemma 5.10:

The output spectrum $\Phi_y^d(\omega)$ of the sampled-data models 5.197 and 5.198 is

$$\Phi_y^\delta(\gamma_\omega) = C(\gamma_\omega I_n - A_i)^{-1} \Omega_i (\gamma_\omega^* I_n - A_i^T)^{-1} C^T \tag{5.209}$$

where $\gamma_\omega = \frac{1}{\Delta}(e^{j\omega\Delta} - 1)$ and $*$ denote complex conjugation. Using Equation 5.205, this spectrum can be equivalently obtained as

$$\Phi_y^q(e^{j\omega\Delta}) = \Delta C(e^{j\omega\Delta}I_n - A_q)^{-1}\Omega_q(e^{-j\omega\Delta}I_n - A_q^T)^{-1}C^T \tag{5.210}$$

We can spectrally factor 5.210 to obtain a model driven by a single noise source. In the sequel, we will use this idea to study asymptotic sampling zeros as was done earlier for deterministic systems.

Next we present examples showing how stochastic sampled-data models can be obtained utilizing the above ideas.

Example 5.13:

We consider the first-order continuous-time auto-regressive (CAR) system

$$\frac{dy(t)}{dt} - a_0 y(t) = b_0 \dot{v}(t) \tag{5.211}$$

where $a_0 < 0$ and $\dot{v}(t)$ is a CTWN process of unitary spectral density, that is, $\sigma_v^2 = 1$. A suitable state-space model can readily be obtained as

$$\frac{dx(t)}{dt} = a_0 x(t) + b_0 \dot{v}(t) \tag{5.212}$$

$$y(t) = x(t) \tag{5.213}$$

An equivalent sampled-data model for this system is readily obtained in terms of the shift operator q or, equivalently, using the *delta* operator:

$$q x_k = e^{a_0 \Delta} x_k + \tilde{v}_k \quad \Longleftrightarrow \quad \delta x_k = \left(\frac{e^{a_0 \Delta} - 1}{\Delta}\right) x_k + v_k \tag{5.214}$$

$$y_k = x_k \tag{5.215}$$

where $\tilde{v}_k$ and v_k are DTWN processes with variance Ω_q and $\frac{\Omega_i}{\Delta}$, respectively. Note that these variances are not very useful when considering the sampling period Δ tending to zero. If we compute them, for example, using Remark 5.13, we can see that they are badly scaled:

$$\Omega_q = \Delta\Omega_i = b_0^2 \frac{(e^{2a_0\Delta} - 1)}{2a_0} \xrightarrow{\Delta \to 0} 0 \tag{5.216}$$

$$\frac{\Omega_i}{\Delta} = b_0^2 \frac{(e^{2a_0\Delta} - 1)}{2a_0\Delta^2} \xrightarrow{\Delta \to 0} \infty \tag{5.217}$$

On the other hand, as noticed in Remark 5.12, the *spectral density* Ω_i converges naturally to its continuous-time counterpart:

$$\Omega_i = b_0^2 \frac{(e^{2a_0\Delta} - 1)}{2a_0\Delta} \xrightarrow{\Delta \to 0} b_0^2 \tag{5.218}$$

$\square$

In the previous example, a stochastic sampled-data model was immediately obtained having a single scalar noise source. For higher-order systems, Lemma 5.9 gives a sampled-data model in terms of a vector input v_k. However, as described above, we can use spectral factorization to obtain an *equivalent* sampled-data model, with a single scalar noise source as input, using spectral factorization. The output of this system has the same second-order statistics, that is, the same discrete-time spectrum 5.210, as the original sampled-data model.

Example 5.14:

Consider the second-order CAR system

$$\frac{d^2y(t)}{dt} + a_1\frac{dy(t)}{dt} + a_0 y(t) = \dot{v}(t) \tag{5.219}$$

where $\dot{v}(t)$ is CTWN process of unitary spectral density, that is, $\sigma_v^2 = 1$. An appropriate state-space model is given by

$$\frac{dx(t)}{dt} = \begin{bmatrix} 0 & 1 \\ -a_0 & -a_1 \end{bmatrix} x(t) + \begin{bmatrix} 0 \\ 1 \end{bmatrix}\dot{v}(t) \tag{5.220}$$

$$y(t) = \begin{bmatrix} 1 & 0 \end{bmatrix} x(t) \tag{5.221}$$

Using Equation 5.210, and after some lengthy calculations, we see that the discrete-time output spectrum has the form

$$\Phi_y^q(z) = K\frac{z(b_2 z^2 + b_1 z + b_0)}{(z - e^{\lambda_1\Delta})(z - e^{\lambda_2\Delta})(1 - e^{\lambda_1\Delta}z)(1 - e^{\lambda_2\Delta}z)} \tag{5.222}$$

where λ_1 and λ_2 are the continuous-time system poles, and

$$b_2 = (\lambda_1 - \lambda_2)\left[e^{(\lambda_1+\lambda_2)\Delta}(\lambda_2 e^{\lambda_1\Delta} - \lambda_1 e^{\lambda_2\Delta}) + \lambda_1 e^{\lambda_1\Delta} - \lambda_2 e^{\lambda_2\Delta} \right] \tag{5.223}$$

$$b_1 = \left[(\lambda_1 + \lambda_2)(e^{2\lambda_1\Delta} - e^{2\lambda_2\Delta}) + (\lambda_1 - \lambda_2)(e^{2(\lambda_1+\lambda_2)\Delta} - 1) \right] \tag{5.224}$$

$$b_0 = b_2 \tag{5.225}$$

$$K = \frac{\Delta}{2\lambda_1\lambda_2(\lambda_1 - \lambda_2)^2(\lambda_1 + \lambda_2)} \tag{5.226}$$

If we perform spectral factorization on the sampled spectrum 5.222 we can obtain a sampled-data model in terms of only one noise source, that is,

$$\Phi_y^q(z) = H_q(z)H_q(z^{-1}) \tag{5.227}$$

where

$$H_q(z) = \frac{\sqrt{K}(c_1 z + c_0)}{(z - e^{\lambda_1\Delta})(z - e^{\lambda_2\Delta})} \tag{5.228}$$

The expression for the numerator coefficients (and, thus, of the only sampling zero) of the latter discrete-time model are involved. However, it is possible to obtain an asymptotic characterization of this sampled-data model as the sampling period goes to zero, in a similar fashion as was done for the deterministic case. We will explore this idea in the next section.

5.3.11 Asymptotic Sampling Zeros for Stochastic Systems

In the previous section we have seen that the output spectrum of the sampled-data model contains *sampling zeros* which have no counterpart in the underlying continuous-time system. Similar to the deterministic case, these zeros can be asymptotically characterized. The following result characterizes the asymptotic sampling zeros of the output spectrum in the case of instantaneous sampling.

Lemma 5.11: [11]

Consider the instantaneous sampling of the continuous-time process (Equation 5.181). We then have that

$$\Phi_y^d(\omega) \xrightarrow{\Delta \to 0} \Phi_y(\omega) \tag{5.229}$$

uniformly in s, on compact subsets. Moreover, let $\pm z_i$, $i = 1, \ldots, m$ *be the 2m zeros of* $\Phi_y(s)$, *and* $\pm p_i$, $i = 1, \ldots, n$ *its 2n poles. Then*

  i. $2m$ *zeros of* $\Phi_y^d(z)$ *will converge to 1 as* $e^{\pm z_i \Delta}$;
 ii. *The remaining* $2(n - m) - 1$ *will converge to the zeros of* $z B_{2(n-m)-1}(z)$ *as* Δ *goes to zero;*
iii. *The 2n poles of* $\Phi_y^d(z)$ *equal* $e^{\pm p_i \Delta}$, *and will hence go to 1 as* Δ *goes to zero.*

Example 5.15:

Consider again the second-order CAR system in Example 5.14. The discrete-time spectrum 5.222 was obtained for the case of instantaneous sampling of the output $y(t)$. Exact expressions for the sampling zeros of this spectrum are quite involved. However, performing a Taylor-series expansion of the numerator we have that

$$Kz(b_2 z^2 + b_1 z + b_0) = \frac{\Delta^4}{3!} z(z^2 + 4z + 1) + \mathcal{O}(\Delta^5) \tag{5.230}$$

which, asymptotically, as Δ goes to zero, is consistent with Lemma 5.11, noting that $B_3(z) = z^2 + 4z + 1$ as in Equation 5.136.

The asymptotic sampled spectrum can be obtained as

$$
\begin{aligned}
\Phi_y^q(z) &= \frac{\Delta^4}{6} \frac{(z + 4 + z^{-1})}{(z - e^{\lambda_1 \Delta})(z - e^{\lambda_2 \Delta})(z^{-1} - e^{\lambda_1 \Delta})(z^{-1} - e^{\lambda_2 \Delta})} \\
&= \frac{\Delta^4}{6(2 - \sqrt{3})} \frac{(z + 2 - \sqrt{3})}{(z - e^{\lambda_1 \Delta})(z - e^{\lambda_2 \Delta})} \frac{(z^{-1} + 2 - \sqrt{3})}{(z^{-1} - e^{\lambda_1 \Delta})(z^{-1} - e^{\lambda_2 \Delta})}
\end{aligned}
\tag{5.231}
$$

Then, the spectrum can be written as $\Phi_y^q(z) = H_q(z) H_q(z^{-1})$, where

$$H_q(z) = \frac{\Delta^2}{3 - \sqrt{3}} \frac{(z + 2 - \sqrt{3})}{(z - e^{\lambda_1 \Delta})(z - e^{\lambda_2 \Delta})} \tag{5.232}$$

The corresponding δ-operator model can be obtained by changing variable $z = 1 + \gamma \Delta$. This yields the following discrete-time model:

$$H_i(\gamma) = \frac{\left(1 + \frac{1}{3 - \sqrt{3}} \Delta \gamma\right)}{\left(\gamma - \frac{e^{\lambda_1 \Delta} - 1}{\Delta}\right)\left(\gamma - \frac{e^{\lambda_2 \Delta} - 1}{\Delta}\right)} \tag{5.233}$$

which clearly converges to the underlying continuous-time system 5.219, as the sampling period goes to zero. $\qquad\square$

5.3.12 Model Error Quantification for Stochastic Systems

In this section we study the stochastic version of model errors results described in Section 5.3.6 for the case of deterministic inputs. We have seen that the discrete-time spectrum is given by an infinite sum (see Equation 5.190). It can be shown (see, e.g. [12]) that the sum in Equation 5.190 can also be calculated by using the Hurwitz-Zeta function given by

$$\zeta(r, a) = \sum_{k=0}^{\infty} \frac{1}{(k + a)^r}, \quad \text{Re}\{r\} > 1, \ a \notin \mathbb{Z}_0^- \tag{5.234}$$

This can be used to develop alternative expressions to those given in Equation 5.138. For simple cases, it is possible to find a closed-form expression for the discrete-time spectrum. Indeed, we can readily establish the following result.

Lemma 5.12:

When the continuous spectrum is given by $\Psi_r^c(s) = 1/s^r$, where r is an arbitrary positive integer, then we see that the corresponding discrete-time spectrum is given by*

$$\Psi_r^d(e^{j\omega\Delta}) = \frac{\Delta^r}{(r-1)!} \frac{z B_{r-1}(z)}{(z-1)^r} \tag{5.235}$$

Lemma 5.12 is of particular interest in analyzing sampled-data models since, for high frequencies, all (finite-dimensional) systems behave as $1/s^r$, where r is the relative degree of the continuous-time spectrum of interest. In fact, for some simple cases, we have that the discrete-time spectrum corresponding to $1/s^r$ is given by

$$\Psi_{r=2}^d(e^{j\omega\Delta}) = \left[\frac{\left(\frac{\Delta}{2}\right)}{\sin\left(\frac{\omega\Delta}{2}\right)} \right]^2 \tag{5.236}$$

$$\Psi_{r=3}^d(e^{j\omega\Delta}) = \cos\left(\frac{\omega\Delta}{2}\right) \left[\frac{\left(\frac{\Delta}{2}\right)}{\sin\left(\frac{\omega\Delta}{2}\right)} \right]^3 \tag{5.237}$$

$$\Psi_{r=4}^d(e^{j\omega\Delta}) = \frac{1}{3}\left[1 + 2\cos^2\left(\frac{\omega\Delta}{2}\right)\right] \left[\frac{\left(\frac{\Delta}{2}\right)}{\sin\left(\frac{\omega\Delta}{2}\right)} \right]^4 \tag{5.238}$$

$$\Psi_{r=5}^d(e^{j\omega\Delta}) = \frac{1}{3}\cos\left(\frac{\omega\Delta}{2}\right)\left[2 + \cos^2\left(\frac{\omega\Delta}{2}\right)\right] \left[\frac{\left(\frac{\Delta}{2}\right)}{\sin\left(\frac{\omega\Delta}{2}\right)} \right]^5 \tag{5.239}$$

The above results give very simple expressions for the discrete spectrum corresponding to $\Phi^c(s) = \Psi_r^c(s) = 1/s^r$. These results raise the question as to whether or not there exist simple relationships between $\Phi^d(e^{j\omega\Delta})$ and $\Phi^c(j\omega)$ in more general cases. The following result gives a surprisingly simple connection, which holds generally when $\Delta \to 0$.

Lemma 5.13: [13]

Assuming that there exists an ω_N such that for $\omega \geq \omega_N$ we have that $|\Phi^c(j\omega)| \leq \frac{\beta}{\omega^2}$, then

$$|\Phi^d(e^{j\omega\Delta}) - \Phi^c(j\omega)| = O(\Delta^2), \quad 0 \leq \omega \leq \frac{\pi}{\Delta} \tag{5.240}$$

The result in Lemma 5.13 is valid for general systems. However, there is a subtle caveat. Specifically, for high frequencies, the continuous-time spectrum also satisfies $\Phi^c(j\omega) = O(\Delta^2)$. This means that the continuous- and discrete-time spectra are not necessarily close to each other at high frequencies, in the sense of the *relative error* being small. To illustrate the difficulty, Figure 5.11 shows the relative error given by

$$R(r, \omega) = \left| \frac{\Phi^d(e^{j\omega\Delta}) - \Phi^c(j\omega)}{\Phi^c(j\omega)} \right| \tag{5.241}$$

for $\Phi^c(s) = \Psi_r^c(s) = 1/s^r$ and different values of r. It can be seen that the relative error between continuous and discrete spectra is certainly not small for frequencies close to $\omega = \pi/\Delta$.

---

* Here and in the sequel, we use Ψ_r^c for a particular spectrum parametrized in terms of r. Ψ_r^d represents the corresponding discrete-time spectrum of Ψ_r^c.

Thus, Lemma 5.13 does not adequately achieve the desired objective of connecting continuous and discrete spectra. This raises a follow-up question as to whether there exist functions, closely related to $\Phi^c(j\omega)$, which are near to $\Phi^d(e^{j\omega\Delta})$ in the sense of small *relative errors*. The following result is immediate.

Lemma 5.14: [13]

Assuming that for $\omega \geq \omega_N$ we have that $\frac{\alpha}{\omega^p} \leq |\Phi^c(j\omega)| \leq \frac{\beta}{\omega^2}$ ($p \geq 2$), then for finite bandwidth ($\omega \leq \omega_B < \pi/\Delta$) we have that

$$\left| \frac{\Phi^d(e^{j\omega\Delta}) - \Phi^c(j\omega)}{\Phi^c(j\omega)} \right| = O(\Delta^2) \tag{5.242}$$

Lemma 5.14 does not hold for frequencies in the vicinity of π/Δ. Indeed, this has already been illustrated for a special case in Figure 5.11. We are thus motivated to ask a further question: Are there spectra related to $\Phi^c(j\omega)$, which are close to $\Phi^d(e^{j\omega\Delta})$ (in the sense of relative error) over the full range ($0 < \omega \leq \pi/\Delta$)? We explore this question below.

Let the continuous-time spectrum be given by

$$\Phi^c(s) = K_0 \frac{\prod_{i=1}^{m}(s - z_i)}{\prod_{i=1}^{n}(s - p_i)}, \quad r = n - m \tag{5.243}$$

Some candidate approximations to the associated discrete-time spectrum are given by

- *Model 1:*

$$\overline{\Phi_1^d}(e^{j\omega\Delta}) = \frac{\Delta^r}{(r-1)!} \frac{e^{j\omega\Delta} B_{r-1}(e^{j\omega\Delta})}{(e^{j\omega\Delta} - 1)^r} \tag{5.244}$$

This model is exact when $\Phi^c(s) = 1/s^r$.

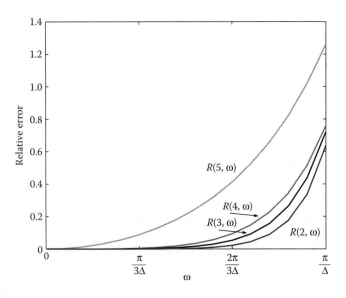

FIGURE 5.11 $R(r, \omega)$ for different values of r.

- *Model 2:*

$$\overline{\Phi_2^d}(e^{j\omega\Delta}) = \Phi^c(j\omega)\frac{e^{j\omega\Delta}B_{r-1}(e^{j\omega\Delta})}{(r-1)!} \tag{5.245}$$

Note that this model includes the continuous spectrum.
- *Model 3:*

$$\overline{\Phi_3^d}(e^{j\omega\Delta}) = \Phi^c\left(\frac{e^{j\omega\Delta}-1}{\Delta}\right)\frac{e^{j\omega\Delta}B_{r-1}(e^{j\omega\Delta})}{(r-1)!} \tag{5.246}$$

Note that this model uses an Euler approximation for the intrinsic poles and zeros.
- *Model 4:*

$$\overline{\Phi_4^d}(e^{j\omega\Delta}) = \Phi^c\left(\frac{e^{j\omega\Delta}-1}{\Delta}\right)\frac{e^{j\omega\Delta}p_{r-1}(e^{j\omega\Delta})}{(r-1)!} \tag{5.247}$$

where $p_{r-1}(z)$ is the polynomial corresponding to the corrected sampling zeros introduced in Section 5.3.6. This latter model is the same as the previous one when the relative degree of the continuous spectrum, r, is an even number (e.g., for the case of auto-spectrum).
- *Model 5:*

$$\overline{\Phi_5^d}(e^{j\omega\Delta}) = K_5\frac{\prod_{i=1}^m(e^{j\omega\Delta}-e^{z_i\Delta})}{\prod_{i=1}^n(e^{j\omega\Delta}-e^{p_i\Delta})}\frac{e^{j\omega\Delta}p_{r-1}(e^{j\omega\Delta})}{(r-1)!} \tag{5.248}$$

where K_5 is such that $\overline{\Phi_5^d}(1) = \Phi^c(0)$. Note that this is a refined form of Model 4.

We present a simple example to illustrate the errors introduced by the various models described above.

Example 5.16:

Let $\Phi^c(s)$ be given by

$$\Phi^c(s) = \frac{1}{(s+10)^2(-s+10)^2} \quad s = j\omega \tag{5.249}$$

We choose $\Delta = 0.01$ and we compare the following models:

- *Model 0:* The true discrete spectrum.
- *Models 1 to 5* as described above. (Note that Model 4 is identical to Model 3 in this case.)
- *Model 6:* The continuous spectrum.

Figure 5.12a shows the various spectra. Note that it is virtually impossible to distinguish Models 0, 2, 3, 5, and 6 on this scale. The only model which shows any discernible difference is Model 1 which is clearly only valid at high frequencies. This is to be expected since no attempt was made to accurately map the intrinsic poles and zeros.

The above observations are consistent with Lemma 5.13. More informative results are shown in Figure 5.12b. This shows the relative error with respect to the true discrete spectrum, that is,

$$\rho_k(e^{j\omega\Delta}) = \left|\frac{\Phi^d(e^{j\omega\Delta}) - \overline{\Phi_k^d}(e^{j\omega\Delta})}{\Phi^d(e^{j\omega\Delta})}\right| \tag{5.250}$$

Note that a relative error of 10^0 implies 100% errors. Again, we see that Model 1 is only useful at high frequencies. We also see that Model 6 only gives small errors over a limited bandwidth. Perhaps, more surprisingly, we see that Model 2 (which corresponds to the continuous spectrum modified

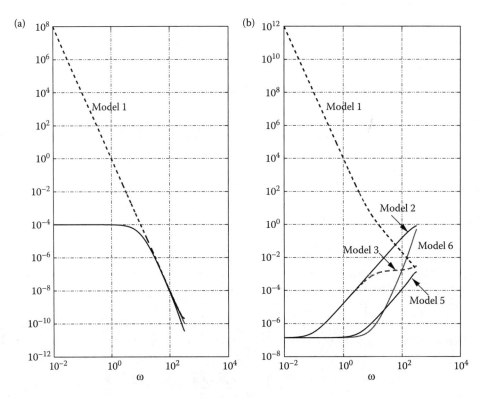

FIGURE 5.12 (a) Magnitude of the spectra. (b) Relative errors.

by the sampling zeros) is also only valid over a limited range. The only models which give small relative errors over the complete frequency range are Model 3 (which maps the poles and zeros using $1 + \zeta\Delta$) and Model 5 (which maps the poles and zeros using $e^{\zeta\Delta}$). Moreover, the performance of these models improves as Δ is decreased. Finally, in terms of the maximal relative error over all frequencies, both Models 3 and 5 perform equally well.

Our conclusion from the previous example is that, in order to obtain an adequate discrete spectrum over the complete frequency range $(0, \pi/\Delta)$, one needs to modify the continuous spectrum in two ways:

i. Map the poles and zeros, either using $e^{\zeta\Delta}$ (Model 5) or using $1 + \zeta\Delta$ (Model 3); and
ii. add appropriate discrete sampling zeros.

If both of these two steps are taken, then a model having relative error of the order of Δ is obtained.

5.3.13 Robustness Issues

The reader will have noticed that the above discussion about sampled-data models is based on capturing the effect of folding. Thus, to obtain approximate models one needs to make hypotheses about the high-frequency behavior of the system. For example, asymptotic sampling zeros follow by applying the assumption that when the sampling frequency is sufficiently high, then the model behaves *above the Nyquist frequency* as $1/s^r$ (where r is the relative degree). Clearly, this begs the question about unmodeled high-frequency poles or zeros. If these are present, then they will clearly destroy the validity of discrete-time models based on the (false) assumption that the continuous-time model is behaving as $1/s^r$.

Thus, one needs to be very careful about the frequency range of validity of models. In particular, sampling zeros correspond to very precise assumptions about how the system behaves above the Nyquist frequency. If one is uncertain about this behavior, then we recommend that one sample as quickly as possible, but then to only use the model up to a fraction of the Nyquist frequency. In this case, one can ignore sampling zeros, folding, etc. Indeed, a very accurate model in terms of either absolute or relative errors, can simply be obtained by replacing s by δ. Of course, this model is *not* accurate (as we have shown in Sections 5.3.6 and 5.3.12) in the vicinity of the Nyquist frequency.

5.3.14 Extensions to Nonlinear Systems

In this section we will make some brief comments about the extension of the results to nonlinear systems. A key departure point from the linear sampled-data models considered in the previous sections, is that, for linear systems, we can, in principle, obtain ESD models, whereas for nonlinear systems, only *approximate* sampled-data models can be obtained (e.g., by using rth-order Runge–Kutta integration). However, the accuracy of these models can be characterized in a precise way.

We have seen above that, in the linear case, the use of Euler integration leads to a discrete-time model having no sampling zeros. If one uses an integration method with an rth-order Taylor' series, where r is the relative degree, then it turns out that this exactly captures the asymptotic sampling zeros in the linear case. This leads to the following conjecture: Say one were to use an rth-order Runge–Kutta method in the nonlinear case, would this reveal anything about *nonlinear sampling zeros*? Actually, this is precisely what happens. Indeed, for deterministic inputs, the resultant approximate sampling *zero dynamics* for a nonlinear system of relative degree r are the same as for a linear system having the same relative degree [7]. Related results hold for stochastic systems [14,15].

5.3.15 Summary

In this chapter we have studied sampled-data models for linear and nonlinear systems. We also discussed the implications, and inherent difficulties, of using sampled-data models, defined at discrete-time instants, to represent real systems evolving in continuous time. The following topics have been covered.

- *Sampling of continuous-time systems:* The sampled-data models obtained were shown to depend, not only on the underlying continuous-time system, but also on the details of the sampling process itself. Specifically, the hold device, used to generate a continuous-time input, and the sampling device, that gives us the output sequence of samples, both play an important role in the sampling process. The effect of these *artifacts* of sampling become negligible when the sampling period goes to zero. However, for any finite sampling rate, their role has to be considered to obtain accurate sampled-data models.

- *Sampling zeros:* Sampled-data models have, in general, more zeros than the underlying continuous-time system. These extra zeros, called *sampling zeros*, have no continuous-time counterpart. For the linear case, their presence can be interpreted as a consequence of the *aliasing* effect of the system frequency response (or spectrum), where high-frequency components are folded back to low frequencies due to the sampling process. We have seen that sampling zeros arise in both deterministic and stochastic systems. They can be asymptotically characterized in terms of the Euler–Fröbenius polynomials.

- *Approximate sampled-data models:* The presence of *sampling zeros* in discrete-time models is an illustration of the inherent difference between continuous- and discrete-time system descriptions. When using δ-operator models, the sampling zeros go to infinity as the sampling period goes to zero; nonetheless they generally have to be taken into account to obtain accurate discrete-time descriptions.

- *Nonlinear systems:* The above ideas can be also extended to the nonlinear case. In fact, the sampled-data models obtained for nonlinear systems contain extra zero dynamics with no counterpart in continuous time. These *sampling zero dynamics* are a consequence of using a more accurate discretization procedure than simple Euler integration. Surprisingly, if an rth-order Taylor series is used for the integration (where r is the relative degree), then the extra *zero dynamics* turn out to be the same as the dynamics associated with the asymptotic sampling zeros in the linear case.

- *Robustness issues:* The use of sampled data taken from continuous-time systems inherently implies a *loss of information*. Even though it is possible to obtain accurate models, there will always exist a gap between the discrete- and continuous-time representations. As a consequence, one needs to rely on *assumptions* on the inter sample behavior of signals or, equivalently, on the characteristics of the system response beyond the sampling frequency. Based on these issues we have stressed the concept of *bandwidth of validity* for continuous-time models, within which assumptions, such as relative degree or *white* noise, can be trusted. We have emphasized the importance of this concept, in particular, when utilizing asymptotic results for sampling zeros.

References

1. R.H. Middleton and G.C. Goodwin. *Digital Control and Estimation. A Unified Approach*. Prentice-Hall, Englewood Cliffs, NJ, 1990.
2. A. Feuer and G.C. Goodwin. *Sampling in Digital Signal Processing and Control*. Birkhäuser, Boston, 1996.
3. K.J. Åström, P. Hagander, and J. Sternby. Zeros of sampled systems. *Automatica*, 20(1):31–38, 1984.
4. B. Mårtensson. Zeros of sampled systems. Master's thesis, Department of Automatic Control, Lund University, Lund, Sweden, 1982. Report TFRT-5266.
5. S.R. Weller, W. Moran, B. Ninness, and A.D. Pollington. Sampling zeros and the Euler–Fröbenius polynomials. *IEEE Transactions on Automatic Control*, 46(2):340–343, February 2001.
6. G.C. Goodwin, R.H. Middleton, and H.V. Poor. High-speed digital signal processing and control. *Proceedings of the IEEE*, 80(2):240–259, 1992.
7. J.I. Yuz and G.C. Goodwin. On sampled-data models for nonlinear systems. *IEEE Transactions on Automatic Control*, 50(10):1477–1489, October 2005.
8. M.J. Báchuta. On zeros of pulse transfer functions. *IEEE Transactions on Automatic Control*, 44(6):1229–1234, 1999.
9. G.C. Goodwin, S.F. Graebe, and M.E. Salgado. *Control System Design*. Prentice-Hall, Englewood Cliffs, NJ, 2001.
10. G.C. Goodwin, J.I. Yuz, and J.C. Agüero. Relative error issues in sampled-data models. In *17th IFAC World Congress*, Seoul, Korea, 2008.
11. B. Wahlberg. Limit results for sampled systems. *International Journal of Control*, 48(3):1267–1283, 1988.
12. V.S. Adamchik. On the Hurwitz function for rational arguments. *Applied Mathematics and Computation*, 187(1):3–12, 2007.
13. J.C. Agüero, G.C. Goodwin, T. Söderström, and J.I. Yuz. Sampled data errors in variables systems. In *15th IFAC Symposium on System Identification—SYSID 2009*, Saint Malo, France, 5–8 July 2009.
14. J.I. Yuz and G.C. Goodwin. Sampled-data models for stochastic nonlinear systems. In *14th IFAC Symposium on System Identification, SYSID 2006*, Newcastle, Australia, March 2006.
15. G.C. Goodwin, J.I. Yuz, and M.E. Salgado. Insights into the zero dynamics of sampled-data models for linear and nonlinear stochastic systems. In *European Control Conference—ECC'07*, Kos, Greece, 2–5 July 2007.

5.4 Behavioral Methods in Control

Harry L. Trentelman

5.4.1 Introduction

In systems and control, traditionally *control* has been almost invariably associated with the concepts of input, output, and feedback. The mechanism of feedback involves observations, is able to adapt to its environment, and decides on the basis of the observed sensor outputs what the actuator inputs should be. In other words, the typical features of feedback are the presence of sensing and generating control inputs on the basis of the sensed observations. This principle, often called *intelligent control*, is depicted in Figure 5.13.

It has been argued in [1] and [2] that for many practical control devices it is hard, or even impossible, to give an interpretation as feedback controllers. Consider, for example, passive vibration-control systems consisting of special bracings and oil dampers to protect large buildings against earthquakes, or passive suspension applied by springs and dampers in automobiles. Such control mechanisms do certainly not act as feedback controllers in the sense that control inputs are generated on the basis of sensed measurement outputs. Rather, from a physical point of view, the action of such passive controllers can best be understood using the concepts of interconnection and variable sharing. Many other devices act as controllers, but not as feedback controllers. Examples are strips mounted on objects to improve aerodynamic properties, stabilizers for ships, etc.

The behavioral point of view provides a natural framework for this more general way of looking at control. In the behavioral approach, control means restricting the behavior of a system, for example the plant, by interconnecting it with another system, the controller. In this paper, we will explain the basic idea of control in the framework of behaviors, which allows a general interconnection stucture between plant and controller. This will be the subject of Section 5.4.3.

Although there is an increasing body of literature on control of multidimensional linear systems, in particular of behaviors represented by constant coefficient, linear, partial differential equations, in the present paper we will restrict ourselves to one-dimensional *linear differential systems*. These are systems represented by ordinary, constant coefficient, linear differential equations. In Section 5.4.4 we will review the basic concepts and results for this class of systems.

In Section 5.4.5, we discuss the implementability problem. This problem may actually be considered as a basic question in control: Given a plant behavior, together with some "desired" behavior, the latter is called *implementable* (sometimes called: Achievable) if it can be achieved as controlled behavior by interconnecting the plant with a suitable controller. In this section, for a given plant, we give a complete characterization of all implementable behaviors. We also discuss the issues of regular interconnection and regular implementability.

Next, in Section 5.4.6, we turn to the most basic of control problems: The problems of finding stabilizing controllers and finding controllers that assign the poles of the controlled behavior. We give behavioral, representation-free, formulations of these problems, and give necessary and sufficient conditions for the existence of stabilizing controllers, and for pole placement. These conditions will turn out to involve the behavioral versions of the notions of controllability, observability, stabilizability, and detectability.

Section 5.4.7 then deals with the natural problem of controller parametrization: Given a plant behavior and an implementable desired behavior, we give a parametrization of all controllers that regularly implement this desired behavior. We also parametrize all controllers that stabilize a given stabilizable plant, thus establishing a behavioral analogue of the well-known Youla parametrization.

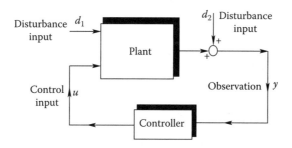

FIGURE 5.13 Intelligent control.

In Section 5.4.8, we return to the stabilization problem. We study the problem of finding, for a given plant, stabilizing controllers with the property that pre specified components of the interconnection variable are free in these controllers. In this problem we embed classical feedback thinking into the behavioral control framework by realizing that the controller should not be allowed to constrain interconnection variables that correspond to, for example, measured plant outputs. These variables should obviously remain free in the controller: The controller should "respect" the values that these variables take.

Finally, in Section 5.4.9, we look at robust control in a behavioral framework, and study the problem of finding, for a given nominal plant, controllers that stabilize (in the behavioral sense) all plants in a given ball around the nominal plant. This problem is closely related to behavioral $\mathcal{H}_\infty$-control, and we formulate the appropriate behavioral "small gain argument" to formalize this.

5.4.2 Notation and Nomenclature

To conclude this section, we will spend a few words on the notation and nomenclature used. We use the standard symbols for the fields of real and complex numbers $\mathbb{R}$ and $\mathbb{C}$. We use $\mathbb{R}^n$, $\mathbb{R}^{n \times m}$, etc. for the real linear spaces of vectors and matrices with components in $\mathbb{R}$. Often, the notation $\mathbb{R}^w$, $\mathbb{R}^d$, $\mathbb{R}^c$ ... is used if w, d, c ... denote typical elements of that vector space, or typical functions taking their values in that vector space. Vectors are understood to be column vectors in equations, in text, however, we sometimes write them as row vectors. Sometimes, we also use the notation $\text{col}(w_1, w_2)$ to represent the column vector formed by stacking w_1 over w_2.

$\mathcal{C}^\infty(\mathbb{R}, \mathbb{R}^w)$ will denote the set of infinitely often differentiable functions from $\mathbb{R}$ to $\mathbb{R}^w$. The space of all measurable functions w from $\mathbb{R}$ to $\mathbb{R}^w$ such that $\int_{-\infty}^{\infty} \|w\|^2 dt < \infty$ is denoted by $\mathfrak{L}_2(\mathbb{R}, \mathbb{R}^w)$. The $\mathfrak{L}_2$-norm of w is $\|w\|_2 := (\int_{-\infty}^{\infty} \|w\|^2 dt)^{1/2}$.

$\mathbb{R}[\xi]$ denotes the ring of polynomials in the indeterminate ξ with real coefficients, and $\mathbb{R}(\xi)$ denotes its quotient field of real rational functions in the indeterminate ξ. A polynomial $r \in \mathbb{R}[\xi]$ is called *monic* if the coefficient of its highest degree monomial is equal to 1. We use $\mathbb{R}^n[\xi]$, $\mathbb{R}^{n \times m}[\xi]$, $\mathbb{R}^n(\xi)$, $\mathbb{R}^{n \times m}(\xi)$, etc. for the spaces of vectors and matrices with components in $\mathbb{R}[\xi]$ and $\mathbb{R}(\xi)$, respectively. Elements of $\mathbb{R}^{n \times m}[\xi]$ are called *real polynomial matrices* and elements of $\mathbb{R}^{n \times m}(\xi)$ are called *real rational matrices*.

$\det(A)$ denotes the determinant of a square matrix A. A square, nonsingular real polynomial matrix R is called *Hurwitz* if all roots of the polynomial $\det(R)$ lie in the open left-half complex plane $\mathbb{C}^-$. A proper real rational matrix G is called *stable* if all its poles are in $\mathbb{C}^-$. If G is a proper stable rational matrix, then its $\mathcal{H}_\infty$ norm is defined as $\|G\|_\infty := \sup_{\lambda \in \bar{\mathbb{C}}^+} \|G(\lambda)\|$.

5.4.3 Control in a Behavioral Setting

In this section, we will first explain the basic elements of control in the context of the behavioral approach to dynamical systems.

In the behavioral approach, a dynamical system is a triple, $\Sigma = (T, W, \mathfrak{B})$, with T a set called the *time axis*, W a set called the *signal space*, and $\mathfrak{B} \subset W^T$ the *behavior*. The behavior consists of a family of admissible functions $w : T \to W$. The dynamical system aimes at specifying which functions of time $t \in T$ of the variable w can occur. This variable is called the *manifest variable* of the system. Since T and W are often apparent from the context we identify the system $\Sigma = (T, W, \mathfrak{B})$ simply with its behavior $\mathfrak{B}$. For a basic introduction to dynamical systems in a behavioral framework, we refer to the textbook [5], and for background information to [3] or [4].

The basic idea of control in this framework is very simple. If $\Sigma_1 = (T, W, \mathfrak{B}_1)$ and $\Sigma_2 = (T, W, \mathfrak{B}_2)$ are two dynamical systems with the same time axis and the same signal space, then the *full interconnection* $\Sigma_1 \wedge \Sigma_2$ of Σ_1 and Σ_2 is defined as the dynamical system $(T, W, \mathfrak{B}_1 \cap \mathfrak{B}_2)$, that is, the system whose behavior is equal to the set-theoretic intersection of the behaviors $\mathfrak{B}_1$ and $\mathfrak{B}_2$. We speak of full interconnection since the entire variable w of $\mathfrak{B}_1$ is shared with $\mathfrak{B}_2$ in the interconnection.

More often, the interconnection can only take place through pre specified components of the manifest variable. In that case, we speak of *partial interconnection*. Let $\Sigma_1 = (T, W_1 \times C, \mathfrak{B}_1)$ and $\Sigma_2 = (T, W_2 \times C, \mathfrak{B}_2)$ be two dynamical systems with the same time axis. We assume that the signal spaces $W_1 \times C$ and $W_2 \times C$ of Σ_1 and Σ_2, respectively, are product spaces, with the factor C in common. Correspondingly, trajectories of $\mathfrak{B}_1$ are denoted by (w_1, c) and trajectories of $\mathfrak{B}_2$ by (w_2, c). We define the *interconnection of Σ_1 and Σ_2 through c* as the dynamical system

$$\Sigma_1 \wedge_c \Sigma_2 := (T, W_1 \times W_2 \times C, \mathfrak{B})$$

with interconnected behavior

$$\mathfrak{B} = \{(w_1, w_2, c) : T \to W_1 \times W_2 \times C \mid (w_1, c) \in \mathfrak{B}_1 \text{ and } (w_2, c) \in \mathfrak{B}_2\}.$$

The behaviors $\mathfrak{B}_1$ and $\mathfrak{B}_2$ in this case only share the variable c, which is called the *interconnection variable*. Often, we denote the interconnected behavior $\mathfrak{B}$ by $\mathfrak{B}_1 \wedge_c \mathfrak{B}_2$. This interconnection is illustrated in Figure 5.14.

In this context, control is formalized as follows. Assume that the *plant*, a dynamical system $\Sigma_p = (T, W \times C, \mathcal{P}_{\text{full}})$ is given. It has two types of terminals: terminals carrying *to-be-controlled variables* w and terminals carrying *interconnection variables* c. Therefore, the signal space of the plant is given as the product space $W \times C$, where W is the space in which w takes its values, and C denotes the space in which c takes its values. The behavior of the plant is denoted by $\mathcal{P}_{\text{full}}$, called the *full plant behavior*, and consists of all $(w, c) : T \to W \times C$ that are compatible with the laws of the plant.

In the classical controller configuration, the to-be-controlled variables combine the exogenous disturbance inputs and the to-be-controlled outputs, while the interconnection variables combine the sensor outputs and the actuator inputs. A *feedback controller* may be viewed as a signal processor that processes the sensor outputs and returns the actuator inputs. It is the synthesis of such feedback processors that is traditionally viewed as control design. However, we will look at control from a somewhat broader perspective, and we consider any law that restricts the behavior of the interconnection variables as a controller.

Thus a *controller* for the plant Σ_p is a dynamical system $\Sigma_c = (T, C, \mathcal{C})$ with controller behavior $\mathcal{C}$. The interconnected system $\Sigma_p \wedge_c \Sigma_c$ is called the *controlled system*. A control problem for the plant Σ_p is now to specify a set of admissible controllers, to describe what desirable properties the controlled system should have, and, finally, to find an admissible controller Σ_c such that $\Sigma_p \wedge_c \Sigma_c$ has the desired properties. In this framework, control is nothing more than general interconnection through the interconnection variables (see Figure 5.15).

The main motivation for this alternative formulation of control is a practical one: many controllers, for example, physical devices such as dampers, heat insulators, matched impedances, etc. simply do not act as signal processors. For a more elaborate discussion of this point of view, we refer to [1]. As

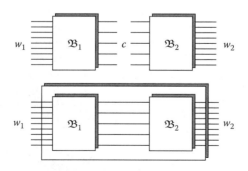

FIGURE 5.14 Interconnection of Σ_1 and Σ_2.

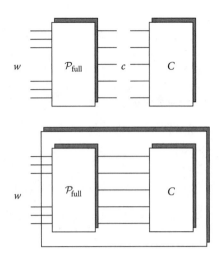

FIGURE 5.15 Σ_p controlled by Σ_c.

a comment related to our motivation to view a controller as any law that restricts the behavior of the control variables, we emphasize our misgivings regarding the omnipresence of signal flow graph thinking in systems modeling. The point of view that system interconnection should, or even can, be viewed as identifying inputs of one system with outputs of another, simply does not agree with physical reality. When interconnecting two systems, certain variables of one system are indeed identified with certain variables of another. There is no reason for these variables to act as inputs, respectively outputs. When connecting two wires of two electrical circuits, we impose the equality of two voltages and the equality of a current into one circuit with the current out of another. Nothing requires that one circuit should be current controlled, and the other voltage controlled. When connecting two pins of two mechanical systems, we impose the equality of two generalized positions and of the generalized force on one system with (the negative of) the generalized force on another. If the intuitive classification of forces as inputs is tenable, then this interconnection results in equating two inputs and two outputs, and not equating inputs with outputs. For thermal connections, we identify temperatures, and the heat flow into one system with that out of another. Typically again, this results in equating inputs *and* (not *with*) outputs. Pressures and flows: Same story.

5.4.4 Linear Differential Systems

In this section we will discuss control in a behavioral framework for *linear differential systems*. We will first review the basic concepts and ideas. For more detailed information we refer to [5].

As explained in Section 5.4.3, in the behavioral approach, a dynamical system is given by a triple $\Sigma = (T, W, \mathfrak{B})$, where T is the time axis, W is the signal space, and the behavior $\mathfrak{B}$ is a subset of W^T, the set of all functions from T to W. A *linear differential system* is a dynamical system with time axis $T = \mathbb{R}$, and whose signal space W is a finite-dimensional Euclidean space, say, $\mathbb{R}^w$. Correspondingly, the manifest variable is then given as $w = \text{col}(w_1, w_2, \ldots, w_w)$. The behavior $\mathfrak{B}$ is a linear subspace of $\mathfrak{C}^\infty(\mathbb{R}, \mathbb{R}^w)$ consisting of all solutions of a set of higher order, linear, constant coefficient differential equations. More precisely, there exists a positive integer g and a polynomial matrix $R \in \mathbb{R}^{g \times w}[\xi]$ such that

$$\mathfrak{B} = \left\{ w \in \mathfrak{C}^\infty(\mathbb{R}, \mathbb{R}^w) \mid R\left(\frac{d}{dt}\right) w = 0 \right\}.$$

The set of linear differential systems with manifest variable w taking its value in $\mathbb{R}^w$ is denoted by $\mathfrak{L}^w$.

We make a clear distinction between the behavior as defined as the space of all solutions of a set of (differential) equations, and the set of equations itself. A set of equations in terms of which the behavior is defined is called a *representation* of the behavior. Let $R \in \mathbb{R}^{g \times w}[\xi]$ be a polynomial matrix. If the behavior $\mathfrak{B}$ is represented by $R(\frac{d}{dt})w = 0$ then we call this a *kernel representation* of $\mathfrak{B}$. Further, a kernel representation is said to be *minimal* if every other kernel representation of $\mathfrak{B}$ has at least g rows. A given kernel representation $R\left(\frac{d}{dt}\right)w = 0$ is minimal if and only if the polynomial matrix R has full row rank.

We speak of a system as the behavior $\mathfrak{B}$, one of whose representations is given by $R\left(\frac{d}{dt}\right)w = 0$ or just $\mathfrak{B} = \ker(R)$. The "$\frac{d}{dt}$" is often suppressed to enhance readability.

We will also encounter behaviors $\mathfrak{B}$ with manifest variable w, which are represented by equations of the form $R(\frac{d}{dt})w = M(\frac{d}{dt})\ell$, in which an auxiliary, *latent variable* ℓ appears. Here, R and M are polynomial matrices with the same number of rows. Through such an equation, we can consider the subspace of all $w \in \mathcal{C}^{\infty}(\mathbb{R}, \mathbb{R}^w)$ for which there exists an $\ell \in \mathcal{C}^{\infty}(\mathbb{R}, \mathbb{R}^l)$ such that the equation holds:

$$\mathfrak{B} = \left\{ w \in \mathcal{C}^{\infty}(\mathbb{R}, \mathbb{R}^w) \mid \exists \, \ell \in \mathcal{C}^{\infty}(\mathbb{R}, \mathbb{R}^l) \text{ such that } R(\frac{d}{dt})w = M(\frac{d}{dt})\ell \right\}.$$

By the elimination theorem (see [5], Chapter 6, Theorem 6.2.6), $\mathfrak{B} \in \mathcal{L}^w$, that is, $\mathfrak{B}$ is again a linear differential system. We call $R(\frac{d}{dt})w = M(\frac{d}{dt})\ell$ a *latent variable representation* of $\mathfrak{B}$.

Let $\mathfrak{B} \in \mathcal{L}^w$. Let $R(\frac{d}{dt})w = 0$ be a kernel representation. Assume rank$(R) < $ w (which also means that it is *under determined*: the number of variables is strictly larger than the number of equations). Then, obviously, some components of $w = \text{col}(w_1, w_2, \ldots, w_w)$ are unconstrained by the requirement $w \in \mathfrak{B}$. These components are said to be *free* in $\mathfrak{B}$. The maximum number of such components is called the *input cardinality* of $\mathfrak{B}$ (denoted as m($\mathfrak{B}$)). Once m($\mathfrak{B}$) free components are chosen, the remaining w $-$ m($\mathfrak{B}$) components are determined up to a finite-dimensional affine subspace of $\mathcal{C}^{\infty}(\mathbb{R}, \mathbb{R}^{w-m(\mathfrak{B})})$. These are called *outputs*, and the number of outputs is denoted by p($\mathfrak{B}$), called the *output cardinality* of $\mathfrak{B}$. Thus, possibly after a permutation of components, $w \in \mathfrak{B}$ can be partitioned as $w = (u, y)$, with the m($\mathfrak{B}$) components of u as inputs, and the p($\mathfrak{B}$) components of y as outputs. We say that (u, y) is an *input–output partition* of $w \in \mathfrak{B}$, with input u and output y.

The input–output structure of $\mathfrak{B} \in \mathcal{L}^w$ is reflected in its kernel representations as follows. Suppose $R(\frac{d}{dt})w = 0$ is a minimal kernel representation of $\mathfrak{B}$. Partition $R = (Q \;\; P)$, and accordingly $w = (w_1, w_2)$. Then $w = (w_1, w_2)$ is an i/o partition (with input w_1 and output w_2) if and only if P is square and nonsingular. In general, there exist many input–output partitions, but the integers m($\mathfrak{B}$) and p($\mathfrak{B}$) are invariants associated with a behavior. It can be verified that p($\mathfrak{B}$) is equal to the rank of the polynomial matrix in any (not necessarily minimal) kernel representation of $\mathfrak{B}$ (for details see [5]).

Definition 5.2:

A behavior whose input cardinality is equal to 0 is called autonomous. An autonomous behavior $\mathfrak{B}$ is said to be stable if for all $w \in \mathfrak{B}$ we have $w(t) \to 0$ as $t \to \infty$.

In the context of stability, we often need to describe regions of the complex plane $\mathbb{C}$. We denote the closed right-half of the complex plane by $\bar{\mathbb{C}}^+$ and the open left-half complex plane by $\mathbb{C}^-$. A polynomial matrix $R \in \mathbb{R}^{w \times w}[\xi]$ is called *Hurwitz* if rank$(R(\lambda)) = $ w for all $\lambda \in \mathbb{C}^+$ (equivalently, det(R) has no roots in $\mathbb{C}^+$). If $\mathfrak{B} \in \mathcal{L}^w$ is represented by the minimal kernel representation $R(\frac{d}{dt})w = 0$ then $\mathfrak{B}$ is stable if and only if R is Hurwitz (see [5], Chapter 7).

For autonomous behaviors, we also speak about *poles of the behavior*. Let $\mathfrak{B} \in \mathcal{L}^w$ be autonomous. Then there exists an $R \in \mathbb{R}^{w \times w}[\xi]$ such that $\mathfrak{B}$ is represented minimally by $R\left(\frac{d}{dt}\right)w = 0$. Obviously, for any nonzero $\alpha \in \mathbb{R}$, αR also yields a kernel representation of $\mathfrak{B}$. Hence, we can choose R such that det(R) is a monic polynomial. This monic polynomial is denoted by $\chi_\mathfrak{B}$ and is called *the characteristic polynomial of* $\mathfrak{B}$. $\chi_\mathfrak{B}$ depends only on $\mathfrak{B}$, and not on the polynomial matrix R we used to define it: if R_1, R_2 both represent $\mathfrak{B}$ minimally then there exists a unimodular U such that $R_2 = UR_1$ (see [5], Chapter 3). Hence, if $\det(R_1)$ and $\det(R_2)$ are monic then $\det(R_1) = \det(R_2)$. The *poles* of $\mathfrak{B}$ are defined as the roots of $\chi_\mathfrak{B}$. Note that $\chi_\mathfrak{B} = 1$ if and only if $\mathfrak{B} = 0$. A behavior is stable if and only if all its poles are in $\mathbb{C}^-$.

Next, we review the concept of controllability in the behavioral approach.

Definition 5.3:

A behavior $\mathfrak{B} \in \mathfrak{L}^w$ *is controllable if for all* $w_1, w_2 \in \mathfrak{B}$, *there exist a* $T \geq 0$ *and a* $w \in \mathfrak{B}$ *such that* $w(t) = w_1(t)$ *for* $t < 0$ *and* $w(t + T) = w_2(t)$ *for* $t \geq 0$.

Often, we encounter behaviors $\mathfrak{B} \in \mathfrak{L}^w$ that are neither autonomous nor controllable. The *controllable part* of a behavior $\mathfrak{B}$ is defined as the largest controllable subbehavior of $\mathfrak{B}$. This is denoted by $\mathfrak{B}_{cont}$. A given $\mathfrak{B} \in \mathfrak{L}^w$ can always be decomposed as $\mathfrak{B} = \mathfrak{B}_{cont} \oplus \mathfrak{B}_{aut}$, where $\mathfrak{B}_{cont}$ is the (unique) controllable part of $\mathfrak{B}$, and $\mathfrak{B}_{aut}$ is a (nonunique) autonomous subbehavior of $\mathfrak{B}$. For details we refer to [5].

Definition 5.4:

A behavior $\mathfrak{B} \in \mathfrak{L}^w$ *is called stabilizable if for all* $w_1 \in \mathfrak{B}$, *there exists a* $w \in \mathfrak{B}$ *such that* $w(t) = w_1(t)$ *for* $t < 0$, *and* $w(t) \to 0$ *as* $t \to \infty$.

Thus every trajectory in a stabilizable behavior $\mathfrak{B}$ can be steered to 0, asymptotically. Conditions for controllability and stabilizability in terms of the polynomial matrix appearing in any kernel representation of $\mathfrak{B}$ are well-known. Indeed, if $\mathfrak{B} = \ker(R)$, then $\mathfrak{B}$ is controllable if and only if $\text{rank}(R(\lambda)) = \text{rank}(R)$ for all $\lambda \in \mathbb{C}$. $\mathfrak{B}$ is stabilizable if and only if $\text{rank}(R(\lambda)) = \text{rank}(R)$ for all $\lambda \in \bar{\mathbb{C}}^+$.

We shall also deal with systems in which the signal space comes as a product space, with the first component viewed as an observed variable, and the second as a to-be-deduced variable. We talk about observability (in such systems).

Definition 5.5:

Given $\mathfrak{B} \in \mathfrak{L}^{w_1+w_2}$ *with manifest variable* $w = (w_1, w_2)$, w_2 *is said to be observable from* w_1 *if* (w_1, w_2'), $(w_1, w_2'') \in \mathfrak{B}$ *implies* $w_2' = w_2''$.

If $R_1\left(\frac{d}{dt}\right)w_1 + R_2\left(\frac{d}{dt}\right)w_2 = 0$ is a kernel representation of $\mathfrak{B}$, then observability of w_2 from w_1 is equivalent to $R_2(\lambda)$ having full column rank for all $\lambda \in \mathbb{C}$. The weaker notion of *detectability* is defined along similar lines:

Definition 5.6:

Given $\mathfrak{B} \in \mathfrak{L}^{w_1+w_2}$, w_2 *is said to be detectable from* w_1 *if* (w_1, w_2'), $(w_1, w_2'') \in \mathfrak{B}$ *implies* $w_2'(t) - w_2''(t) \to 0$ *as* $t \to \infty$.

In the above kernel representation, detectability of w_2 from w_1 is equivalent to $R_2(\lambda)$ having full column rank for all $\lambda \in \bar{\mathbb{C}}^+$. For details, we refer to [5].

To conclude this section, we review some facts on elimination of variables. Let $\mathfrak{B} \in \mathcal{L}^{w_1 + w_2}$ with system variable $w = (w_1, w_2)$. Let P_{w_1} denote the projection onto the w_1-component. Then the set $P_{w_1}\mathfrak{B}$ consisting of all w_1 for which there exists w_2 such that $(w_1, w_2) \in \mathfrak{B}$ is again a linear differential system. We denote $P_{w_1}\mathfrak{B}$ by $\mathfrak{B}_{w_1}$, and call it the behavior obtained by eliminating w_2 from $\mathfrak{B}$.

If $\mathfrak{B} = \ker(R_1 \; R_2)$, then a representation for $\mathfrak{B}_{w_1}$ is obtained as follows: choose a unimodular matrix U such that $UR_2 = \begin{pmatrix} R_{12} \\ 0 \end{pmatrix}$, with R_{12} full row rank, and conformably partition $UR_1 = \begin{pmatrix} R_{11} \\ R_{21} \end{pmatrix}$. Then $\mathfrak{B}_{w_1} = \ker(R_{21})$ (see [5], Section 6.2.2).

5.4.5 Implementability

We now turn to the question what controlled behaviors can be achieved by interconnecting a given plant with a controller. This problem may actually be considered as a basic question in engineering design: a behavior is prescribed, and the question is whether this "desired" behavior can be achieved by inserting a suitably designed subsystem into the over all system. Details on the implementability problem can be found in [6].

We first consider the full interconnection case, in which the interconnection variable c coincides with the to be controlled variable w. In that case we have a plant behavior $\mathcal{P} \in \mathfrak{L}^w$, and a controller for $\mathcal{P}$ is also a behavior $\mathcal{C} \in \mathfrak{L}^w$. The full interconnection of $\mathcal{P}$ and $\mathcal{C}$ is the system whose behavior is the intersection $\mathcal{P} \cap \mathcal{C}$. This controlled behavior is again a linear differential system. Indeed, if $\mathcal{P} = \ker(R)$ and $\mathcal{C} = \ker(C)$, then $\mathcal{P} \cap \mathcal{C} = \ker \begin{pmatrix} R \\ C \end{pmatrix} \in \mathfrak{L}^w$.

Definition 5.7:

Let $\mathcal{K} \in \mathfrak{L}^w$ be a given behavior, to be interpreted as a desired behavior. If $\mathcal{K}$ can be achieved as controlled behavior, that is, if there exists $\mathcal{C} \in \mathfrak{L}^w$ such that $\mathcal{K} = \mathcal{P} \cap \mathcal{C}$, then we call $\mathcal{K}$ implementable by full interconnection (with respect to $\mathcal{P}$).

Obviously, a given $\mathcal{K} \in \mathfrak{L}^w$ is implementable by full interconnection with respect to $\mathcal{P}$ if and only if $\mathcal{K} \subset \mathcal{P}$. Indeed, if $\mathcal{K} \subset \mathcal{P}$, then with "controller" $\mathcal{C} = \mathcal{K}$ we have $\mathcal{K} = \mathcal{P} \cap \mathcal{C}$.

Next we consider the case that not all variables are available for interconnection, but in which interconnection can only take place through prespecified interconnection variables c, that is, the case of partial interconnection.

Before the controller acts, there are two behaviors of the plant that are relevant: the behavior $\mathcal{P}_{\text{full}} \in \mathfrak{L}^{w+c}$ (the full plant behavior) of the variables w and c combined, and the behavior $(\mathcal{P}_{\text{full}})_w$ of the to-be-controlled variables w (with the interconnection variable c eliminated). Hence,

$$\mathcal{P}_{\text{full}} = \{(w, c) \in \mathcal{C}^\infty(\mathbb{R}, \mathbb{R}^{w \times c}) \mid (w, c) \text{ satisfies the plant equations}\},$$

$$(\mathcal{P}_{\text{full}})_w = \{w \in \mathcal{C}^\infty(\mathbb{R}, \mathbb{R}^w) \mid \exists \, c \in \mathcal{C}^\infty(\mathbb{R}, \mathbb{R}^c) \text{ such that } (w, c) \in \mathcal{P}_{\text{full}}\}.$$

By the elimination theorem, $(\mathcal{P}_{\text{full}})_w \in \mathfrak{L}^w$. The controller restricts the interconnection variables c and (assuming it is a linear differential system) is described by a *controller behavior* $\mathcal{C} \in \mathfrak{L}^c$. Hence,

$$\mathcal{C} = \{c \in \mathcal{C}^\infty(\mathbb{R}, \mathbb{R}^c) \mid c \text{ satisfies the controller equations}\}.$$

The *full controlled behavior* $\mathcal{P}_{\text{full}} \wedge_c \mathcal{C}$ is obtained by the interconnection of $\mathcal{P}_{\text{full}}$ and $\mathcal{C}$ through the variable c and is defined as

$$\mathcal{P}_{\text{full}} \wedge_c \mathcal{C} = \{(w, c) \mid (w, c) \in \mathcal{P}_{\text{full}} \text{ and } c \in \mathcal{C}\}.$$

Eliminating c from the full controlled behavior, we obtain its restriction $(\mathcal{P}_{\text{full}} \wedge_c \mathcal{C})_w$ to the behavior of the to-be-controlled variable w, defined by

$$(\mathcal{P}_{\text{full}} \wedge_c \mathcal{C})_w = \{w \in \mathfrak{C}^\infty(\mathbb{R}, \mathbb{R}^w) \mid \exists\, c \in \mathcal{C} \text{ such that } (w, c) \in \mathcal{P}_{\text{full}}\}.$$

Note that, again by the elimination theorem, $(\mathcal{P}_{\text{full}} \wedge_c \mathcal{C})_w \in \mathfrak{L}^w$.

Definition 5.8:

Given $\mathcal{P}_{\text{full}} \in \mathfrak{L}^{w+c}$, we say that $\mathcal{C} \in \mathfrak{L}^c$ implements $\mathcal{K} \in \mathfrak{L}^w$ through c if $\mathcal{K} = (\mathcal{P}_{\text{full}} \wedge_c \mathcal{C})_w$.

We now discuss the following problem:

> *The implementability problem:* Given $\mathcal{P}_{\text{full}} \in \mathfrak{L}^{w+c}$, give a characterization of all $\mathcal{K} \in \mathfrak{L}^w$ for which there exists a $\mathcal{C} \in \mathfrak{L}^c$ that implements $\mathcal{K}$ through c.

This problem has a very simple and elegant solution: it depends only on the projected full plant behavior $(\mathcal{P}_{\text{full}})_w$ and on the behavior consisting of the plant trajectories with the interconnection variables put equal to zero. This behavior is denoted by $\mathbb{N}_w(\mathcal{P}_{\text{full}})$, and is called the *hidden behavior*. It is defined as

$$\mathcal{N}_w(\mathcal{P}_{\text{full}}) = \{w \mid (w, 0) \in \mathcal{P}_{\text{full}}\}.$$

Theorem 5.4: Sandwich Theorem

Let $\mathcal{P}_{\text{full}} \in \mathfrak{L}^{w+c}$ be the full plant behavior. Then $\mathcal{K} \in \mathfrak{L}^w$ is implementable by a controller $\mathcal{C} \in \mathfrak{L}^c$ acting on the interconnection variable c if and only if

$$\mathcal{N}_w(\mathcal{P}_{\text{full}}) \subset \mathcal{K} \subset (\mathcal{P}_{\text{full}})_w.$$

Theorem 5.4 shows that $\mathcal{K}$ can be *any* behavior that is wedged in between the given behaviors $\mathcal{N}_w(\mathcal{P}_{\text{full}})$ and $(\mathcal{P}_{\text{full}})_w$. The necessity of this condition is quite intuitive: $\mathcal{K} \subset (\mathcal{P}_{\text{full}})_w$ states that the controlled behavior must be part of the plant behavior. Logical, since the controller merely restricts what can happen. The condition $\mathcal{K} \supset \mathcal{N}_w(\mathcal{P}_{\text{full}})$ states that the behavior $\mathcal{N}_w(\mathcal{P}_{\text{full}})$ must remain possible, whatever be the controller. This is quite intuitive also, since the subbehavior of the plant behavior that is compatible with $c = 0$, hence when the controller receives no information on what is happening in the plant, must remain possible in the controlled behavior, whatever controller is chosen. This observation has important consequences in control: in order for there to exist a controller that achieves acceptable performance, the hidden behavior must already meet the specifications, since *there is simply no way to eliminate it by means of control*. The fact that the hidden behavior must meet the control specifications has been observed before in a $\mathcal{H}_\infty$-control context for example in [7–9].

Theorem 5.4 reduces control problems to finding the controlled behavior $\mathcal{K}$ directly. Of course, the problem of how to actually implement $\mathcal{K}$ needs to be addressed at some point. This problem was studied in [10–12]. In particular, the question when a particular controlled behavior can be implemented by a feedback processor remains a very important one, and is discussed, for example, in [1] and [13].

5.4.5.1 Regular Implementability

As discussed above, the implementability problem is to characterize all behaviors that can be achieved as controlled behavior by interconnecting the plant with some controller. It turns out that in control problems in a behavioral framework one often has to require that the interconnection of plant and controller is a *regular interconnection*. In this section we will discuss the issue of implementability by regular interconnection. Detailed material can be found in [14]. Furthermore, in Section 5.4.6.3 some remarks on the relevance of regular interconnections can be found. We now first deal with the full interconnection case.

Let $\mathcal{P} \in \mathfrak{L}^w$ be a plant behavior, and let $\mathcal{C} \in \mathfrak{L}^w$ be a controller.

Definition 5.9:

The interconnection of $\mathcal{P}$ and $\mathcal{C}$ is called regular if

$$\mathrm{p}(\mathcal{P}) + \mathrm{p}(\mathcal{C}) = \mathrm{p}(\mathcal{P} \cap \mathcal{C}),$$

in other words, if the output cardinalities of the plant and the controller add up to the output cardinality of the controlled behavior. In that case, we also call the controller $\mathcal{C}$ regular (with respect to $\mathcal{P}$).

In terms of kernel representations this condition can be expressed as follows. Let $\mathcal{P} = \ker(R)$ and $\mathcal{C} = \ker(C)$ be minimal kernel representations of the plant and the controller, respectively. Then $\mathcal{P} \cap \mathcal{C} = \ker \begin{pmatrix} R \\ C \end{pmatrix}$ is a kernel representation of the controlled behavior. Since the output cardinality of a behavior is equal to the rank of the polynomial matrix in any of its kernel representations, the interconnection of $\mathcal{P}$ and $\mathcal{C}$ is regular if and only if $\begin{pmatrix} R \\ C \end{pmatrix}$ has full row rank, equivalently yields a minimal kernel representation of $\mathcal{P} \cap \mathcal{C}$.

Definition 5.10:

Given $\mathcal{P} \in \mathfrak{L}^w$, a given behavior $\mathcal{K} \in \mathfrak{L}^w$ is called regularly implementable with respect to $\mathcal{P}$ by full interconnection if there exists a regular controller $\mathcal{C} \in \mathfrak{L}^w$ that implements $\mathcal{K}$.

We now formulate the problem of regular implementability by full interconnection:

> *The problem of regular implementability by full interconnection:* Given $\mathcal{P} \in \mathfrak{L}^w$, give a characterization of all $\mathcal{K} \in \mathfrak{L}^w$ for which there exists a regular controller $\mathcal{C} \in \mathfrak{L}^w$ that implements $\mathcal{K}$ by full interconnection.

The following theorem from [14] gives such characterization. Recall from Section 5.4.4 the definition of controllable part of a behavior.

Theorem 5.5:

Let $\mathcal{P} \in \mathfrak{L}^w$. Let $\mathcal{P}_{cont}$ be its controllable part. Let $\mathcal{K} \in \mathfrak{L}^w$. Then $\mathcal{K}$ is regularly implementable with respect to $\mathcal{P}$ by full interconnection if and only if $\mathcal{K} + \mathcal{P}_{cont} = \mathcal{P}$.

Next, we turn to the partial interconnection case. Here, the problem is to find, for a given $\mathcal{P}_{full}$ and a given desired behavior $\mathcal{K}$, a controller $\mathcal{C}$ such that the manifest controlled behavior is equal to $\mathcal{K}$. Again, often we shall restrict ourselves to $\mathcal{C}$'s such that the interconnection of $\mathcal{P}_{full}$ and $\mathcal{C}$ is *regular*. A motivation for this is provided in Section 5.4.6.3.

Definition 5.11:

Let $\mathcal{P}_{full} \in \mathfrak{L}^{w+c}$ *and* $C \in \mathfrak{L}^w$. *The interconnection of* $\mathcal{P}_{full}$ *and* C *through* c *is called regular if*

$$p(\mathcal{P}_{full} \wedge_c C) = p(\mathcal{P}_{full}) + p(C),$$

that is, the output cardinalities of $\mathcal{P}_{full}$ *and* C *add up to that of the full controlled behavior* $\mathcal{P}_{full} \wedge_c C$. *In that case we also call the controller* C *regular.*

Definition 5.12:

A given $\mathcal{K} \in \mathfrak{L}^w$ *is called regularly implementable if there exists a* $C \in \mathfrak{L}^c$ *such that* $\mathcal{K}$ *is implemented by* C, *and the interconnection of* $\mathcal{P}_{full}$ *and* C *is regular.*

Similar to plain implementability, an important question is under what conditions a given subbehavior $\mathcal{K}$ of $\mathcal{P}$ is regularly implementable:

> *The problem of regular implementability by partial interconnection:* Given $\mathcal{P}_{full} \in \mathfrak{L}^{w+c}$, give a characterization of all $\mathcal{K} \in \mathfrak{L}^w$ for which there exists a controller $C \in \mathfrak{L}^c$ that implements $\mathcal{K}$ by regular interconnection through c.

The following theorem from [14] provides a solution to this problem:

Theorem 5.6:

Let $\mathcal{P}_{full} \in \mathfrak{L}^{w+c}$. *Let* $(\mathcal{P}_{full})_w$ *and* $\mathcal{N}_w(\mathcal{P}_{full})$ *be the corresponding projected plant behavior and hidden behavior, respectively. Let* $(\mathcal{P}_{full})_{w,cont}$ *be the controllable part of* $(\mathcal{P}_{full})_w$. *Let* $\mathcal{K} \in \mathfrak{L}^w$. *Then* $\mathcal{K}$ *is implementable with respect to* $\mathcal{P}_{full}$ *by regular interconnection through* c *if and only if the following two conditions are satisfied:*

- $\mathcal{N}_w(\mathcal{P}_{full}) \subset \mathcal{K} \subset (\mathcal{P}_{full})_w$
- $\mathcal{K} + (\mathcal{P}_{full})_{w,cont} = (\mathcal{P}_{full})_w$

The above theorem has two conditions. The first one is exactly the condition for implementability through c (as in the Sandwich theorem). The second condition formalizes the notion that the autonomous part of $(\mathcal{P}_{full})_w$ is taken care of by $\mathcal{K}$. While the autonomous part of $(\mathcal{P}_{full})_w$ is not unique, $(\mathcal{P}_{full})_{w,cont}$ is. This makes verifying the regular implementability of a given $\mathcal{K}$ computable. As a consequence of this theorem, note that if $(\mathcal{P}_{full})_w$ is controllable, then $\mathcal{K} \in \mathfrak{L}^w$ is regularly implementable if and only if it is implementable.

We conclude this section with an example.

Example 5.17:

Consider the plant behavior $\mathcal{P}_{full}$ with manifest variable $w = (w_1, w_2)$ and control variable $c = (c_1, c_2)$ represented by

$$w_1 + \dot{w}_2 + \dot{c}_1 + c_2 = 0$$
$$c_1 + c_2 = 0$$

Clearly, the projected plant behavior $(\mathcal{P}_{\text{full}})_w$ is equal to $\mathfrak{C}^{\infty}(\mathbb{R}, \mathbb{R}^2)$. For the desired behavior $\mathcal{K}$ we take $\mathcal{K} = \{(w_1, w_2) \mid w_1 + \dot{w}_2 = 0\}$. The following controller regularly implements $\mathcal{K}$ through c: $\mathcal{C} = \{(c_1, c_2) \mid \dot{c}_1 + c_2 = 0\}$. Also every controller represented by $kc_1 + c_2 = 0$, with $k \neq 1$, regularly implements $\mathcal{K}$.

5.4.6 Pole Placement and Stabilization

In this section we will discuss the problems of pole placement and stabilization from a behavioral point of view. Given a plant behavior, the stabilization problem is to find a regular controller such that the controlled behavior is stable. In the pole placement problem it is required to find a regular controller such that the controlled behavior is autonomous and has a given desired polynomial as its characteristic polynomial.

Again, we distinguish between the full interconnection and the partial interconnection case. A detailed treatment of pole placement and stabilization in a behavioral framework for the full interconnection case can be found in [1]. The partial interconnection case has been described extensively in [14].

5.4.6.1 Full Interconnection

We first introduce the pole placement problem. Given a plant behavior, the problem is to find conditions under which for every a "desired" real monic polynomial, there exists a regular controller such that the controlled behavior is autonomous and has the desired polynomial as its characteristic polynomial:

> *Pole placement by full interconnection:* Given $\mathcal{P} \in \mathfrak{L}^{w}$, find conditions under which there exists, and compute, for every monic $r \in \mathbb{R}[\xi]$, a $\mathcal{C} \in \mathfrak{L}^{w}$ such that:
>
> - The interconnection of $\mathcal{P}$ and $\mathcal{C}$ is regular.
> - The controlled behavior $\mathcal{P} \cap \mathcal{C}$ is autonomous and has r as its characteristic polynomial.

Suppressing the controller $\mathcal{C}$ from the problem formulation, the problem can be stated alternatively as: Given $\mathcal{P} \in \mathfrak{L}^{w}$, find conditions under which there exists, and compute, for every monic $r \in \mathbb{R}[\xi]$, an autonomous behavior $\mathcal{K} \in \mathfrak{L}^{w}$ that is regularly implementable by full interconnection, and such that $\chi_{\mathcal{K}} = r$.

A solution to this problem is given below.

Theorem 5.7:

Let $\mathcal{P} \in \mathfrak{L}^{w}$. For every monic $r \in \mathbb{R}[\xi]$, there exists a regular controller $\mathcal{C} \in \mathfrak{L}^{w}$ such that $\mathcal{P} \cap \mathcal{C}$ is autonomous and its characteristic polynomial $\chi_{\mathcal{P} \cap \mathcal{C}}$ is equal to r if and only $\mathcal{P}$ is controllable and $\mathrm{m}(\mathcal{P}) \geq 1$ (i.e., $\mathcal{P}$ has at least one input component).

Next, we consider the problem of stabilization by full interconnection.

> *Stabilization by full interconnection:* Given $\mathcal{P} \in \mathfrak{L}^{w}$, find conditions for the existence of, and compute $\mathcal{C} \in \mathfrak{L}^{w}$ such that
>
> - The interconnection of $\mathcal{P}$ and $\mathcal{C}$ is regular.
> - The controlled behavior $\mathcal{P} \cap \mathcal{C}$ is autonomous and stable.

Again, suppressing the controller $\mathcal{C}$ from the formulation, the stabilization problem can be restated as: given $\mathcal{P}$, find conditions for the existence of, and compute a behavior $\mathcal{K} \in \mathfrak{L}^{w}$ that is autonomous, stable and regularly implementable by full interconnection.

A necessary and sufficient condition for the existence of a regular stabilizing controller for $\mathcal{P}$ is stabilizability of $\mathcal{P}$ as defined in Section 5.4.4:

Theorem 5.8:

Let $\mathcal{P} \in \mathfrak{L}^{w}$. There exists a regular controller $\mathcal{C} \in \mathfrak{L}^{w}$ such that $\mathcal{P} \cap \mathcal{C}$ is autonomous and stable if and only $\mathcal{P}$ is stabilizable.

A regular controller that stabilizes a given plant is said to *regularly stabilize* this plant. In Section 5.4.7 we will deal with the problem how to compute regularly stabilizing controllers. In fact, there we will establish a parametrization of all such controllers.

5.4.6.2 Partial Interconnection

Again, we first introduce the pole placement problem. Given a full plant behavior the problem is to find conditions under which for every monic real polynomial there exists a regular controller such that the projected controlled behavior has this polynomial as its characteristic polynomial:

Pole placement by partial interconnection: Given $\mathcal{P}_{\text{full}} \in \mathfrak{L}^{w+c}$, find conditions under which there exists, and compute, for every monic $r \in \mathbb{R}[\xi]$, a $\mathcal{C} \in \mathfrak{L}^{c}$ such that:

- The interconnection of $\mathcal{P}_{\text{full}}$ and $\mathcal{C}$ is regular.
- The projected full controlled behavior $(\mathcal{P}_{\text{full}} \wedge_c \mathcal{C})_w$ is autonomous and has r as its characteristic polynomial.

Suppressing the controller $\mathcal{C}$ from the problem formulation, the problem can alternatively be stated as: given $\mathcal{P}_{\text{full}}$, find conditions under which there exists, and compute, for every monic $r \in \mathbb{R}[\xi]$ a regularly implementable, autonomous $\mathcal{K} \in \mathfrak{L}^{w}$ such that $\chi_\mathcal{K} = r$.

Necessary and sufficient conditions for pole placement are given in the following theorem, and involve observability of the to-be-controlled variable from the interconnection variable, and controllability of the projected full plant behavior:

Theorem 5.9:

Let $\mathcal{P}_{\text{full}} \in \mathfrak{L}^{w+c}$. For every monic $r \in \mathbb{R}[\xi]$, there exists a regular controller $\mathcal{C} \in \mathfrak{L}^{c}$ such that the characteristic polynomial of $(\mathcal{P}_{\text{full}} \wedge_c \mathcal{C})_w$ is equal to r if and only

- *In $\mathcal{P}_{\text{full}}$, w is observable from c.*
- *$(\mathcal{P}_{\text{full}})_w$ is controllable and $\mathrm{m}((\mathcal{P}_{\text{full}})_w) \geq 1$.*

Note that, by definition, observability of w from c means that if $(w_1, c), (w_2, c) \in \mathcal{P}_{\text{full}}$, then $w_1 = w_2$, or equivalently, $(w, 0) \in \mathcal{P}_{\text{full}}$ implies $w = 0$. Thus w is observable from c in $\mathcal{P}_{\text{full}}$ if and only if the hidden behavior $\mathbb{N}_w(\mathcal{P}_{\text{full}})$ is equal to $\{0\} \in \mathfrak{L}^{w}$.

Next, we formulate the stabilization problem, which deals with finding a regular controller for the full plant such that the projected behavior is autonomous and stable:

> *Stabilization by partial interconnection:* Given $\mathcal{P}_{\text{full}} \in \mathfrak{L}^{w+c}$, find conditions for the existence of, and compute $C \in \mathfrak{L}^c$ such that
>
> - The interconnection of $\mathcal{P}_{\text{full}}$ and C is regular.
> - The projected full controlled behavior $(\mathcal{P}_{\text{full}} \wedge_c C)_w$ is autonomous and stable.

Again, suppressing the controller C from the formulation, the stabilization problem can be restated as: Given $\mathcal{P}_{\text{full}}$, find conditions for the existence of, and compute a behavior $\mathcal{K} \in \mathfrak{L}^w$ that is autonomous, stable, and regularly implementable.

A solution to this problem involves both the notions of stabilizability and detectability in a behavioral framework (see Section 5.4.4):

Theorem 5.10:

Let $\mathcal{P}_{\text{full}} \in \mathfrak{L}^{w+c}$. *There exists a regular controller $C \in \mathfrak{L}^c$ such that the projected behavior $(\mathcal{P}_{\text{full}} \wedge_c C)_w$ is autonomous and stable if and only if*

- *in $\mathcal{P}_{\text{full}}$, w is detectable from c.*
- *$(\mathcal{P}_{\text{full}})_w$ is stabilizable.*

Since, by linearity, detectability of w from c in $\mathcal{P}_{\text{full}}$ is equivalent with: $(w, 0) \in \mathcal{P}_{\text{full}}$ implies $w(t) \to 0$ $(t \to \infty)$, detectability is equivalent with—the hidden behavior $\mathbb{N}_w(\mathcal{P}_{\text{full}})$ is stable.

Example 5.18:

Consider the full plant behavior $\mathcal{P}_{\text{full}}$ with to-be-controlled variable (w_1, w_2) and interconnection variable (c_1, c_2), represented by

$$w_1 + \dot{w}_2 + \dot{c}_1 + c_2 = 0$$
$$w_2 + c_1 + c_2 = 0$$
$$\dot{c}_1 + c_1 + \dot{c}_2 + c_2 = 0$$

A stabilizing regular controller is given by $C = \{(c_1, c_2) \mid \dot{c}_2 + 2c_1 + c_2 = 0\}$. Indeed, by eliminating c from the full controlled behavior $\mathcal{P}_{\text{full}} \wedge_c C$ we find that $(\mathcal{P}_{\text{full}} \wedge_c C)_w = \ker(R)$, with

$$R(\xi) = \begin{pmatrix} 0 & \xi + 1 \\ -1 & 2 \end{pmatrix}$$

which is Hurwitz. Yet another class of stabilizing controllers is represented by $C(\xi) = (\xi(\xi + 1) + k, \xi + 1 + k)$, $k \in \mathbb{R}$. In Section 5.4.7 we will find a parametrization of *all* 1×2 polynomial matrices $C(\xi)$ such that $\ker(C)$ is a stabilizing controller.

Neither in the problem formulations nor in the conditions of Theorems 5.9 and 5.10, representations of the given plant appear. Indeed, our problem formulations and their resolutions are completely *representation free*, and are formulated purely in terms of properties of the *behavior* $\mathcal{P}_{\text{full}}$. Thus, our treatment of the pole placement and stabilization problems is genuinely behavioral. Of course, Theorems 5.9 and 5.10 are applicable to any particular representation of $\mathcal{P}_{\text{full}}$ as well. For example, in [14], it was illustrated how the classical results on pole placement and stabilization by dynamic output feedback can be derived from the results in this section. Indeed, this can be done starting with $\mathcal{P}_{\text{full}}$ represented in input–state–output representation.

In both the stabilization problem and the pole placement problem, we have restricted ourselves to regular interconnections. We give an explanation for this in Section 5.4.6.3. At this point we note that if in the above problem formulations we omit the requirement that the interconnection should be regular, then in the stabilization problem a necessary and sufficient condition for the existence of a stabilizing controller is that $\mathcal{N}_w(\mathcal{P}_{\text{full}})$ is stable (equivalently: in $\mathcal{P}_{\text{full}}$, w is detectable from c). In the pole placement problem, necessary and sufficient conditions are that $\mathcal{N}_w(\mathcal{P}_{\text{full}}) = \{0\}$ (i.e., in $\mathcal{P}_{\text{full}}$, w is observable from c) and that $\mathcal{P}$ is not autonomous.

5.4.6.3 Disturbances and Regular Interconnection

In this section we have formulated the problems of stabilization and pole placement for a given plant $\mathcal{P}_{\text{full}}$ with to-be-controlled variable w and control variable c. In most system models, an unknown external disturbance variable, d, also occurs. The stabilization problem is then to find a controller acting on c such that whenever $d(t) = 0$ ($t \geq 0$), we have $w(t) \to 0$ ($t \to \infty$). Typically, the disturbance d is assumed to be free, in the sense that *every* $\mathfrak{C}^\infty$ function d is compatible with the equations of the model. As an example, think of a model of a car suspension system given by $R_1\left(\frac{d}{dt}\right)w + R_2\left(\frac{d}{dt}\right)c + R_3\left(\frac{d}{dt}\right)d = 0$, where d is the road profile as a function of time. In the stabilization problem, one puts $d = 0$ and solves the stabilization problem for the full plant $\mathcal{P}_{\text{full}}$ represented by $R_1\left(\frac{d}{dt}\right)w + R_2\left(\frac{d}{dt}\right)c = 0$. In doing this, one should make sure that the stabilizing controller C: $C\left(\frac{d}{dt}\right)c = 0$, when connected to the actual model, *does not put restrictions on d*. The notion of regular interconnection captures this, as explained below.

Consider the full plant behavior $\mathcal{P}_{\text{full}} \in \mathfrak{L}^{w+c}$. An *extension* of $\mathcal{P}_{\text{full}}$ is a behavior $\mathcal{P}_{\text{full}}^{\text{ext}} \in \mathfrak{L}^{w+c+d}$ (with d an arbitrary positive integer), with variables (w, c, d), such that

1. d is free in $\mathcal{P}_{\text{full}}^{\text{ext}}$.
2. $\mathcal{P}_{\text{full}} = \{(w, c) \mid \text{ such that } (w, c, 0) \in \mathcal{P}_{\text{full}}^{\text{ext}}\}$.

Thus, $\mathcal{P}_{\text{full}}^{\text{ext}}$ being an extension of $\mathcal{P}_{\text{full}}$ formalizes that $\mathcal{P}_{\text{full}}$ has exactly those signals (w, c) that are compatible with the disturbance $d = 0$ in $\mathcal{P}_{\text{full}}^{\text{ext}}$. Of course, a given full behavior $\mathcal{P}_{\text{full}}$ has many extensions.

For a given extension $\mathcal{P}_{\text{full}}^{\text{ext}}$ and a given controller $C \in \mathfrak{L}^c$, we consider the *extended controlled behavior* given by

$$\mathcal{P}_{\text{full}}^{\text{ext}} \wedge_c C = \{(w, c, d) \mid (w, c, d) \in \mathcal{P}_{\text{full}}^{\text{ext}} \text{ and } c \in C\}.$$

A controller C shall be acceptable only if the disturbance d remains free in $\mathcal{P}_{\text{full}}^{\text{ext}} \wedge_c C$, *for any possible extension* $\mathcal{P}_{\text{full}}^{\text{ext}}$. It turns out that this is guaranteed exactly, by the regularity of the interconnection of $\mathcal{P}_{\text{full}}$ and C ! Indeed, the following was proven in [14], Theorem 7.1:

Theorem 5.11:

The following two conditions are equivalent.

1. *The interconnection of $\mathcal{P}_{\text{full}}$ and C is regular.*
2. *For any extension $\mathcal{P}_{\text{full}}^{\text{ext}}$ of $\mathcal{P}_{\text{full}}$, d is free in $\mathcal{P}_{\text{full}}^{\text{ext}}$.*

5.4.7 Parametrization of Stabilizing Controllers

In Section 5.4.5, both for the full information as well as for the partial interconnection case, necessary and sufficient conditions have been given for a desired behavior to be regularly implementable. In Section 5.4.6, conditions have been given for the existence of regular, stabilizing controllers. The present section deals with the issue on how to find the required controllers. Both for the full interconnection case as well as for the partial interconnection case, the corresponding parametrization problems can be formulated and have been resolved in [12]. We will however restrict ourselves here to the full interconnection case. We will only briefly state the main results. Details can be found in [12].

The first problem that we formulate is the problem of parametrizing all controllers that regularly implement a given desired behavior.

Parametrization of regularly implementing controllers: Let $\mathcal{P} \in \mathfrak{L}^{\mathrm{w}}$ be the plant behavior, and let $\mathcal{K} \in \mathfrak{L}^{\mathrm{w}}$ be a desired behavior. Let $\mathcal{P} = \ker(R)$ and $\mathcal{K} = \ker(K)$ be minimal representations of the plant and desired behavior, respectively. Find a parametrization, in terms of the polynomial matrices R and K, of *all* polynomial matrices C such that the controller $\ker(C)$ regularly implements $\mathcal{K}$ by full interconnection.

The next problem is to parametrize all regular, stabilizing controllers.

Parametrization of regular and stabilizing controllers: Let $\mathcal{P} \in \mathfrak{L}^{\mathrm{w}}$ be a plant behavior. Let $\mathcal{P} = \ker(R)$ be a minimal kernel representation. Find a parametrization, in terms of the polynomial matrix R, of *all* polynomial matrices C such that the controller $\ker(C)$ is regular and $\mathcal{P} \cap \ker(C)$ is stable.

First, we will establish a parametrization of all controllers that regularly implement a given behavior. We make use of the following lemma from [12], which gives conditions in terms of the representations for regular implementability.

Lemma 5.15:

Let $\mathcal{P}, \mathcal{K} \in \mathfrak{L}^{\mathrm{w}}$. Let $\mathcal{P} = \ker(R)$ and $\mathcal{K} = \ker(K)$ be minimal kernel representations. Then $\mathcal{K}$ is regularly implementable with respect to $\mathcal{P}$ by full interconnection if and only if there exists a polynomial matrix F with $F(\lambda)$ full row rank for all $\lambda \in \mathbb{C}$, such that $R = FK$.

From the above, for a given regularly implementable $\mathcal{K}$ it is easy to obtain a controller, which regularly implements it. Indeed, if $R = FK$ with $F(\lambda)$ full row rank for all λ, let V be such that $\mathrm{col}(F, V)$ is unimodular. Then clearly the controller $\ker(VK)$ does the job. Indeed, the corresponding controlled behavior is given by

$$\mathcal{P} \cap \ker(VK) = \ker \begin{pmatrix} R \\ VK \end{pmatrix} = \ker \begin{pmatrix} F \\ V \end{pmatrix} K = \ker(K) = \mathcal{K},$$

and the interconnection is regular since $\begin{pmatrix} R \\ VK \end{pmatrix}$ has full row rank. A parametrization of all controllers that regularly implement $\mathcal{K}$ is described in the following theorem:

Theorem 5.12:

Let $\mathcal{P} \in \mathfrak{L}^{\mathrm{w}}$, with minimal kernel representation $\mathcal{P} = \ker(R)$. Let $\mathcal{K} \in \mathfrak{L}^{\mathrm{w}}$ be regularly implementable by full interconnection, and let $\mathcal{K} = \ker(K)$ be a minimal kernel representation. Let F be as in Lemma 5.15 and let V be such that $\mathrm{col}(F, V)$ is unimodular. Then for any $\mathcal{C} \in \mathfrak{L}^{\mathrm{w}}, \mathcal{C} = \ker(C)$, the following statements are equivalent:

1. *$\mathcal{C} = \ker(C)$ is a minimal kernel representation, and $\mathcal{C}$ regularly implements $\mathcal{K}$.*
2. *There exist a polynomial matrix G and a unimodular polynomial matrix U such that $C = GR + UVK$.*

Thus a parametrization is given by $C = GR + UVK$, where G ranges over *all* polynomial matrices, and U ranges over all unimodular polynomial matrices.

We now turn to parametrizing all regular controllers that stabilize a given plant behavior. Let $\mathcal{P} = \ker(R)$ be a minimal kernel representation. Assume that $\mathcal{P}$ is stabilizable, equivalently, $R(\lambda)$ has full row rank for all $\lambda \in \bar{\mathbb{C}}^+$. The following theorem yields a parametrization of all stabilizing controllers.

Theorem 5.13:

Let $\mathcal{P} \in \mathfrak{L}^w$ be stabilizable. Let $\mathcal{P} = \ker(R)$ be a minimal kernel representation, and let R_1 be such that $\ker(R_1)$ is a minimal kernel representation of the controllable part $\mathcal{P}_{\text{cont}}$ of $\mathcal{P}$. Let C_0 be such that $\text{col}(R_1, C_0)$ is unimodular. Then for any $C \in \mathfrak{L}^w$ with $C = \ker(C)$, the following statements are equivalent:

1. *$\mathcal{P} \cap C$ is autonomous and stable, the interconnection is regular, and the representation $C = \ker(C)$ is minimal.*
2. *There exist a polynomial matrix G and a Hurwitz polynomial matrix D such that $C = GR + DC_0$.*

Thus, a parametrization of all regular stabilizing controllers $\ker(C)$ is given by $C = GR + DC_0$, where G ranges over all polynomial matrices, and D ranges over all Hurwitz polynomial matrices. We also refer to [15].

Remark 5.13

In the special case that the plant $\mathcal{P}$ to be stabilized is given together with an input–output partition $w = (y, u)$, our parametrization result of Theorem 5.13 specializes to the well-known Youla parametrization of all stabilizing controllers (see [16,17]). For simplicity, assume that $\mathcal{P}$ is controllable. Assume that, in $\mathcal{P}$, G is the transfer matrix from u to y. Let $P^{-1}Q$ be a left coprime factorization of G. Then $\mathcal{P} = \ker(P \ - Q)$. Choose polynomial matrices X and Y such that

$$\begin{pmatrix} P & -Q \\ X & Y \end{pmatrix}$$

is unimodular. According to Theorem 5.13, a parametrization of all stabilizing controllers $\ker(Q_c \ P_c)$ is given by $(Q_c \ P_c) = F(P - Q) + D(X \ Y)$, where F is arbitrary polynomial and D is Hurwitz. In transfer matrix form this yields $C := -P_c^{-1}Q_c = -(DY - FQ)^{-1}(DX + FP) = -(Y - D^{-1}FQ)^{-1}(X + D^{-1}FP)$. Finally, denote $D^{-1}F$ by T, and let T vary over all proper stable rational matrices to obtain the original Youla parametrization $C = -(Y - TQ)^{-1}(X + TP)$ (see [16]).

5.4.8 Stabilization Using Controllers with *a Priori* Input–Output Structure

Often, certain components of the plant interconnection variables represent plant *sensor measurements*, or *unknown disturbance inputs* to the plant. As argued in [18–20], in these cases, by physical considerations, not all regular controllers are admissible anymore, since only those controllers are allowed that do not put constraints on these particular plant interconnection variables: the controller should respect the values that the plant has given to these variables. In the behavioral framework this is formalized by requiring these plant interconnection variables to be *free* in the controllers that are allowed.

Therefore, in this section we deal with the problems of finding necessary and sufficient conditions for a behavior to be stabilizable using regular controllers in which an *a priori* given subset of the plant interconnection variables is free or maximally free, respectively. In other words, we require *a priori* given components of the plant interconnection variable to be part of the controller input, or even to be the controller input. The complementary subset in the set of all interconnection variables then necessarily contains the controller output, or is equal to the controller output. We study these problems in both the full and the partial interconnection case. The material of this section can be found in full detail in [20]. Related results can be found in [19].

5.4.8.1 Full Interconnection

As usual, we first consider the full interconnection case. The problems we want to deal with are formulated as follows.

Stabilization with pre specified free variables: Let $\mathcal{P} \in \mathcal{L}^{w_1 + w_2}$ with plant variable (w_1, w_2). Find necessary and sufficient conditions for the existence of, and compute a regular, stabilizing controller $\mathcal{C} \in \mathfrak{L}^{w_1 + w_2}$ in which w_2 is free.

Stabilization with prespecified i/o structure: Let $\mathcal{P} \in \mathcal{L}^{w_1 + w_2}$ with plant variable (w_1, w_2). Find necessary and sufficient conditions for the existence of, and compute a regular, stabilizing controller $\mathcal{C} \in \mathfrak{L}^{w_1 + w_2}$ in which w_2 is input and w_1 is output.

Theorem 5.14:

Let $\mathcal{P} \in \mathcal{L}^{w_1 + w_2}$ with plant variable (w_1, w_2). Then we have the following:

1. *There exists a stabilizing controller $\mathcal{C} \in \mathcal{L}^{w_1 + w_2}$ in which w_2 is free if and only if $\mathcal{P}$ is stabilizable and* $w_2 \leq \mathrm{p}(\mathcal{P})$,
2. *There exists a stabilizing controller $\mathcal{C} \in \mathcal{L}^{w_1 + w_2}$ in which w_2 is input and w_1 is output if and only if $\mathcal{P}$ is stabilizable and* $w_2 = \mathrm{p}(\mathcal{P})$.

In other words, necessary and sufficient conditions for the existence of a regular, stabilizing controller in which the given plant variable w_2 is free are that the plant is stabilizable, and the size of w_2 does not exceed the output cardinality of the plant. If we require w_2 to be input to the controller (so, consequently, w_1 output), then the size of w_2 should be equal to the plant output cardinality. It follows from this theorem that the actual choice of components that we want to be free in the controller does not matter, in the sense that if the plant is stabilizable and if $w_2 \leq \mathrm{p}(\mathcal{P})$, then for *any choice* of w_2 components of the plant variable there exists a regular, stabilizing controller in which these components are free. A similar statement holds for the input–output assignment in the controller under the condition $w_2 = \mathrm{p}(\mathcal{P})$.

5.4.8.2 Partial Interconnection

Next, we study the above problems in the case of partial interconnection. The exact statement of these problems is as follows:

Stabilization with prespecified free variables: Let $\mathcal{P}_{\text{full}} \in \mathcal{L}^{w + c_1 + c_2}$ with system variable (w, c), where $c = (c_1, c_2)$. Find necessary and sufficient conditions for the existence of, and compute a regular, stabilizing controller $\mathcal{C} \in \mathfrak{L}^{c_1 + c_2}$ in which c_2 is free.

Stabilization with prespecified i/o structure: Let $\mathcal{P}_{\text{full}} \in \mathcal{L}^{w + c_1 + c_2}$ with system variable (w, c), where $c = (c_1, c_2)$. Find necessary and sufficient conditions for the existence of, and compute a regular, stabilizing controller $\mathcal{C} \in \mathfrak{L}^{c_1 + c_2}$ in which c_2 is input and c_1 is output.

Recall the following definitions of hidden and projected behaviors from Section 5.4.5:

$$\mathcal{N}_c(\mathcal{P}_{\text{full}}) = \{(c_1, c_2) \mid (c_1, c_2, 0) \in \mathcal{P}_{\text{full}}\},$$
$$\mathcal{N}_{c_1}(\mathcal{P}_{\text{full}}) = \{c_1 \mid (c_1, 0, 0) \in \mathcal{P}_{\text{full}}\},$$
$$(\mathcal{P}_{\text{full}})_c = \{(c_1, c_2) \mid \exists w \text{ such that } (c_1, c_2, w) \in \mathcal{P}_{\text{full}}\}.$$

In the following, we assume that c_1 has size $\mathtt{c}_2$ and c_2 has size $\mathtt{c}_2$, $\mathtt{c}_1 + \mathtt{c}_2 = \mathtt{c}$.

Theorem 5.15:

Let $\mathcal{P}_{\text{full}} \in \mathcal{L}^{w+c_1+c_2}$ with system variable (w, c), with $c = (c_1, c_2)$. There exists a regular, stabilizing controller $C \in \mathcal{L}^{c_1+c_2}$ in which c_2 is free if and only if

1. $(\mathcal{P}_{\text{full}})_w$ *is stabilizable and w is detectable from c in $\mathcal{P}_{\text{full}}$.*
2. $\mathrm{p}(\mathcal{N}_c(\mathcal{P}_{\text{full}})) - \mathrm{p}(\mathcal{N}_{c_1}(\mathcal{P}_{\text{full}})) \leq \mathrm{p}((\mathcal{P}_{\text{full}})_c).$

Thus, in addition to the obvious condition (1) for the existence of a regular stabilizing controller, the theorem requires that the difference between the output cardinalities of the hidden behaviors $\mathcal{N}_c(\mathcal{P}_{\text{full}})$ and $\mathcal{N}_{c_1}(\mathcal{P}_{\text{full}})$ does not exceed the output cardinality of the projected behavior $(\mathcal{P}_{\text{full}})_c$.

If we want to assign the input–output structure of the controller, then the following theorem holds:

Theorem 5.16:

Let $\mathcal{P}_{\text{full}} \in \mathcal{L}^{w+c_1+c_2}$ with system variable (w, c), where $c = (c_1, c_2)$. Consider the following conditions

1. $(\mathcal{P}_{\text{full}})_w$ *is stabilizable and w is detectable from c in $\mathcal{P}_{\text{full}}$.*
2. $\mathrm{p}(\mathcal{N}_c(\mathcal{P}_{\text{full}})) - \mathrm{p}(\mathcal{N}_{c_1}(\mathcal{P}_{\text{full}})) = \mathrm{p}((\mathcal{P}_{\text{full}})_c).$
3. $\mathrm{p}(\mathcal{N}_c(\mathcal{P}_{\text{full}})) = \mathtt{c}_1 + \mathrm{p}((\mathcal{P}_{\text{full}})_c).$

If conditions 1, 2, and 3 hold, then there exists a stabilizing controller $C \in \mathcal{L}^{c_1+c_2}$ for which c_2 is input and c_1 is output. If $\mathcal{N}_c(\mathcal{P}_{\text{full}})$ is autonomous, then these conditions are also necessary, and conditions 2 and 3 reduce to the single condition $\mathrm{p}((\mathcal{P}_{\text{full}})_c) = \mathtt{c}_2$.

We will illustrate the above by means of a worked-out example.

Example 5.19:

Let $\mathcal{P}_{\text{full}} \in \mathfrak{L}^5$ with manifest variable $w = (w_1, w_2)$ and interconnection variable $c = (c_1, c_2, c_3)$ be represented by

$$w_1 + \dot{w}_2 + \dot{c}_3 = 0,$$
$$w_2 + c_1 + c_2 + c_3 = 0.$$

Clearly, $(\mathcal{P}_{\text{full}})_w = \mathfrak{C}^\infty(\mathbb{R}, \mathbb{R}^2)$ and $(\mathcal{P}_{\text{full}})_c = \mathfrak{C}^\infty(\mathbb{R}, \mathbb{R}^2)$. $(\mathcal{P}_{\text{full}})_w$ is trivially stabilizable, and w is detectable from c in $\mathcal{P}_{\text{full}}$. Clearly, $\mathrm{p}((\mathcal{P}_{\text{full}})_c) = 0$. We compute

$$\mathcal{N}_c(\mathcal{P}_{\text{full}}) = \ker\left(N\left(\frac{d}{dt}\right)\right), \quad \mathcal{N}_{(c_1, c_2)}(\mathcal{P}_{\text{full}}) = \ker\left(N_{12}\left(\frac{d}{dt}\right)\right)$$

and

$$\mathcal{N}_{(c_2,c_3)}(\mathcal{P}_{\text{full}}) = \ker\left(N_{23}\left(\frac{d}{dt}\right)\right).$$

where $N(\xi) = \begin{pmatrix} 0 & 0 & \xi \\ 1 & 1 & 1 \end{pmatrix}$, $N_{12}(\xi) = \begin{pmatrix} 0 & 0 \\ 1 & 1 \end{pmatrix}$, and $N_{23}(\xi) = \begin{pmatrix} 0 & \xi \\ 1 & 1 \end{pmatrix}$. As a consequence, $\text{p}(\mathcal{N}_c(\mathcal{P}_{\text{full}})) = \text{rank}(N) = 2$, $\text{p}(\mathcal{N}_{(c_2,c_3)}(\mathcal{P}_{\text{full}})) = \text{rank}(N_{23}) = 2$. Furthermore, $\text{p}(\mathcal{N}_{(c_1,c_2)}(\mathcal{P}_{\text{full}})) = \text{rank}(N_{12}) = 1$. From these calculations it is evident that $\text{p}(\mathcal{N}_c(\mathcal{P}_{\text{full}})) - \text{p}(\mathcal{N}_{(c_2,c_3)}(\mathcal{P}_{\text{full}})) = \text{p}((\mathcal{P}_{\text{full}})_c)$ and $\text{p}(\mathcal{N}_c(\mathcal{P}_{\text{full}})) - \text{p}(\mathcal{N}_{(c_1,c_2)}(\mathcal{P}_{\text{full}})) > \text{p}((\mathcal{P}_{\text{full}})_c)$. Therefore, from Theorem 5.15 we conclude that the plant is stabilizable using a controller in which c_1 is free. We also conclude that there does not exist a controller which stabilizes the plant and in which c_3 is free.

5.4.9 Rational Representations

Recently, in [21], representations of linear differential systems using *rational matrices* instead of polynomial matrices were introduced. In [21], a meaning was given to the equation $R\left(\frac{d}{dt}\right)w = 0$, where $R(\xi)$ is a given real *rational* matrix. In order to do this, we need the concept of *left coprime factorization*. Let R be a real rational matrix R. Then a factorization $R = P^{-1}Q$ is called a left coprime factorization of R over $\mathbb{R}[\xi]$ if P and Q are real polynomial matrices with P nonsingular, and the complex matrix $\big(P(\lambda)\ Q(\lambda)\big)$ has full row rank for all $\lambda \in \mathbb{C}$.

Definition 5.13:

Let R be a real rational matrix, and let $R = P^{-1}Q$ be a left coprime factorization of R over $\mathbb{R}[\xi]$. Let $w \in \mathcal{C}^\infty(\mathbb{R}, \mathbb{R}^w)$. Then we define w to be a solution to $R\left(\frac{d}{dt}\right)w = 0$ if $Q\left(\frac{d}{dt}\right)w = 0$.

It can be proven that the space of solutions of $R\left(\frac{d}{dt}\right)w = 0$ defined in this way is independent of the particular left coprime factorization. Hence $R\left(\frac{d}{dt}\right)w = 0$ represents the linear differential system $\Sigma = (\mathbb{R}, \mathbb{R}^w, \ker(Q)) \in \mathcal{L}^w$.

If a behavior $\mathfrak{B}$ is represented by $R\left(\frac{d}{dt}\right)w = 0$ (or: $\mathfrak{B} = \ker(R)$), with $R(\xi)$ a real rational matrix, then we call this a *rational kernel representation of* $\mathfrak{B}$. If R has g rows, then the rational kernel representation is called *minimal* if every rational kernel representation of $\mathfrak{B}$ has at least g rows. It can be shown that a given rational kernel representation $\mathfrak{B} = \ker(R)$ is minimal if and only if the rational matrix R has full row rank. As in the polynomial case, every $\mathfrak{B} \in \mathcal{L}^w$ admits a minimal rational kernel representation. The number of rows in any minimal rational kernel representation of $\mathfrak{B}$ is equal to the number of rows in any minimal polynomial kernel representation of $\mathfrak{B}$, and therefore equal to $\text{p}(\mathfrak{B})$, the output cardinality of $\mathfrak{B}$. In general, if $\mathfrak{B} = \ker(R)$ is a rational kernel representation, then $\text{p}(\mathfrak{B}) = \text{rank}(R)$. This follows immediately from the corresponding result for polynomial kernel representations (see [5]). The following was proven in [21]:

Lemma 5.16:

For every behavior $\mathfrak{B} \in \mathcal{L}^w$ there exists a stable, proper real rational matrix R such that $\mathfrak{B} = \ker(R)$.

In other words, every linear differential system $\mathfrak{B}$ admits a kernel representation with a stable proper rational matrix. We will apply this useful result in the next section on robust stabilization.

5.4.10 Optimal Robust Stabilization and $\mathcal{H}_\infty$-Control

Given a nominal plant, together with a fixed neighborhood of this plant, the problem of robust stabilization is to find a controller that stabilizes all plants in that neighborhood (in an appropriate sense). If a controller achieves this design objective, we say that it robustly stabilizes the nominal plant. In this section we formulate the robust stabilization problem in a behavioral framework. Details on the material in this section can be found in [22].

Let $\mathcal{P} \in \mathfrak{L}^{\mathrm{w}}$ be a stabilizable linear differential system, to be interpreted as the *nominal plant*. In addition, we consider a fixed neighborhood of this plant. Of course, the concept of neighborhood should be made precise. We do this in the following way. Assume that the nominal plant $\mathcal{P}$ is represented in rational kernel representation by $R\left(\frac{d}{dt}\right) w = 0$, where R is a proper, stable real rational matrix. As noted in Lemma 5.16, for a given $\mathcal{P}$ such R exists. For a given $\gamma > 0$, we define the ball $B(\mathcal{P}, \gamma)$ with radius γ around $\mathcal{P}$ as follows:

$$
\begin{aligned}
B(\mathcal{P}, \gamma) := \{ \mathcal{P}_\Delta \in \mathfrak{L}^{\mathrm{w}} \mid \text{ there exists a proper, stable, real rational} \\
R_\Delta \text{ of full row rank such that } \mathcal{P}_\Delta = \ker(R_\Delta) \\
\text{and } \quad \|R - R_\Delta\|_\infty \leq \gamma \}.
\end{aligned}
\tag{5.251}
$$

Then we define the robust stabilization problem as follows:

> *Robust stabilization by full interconnection:* Find conditions for the existence of, and compute a controller $\mathcal{C} \in \mathfrak{L}^{\mathrm{w}}$ that regularly stabilizes all plants $\mathcal{P}_\Delta$ in the ball with radius γ around $\mathcal{P}$, that is, for all $\mathcal{P}_\Delta \in B(\mathcal{P}, \gamma)$, $\mathcal{P}_\Delta \cap \mathcal{C}$ is stable and $\mathcal{P}_\Delta \cap \mathcal{C}$ is a regular interconnection.

It turns out that a controller achieves robust stability in the above sense if and only if it solves a given $\mathcal{H}_\infty$-control problem for an auxiliary system associated with the nominal plant. This is a behavioral version of the small gain theorem. Therefore, we now first formulate an appropriate behavioral version of the $\mathcal{H}_\infty$-control problem. Such problems were studied also in [6,13,23] and [24], see also [8,25,26].

We start with a full plant behavior $\mathcal{P}_{\mathrm{full}} \in \mathfrak{L}^{\mathrm{w+d+c}}$, with system variable (w, d, c). The variable c is, as before, the interconnection variable. The variable to be controlled consists of two components, that is, is given as (w, d), with w a variable that should be kept "small" regardless of d, which should be interpreted as an unknown disturbance. The fact that the variable d represents an unknown disturbance is formalized by assuming d to be free in $\mathcal{P}_{\mathrm{full}}$. As d is interpreted as unknown disturbance, it should be free also after interconnecting the plant with a controller. Furthermore, in the context of $\mathcal{H}_\infty$-control, a controller is called stabilizing if, whenever the disturbance d is zero, the to be controlled variable w tends to zero as time runs off to infinity. Therefore, we define:

Definition 5.14:

Let $\mathcal{P}_{\mathrm{full}} \in \mathfrak{L}^{\mathrm{w+d+c}}$, with d free. A controller $\mathcal{C} \in \mathfrak{L}^{\mathrm{c}}$ is called *disturbance-free* if d is free in $\mathcal{P}_{\mathrm{full}} \wedge_c \mathcal{C}$. A disturbance-free controller $\mathcal{C} \in \mathfrak{L}^{\mathrm{c}}$ is called *stabilizing* if $[(w, 0, c) \in \mathcal{P}_{\mathrm{full}} \wedge_c \mathcal{C}] \Rightarrow [\lim_{t \to \infty} w(t) = 0]$.

For a given controller $\mathcal{C}$, let $(\mathcal{P}_{\mathrm{full}} \wedge_c \mathcal{C})_{(w,d)}$ be the projection of the full controlled behavior onto (w, d). $\mathcal{H}_\infty$-control deals with finding controllers that make this projected behavior (strictly) contractive in the following sense.

Definition 5.15:

Let $\mathcal{P}_{\text{full}} \in \mathfrak{L}^{w+d+c}$. Let $\gamma > 0$. A controller $\mathcal{C} \in \mathfrak{L}^c$ is called strictly γ-contracting if there exists $\epsilon > 0$ such that for all $(w,d) \in (\mathcal{P}_{\text{full}} \wedge_c \mathcal{C})_{(w,d)} \cap \mathfrak{L}_2(\mathbb{R}, \mathbb{R}^{w+d})$ we have $\|w\|_2 \le (\gamma - \epsilon)\|d\|_2$. The projected controlled behavior is then called strictly γ-contractive.

We now formulate the $\mathcal{H}_\infty$ control problem that will be instrumental in our solution of the behavioral robust stabilization problem.

> *The $\mathcal{H}_\infty$-control problem:* Let $\mathcal{P}_{\text{full}} \in \mathfrak{L}^{w+d+c}$. Assume that d is free. Let $\gamma > 0$. Find a disturbance-free, stabilizing, regular and strictly γ-contracting controller $\mathcal{C} \in \mathfrak{L}^c$ for $\mathcal{P}_{\text{full}}$.

In [23], necessary and sufficient conditions for the existence of a required controller were established under the conditions that in $\mathcal{P}_{\text{full}}$ the variable c is observable from (w,d), and the variable (w,d) is detectable from c. The latter condition is sometimes referred to as the *full information* condition. In the present section we will not explicitly state these necessary and sufficient conditions, since they require the introduction of the behavioral theory of dissipative systems, quadratic differential forms and storage functions as treated in [27] and [6]. Instead, we will directly turn to the connection between the robust stabilization problem and the $\mathcal{H}_\infty$-control problem.

Thus, we return to the stabilizable linear differential system $\mathcal{P} \in \mathfrak{L}^w$, to be interpreted as the nominal plant. Associated with this nominal plant $\mathcal{P} \in \mathfrak{L}^w$, we define the auxiliary system $\mathcal{P}_{\text{aux}} \in \mathfrak{L}^{w+d+w}$ by

$$\mathcal{P}_{\text{aux}} := \{(w,d,c) \mid R\left(\frac{d}{dt}\right)w + d = 0, \ c = w\}. \tag{5.252}$$

Let $R(\xi) = P^{-1}(\xi)Q(\xi)$ be a left coprime factorization over $\mathbb{R}[\xi]$, with P Hurwitz. Then by definition

$$\mathcal{P}_{\text{aux}} = \{(w,d,c) \mid Q\left(\frac{d}{dt}\right)w + P\left(\frac{d}{dt}\right)d = 0, \ c = w\}. \tag{5.253}$$

The following lemma formulates a behavioral version of the "small gain theorem":

Lemma 5.17:

Let $\mathcal{P}_{\text{aux}}$ be the auxiliary system represented by Equation 5.252. Let $\mathcal{C} \in \mathfrak{L}^w$ be represented in minimal rational kernel representation by $C\left(\frac{d}{dt}\right)c = 0$. Let $\gamma > 0$. Then the following statements are equivalent:

1. $\mathcal{C}$ regularly stabilizes $\mathcal{P}_\Delta$ for all $\mathcal{P}_\Delta \in B(\mathcal{P}, \gamma)$, that is, $\mathcal{P}_\Delta \cap \mathcal{C}$ is stable and $\mathcal{P}_\Delta \cap \mathcal{C}$ is a regular interconnection for all $\mathcal{P}_\Delta \in B(\mathcal{P}, \gamma)$.
2. $\mathcal{C}$ is a disturbance-free, stabilizing, regular and strictly $\frac{1}{\gamma}$-contracting controller for $\mathcal{P}_{\text{aux}}$.

In other words, given the nominal plant $\mathcal{P} \in \mathfrak{L}^w$, and given $\gamma > 0$, a controller $\mathcal{C} \in \mathfrak{L}^w$ regularly stabilizes all plants $\mathcal{P}_\Delta$ in the ball $B(\mathcal{P}, \gamma)$ around $\mathcal{P}$ if and only if $\mathcal{C}$ solves the $\mathcal{H}_\infty$-control problem for the auxiliary full plant behavior $\mathcal{P}_{\text{aux}}$. This is a full information control problem since, in Equation 5.253, $c = 0$ implies $w = 0$, and therefore (since P is Hurwitz) $d(t) \to 0$ ($t \to \infty$). Without going into the details, we now state necessary and sufficient conditions for the existence of, and outline how to compute such controller $\mathcal{C}$ from the representation of $\mathcal{P}$ and the given tolerance γ. The conditions involve the existence of a suitable

J-spectral factorization. In the following, let Σ_γ be given by

$$\Sigma_\gamma := \begin{pmatrix} -I_w & 0 \\ 0 & \frac{1}{\gamma^2}I_d \end{pmatrix}$$

Denote $R^\sim(\xi) := R^\top(-\xi)$.

Theorem 5.17:

Let $\mathcal{P} \in \mathcal{L}^w$ be stabilizable, and let $R\left(\frac{d}{dt}\right) w = 0$ be a minimal kernel representation, with R real rational, proper and stable. Let $\gamma > 0$. Let $R = DW^{-1}$ be a right coprime factorization of R over $\mathbb{R}[\xi]$. Then there exists a controller $C \in \mathcal{L}^w$ that regularly stabilizes all $\mathcal{P}_\Delta$ in the ball $B(\mathcal{P}, \gamma)$ if and only if there exists a square nonsingular Hurwitz polynomial matrix $F = \begin{pmatrix} F_+ \\ F_- \end{pmatrix}$ such that

1. $-\begin{pmatrix} W \\ D \end{pmatrix}^\sim \Sigma_\gamma \begin{pmatrix} W \\ D \end{pmatrix} = \begin{pmatrix} F_+ \\ F_- \end{pmatrix}^\sim \begin{pmatrix} I_{w-d} & 0 \\ 0 & -I_d \end{pmatrix} \begin{pmatrix} F_+ \\ F_- \end{pmatrix}.$

2. $\begin{pmatrix} W \\ D \end{pmatrix} \begin{pmatrix} F_+ \\ F_- \end{pmatrix}^{-1}$ *is proper.*

3. $\begin{pmatrix} D \\ F_+ \end{pmatrix}$ *is Hurwitz.*

If such F exists, then a suitable controller is computed as follows:

  a. *Factorize: $F_+ W^{-1} = P_1^{-1} C$ with P_1, C polynomial matrices, P_1 Hurwitz.*
  b. *Define $\mathcal{C} \in \mathcal{L}^w$ by $\mathcal{C} := \ker(C)$.*

The controller $\mathcal{C}$ is then regular, disturbance-free, stabilizing and strictly $\frac{1}{\gamma}$-contracting for $\mathcal{P}_{aux}$; so by the small gain argument of Lemma 5.17, it regularly stabilizes all $\mathcal{P}_\Delta$ in the ball $B(\mathcal{P}, \gamma)$.

Of course, for a given nominal plant $\mathcal{P}$, we would like to know the *smallest upper bound* (if it exists) of those γ's for which there exists a controller $\mathcal{C}$ that regularly stabilizes all perturbed plants $\mathcal{P}_\Delta$ in the ball with radius γ around $\mathcal{P}$. This is the problem of *optimal robust stabilization*.

Computation of the optimal stability radius: Find the optimal stability radius

$$\gamma^* := \sup\{\gamma > 0 \mid \exists \mathcal{C} \in \mathcal{L}^w \text{ that regularly stabilizes all } \mathcal{P}_\Delta \in B(\mathcal{P}, \gamma)\}.$$

In [22], a complete solution to this problem was given. Again, for this we would have to introduce the behavioral theory of dissipative systems and two-variable polynomial matrices, which goes beyond the scope of this study. It turns out that the optimal stability radius γ^* can be computed in terms of certain two-variable polynomial matrices obtained after polynomial spectral factorization.

References

1. J.C. Willems, On interconnection, control, and feedback. *IEEE Transactions on Automatic Control,* 42:326–339, 1997.
2. J.C. Willems, The behavioral approach to open and interconnected systems. *IEEE Control Systems Magazine,* December 2007:46–99, 2007.
3. J.C. Willems, Models for dynamics. *Dynamics Reported,* 2:171–269, 1989.
4. J.C. Willems, Paradigms and puzzles in the theory of dynamical systems. *IEEE Transactions on Automatic Control,* 36:259–294, 1991.

5. J.W. Polderman and J.C. Willems, *Introduction to Mathematical Systems Theory: A Behavioral Approach*, Springer-Verlag, Berlin, 1997.
6. J.C. Willems and H.L. Trentelman, Synthesis of dissipative systems using quadratic differential forms—Part I. *IEEE Transactions on Automatic Control*, 47(1):53–69, 2002.
7. P.P. Khargonekar, State-space $\mathcal{H}_\infty$ control theory. In A.C. Antoulas (ed.) *Mathematical System Theory: The Influence of R.E. Kalman*. Springer-Verlag, pp. 159–176, 1991.
8. G. Meinsma, Frequency domain methods in H_∞ control, Doctoral dissertation, University of Twente, Enschede, The Netherlands, 1993.
9. G. Meinsma and M. Green, On strict passivity and its application to interpolation and $\mathcal{H}_\infty$ control. *Proceedings IEEE Conference on Decision and Control*, Tucson, AZ, 1992.
10. A.A. Julius, J.C. Willems, M.N. Belur, and H.L. Trentelman, The canonical controller and regular interconnection. *Systems and Control Letters*, 54(8):787–797, 2005.
11. A.J. van der Schaft, Achievable behavior of general systems. *Systems and Control Letters*, 49:141–149, 2003.
12. C. Praagman, H.L. Trentelman, and R. Zavala Yoe, On the parametrization of all regularly implementing and stabilizing controllers. *SIAM Journal of Control and Optimization*, 45(6):2035–2053, 2007.
13. H.L. Trentelman and J.C. Willems, Synthesis of dissipative systems using quadratic differential forms—Part II. *IEEE Transactions on Automatic Control*, 47(1):70–86, 2002.
14. M.N. Belur and H.L. Trentelman, Stabilization, pole placement and regular implementability. *IEEE Transactions on Automatic Control*, 47(5):735–744, 2002.
15. M. Kuijper, Why do stabilizing controllers stabilize? *Automatica*, 31:621–625, 1995.
16. D.C. Youla, H.A. Jabr, and J.J. Bongiorno, Modern Wiener-Hopf design of optimal controllers. II. The multivariable case. *IEEE Transaction Automation Control*, 21:319–338, 1976.
17. M. Vidyasagar, *Control System Synthesis, A Factorization Approach*, The MIT Press, Cambridge, MA, 1985.
18. A.A. Julius, On interconnection and equivalence of continuous and discrete systems: A behavioral perspective, Doctoral dissertation, University of Twente, The Netherlands, 2005.
19. A.A. Julius, J.W. Polderman, and A.J. van der Schaft, Parametrization of the regular equivalences of the canonical controller. *IEEE Transactions on Automatic Control*, 53(4):1032–1036, 2008.
20. S. Fiaz and H.L. Trentelman, Regular implementability and stabilization using controllers with prespecified input/output partition. *IEEE Transactions on Automatic Control*, 54(7):1562–1568, 2009.
21. J.C Willems and Y. Yamamoto, Behaviors defined by rational functions. *Linear Algebra and Its Applications*, 425:226–241, 2007.
22. H.L. Trentelman, S. Fiaz, and K. Takaba, Optimal robust stabilization and dissipativity synthesis by behavioral interconnection, Manuscript 2009, submitted for publication.
23. H.L. Trentelman and J.C. Willems, $\mathcal{H}_\infty$ control in a behavioral context: the full information case. *IEEE Transactions on Automatic Control*, 44:521–536, 1999.
24. M.N. Belur and H.L. Trentelman, The strict dissipativity synthesis problem and the rank of the coupling QDF. *Systems and Control Letters*, 51(3-4):247–258, 2004.
25. G. Meinsma, Polynomial solutions to $\mathcal{H}_\infty$ problems. *International Journal of Robust and Nonlinear Control*, 4:323–351, 1994.
26. S. Weiland, A.A. Stoorvogel, and B. de Jager, A behavioral approach to the $\mathcal{H}_\infty$ optimal control problem. *Systems and Control Letters*, 32:323–334, 1997.
27. J.C. Willems and H.L. Trentelman, On quadratic differential forms. *SIAM Journal of Control and Optimization*, 36(5):1703–1749, 1998.

5.5 Discrete Event Systems

Richard Hill

5.5.1 Introduction

Discrete event systems (DES) are dynamic systems characterized by discrete states and event-driven evolution. Care is taken to distinguish DES from digital or discrete-time systems. Whereas discrete-time systems are continuous systems sampled at discrete intervals of time, DES are fundamentally discrete. A state of a DES could be a buffer being empty or full, a machine being idle or busy, or a transmission being in second or third gear. DES also evolve according to events, such as a part arriving at a machine or the

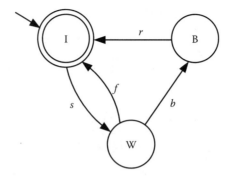

FIGURE 5.16 Simple DES example, **M**.

value of a continuous signal crossing some threshold. For the purposes of modeling a DES, these events are considered to occur instantaneously and asynchronously.

Figure 5.16 shows an example of a generic, isolated machine modeled as a DES using a finite state automaton [1]. The modeled states represent the machine being *Idle* (I), *Working* (W) or *Broken* (B), as opposed to perhaps more customary continuous states such as the position or cutting force of a tool bit. Transitions between these discrete states are indicated by the events *start* (*s*), *finish* (*f*), *break* (*b*), and *repair* (*r*), as opposed to evolving according to time.

A DES model is well suited to capturing important characteristics of many inherently discrete systems including computer or manufacturing systems, as well as for the high-level modeling of complex systems such as a fleet of autonomous vehicles. These models then can be employed for verifying certain system properties and for designing the discrete control logic.

The discrete event control of these systems at its most fundamental level is concerned with the ordering and synchronization of discrete events and the prevention of forbidden events. For example, discrete logic is needed for routing parts through a factory, for choosing between a set of sensors in an airplane control system, and for tasking a fleet of autonomous vehicles. Traditionally, the analysis and design of discrete control logic has been handled in a rather *ad hoc* manner. The programming of a Programmable Logic Controller (PLC) for factory automation is often described as being as much an art as it is a science. As systems become increasingly complex and design cycle times become increasingly short, the need for formal analysis and design techniques has become readily apparent to engineers and users alike. Anyone who has had their desktop PC crash or their car's "check engine" light come on for no apparent reason can begin to appreciate the difficulty of designing DES and their control. Purely logical models also can be appended with timing and probabilistic information for answering questions about when and with what frequency certain events occur and for answering questions about the average behavior of a system. Additionally, discrete event models can be combined with continuous time differential equation models for generating what are referred to as *hybrid system models*.

Much of the body of DES theory has its foundations in computer science research on automata theory and formal verification. More recently, the control engineering community has extended these ideas so that not only can correct behavior be verified, but it is now further possible that control logic can be synthesized to guarantee correct behavior. The basis for much of the theory for the control of DES was developed in the 1980s and has its origins in the seminal work of Ramadge and Wonham [2]. This broad approach to DES control is generally referred to as *supervisory control* and remains an active area of research. Supervisory control is a feedback approach to control that has many analogs with more commonly applied continuous, time-based control techniques and will be the focus of the material presented here. Specifically, Section 5.5.2 discusses logical DES modeling, Section 5.5.3 presents results on supervisory controller existence and synthesis, Section 5.5.4 discusses state estimation and fault diagnosis, Section 5.5.5 introduces more sophisticated approaches to supervisory control involving hierarchy and modularity, Section 5.5.6 introduces some DES models that include timing and probabilistic information, and the chapter concludes with a summary in Section 5.5.7. The discussion of DES in this chapter owes much to the presentation in [1] and [3] and both references provide the reader sources from which to seek further details.

5.5.2 Logical Discrete Event Models

5.5.2.1 Finite State Automata and Languages

As stated previously, DES models are characterized by discrete states and event-driven dynamics. One of the most common types of DES models and the one that will be primarily employed here is the *finite state automaton*, otherwise known as the finite state machine. The model shown in Figure 5.16 is an example of a finite state automaton. This is a strictly logical model where the system remains in each state for an unspecified amount of time and the transitions between states are triggered by events that occur instantaneously. The logic by which these events are triggered is not necessarily specified within the model and any analysis or control synthesis must assume that any transition that is possible in a given state can occur at any time.

The current state of a *deterministic* automaton model can be completely determined by knowledge of the events that have occurred in the model's past. Specifically, the automaton begins in its initial state, denoted graphically here by a short arrow (see Figure 5.16). When an event occurs, the transition structure of the model indicates the next state of the model that is entered. With a deterministic model, the occurrence of an event completely determines the successive state. In other words, an event may not lead to multiple states from a given originating state. In the graphical model, states with double circles are *marked* to indicate successful termination of a process. Mathematically, a deterministic automaton can be denoted by the five-tuple $\mathbf{G} = (Q, \Sigma, \delta, q_0, Q_m)$, where Q is the set of states, Σ is the set of events, $\delta : Q \times \Sigma \to Q$ is the state transition function, $q_0 \in Q$ is the initial state, and $Q_m \subseteq Q$ is the set of marked states. The notation Σ^* will represent the set of all finite strings of elements of Σ, including the empty string ε, and is called the *Kleene-closure* of the set Σ. The empty string ε is a null event for which the system does not change state. In this presentation the function δ is also extended to $\delta : Q \times \Sigma^* \to Q$. The notation $\delta(q, s)!$ for any $q \in Q$ and any $s \in \Sigma^*$ denotes that $\delta(q, s)$ is defined. An event σ is said to be *feasible* at a state q if $\delta(q, \sigma)!$. It is sometimes appropriate to employ a nondeterministic automaton model where there may be multiple initial states and an event σ may transition the model to any one of several different states from the same originating state q. All automata employed here may be assumed to be deterministic unless otherwise noted.

Another type of representation of a DES that will be employed here is that of a *language*. A language representation consists of the set of all possible strings of events $\sigma_1 \sigma_2 \sigma_3 \cdots$ that can be generated by a DES. The language representation of a DES can be formally defined in terms of the automaton model of the DES. For example, the language generated by the deterministic automaton model $\mathbf{G}$ is defined by $\mathcal{L}(\mathbf{G}) = \{s \in \Sigma^* \mid \delta(q_0, s)!\}$. Furthermore, the *marked language* generated by a DES represents the subset of strings in $\mathcal{L}(\mathbf{G})$ that end in the successful termination of a process. Otherwise stated, $\mathcal{L}_m(\mathbf{G}) = \{s \in \Sigma^* \mid \delta(q_0, s) \in Q_m\}$. For the string $s = ru \in \Sigma^*$ formed from the catenation of the strings r and u, r is called a prefix of s and is denoted $r \leq s$. The notation $\overline{K}$ represents the set of all prefixes of strings in the language K, and is referred to as the *prefix-closure* of K.

Let the automaton of Figure 5.16 be named $\mathbf{M}$. Therefore, the languages generated and marked by this automaton are $\mathcal{L}(\mathbf{M}) = \{\varepsilon, s, sf, sb, sbr, sfs, sbrs, \ldots\}$ and $\mathcal{L}_m(\mathbf{M}) = \{\varepsilon, sf, sbr, sfsf, sbrsf, \ldots\}$ where the sets are countably infinite since they contain strings of arbitrarily long length. The fact that these sets are countably infinite makes it impossible to list all of the strings that they contain. These languages, however, have an obvious structure that allows them to be represented by a (nonunique) automaton that generates and marks the given languages. Languages that can be generated by a finite-state automaton are said to be *regular*. Note that there exist *nonregular* languages that cannot be represented by automata with a finite state space. Other formalisms, however, do exist that can represent some nonregular languages with a finite transition structure.

Complex systems often consist of many individual subsystems. It is generally easiest to model each component system by its own automaton. In order to model the synchronous operation of these component models, a *synchronous composition* (or parallel composition) operator denoted ∥ may be employed. With this operator, an event that is common to multiple automata can only occur if that event is able to occur synchronously in each of the automata that share the given event. If a component automaton employs an event that is not shared with any other automata, it may then enact the event without participation of any of the other automata. A formal definition for the synchronous composition of two automata is given below.

Definition 5.16:

The synchronous composition of two automata G_1 and G_2, where $G_1 = (Q_1, \Sigma_1, \delta_1, q_{01}, Q_{m1})$ and $G_2 = (Q_2, \Sigma_2, \delta_2, q_{02}, Q_{m2})$ is the automaton

$$G_1 \| G_2 = (Q_1 \times Q_2, \Sigma_1 \cup \Sigma_2, \delta, (q_{01}, q_{02}), Q_{m1} \times Q_{m2})$$

where the transition function $\delta : (Q_1 \times Q_2) \times (\Sigma_1 \cup \Sigma_2) \to (Q_1 \times Q_2)$ is defined for $q_1 \in Q_1, q_2 \in Q_2$ and $\sigma \in (\Sigma_1 \cup \Sigma_2)$ as

$$\delta((q_1, q_2), \sigma) := \begin{cases} (\delta_1(q_1, \sigma), \delta_2(q_2, \sigma)) & \text{if } \delta_1(q_1, \sigma)! \text{ and } \delta_2(q_2, \sigma)! \\ (\delta_1(q_1, \sigma), q_2) & \text{if } \delta_1(q_1, \sigma)! \text{ and } \sigma \notin \Sigma_2 \\ (q_1, \delta_2(q_2, \sigma)) & \text{if } \sigma \notin \Sigma_1 \text{ and } \delta_2(q_2, \sigma)! \\ \text{undefined} & \text{otherwise.} \end{cases}$$

As an example, the synchronous composition of two instances of the automaton shown in Figure 5.16 can be considered where the event labels b, r, and s are appended by a "1" for the first machine M_1 and a "2" for the second machine M_2. The finish event f for both machines remains unchanged meaning that this event must occur simultaneously in the two machines. Figure 5.17 represents the resulting synchronous composition $M_1 \| M_2$ where each state is given by a pair (q_1, q_2) with the first element representing the state of the first machine and the second element representing the state of the second machine. Note that each of the individual machines are modeled by three states each, while the concurrent operation of the machines requires $3 \times 3 = 9$ states. From this example, one can see that in the worst case the state space of a synchronous composition grows exponentially with the number of components in the system. This "explosion" of the state space represents one of the largest challenges to the application of DES theory.

In terms of the synchronous composition of languages, if G_1 and G_2 possess the same event set Σ, then $\mathcal{L}(G_1) \| \mathcal{L}(G_2) = \mathcal{L}(G_1) \cap \mathcal{L}(G_2)$. In order to precisely define the synchronous composition for languages possessing different event sets, the following *natural projection* operator $P_i : \Sigma^* \to \Sigma_i^*$ needs to be defined:

$$P_i(\varepsilon) := \varepsilon$$

$$P_i(\sigma) := \begin{cases} \sigma, & \sigma \in \Sigma_i \subseteq \Sigma \\ \varepsilon, & \sigma \notin \Sigma_i \subseteq \Sigma \end{cases} \tag{5.254}$$

$$P_i(s\sigma) := P_i(s)P_i(\sigma), s \in \Sigma^*, \sigma \in \Sigma.$$

Given a string $s \in \Sigma^*$, the projection P_i erases those events in the string that are in the alphabet Σ but not in the subset alphabet Σ_i. The inverse projection operation can also be defined:

$$P_i^{-1}(t) := \{s \in \Sigma^* \mid P_i(s) = t\}. \tag{5.255}$$

The effect of the inverse projection P_i^{-1} is to extend the local event set Σ_i to Σ. In terms of automata, this is represented by adding self-loops at every state for each event in the set $(\Sigma \setminus \Sigma_i)$. Formally, a self-loop for the event σ means that $q = \delta(q, \sigma)$. These self-looped events are in essence enabled at every state and as such do not meaningfully restrict the behavior of the system.

The projection definitions given by Equations 5.254 and 5.255 can be extended naturally from strings to languages and then applied to give a formal definition of the synchronous composition for languages defined over different event sets. In the following, $P_i : \Sigma^* \rightarrow \Sigma_i^*$, where $\Sigma = \Sigma_1 \cup \Sigma_2$:

$$L_1 \| L_2 := P_1^{-1}(L_1) \cap P_2^{-1}(L_2).$$

5.5.2.2 Petri Nets

An alternative and widely employed representation of DES is the Petri net [4]. Like automata, Petri nets represent the structure of a DES graphically. Specifically, a Petri net is a *bipartite graph* with two different types of nodes, *transitions* and *places* connected by *arcs* that only connect nodes of different types. Referring to Figure 5.18, transitions represent events and are shown graphically by bars, and places represent some condition being met and are represented by circles. When a *token*, represented by a black dot, is in a place, it means that the condition represented by the place is currently satisfied. The arrangement of tokens in a Petri net graph is referred to as the net's *marking* and corresponds to the state of the DES. When all of the places that have arcs directed toward a particular transition are filled, it means that the conditions for that transition to occur have been met and the transition can "fire." When a transition fires, it consumes the tokens in the places with arcs directed toward the transition and generates new tokens in the places that are reached by arcs directed away from the given transition. The event represented by this transition is again modeled as occurring instantaneously. An arc may generate or consume multiple tokens at once if it possesses a nonunity *weighting*. Mathematically, the Petri net structure is represented (P, T, A, w), where P is the finite set of places, T is the finite set of transitions, $A \subseteq (P \times T) \cup (T \times P)$ is the set of arcs directed from places to transitions and directed from transitions to places, and w is the weighting associated with each arc. The state, or marking, of the Petri net may be represented by a vector of integers equal to the number of tokens in each corresponding place.

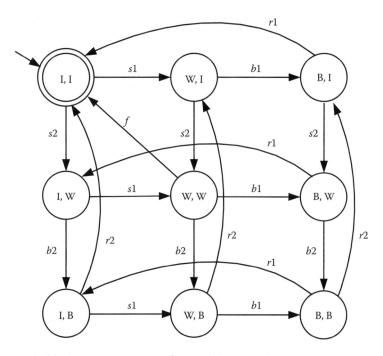

FIGURE 5.17 Model of the concurrent operation of two machines, $\mathbf{M_1} \| \mathbf{M_2}$.

The example in Figure 5.18 can be interpreted as representing two machines operating together with an assembly station. For example, places $\{p1, p2, p3\}$ together with transitions $\{t2, t3, t4, t9\}$ model a simple machine like the one modeled by the automaton in Figure 5.16. A token in place $p1$ indicates that the machine is *idle* and a workpiece is available to be machined. A firing of transition $t2$ then represents the machine beginning operation and causes the token to be removed from place $p1$ and a token to be generated in place $p2$ representing that the machine is *working*. A firing of transition $t3$ represents the machine has broken down and a subsequent firing of transition $t4$ means the machine has been repaired. The second machine is similarly represented by places $\{p4, p5, p6\}$ and transitions $\{t5, t6, t7, t8, t9\}$. Transition $t9$ represents the assembly of the finished parts from the individual machines and place $p7$ represents a buffer containing the finished assemblies.

One of the primary advantages of Petri nets as compared to automata is that they are able to represent concurrency more compactly. As seen in Figure 5.18, the complexity of the graph grew linearly with the addition of the second machine, while the complexity of the automaton representation shown in Figure 5.17 grew exponentially. Despite the compactness of the Petri net representation, the underlying size of the state space is essentially unchanged. This can be seen by considering the different combinations of token placements (presuming each machine employs only a single token at a time).

The number of tokens in place $p7$ represents the total number of assemblies that have been completed and stored in a buffer. The fact that there is no limit to the number of tokens that can be collected in the places of this example illustrates another advantage of the Petri net modeling formalism; Petri nets can

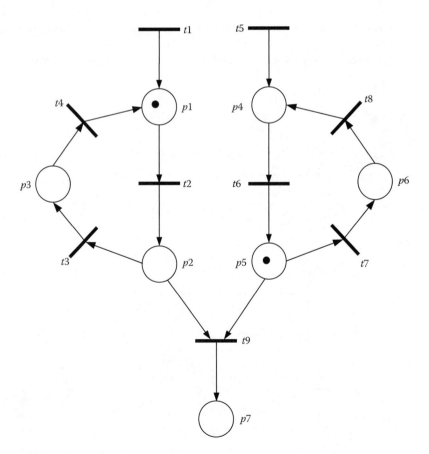

FIGURE 5.18   Petri net example.

represent systems that have an infinite state space. Otherwise stated, Petri nets are able to represent a larger class of behaviors than the class of regular languages that can be represented by finite state automata. This class of Petri net languages, however, presents difficulties for implementing some verification and synthesis algorithms that are readily implementable for regular languages using finite state automata.

Another useful aspect of Petri nets is that their structure and evolution can be represented by linear-algebraic state equations. Specifically, the structure of the Petri net is captured by an $n \times m$ incidence matrix A, where n is the number of transitions, m is the number of places, and the jth element of A is equal to the difference between the weighting of the arc directed from the jth transition toward the ith place and the arc directed from the ith place toward the jth transition. Mathematically, the jth element of A is calculated:

$$a_{j,i} = w(t_j, p_i) - w(p_i, t_j)$$

The (2,1) element of the incidence matrix for the Petri net given in Figure 5.18 is, therefore, equal to -1 since there is an arc of weight one directed from place $p1$ to transition $t2$ and there is no arc directed from transition $t2$ to place $p1$. As mentioned previously, the marking of a Petri net corresponds to its state and can be represented by an m-element vector where the ith element of the vector is equal to the number of tokens in the place p_i. For the current marking of the example in Figure 5.18, the state vector is $\mathbf{x} = [1\ 0\ 0\ 0\ 1\ 0\ 0]^T$. The evolution of the Petri net from the state at step $k-1$ to the state at step k is then given by the following state equation

$$\mathbf{x}_k = \mathbf{x}_{k-1} + A^T \mathbf{u}_k, \ k = 1, 2, 3, \ldots$$

where $\mathbf{u}_k$ is referred to as the *firing vector* and indicates the transition(s) being fired at step k. The current state of the Petri net in Figure 5.18 indicates that transition $t2$ and transition $t7$ could be fired. Petri nets differ from automata in that multiple events can occur concurrently. If transition $t2$ and $t7$ were both fired, then the firing vector would equal $\mathbf{u} = [0\ 1\ 0\ 0\ 0\ 0\ 1\ 0\ 0]^T$.

5.5.2.3 Other Model Types

Other types of discrete event models exist and have found acceptance besides finite state automata and Petri nets; prominent examples include *statecharts* and a variety of *process algebras*. Statecharts are a powerful extension of automata (state machines) that, much like Petri nets, are able to represent concurrency much more compactly than traditional automata can. Process algebras use a small set of primitives to generate traces that describe the desired and actual behavior of a system. With process algebras traces are generated in a language without explicitly representing the system's "state." Operators are also defined for combining these expressions. The *Communicating Sequential Processes* (CSP) formalism [5] is one prominent example of a process algebra. Despite the compactness of some of these representations, if the algorithms for verification or synthesis depend on a search of the state space, there are still challenges with computational complexity. Some formalisms also exist for representing the state space and transition structure of a DES, such as *Binary Decision Diagrams* (BDDs), that offer significant computational advantages. In general, the specific modeling formalism that should be employed depends upon the particular application.

5.5.3 Supervisory Control

The supervisory control framework of Ramadge and Wonham [2] employs a feedback-type architecture similar to that employed in common control engineering practice. Figure 5.19 illustrates the supervisory control architecture where both the plant G and the supervisory controller S are modeled as DES, for example, as automata. In this framework, the supervisor observes the events generated by the plant and makes its control decisions based upon these observations. One aspect of supervisory control that is different from most control techniques is that a supervisor typically prevents rather than forces events

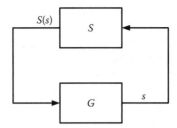

FIGURE 5.19 The supervisory control architecture.

to occur. Furthermore, the logic that the "plant" uses for the triggering of events is not usually explicitly included in its DES model. The conditions by which the transmission of an automobile changes gears, for example, may depend on the engine speed, or load, or driver inputs, but this information would not necessarily be included in a DES model of the transmission. The supervisor does not necessarily need any knowledge of the triggering information to make its control decisions; it is concerned only with preventing an event if it will lead to an "illegal" state or if it will occur in the wrong sequence as compared to other events. In other words, it is the goal of the supervisory controller to restrict the behavior of the uncontrolled plant to meet some given specification. For example, a supervisory controller may prevent a transmission from entering reverse if it has observed events that indicate that the automobile is traveling forward at greater than 15 miles per hour.

While the supervisor can be modeled as an automaton **S**, it is fundamentally a mapping that upon observation of a string generated by a plant automaton **G** outputs a list of events to be enabled, $S : \mathcal{L}(\mathbf{G}) \to 2^{\Sigma}$. This type of language-based formulation of supervisory control is a dynamic event-feedback law characterized by the fact that the control action the supervisor generates depends not on the current state of the plant **G**, but rather on the history of events that brought the system to that state. The resulting closed-loop behavior $S/\mathbf{G}$ can be expressed as a prefix-closed language $\mathcal{L}(S/\mathbf{G})$. It then follows that the closed-loop marked language is defined as $\mathcal{L}_m(S/\mathbf{G}) = \mathcal{L}(S/\mathbf{G}) \cap \mathcal{L}_m(\mathbf{G})$. The closed-loop behavior also can be represented by the synchronous composition $\mathbf{S} \| \mathbf{G}$. A less common formulation of supervisory control implements a *state-feedback law* that bases its control only on the plant's current state.

5.5.3.1 Supervisor Existence

Given a specification language K representing the set of allowed behaviors for a given system, it is desired that a supervisor be synthesized so that the closed-loop behavior $S/\mathbf{G}$ satisfies certain properties. A complicating factor is that generally not all events can be disabled by the supervisor. For example, once a machine begins a process it may not be possible to prevent the process from finishing. Other examples include those events that come from the external environment, such as a human operator hitting a button or opening a door. These events are defined to be uncontrollable and the event set of a DES is thus partitioned into *controllable* and *uncontrollable* events, $\Sigma = \Sigma_c \dot{\cup} \Sigma_{uc}$. The set of events enabled by a supervisor S, therefore, must implicitly include all uncontrollable events.

The most fundamental goal of a supervisor is to achieve the property of *safety*. Safety means that the system's behavior remains within the allowed set of behaviors, $\mathcal{L}(S/\mathbf{G}) \subseteq \overline{K}$. Another desirable property is that the closed-loop system be *nonblocking*. An automaton is said to be nonblocking when from all of its reachable states a marked state can be reached. From a language point of view, this is defined as $\overline{\mathcal{L}_m(S/\mathbf{G})} = \mathcal{L}(S/\mathbf{G})$. In other words, a system is nonblocking if its processes can always run to "completion." A secondary goal of a supervisor is that it be *optimal*. In this context, optimality is defined as the supervisor allowing the largest set of behaviors as determined by set inclusion.

Before a supervisor that provides safe, nonblocking behavior can be constructed, it must be determined if such a supervisor even exists. Specifically, the property of *controllability* guarantees that a supervisor exists that can restrict the behavior of the plant G to exactly achieve the set of allowed behaviors $\overline{K}$. This is stated formally as follows where the language L can be thought of as representing the behavior of the DES G.

Definition 5.17:

Let K and $L = \overline{L}$ be languages defined over the event set Σ. Let $\Sigma_{uc} \subseteq \Sigma$ be the set of uncontrollable events. The language K is said to be controllable with respect to L and Σ_{uc} if

$$\overline{K}\Sigma_{uc} \cap L \subseteq \overline{K}. \tag{5.256}$$

The following theorem [3] employs this controllability property. □

Theorem 5.18:

Let the DES G be represented by the automaton $\mathbf{G} = (Q, \Sigma, \delta, q_0, Q_m)$, where $\Sigma_{uc} \subseteq \Sigma$ is the set of uncontrollable events. Also, let $K \subseteq \mathcal{L}(\mathbf{G})$, where $K \neq \emptyset$. Then there exists supervisor S such that $\mathcal{L}(S/\mathbf{G}) = \overline{K}$ if and only if K is controllable with respect to $\mathcal{L}(\mathbf{G})$ and Σ_{uc}. □

An alternative representation of the controllability condition is

$$\text{for all } s \in \overline{K}, \quad \text{for all } \sigma \in \Sigma_{uc}, \quad s\sigma \in L \ \Rightarrow \ s\sigma \in \overline{K}.$$

A string s that does not satisfy the above expression will be referred to as an uncontrollable string. The controllability condition requires that an uncontrollable event cannot lead the DES G outside of the allowed set of behaviors $\overline{K}$ because such an event cannot be prevented by a supervisor S. Note that controllability is fundamentally a property of a language's prefix-closure $\overline{K}$.

In the case that K is a regular language and $\mathbf{G}$ is a finite state automaton, controllability can be verified by comparing the transition functions of the automata $\mathbf{G}$ and $\mathbf{H}\|\mathbf{G}$ where $\mathbf{H}$ is the automaton generator for $\overline{K}$. More specifically, if an uncontrollable event is defined at a state q in the automaton $\mathbf{G}$, then the same uncontrollable event must be defined in the corresponding state of $\mathbf{H}\|\mathbf{G}$, that is, for those states (p, q) that share the same second element. In the worst case, the computational complexity of this test is $\mathcal{O}(|\Sigma|mn)$, where $|\Sigma|$ is the number of events in the set Σ, m is the number of states in $\mathbf{H}$, and n is the number of states in $\mathbf{G}$.

If it is desired that a supervisor additionally provide nonblocking behavior, then a second requirement denoted $\mathcal{L}_m(\mathbf{G})$-*closure* is needed in addition to controllability. This result is captured in the following theorem [3].

Theorem 5.19:

Let the DES G be represented by the automaton $\mathbf{G} = (Q, \Sigma, \delta, q_0, Q_m)$, where $\Sigma_{uc} \subseteq \Sigma$ is the set of uncontrollable events. Also, let $K \subseteq \mathcal{L}_m(\mathbf{G})$, where $K \neq \emptyset$. Then there exists a nonblocking supervisor S such that $\mathcal{L}_m(S/\mathbf{G}) = K$ and $\mathcal{L}(S/\mathbf{G}) = \overline{K}$ if and only if the following two conditions hold:

1. *Controllability: $\overline{K}\Sigma_{uc} \cap \mathcal{L}(\mathbf{G}) \subseteq \overline{K}$*
2. *$\mathcal{L}_m(\mathbf{G})$-closure: $K = \overline{K} \cap \mathcal{L}_m(\mathbf{G})$.* □

The $\mathcal{L}_m(\mathbf{G})$-closure condition requires that the specification language K has a marking that is consistent with the marking of the uncontrolled plant $\mathbf{G}$. If it is desired to specify the marking of the system through K, then only the controllability condition is needed to guarantee safe, nonblocking behavior. In this instance, the associated S may be referred to as a *marking supervisor*.

In addition to some events being uncontrollable, an additional complication to the DES control problem is that some events may be *unobservable*. For example, a sensor needed to identify the occurrence of a particular event may not be included in a system in order to reduce cost. Another common type of unobservable event is a fault event. For example, the malfunction of an actuator may not be directly observable even though events resulting from its failure could be observed. There are also certain conditions under which an event may be intentionally hidden in a model in order to reduce its complexity. Therefore, in addition to partitioning a system's event set Σ into controllable and uncontrollable events, the event set can also be partitioned into observable and unobservable events, $\Sigma = \Sigma_o \dot{\cup} \Sigma_{uo}$.

The problem of control under partial observation [6] requires that the supervisor S make its control decision based on the natural projection of the string s generated by the plant, where $P : \Sigma^* \to \Sigma_o^*$. Recall from its previous definition that in this instance the natural projection P erases from the string s those events that are unobservable. A supervisor acting under partial observation will, in essence, hold its previous control action until the next observable event is generated by the plant. Such a supervisor will be identified here as a P-supervisor $S_P : P(\mathcal{L}(\mathbf{G})) \to 2^{\Sigma}$. The behavior generated under P-supervision $S_P/\mathbf{G}$ can be represented by the language $\mathcal{L}(S_P/\mathbf{G})$. Under conditions of partial observation it is necessary that two strings that have the same observation not require conflicting control actions. In other words, if an event must be disabled following a string s then it must be disabled following all strings s' with the same observation as s, $P(s') = P(s)$. This property is defined formally as *observability* in the following.

Definition 5.18:

Let K and $L = \overline{L}$ be languages defined over the event set Σ. Let Σ_c and Σ_o be subsets of Σ and let the natural projection P be defined $P : \Sigma^* \to \Sigma_o^*$. The language K is said to be observable with respect to L, P, and Σ_c if for all $s \in \overline{K}$ and for all $\sigma \in \Sigma_c$,

$$(s\sigma \notin \overline{K}) \quad and \quad (s\sigma \in L) \;\Rightarrow\; P^{-1}[P(s)]\sigma \cap \overline{K} = \emptyset \qquad \square$$

The property of observability technically means that if a continuation $\sigma \in \Sigma_c$ can occur in the plant language L following the string s, but is not allowed in the specification language $\overline{K}$ following s, then the continuation should not be allowed following any string that has the same projection as s, that is, for any string $s' \in P^{-1}[P(s)]$. As was the case with controllability, the observability of a language is fundamentally a property of its prefix-closure. In the above definition, the event set Σ_c is arbitrary, but often represents the set of controllable events because if the controllability requirement is satisfied then the above definition is implicitly satisfied for all $\sigma \in \Sigma_{uc}$. The property of observability can be verified with complexity $\mathcal{O}(m^2 n)$ where the term $|\Sigma|$ has been absorbed into the constants of $\mathcal{O}(\cdot)$ and again m is the number of states in the automaton generator of $\overline{K}$ and n is the number of states in $\mathbf{G}$ where $L = \mathcal{L}(\mathbf{G})$. Verification of the observability property will be discussed further when the topic of state estimation is introduced.

This observability property then can be used in conjunction with previously defined properties to guarantee the existence of a safe, nonblocking supervisor that exactly achieves the set of allowed behaviors K under the condition that some events cannot be observed. This result is captured in the following theorem [3].

Theorem 5.20:

Let the DES G be represented by the automaton $G = (Q, \Sigma, \delta, q_0, Q_m)$, *where* $\Sigma_{uc} \subseteq \Sigma$ *is the set of uncontrollable events and* $\Sigma_{uo} \subseteq \Sigma$ *is the set of unobservable events. Also, let* $K \subseteq \mathcal{L}_m(G)$, *where* $K \neq \emptyset$. *Then there exists a nonblocking supervisor* S *such that* $\mathcal{L}_m(S/G) = K$ *and* $\mathcal{L}(S/G) = \overline{K}$ *if and only if the following three conditions hold:*

1. *K is controllable with respect to $\mathcal{L}(G)$ and Σ_{uc}.*
2. *K is observable with respect to $\mathcal{L}(G)$, P, and Σ_c.*
3. *K is $\mathcal{L}_m(G)$-closed.* □

5.5.3.2 Supervisor Synthesis

In the previous section, requirements were presented to provide for the existence of a safe, nonblocking supervisor to exactly achieve the set of allowed behaviors K. These requirements not only demonstrate the existence of a supervisor, but they are also constructive in that if the language K is regular and the plant G is represented by a finite state automaton G, then an automaton that generates $\overline{K}$ and marks K represents a supervisor function that provides safe, nonblocking behavior. For the full observation case, the set of events that are defined in the supervisor automaton $S = (Q_S, \Sigma, \delta_S, q_{S0}, Q_{Sm})$ at a given state q reached by a string of events s defines the output of the function S. More formally, $S(s) = \{\sigma \in \Sigma | \delta_S(q, \sigma)!$ where $q = \delta_S(q_{S0}, s)\}$. As stated previously, the supervised behavior of the system then can be represented by the synchronous composition $S \| G$.

In general, the existence conditions of the previous section provide a means for testing whether or not a given automaton representing the set of allowed behaviors may serve as a supervisor for the uncontrolled plant. The theory of supervisory control is more developed, however, in that it can address the situation where the conditions of Theorem 5.20 are not met. When a supervisor that is able to exactly achieve the set of allowed behaviors K does not exist, it is often desirable to find the supervisor S (or S_P) that can provide the behavior that most closely approximates K. Since the specification language K represents the set of allowed behaviors, it is generally the goal to find the least restrictive supervisor that keeps the system behavior within this set of legal traces. Such a supervisor, if it exists, is considered optimal with respect to set inclusion in that it provides behavior that is equal to the largest achievable sublanguage of K.

The case where all events are observable will be considered first; in other words, consider that the language K is not controllable with respect to $\mathcal{L}(G)$ and Σ_{uc}. It is, therefore, desired to find the largest sublanguage of K (if possible) that is controllable with respect to $\mathcal{L}(G)$. Let $C(K, L)$ represent the set of sublanguages of K that are controllable with respect to the prefix-closed language L and the set of uncontrollable events Σ_{uc}:

$$C(K, L) := \{K' \subseteq K \mid \overline{K}' \Sigma_{uc} \cap L \subseteq \overline{K}'\}$$

It can be shown that the property of controllability is preserved under arbitrary unions. Therefore, the set $C(K, L)$ has a supremal element sup $C(K, L)$ that is equal to the union of all controllable sublanguages of K:

$$\sup C(K, L) := \bigcup_{K' \in C(K,L)} K'$$

Furthermore, the sup C operation preserves $\mathcal{L}_m(G)$-closure. Therefore, based on Theorem 5.19, the supremal controllable sublanguage of K for a given plant G, sup $C(K, \mathcal{L}(G))$, represents the largest set of behaviors that can be achieved by a safe, nonblocking supervisor and the automaton generator for this language represents an optimal supervisor function. If K is controllable with respect to $\mathcal{L}(G)$, it logically follows that $\sup \mathcal{C}(K, \mathcal{L}(\mathbf{G})) = K$. Note also that in the worst case the supremal controllable sublanguage may be empty.

The supremal controllable sublanguage $\sup \mathcal{C}(K, \mathcal{L}(\mathbf{G}))$ can be constructed in a rather intuitive manner. From a language point of view, one can start with the language $\overline{K}$ and remove all strings that violate the controllability condition of Equation 5.256. The resulting prefix-closed language $K^{\uparrow}$ is then controllable, but may be blocking. In other words, there may be strings in $K^{\uparrow}$ whose continuations in the marked language K are not in $K^{\uparrow}$. The intersection of $K^{\uparrow}$ with K produces a language $K^{\uparrow} \cap K$ that is nonblocking, but that may no longer be controllable. Therefore, the whole process starts over again with the language $K^{\uparrow} \cap K$. This process is repeated until the resulting language is both controllable and nonblocking or the language is empty.

The automaton generator for the supremal controllable sublanguage $\sup \mathcal{C}(K, \mathcal{L}(\mathbf{G}))$ can be constructed in much the same manner as the outline presented from a language point of view. Consider the following example.

Example 5.20:

Consider the automata **G** and **H** shown in Figure 5.20 that generate and mark the plant and specification languages, respectively. Also, let the event set Σ be partitioned into controllable and uncontrollable event sets $\Sigma_c = \{a, c\}$ and $\Sigma_{uc} = \{b\}$.

The automaton generator for the supremal controllable sublanguage $\sup \mathcal{C}(K, \mathcal{L}(\mathbf{G}))$ can be constructed by beginning with the parallel composition **H‖G** which is shown in Figure 5.21. Comparing

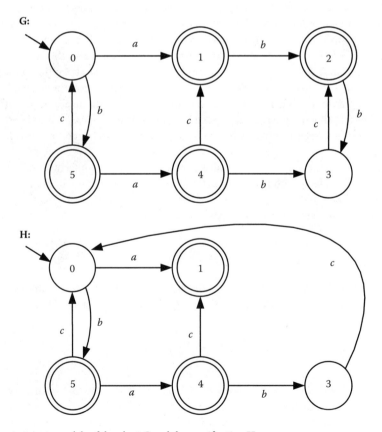

FIGURE 5.20   Automata models of the plant **G** and the specification **H**.

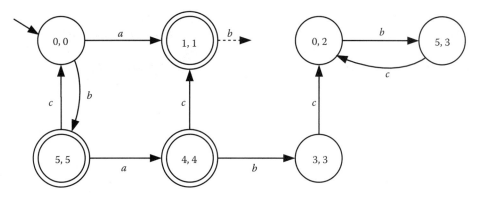

FIGURE 5.21 Automaton model of **H**||**G**.

this automaton to the uncontrolled plant **G**, it is apparent that the specification language K is not controllable with respect to $\mathcal{L}(\mathbf{G})$ and Σ_{uc} since the specification requires that the uncontrollable continuation b be disabled following the occurrence of the string a.

If the state $(1, 1)$ is removed from the automaton **H**||**G**, then the generated language is controllable, but it is still blocking since from states $(3, 3)$, $(0, 2)$, and $(5, 3)$ a marked state cannot be reached. Removing these states makes the automaton nonblocking, but in turn also makes the generated language uncontrollable since it would require that the uncontrollable continuation b be disabled following an occurrence of the string ba. Removing the state $(4, 4)$ leaves an automaton that generates and marks a controllable and nonblocking sublanguage of K. Otherwise stated, the automaton generator for $\sup \mathcal{C}(K, \mathcal{L}(\mathbf{G}))$ consists of the states $(0, 0)$ and $(5, 5)$ and the transitions between them.

To summarize the procedure of the above example, the state space of **H**||**G** is searched and states are alternately removed that are either reached by strings that violate controllability or are blocking. This process continues until the generated language is controllable and nonblocking or the resulting automaton is empty. The maximum number of iterations of this naive algorithm is nm since **H**||**G** has at most nm states and at least one state is removed each time through the process. Note that there are multiple academically available software packages to implement the various verification and synthesis algorithms mentioned so far. Two widely employed software tools are DESUMA [7] and Supremica [8]. Furthermore, the development of increasingly efficient algorithms and data structures is an active area of research; even the polynomial complexity algorithms mentioned thus far can be overwhelmed by systems with very large state spaces.

5.5.3.3 Supervisor Synthesis under Partial Observation

The presence of events that cannot be observed makes the problem of supervisor synthesis much more difficult. Recalling Theorem 5.20, under partial observation the existence of a supervisor that can provide safety requires the property of observability. As stated previously, the property of observability can be verified with polynomial complexity in the number of states of the automaton representations of the languages. Despite this, and the fact that a supremal controllable sublanguage can also be constructed with polynomial complexity, the construction of a supervisor under partial observation to achieve a prescribed behavior K has, in the worst case, exponential complexity in the number of states. This fact will become clearer when the problem of state estimation is discussed.

The problem of constructing a supervisor when the legal language K is not observable is also much more difficult. Unlike controllability, the property of observability is not closed under union. Therefore, when a language K is not observable there does not generally exist a supremal observable sublanguage.

The implication of this fact is that it is, in general, not possible to find the supervisor under partial observation that is "optimal" with respect to set inclusion. For example, there may exist multiple observable sublanguages of K that are not contained in any other observable sublanguage of K. Otherwise stated, there may be multiple *maximal* observable sublanguages of K that are incomparable. While it may not always be possible to construct a supremal observable sublanguage, techniques do exist for constructing maximal observable sublanguages.

One approach for dealing with unobservability is to try to achieve a stronger property that implies observability and is closed under union. The property of *normality* is such a condition and is defined in the following.

Definition 5.19:

Let K and $L = \overline{L}$ be languages defined over the event set Σ. Let the natural projection P be defined for the subset of events $\Sigma_o \subseteq \Sigma$ such that $P : \Sigma^* \rightarrow \Sigma_o^*$. The language K is said to be normal with respect to L and P if

$$\overline{K} = P^{-1}[P(\overline{K})] \cap L$$

$\square$

Normality means that the language $\overline{K}$ (with unobservable events) can be completely reconstructed from knowledge of its observation $P(\overline{K})$ and the language L. From a practical point of view, L may represent the behavior of the plant $\mathcal{L}(\mathbf{G})$ and hence the legal behavior $\overline{K}$ that cannot be performed by the plant does not need to be reconstructed. The fact that the property of normality is closed under union provides that a supremal normal sublanguage of K with respect to L does exist and is denoted $\sup \mathcal{N}(K, L)$. Furthermore, a supremal controllable and normal sublanguage of K, $\sup \mathcal{CN}(K, L)$, also exists and the automaton generator of such a language can represent a supervisor $\mathcal{S}_P$ that can be employed under partial observation. Note that the construction of such a supervisor automaton has, in the worst case, exponential complexity in the number of states of the automaton representations of the languages of interest. A special case when a supremal control and observable sublanguage does exist is when all controllable events are observable $\Sigma_c \subseteq \Sigma_o$.

An approach to control under partial observation that can be constructed with reduced complexity is to employ a static *state-feedback* approach to control where the control is based on the current state of the automaton rather than the history of events that brought the system to that state [1]. Under partial observation, the state space is observed through a mask that effectively partitions the state space into sets of indistinguishable states. Since the control action must be the same for different strings that take the system to the same state (or partition), a *state-feedback* approach to control is, in general, more restrictive than the dynamic event-feedback laws that are the focus of the discussion presented here. There may be, however, instances where this trade-off of permissiveness for reduced computational complexity is acceptable.

5.5.3.4 Control of Petri Nets

In the preceding section, the supervisory control framework was explored using the modeling formalism of automata. Petri nets can also be employed for verifying system properties and synthesizing control. Even though the representation of the state space differs with Petri nets, the underlying size of the state space is essentially unchanged. Therefore, the necessary computation for verification or synthesis is comparable if it depends on the number of states of the system. Furthermore, supervisory control in the sense of Ramadge and Wonham is more difficult to implement using Petri nets because it is not as straightforward to remove blocking states or implement a feedback architecture.

There are, however, techniques for control that are specific to Petri nets that do not require the entire explicit state space to be built [9]. Additionally, some properties of some classes of systems can be verified without exploring the entire state space. Some properties, for example, can be determined based on the Petri net's incidence matrix.

5.5.4 Estimation and Diagnosis

When an automaton **G** has unobservable events or is nondeterministic, the same observed string can lead the DES to different states; therefore, it may not be possible to determine the current state of the system. Under these conditions, an observer automaton $\mathbf{G_{obs}}$ can be constructed with a transition structure defined such that each string s leads to a single state that is the catenation of all states in the original automaton **G** that are reached by strings bearing the observation s. This observer automaton $\mathbf{G_{obs}}$ is useful in that it is deterministic, but generates the same observed language as **G**, that is, $\mathcal{L}(\mathbf{G_{obs}}) = P(\mathcal{L}(\mathbf{G}))$. The observed marked languages are also equal, $\mathcal{L}_m(\mathbf{G_{obs}}) = P(\mathcal{L}_m(\mathbf{G}))$. The observer automaton can, therefore, be used to convert a nondeterministic automaton into a deterministic one. The observer automaton also can be used to construct a supervisor for a partially observed plant; consider the following example.

Example 5.21:

Consider the nondeterministic automaton **G** shown in Figure 5.22 where the set of observable and unobservable events are $\Sigma_o = \{b, c, d\}$ and $\Sigma_{uo} = \{a, e\}$, respectively. The automaton **G** is considered nondeterministic because after the occurrence of event b at state 1, it is unclear whether the system is in state 0 or state 1. This type of uncertainty also arises because events a and e are unobservable. For example, the transition of the automaton from state 0 to state 1 by event a cannot be observed.

The corresponding observer automaton $\mathbf{G_{obs}}$ can be constructed in the manner of [3] and is shown in Figure 5.23.

As expected, the resulting observer automaton $\mathbf{G_{obs}}$ is deterministic and generates the same observed language as **G**. Inspection of the observer automaton also provides some intuition for the observability property introduced in Definition 5.18. For example, let the automaton **G** be the plant and assume that there exists a specification language K that is generated and marked by an automaton **H** that is equivalent to **G** except that the d transition from state 2 to state 0 has been removed. In other words, the specification language K includes the string $aced$, but not the string acd. The synchronous composition $\mathbf{H} \| \mathbf{G}$ equals **H** and indicates that disablement of event d is desired

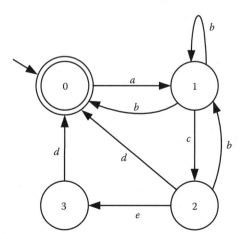

FIGURE 5.22 Model of a nondeterministic automaton with unobservable events, **G**.

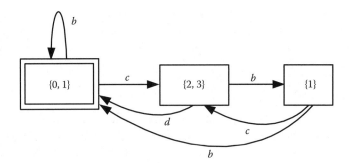

FIGURE 5.23   Example 5.21 observer automaton, $\mathbf{G_{obs}}$.

at state 2 of **G**, but not at state 3. The observer automaton for **H**∥**G** would in this case be equal to $\mathbf{G_{obs}}$ and demonstrates that states 2 and 3 cannot be distinguished following an observation of the string c where $P(ac) = P(ace) = c$. Therefore, the language K requires conflicting control actions for an observation of the string c and hence is not observable.

An issue that arises with building an observer automaton is that in the worst case it has exponential complexity in the number of states of the original automaton because the state space of the observer can be as large as the power set 2^Q of the original automaton's state set Q. This gives some intuition as to why an event-feedback supervisor for a partially observed system cannot be guaranteed to be constructed with polynomial complexity.

The observer automaton also can be used to provide an estimate of what state a partially observed or nondeterministic system is in. For example, following an observation of the string cd the observer automaton $\mathbf{G_{obs}}$ indicates that the automaton **G** is in either state 0 or state 1. The concept of an observer can, therefore, be modified for the purposes of fault diagnosis where the unobservable events represent faults. Identification of the occurrence of a fault may be desirable for several reasons and can be inferred from the occurrence of other events that can be observed. A diagnoser automaton has a similar structure to an observer automaton, except that the states in the observer automaton are appended with a label to indicate whether or not a specific type of fault has occurred or not, for example, $\{2F1, 3NF\}$ where the label $F1$ indicates that a fault of type 1 has occurred and the label NF indicates that no fault has occurred. If all of the labels in a diagnoser automaton state are consistent, then a conclusion can be made as to whether or not a certain type of fault has occurred. See [10] for further details.

5.5.5 Hierarchical and Modular Supervisory Control

The approach to supervisory control that has been discussed so far builds a single monolithic supervisor to control a single plant. As mentioned previously, if a system is complicated and involves multiple subsystems operating concurrently subject to multiple specifications, the complexity of verification and supervisor design can become unmanageable very quickly. An approach for dealing with this complexity is to construct a series of smaller modular supervisors that each attempt to satisfy a single component specification [11]. The control of the monolithic system is then achieved by the conjunction of the control actions of a series of modular supervisors. Specifically, in order for an event to be enabled for the monolithic plant at a given instant that event must be enabled by each of the modular supervisors simultaneously. The advantage of *modular supervisory control* is that it avoids building the full monolithic system, thereby mitigating the state-space explosion problem. Modular control is able to provide safety, but does not guarantee nonblocking unless the modular supervised languages are shown *a priori* to be *nonconflicting.* Two languages K_1 and K_2 are formally defined to be nonconflicting if

$$\overline{K_1 \cap K_2} = \overline{K_1} \cap \overline{K_2}$$

Unfortunately, verifying nonconflict is typically as computationally expensive as building the monolithic system. Therefore, these approaches provide complexity savings in terms of implementation, but not in terms of design. If the modular supervisors are nonconflicting, then the behavior achieved by their conjunction is also optimal.

Another approach for dealing with complexity is to employ abstraction. This approach is classified as *hierarchical supervisory control* and is able to reduce the complexity of verification and controller synthesis by performing the analysis on the abstracted system, which, in general, has fewer states and fewer transitions [12]. In order for the verification to be valid for the real unabstracted system, or for the hierarchical supervisor to provide certain properties like safety, nonblocking, and optimality when applied to the unabstracted system, the abstraction may need to satisfy specific properties. For example, abstractions meeting requirements for *conflict equivalence* can be employed to verify nonconflict, and abstractions that meet requirements for *observational equivalence* (the observer property) can guarantee that a hierarchical supervisor provides nonblocking.

More recent approaches to supervisory control combine elements of modular and hierarchical approaches to control [13]. One class of approach is to build up the global system incrementally employing abstraction, for example. This approach is hierarchical in that different amounts of abstraction define different levels of the hierarchy. The approach is also modular in that each "level" has its own control and the component supervisors work in conjunction to meet the global desired behavior of the overall system. A second class of approach builds modular supervisors that provide safety and nonblocking locally, then employ another level of supervision to resolve conflict between these local supervisors in order to provide global nonblocking. This higher level of control is generally constructed from an abstraction of the local modular subsystems. It is often the case that there is a trade-off between the amount of complexity reduction that can be achieved and the permissiveness of the resulting control.

A related type of DES control referred to as *decentralized supervisory control* involves a plant being controlled by a series of component supervisors, each of which base their control decisions on a different abstraction of the system. In a distributed control architecture, for example, one supervisor may be able to observe event a because it is local, but not event b because it originates in another subsystem, while a different supervisor can observe event b, but not event a. Different approaches to the control and verification of such systems have been devised.

5.5.6 Timed and Stochastic Discrete Event Systems

So far, only purely logical discrete event models have been considered. These models can be appended with timing and probabilistic information to answer questions about when and with what frequency events occur and about the average behavior of systems [3].

One approach for modeling the timed behavior of DES is to append an automaton model with a *clock structure*. In this framework the operation of a DES is no longer indicated by strings of events only $\{\sigma_1, \sigma_2, \ldots\}$, but by sequences where the time of occurrence of each event is also indicated $\{(\sigma_1, t_1), (\sigma_2, t_2), \ldots\}$. An automaton with clock structure is referred to as a *timed automaton* and defined by the five-tuple $\mathbf{G_V} = (Q, \Sigma, \delta, q_0, V)$, where V is the clock structure and $\mathbf{G} = (Q, \Sigma, \delta, q_0)$ is the underlying logical automaton. The clock structure V is comprised of *clock sequences* v_i, one for each corresponding event σ_i in the alphabet Σ. Each clock sequence $v_i = \{v_{i,1}, v_{i,2}, \ldots\}$ indicates the time at which the corresponding event σ_i will occur if it is feasible. In a timed automaton $\mathbf{G_V}$ for example, if there are multiple feasible events at a given state, the transition that actually will be taken is determined by the clock sequence of the event whose next possible occurrence comes first. A consequence of this structure is that the behavior of a timed automaton $\mathbf{G_V}$ is completely predetermined and the language generated by the timed automaton $\mathcal{L}(\mathbf{G_V})$ consists of a single trace.

The stochastic nature of a DES can be modeled by appending probabilistic information to the clock structure of a timed automaton. Specifically, each clock sequence v_i corresponding to an event σ_i can be redefined as a stochastic clock sequence $V_i = \{V_{i,1}, V_{i,2}, \ldots\}$, where the sequence V_i is characterized by a distribution function $F_i(t) = Prob[V_i \leq t]$. A *stochastic timed automaton* is then defined by the five-tuple $\mathbf{G_F} = (Q, \Sigma, p, p_0, F)$, where p is the probabilistic transition function such that $p(q'; q, \sigma)$ defines the probability that q' is reached by the event σ from the state q and $p(q'; q, \sigma) = 0$ if $q' \notin \delta(q, \sigma)$, p_0 is the probability mass function of the initial state such that $p_0(q)$ is the probability that q is the initial state, and F is the stochastic clock structure that consists of the individual distribution functions F_i. The language generated by an untimed automaton corresponds to the set of traces defined by a timed automaton over all possible clock structures; when a single clock structure is chosen this language then reduces again to a single trace.

A limitation of the deterministic timed automaton model presented above is that it may be too restrictive, while a limitation of the stochastic timed automaton model presented above is that its complexity makes it difficult to analyze and control. In cases where the DES is too complicated to analyze, *discrete event simulation* can be a powerful tool for analysis. Another approach is to employ a different type of automaton model referred to as a *timed discrete event system* that introduces time and allows for non-determinism. This type of model defines intervals of time for each transition in the automaton without explicitly attaching probability distribution functions to the timing of the transitions. The timed DES model allows for nondeterminism in that the exact sequence of events is not completely determined and the model is simple enough that a theory of supervisory control has been developed based on them [14]. A timed DES model $\mathbf{G} = (Q, \Sigma, \delta, q_0, Q_m)$ has the same format as a traditional automaton except that the event set Σ includes the event *tick* which represents the passage of time of the global clock. This type of model allows for control to be formulated that meets temporal specifications, but can also significantly increase the required complexity.

Alternative models that include timing information other than the automata-based models introduced above also exist. One example is the *timed Petri net*, which attaches a clock structure to the Petri net in an analogous manner to timed automata. Here each clock sequence is associated with a transition. Another class of model based on the algebraic structure of timed Petri nets is called *max-plus algebra*. This formalism is a diod algebra based on two operations, the "max" operation written $\oplus$ and the "+" operation written $\otimes$. The max-plus algebra allows questions to be formally answered about whether or not events occur within some time window, and provides information about the steady-state behavior of a system. Finally, process algebra type models can also capture the notion of time without explicitly employing a clock. For example, such concepts as "something *eventually* being true" or commanding an action "*until* something else is achieved" are concepts that can be expressed employing *temporal logic*.

5.5.7 Conclusion

The same underlying real-world system can be modeled in many different ways. An alternative to commonly employed time-driven models like differential and difference equations are DES models. DES models have discrete state spaces and event-driven evolution and can be very useful for answering many important equations about system behavior and for designing control systems to achieve desired behavior. Three levels of abstraction for DES models are models that capture (1) purely logical behavior, (2) timed logical behavior, and (3) stochastic timed logical behavior. Each level of abstraction can be used to answer different questions and achieve different behavior, from guaranteeing a specific ordering of events, to achieving a task in a desired time frame, to assuring a task is completed with some level of certainty. The focus of this chapter has been on the analysis of automata models and the design of supervisory control in the sense of Ramadge and Wonham [2]. Other logical, timed, and timed stochastic models have been introduced including Petri net and process algebra models. Further details about DES theory can be found in many sources, with [1,3] and [15] being three particularly good references.

References

1. W. M. Wonham, *Supervisory Control of Discrete-Event Systems*. ECE Dept., University of Toronto, current update 2009.07.01, available at http://www.control.utoronto.ca/DES.
2. P. J. Ramadge and W. M. Wonham, The control of discrete event systems, *Proceedings of the IEEE*, Special issue on discrete event dynamic systems, 77(1):81–98, 1989.
3. C. G. Cassandras and S. Lafortune, *Introduction to Discrete Event Systems*, 2nd edn, New York, NY: Springer, 2007.
4. T. Murata, Petri nets: Properties, analysis, applications, *Proceedings of the IEEE*, 77(4):541–580, 1989.
5. C. A. R. Hoare, *Communicating Sequential Processes*. London: Prentice-Hall, Inc, 1985.
6. F. Lin and W. M. Wonham, On observability of discrete event systems, *Information Sciences*, 44:173–198, 1988.
7. *DESUMA Software Package*, available: http://www.eecs.umich.edu/umdes/toolboxes.html.
8. *Supremica Software Package*, available: http://www.supremica.org.
9. J. O. Moody and P. Antsaklis, *Supervisory Control of Discrete Event Systems Using Petri Nets*. Boston: Kluwer Academic Publishers, 1998.
10. M. Sampath, R. Sengupta, S. Lafortune, K. Sinnamohideen, and D. Teneketzis, Diagnosability of discrete-event systems, *IEEE Transactions on Automatic Control*, 40(9):1555–1575, September 1995.
11. M. H. de Queiroz and J. E. R. Cury, Modular supervisory control of composed systems, in *Proceedings of the American Control Conference*, Chicago, USA, 2000, 4051–4055.
12. H. Zhong and W. M. Wonham, On the consistency of hierarchical supervision in discrete-event systems, *IEEE Transactions on Automatic Control*, 35(10):1125–1134, 1990.
13. K. C. Wong and W. M. Wonham, Modular control and coordination of discrete-event systems, *Discrete Event Dynamic Systems: Theory and Applications*, 8:247–297, 1998.
14. B. A. Brandin and W. M. Wonham, Supervisory control of timed discrete-event systems, *IEEE Transactions on Automatic Control*, 39(2):329–442, 1994.
15. B. Hruz and M. C. Zhou, *Modeling and Control of Discrete-Event Dynamic Systems*. New York, NY: Springer, 2007.

6

Graphical Models

Dean K. Frederick
Rensselaer Polytechnic Institute

Charles M. Close
Rensselaer Polytechnic Institute

Norman S. Nise
California State Polytechnic University

6.1 Block Diagrams

Dean K. Frederick and Charles M. Close

6.1.1 Introduction

A block diagram is an interconnection of symbols representing certain basic mathematical operations in such a way that the overall diagram obeys the system's mathematical model. In the diagram, the lines interconnecting the blocks represent the variables describing the system behavior, such as the input and state variables. Inspecting a block diagram of a system may provide new insight into the system's structure and behavior beyond that available from the differential equations themselves.

Throughout most of this chapter, we restrict the discussion to fixed linear systems that contain no initial stored energy. After we transform the equations describing such a system, the variables that we use are the Laplace transforms of the corresponding functions of time. The parts of the system can then be described by their transfer functions. Recall that transfer functions give only the zero-state response. However, the steady-state response of a stable system does not depend on the initial conditions, so in that case there is no loss of generality in using only the zero-state response.

After defining the components to be used in our diagrams, we develop rules for simplifying block diagrams, emphasizing those diagrams that represent feedback systems. The chapter concludes by pointing out that graphical models can be used for more general systems than those considered here, including nonlinear blocks, multi-input/multi-output (MIMO) systems, and discrete-time systems.

Computer programs for the analysis and design of control systems exist that allow the entry of block diagram models in graphical form. These programs are described in another section of this book.

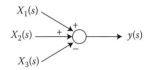

FIGURE 6.1 Summer representing $Y(s) = X_1(s) + X_2(s) - X_3(s)$.

6.1.2 Diagram Blocks

The operations that we generally use in block diagrams are summation, gain, and multiplication by a transfer function. Unless otherwise stated, all variables are Laplace-transformed quantities.

6.1.2.1 Summer

The addition and subtraction of variables is represented by a *summer*, or *summing junction*. A summer is represented by a circle that has any number of arrows directed toward it (denoting inputs) and a single arrow directed away from it (denoting the output). Next to each entering arrowhead is a plus or minus symbol indicating the sign associated with the variable that the particular arrow represents. The output variable, appearing as the one arrow leaving the circle, is defined to be the sum of all the incoming variables, with the associated signs taken into account. A summer having three inputs $X_1(s)$, $X_2(s)$, and $X_3(s)$ appears in Figure 6.1.

6.1.2.2 Gain

The multiplication of a single variable by a constant is represented by a *gain* block. We place no restriction on the value of the gain, which may be positive or negative. It may be an algebraic function of other constants and/or system parameters. Several self-explanatory examples are shown in Figure 6.2.

6.1.2.3 Transfer Function

For a fixed linear system with no initial stored energy, the transformed output $Y(s)$ is given by

$$Y(s) = H(s)U(s)$$

where $H(s)$ is the transfer function and $U(s)$ is the transformed input. When dealing with parts of a larger system, we often use $F(s)$ and $X(s)$ for the transfer function and transformed input, respectively, of an individual part. Then

$$Y(s) = F(s)X(s) \tag{6.1}$$

Any system or combination of elements can be represented by a block containing its transfer function $F(s)$, as indicated in Figure 6.3a. For example, the first-order system that obeys the input–output equation

$$\dot{y} + \frac{1}{\tau}y = Ax(t)$$

(a)   (b)   (c)

$$X(s) \xrightarrow{\quad} \boxed{A} \xrightarrow{\quad Y(s)} \qquad X(s) \xrightarrow{\quad} \boxed{-5} \xrightarrow{\quad Y(s)} \qquad X(s) \xrightarrow{\quad} \boxed{\dfrac{K}{M}} \xrightarrow{\quad Y(s)}$$

FIGURE 6.2 Gains. (a) $Y(s) = AX(s)$; (b) $Y(s) = -5X(s)$; and (c) $Y(s) = (K/M)X(s)$.

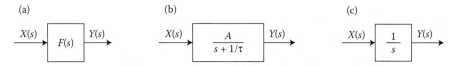

FIGURE 6.3 Basic block diagrams. (a) Arbitrary transfer function; (b) first-order system; and (c) integrator.

has as its transfer function

$$F(s) = \frac{A}{s + \dfrac{1}{\tau}}$$

Thus, it could be represented by the block diagram shown in Figure 6.3b. Note that the gain block in Figure 6.2a can be considered as a special case of a transfer function block, with $F(s) = A$.

6.1.2.4 Integrator

Another important special case of a general transfer function block, one that appears frequently in our diagrams, is the *integrator* block. An integrator that has an input $x(t)$ and an output $y(t)$ obeys the relationship

$$y(t) = y(0) + \int_0^t x(\lambda)\,d\lambda$$

where λ is the dummy variable of integration. Setting $y(0)$ equal to 0 and transforming the equation give

$$Y(s) = \frac{1}{s}X(s)$$

Hence, the transfer function of the integrator is $Y(s)/X(s) = 1/s$, as shown in Figure 6.3c.

Because a block diagram is merely a pictorial representation of a set of algebraic Laplace-transformed equations, it is possible to combine blocks by calculating equivalent transfer functions and thereby to simplify the diagram. We now present procedures for handling series and parallel combinations of blocks. Methods for simplifying diagrams containing feedback paths are discussed in the next section.

6.1.2.5 Series Combination

Two blocks are said to be in *series* when the output of one goes only to the input of the other, as shown in Figure 6.4a. The transfer functions of the individual blocks in the figure are $F_1(s) = V(s)/X(s)$ and $F_2(s) = Y(s)/V(s)$.

When we evaluate the individual transfer functions, it is essential that we take any *loading effects* into account. This means that $F_1(s)$ is the ratio $V(s)/X(s)$ when the two subsystems are connected, so any effect the second subsystem has on the first is accounted for in the mathematical model. The same statement holds for calculating $F_2(s)$. For example, the input–output relationship for a linear potentiometer loaded by a resistor connected from its wiper to the ground node differs from that of the unloaded potentiometer.

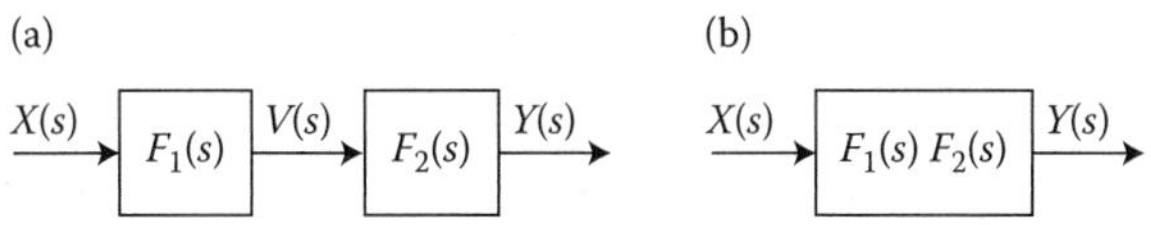

FIGURE 6.4 (a) Two blocks in series and (b) equivalent diagram.

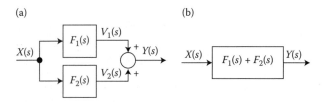

FIGURE 6.5　(a) Two blocks in parallel and (b) equivalent diagram.

In Figure 6.4a, $Y(s) = F_2(s)V(s)$ and $V(s) = F_1(s)X(s)$. It follows that

$$Y(s) = F_2(s)[F_1(s)X(s)]$$
$$= [F_1(s)F_2(s)]X(s)$$

Thus, the transfer function relating the input transform $X(s)$ to the output transform $Y(s)$ is $F_1(s)$ $F_2(s)$, the product of the individual transfer functions. The equivalent block diagram is shown in Figure 6.4b.

6.1.2.6　Parallel Combination

Two systems are said to be in *parallel* when they have a common input and their outputs are combined by a summing junction. If, as indicated in Figure 6.5a, the individual blocks have the transfer functions $F_1(s)$ and $F_2(s)$ and the signs at the summing junction are both positive, the overall transfer function $Y(s)/X(s)$ is the sum $F_1(s) + F_2(s)$, as shown in Figure 6.5b. To prove this statement, we note that

$$Y(s) = V_1(s) + V_2(s)$$

where $V_1(s) = F_1(s)X(s)$ and $V_2(s) = F_2(s)X(s)$. Substituting for $V_1(s)$ and $V_2(s)$, we have

$$Y(s) = [F_1(s) + F_2(s)]X(s)$$

If either of the summing-junction signs associated with $V_1(s)$ or $V_2(s)$ is negative, we must change the sign of the corresponding transfer function in forming the overall transfer function. The following example illustrates the rules for combining blocks that are in parallel or in series.

Example 6.1:

Evaluate the transfer functions $Y(s)/U(s)$ and $Z(s)/U(s)$ for the block diagram shown in Figure 6.6, giving the results as rational functions of s.

Solution

Because $Z(s)$ can be viewed as the sum of the outputs of two parallel blocks, one of which has $Y(s)$ as its output, we first evaluate the transfer function $Y(s)/U(s)$. To do this, we observe that $Y(s)$ can be considered the output of a series combination of two parts, one of which is a parallel combination of two

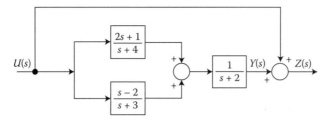

FIGURE 6.6 Block diagram for Example 6.1.

blocks. Starting with this parallel combination, we write

$$\frac{2s+1}{s+4} + \frac{s-2}{s+3} = \frac{3s^2+9s-5}{s^2+7s+12}$$

and redraw the block diagram as shown in Figure 6.7a. The series combination in this version has the transfer function

$$\frac{Y(s)}{U(s)} = \frac{3s^2+9s-5}{s^2+7s+12} \cdot \frac{1}{s+2}$$

$$= \frac{3s^2+9s-5}{s^3+9s^2+26s+24}$$

which leads to the diagram shown in Figure 6.7b. We can reduce the final parallel combination to the single block shown in Figure 6.7c by writing

$$\frac{Z(s)}{U(s)} = 1 + \frac{Y(s)}{U(s)}$$

$$= 1 + \frac{3s^2+9s-5}{s^3+9s^2+26s+24}$$

$$= \frac{s^3+12s^2+35s+19}{s^3+9s^2+26s+24}$$

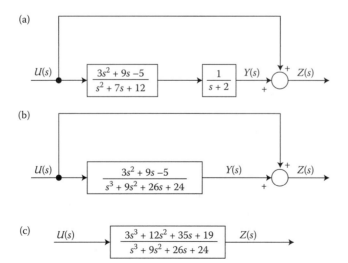

FIGURE 6.7 Equivalent block diagrams for the diagram shown in Figure 6.6.

In general, it is desirable to reduce the transfer functions of combinations of blocks to rational functions of s in order to simplify the subsequent analysis. This will be particularly important in the following section when we are reducing feedback loops to obtain an overall transfer function.

6.1.3 Block Diagrams of Feedback Systems

Figure 6.8a shows the block diagram of a general feedback system that has a forward path from the summing junction to the output and a feedback path from the output back to the summing junction. The transforms of the system's input and output are $U(s)$ and $Y(s)$, respectively. The transfer function $G(s) = Y(s)/V(s)$ is known as the *forward transfer function*, and $H(s) = Z(s)/Y(s)$ is called the *feedback transfer function*. We must evaluate both of these transfer functions with the system elements connected in order properly to account for the loading effects of the interconnections. The product $G(s)H(s)$ is referred to as the *open-loop transfer function*. The sign associated with the feedback signal from the block $H(s)$ at the summing junction is shown as minus because a minus sign naturally occurs in the majority of feedback systems, particularly in control systems.

Given the model of a feedback system in terms of its forward and feedback transfer functions $G(s)$ and $H(s)$, it is often necessary to determine the *closed-loop transfer function* $T(s) = Y(s)/U(s)$. We do this by writing the algebraic transform equations corresponding to the block diagram shown in Figure 6.8a and solving them for the ratio $Y(s)/U(s)$. We can write the following transform equations directly from the block diagram.

$$V(s) = U(s) - Z(s)$$
$$Y(s) = G(s)V(s)$$
$$Z(s) = H(s)Y(s)$$

If we combine these equations in such a way as to eliminate $V(s)$ and $Z(s)$, we find that

$$Y(s) = G(s)[U(s) - H(s)Y(s)]$$

which can be rearranged to give

$$[1 + G(s)H(s)]Y(s) = G(s)U(s)$$

Hence, the closed-loop transfer function $T(s) = Y(s)/U(s)$ is

$$T(s) = \frac{G(s)}{1 + G(s)H(s)} \tag{6.2}$$

where it is implicit that the sign of the feedback signal at the summing junction is negative. It is readily shown that when a plus sign is used at the summing junction for the feedback signal, the closed-loop

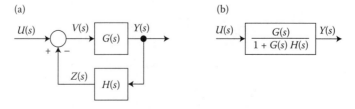

FIGURE 6.8   (a) Block diagram of a feedback system and (b) equivalent diagram.

transfer function becomes

$$T(s) = \frac{G(s)}{1 - G(s)H(s)} \tag{6.3}$$

A commonly used simplification occurs when the feedback transfer function is unity, that is, when $H(s) = 1$. Such a system is referred to as a *unity-feedback system*, and Equation 6.2 reduces to

$$T(s) = \frac{G(s)}{1 + G(s)} \tag{6.4}$$

We now consider three examples that use Equations 6.2 and 6.3. The first two illustrate determining the closed-loop transfer function by reducing the block diagram. They also show the effects of feedback gains on the closed-loop poles, time constant, damping ratio, and undamped natural frequency. In Example 6.4, a block diagram is drawn directly from the system's state-variable equations and then reduced to give the system's transfer functions.

Example 6.2:

Find the closed-loop transfer function for the feedback system shown in Figure 6.9a, and compare the locations of the poles of the open-loop and closed-loop transfer functions in the *s*-plane.

Solution

By comparing the block diagram shown in Figure 6.9a with that shown in Figure 6.8a, we see that $G(s) = 1/(s + \alpha)$ and $H(s) = \beta$. Substituting these expressions into Equation 6.2 gives

$$T(s) = \frac{\dfrac{1}{s + \alpha}}{1 + \left(\dfrac{1}{s + \alpha}\right)\beta}$$

which we can write as a rational function of *s* by multiplying the numerator and denominator by $s + \alpha$. Doing this, we obtain the closed-loop transfer function

$$T(s) = \frac{1}{s + \alpha + \beta}$$

This result illustrates an interesting and useful property of feedback systems: the fact that the poles of the closed-loop transfer function differ from the poles of the open-loop transfer function $G(s)H(s)$. In this case, the single open-loop pole is at $s = -\alpha$, whereas the single closed-loop pole is at $s = -(\alpha + \beta)$.

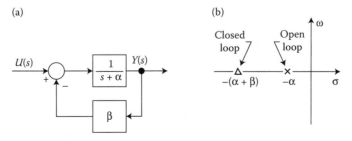

FIGURE 6.9 Single-loop feedback system for Example 6.2.

These pole locations are indicated in Figure 6.9b for positive α and β. Hence, in the absence of feedback, the pole of the transfer function $Y(s)/U(s)$ is at $s = -\alpha$, and the free response is of the form $\epsilon^{-\alpha t}$. With feedback, however, the free response is $\epsilon^{-(\alpha+\beta)t}$. Thus, the time constant of the open-loop system is $1/\alpha$, whereas that of the closed-loop system is $1/(\alpha + \beta)$.

Example 6.3:

Find the closed-loop transfer function of the two-loop feedback system shown in Figure 6.10. Also express the damping ratio ζ and the undamped natural frequency ω_n of the closed-loop system in terms of the gains a_0 and a_1.

Solution

Because the system's block diagram contains one feedback path inside another, we cannot use Equation 6.2 directly to evaluate $Y(s)/U(s)$. However, we can redraw the block diagram such that the summing junction is split into two summing junctions, as shown in Figure 6.11a. Then it is possible to use Equation 6.2 to eliminate the inner loop by calculating the transfer function $W(s)/V(s)$. Taking $G(s) = 1/s$ and $H(s) = a_1$ in Equation 6.2, we obtain

$$\frac{W(s)}{V(s)} = \frac{\frac{1}{s}}{1 + \frac{a_1}{s}} = \frac{1}{s + a_1}$$

Redrawing Figure 6.11a with the inner loop replaced by a block having $1/(s+a_1)$ as its transfer function gives Figure 6.11b. The two blocks in the forward path of this version are in series and can be combined by multiplying their transfer functions, which gives the block diagram shown in Figure 6.11c. Then we can apply Equation 6.2 again to find the overall closed-loop transfer function $T(s) = Y(s)/U(s)$ as

$$T(s) = \frac{\frac{1}{s(s+a_1)}}{1 + \frac{1}{s(s+a_1)} \cdot a_0} = \frac{1}{s^2 + a_1 s + a_0} \tag{6.5}$$

The block diagram representation of the feedback system corresponding to Equation 6.5 is shown in Figure 6.11d.

The poles of the closed-loop transfer function are the roots of the equation

$$s^2 + a_1 s + a_0 = 0 \tag{6.6}$$

which we obtain by setting the denominator of $T(s)$ equal to zero and which is the characteristic equation of the closed-loop system. Equation 6.6 has two roots, which may be real or complex, depending on the

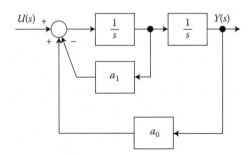

FIGURE 6.10 System with two feedback loops for Example 6.3.

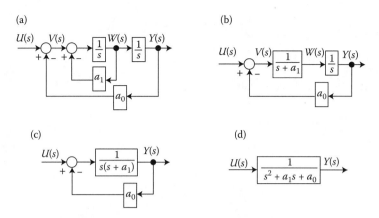

FIGURE 6.11   Equivalent block diagrams for the system shown in Figure 6.10.

sign of the quantity $a_1^2 - 4a_0$. However, the roots of Equation 6.6 will have negative real parts and the closed-loop system will be stable provided that a_0 and a_1 are both positive.

If the poles are complex, it is convenient to rewrite the denominator of $T(s)$ in terms of the damping ratio ζ and the undamped natural frequency ω_n. By comparing the left side of Equation 6.6 with the characteristic polynomial of a second-order system written as $s^2 + 2\zeta\omega_n s + \omega_n^2$, we see that

$$a_0 = \omega_n^2 \quad \text{and} \quad a_1 = 2\zeta\omega_n$$

Solving the first of these equations for ω_n and substituting it into the second gives the damping ratio and the undamped natural frequency of the closed-loop system as

$$\zeta = \frac{a_1}{2\sqrt{a_0}} \quad \text{and} \quad \omega_n = \sqrt{a_0}$$

We see from these expressions that a_0, the gain of the outer feedback path in Figure 6.10, determines the undamped natural frequency ω_n and that a_1, the gain of the inner feedback path, affects only the damping ratio. If we can specify both a_0 and a_1 at will, then we can attain any desired values of ζ and ω_n for the closed-loop transfer function.

Example 6.4:

Draw a block diagram for the translational mechanical system shown in Figure 6.12, whose state-variable equations can be written as

$$\dot{x}_1 = v_1$$

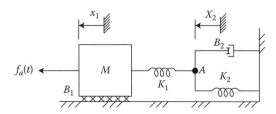

FIGURE 6.12   Translational system for Example 6.4.

$$\dot{v}_1 = \frac{1}{M}[-K_1 x_1 - B_1 v_1 + K_1 x_2 + f_a(t)]$$

$$\dot{x}_2 = \frac{1}{B_2}[K_1 x_1 - (K_1 + K_2)x_2]$$

Reduce the block diagram to determine the transfer functions $X_1(s)/F_a(s)$ and $X_2(s)/F_a(s)$ as rational functions of s.

Solution

Transforming the three differential equations with zero initial conditions, we have

$$sX_1(s) = V_1(s)$$

$$MsV_1(s) = -K_1 X_1(s) - B_1 V_1(s) + K_1 X_2(s) + F_a(s)$$

$$B_2 sX_2(s) = K_1 X_1(s) - (K_1 + K_2)X_2(s)$$

We use the second of these equations to draw a summing junction that has $MsV_1(s)$ as its output. After the summing junction, we insert the transfer function $1/Ms$ to get $V_1(s)$, which, from the first equation, equals $sX_1(s)$. Thus, an integrator whose input is $V_1(s)$ has $X_1(s)$ as its output. Using the third equation, we form a second summing junction that has $B_2 sX_2(s)$ as its output. Following this summing junction by the transfer function $1/B_2 s$, we get $X_2(s)$ and can complete the four feedback paths required by the summing junctions. The result of these steps is the block diagram shown in Figure 6.13a.

To simplify the block diagram, we use Equation 6.2 to reduce each of the three inner feedback loops, obtaining the version shown in Figure 6.13b. To evaluate the transfer function $X_1(s)/F_a(s)$, we can apply Equation 6.3 to this single-loop diagram because the sign associated with the feedback signal at the summing junction is positive rather than negative. Doing this with

$$G(s) = \frac{1}{Ms^2 + B_1 s + K_1} \quad \text{and} \quad H(s) = \frac{K_1^2}{B_2 s + K_1 + K_2}$$

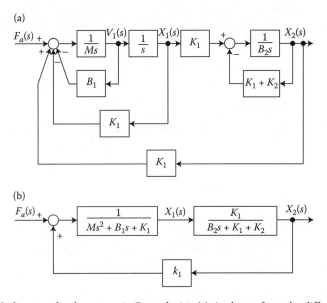

FIGURE 6.13 Block diagrams for the system in Example 6.4. (a) As drawn from the differential equations and (b) with the three inner feedback loops eliminated.

we find

$$
\begin{aligned}
\frac{X_1(s)}{F_a(s)} &= \frac{\dfrac{1}{Ms^2 + B_1 s + K_1}}{1 - \dfrac{1}{Ms^2 + B_1 s + K_1} \cdot \dfrac{K_1^2}{B_2 s + K_1 + K_2}} \\
&= \frac{B_2 s + K_1 + K_2}{(Ms^2 + B_1 s + K_1)(B_2 s + K_1 + K_2) - K_1^2} \\
&= \frac{B_2 s + K_1 + K_2}{P(s)}
\end{aligned}
\tag{6.7}
$$

where

$$
P(s) = MB_2 s^3 + [(K_1 + K_2)M + B_1 B_2]s^2 + [B_1(K_1 + K_2) + B_2 K_1]s + K_1 K_2
$$

To obtain $X_2(s)/F_a(s)$, we can write

$$
\frac{X_2(s)}{F_a(s)} = \frac{X_1(s)}{F_a(s)} \cdot \frac{X_2(s)}{X_1(s)}
$$

where $X_1(s)/F_a(s)$ is given by Equation 6.7 and, from Figure 6.13b,

$$
\frac{X_2(s)}{X_1(s)} = \frac{K_1}{B_2 s + K_1 + K_2}
\tag{6.8}
$$

The result of multiplying Equations 6.7 and 6.8 is a transfer function with the same denominator as Equation 6.7 but with a numerator of K_1.

In the previous examples, we used the rules for combining blocks that are in series or in parallel, as shown in Figures 6.4 and 6.5. We also repeatedly used the rule for simplifying the basic feedback configuration given in Figure 6.8a. A number of other operations can be derived to help simplify block diagrams. To conclude this section, we present and illustrate two of these additional operations.

Keep in mind that a block diagram is just a means of representing the algebraic Laplace-transformed equations that describe a system. Simplifying or reducing the diagram is equivalent to manipulating the equations. In order to prove that a particular operation on the block diagram is valid, we need only show that the relationships among the transformed variables of interest are left unchanged.

6.1.3.1 Moving a Pick-Off Point

A *pick-off point* is a point where an incoming variable in the diagram is directed into more than one block. In the partial diagram of Figure 6.14a, the incoming signal $X(s)$ is used not only to provide the output $Q(s)$ but also to form the signal $W(s)$, which in practice might be fed back to a summer that appears earlier in the complete diagram. The pick-off point can be moved to the right of $F_1(s)$ if the transfer function of the block leading to $W(s)$ is modified as shown in Figure 6.14b. Both parts of the figure give the same equations:

$$
Q(s) = F_1(s)X(s)
$$
$$
W(s) = F_2(s)X(s)
$$

Example 6.5:

Use Figure 6.14 to find the closed-loop transfer function for the system shown in Figure 6.10.

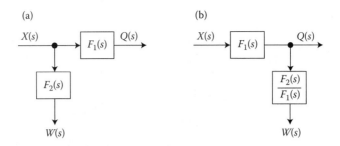

FIGURE 6.14 Moving a pick-off point.

Solution

The pick-off point leading to the gain block a_1 can be moved to the output $Y(s)$ by replacing a_1 by a_1s, as shown in Figure 6.15a. Then the two integrator blocks, which are now in series, can be combined to give the transfer function $G(s) = 1/s^2$. The two feedback blocks are now in parallel and can be combined into the single transfer function $a_1s + a_0$, as shown in Figure 6.15b. Finally, by Equation 6.2,

$$T(s) = \frac{Y(s)}{U(s)} = \frac{1/s^2}{1 + (a_1s + a_0)/s^2} = \frac{1}{s^2 + a_1s + a_0}$$

which agrees with Equation 6.5, as found in Example 6.3.

6.1.3.2 Moving a Summing Junction

Suppose that, in the partial diagram of Figure 6.16a, we wish to move the summing junction to the left of the block that has the transfer function $F_2(s)$. We can do this by modifying the transfer func-

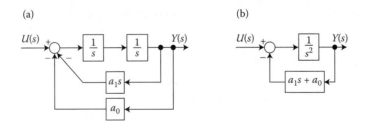

FIGURE 6.15 Equivalent block diagrams for the system shown in Figure 6.10.

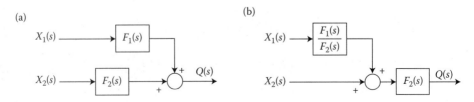

FIGURE 6.16 Moving a summing junction.

tion of the block whose input is $X_1(s)$, as shown in part (b) of the figure. For each part of the figure,

$$Q(s) = F_1(s)X_1(s) + F_2(s)X_2(s)$$

Example 6.6:

Find the closed-loop transfer function $T(s) = Y(s)/U(s)$ for the feedback system shown in Figure 6.17a.

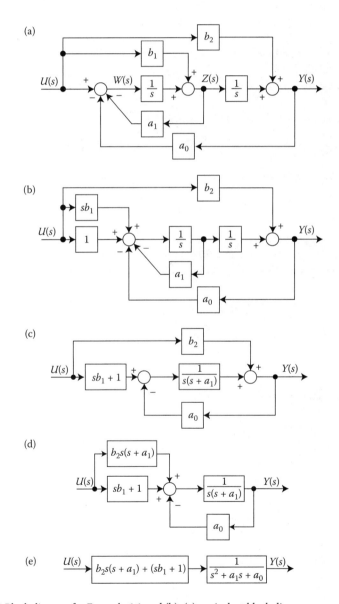

FIGURE 6.17 (a) Block diagram for Example 6.6; and (b)–(e) equivalent block diagrams.

Solution

We cannot immediately apply Equation 6.2 to the inner feedback loop consisting of the first integrator and the gain block a_1 because the output of block b_1 enters a summer within that loop. We therefore use Figure 6.16 to move this summer to the left of the first integrator block, where it can be combined with the first summer. The resulting diagram is given in Figure 6.17b.

Now Equation 6.2 can be applied to the inner feedback loop to give the transfer function

$$G_1(s) = \frac{1/s}{1 + a_1/s} = \frac{1}{s + a_1}$$

The equivalent block with the transfer function $G_1(s)$ is then in series with the remaining integrator, which results in a combined transfer function of $1/[s(s + a_1)]$. Also, the two blocks with gains of sb_1 and 1 are in parallel and can be combined into a single block. These simplifications are shown in Figure 6.17c.

We can now repeat the procedure and move the right summer to the left of the block labeled $1/[s(s + a_1)]$, where it can again be combined with the first summer. This is done in part (d) of the figure. The two blocks in parallel at the left can now be combined by adding their transfer functions, and Equation 6.2 can be applied to the right part of the diagram to give

$$\frac{\dfrac{1}{s(s + a_1)}}{1 + \dfrac{a_0}{s(s + a_1)}} = \frac{1}{s^2 + a_1 s + a_0}$$

These steps yield Figure 6.17e, from which we see that

$$T(s) = \frac{b_2 s^2 + (a_1 b_2 + b_1)s + 1}{s^2 + a_1 s + a_0} \tag{6.9}$$

Because performing operations on a given block diagram is equivalent to manipulating the algebraic equations that describe the system, it may sometimes be easier to work with the equations themselves. As an alternative solution to the last example, suppose that we start by writing the equations for each of the three summers in Figure 6.17a:

$$W(s) = U(s) - a_0 Y(s) - a_1 Z(s)$$

$$Z(s) = \frac{1}{s} W(s) + b_1 U(s)$$

$$Y(s) = \frac{1}{s} Z(s) + b_2 U(s)$$

Substituting the expression for $W(s)$ into the second equation, we see that

$$Z(s) = \frac{1}{s} [U(s) - a_0 Y(s) - a_1 Z(s)] + b_1 U(s)$$

from which

$$Z(s) = \frac{1}{s + a_1} [-a_0 Y(s) + (b_1 s + 1)U(s)] \tag{6.10}$$

Substituting Equation 6.10 into the expression for $Y(s)$ gives

$$Y(s) = \frac{1}{s(s + a_1)} [-a_0 Y(s) + (b_1 s + 1)U(s)] + b_2 U(s)$$

Rearranging this equation, we find that

$$\frac{Y(s)}{U(s)} = \frac{b_2 s^2 + (a_1 b_2 + b_1)s + 1}{s^2 + a_1 s + a_0}$$

which agrees with Equation 6.9, as found in Example 6.6.

6.1.4 Summary

Block diagrams are an important way of representing the structure and properties of fixed linear systems. We start the construction of a block diagram by transforming the system equations, assuming zero initial conditions.

The blocks used in diagrams for feedback systems may contain transfer functions of any degree of complexity. We developed a number of rules, including those for series and parallel combinations and for the basic feedback configuration, for simplifying block diagrams.

6.1.5 Block Diagrams for Other Types of Systems

So far, we have considered block diagrams of linear, continuous-time models having one input and one output. Because of this restriction, we can use the system's block diagram to develop an overall transfer function in terms of the complex Laplace variable s. For linear discrete-time models we can draw block diagrams and do corresponding manipulations to obtain the closed-loop transfer function in terms of the complex variable z, as used in the z-transform. We can also construct block diagrams for MIMO systems, but we must be careful to obey the rules of matrix algebra when manipulating the diagrams to obtain equivalent transfer functions.

System models that contain nonlinearities, such as backlash, saturation, or dead band, can also be represented in graphical form. Typically, one uses transfer-function blocks for the linear portion of the system and includes one or more special blocks to represent the specific nonlinear operations. While such diagrams are useful for representing the system and for preparing computer simulations of the nonlinear model, they cannot be analyzed in the ways that we have done in the previous two sections. Similar comments can be made for time-varying elements.

Acknowledgments

Section 6.1 is closely based on Sections 13.1 and 13.5 of *Modeling and Analysis of Dynamic Systems*, Second Edition, 1993, by Charles M. Close and Dean K. Frederick. It is used with permission of the publisher, Houghton Mifflin Company.

6.2 Signal-Flow Graphs*

Norman S. Nise

6.2.1 Introduction

Signal-flow graphs are an alternate system representation. Unlike *block diagrams* of *linear systems*, which consist of blocks, signals, summing junctions, and pickoff points, signal-flow graphs of linear systems consist of only *branches*, which represent systems, and *nodes*, which represent signals. These elements are shown in Figure 6.18a and Figure 6.18b respectively. A system (Figure 6.18a) is represented by a line with an arrow showing the direction of signal flow through the system. Adjacent to the line we write the *transfer function*. A signal (Figure 6.18b) is a node with the signal name written adjacent to the node.

Figure 6.18c shows the interconnection of the systems and the signals. Each signal is the sum of signals flowing into it. For example, in Figure 6.18c the signal $X(s) = R_1(s)G_1(s) - R_2(s)G_2(s) + R_3(s)G_3(s)$. The signal, $C_3(s) = -X(s)G_6(s) = -R_1(s)G_1(s)G_6(s) + R_2(s)G_2(s)G_6(s) - R_3(s)G_3(s)G_6(s)$.

---

* Reprinted from Nise, N. S., *Control Systems Engineering*, 5th ed., John Wiley & Sons, Inc., Hoboken, NJ. With permission. © 2008.

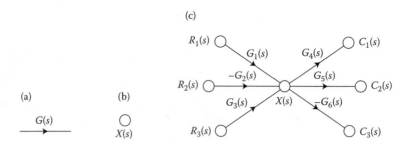

FIGURE 6.18 Signal-flow graph component parts: (a) system; (b) signal; and (c) interconnection of systems and signals.

Notice that the summing of negative signals is handled by associating the negative sign with the system and not with a summing junction as in the case of block diagrams.

6.2.2 Relationship between Block Diagrams and Signal-Flow Graphs

To show the parallel between block diagrams and signal-flow graphs, we convert some block diagram forms to signal-flow graphs. In each case, we first convert the signals to nodes and then interconnect the nodes with systems.

Let us convert the *cascaded, parallel*, and *feedback* forms of the block diagrams shown in Figures 6.19 through 6.21, respectively, into signal-flow graphs.

In each case, we start by drawing the signal nodes for that system. Next, we interconnect the signal nodes with system branches. Figures 6.22a, 6.22c, and 6.22e show the signal nodes for the cascaded, parallel, and feedback forms, respectively. Next, interconnect the nodes with branches that represent the subsystems. This is done in Figures 6.22b, 6.22d, and 6.22f for the cascaded, parallel, and feedback forms, respectively. For the parallel form, positive signs are assumed at all inputs to the summing junction; for the feedback form, a negative sign is assumed for the feedback.

In the next example we start with a more complicated block diagram and end with the equivalent signal-flow graph. Convert the block diagram of Figure 6.23 to a signal-flow graph. Begin by drawing the signal nodes as shown in Figure 6.24a.

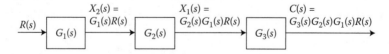

FIGURE 6.19 Cascaded subsystems.

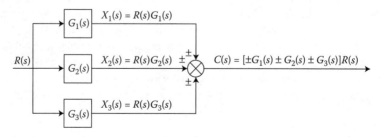

FIGURE 6.20 Parallel subsystems.

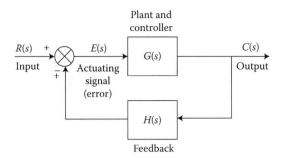

FIGURE 6.21 Feedback control system.

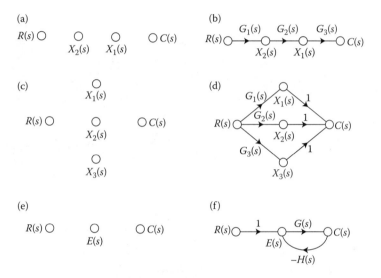

FIGURE 6.22 Building signal-flow graphs: (a) cascaded system: nodes (from Figure 6.19); (b) cascaded system: signal-flow graph; (c) parallel system: nodes (from Figure 6.20); (d) parallel system: signal-flow graph; (e) feedback system: nodes (from Figure 6.21); and (f) feedback system: signal-flow graph.

Next, interconnect the nodes showing the direction of signal flow and identifying each transfer function. The result is shown in Figure 6.24b. Notice that the negative signs at the summing junctions of the block diagram are represented by the negative transfer functions of the signal-flow graph.

Finally, if desired, simplify the signal-flow graph to the one shown in Figure 6.24c by eliminating signals that have a single flow in and a single flow out such as $V_2(s)$, $V_6(s)$, $V_7(s)$, and $V_8(s)$.

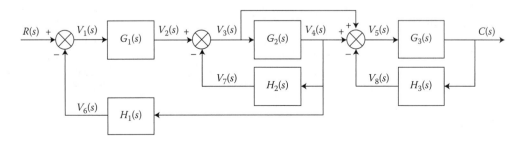

FIGURE 6.23 Block diagram.

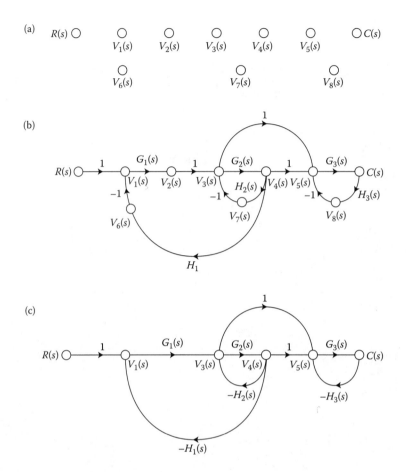

FIGURE 6.24 Signal-flow graph development: (a) signal nodes; (b) signal-flow graph; and (c) simplified signal-flow graph.

6.2.3 Mason's Rule

Block diagram reduction requires successive application of fundamental relationships in order to arrive at the system transfer function. On the other hand, Mason's Rule for the reduction of signal-flow graphs to a transfer function relating the output to the input requires the application of a single formula. The formula was derived by S. J. Mason when he related the signal-flow graphs to the simultaneous equations that can be written from the graph ([1,2]).

In general, it can be complicated to implement the formula without making mistakes. Specifically, the existence of what we later call nontouching loops increases the complexity of the formula. However, many systems do not have nontouching loops and thus lend themselves to the easy application of Mason's gain formula. For these systems, you may find Mason's Rule easier to use than block diagram reduction.

The formula has several component parts that must be evaluated. We must first be sure that the definitions of the component parts are well understood. Then, we must exert care in evaluating the component parts of Mason's formula. To that end, we now discuss some basic definitions applicable to signal-flow graphs. Later we will state Mason's Rule and show an example.

6.2.3.1 Definitions

Loop Gain: Loop gain is the product of branch gains found by traversing a path that starts at a node and ends at the same node without passing through any other node more than once and following

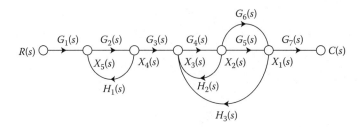

FIGURE 6.25 Sample signal-flow graph for demonstrating Mason's Rule.

the direction of the signal flow. For examples of loop gains, look at Figure 6.25. There are four loop gains as follows:

$$1. \quad G_2(s)H_1(s) \tag{6.11a}$$

$$2. \quad G_4(s)H_2(s) \tag{6.11b}$$

$$3. \quad G_4(s)G_5(s)H_3(s) \tag{6.11c}$$

$$4. \quad G_4(s)G_6(s)H_3(s) \tag{6.11d}$$

Forward-Path Gain: Forward-path gain is the product of gains found by traversing a path from the input node to the output node of the signal-flow graph in the direction of signal flow. Examples of forward-path gains are also shown in Figure 6.25. There are two forward-path gains as follows:

$$G_1(s)G_2(s)G_3(s)G_4(s)G_5(s)G_7(s) \tag{6.12a}$$

$$G_1(s)G_2(s)G_3(s)G_4(s)G_6(s)G_7(s) \tag{6.12b}$$

Nontouching Loops: Nontouching loops are loops that do not have any nodes in common. In Figure 6.25, loop $G_2(s)H_1(s)$ does not touch loop $G_4(s)H_2(s)$, loop $G_4(s)G_5(s)H_3(s)$, or loop $G_4(s)G_6(s)H_3(s)$.

Nontouching-Loop Gain: Nontouching-loop gain is the product of loop gains from nontouching loops taken two, three, four, etc. at a time. In Figure 6.25, the product of loop gain $G_2(s)H_1(s)$ and loop gain $G_4(s)H_2(s)$ is a nontouching-loop gain taken two at a time. In summary, all three of the nontouching-loop gains taken two at a time are

$$1. \quad [G_2(s)H_1(s)][G_4(s)H_2(s)] \tag{6.13a}$$

$$2. \quad [G_2(s)H_1(s)][G_4(s)G_5(s)H_3(s)] \tag{6.13b}$$

$$3. \quad [G_2(s)H_1(s)][G_4(s)G_6(s)H_3(s)] \tag{6.13c}$$

The product of loop gains $[G_4(s)G_5(s)H_3(s)][G_4(s) G_6(s)H_3(s)]$ is not a nontouching-loop gain since these two loops have nodes in common. In our example there are no nontouching-loop gains taken three at a time since three nontouching loops do not exist in the example.

We are now ready to state Mason's Rule.

Mason's Rule: The transfer function, $C(s)/R(s)$, of a system represented by a signal-flow graph is

$$G(s) = \frac{C(s)}{R(s)} = \frac{\sum_k T_k \Delta_k}{\Delta} \tag{6.14}$$

where $\sum$ denotes summation; $k =$ number of forward paths; $T_k =$ the kth forward-path gain; $\Delta = 1 - \sum$ loop gains $+ \sum$ nontouching-loop gains taken two at a time $- \sum$ nontouching-loop gains taken three at a time $+ \sum$ nontouching-loop gains taken four at a time $- \cdots$; $\Delta_k = \Delta - \sum$ loop gain terms in Δ touching the kth forward path. In other words, Δ_k is formed by eliminating from Δ those loop gains that touch the kth forward path.

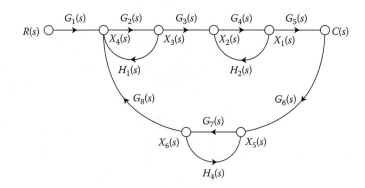

FIGURE 6.26 Signal-flow graph.

Notice the alternating signs for each component part of Δ. The following example will help clarify Mason's Rule.

Find the transfer function, $C(s)/R(s)$, for the signal-flow graph in Figure 6.26. First, identify the *forward-path gains*. In this example, there is only one:

$$G_1(s)G_2(s)G_3(s)G_4(s)G_5(s) \tag{6.15}$$

Second, identify the *loop gains*. There are four loops as follows:

$$
\begin{array}{lll}
1. & G_2(s)H_1(s) & (6.16\text{a})\\[4pt]
2. & G_4(s)H_2(s) & (6.16\text{b})\\[4pt]
3. & G_7(s)H_4(s) & (6.16\text{c})\\[4pt]
4. & G_2(s)G_3(s)G_4(s)G_5(s)G_6(s)G_7(s)G_8(s) & (6.16\text{d})
\end{array}
$$

Third, identify the *nontouching-loop gains taken two at a time*. From Equations 6.16a to 6.16d and Figure 6.26, we see that loop 1 does not touch loop 2, loop 1 does not touch loop 3, and loop 2 does not touch loop 3. Notice that loops 1, 2, and 3 all touch loop 4. Thus, the combinations of nontouching-loop gains taken two at a time are as follows:

$$
\begin{array}{lll}
\text{Loop 1 and loop 2:} & G_2(s)H_1(s)G_4(s)H_2(s) & (6.17\text{a})\\[4pt]
\text{Loop 1 and loop 3:} & G_2(s)H_1(s)G_7(s)H_4(s) & (6.17\text{b})\\[4pt]
\text{Loop 2 and loop 3:} & G_4(s)H_2(s)G_7(s)H_4(s) & (6.17\text{c})
\end{array}
$$

Finally, the *nontouching-loop gains taken three at a time* are as follows:

$$
\text{Loops 1, 2, and 3:} \quad G_2(s)H_1(s)G_4(s)H_2(s)G_7(s)H_4(s) \tag{6.18}
$$

Now, from Equation 6.14 and its definitions, we form Δ and Δ_k:

$$
\begin{aligned}
\Delta = 1 - [&G_2(s)H_1(s) + G_4(s)H_2(s) + G_7(s)H_4(s)\\
+ &G_2(s)G_3(s)G_4(s)G_5(s)G_6(s)G_7(s)G_8(s)] + [G_2(s)H_1(s)G_4(s)H_2(s)\\
+ &G_2(s)H_1(s)G_7(s)H_4(s) + G_4(s)H_2(s)G_7(s)H_4(s)] - [G_2(s)H_1(s)G_4(s)H_2(s)G_7(s)H_4(s)]
\end{aligned} \tag{6.19}
$$

We form Δ_k by eliminating from Δ those loop gains that touch the kth forward path:

$$\Delta_1 = 1 - G_7(s)H_4(s) \tag{6.20}$$

Equations 6.15, 6.19, and 6.20 are substituted into Equation 6.14 yielding the transfer function

$$G(s) = \frac{T_1 \Delta_1}{\Delta} = \frac{[G_1(s)G_2(s)G_3(s)G_4(s)G_5(s)][1 - G_7(s)H_4(s)]}{\Delta} \qquad (6.21)$$

Since there is only one forward path, $G(s)$ consists only of one term rather than the sum of terms each coming from a forward path.

6.2.4 Signal-Flow Graphs of Differential Equations and State Equations

In this section, we show how to convert a differential equation or state-space representation to a signal-flow graph. Consider the differential equation

$$\frac{d^3c}{dt^3} + 9\frac{d^2c}{dt^2} + 26\frac{dc}{dt} + 24c = 24r \qquad (6.22)$$

Converting to the *phase-variable* representation in state-space*

$$\dot{x}_1 = x_2 \qquad (6.23a)$$
$$\dot{x}_2 = x_3 \qquad (6.23b)$$
$$\dot{x}_3 = -24x_1 - 26x_2 - 9x_3 + 24r \qquad (6.23c)$$
$$y = x_1 \qquad (6.23d)$$

To draw the associated signal-flow graph, first identify three nodes, as in Figure 6.27a, to be the three state variables, $x_1, x_2,$ and x_3. Also identify a node as the input, r, and another node as the output, y. The first of the three state equations, $\dot{x}_1 = x_2$, is modeled in Figure 6.27b by realizing that the derivative of state variable x_1, which is x_2, would appear to the left at the input to an integrator. Remember, division by s in the frequency domain is equivalent to integration in the time domain. Similarly, the second equation, $\dot{x}_2 = x_3$, is added in Figure 6.27c. The last of the state equations, $\dot{x}_3 = -24x_1 - 26x_2 - 9x_3 + 24r$, is added in Figure 6.27d by forming $\dot{x}_3$ at the input of an integrator whose output is x_3. Finally, the output, $y = x_1$, is also added in Figure 6.27d completing the signal-flow graph. Notice that the state variables are outputs of the integrators.

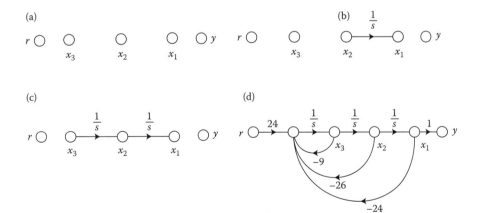

FIGURE 6.27 Stages in the development of a signal-flow graph in phase-variable form for the system of Equations 6.23. (a) Place nodes; (b) form dx_1/dt; (c) form dx_2/dt; (d) form dx_3/dt and output, y.

---

* See [3] for description of how to convert differential equations into the phase-variable representation in state space.

6.2.5 Signal-Flow Graphs of Transfer Functions

To convert transfer functions to signal-flow graphs, we first convert the transfer function to a state-space representation. The signal-flow graph then follows from the state equations as in the preceding section. Consider the transfer function

$$G(s) = \frac{s^2 + 7s + 2}{s^3 + 9s^2 + 26s + 24} \tag{6.24}$$

Converting to the phase-variable representation in state-space[*]

$$\dot{x}_1 = x_2 \tag{6.25a}$$
$$\dot{x}_2 = x_3 \tag{6.25b}$$
$$\dot{x}_3 = -24x_1 - 26x_2 - 9x_3 + 24r \tag{6.25c}$$
$$y = 2x_1 + 7x_2 + x_3 \tag{6.25d}$$

Following the same procedure used to obtain Figure 6.27d, we arrive at Figure 6.28. Notice that the denominator of the transfer function is represented by the feedback paths, while the numerator of the transfer function is represented by the *linear combination* of state variables forming the output.

6.2.6 A Final Example

We conclude this chapter with an example that demonstrates the application of signal-flow graphs and the previously discussed forms to represent in state space the feedback control system shown in Figure 6.29. We first draw the signal-flow diagram for the forward transfer function, $G(s) = 100(s + 5)/[(s + 2)(s + 3)]$, and then add the feedback path. In many physical systems, the forward transfer function consists of several systems in cascade. Thus, for this example, instead of representing $G(s)$ in phase-variable form using the methods previously described, we arrive at the signal-flow graph by considering $G(s)$ to be the following terms in cascade:

$$G(s) = 100 * \frac{1}{(s + 2)} * \frac{1}{(s + 3)} * (s + 5) \tag{6.26}$$

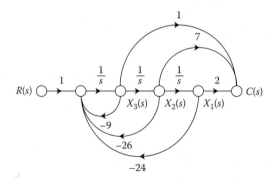

FIGURE 6.28 Signal-flow graph in phase-variable form for Equation 6.24.

---

[*] See [3] for a description of how to convert transfer functions into the phase-variable representation in state space.

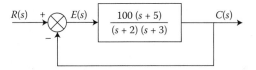

FIGURE 6.29 Feedback control system.

Each first-order term is of the form

$$\frac{C_i(s)}{R_i(s)} = \frac{1}{(s + a_i)} \tag{6.27}$$

Cross multiplying,

$$(s + a_i)C_i(s) = R_i(s) \tag{6.28}$$

Taking the inverse Laplace transform with zero initial conditions,

$$\frac{dc_i(t)}{dt} + a_i c_i(t) = r_i(t) \tag{6.29}$$

Solving for $\dfrac{dc_i(t)}{dt}$,

$$\frac{dc_i(t)}{dt} = -a_i c_i(t) + r_i(t) \tag{6.30}$$

Figure 6.30 shows Equation 6.30 as a signal-flow graph. Here again, a node is assumed for $C_i(s)$ at the output of an integrator, and its derivative formed at the input. Using Figure 6.30 as a model, we represent the first three terms on the right of Equation 6.26 as shown in Figure 6.31a. To cascade the zero, $(s + 5)$, we identify the output of $100/[(s + 2)(s + 3)]$ as $X_1(s)$. Cascading $(s + 5)$ yields

$$C(s) = (s + 5)X_1(s) = sX_1(s) + 5X_1(s) \tag{6.31}$$

Thus, $C(s)$ can be formed from a linear combination of previously derived signals as shown in Figure 6.31a. Finally, add the feedback and input paths as shown in Figure 6.31b.

Now, by inspection, write the state equations.

$$\dot{x}_1 = -3x_1 + x_2 \tag{6.32a}$$
$$\dot{x}_2 = -2x_2 + 100(r - c) \tag{6.32b}$$

But from Figure 6.31b,

$$c = 5x_1 + (x_2 - 3x_1) = 2x_1 + x_2 \tag{6.32c}$$

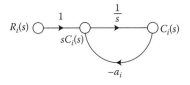

FIGURE 6.30 First-order subsystem.

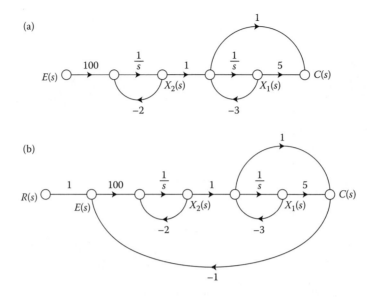

FIGURE 6.31 Steps in drawing the signal-flow graph for the feedback system of Figure 6.29: (a) forward transfer function and (b) closed-loop system.

Substituting Equation 6.32c into Equation 6.32b, the state equations for the system are

$$\dot{x}_1 = -3x_1 + x_2 \tag{6.33a}$$

$$\dot{x}_2 = -200x_1 - 102x_2 + 100r \tag{6.33b}$$

The output equation is the same as Equation 6.32c, or

$$y = c = 2x_1 + x_2 \tag{6.33c}$$

In vector-matrix form,

$$\dot{x} = \begin{bmatrix} -3 & 1 \\ -200 & -102 \end{bmatrix} x + \begin{bmatrix} 0 \\ 100 \end{bmatrix} r \tag{6.34a}$$

$$y = \begin{bmatrix} 2 & 1 \end{bmatrix} x \tag{6.34b}$$

In this chapter, we discussed signal-flow graphs. We defined them and related them to block diagrams. We showed that signals are represented by nodes, and systems by branches. We showed how to draw signal-flow graphs from state equations and how to use signal-flow graphs as an aid to obtaining state equations.

6.2.7 Defining Terms

Block diagram: A representation of the interconnection of subsystems. In a linear system, the block diagram consists of blocks representing subsystems; arrows representing signals; summing junctions, which show the algebraic summation of two or more signals; and pickoff points, which show the distribution of one signal to multiple subsystems.

Branches: Lines that represent subsystems in a signal-flow diagram.

Cascaded form: An interconnection of subsystems, where the output of one subsystem is the input of the next. For linear systems with real and distinct poles (eigenvalues), this model leads to a triangular system matrix in state-space with the poles along the diagonal.

Companion matrix: A system matrix that contains the coefficients of the characteristic equation along a row or column. If the first row contains the coefficients, the matrix is an upper companion matrix; if the last row contains the coefficients, the matrix is a lower companion matrix; if the first column contains the coefficients, the matrix is a left companion matrix; and if the last column contains the coefficients, the matrix is a right companion matrix.

Feedback form: An interconnection of two subsystems: forward-path and feedback. The input to the forward-path subsystem is the algebraic sum of two signals: (1) the system input and (2) the system output operated on by the feedback subsystem.

Forward-path gain: The product of gains found by traversing a path from the input node to the output node of a signal-flow graph in the direction of signal flow.

Linear combination: A linear combination of n variables, x_i, for $i = 1$ to n, is given by the sum, $S = K_n X_n + K_{n-1} X_{n-1} + \cdots K_1 X_1$, where each K_i is a constant.

Linear system: A system possessing the properties of superposition and homogeneity.

Loop gain: The product of branch gains found by traversing a path that starts at a node and ends at the same node without passing through any other node more than once and following the direction of the signal flow.

Mason's Rule: The transfer function, $C(s)/R(s)$, of a system represented by a signal-flow graph is

$$G(s) = \frac{C(s)}{R(s)} = \frac{\sum_k T_k \Delta_k}{\Delta}$$

where $\sum$ denotes summation; $k =$ number of forward paths; $T_k =$ the kth forward-path gain; $\Delta = 1 - \sum$ loop gains $+ \sum$ nontouching-loop gains taken two at a time $- \sum$ nontouching-loop gains taken three at a time $+ \sum$ nontouching-loop gains taken four at a time $- \ldots$; $\Delta_k = \Delta - \sum$ loop gain terms in Δ touching the kth forward path. In other words, Δ_k is formed by eliminating from Δ those loop gains that touch the kth forward path.

Nodes: Points in a signal-flow diagram that represent signals.

Nontouching-loop gain: The product of loop gains from nontouching loops taken two, three, four, etc. at a time.

Nontouching loops: Loops that do not have any nodes in common.

Parallel form: An interconnection of subsystems, where the input is common to all subsystems and the output is an algebraic sum of the outputs of the subsystems.

Phase-variable form: A system representation where the state variables are successive derivatives and the system matrix is a lower *companion matrix*.

Signal-flow graphs: A representation of the interconnection of subsystems that form a system. The representation consists of nodes representing signals, and lines with arrows representing subsystems.

Transfer function: The ratio of the Laplace transform of the output of a system to the Laplace transform of the input.

References

1. Mason, S., Feedback theory—some properties of signal-flow graphs, *Proc. IRE,* September 1953, 1144–1156.
2. Mason, S., Feedback theory—further properties of signal-flow graphs, *Proc. IRE,* July 1956, 920–926.
3. Nise, N. S., *Control Systems Engineering,* 5th ed., John Wiley & Sons, Inc., Hoboken, NJ, 2008, 224–229, 236–254.

Further Reading

1. Dorf, R. and Bishop, R. 2008. *Modern Control Systems*, 11th ed., Pearson Education, Inc., Upper Saddle River, NJ.
2. Hostetter, G., Savant, C., Jr., and Stefani, R. 1994. *Design of Feedback Control Systems,* 3rd ed., Holt Rinehart and Winston, New York.
3. Kuo, B. C. and Golnaraghi, F. 2003. *Automatic Control Systems*, 8th ed., John Wiley & Sons, Inc., Hoboken, NJ.
4. Timothy, L. and Bona, B. 1968. *State Space Analysis: An Introduction*, McGraw-Hill, New York.

III

Analysis and Design Methods for Continuous-Time Systems

7

Analysis Methods

Raymond T. Stefani
California State University

William A. Wolovich
Brown University

7.1 Time Response of Linear Time-Invariant Systems

Raymond T. Stefani

7.1.1 Introduction

Linear time-invariant systems* may be described by either a scalar *n*th-order linear differential equation with constant coefficients or a coupled set of *n* first-order linear differential equations with constant coefficients using state variables. The solution in either case may be separated into two components: the *zero-state response*, found by setting the initial conditions to zero; and *zero-input response*, found by setting the input to zero. Another division is into the *forced response* (having the form of the input) and the *natural response* due to the characteristic polynomial.

7.1.2 Scalar Differential Equation

Suppose the system has an input (forcing function) $r(t)$ with the resulting output being $y(t)$. As an example of a second-order linear differential equation with constant coefficients

$$\frac{d^2y}{dt} + 6\frac{dy}{dt} + 8y = \frac{dr}{dt} + 8r \tag{7.1}$$

---

*   This section includes excerpts from *Design of Feedback Control Systems,* Third Edition by Raymond T. Stefani, Clement J. Savant, Barry Shahian, and Gene H. Hostetter, copyright 1994 by Saunders College Publishing, reprinted by permission of the publisher.

The Laplace transform may be evaluated where the initial conditions are taken at time 0^- and $y'(0^-)$ means the value of dy/dt at time 0^-.

$$s^2 Y(s) - sy(0^-) - y'(0^-) + 6[sY(s) - y(0^-)] + 8Y(s) = sR(s) - r(0^-) + 8R(s)$$

$$Y(s)[s^2 + 6s + 8] = (s+8)R(s) + sy(0^-) + y'(0^-) + 6y(0^-) - r(0^-)$$

(7.2)

In Equation 7.2, the quadratic $s^2 + 6s + 8$ is the *characteristic polynomial* while the first term on the right-hand side is due to the input $R(s)$ and the remaining terms are due to initial conditions (the initial state). Solving for $Y(s)$

$$Y(s) = \underbrace{\left[\frac{s+8}{s^2+6s+8}\right]R(s)}_{\text{zero-state response}} + \underbrace{\frac{sy(0^-) + y'(0^-) + 6y(0^-) - r(0^-)}{s^2+6s+8}}_{\text{zero-input response}}$$

(7.3)

In Equation 7.3, the zero-state response results by setting the initial conditions to zero while the zero-input response results from setting the input $R(s)$ to zero. The system transfer function results from

$$T(s) = \left.\frac{Y(s)}{R(s)}\right|_{\text{initial conditions}=0} = \frac{s+8}{s^2+6s+8}$$

(7.4)

Thus, the zero-state response is simply $T(s)R(s)$. The denominator of $T(s)$ is the characteristic polynomial, which, in this case, has roots at -2 and -4.

To solve Equation 7.4, values must be established for the input and the initial conditions. With a unit-step input and choices for $y(0^-)$ and $y'(0^-)$

$$r(0^-) = 0 \quad R(s) = 1/s \quad y(0^-) = 10 \quad y'(0^-) = -4$$

$$Y(s) = \underbrace{\left[\frac{s+8}{s^2+6s+8}\right]\frac{1}{s}}_{\text{zero-state response}} + \underbrace{\frac{10s+56}{s^2+6s+8}}_{\text{zero-input response}}$$

(7.5)

Next, the zero-state and zero-input responses can be expanded into partial fractions for the poles of $T(s)$ at -2 and -4 and for the pole of $R(s)$ at $s = 0$.

$$Y(s) = \overbrace{\frac{1}{s}}^{\substack{\text{forced}\\\text{response}}} \underbrace{- \frac{1.5}{s+2} + \frac{0.5}{s+4}}_{\text{zero-state response}} + \overbrace{\underbrace{\frac{18}{s+2} - \frac{8}{s+4}}_{\text{zero-input response}}}^{\text{natural response}}$$

(7.6)

In this case, the forced response is the term with the pole of $R(s)$, and the natural response contains the terms with the poles of $T(s)$, since there are no common poles between $R(s)$ and $T(s)$. When there are common poles, those multiple poles are usually assigned to the forced response.

The division of the total response into zero-state and zero-input components is a rather natural and logical division, because these responses can easily be obtained empirically by setting either the initial conditions or the input to zero and then obtaining each response. The forced response cannot be obtained separately in most cases; so the division into forced and natural components is more mathematical than practical.

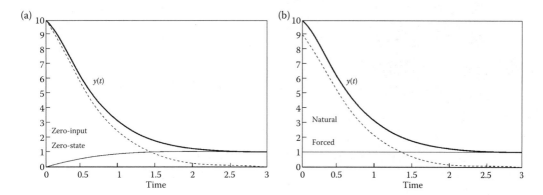

FIGURE 7.1 $y(t)$ and components. (a) Zero-state and zero-input components. (b) Forced and natural components.

The inverse transform of Equation 7.6 is

$$Y(s) = \underbrace{\overbrace{1 - 1.5e^{-2t}}^{\text{forced response}} + 0.5e^{-4t}}_{\text{zero-state response}} + \underbrace{\overbrace{18e^{-2t} - 8e^{-4t}}^{\text{natural response}}}_{\text{zero-input response}} \tag{7.7}$$

The total response is therefore

$$y(t) = 1 + 16.5e^{-2t} - 7.5e^{-4t} \tag{7.8}$$

Figure 7.1 contains a plot of $y(t)$ and its components.

7.1.3 State Variables

A linear system may also be described by a coupled set of n first-order linear differential equations, in this case having constant coefficients.

$$\frac{dx}{dt} = Ax + Br$$
$$y = Cx + Dr \tag{7.9}$$

where x is an $n \times 1$ column vector. If r is a scalar and there are m outputs, then A is $n \times n$, B is $n \times 1$, C is $m \times n$, and D is $m \times 1$. In most practical systems, D is zero because there is usually no instantaneous output response due to an applied input. The Laplace transform can be obtained in vector form, where I is the identity matrix:

$$sIX(s) - x(0^-) = AX(s) + BR(s)$$
$$Y(s) = CX(s) + DR(s) \tag{7.10}$$

Solving

$$X(s) = \underbrace{(sI - A)^{-1}BR(s)}_{\text{zero-state response}} + \underbrace{(sI - A)^{-1}x(0^-)}_{\text{zero-input response}}$$

$$Y(s) = \underbrace{[C(sI - A)^{-1}B + D]R(s)}_{\text{zero-state response}} + \underbrace{C(sI - A)^{-1}x(0^-)}_{\text{zero-input response}} \tag{7.11}$$

Thus, the transfer function $T(s)$ becomes $[C(sI - A)^{-1}B + D]$.

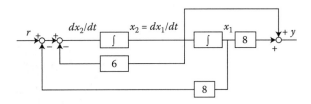

FIGURE 7.2 Second-order system in state variable form.

The time response can be found in two ways. The inverse Laplace transform of Equation 7.11 can be taken, or the response can be calculated by using the *state transition matrix* $\Phi(t)$, which is the inverse Laplace transform of the *resolvant matrix* $\Phi(s)$.

$$\Phi(s) = (sI - A)^{-1} \quad \Phi(t) = L^{-1}\{\Phi(s)\} \tag{7.12}$$

Figure 7.2 contains a second-order system in state-variable form. The system is chosen to have the dynamics of Equation 7.1. From Figure 7.2

$$\begin{bmatrix} dx_1/dt \\ dx_2/dt \end{bmatrix} = \begin{bmatrix} 0 & 1 \\ -8 & -6 \end{bmatrix} \begin{bmatrix} x_1 \\ x_2 \end{bmatrix} + \begin{bmatrix} 0 \\ 1 \end{bmatrix}$$

$$y = \begin{bmatrix} 8 & 1 \end{bmatrix} \begin{bmatrix} x_1 \\ x_2 \end{bmatrix} \tag{7.13}$$

Thus, D is zero. The resolvant matrix is

$$\Phi(s) = (sI - A)^{-1} = \begin{bmatrix} s & -1 \\ 8 & s+6 \end{bmatrix}^{-1} = \frac{1}{s^2 + 6s + 8} \begin{bmatrix} s+6 & 1 \\ -8 & s \end{bmatrix} \tag{7.14}$$

7.1.4 Inverse Laplace Transform Approach

First, the time response for $y(t)$ is calculated using the inverse Laplace transform approach. The transfer function is as before since D is zero and

$$C(sI - A)^{-1}B = \frac{\begin{bmatrix} 8 & 1 \end{bmatrix}}{s^2 + 6s + 8} \begin{bmatrix} s+6 & 1 \\ -8 & s \end{bmatrix} \begin{bmatrix} 0 \\ 1 \end{bmatrix} = \frac{s+8}{s^2 + 6s + 8} \tag{7.15}$$

Suppose a unit-step input is chosen so that $R(s)$ is $1/s$. It follows that the zero-state response of Equation 7.11 is the same as in Equations 7.5 through 7.7 since both $T(s)$ and $R(s)$ are the same. Suppose $x_1(0^-)$ is 1 and $x_2(0^-)$ is 2. The zero-input response becomes

$$C(sI - A)^{-1}x(0^-) = \frac{\begin{bmatrix} 8 & 1 \end{bmatrix}}{s^2 + 6s + 8} \begin{bmatrix} s+6 & 1 \\ -8 & s \end{bmatrix} \begin{bmatrix} 1 \\ 2 \end{bmatrix} = \frac{10s + 56}{s^2 + 6s + 8} \tag{7.16}$$

The zero-input response is also the same as in Equations 7.5 through 7.7 because the initial conditions on the state variables cause the same initial conditions as were used for y; that is,

$$y(0^-) = 8x_1(0^-) + x_2(0^-) = 8 + 2 = 10$$
$$y'(0^-) = 8dx_1/dt(0^-) + dx_2/dt(0^-)$$
$$= 8x_2(0^-) + [-8x_1(0^-) - 6x_2(0^-) + r(0^-)]$$
$$= 16 + [-8 - 12 + 0] = -4$$

7.1.5 State Transition Matrix Approach

The second procedure for calculating the time response is to use the state transition matrix. $\Phi(s)$ in Equation 7.14 may be expanded in partial fractions to obtain $\Phi(t)$

$$
\begin{aligned}
\Phi(s) &= \begin{bmatrix} (2/s+2-1/s+4) & (0.5/s+2-0.5/s+4) \\ (-4/s+2+4/s+4) & (-1/s+2+2/s+4) \end{bmatrix} \\
\Phi(t) &= \begin{bmatrix} (2e^{-2t}-e^{-4t}) & (0.5e^{-2t}-0.5e^{-4t}) \\ (-4e^{-2t}+4e^{-4t}) & (-e^{-2t}+2e^{-4t}) \end{bmatrix}
\end{aligned}
\tag{7.17}
$$

Using $\Phi(t)$, the solution to Equation 7.11 is

$$
x(t) = \underbrace{\int_0^t \Phi(t-\tau)Br(\tau)\,d\tau}_{\text{zero-state response}} + \underbrace{\Phi(t)x(0^-)}_{\text{zero-input response}}
$$

$$
y(t) = \underbrace{\int_0^t C\Phi(t-\tau)Br(\tau)\,dt + Dr(t)}_{\text{zero-state response}} + \underbrace{C\Phi(t)x(0^-)}_{\text{zero-input response}}
\tag{7.18}
$$

For this example, the zero-input response for $y(t)$ is

$$
\begin{aligned}
y_{zi}(t) &= \begin{bmatrix} 8 & 1 \end{bmatrix} \begin{bmatrix} (2e^{-2t}-e^{-4t}) & (0.5e^{-2t}-0.5e^{-4t}) \\ (-4e^{-2t}+4e^{-4t}) & (-e^{-2t}+2e^{-4t}) \end{bmatrix} \begin{bmatrix} 1 \\ 2 \end{bmatrix} \\
&= 18e^{-2t} - 8e^{-4t},
\end{aligned}
\tag{7.19}
$$

which agrees with Equation 7.7. The form for the zero-state response in Equation 7.18 is called a convolution integral. It is therefore necessary to evaluate the integral of

$$
\begin{aligned}
C\Phi(t-\tau)Bu(\tau) &= \begin{bmatrix} 8 & 1 \end{bmatrix} \begin{bmatrix} \Phi_{11}(t-\tau) & \Phi_{12}(t-\tau) \\ \Phi_{21}(t-\tau) & \Phi_{22}(t-\tau) \end{bmatrix} \begin{bmatrix} 0 \\ 1 \end{bmatrix} \tag{1} \\
&= 8\Phi_{12}(t-\tau) + \Phi_{22}(t-\tau) \\
&= 3e^{-2(t-\tau)} - 2e^{-4(t-\tau)} \\
&= 3e^{-2t}e^{2\tau} - 2e^{-4t}e^{4\tau}
\end{aligned}
\tag{7.20}
$$

After integrating with respect to τ

$$
\begin{aligned}
y_{zs}(t) &= (3/2)e^{-2t}[e^{2\tau}]\big|_0^t - (2/4)e^{-4t}[e^{4\tau}]\big|_0^t \\
&= 1.5e^{-2t}[e^{2t}-1] - 0.5e^{-4t}[e^{4t}-1] \\
&= 1.5[1-e^{-2t}] - 0.5[1-e^{-4t}] \\
&= 1 - 1.5e^{-2t} + 0.5e^{-4t}
\end{aligned}
\tag{7.21}
$$

which agrees with Equation 7.7.

7.1.6 Use of MATLAB

MATLAB® software produced by The Math Works Inc. provides a platform for calculating the inverse Laplace transform and for calculating and plotting the various time response components of a linear time-invariant system, such as those covered in this section. To use MATLAB in this context, it is necessary to define *s* and *t* as variables and then to use the appropriate MATLAB commands. For example, the zero-state response of the transfer function in Equation 7.6, as plotted in Figure 7.1, can be obtained by defining the Laplace transform in terms of *s*, using the *ilaplace* command to get the inverse Laplace transform in terms of *t* and then plotting the time response using the ezplot command. The latter command requires a vector giving minimum and maximum times, which would be coded as [0 3] to reproduce the zero-state response of Figure 7.1.

$$Syms \ s \ t$$
$$zst = ilaplace((s+8)/(s \wedge 3 + 6*s \wedge 2 + 8*s)) \tag{7.22}$$
$$ezplot(zst, \ [0 \ 3])$$

Similarly, the inverse Laplace transform and time response for the zero-input component may be found by

$$zit = ilaplace((10*s+6)/(s \wedge 2 + 6*s + 8))$$
$$ezplot(zit, \ [0 \ 3]) \tag{7.23}$$

The two components may be combined using *ezplot(zst + zit, [0 3])*.

When working with the state space form, the system matrices are input row by row. The system of Equation 7.13 and Figure 7.2 may be input to MATLAB and then the transfer function of Equation 7.15 can be found.

$$a = [0 \ 1; -8 \ -6], \quad b = [0 \ 1]'$$
$$c = [8 \ 1], \quad d = 0 \tag{7.24}$$
$$tfn = c * inv(s * eye(2) - a) * b$$

The command *eye(n)* creates an $n \times n$ identity matrix, *inv* creates a matrix inverse and the prime symbol used to find *b* creates a transpose. The inverse Laplace transform of Equation 7.21 for the zero-state component would be found by *ilaplace (tfn/s)*. The time response could be obtained using *ezplot*.

For the zero-input transfer function of Equation 7.16, inverse Laplace transform of Equation 7.19 and time response, one could use

$$x0 = [1 \ 2]'$$
$$zis = c * inv(s * eye(2) - a) * x0 \tag{7.25}$$
$$zit = ilplace(\ zis), \quad ezplot(zit, \ [0 \ 3])$$

The $\Phi(s)$ and $\Phi(t)$ of Equation 7.17 may be found by

$$phis = inv(s * eye(2) - a)$$
$$phit = ilaplace(phis) \tag{7.26}$$

7.1.7 Eigenvalues, Poles, and Zeros

It has been noted that the denominator of a transfer function $T(s)$ is the characteristic polynomial. In the previous examples, that denominator was $s^2 + 6s + 8$ with roots at -2 and -4. The characteristic polynomial may also be found from the system matrix since the characteristic polynomial is $|sI - A|$, where $|.|$ means the determinant. The eigenvalues of a matrix are those *s* values satisfying $|sI - A| = 0$; hence, the eigenvalues of *A* are the same as the poles of the transfer function. As in Equations 7.6 and 7.7, the poles of $T(s)$ establish terms present in the natural response. The coefficients of the partial fraction expansion (and, thus, the shape of the response) depend on the numerator of $T(s)$ which, in turn, depends on the zeros of $T(s)$. In fact, some zeros of $T(s)$ can cancel poles of $T(s)$, eliminating terms from the natural response.

To simplify this discussion, suppose the initial conditions are set to zero, and interest is focused on the zero-state response for a unit-step input. As one example, consider a system with a closed-loop transfer function $T_1(s)$:

$$T_1(s) = \frac{6s^2 + 10s + 2}{(s+1)(s+2)} = \frac{6(s+0.232)(s+1.434)}{(s+1)(s+2)}$$

The zero-state response for a unit-step input is

$$Y_1(s) = \frac{6s^2 + 10s + 2}{s(s+1)(s+2)} = \frac{1}{s} + \frac{2}{s+1} + \frac{3}{s+2}$$

If the denominator of the transfer function (the characteristic polynomial) remains the same but the numerator changes to $3s^2 + 7s + 2$, the zero-state response changes considerably due to a cancellation of the pole at -2 by a zero at the same location:

$$T_2(s) = \frac{3s^2 + 7s + 2}{(s+1)(s+2)} = \frac{3(s+1/3)(s+2)}{(s+1)(s+2)}$$

$$Y_2(s) = \frac{3s^2 + 7s + 2}{s(s+1)(s+2)} = \frac{1}{s} + \frac{2}{s+1}$$

Similarly, if the numerator changes to $4s^2 + 6s + 2$, there is a cancellation of the pole at -1 by a zero at -1.

$$T_3(s) = \frac{4s^2 + 6s + 2}{(s+1)(s+2)} = \frac{4(s+0.5)(s+1)}{(s+1)(s+2)}$$

$$Y_3(s) = \frac{4s^2 + 6s + 2}{s(s+1)(s+2)} = \frac{1}{s} + \frac{3}{s+2}$$

The time responses $y_1(t), y_2(t)$, and $y_3(t)$ are shown in Figure 7.3.

In summary, the terms in the time response are determined by the poles of the transfer function (eigenvalues of the system matrix), while the relative excitation of each term is dictated by the zeros of the transfer function.

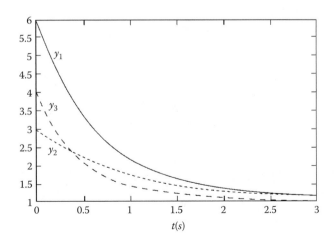

FIGURE 7.3 Time response example.

7.1.8 Second-Order Response

Many systems are describable by a second-order linear differential equation with constant coefficients while many other higher-order systems have complex conjugate dominant roots that cause the response to be nearly second order. Thus, the study of second-order response characteristics is important in understanding system behavior. The standard form for a second-order transfer function is

$$T(s) = \frac{\omega_n^2}{s^2 + 2\zeta\omega_n + \omega_n^2} \tag{7.27}$$

where ζ is called the damping ratio and ω_n is called the undamped natural frequency. The roots of the characteristic polynomial depend on ζ as shown in Table 7.1. When the damping ratio exceeds one (overdamped response), there are two real roots, which are distinct; hence, the natural response contains two exponentials with differing time constants. When the damping ratio equals one (critically damped response), there are two equal real roots and the natural response contains one term $K_1 \exp(-\omega_n t)$ and a second term $K_2 t \exp(-\omega_n t)$. When $0 \leq \zeta < 1$, the resulting oscillatory response is called underdamped. The zero-state response for a unit-step input is

$$y(t) = 1 - (1/k)e^{-at}\cos(\omega t - \Theta)$$
$$k = (1 - \zeta^2)^{1/2} \quad \Theta = \tan^{-1}(\zeta/k) \tag{7.28}$$
$$a = \zeta\omega_n \quad \omega = \omega_n(1 - \zeta^2)^{1/2}$$

When the damping ratio ζ is zero (undamped behavior), the system is marginally stable and there are complex roots $\pm j\omega_n$; hence, the radian frequency of the sinusoid becomes ω_n, explaining the term undamped natural frequency. When $0 < \zeta < 1$, the system is underdamped-stable and the radian frequency becomes $\omega = \omega_n(1 - \zeta^2)^{1/2}$, called the damped natural frequency.

Figure 7.4 shows the unit-step response for various values of ζ from 0 to 1. To normalize (generalize) these plots, the horizontal axis is $\omega_n t$. Associated with this type of response are three figures of merit: percent overshoot, rise time, and settling time.

Percent overshoot is defined by

$$\% \text{ overshoot} = 100 \frac{\text{max value-steady-state value}}{\text{steady-state value}}$$

From Figure 7.4, note that percent overshoot varies from 0% to 100%.

Rise time, Tr, is defined as the time required for the unit-step response to rise from 10% of the steady-state value to 90% of the steady-state value. Alternatively, rise time may be defined from 5% of the steady-state value to 95% of the steady-state value, but the 10–90% range is used here.

Settling time, Ts, is defined as the minimum time after which the response remains within $\pm 5\%$ of the steady-state value.

Figure 7.5a shows the product $\omega_n T_r$ versus damping ratio ζ. Figure 7.5b shows percent overshoot versus ζ. Figure 7.5c shows the product $\omega_n T_s$ versus ζ. Note that Figures 7.5a and 7.5b have opposite

TABLE 7.1   Roots of a Second-Order Characteristic Polynomial

Range for ζ	Type of Response	Root Locations
$\zeta > 1$	Overdamped	$-\zeta\omega_n \pm \omega_n(\zeta^2 - 1)^{1/2}$
$\zeta = 1$	Critically damped	$-\omega_n, \quad -\omega_n$
$0 \leq \zeta < 1$	Underdamped	$-\zeta\omega_n \pm j\omega_n(1 - \zeta^2)^{1/2}$

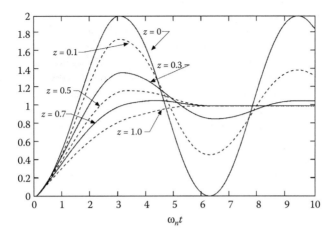

FIGURE 7.4 Second-order zero-state unit step responses.

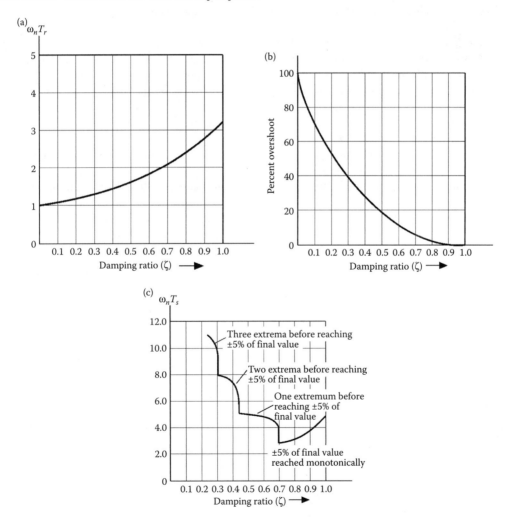

FIGURE 7.5 Figures of merit for second-order zero-state unit step responses. (a) Rise time. (b) Percent overshoot. (c) Settling time.

slopes. As ζ diminishes from 1 to 0, the rise time drops while the percent overshoot increases. The settling time curve of Figure 7.5c provides a trade-off between percent overshoot and rise time. As ζ drops from 1 to 0.7, the product $\omega_n T_s$ drops monotonically toward a minimum value. For that range of ζ values, the time response enters a value 5% below the steady-state value at $\omega_n T_s$ and does not exceed the upper limit 5% above the steady-state value after $\omega_n T_s$. Near $\zeta = 0.7$, the percent overshoot is near 5%; so a small reduction in ζ below 0.7 causes $\omega_n T_s$ to jump upward to a value at which the response peaks and then enters the $\pm 5\%$ boundary. The segment of Figure 7.5c, as ζ drops from about 0.7 to about 0.43, is a plot of increasing values of $\omega_n T_s$, corresponding to response curves in Figure 7.4, which go through a peak value where the derivative is zero (extremum point) prior to entering and staying within the $\pm 5\%$ boundary. Additional $\omega_n T_s$ curve segments correspond to regions where the unit-step response curve goes through an integer number of extrema prior to entering the $\pm 5\%$ boundary, which is entered alternatively from above and below. For a damping ratio of zero, the value of $\omega_n T_s$ is infinite since the peak values are undamped.

7.1.9 Defining Terms

Characteristic polynomial: Denominator of the transfer function. The roots determine terms in the *natural response*.

Forced response: The part of the response of the form of the forcing function.

MATLAB: A software package produced by The Math Works Inc which facilitates calculating and plotting the time response, among many other options, using the Controls Toolbox.

Natural response: The part of the response whose terms follow from the roots of the *characteristic polynomial*.

Percent overshoot: 100 (max. value–steady-state value)/steady-state value.

Resolvant matrix: $\Phi(s) = [sI - A]^{-1}$.

Rise time: The time required for the unit-step response to rise from 10% of the steady-state value to 90% of the steady-state value.

Settling time: The minimum time after which the response remains within $\pm 5\%$ of the steady-state value.

State transition matrix: $\Phi(t) =$ The inverse Laplace transform of the *resolvant matrix*.

Zero-input response: The part of the response found by setting the input to zero.

Zero-state response: The part of the response found by setting the initial conditions to zero.

Reference

1. Stefani, R.T., Shahian, B., Savant, C.J., and Hostetter, G.H., *Design of Feedback Control Systems*, 4th edn., Oxford University Press, Oxford, New York, 2002.

Further Reading

Additional information may be found in *IEEE Control Systems Mag.*; *IEEE Trans. Autom. Control*; and *IEEE Trans. Systems, Man, and Cybern.*

7.2 Controllability and Observability

William A. Wolovich

7.2.1 Introduction

The ultimate objective of any control system* is to improve and often to optimize the performance of a given dynamical system. Therefore, an obvious question that should be addressed is *how* do we design an appropriate controller? Before we can resolve this question, however, it is usually necessary to determine whether or not an appropriate controller exists; i.e., *can* we design a satisfactory controller?

Most physical systems are designed so that the control input does affect the complete system and, as a consequence, an appropriate controller does exist. However, this is not always the case. Moreover, in the *multi-input/multi-output* (MIMO) cases, certain control inputs may affect only part of the dynamical behavior. For example, the steering wheel of an automobile does not affect its speed, nor does the accelerator affect its heading; i.e., the speed of an automobile is uncontrollable via the steering wheel, and the heading is uncontrollable via the accelerator. In certain cases, it is important to determine whether or not complete system control is possible if one or more of the inputs (actuators) or outputs (sensors) fails to perform as expected.

The primary purpose of this chapter is to introduce two fundamental concepts associated with dynamical systems, namely controllability and observability, which enable us to resolve the "can" question for a large class of dynamical systems. These dual concepts, which were first defined by R. E. Kalman [1] using state-space representations, are by no means restricted to systems described in state-space form. Indeed, problems associated with analyzing and/or controlling systems that were either uncontrollable or unobservable were encountered long before the state-space approach to control system analysis and design was popularized in the early 1960s.

In those (many) cases where state-space equations can be employed to define the behavior of a dynamical system, controllability implies an ability to transfer the entire state of the system from any initial state $x(t_0)$ to any final state $x(t_f)$ over any arbitrary time interval $t_f - t_0 > 0$ through the employment of an appropriate control input $u(t)$ defined over the time interval. The concept of observability implies an ability to determine the entire initial state of the system from knowledge of the input and the output $y(t)$ over any arbitrary time interval $t_f - t_0 > 0$. These dual concepts play a crucial role in many of the control system design methodologies that have evolved since the early 1960s, such as pole placement, LQG, H_∞ and minimum time optimization, realization theory, adaptive control, and system identification.

These dual concepts are not restricted to linear and/or time-invariant systems, and numerous technical papers have been directed at extending controllability and observability to other classes of systems, such as nonlinear, distributed-parameter, discrete-event, and behavioral systems. This chapter, however, focuses on the class of linear time-invariant systems whose dynamical behavior can be described by finite dimensional state-space equations or (equivalently) by one or more ordinary linear differential equations, since a fundamental understanding of these two concepts should be obtained before any extensions can be undertaken.

7.2.2 State-Space Controllability and Observability

This section deals with the controllability and observability properties of systems described by linear, time-invariant state-space representations. In particular, consider a *single-input/single-output* (SISO) linear, time-invariant system defined by the *state-space representation*:

$$\dot{\mathbf{x}}(t) = A\mathbf{x}(t) + Bu(t); \quad y(t) = C\mathbf{x}(t) + Eu(t) \tag{7.29}$$

---

* Excerpts and figures from *Automatic Control Systems, Basic Analysis and Design,* by William A. Wolovich, copyright ©1994 by Saunders College Publishing, reproduced by permission of the publisher.

whose state matrix A has (n) distinct eigenvalues, $\lambda_1, \lambda_2, \ldots, \lambda_n$, which define the *poles* of the system and the corresponding *modes* $e^{\lambda_i t}$. Such an A can be diagonalized by any one of its eigenvector matrices V. More specifically, there exists a state transformation matrix $Q = V^{-1}$ that diagonalizes A, so that if

$$\hat{\mathbf{x}}(t) = Q\mathbf{x}(t) = V^{-1}\mathbf{x}(t) \tag{7.30}$$

the dynamical behavior of the equivalent system in *modal canonical form* then is defined by the state-space representation:

$$\dot{\hat{\mathbf{x}}}(t) = \begin{bmatrix} \dot{\hat{x}}_1(t) \\ \dot{\hat{x}}_2(t) \\ \vdots \\ \dot{\hat{x}}_n(t) \end{bmatrix} = \underbrace{\begin{bmatrix} \lambda_1 & 0 & 0 & \cdots \\ 0 & \lambda_2 & 0 & \cdots \\ \vdots & & \ddots & \\ 0 & 0 & \cdots & \lambda_n \end{bmatrix}}_{QAQ^{-1} = \hat{A}} \begin{bmatrix} \hat{x}_1(t) \\ \hat{x}_2(t) \\ \vdots \\ \hat{x}_n(t) \end{bmatrix} + \underbrace{\begin{bmatrix} \hat{B}_{11} \\ \hat{B}_{21} \\ \vdots \\ \hat{B}_{n1} \end{bmatrix}}_{QB = \hat{B}} u(t)$$

$$y(t) = \underbrace{[\hat{C}_{11}, \hat{C}_{12}, \ldots \hat{C}_{1n}]}_{CQ^{-1} = \hat{C}} \begin{bmatrix} \hat{x}_1(t) \\ \hat{x}_2(t) \\ \vdots \\ \hat{x}_n(t) \end{bmatrix} + Eu(t) \tag{7.31}$$

as depicted in Figure 7.6.

7.2.2.1 Controllability

If $\hat{B}_{k1} = 0$ for any $k = 1, 2, \ldots, n$, then the state $\hat{x}_k(t)$ is *uncontrollable* by the input $u(t) = u_1(t)$, since its time behavior is characterized by the mode $e^{\lambda_k t}$, independent of $u(t)$; i.e.,

$$\hat{x}_k(t) = e^{\lambda_k(t-t_0)}\hat{x}_k(t_0) \tag{7.32}$$

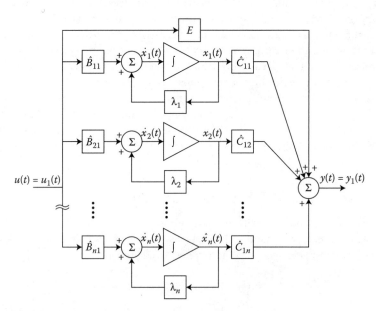

FIGURE 7.6   A state-space system in modal canonical form. (Reproduced from Wolovich, W. A., *Automatic Control Systems, Basic Analysis and Design,* © 1994 by Saunders College Publishing. With permission.)

The lack of controllability of the state $\hat{x}_k(t)$ (or the mode $e^{\lambda_k t}$) by $u(t)$ is reflected by a completely zero kth row of the so-called *controllability matrix* of the system, namely the $(n \times n)$ matrix

$$\hat{C} \stackrel{\text{def}}{=} [\hat{B}, \hat{A}\hat{B}, \ldots, \hat{A}^{n-1}\hat{B}]$$

$$= \begin{bmatrix} \hat{B}_{11} & \lambda_1 \hat{B}_{11} & \cdots & \lambda_1^{n-1} \hat{B}_{11} \\ \hat{B}_{21} & \lambda_2 \hat{B}_{21} & \cdots & \lambda_2^{n-1} \hat{B}_{21} \\ \vdots & \vdots & \ddots & \vdots \\ \hat{B}_{n1} & \lambda_n \hat{B}_{n1} & \cdots & \lambda_n^{n-1} \hat{B}_{n1} \end{bmatrix} \tag{7.33}$$

because $\hat{A}^m = \Lambda^m = diag[\lambda_i{}^m]$, a diagonal matrix for all integers $m \geq 0$. Therefore, each zero kth row element $\hat{B}_{k1}$ of $\hat{B}$ implies an *uncontrollable state* $\hat{x}_k(t)$, whose time behavior is characterized by the *uncontrollable mode* $e^{\lambda_k t}$, as well as a completely zero kth row of the controllability matrix $\hat{C}$.*

On the other hand, each nonzero kth row element of $\hat{B}$ implies a direct influence of $u(t)$ on $\hat{x}_k(t)$, and hence a *controllable* state $\hat{x}_k(t)$ (or mode $e^{\lambda_k t}$) and a corresponding nonzero kth row of $\hat{C}$ defined by $\hat{B}_{k1}[1, \lambda_k, \lambda_k^2, \ldots, \lambda_k^{n-1}]$. In the case (assumed here) of distinct eigenvalues, each such nonzero row of $\hat{B}$ increases the rank of $\hat{C}$ by one. Therefore, the rank of $\hat{C}$ corresponds to the total number of states or modes that are controllable by the input $u(t)$, which is termed the *controllability rank* of the system.

Fortunately, it is not necessary to transform a given state-space system to modal canonical form in order to determine its controllability rank. In particular, Equation 7.31 implies that $B = Q^{-1}\hat{B}$, $AB = Q^{-1}\hat{A}QQ^{-1}\hat{B} = Q^{-1}\hat{A}\hat{B}$, or that $A^m B = Q^{-1}\hat{A}^m\hat{B}$ in general, which defines the *controllability matrix* of the system defined by Equation 7.29, namely,

$$C \stackrel{\text{def}}{=} [B, AB, \ldots, A^{n-1}B] = Q^{-1}\hat{C} \tag{7.34}$$

with $Q^{-1} = V$ nonsingular. Therefore, *the rank of* C (which is equal to the rank of $\hat{C}$) *equals the controllability rank of the system.* It is important to note that this result holds in the case of nondistinct eigenvalues, as well the multi-input case where B has m columns, so that

$$B = [B_1, B_2, \ldots, B_m] \tag{7.35}$$

and the controllability matrix C, as defined by Equation 7.34, is an $(n \times nm)$ matrix.

In light of the preceding, any state-space system defined by Equation 7.29 is said to be *completely (state or modal) controllable* if its $(n \times nm)$ controllability matrix C has full rank n. Otherwise, the system is said to be *uncontrollable*, although some $(<n)$ of its states generally will be controllable. Note that for a general state-space system, the rank of C tells us only the *number* of controllable (and uncontrollable) modes, and not their identity, an observation that holds relative to the observability properties of a system as well.

We finally observe that there are several alternative ways of establishing state-space controllability. In particular, it is well known [2], [3] that *a state-space system defined by* Equation 7.29 *is controllable if and only if any one of the following (equivalent) conditions is satisfied:*

- The (n) rows of $e^{At}B$, where e^{At} represents the (unique) *state transition matrix* of the system, are linearly independent over the real field $\mathcal{R}$ for all t.
- The *controllability grammian*

$$G_c(t_0, t_f) \stackrel{\text{def}}{=} \int_{t_0}^{t_f} e^{-A\tau} BB^T e^{-A^T\tau} d\tau$$

  is nonsingular for all $t_f > t_0$.

---

* The reader should be careful not to confuse the controllability matrix $\hat{C}$ of Equation 7.33 with the output matrix $\hat{C}$ of Equation 7.31.

- The controllability matrix C defined by Equation 7.34 has full rank n.
- The $n \times (n+m)$ matrix $[\lambda I - A, \ B]$ has rank n at all eigenvalues λ_i of A or, equivalently, $\lambda I - A$ and B are *left coprime** polynomial matrices.

Since the solution to Equation 7.29 is given by

$$\mathbf{x}(t) = e^{A(t-t_0)} \mathbf{x}(t_0) + \int_{t_0}^{t} e^{A(t-\tau)} B\mathbf{u}(\tau) \, d\tau \tag{7.36}$$

it follows that the controllability grammian-based control input

$$\mathbf{u}(t) = B^T e^{-A^T t} G_c^{-1}(t_0, t_f) \left[e^{-At_f} \mathbf{x}(t_f) - e^{-At_0} \mathbf{x}(t_0) \right] \tag{7.37}$$

transfers any initial state $\mathbf{x}(t_0)$ to any arbitrary final state $\mathbf{x}(t_f)$ at any arbitrary $t_f > t_0$, an observation that is consistent with the more traditional definition of controllability.

7.2.2.2 Observability

We next note, in light of Figure 7.6, that if $\hat{C}_{1i} = 0$ for any $i = 1, 2, \ldots n$, then the state $\hat{x}_i(t)$ is *unobservable* at the output $y(t) = y_1(t)$, in the sense that the mode $e^{\lambda_i t}$, which defines the time behavior of

$$\hat{x}_i(t) = e^{\lambda_i(t-t_0)} \hat{x}_i(t_0) \tag{7.38}$$

does not appear at the output $y(t)$. This lack of observability of the state $\hat{x}_i(t)$ (or the mode $e^{\lambda_i t}$) at $y(t)$ is reflected by a completely zero (ith) column of the so-called *observability matrix* of the system, namely, the ($n \times n$) matrix

$$\hat{\mathcal{O}} \stackrel{\text{def}}{=} \begin{bmatrix} \hat{C} \\ \hat{C}\hat{A} \\ \vdots \\ \hat{C}\hat{A}^{n-1} \end{bmatrix} = \begin{bmatrix} \hat{C}_{11} & \hat{C}_{12} & \cdots & \hat{C}_{1n} \\ \lambda_1 \hat{C}_{11} & \lambda_2 \hat{C}_{12} & \cdots & \lambda_n \hat{C}_{1n} \\ \vdots & \vdots & \ddots & \vdots \\ \lambda_1^{n-1} \hat{C}_{11} & \lambda_2^{n-1} \hat{C}_{12} & \cdots & \lambda_n^{n-1} \hat{C}_{1n} \end{bmatrix} \tag{7.39}$$

analogous to a completely zero (kth) row of $\hat{C}$ in Equation 7.33.

On the other hand, each nonzero ith-column element $\hat{C}_{1i}$ of $\hat{C}$ implies a direct influence of $\hat{x}_i(t)$ on $y(t)$, hence an *observable* state $\hat{x}_i(t)$ or mode $e^{\lambda_i t}$, and a corresponding nonzero ith column of $\hat{\mathcal{O}}$ defined by $[1, \lambda_i, \lambda_i^2, \ldots, \lambda_i^{n-1}]^T \hat{C}_{1i}$. In the case (assumed here) of distinct eigenvalues, each such nonzero element of $\hat{C}$ increases the rank of $\hat{\mathcal{O}}$ by one. Therefore, the rank of $\hat{\mathcal{O}}$ corresponds to the total number of states or modes that are observable at the output $y(t)$, which is termed the *observability rank* of the system.

As in the case of controllability, it is not necessary to transform a given state-space system to modal canonical form in order to determine its observability rank. In particular, Equation 7.31 implies that $C = \hat{C}Q, CA = \hat{C}QQ^{-1}\hat{A}Q = \hat{C}\hat{A}Q$, or that $CA^m = \hat{C}\hat{A}^m Q$ in general, which defines the *observability matrix* of the system defined by Equation 7.29, namely

$$\mathcal{O} \stackrel{\text{def}}{=} \begin{bmatrix} C \\ CA \\ \vdots \\ CA^{n-1} \end{bmatrix} = \hat{\mathcal{O}}Q \tag{7.40}$$

with $Q = V^{-1}$ nonsingular. Therefore, *the rank of $\mathcal{O}$* (which is equal to the rank of $\hat{\mathcal{O}}$) *equals the observability rank of the system.* It is important to note that this result holds in the case of nondistinct eigenvalues,

---

* Two polynomials are called *coprime* if they have no common roots. Two polynomial matrices $P(\lambda)$ and $R(\lambda)$, which have the same number of rows, are left coprime if the rank of the composite matrix $[P(\lambda), \ R(\lambda)]$ remains the same for all (complex) values of λ. Right coprime polynomial matrices, which have the same number of columns, are defined in an analogous manner.

as well as the multi-output case where C has p rows, so that

$$C = \begin{bmatrix} C_1 \\ C_2 \\ \vdots \\ C_p \end{bmatrix} \tag{7.41}$$

and the observability matrix $\mathcal{O}$, as defined by Equation 7.40, is a $(pn \times n)$ matrix. In view of the preceding, a state-space system defined by Equation 7.29 is said to be *completely (state or modal) observable* if its $(pn \times n)$ observability matrix $\mathcal{O}$ has full rank n. Otherwise, the system is said to be *unobservable*, although some $(<n)$ of its states generally are observable.

As in the case of controllability, there are several alternative ways of establishing state-space observability. In particular, it is well known [2,3] that *a state-space system defined by* Equation 7.29 *is observable if and only if any one of the following (equivalent) conditions is satisfied*:

- The (n) columns of Ce^{At} are linearly independent over $\mathcal{R}$ for all t.
- The *observability grammian*

$$G_o(t_0, t_f) \overset{\text{def}}{=} \int_{t_0}^{t_f} e^{A^T \tau} C^T C e^{A\tau} \, d\tau$$

  is nonsingular for all $t_f > t_0$.
- The observability matrix $\mathcal{O}$ defined by Equation 7.40 has full rank n.
- The $(n+p) \times n$ matrix $\begin{bmatrix} \lambda I - A \\ C \end{bmatrix}$ has rank n at all eigenvalues λ_i of A or, equivalently, $\lambda I - A$ and C are *right coprime* polynomial matrices.

If a state-space system is observable, and if

$$f(t) \overset{\text{def}}{=} y(t) - C \int_{t_0}^{t} e^{A(t-\tau)} Bu(\tau) \, d\tau - Eu(t) \tag{7.42}$$

it then follows that its initial state can be determined via the relation

$$\mathbf{x}(t_0) = e^{At_0} G_o^{-1}(t_0, t_f) e^{A^T t_0} \int_{t_0}^{t_f} e^{A^T(t-t_0)} C^T f(t) \, dt \tag{7.43}$$

which is consistent with the more traditional definition of observability.

7.2.2.3 Component Controllability and Observability

In the multi-input and/or multi-output cases, it often is useful to determine the controllability and observability rank of a system relative to the individual components of its input and output. Such a determination would be important, for example, if one or more of the actuators or sensors were to fail.

In particular, suppose the system defined by Equation 7.29 has $m > 1$ inputs, $u_1(t), u_2(t), \ldots, u_m(t)$, so that the input matrix B has m columns, as in Equation 7.35. If we disregard all inputs except for $u_j(t)$, the resulting *controllability matrix associated with input* $u_j(t)$ is defined as the $(n \times n)$ matrix

$$\mathcal{C}_j \overset{\text{def}}{=} [B_j, AB_j, \ldots, A^{n-1} B_j] \tag{7.44}$$

The rank of each such $\mathcal{C}_j$ would determine the number of states or modes that are controllable by input component $u_j(t)$.

In a dual manner, suppose the given state-space system has $p > 1$ outputs, $y_1(t), y_2(t), \ldots y_p(t)$, so that the output matrix C has p rows, as in Equation 7.41. If we disregard all outputs except $y_q(t)$, the resulting *observability matrix associated with output $y_q(t)$* is defined as the $(n \times n)$ matrix

$$\mathcal{O}_q \overset{\text{def}}{=} \begin{bmatrix} C_q \\ C_q A \\ \vdots \\ C_q A^{n-1} \end{bmatrix} \tag{7.45}$$

As in the case of controllability, the rank of each such $\mathcal{O}_q$ determines the number of states or modes that are observable by output component $y_q(t)$.

Example 7.1:

To illustrate the preceding, we next note that the linearized equations of motion of an orbiting satellite can be defined by the state-space representation [4]

$$\underbrace{\begin{bmatrix} \dot{x}_1(t) \\ \dot{x}_2(t) \\ \dot{x}_3(t) \\ \dot{x}_4(t) \end{bmatrix}}_{\dot{\mathbf{x}}(t)} = \underbrace{\begin{bmatrix} 0 & 1 & 0 & 0 \\ 3\omega^2 & 0 & 0 & 2\omega \\ 0 & 0 & 0 & 1 \\ 0 & -2\omega & 0 & 0 \end{bmatrix}}_{A} \underbrace{\begin{bmatrix} x_1(t) \\ x_2(t) \\ x_3(t) \\ x_4(t) \end{bmatrix}}_{\mathbf{x}(t)} + \underbrace{\begin{bmatrix} 0 & 0 \\ 1 & 0 \\ 0 & 0 \\ 0 & 1 \end{bmatrix}}_{B} \underbrace{\begin{bmatrix} u_1(t) \\ u_2(t) \end{bmatrix}}_{\mathbf{u}(t)}$$

with a defined output

$$\underbrace{\begin{bmatrix} y_1(t) \\ y_2(t) \end{bmatrix}}_{\mathbf{y}(t)} = \underbrace{\begin{bmatrix} 1 & 0 & 0 & 0 \\ 0 & 0 & 1 & 0 \end{bmatrix}}_{C} \mathbf{x}(t)$$

The reader can verify that the $(n \times nm = 4 \times 8)$ controllability matrix $C = [B, AB, A^2B, A^3B]$ has full rank $4 = n$ in this case, so that the entire state is controllable using both inputs. However, since

$$C_1 = [B_1, AB_1, A^2B_1, A^3B_1] = \begin{bmatrix} 0 & 1 & 0 & -\omega^2 \\ 1 & 0 & -\omega^2 & 0 \\ 0 & 0 & -2\omega & 0 \\ 0 & -2\omega & 0 & 2\omega^3 \end{bmatrix}$$

is singular (i.e., the determinant of C_1, namely $|\,C_1\,| = 4\omega^4 - 4\omega^4 = 0$, with rank $C_1 = 3 < 4 = n$,) it follows that one of the "states" cannot be controlled by the radial thruster $u_1(t)$ alone, which would be unfortunate if the tangential thruster $u_2(t)$ were to fail.

We next note that

$$C_2 = [B_2, AB_2, A^2B_2, A^3B_2] = \begin{bmatrix} 0 & 0 & 2\omega & 0 \\ 0 & 2\omega & 0 & -2\omega^3 \\ 0 & 1 & 0 & -4\omega^2 \\ 1 & 0 & -4\omega^2 & 0 \end{bmatrix}$$

is nonsingular, since $|\,C_2\,| = 4\omega^4 - 16\omega^4 = -12\omega^4 \neq 0$, so that complete state control is possible by the tangential thruster $u_2(t)$ alone if the radial thruster $u_1(t)$ were to fail.

Insofar as observability is concerned, $y_1(t) = C_1\mathbf{x}(t) = x_1(t) = r(t) - d$ represents the radial deviation of $r(t)$ from a nominal radius $d = 1$, while output $y_2(t) = C_2\mathbf{x}(t) = x_3(t) = \alpha(t) - \omega t$ represents the tangential deviation of $\alpha(t)$ from a nominal angular position defined by ωt. The reader can verify that the $(pn \times n = 8 \times 4)$ observability matrix $\mathcal{O}$ has full rank $n = 4$ in this case, so that the entire state is observable using both outputs. However, since

$$\mathcal{O}_1 = \begin{bmatrix} C_1 \\ C_1 A \\ C_1 A^2 \\ C_1 A^3 \end{bmatrix} = \begin{bmatrix} 1 & 0 & 0 & 0 \\ 0 & 1 & 0 & 0 \\ 3\omega^2 & 0 & 0 & 2\omega \\ 0 & -\omega^2 & 0 & 0 \end{bmatrix}$$

is clearly singular (because its third column is zero), with rank $\mathcal{O}_1 = 3 < 4 = n$, it follows that one of the "states" cannot be observed by $y_1(t)$ alone.

We finally note that

$$\mathcal{O}_2 = \begin{bmatrix} C_2 \\ C_2 A \\ C_2 A^2 \\ C_2 A^3 \end{bmatrix} = \begin{bmatrix} 0 & 0 & 1 & 0 \\ 0 & 0 & 0 & 1 \\ 0 & -2\omega & 0 & 0 \\ -6\omega^3 & 0 & 0 & -4\omega^2 \end{bmatrix}$$

is nonsingular, since $|\mathcal{O}_2| = -12\omega^4 \neq 0$, so that the entire state can be observed by $y_2(t)$ alone.

7.2.2.4 MIMO Case

In the general MIMO case, the explicit modal controllability and observability properties of a system with distinct eigenvalues can be determined by transforming the system to modal canonical form. In particular, a zero in any kth row of column $\hat{B}_j$ of the input matrix $\hat{B}$ implies the uncontrollability of state $\hat{x}_k(t)$ (or the mode $e^{\lambda_k t}$) by $u_j(t)$. Furthermore, a completely zero kth row of $\hat{B}$ implies the complete uncontrollability of state $\hat{x}_k(t)$ (or the mode $e^{\lambda_k t}$) with respect to the entire vector input $\mathbf{u}(t)$. Each such zero row of $\hat{B}$ implies a corresponding zero row of $\hat{\mathcal{C}}$, thereby reducing the rank of the $(n \times nm)$ controllability matrices $\hat{\mathcal{C}}$ and $\mathcal{C}$ by one. The number of controllable modes therefore is given by the rank of $\hat{\mathcal{C}}$ or $\mathcal{C}$, the controllability rank of the system.

Dual results hold with respect to the observability properties of a system. In particular, a zero in any ith column of row $\hat{C}_q$ of the output matrix $\hat{C}$ implies the unobservability of state $\hat{x}_i(t)$ (or the mode $e^{\lambda_i t}$) by $y_q(t)$. Furthermore, a completely zero ith column of $\hat{C}$ implies the complete unobservability of state $\hat{x}_i(t)$ (or the mode $e^{\lambda_i t}$) with respect to the entire vector output $\mathbf{y}(t)$. Each such zero column of $\hat{C}$ implies a corresponding zero column of $\hat{\mathcal{O}}$, thereby reducing the rank of the $(pn \times n)$ observability matrices $\hat{\mathcal{O}}$ and $\mathcal{O}$ by one. The number of observable modes therefore is given by the rank of $\hat{\mathcal{O}}$ or $\mathcal{O}$, the observability rank of the system. Section 2.6 of [5] contains a MIMO example that illustrates the preceding.

7.2.3 Differential Operator Controllability and Observability

Suppose the defining differential equations of a dynamical system are in the *differential operator form*

$$a(D)z(t) = b(D)u(t)$$
$$y(t) = c(D)z(t) + e(D)u(t) \tag{7.46}$$

where $a(D)$, $b(D)$, $c(D)$, and $e(D)$ are polynomials* in the *differential operator* $D = \frac{d}{dt}$, with $a(D)$ a monic polynomial of degree n, which defines the *order* of this representation, and $z(t)$ is a single-valued function of time called the *partial state*. We often find it convenient to "abbreviate" Equation 7.46 by the polynomial quadruple $\{a(D), b(D), c(D), e(D)\}$; i.e., $\{a(D), b(D), c(D), e(D)\} \iff$ Equation 7.46.

---

* We will later allow both $b(D)$ and $c(D)$ to be polynomial vectors, thereby enlarging the class of systems considered beyond the SISO case defined by Equation 7.46.

7.2.3.1 An Equivalent State-Space Representation

We first show that Equation 7.46 has an *equivalent state-space representation* that can be determined directly by inspection of $a(D)$ and $b(D)$ when $deg[b(D)] < n = deg[a(D)]$. In particular, suppose we employ the coefficients of

$$a(D) = D^n + a_{n-1}D^{n-1} + \cdots + a_1 D + a_0$$

and

$$b(D) = b_{n-1}D^{n-1} + \cdots + b_1 D + b_0$$

in order to define the following state-space system:

$$\underbrace{\begin{bmatrix} \dot{x}_1(t) \\ \dot{x}_2(t) \\ \vdots \\ \dot{x}_n(t) \end{bmatrix}}_{\dot{\mathbf{x}}(t)} = \underbrace{\begin{bmatrix} 0 & 0 & 0 & \cdots & -a_0 \\ 1 & 0 & 0 & \cdots & -a_1 \\ 0 & 1 & 0 & \cdots & \\ \vdots & \vdots & & \ddots & \vdots \\ 0 & 0 & \cdots & 1 & -a_{n-1} \end{bmatrix}}_{A} \underbrace{\begin{bmatrix} x_1(t) \\ x_2(t) \\ \vdots \\ x_n(t) \end{bmatrix}}_{\mathbf{x}(t)} + \underbrace{\begin{bmatrix} b_0 \\ b_1 \\ \vdots \\ b_{n-1} \end{bmatrix}}_{B} u(t) \tag{7.47}$$

with

$$z(t) \stackrel{\text{def}}{=} x_n(t) = [0\ 0\ \ldots 0\ 1]\mathbf{x}(t) \stackrel{\text{def}}{=} C_z \mathbf{x}(t) \tag{7.48}$$

Since A is a (right column) companion matrix, the *characteristic polynomial of A* is given by

$$|\lambda I - A| = \lambda^n + a_{n-1}\lambda^{n-1} + \cdots + a_1\lambda + a_0 = a(\lambda) \tag{7.49}$$

Therefore, *the n zeros of $a(\lambda)$ correspond to the n eigenvalues λ_i of A, which define the system modes $e^{\lambda_i t}$.* As in the previous section, we assume that these n eigenvalues of A are distinct.

In terms of the differential operator D, Equation 7.47 can be written as

$$\underbrace{\begin{bmatrix} D & 0 & 0 & \cdots & a_0 \\ -1 & D & 0 & \cdots & a_1 \\ 0 & -1 & D & \cdots & a_2 \\ \vdots & & \ddots & & \vdots \\ 0 & 0 & \cdots & -1 & D+a_{n-1} \end{bmatrix}}_{(DI-A)} \underbrace{\begin{bmatrix} x_1(t) \\ x_2(t) \\ \vdots \\ x_n(t) \end{bmatrix}}_{\mathbf{x}(t)} = \underbrace{\begin{bmatrix} b_0 \\ b_1 \\ \vdots \\ b_{n-1} \end{bmatrix}}_{B} u(t) \tag{7.50}$$

If we now premultiply Equation 7.50 by the row vector $[1\ D\ D^2\ \ldots\ D^{n-1}]$, noting that $x_n(t) = z(t)$, we obtain the relation

$$[0\ 0\ \ldots\ 0\ a(D)] \begin{bmatrix} x_1(t) \\ x_2(t) \\ \vdots \\ x_n(t) \end{bmatrix} = a(D)z(t)$$

$$= b(D)u(t) \tag{7.51}$$

thereby establishing the equivalence of the state-space system defined by Equation 7.47 and the partial state/input relation $a(D)z(t) = b(D)u(t)$ of Equation 7.46.

Since $x_n(t) = z(t)$, in light of Equation 7.48, Equation 7.47 implies that

$$Dz(t) = \dot{x}_n(t) = x_{n-1}(t) - a_{n-1}x_n(t) + b_{n-1}u(t)$$

$$D^2 z(t) = \dot{x}_{n-1}(t) - a_{n-1}\dot{x}_n(t) + b_{n-1}\dot{u}(t)$$

$$= x_{n-2}(t) - a_{n-2}x_n(t) + b_{n-2}u(t) - a_{n-1}[x_{n-1}(t) - a_{n-1}x_n(t) + b_{n-1}u(t)] + b_{n-1}\dot{u}(t)$$

etc., which enables us to express the output relation of Equation 7.46, namely,

$$y(t) = c(D)z(t) + e(D)u(t) = c(D)x_n(t) + e(D)u(t)$$

as a function of $\mathbf{x}(t)$ and $u(t)$ and its derivatives. As a consequence,

$$y(t) = C\mathbf{x}(t) + E(D)u(t) \tag{7.52}$$

for some constant $(1 \times n)$ vector C and a corresponding polynomial $E(D)$.

We have therefore established a *complete equivalence relationship* between the differential operator representation of Equation 7.46 and the state-space representation defined by Equations 7.47 and 7.52, with E expanded to $E(D)$ (if necessary) to include derivatives of the input. We denote this equivalence relationship as

$$\underbrace{\{A, B, C, E(D)\}}_{\text{of Equations 7.47 and 7.52}} \overset{\text{equiv}}{\Longleftrightarrow} \underbrace{\{a(D), b(D), c(D), e(D)\}}_{\text{of Equation 7.46}} \tag{7.53}$$

7.2.3.2 Observable Canonical Forms

If $c(D) = 1$ in Equation 7.46, so that

$$a(D)z(t) = b(D)u(t); \quad y(t) = z(t) + e(D)u(t) \tag{7.54}$$

the equivalent state-space system defined by Equations 7.47 and 7.52 is characterized by an output matrix $C = C_z = [0\ 0\ \dots 0\ 1]$ and an $E(D) = e(D)$ in Equation 7.52; i.e.,

$$y(t) = \underbrace{[0\ 0\ \dots 0\ 1]}_{C = C_z}\mathbf{x}(t) + \underbrace{E(D)}_{e(D)}u(t) \tag{7.55}$$

Therefore, Equations 7.47 and 7.55 represent a state-space system equivalent to the differential operator system defined by Equation 7.54. We denote this equivalence relationship as

$$\underbrace{\{A, B, C, E(D)\}}_{\text{of Equations 7.47 and 7.55}} \overset{\text{equiv}}{\Longleftrightarrow} \underbrace{\{a(D), b(D), c(D) = 1, e(D)\}}_{\text{of Equation 7.54}} \tag{7.56}$$

Moreover, *both of these representations are completely observable*. In particular, the differential operator representation is observable because $a(D)$ and $c(D) = 1$ are coprime,[*] and the state-space representation is observable because its observability matrix

$$\mathcal{O} = \begin{bmatrix} C \\ CA \\ CA^2 \\ \vdots \\ CA^{n-1} \end{bmatrix} = \begin{bmatrix} 0 & \dots & 0 & 1 \\ 0 & \dots & 1 & * \\ \vdots & & * & \vdots \\ 0 & 1 & \dots & * \\ 1 & * & \dots & * \end{bmatrix} \tag{7.57}$$

(where $*$ denotes an irrelevant scalar) is nonsingular.

---

[*] We formally establish this condition for differential operator observability later in this section.

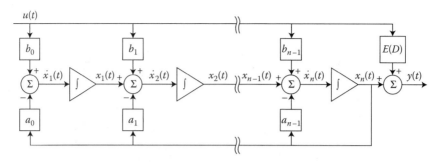

FIGURE 7.7   A state-space system in observable canonical form. (Reproduced from Wolovich, W. A., *Automatic Control Systems, Basic Analysis and Design,* © 1994 by Saunders College Publishing. With permission.)

Note further that the $\{A, C\}$ pair of Equations 7.47 and 7.55 is in a special canonical form. In particular, A is a right column companion matrix and C is identically zero except for a 1 in its right-most column. In light of these observations, we say that both of the representations defined by Equation 7.56 are in *observable canonical form.* Figure 7.7 depicts a block diagram of a state-space system in observable canonical form, as defined by Equations 7.47 and 7.55.

7.2.3.3 Differential Operator Controllability

Because of the right-column companion form structure of A in Equation 7.47, it follows that in the case (assumed here) of distinct eigenvalues,* the vector $[1\ \lambda_i\ \lambda_i^2 \ldots \lambda_i^{n-1}]$ is a row eigenvector of A in the sense that

$$[1\ \lambda_i\ \lambda_i^2 \ldots \lambda_i^{n-1}]A = \lambda_i[1\ \lambda_i\ \lambda_i^2 \ldots \lambda_i^{n-1}] \tag{7.58}$$

for each $i = 1, 2, \ldots n$. Therefore, the transpose of a Vandermonde matrix V of n column eigenvectors of A, namely,

$$V^T = \begin{bmatrix} 1 & 1 & \cdots & 1 \\ \lambda_1 & \lambda_2 & \cdots & \lambda_n \\ \lambda_1^2 & \lambda_2^2 & \cdots & \lambda_n^2 \\ \vdots & \vdots & & \vdots \\ \lambda_1^{n-1} & \lambda_2^{n-1} & \cdots & \lambda_n^{n-1} \end{bmatrix}^T = \begin{bmatrix} 1 & \lambda_1 & \lambda_1^2 & \cdots & \lambda_1^{n-1} \\ 1 & \lambda_2 & \lambda_2^2 & \cdots & \lambda_2^{n-1} \\ \vdots & \vdots & \vdots & & \vdots \\ 1 & \lambda_n & \lambda_n^2 & \cdots & \lambda_n^{n-1} \end{bmatrix} \tag{7.59}$$

diagonalizes A. Otherwise stated, a transformation of state defined by $\hat{x}(t) = V^T\mathbf{x}(t)$ reduces the state-space system defined by Equation 7.47 to the modal canonical form

$$\begin{bmatrix} \dot{\hat{x}}_1(t) \\ \dot{\hat{x}}_2(t) \\ \vdots \\ \dot{\hat{x}}_n(t) \end{bmatrix} = \underbrace{\begin{bmatrix} \lambda_1 & 0 & 0 & \cdots \\ 0 & \lambda_2 & 0 & \cdots \\ \vdots & & \ddots & \\ 0 & 0 & \cdots & \lambda_n \end{bmatrix}}_{V^T A V^{-T} = \hat{A} = diag[\lambda_i]} \begin{bmatrix} \hat{x}_1(t) \\ \hat{x}_2(t) \\ \vdots \\ \hat{x}_n(t) \end{bmatrix} + \underbrace{\begin{bmatrix} b(\lambda_1) \\ b(\lambda_2) \\ \vdots \\ b(\lambda_n) \end{bmatrix}}_{V^T B} u(t) \tag{7.60}$$

with the elements of $V^T B$ given by $b(\lambda_i)$ because

$$[1\ \lambda_i\ \lambda_i^2 \ldots \lambda_i^{n-1}]B = b(\lambda_i), \ \text{for } i = 1, 2, \ldots, n. \tag{7.61}$$

In light of the results presented in the previous section, and Figure 7.6 in particular, each $\hat{x}_i(t)$ is controllable if and only if $b(\lambda_i) \neq 0$ when $a(\lambda_i) = 0$. Therefore, *the state-space system defined by*

---

* Although the results presented hold in the case of nondistinct eigenvalues as well.

Equation 7.47 *is completely (state or modal) controllable if and only if the polynomials $a(\lambda)$ and $b(\lambda)$, or the differential operator pair $a(D)$ and $b(D)$, are coprime.*

When this is not the case, every zero λ_k of $a(\lambda)$, which also is a zero of $b(\lambda)$, implies an uncontrollable state $\hat{x}_k(t) = [1 \ \lambda_k \ \lambda_k^2 \ \ldots \ \lambda_k^{n-1}]\mathbf{x}(t)$, characterized by an uncontrollable mode $e^{\lambda_k t}$. Moreover, each such λ_k reduces the controllability rank of the system by one. The controllability properties of a dynamical system in differential operator form therefore can be completely specified by the (zeros of the) polynomials $a(D)$ and $b(D)$ of Equation 7.46, independent of any state-space representation.

7.2.3.4 Controllable Canonical Forms

When $b(D) = 1$ and $deg[c(D)] < n = deg[a(D)]$, the differential operator system defined by Equation 7.46, namely,

$$\underbrace{(D^n + a_{n-1}D^{n-1} + \cdots + a_1 D + a_0)}_{a(D)} z(t) = u(t)$$

$$y(t) = \underbrace{(c_{n-1}D^{n-1} + \cdots + c_1 D + c_0)}_{c(D)} z(t) + e(D)u(t) \tag{7.62}$$

has an alternative, *equivalent state-space representation,* which can be determined directly by inspection of $a(D)$ and $c(D)$.

In particular, suppose we employ the coefficients of $a(D)$ and $c(D)$ to define the following state-space system:

$$\underbrace{\begin{bmatrix} \dot{x}_1(t) \\ \dot{x}_2(t) \\ \vdots \\ \dot{x}_n(t) \end{bmatrix}}_{\dot{\mathbf{x}}(t)} = \underbrace{\begin{bmatrix} 0 & 1 & 0 & \cdots & 0 \\ 0 & 0 & 1 & \cdots & 0 \\ \vdots & \vdots & & \ddots & \vdots \\ -a_0 & -a_1 & & \cdots & -a_{n-1} \end{bmatrix}}_{A} \underbrace{\begin{bmatrix} x_1(t) \\ x_2(t) \\ \vdots \\ x_n(t) \end{bmatrix}}_{\mathbf{x}(t)} + \underbrace{\begin{bmatrix} 0 \\ \vdots \\ 0 \\ 1 \end{bmatrix}}_{B} u(t)$$

$$y(t) = \underbrace{[c_0 \ c_1 \ \cdots \ c_{n-1}]}_{C} \begin{bmatrix} x_1(t) \\ x_2(t) \\ \vdots \\ x_n(t) \end{bmatrix} + \underbrace{E(D)}_{e(D)} u(t) \tag{7.63}$$

Since A is a (bottom row) companion matrix, the characteristic polynomial of A is given by

$$|\lambda I - A| = \lambda^n + a_{n-1}\lambda^{n-1} + \cdots + a_1\lambda + a_0 = a(\lambda) \tag{7.64}$$

as in Equation 7.49. Therefore, the n zeros of $a(\lambda)$ correspond to the n eigenvalues λ_i of A, which define the system modes $e^{\lambda_i t}$.

If $z(t) \overset{\text{def}}{=} x_1(t)$ in Equation 7.63, it follows that $Dz(t) = \dot{x}_1(t) = x_2(t)$, $D^2 z(t) = \dot{x}_2(t) = x_3(t), \ldots D^{n-1}z(t) = \dot{x}_{n-1}(t) = x_n(t)$, or that

$$\begin{bmatrix} 1 \\ D \\ \vdots \\ D^{n-1} \end{bmatrix} z(t) = \begin{bmatrix} x_1(t) \\ x_2(t) \\ \vdots \\ x_n(t) \end{bmatrix} = \mathbf{x}(t) \tag{7.65}$$

The substitution of Equation 7.65 for $\mathbf{x}(t)$ in Equation 7.63 therefore implies that

$$
\begin{bmatrix}
D & -1 & 0 & \dots & 0 \\
0 & D & -1 & \dots & 0 \\
\vdots & \vdots & & \ddots & \vdots \\
a_0 & a_1 & \dots & & D+a_{n-1}
\end{bmatrix}
\begin{bmatrix}
1 \\ D \\ \vdots \\ D^{n-1}
\end{bmatrix}
z(t) =
\begin{bmatrix}
0 \\ \vdots \\ 0 \\ 1
\end{bmatrix}
u(t)
$$

or that

$$
a(D)z(t) = u(t)
$$
$$
y(t) = C\mathbf{x}(t) + E(D)u(t) = c(D)z(t) + e(D)u(t) \tag{7.66}
$$

thus establishing the equivalence of the two representations. We denote this *equivalence relationship* as

$$
\underbrace{\{A, B, C, E(D)\}}_{\text{of Equation 7.63}} \overset{\text{equiv}}{\Longleftrightarrow} \underbrace{\{a(D), b(D) = 1, c(D), e(D)\}}_{\text{of Equation 7.62}} \tag{7.67}
$$

Note that *both of the representations defined by* Equation 7.67 *are completely controllable.* In particular, the differential operator representation is controllable because $a(D)$ and $b(D) = 1$ clearly are coprime, and the state-space representation is controllable because its controllability matrix, namely,

$$
\mathcal{C} = [B, AB, \dots, A^{n-1}B] =
\begin{bmatrix}
0 & \dots & 0 & 1 \\
0 & \dots & 1 & * \\
\vdots & & * & \vdots \\
0 & 1 & \dots & * \\
1 & * & \dots & *
\end{bmatrix} \tag{7.68}
$$

(where $*$ denotes an irrelevant scalar) is nonsingular.

Furthermore, the $\{A, B\}$ pair of Equation 7.63 is in a special canonical form. In particular, A is a bottom row companion matrix and B is identically zero except for the 1 in its bottom row. In light of these observations, we say that both of the representations defined by Equation 7.67 are in *controllable canonical form.* Figure 7.8 depicts a block diagram of a state-space system in controllable canonical form, as defined by Equation 7.63.

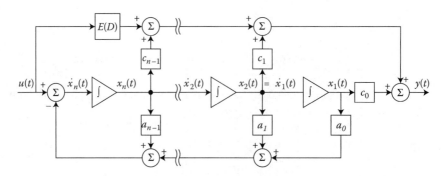

FIGURE 7.8 A state-space system in controllable canonical form. (Reproduced from Wolovich, W. A., *Automatic Control Systems, Basic Analysis and Design,* Saunders College Publishing, Boston, MA. With permission. © 1994.)

7.2.3.5 Differential Operator Observability

Because of the bottom row companion form structure of A in Equation 7.63, it follows that for each

$i = 1, 2, \ldots n,$ $\begin{bmatrix} 1 \\ \lambda_i \\ \vdots \\ \lambda_i^{n-1} \end{bmatrix}$ is a column eigenvector of A in the sense that

$$A \begin{bmatrix} 1 \\ \lambda_i \\ \vdots \\ \lambda_i^{n-1} \end{bmatrix} = \begin{bmatrix} 1 \\ \lambda_i \\ \vdots \\ \lambda_i^{n-1} \end{bmatrix} \lambda_i \qquad (7.69)$$

Therefore, if V is a Vandermonde matrix of n column eigenvectors of A, as in Equation 7.59, it follows that its inverse V^{-1} diagonalizes A. Otherwise stated, a transformation of state defined by $\hat{\mathbf{x}}(t) = V^{-1}\mathbf{x}(t)$ reduces the state-space system defined by Equation 7.63 to the following modal canonical form:

$$\begin{bmatrix} \dot{\hat{x}}_1(t) \\ \dot{\hat{x}}_2(t) \\ \vdots \\ \dot{\hat{x}}_n(t) \end{bmatrix} = \underbrace{\begin{bmatrix} \lambda_1 & 0 & 0 & \cdots \\ 0 & \lambda_2 & 0 & \cdots \\ \vdots & & \ddots & \\ 0 & 0 & \cdots & \lambda_n \end{bmatrix}}_{V^{-1}AV = \hat{A} = diag[\lambda_i]} \begin{bmatrix} \hat{x}_1(t) \\ \hat{x}_2(t) \\ \vdots \\ \hat{x}_n(t) \end{bmatrix} + \underbrace{\begin{bmatrix} \hat{b}_0 \\ \hat{b}_1 \\ \vdots \\ \hat{b}_{n-1} \end{bmatrix}}_{V^{-1}B} u(t)$$

$$y(t) = \underbrace{[c(\lambda_1), \ c(\lambda_2), \ \ldots c(\lambda_n)]}_{CV} \begin{bmatrix} \hat{x}_1(t) \\ \hat{x}_2(t) \\ \vdots \\ \hat{x}_n(t) \end{bmatrix} + E(D)u(t) \qquad (7.70)$$

with the elements of CV given by $c(\lambda_i)$ because

$$C \begin{bmatrix} 1 \\ \lambda_i \\ \lambda_i^2 \\ \vdots \\ \lambda_i^{n-1} \end{bmatrix} = c(\lambda_i), \quad \text{for } i = 1, 2, \ldots n \qquad (7.71)$$

In light of the results presented in the previous section, and Figure 7.6 in particular, each $\hat{x}_i(t)$ is observable if and only if $c(\lambda_i) \neq 0$. Therefore, *the state-space system defined by* Equations 7.47 and 7.52 *is completely (state or modal) observable if and only if the polynomials* $a(\lambda)$ *and* $c(\lambda)$, *or the differential operator pair* $a(D)$ *and* $c(D)$, *are coprime.*

When this is not the case, every zero λ_k of $a(\lambda)$, which is also a zero of $c(\lambda)$, implies an unobservable state $\hat{x}_k(t)$, characterized by an uncontrollable mode $e^{\lambda_k t}$. Moreover, each such λ_k reduces the observability rank of the system by one. The observability properties of a dynamical system in differential operator form therefore can be completely specified by the (zeros of the) polynomials $a(D)$ and $c(D)$ of Equation 7.46, independent of any state-space representation.

7.2.3.6 The MIMO Case

Although we initially assumed that Equation 7.46 defines a SISO system, it can be modified to include certain MIMO systems as well. In particular, a vector input

$$\mathbf{u}(t) = \begin{bmatrix} u_1(t) \\ u_2(t) \\ \vdots \\ u_m(t) \end{bmatrix} \tag{7.72}$$

can be accommodated by allowing the polynomial $b(D)$ in Equation 7.46 to be a row vector of polynomials, namely,

$$b(D) = [b_1(D), b_2(D), \dots, b_m(D)] \tag{7.73}$$

Each polynomial element of $b(D)$ then defines a corresponding real $(n \times 1)$ column of the input matrix B of an equivalent state-space system, analogous to that defined by Equation 7.47.

In a dual manner, a vector output

$$\mathbf{y}(t) = \begin{bmatrix} y_1(t) \\ y_2(t) \\ \vdots \\ y_p(t) \end{bmatrix} \tag{7.74}$$

can be accommodated by allowing the polynomial $c(D)$ in Equation 7.46 to be a column vector of polynomials, namely,

$$c(D) = \begin{bmatrix} c_1(D) \\ c_2(D) \\ \vdots \\ c_p(D) \end{bmatrix} \tag{7.75}$$

Of course, $e(D)$ also is a vector or matrix of polynomials in these cases. Each polynomial element of $c(D)$ then defines a corresponding real $(1 \times n)$ row of the output matrix C of an equivalent state-space system, analogous to that defined by Equation 7.52. A block diagram of such a MIMO system is depicted in Figure 7.9.

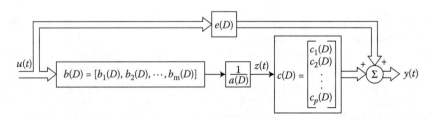

FIGURE 7.9 A MIMO differential operator system. (Reproduced from Wolovich, W. A., *Automatic Control Systems, Basic Analysis and Design,* Saunders College Publishing, Boston, MA. With permission. © 1994.)

Example 7.2:

To illustrate the preceding, consider a dynamical system defined by the (two-input/two-output) differential equation

$$\frac{d^4z(t)}{dt^4} + 2\frac{d^3z(t)}{dt^3} - 6\frac{d^2z(t)}{dt^2} - 22\frac{dz(t)}{dt} - 15z(t) = \frac{d^2u_1(t)}{dt^2} + 4\frac{du_1(t)}{dt} + 5u_1(t) + \frac{d^2u_2(t)}{dt^2} - u_2(t)$$

with

$$y_1(t) = -2\frac{dz(t)}{dt} + 6z(t)$$

and

$$y_2(t) = -\frac{dz(t)}{dt} - z(t)$$

This system can readily be placed in a MIMO differential operator form analogous to that defined by Equation 7.46, namely,

$$\underbrace{(D^4 + 2D^3 - 6D^2 - 22D - 15)}_{a(D)} z(t) = \underbrace{[D^2 + 4D + 5, \ D^2 - 1]}_{b(D) = [b_1(D), b_2(D)]} \underbrace{\begin{bmatrix} u_1(t) \\ u_2(t) \end{bmatrix}}_{\mathbf{u}(t)}$$

$$\mathbf{y}(t) = \begin{bmatrix} y_1(t) \\ y_2(t) \end{bmatrix} = \begin{bmatrix} c_1(D) \\ c_2(D) \end{bmatrix} z(t) = \underbrace{\begin{bmatrix} -2D + 6 \\ -D - 1 \end{bmatrix}}_{c(D)} z(t)$$

Since $a(D)$ can be factored as

$$a(D) = (D + 1)(D - 3)(D^2 + 4D + 5)$$
$$= (D + 1)(D - 3)(D + 2 - j)(D + 2 + j)$$

the system modes are defined by the ($n = 4$) zeros of $a(D)$, namely, $-1, +3$, and $-2 \pm j$.

We next note that $b_1(D) = D^2 + 4D + 5$, which is a factor of $a(D)$ as well. Therefore, the modes $e^{(-2+j)t}$ and $e^{(-2-j)t}$, which imply the real-valued modes $e^{-2t} \sin t$ and $e^{-2t} \cos t$, are uncontrollable by $u_1(t)$. Moreover, since $b_2(D) = (D + 1)(D - 1)$, the mode e^{-t} is uncontrollable by $u_2(t)$. Therefore, the remaining mode e^{3t} is the only one that is controllable by both inputs. Since all of the modes are controllable by at least one of the inputs, the system is completely (state or modal) controllable by the vector input $\mathbf{u}(t)$. This latter observation also holds because $a(D)$ and the polynomial vector $b(D) = [b_1(D), b_2(D)]$ are coprime; i.e., none of the zeros of $a(D)$ are also zeros of *both* $b_1(D)$ and $b_2(D)$.

We further note that $c_1(D) = -2(D - 3)$ while $c_2(D) = -(D + 1)$. Therefore, the mode e^{3t} is unobservable by $y_1(t)$, while e^{-t} is unobservable by $y_2(t)$. Since all of the modes are observable by at least one of the outputs, the system is completely (state or modal) observable by the vector output $\mathbf{y}(t)$. This latter observation also holds because $a(D)$ and the polynomial vector $c(D)$ are coprime; i.e., none of the zeros of $a(D)$ are also zeros of *both* $c_1(D)$ and $c_2(D)$.

In the general p-output, m-input differential operator case, $a(D)$ in Equation 7.46 could be a $(q \times q)$ polynomial matrix, with $\mathbf{z}(t)$ a q-dimensional partial state vector [2] [3]. The zeros of the determinant of $a(D)$ would then define the (n) poles of the MIMO system and its corresponding modes. Moreover, $b(D)$, $c(D)$ and $e(D)$ would be polynomial matrices in D of dimensions $(q \times m)$, $(p \times q)$ and $(p \times m)$, respectively. In such cases, the controllability and observability properties of the system can be determined directly in terms of the defining polynomial matrices.

In particular, as shown in [2], $a(D)$ and $b(D)$ always have a *greatest common left divisor*, namely a nonsingular polynomial matrix $g_l(D)$ that is a left divisor of both $a(D)$ and $b(D)$ in the sense that $a(D) = g_l(D)\hat{a}(D)$ and $b(D) = g_l(D)\hat{b}(D)$, for some appropriate pair of polynomial matrices, $\hat{a}(D)$ and $\hat{b}(D)$. Furthermore, *the determinant of $g_l(D)$ is a polynomial of maximum degree whose zeros define all of the uncontrollable modes of the system.*

If the degree of $|g_l(D)|$ is zero, then the following (equivalent) conditions hold:

- $g_l(D)$ is a *unimodular matrix**.
- $a(D)$ and $b(D)$ are *left coprime*.
- The differential operator system is controllable.

The astute reader will note that a nonunimodular $g_l(D)$ implies a lower-order differential operator representation between $\mathbf{z}(t)$ and $\mathbf{u}(t)$ than that defined by Equation 7.46, namely, $\hat{a}(D)\mathbf{z}(t) = \hat{b}(D)\mathbf{u}(t)$, which implies a corresponding pole-zero "cancellation" relative to the transfer function matrix relationship between the partial state and the input.

By duality, $a(D)$ and $c(D)$ always have a *greatest common right divisor $g_r(D)$, whose determinant defines all of the unobservable modes of the system.* If the degree of $|g_r(D)|$ is zero, then the following (equivalent) conditions hold:

- $g_r(D)$ is a *unimodular matrix*.
- $a(D)$ and $c(D)$ are *right coprime*.
- The differential operator system is observable.

The astute reader will note that a nonunimodular $g_l(D)$ or $g_r(D)$ implies a lower-order differential operator representation between $\mathbf{z}(t)$ and $\mathbf{u}(t)$ or $\mathbf{y}(t)$ than that defined by Equation 7.46. For example, if $g_l(D)$ is nonunimodular, then $\hat{a}(D)\mathbf{z}(t) = \hat{b}(D)\mathbf{u}(t)$, which implies a corresponding pole-zero "cancellation" relative to the transfer function matrix relationship between the partial state and the input. A dual observation holds when $g_r(D)$ is nonunimodular.

The preceding observations, which extend the notions of controllability and observability to a more general class of differential operator systems, are fully developed and illustrated in a number of references, such as [2] and [3].

References

1. Kalman, R. E., Contributions to the theory of optimal control, *Bol. Sociedad Mat. Mex.,* 1960.
2. Wolovich, W. A., *Linear Multivariable Systems,* Springer-Verlag, 1974.
3. Chen, C.-T., *Linear System Theory and Design,* Holt, Rinehart & Winston, New York, 1984.
4. Brockett, Roger W., *Finite Dimensional Linear Systems,* John Wiley & Sons, New York, 1970.
5. Wolovich, W. A., *Automatic Control Systems, Basic Analysis and Design,* Saunders College Publishing, Boston, MA, 1994.
6. Athans, M. and Falb, P. L., *Optimal Control: An Introduction to the Theory and Its Applications,* McGraw-Hill Book Company, New York, 1966.
7. *The MATLAB User's Guide,* The Math Works, Inc., South Natick, MA.

---

* A polynomial matrix whose determinant is a real scalar, independent of D, so that its inverse is also a unimodular matrix.

8

Stability Tests

Robert H. Bishop
The University of Texas at Austin

Richard C. Dorf
University of California, Davis

Charles E. Rohrs
Rohrs Consulting

Mohamed Mansour
Swiss Federal Institute of Technology

Raymond T. Stefani
California State University

8.1 The Routh–Hurwitz Stability Criterion

Robert H. Bishop and Richard C. Dorf

8.1.1 Introduction

In terms of linear systems, we recognize that the stability requirement may be defined in terms of the location of the poles of the closed-loop transfer function. Consider a single-input, single-output closed-loop system transfer function given by

$$T(s) = \frac{p(s)}{q(s)} = \frac{K \prod_{i=1}^{M}(s + z_i)}{\prod_{k=1}^{Q}(s + \sigma_k) \prod_{m=1}^{R}\left(s^2 + 2\alpha_m s + \alpha_m^2 + \omega_m^2\right)}, \qquad (8.1)$$

where $q(s)$ is the characteristic equation whose roots are the poles of the closed-loop system. The output response for an impulse function input is then

$$c(t) = \sum_{k=1}^{Q} A_k e^{-\sigma_k t} + \sum_{m=1}^{R} B_m \left(\frac{1}{\omega_m}\right) e^{-\alpha_m t} \sin \omega_m t. \qquad (8.2)$$

To obtain a bounded response to a bounded input, the poles of the closed-loop system must be in the left-hand portion of the s-plane (i.e., $\sigma_k > 0$ and $\alpha_m > 0$). *A necessary and sufficient condition that a feedback system be stable is that all the poles of the system transfer function have negative real parts.* We will call a system not stable if not all the poles are in the left half-plane. If the characteristic equation has simple roots on the imaginary axis ($j\omega$-axis) with all other roots in the left half-plane, the steady-state output is sustained oscillations for a bounded input, unless the input is a sinusoid (which is bounded) whose frequency is equal to the magnitude of the $j\omega$-axis roots. For this case, the output becomes unbounded. Such a system is called marginally stable, since only certain bounded inputs (sinusoids of the frequency of the poles) cause the output to become unbounded. For an unstable system, the characteristic equation has at least one root in the right half of the s-plane or repeated $j\omega$-axis roots; for this case, the output becomes unbounded for any input.

8.1.2 The Routh–Hurwitz Stability Criterion

The discussion and determination of *stability* has occupied the interest of many engineers. Maxwell and Vishnegradsky first considered the question of stability of dynamic systems. In the late 1800s, A. Hurwitz and E. J. Routh published independently a method of investigating the stability of a linear system [1] and [2]. The Routh–Hurwitz stability method provides an answer to the question of stability by considering the characteristic equation of the system. The characteristic equation in Equation 8.1 can be written as

$$q(s) = a_n s^n + a_{n-1} s^{n-1} + \cdots + a_1 s + a_0 = 0. \tag{8.3}$$

We require that all the coefficients of the polynomial must have the same sign if all the roots are in the left half-plane. Also, it is necessary that all the coefficients for a stable system be nonzero. However, although necessary, these requirements are not sufficient. That is, we immediately know the system is unstable if they are not satisfied; yet if they are satisfied, we must proceed to ascertain the stability of the system. The *Routh–Hurwitz criterion* is a necessary and sufficient criterion for the stability of linear systems. The method was originally developed in terms of determinants, but here we utilize the more convenient array formulation. The Routh–Hurwitz criterion is based on ordering the coefficients of the characteristic equation in Equation 8.3 into an array or schedule as follows [3]:

$$
\begin{array}{c|cccc}
s^n & a_n & a_{n-2} & a_{n-4} & \cdots \\
s^{n-1} & a_{n-1} & a_{n-3} & a_{n-5} & \cdots
\end{array}
$$

Further rows of the schedule are then completed as follows:

$$
\begin{array}{c|cccc}
s^n & a_n & a_{n-2} & a_{n-4} \\
s^{n-1} & a_{n-1} & a_{n-3} & a_{n-5} \\
s^{n-2} & b_{n-1} & b_{n-3} & b_{n-5} \\
s^{n-3} & c_{n-1} & c_{n-3} & c_{n-5} \\
\cdot & \cdot & \cdot & \cdot \\
\cdot & \cdot & \cdot & \cdot \\
\cdot & \cdot & \cdot & \cdot \\
s^0 & h_{n-1} & h_{n-3}
\end{array}
$$

where

$$b_{n-1} = \frac{(a_{n-1})(a_{n-2}) - a_n(a_{n-3})}{a_{n-1}} = -\frac{1}{a_{n-1}} \begin{vmatrix} a_n & a_{n-2} \\ a_{n-1} & a_{n-3} \end{vmatrix},$$

$$b_{n-3} = -\frac{1}{a_{n-1}} \begin{vmatrix} a_n & a_{n-4} \\ a_{n-1} & a_{n-5} \end{vmatrix},$$

and

$$c_{n-1} = -\frac{1}{b_{n-1}} \begin{vmatrix} a_{n-1} & a_{n-3} \\ b_{n-1} & b_{n-3} \end{vmatrix},$$

and so on. The algorithm for calculating the entries in the array can be followed on a determinant basis or by using the form of the equation for b_{n-1}.

The Routh–Hurwitz criterion states that the number of roots of $q(s)$ with positive real parts is equal to the number of changes of sign in the first column of the array. This criterion requires that there be no changes in sign in the first column for a stable system. This requirement is both necessary and sufficient.

Four distinct cases must be considered and each must be treated separately:

1. No element in the first column is zero.
2. There is a zero in the first column, but some other elements of the row containing the zero in the first column are nonzero.
3. There is a zero in the first column, and the other elements of the row containing the zero are also zero.
4. As in case 3 with *repeated* roots on the $j\omega$-axis.

Case 1

No element in the first column is zero.

Example

The characteristic equation of a third-order system is

$$q(s) = a_3 s^3 + a_2 s^2 + a_1 s + a_0. \tag{8.4}$$

The array is written as

$$\begin{array}{c|cc}
s^3 & a_3 & a_1 \\
s^2 & a_2 & a_0 \\
s^1 & b_1 & 0 \\
s^0 & c_1 & 0
\end{array}$$

where

$$b_1 = \frac{a_2 a_1 - a_0 a_3}{a_2} \quad \text{and} \quad c_1 = \frac{b_1 a_0}{b_1} = a_0.$$

For the third-order system to be stable, it is necessary and sufficient that the coefficients be positive and $a_2 a_1 > a_0 a_3$. The condition $a_2 a_1 = a_0 a_3$ results in a marginal stability case, and one pair of roots lies on the imaginary axis in the s-plane. This marginal stability case is recognized as Case 3 because there is a zero in the first column when $a_2 a_1 = a_0 a_3$. It is discussed under Case 3.

Case 2

Zeros in the first column while some other elements of the row containing a zero in the first column are nonzero. If only one element in the array is zero, it may be replaced with a small positive number ϵ that is allowed to approach zero after completing the array.

Example

Consider the characteristic equation

$$q(s) = s^4 + s^3 + s^2 + s + K, \tag{8.5}$$

where it is desired to determine the gain K that results in marginal stability. The Routh–Hurwitz array is then

s^4	1	1	K
s^3	1	1	0
s^2	ϵ	K	0
s^1	c_1	0	0
s^0	K	0	0

where

$$c_1 = \frac{\epsilon - K}{\epsilon} \rightarrow \frac{-K}{\epsilon} \quad \text{as } \epsilon \rightarrow 0.$$

Therefore, for any value of K greater than zero, the system is unstable (with $\epsilon > 0$). Also, because the last term in the first column is equal to K, a negative value of K results in an unstable system. Therefore, the system is unstable for all values of gain K.

Case 3

Zeros in the first column, and the other elements of the row containing the zero are also zero. Case 3 occurs when all the elements in one row are zero or when the row consists of a single element that is zero. This condition occurs when the characteristic polynomial contains roots that are symmetrically located about the origin of the s-plane. Therefore, Case 3 occurs when factors such as $(s + \sigma)(s - \sigma)$ or $(s + j\omega)(s - j\omega)$ occur. This problem is circumvented by utilizing the *auxiliary* polynomial, $U(s)$, which is formed from the row that immediately precedes the zero row in the Routh array. The order of the auxiliary polynomial is always even and indicates the number of symmetrical root pairs.

To illustrate this approach, let us consider a third-order system with a characteristic equation

$$q(s) = s^3 + 2s^2 + 4s + K, \tag{8.6}$$

where K is an adjustable loop gain. The Routh array is then

s^3	1	4
s^2	2	K
s^1	$\frac{8-K}{2}$	0
s^0	K	0

Therefore, for a stable system, we require that

$$0 < K < 8.$$

When $K = 8$, we have two roots on the $j\omega$-axis and a marginal stability case. Note that we obtain a row of zeros (Case 3) when $K = 8$. The auxiliary polynomial, $U(s)$, is formed from the row preceding the row of zeros which, in this case, is the s^2 row. We recall that this row contains the coefficients of the even powers of s and therefore, in this case, we have

$$U(s) = 2s^2 + Ks^0 = 2s^2 + 8 = 2(s^2 + 4) = 2(s + j2)(s - j2). \tag{8.7}$$

Case 4

Repeated roots of the characteristic equation on the $j\omega$-axis. If the roots of the characteristic equation on the $j\omega$-axis are simple, the system is neither stable nor unstable; it is instead called marginally stable, since it has an undamped sinusoidal mode. If the $j\omega$-axis roots are repeated, the system response will be unstable, with a form $t\left(\sin(\omega t + f)\right)$. The Routh–Hurwitz criterion does not reveal this form of instability [4]. Consider the system with a characteristic equation

$$q(s) = (s + 1)(s + j)(s - j)(s + j)(s - j) = s^5 + s^4 + 2s^3 + 2s^2 + s + 1. \tag{8.8}$$

The Routh array is

$$
\begin{array}{c|ccc}
s^5 & 1 & 2 & 1 \\
s^4 & 1 & 2 & 1 \\
s^3 & \epsilon & \epsilon & 0 \\
s^2 & 1 & 1 & \\
s^1 & \epsilon & 0 & \\
s^0 & 1 & & \\
\end{array}
$$

where $\epsilon \to 0$. Note the absence of sign changes, a condition that falsely indicates that the system is marginally stable. The impulse response of the system increases with time as $t \sin(t + f)$. The auxiliary equation at the s^2 line is $(s^2 + 1)$ and the auxiliary equation at the s^4 line is $(s^4 + 2s^2 + 1) = (s^2 + 1)^2$, indicating the repeated roots on the $j\omega$-axis.

8.1.3 Design Example: Tracked Vehicle Turning Control

Using Routh–Hurwitz methods, the design of a turning control system for a tracked vehicle (which can be modeled as a two-input, two-output system [5]) is considered. As shown in Figure 8.1a, the system has throttle and steering inputs and vehicle heading and track speed differences as outputs. The two vehicle tracks are operated at different speeds in order to turn the vehicle. The two-input, two-output system model can be simplified to two independent single-input, single-output systems for use in the control design phase. The single-input, single-output vehicle heading feedback control system is shown in Figure 8.1b. For purposes of discussion, the control problem is further simplified to the selection of two parameters. Our objective is to select the parameters K and a so that the system is stable and the steady-state error for a ramp command is less than or equal to 24% of the magnitude of the command. The characteristic equation of the feedback system is

$$
1 + G_c G(s) = 0 \tag{8.9}
$$

or

$$
1 + \frac{K(s+a)}{s(s+1)(s+2)(s+5)} = 0. \tag{8.10}
$$

Therefore, we have

$$
q(s) = s(s+1)(s+2)(s+5) + K(s+a) = 0 \tag{8.11}
$$

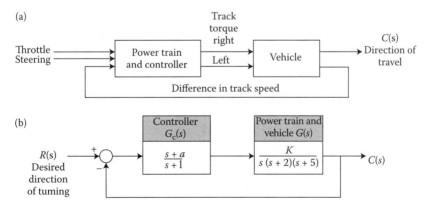

FIGURE 8.1 (a) Turning control for a two-track vehicle; (b) block diagram. (From Dorf, R. C. and Bishop, R. H., *Modern Control Systems*, 7th ed., Addison-Wesley, Reading, MA, 293, 1995. With permission.)

or

$$s^4 + 8s^3 + 17s^2 + (K + 10)s + Ka = 0. \tag{8.12}$$

To determine the stable region for K and a, we establish the Routh array as

s^4	1	17	Ka
s^3	8	$K + 10$	0
s^2	b_3	Ka	
s^1	c_3		
s^0	Ka		

where

$$b_3 = \frac{126 - K}{8} \quad \text{and} \quad c_3 = \frac{b_3(K + 10) - 8Ka}{b_3}.$$

For the elements of the first column to be positive, we require that Ka, b_3, and c_3 be positive. We therefore require

$$K < 126$$
$$Ka > 0 \tag{8.13}$$
$$(K + 10)(126 - K) - 64Ka > 0.$$

The region of stability for $K > 0$ is shown in Figure 8.2. The steady-state error to a ramp input $r(t) = At$, $t > 0$ is

$$e_{ss} = \frac{A}{K_v}, \tag{8.14}$$

where K_v is the *velocity error constant*, and in this case $K_v = Ka/10$. Therefore, we have

$$e_{ss} = \frac{10A}{Ka}. \tag{8.15}$$

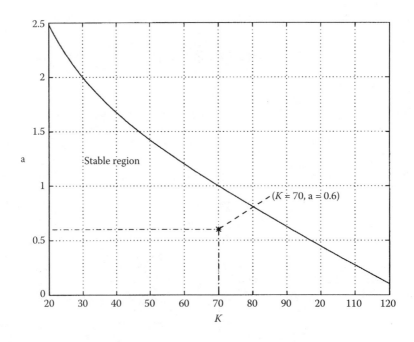

FIGURE 8.2 The stability region. (From Dorf, R. C. and Bishop, R. H., *Modern Control Systems,* 7th ed., Addison-Wesley, Reading, MA, 299, 1995. With permission.)

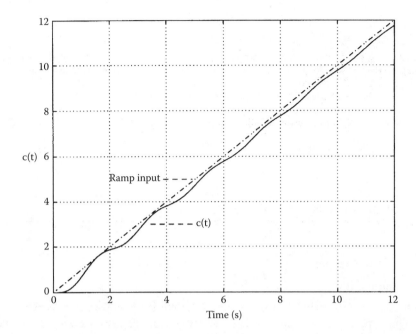

FIGURE 8.3 Ramp response for $a = 0.6$ and $K = 70$ for two-track vehicle turning control. (From Dorf, R. C. and Bishop, R. H., *Modern Control Systems*, 7th ed., Addison-Wesley, Reading, MA, 300, 1995. With permission.)

When e_{ss} is equal to 23.8% of A, we require that $Ka = 42$. This can be satisfied by the selected point in the stable region when $K = 70$ and $a = 0.6$, as shown in Figure 8.2. Of course, another acceptable design is attained when $K = 50$ and $a = 0.84$. We can calculate a series of possible combinations of K and a that can satisfy $Ka = 42$ and that lie within the stable region, and all will be acceptable design solutions. However, not all selected values of K and a will lie within the stable region. Note that K cannot exceed 126.

The corresponding unit ramp input response is shown in Figure 8.3. The steady-state error is less than 0.24, as desired.

8.1.4 Conclusions

In this chapter, we have considered the concept of the stability of a feedback control system. A definition of a stable system in terms of a bounded system response to a bounded input was outlined and related to the location of the poles of the system transfer function in the s-plane.

The Routh–Hurwitz stability criterion was introduced, and several examples were considered. The relative stability of a feedback control system was also considered in terms of the location of the poles and zeros of the system transfer function in the s-plane.

8.1.5 Defining Terms

Stability: A performance measure of a system. A system is stable if all the poles of the transfer function have negative real parts.

Routh–Hurwitz criterion: A criterion for determining the stability of a system by examining the characteristic equation of the transfer function.

References

1. Hurwitz, A., On the conditions under which an equation has only roots with negative real parts, *Mathematische Annalen*, 46, 273–284, 1895. Also in *Selected Papers on Mathematical Trends in Control Theory*, Dover, New York, 70–82, 1964.
2. Routh, E. J., *Dynamics of a System of Rigid Bodies*, Macmillan, New York, 1892.
3. Dorf, R. C. and Bishop, R. H., *Modern Control Systems*, 7th ed., Addison-Wesley, Reading, MA, 1995.
4. Clark, R. N., The Routh–Hurwitz stability criterion, revisited, *IEEE Control Syst. Mag.*, 12 (3), 119–120, 1992.
5. Wang, G. G., Design of turning control for a tracked vehicle, *IEEE Control Syst. Mag.*, 10 (3), 122–125, 1990.

8.2 The Nyquist Stability Test*

Charles E. Rohrs

8.2.1 The Nyquist Criterion

8.2.1.1 Development of the Nyquist Theorem

The Nyquist criterion is a graphical method and deals with the loop gain transfer function, i.e., the open-loop transfer function. The graphical character of the Nyquist criterion is one of its most appealing features.

Consider the controller configuration shown in Figure 8.4. The loop gain transfer function is given simply by $G(s)$. The closed-loop transfer function is given by

$$\frac{Y(s)}{R(s)} = \frac{G(s)}{1+G(s)} = \frac{K_G N_G(s)/D_G(s)}{1+K_G N_G(s)/D_G(s)}$$

$$= \frac{K_G N_G(s)}{D_G(s)+K_G N_G(s)} = \frac{K_G N_G(s)}{D_k(s)}$$

where $N_G(s)$ and $D_G(s)$ are the numerator and denominator, respectively, of $G(s)$, K_G is a constant gain, and $D_k(s)$ is the denominator of the closed-loop transfer function. The closed-loop poles are equal to the zeros of the function

$$1+G(s) = 1 + \frac{K_G N_G(s)}{D_G(s)} = \frac{D_G(s)+K_G N_G(s)}{D_G(s)} \qquad (8.16)$$

Of course, the numerator of Equation 8.16 is just the closed-loop denominator polynomial, $D_k(s)$, so that

$$1+G(s) = \frac{D_k(s)}{D_G(s)} \qquad (8.17)$$

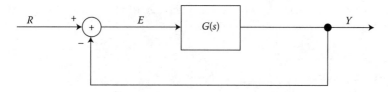

FIGURE 8.4   A control loop showing the loop gain $G(s)$.

*   Much of the material of this section is taken from Rohrs, Charles E., Melsa, James L., and Schultz, Donald G., *Linear Control Systems*, McGraw-Hill, New York, 1993. It is used with permission.

In other words, we can determine the stability of the closed-loop system by locating the zeros of $1 + G(s)$. This result is of prime importance in the following development.

For the moment, let us assume that $1 + G(s)$ is known in factored form so that we have

$$1 + G(s) = \frac{(s + \lambda_{k1})(s + \lambda_{k2}) \cdots (s + \lambda_{kn})}{(s + \lambda_1)(s + \lambda_2) \cdots (s + \lambda_n)} \tag{8.18}$$

Obviously, if $1 + G(s)$ were known in factored form, there would be no need for the use of the Nyquist criterion, since we could simply observe whether any of the zeros of $1 + G(s)$ [which are the poles of $Y(s)/R(s)$], lie in the right half of the s plane. In fact, the primary reason for using the Nyquist criterion is to avoid this factoring. Although it is convenient to think of $1 + G(s)$ in factored form at this time, no actual use is made of that form.

Let us suppose that the pole–zero plot of $1 + G(s)$ takes the form shown in Figure 8.5a. Consider next an arbitrary closed contour, such as that labeled Γ in Figure 8.5a, which encloses one and only one zero of $1 + G(s)$ and none of the poles. Associated with each point on this contour is a value of the complex function $1 + G(s)$. The value of $1 + G(s)$ for any value of s on Γ may be found analytically by substituting the appropriate complex value of s into the function. Alternatively, the value

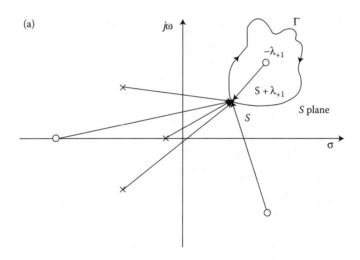

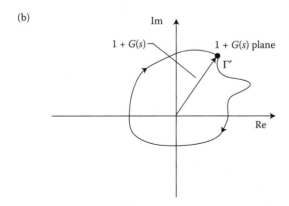

FIGURE 8.5 (a) Pole–zero plot of $1 + G(s)$ in the s plane; (b) plot of the Γ contour in the $1 + G(s)$ plane.

may be found graphically by considering the distances and angles from s on Γ to the zeros and poles.

If the complex value of $1 + G(s)$ associated with every point on the contour Γ is plotted, another closed contour Γ' is created in the complex $1 + G(s)$ plane, as shown in Figure 8.5b. The function $1 + G(s)$ is said to map the contour Γ in the s plane into the contour Γ' in the $1 + G(s)$ plane. What we wish to demonstrate is that, if a zero is enclosed by the contour Γ, as in Figure 8.5a, the contour Γ' encircles the origin of the $1 + G(s)$ plane in the same sense that the contour Γ encircles the zero in the s plane. In the s plane, the zero is encircled in the clockwise direction; hence we must show that the origin of the $1 + G(s)$ plane is also encircled in the clockwise direction. This result is known as the *Principle of the Argument*.

The key to the Principle of the Argument rests in considering the value of the function $1 + G(s)$ at any point s as simply a complex number. This complex number has a magnitude and a phase angle. Since the contour Γ in the s plane does not pass through a zero, the magnitude is never zero. Now we consider the phase angle by rewriting Equation 8.18 in polar form:

$$
\begin{aligned}
1 + G(s) &= \frac{|s + \lambda_{k1}| \, \underline{/\arg(s + \lambda_{k1})} \cdots |s + \lambda_{kn}| \, \underline{/\arg(s + \lambda_{kn})}}{|s + \lambda_1| \, \underline{/\arg(s + \lambda_1)} \cdots |s + \lambda_n| \, \underline{/\arg(s + \lambda_n)}} \\
&= \frac{|s + \lambda_{k1}| \cdots |s + \lambda_{kn}|}{|s + \lambda_1| \cdots |s + \lambda_n|} \, \underline{/\arg(s + \lambda_{k1}) + \cdots} \\
&\quad + \underline{\arg(s + \lambda_{kn}) - \arg(s + \lambda_1) - \cdots - \arg(s + \lambda_n)}
\end{aligned} \tag{8.19}
$$

We assume that the zero encircled by Γ is at $s = -\lambda_{k1}$. Then the phase angle associated with this zero changes by a full $-360°$ as the contour Γ is traversed clockwise in the s plane. Since the argument or angle of $1 + G(s)$ includes the angle of this zero, the argument of $1 + G(s)$ also changes by $-360°$. As seen from Figure 8.5a, the angles associated with the remaining poles and zeros make no net change as the contour Γ is traversed. For any fixed value of s, the vector associated with each of these other poles and zeros has a particular angle associated with it. Once the contour has been traversed back to the starting point, these angles return to their original value; they have not been altered by plus or minus 360° simply because these poles and zeros are not enclosed by Γ.

In a similar fashion, we could show that, if the Γ contour were to encircle two zeros of $1 + G(s)$ in the clockwise direction on the s plane, the Γ' contour would encircle the origin of the $1 + G(s)$ plane twice in the clockwise direction. On the other hand, if the Γ contour were to encircle only one pole and no zero of $1 + G(s)$ in the *clockwise* direction, then the contour Γ' would encircle the origin of the $1 + G(s)$ plane once in the *counterclockwise* direction. This change in direction comes about because angles associated with poles are accompanied by negative signs in the evaluation of $1 + G(s)$, as indicated by Equation 8.19. In general, the following conclusion can be drawn: *The net number of clockwise encirclements by Γ' of the origin in the $1 + G(s)$ plane is equal to the difference between the number of zeros n_z and the number of poles n_p of $1 + G(s)$ encircled in the clockwise direction by Γ.*

This result means that the difference between the number of zeros and the number of poles enclosed by *any* closed contour Γ may be determined simply by counting the net number of clockwise encirclements of the origin of the $1 + G(s)$ plane by Γ'. For example, if we find that Γ' encircles the origin three times in the clockwise direction and once in the counterclockwise direction, then $n_z - n_p$ must be equal to $3 - 1 = 2$. Therefore, in the s plane, Γ must encircle two zeros and no poles, three zeros and one pole, or any other combination such that $n_z - n_p$ is equal to 2.

In terms of stability analysis, the problem is to determine the number of zeros of $1 + G(s)$, i.e., the number of poles of $Y(s)/R(s)$, that lie in the right half of the s–plane. Accordingly, the contour Γ is chosen as the entire $j\omega$ axis and an infinite semicircle enclosing the right half-plane as shown in Figure 8.6a. This contour is known as the *Nyquist D contour* as it resembles the capital letter D.

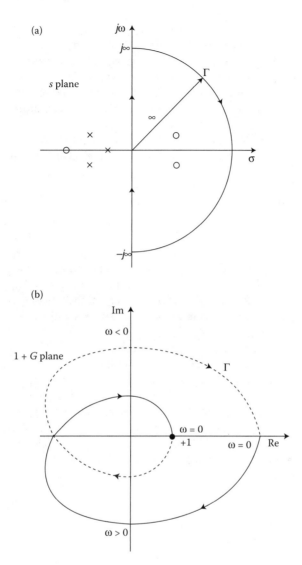

FIGURE 8.6 (a) Γ contour in the *s* plane—the Nyquist contour; (b) Γ′ contour in the 1 + *G* plane.

In order to avoid any problems in plotting the values of 1 + *G*(*s*) along the infinite semicircle, let us assume that

$$\lim_{|s| \to \infty} G(s) = 0$$

This assumption is justified since, in general, the loop gain transfer function *G*(*s*) is strictly proper. With this assumption, the entire infinite semicircle portion of Γ maps into the single point *s* = +1 + *j*0 on the 1 + *G*(*s*) plane.

The mapping of Γ therefore involves simply plotting the complex values of 1 + *G*(*s*) for *s* = *j*ω as ω varies from −∞ to +∞. For ω ≥ 0, Γ′ is nothing more than the polar plot of the frequency response of the function 1 + *G*(*s*). The values of 1 + *G*(*j*ω) for negative values of ω are the mirror image of the values of 1 + *G*(*j*ω) for positive values of ω reflected about the real axis. The Γ′ contour may therefore be found by plotting the frequency response 1 + *G*(*s*) for positive ω and then reflecting this plot about the real axis to find the plot for negative ω. The Γ′ plot is always symmetrical about the real axis of the 1 + *G* plane.

Care must be taken to establish the direction that the Γ' plot is traced as the D-contour moves up the $j\omega$ axis, around the infinite semicircle and back up the $j\omega$ axis from $-\infty$ towards 0.

From the Γ' contour in the $1 + G(s)$ plane, as shown in Figure 8.6b, the number of zeros of $1 + G(s)$ in the right half of the s−plane may be determined by the following procedure. The net number of clockwise encirclements of the origin by Γ' is equal to the number of zeros minus the number of poles of $1 + G(s)$ in the right half of the s−plane. Note that we must know the number of poles of $1 + G(s)$ in the right half-plane if we are to be able to ascertain the exact number of zeros in the right half-plane and therefore determine stability. This requirement usually poses no problem since the poles of $1 + G(s)$ correspond to the poles of the loop gain transfer function. In Equation 8.17 the denominator of $1 + G(s)$ is just $D_G(s)$, which is usually described in factored form. Hence, the number of zeros of $1 + G(s)$ or, equivalently, the number of poles of $Y(s)/R(s)$ in the right half-plane may be found by determining the net number of clockwise encirclements of the origin by Γ' and then adding the number of poles of the loop gain located in the right-half s−plane.

At this point the reader may revolt. Our plan for finding the number of poles of $Y(s)/R(s)$ in the right-half s−plane involves counting encirclements in the $1 + G(s)$ plane and observing the number of loop gain poles in the right-half s−plane. Yet we were forced to start with the assumption that all the poles and zeros of $1 + G(s)$ are known, so that the Nyquist contour can be mapped by the function of $1 + G(s)$. Admittedly, we know the poles of this function because they are the poles of the loop gain, but we do not know the zeros; in fact, we are simply trying to find how many of these zeros lie in the right-half s−plane.

What we do know are the poles and zeros of the loop gain transfer function $G(s)$. Of course, this function differs from $1 + G(s)$ only by unity. Any contour that is chosen in the s−plane and mapped through the function $G(s)$ has exactly the same shape as if the contour were mapped through the function $1 + G(s)$ except that it is displaced by one unit. Figure 8.7 is typical of such a situation. In this diagram the -1 point of the $G(s)$ plane is the origin of the $1 + G(s)$ plane. If we now map the boundary of the right-half s-plane through the mapping function $G(s)$, which we often know in pole–zero form, information concerning the zeros of $1 + G(s)$ may be obtained by counting the encirclements of the -1 point. The important point is that, by plotting the open-loop frequency-response information, we may reach stability conclusions regarding the closed-loop system.

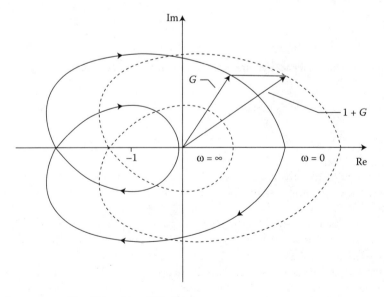

FIGURE 8.7   Comparison of the $G(s)$ and $1 + G(s)$ plots.

As mentioned previously contour Γ of Figure 8.6a is referred to as the Nyquist D-contour. The map of the Nyquist D-contour through $G(s)$ is called the *Nyquist diagram of $G(s)$*. There are three parts to the Nyquist diagram. The first part is the polar plot of the frequency response of $G(s)$ from $\omega = 0$ to $\omega = \infty$. The second part is the mapping of the infinite semicircle around the right half-plane. If $G(s)$ is strictly proper, this part maps entirely into the origin of the $G(s)$ plane. The third part is the polar plot of the negative frequencies, $\omega = -\infty$ to $\omega = 0$. The map of these frequencies forms a mirror image in the $G(s)$ plane about the real axis of the first part.

In terms of the Nyquist diagram of $G(s)$, the Nyquist stability criterion may be stated as follows:

Theorem 8.1: The Nyquist Theorem

The closed-loop system is stable if and only if the net number of clockwise encirclements of the points $s = -1 + j0$ by the Nyquist diagram of $G(s)$ plus the number of poles of $G(s)$ in the right half-plane is zero.

Notice that while the *net* number of clockwise encirclements are counted in the first part of the Nyquist criterion, only the number of right half-plane *poles* of $G(s)$ are counted in the second part. Right half-plane zeros of $G(s)$ are not part of the formula in determining stability using the Nyquist criterion.

Because the Nyquist diagram involves the loop gain transfer function $G(s)$, a good approximation of the magnitude and phase of the frequency response plot can be obtained by using the Bode diagram straight-line approximations for the magnitude and for the phase. The Nyquist plot can be obtained by transferring the magnitude and phase information to a polar plot. If a more accurate plot is needed, the exact magnitude and phase may be determined for a few values of ω in the range of interest. However, in most cases, the approximate plot is accurate enough for practical problems.

An alternative procedure for obtaining the Nyquist diagram is to plot accurately the poles and zeros of $G(s)$ and obtain the magnitude and phase by graphical means. In either of these methods, the fact that $G(s)$ is known in factored form is important. Even if $G(s)$ is not known in factored form, the frequency-response plot can still be obtained by simply substituting the values $s = j\omega$ into $G(s)$ or by frequency-response measurements on the actual system.

Of course, computer programs that produce Nyquist plots are generally available. However, the ability to plot Nyquist plots by hand helps designers know how they can affect such plots by adjusting compensators.

It is also important to note that the information required for a Nyquist plot may be obtainable by measuring the frequency response of a stable plant directly and plotting this information. Thus, Nyquist ideas can be applied even if the system is a "black box" as long as it is stable.

8.2.1.2 Examples of the Nyquist Theorem

Example 8.1:

To illustrate the use of the Nyquist* criterion, let us consider the simple first-order system shown in Figure 8.8a. For this system the loop gain transfer function takes the form

$$G(s) = KG_P(s) = \frac{K}{s+10} = \frac{50}{s+10}$$

The magnitude and phase plots of the frequency response of $KG_P(s)$ are shown. From these plots the Nyquist diagram $KG_P(s)$ may be easily plotted, as shown in Figure 8.9. For example, the point

---

*   Throughout the examples of this section, we assume that the gain $K > 0$. If $K < 0$, all the theory holds with the critical point shifted to $+1 + j0$.

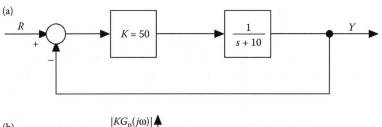

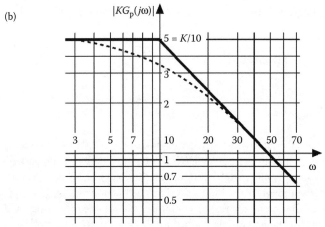

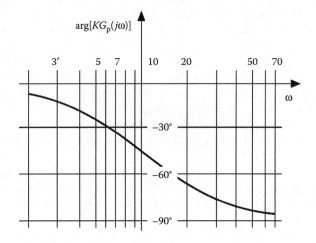

FIGURE 8.8 Simple first-order example. (a) Block diagram; (b) magnitude and phase plots.

associated with $\omega = 10$ rad/s is found to have a magnitude of $K/(10\sqrt{2})$ and a phase angle of $-45°$. The point at $\omega = -10$ rad/s is just the mirror image of the value at $\omega = 10$ rad/s.

From Figure 8.9 we see that the Nyquist diagram can never encircle the $s = -1 + j0$ point for any positive value of K, and therefore the closed-loop system is stable for all positive values of K. In this simple example, it is easy to see that this result is correct since the closed-loop transfer function is given by

$$\frac{Y(s)}{R(s)} = \frac{K}{s + 10 + K}$$

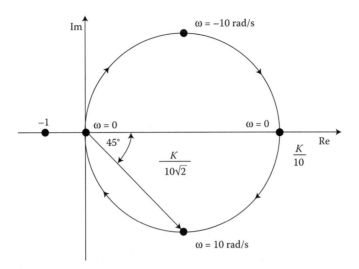

FIGURE 8.9 Nyquist diagram for Example 8.3.

For all positive values of K, the pole of $Y(s)/R(s)$ is in the left half-plane.

In this example, $G(s)$ remains finite along the entire Nyquist contour. This is not always the case even though we have assumed that $G(s)$ approaches zero as $|s|$ approaches infinity. If a pole of $G(s)$ occurs on the $j\omega$ axis, as often happens at the origin because of an integrator in the plant, a slight modification of the Nyquist contour is necessary. The method of handling the modification is illustrated in the Example 8.2.

Example 8.2:

Consider a system whose loop transfer function is given by

$$G(s) = \frac{(2K/7)\left[(s+3/2)^2 + \left(\sqrt{5/2}\right)^2\right]}{s(s+2)(s+3)}$$

The pole–zero plot of $G(s)$ is shown in Figure 8.10a. Since a pole occurs on the standard Nyquist contour at the origin, it is not clear how this problem should be handled. As a beginning, let us plot the Nyquist diagram for $\omega = +\varepsilon$ to $\omega = -\varepsilon$, including the infinite semicircle; when this is done, the small area around the origin is avoided. The resulting plot is shown as the solid line in Figure 8.10b with corresponding points labeled.

From Figure 8.10b we cannot determine whether the system is stable until the Nyquist diagram is completed by joining the points at $\omega = -\varepsilon$ and $\omega = +\varepsilon$. In order to join these points, let us use a semicircle of radius ε to the right of the origin, as shown in Figure 8.10a. Now $G(s)$ is finite at all points on the contour in the s plane, and the mapping to the G plane can be completed as shown by the dashed line in Figure 8.10b. The small semicircle used to avoid the origin in the s plane maps into a large semicircle in the G plane.

It is important to know whether the large semicircle in the G plane swings to the right around positive real values of s or to the left around negative real values of s. There are two ways to determine this. The first way borrows a result from complex variable theory, which says that the Nyquist diagram is a *conformal map* and for a conformal map right turns in the s plane correspond to right turns in the $G(s)$ plane. Likewise, left turns in the s plane correspond to left turns in the $G(s)$ plane.

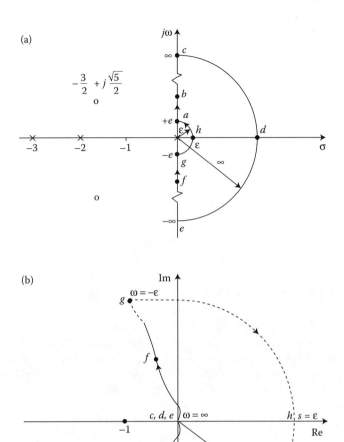

FIGURE 8.10 Example 8.2. (a) Pole–zero plot; (b) Nyquist diagram.

The second method of determining the direction of the large enclosing circle on the $G(s)$ plane comes from a graphical evaluation of $G(s)$ on the circle of radius ε in the s plane. The magnitude is very large here due to the proximity of the pole. The phase at $s = -\varepsilon$ is slightly larger than $+90°$ as seen from the solid line of the Nyquist plot. The phase contribution from all poles and zeros except the pole at the origin does not change appreciably as the circle of the radius ε is traversed. The angle from the pole at the origin changes from $-90°$ through $0°$ to $+90°$. Since angles from poles contribute in a negative manner, the contribution from the pole goes from $+90°$ through $0°$ to $-90°$. Thus, as the semicircle of radius ε is traversed in the s–plane, a semicircle moving in a clockwise direction through about $180°$ is traversed in the $G(s)$ plane. The semicircle is traced in the clockwise direction as the angle associated with $G(s)$ becomes more negative. Notice that this is consistent with the conformal mapping rule, which matches right turns of $90°$ at the top and bottom of both circles.

In order to ensure that no right half-plane zeros of $1 + G(s)$ can escape discovery by lying in the ε-radius semicircular indentation in the s plane, ε is made arbitrarily small, with the result that the radius of the large semicircle in the *G* plane approaches infinity. As $\varepsilon \to 0$, the shape of the Nyquist diagram remains unchanged, and we see that there are no encirclements of the $s = -1 + j0$ point. Since there are no poles of $G(s)$ in the right half-plane, the system is stable. In addition, since changing the magnitude of *K* can never cause the Nyquist diagram to encircle the -1 point, the closed-loop system must be stable for all values of positive *K*.

We could just as well close the contour with a semicircle of radius ε into the left half-plane. Note that if we do this, the contour encircles the pole at the origin and this pole is counted as a right half-plane pole of $G(s)$. In addition, by applying either the conformal mapping with left turns or the graphical evaluation, we close the contour in the $G(s)$ plane by encircling the negative real axis. There is 1 counterclockwise encirclement (-1 clockwise encirclement) of the -1 point. The Nyquist criterion says that -1 clockwise encirclement plus 1 right half-plane pole of $G(s)$ yield zero closed-loop right half-plane poles. The result that the closed-loop system is stable for all positive values of *K* remains unchanged, as it must. The two approaches are equally good although philosophically the left turn contour, which places the pole on the $j\omega$ axis in the right half-plane, is more in keeping with the convention of poles on the $j\omega$ axis being classified as unstable.

In each of the two preceding examples, the system was open-loop stable; that is, all the poles of $G(s)$ were in the left half-plane. The next example illustrates the use of the Nyquist criterion when the system is open-loop unstable.

Example 8.3:

This example is based on the system shown in Figure 8.11. The loop gain transfer function for this system is

$$G(s) = \frac{K(s+1)}{(s-1)(s+2)}$$

We use the Bode diagrams of magnitude and phase as an assistance in plotting the Nyquist diagram. The magnitude and phase plots are shown in Figure 8.12a.

The Nyquist diagram for this system is shown in Figure 8.12b. Note that the exact shape of the plot is not very important since the only information we wish to obtain at this time is the number of encirclements of the $s = -1 + j0$ point. It is easy to see that the Nyquist diagram encircles the -1 point once in the *counterclockwise* direction if $K > 2$ and has no encirclements if $K < 2$. Since this system has one right half-plane pole in $G(s)$, it is necessary that there be one counterclockwise encirclement if the system is to be stable. Therefore, this system is stable if and only if $K > 2$.

Besides providing simple yes/no information about whether a closed-loop system is stable, the Nyquist diagram also provides a clear graphical image indicating how close to instability a system may be. If the Nyquist diagram passes close to the -1 point and there is some mismodeling of the plant so that the characteristics of the plant are slightly different from those plotted in the Nyquist plot, then the true Nyquist characteristic may encircle the -1 point more or fewer times than the nominal Nyquist plot. The actual closed-loop system may be unstable. The Nyquist plot gives direct visual evidence of the frequencies

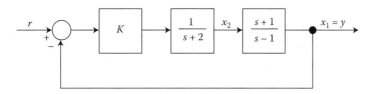

FIGURE 8.11   Example 8.3.

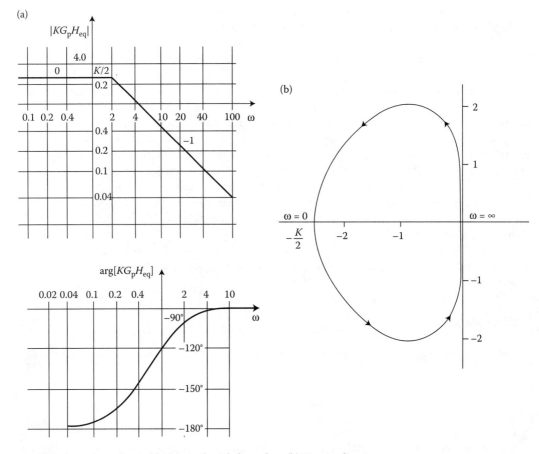

FIGURE 8.12   Example 8.3. (a) Magnitude and phase plots; (b) Nyquist diagram.

where the plant's nominal Nyquist plot passes near the -1 point. At these frequencies great care must be taken to be sure that the nominal Nyquist plot accurately represents the plant transfer characteristic, or undiagnosed instability may result. These ideas are formalized by a theory that goes under the name of stability robustness theory.

8.2.2 Closed-Loop Response and Nyquist Diagrams

The Nyquist diagram has another important use. There are many possible designs that result in closed-loop systems that are stable but have highly oscillatory and thus unsatisfactory responses to inputs and disturbances. Systems that are oscillatory are often said to be relatively less stable than systems that are more highly damped. The Nyquist diagram is very useful in determining the relative stability of a closed-loop system.

For this development we must start with a system in the G configuration (Figure 8.4). The key for extracting information about the closed-loop system is to determine the frequency-response function of the closed-loop system, often referred to as the $M-$curve. The $M-$curve is, of course, a function of frequency and may be determined analytically as

$$M(j\omega) = \frac{Y(j\omega)}{R(j\omega)} = \frac{G(j\omega)}{1 + G(j\omega)} \tag{8.20}$$

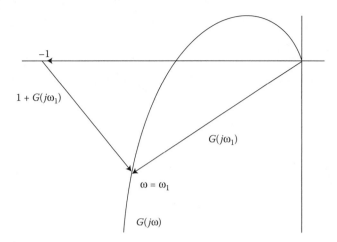

FIGURE 8.13 Graphical determination of $M(j\omega)$.

Figure 8.13 illustrates how the value of $M(j\omega_1)$ may be determined directly from the Nyquist diagram of $G(j\omega)$ at one particular frequency, ω_1. In this figure the vectors -1 and $G(j\omega_1)$ are indicated, as is the vector $(G(j\omega_1) - (-1)) = 1 + G(j\omega_1)$. The length of the vector $G(j\omega_1)$ divided by the length of $1 + G(j\omega_1)$ is thus the value of the magnitude $M(j\omega_1)$. The $\arg M(j\omega_1)$ is determined by subtracting the angle associated with the $1 + G(j\omega_1)$ vector from that of $G(j\omega_1)$. The complete $M-$curve may be found by repeating this procedure over the range of frequencies of interest.

In terms of the magnitude portion of the $M(j\omega)$ plot, the point-by-point procedure illustrated above may be considerably simplified by plotting contours of constant $|M(j\omega)|$ on the Nyquist plot of $G(s)$. The magnitude plot of $M(j\omega)$ can then be read directly from the Nyquist diagram of $G(s)$. Fortunately, these contours of constant $|M(j\omega)|$ have a particularly simple form. For $|M(j\omega)| = M$, the contour is simply a circle. These circles are referred to as constant M-circles or simply M-circles.

If these constant M-circles are plotted together with the Nyquist diagram of $G(s)$, as shown in Figure 8.14 for the system $G(s) = \frac{42}{s(s+2)(s+15)}$, the values of $|M(j\omega)|$ may be read directly from the plot. Note that the $M = 1$ circle degenerates to the straight line $X = -0.5$. For $M < 1$, the constant M-circles lie to the right of this line, whereas, for $M > 1$, they lie to the left. In addition, the $M = 0$ circle is the point $0 + j0$, and $M = \infty$ corresponds to the point $-1.0 + j0$.

In an entirely similar fashion, the contours of constant $\arg(M(j\omega))$ can be found. These contours turn out to be segments of circles. The circles are centered on the line $X = -\frac{1}{2}$. The contour of the $\arg(M(j\omega)) = \beta$ for $0 < \beta < 180°$ is the upper half-plane portion of the circle centered at $-\frac{1}{2} + j1/(2 \tan \beta)$ with a radius $|1/(2 \sin \beta)|$. For β in the range $-180° < \beta < 0°$, the portions of the same circles in the lower half-plane are used. Figure 8.15 shows the plot of the constant-phase contours for some values of β. Notice that one circle represents $\beta = 45°$ above the real axis while the same circle represents $\beta = -135°$ below the real axis.

By using these constant-magnitude and constant-phase contours, it is possible to read directly the complete closed-loop frequency response from the Nyquist diagram of $G(s)$. In practice it is common to dispense with the constant-phase contours, since it is the magnitude of the closed-loop frequency response that provides the most information about the transient response of the closed-loop system. In fact, it is common to simplify the labor further by considering only one point on the magnitude plot, namely, the point at which M is maximum. This point of peak magnitude is referred to as M_p, and the frequency at which the peak occurs is ω_p. The point M_p may be easily found by considering the contours of larger and larger values of M until the contour is found that is just tangent to the plot of $G(s)$. The value associated with this contour is then M_p, and the frequency at which the M_p contour and $G(s)$ touch is

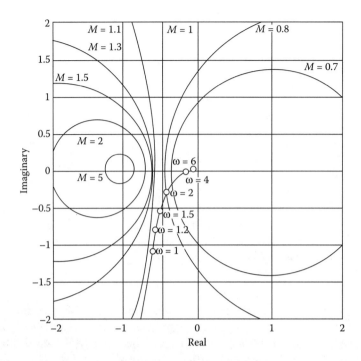

FIGURE 8.14   Constant M-contours.

ω_p. In the plot of $G(s)$ shown in Figure 8.14, for example, the value of M_p is 1.1 at the frequency $\omega_p \approx 1.1$ rad/s.

One of the primary reasons for determining M_p and ω_p, in addition to the obvious saving of labor as compared with the determination of the complete frequency response, is the close correlation of these quantities with the behavior of the closed-loop system. In particular, for the simple second-order closed-loop system,

$$\frac{Y(s)}{R(s)} = \frac{\omega_n^2}{s^2 + 2s\zeta\omega_n + \omega_n^2} \tag{8.21}$$

the values of M_p and ω_p completely characterize the system. In other words, for this second-order system, M_p and ω_p specify the damping, ζ, and the natural frequency, ω_n, the only parameters of the system. The following equations relate the maximum point of the frequency response of Equation 8.21 to the values of ζ and ω_n;

$$\omega_p = \omega_n\sqrt{1 - 2\zeta^2} \tag{8.22}$$

$$M_p = \frac{1}{2\zeta\sqrt{1 - \zeta^2}} \quad \text{for } \zeta \leq 0.707 \tag{8.23}$$

From these equations one may determine ζ and ω_n if M_p and ω_p are known, and vice versa. Figure 8.16 graphically displays the relations between M_p and ω_p and ζ and ω_n for a second-order system. Once ζ and ω_n are known, we may determine the time behavior of this second-order system.

Not all systems are of a simple second-order form. However, it is common practice to assume that the behavior of many high-order systems is closely related to that of a second-order system with the same M_p and ω_p.

Two other measures of the qualitative nature of the closed-loop response that may be determined from the Nyquist diagram of $G(s)$ are the phase margin and crossover frequency. The crossover frequency ω_c

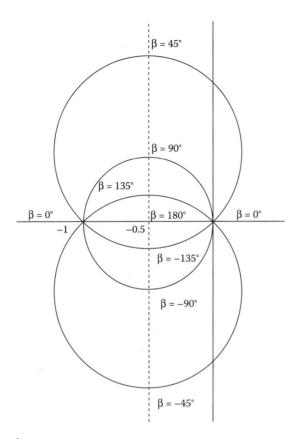

FIGURE 8.15   Constant-phase contours.

is the positive value of ω for which the magnitude of $G(j\omega)$ is equal to unity, that is,

$$|G(j\omega_c)| = 1 \tag{8.24}$$

The phase margin ϕ_m is defined as the difference between the argument of $G(j\omega_c)$ (evaluated at the crossover frequency) and $-180°$. In other words, if we define β_c as

$$\beta_c = \arg(G(j\omega_c)) \tag{8.25}$$

the phase margin is given by

$$\phi_m = \beta_c - (-180°) = 180° + \beta_c \tag{8.26}$$

While it is possible for a complicated system to possess more than one crossover frequency, most systems are designed to possess just one. The phase margin takes on a particularly simple and graphic meaning in the Nyquist diagram of $G(s)$. Consider, for example, the Nyquist diagram shown in Figure 8.17. In that diagram, we see that the phase margin is simply the angle between the negative real axis and the vector $G(j\omega_c)$. The vector $G(j\omega_c)$ may be found by intersecting the $G(s)$ locus with the unit circle. The frequency associated with the point of intersection is ω_c.

It is possible to determine ϕ_m and ω_c more accurately directly from the Bode plots of the magnitude and phase of $G(s)$. The value of ω for which the magnitude crosses unity is ω_c. The phase margin is then determined by inspection from the phase plot by noting the difference between the phase shift at ω_c and $-180°$. Consider, for example, the Bode magnitude and phase plots shown in Figure 8.18 for the $G(s)$

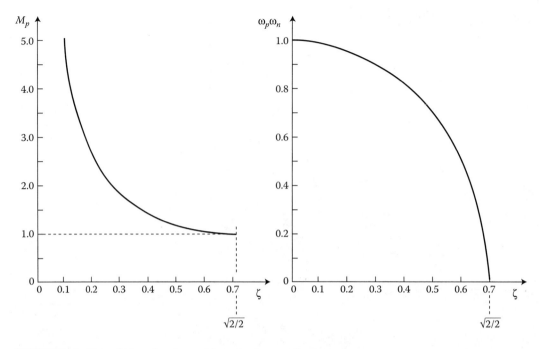

FIGURE 8.16   Plots of M_p and ω_p/ω_n vs. ζ for a simple second-order system.

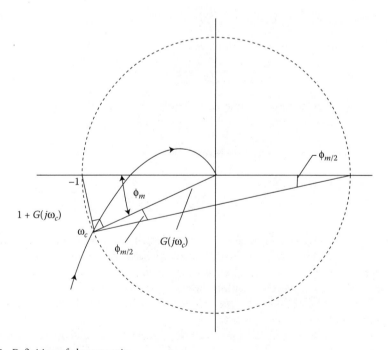

FIGURE 8.17   Definition of phase margin.

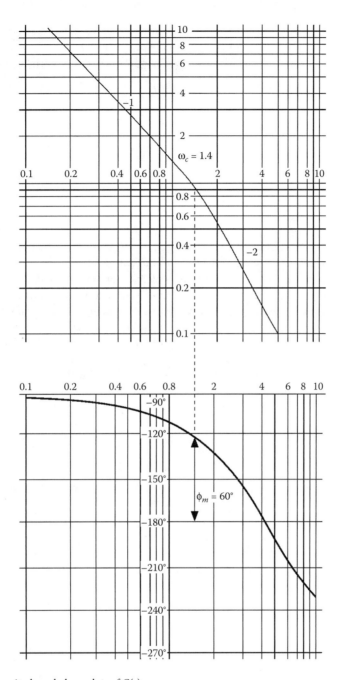

FIGURE 8.18 Magnitude and phase plots of $G(s)$.

function of Figure 8.14. In time-constant form this transfer function is

$$G(s) = \frac{1.4}{s(1 + s/2)(1 + s/15)}$$

From this figure we see that $\omega_c = 1.4$ and $\phi_m = 60°$.

The value of the magnitude of the closed-loop frequency response at ω_c can be derived from ϕ_m. We shall call this value M_c. Often the closest point to the -1 point on a Nyquist plot occurs at a frequency that is close to ω_c. This means that M_c is often a good approximation to M_p. A geometric construction shown in Figure 8.17 shows that a right triangle exists with a hypotenuse of 2, one side of length $|1 + G(j\omega_c)|$, and the opposite angle of $\phi_m/2$ where ϕ_m is the phase margin. From this construction, we see

$$\frac{\sin \phi_m}{2} = \frac{|1 + G(j\omega_c)|}{2} \tag{8.27}$$

Since at $\omega = \omega_{c'}$

$$|G(j\omega_c)| = 1 \tag{8.28}$$

$$M_c = \frac{|G(j\omega_c)|}{|1 + G(j\omega_c)|} = \frac{1}{2 \sin \phi_m/2} \tag{8.29}$$

An oscillatory characteristic in the closed-loop time response can be identified by a large peak in the closed-loop frequency response which, in turn, can be identified by a small phase margin and the corresponding large value of M_c. Unfortunately, the correlation between response and phase margin is somewhat poorer than the correlation between the closed-loop time response and the peak M. This lower reliability of the phase margin measure is a direct consequence of the fact that ϕ_m is determined by considering only one point, ω_c on the G plot, whereas M_p is found by examining the entire plot, to find the maximum M. Consider, for example, the two Nyquist diagrams shown in Figure 8.19. The phase margin for these two diagrams is identical; however, it can be seen from the above discussion that the closed-loop step response resulting from closing the loop gain of Figure 8.19b is far more oscillatory and underdamped then the closed-loop step response resulting from closing the loop gain of Figure 8.19a.

In other words, the relative ease of determining ϕ_m as compared with M_p has been obtained only by sacrificing some of the reliability of M_p. Fortunately, for many systems the phase margin provides a simple and effective means of estimating the closed-loop response from the $G(j\omega)$ plot.

A system such as that shown in Figure 8.19b can be identified as a system having a fairly large M_p by checking another parameter, the gain margin. The gain margin is easily determined from the Nyquist plot of the system. The gain margin is defined as the ratio of the maximum possible gain for stability to the actual system gain. If a plot of $G(s)$ for $s = j\omega$ intercepts the negative real axis at a point $-a$ between the origin and the critical -1 point, then the gain margin is simply

$$\text{Gain margin} = GM = \frac{1}{a}$$

If a gain greater than or equal to $1/a$ were placed in series with $G(s)$, the closed-loop system would be unstable.

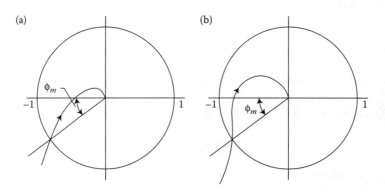

FIGURE 8.19   Two systems with the same phase margin but different M_p.

While the gain margin does not provide very complete information about the response of the closed-loop system, a small gain margin indicates a Nyquist plot that approaches the −1 point closely at the frequency where the phase shift is 180°. Such a system has a large M_p and an oscillatory closed-loop time response independent of the phase margin of the system.

While a system may have a large phase margin and a large gain margin and still get close enough to the critical point to create a large M_p, such phenomena can occur only in high-order loop gains. However, one should never forget to check any results obtained by using phase margin and gain margin as indicators of the closed-loop step response, lest an atypical system slip by. A visual check to see if the Nyquist plot approaches the critical −1 point too closely should be sufficient to determine if the resulting closed-loop system may be too oscillatory.

Using the concepts that give rise to the M-circles, a designer can arrive at a pretty good feel for the nature of the closed-loop transient response by examining the loop gain Bode plots. The chain of reasoning is as follows: From the loop gain Bode plots, the shape of the Nyquist plot of the loop gain can be envisioned. From the shape of the loop gain Nyquist plot, the shape of the Bode magnitude plot of the closed-loop system can be envisioned using the concepts of this section. Certain important points are evaluated by returning to the loop gain Bode plots. From the shape of the Bode magnitude plot of the closed-loop system, the dominant poles of the closed-loop transfer function are identified. From the knowledge of the dominant poles, the shape of the step response of the closed-loop system is determined. Example 8.4 illustrates this chain of thought.

Example 8.4:

Consider the loop gain transfer function

$$G(s) = \frac{80}{s(s+1)(s+10)}$$

The Bode plots for this loop gain are given in Figure 8.20.

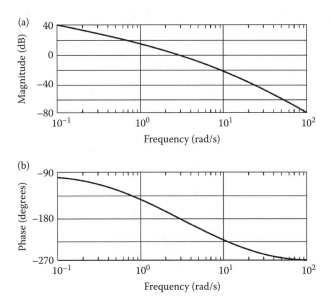

FIGURE 8.20 Bode plots of loop gain. (a) Magnitude plot; (b) Phase plot.

From the Bode plots the Nyquist plot can be envisioned. The Nyquist plot begins far down the negative imaginary axis since the Bode plot has large magnitude and $-90°$ phase at low frequency. It swings to the left as the phase lag increases and then spirals clockwise towards the origin, cutting the negative real axis and approaching the origin from the direction of the positive imaginary axis, i.e., from the direction associated with $-270°$ phase. From the Bode plot it is determined that the Nyquist plot does not encircle the -1 point since the Bode plot shows that the magnitude crosses unity ($0dB$) before the phase crosses $-180°$.

From the Bode plot it can be seen that the Nyquist plot passes very close to the -1 point near the crossover frequency. In this case $\omega_p \approx \omega_c$ and the phase margin is a key parameter to establish how large M_p, the peak in the closed-loop frequency magnitude plot, is. The crossover frequency is read from the Bode magnitude plot as $\omega_c = 2.5$ rad/s and the phase margin is read from the Bode phase plot as $\phi_m = 6°$.

Our visualization of the Nyquist plot is confirmed by the diagram of the actual Nyquist plot shown in Figure 8.21.

The magnitude of the closed-loop frequency response for this system can be envisioned using the techniques learned in this section. At low frequencies $G(s)$ is very large; the distance from the origin to the Nyquist plot is very nearly the same as the distance from the -1 point to the Nyquist plot and the closed-loop frequency response has magnitude near one. As the Nyquist plot of the loop gain approaches -1, the magnitude of the closed-loop frequency-response function increases to a peak. At higher frequencies the loop gain becomes small and the closed-loop frequency response decreases with the loop gain since the distance from -1 point to the loop gain Nyquist plot approaches unity. Thus, the closed-loop frequency response starts near $0dB$, peaks as the loop gain approaches -1 and then falls off.

The key point occurs when the loop gain approaches the -1 point and the closed-loop frequency response peaks. The closest approach to the -1 point occurs at a frequency very close to the crossover frequency, which has been established as $\omega_c = 2.5$ rad/s. The height of the peak can be established using the phase margin which has been established as $\phi_m = 6°$, and Equation 8.29. The height of the peak should be very close to $(2\sin(\phi_m/2))^{-1} = 9.5 = 19.6$ dB.

Our visualization of the magnitude of the closed-loop frequency response is confirmed by the actual plot shown in Figure 8.22.

From the visualization of the closed-loop frequency response function and the information about the peak of the frequency response, it is possible to identify the dominant closed-loop poles. The frequency of the peak identifies the natural frequency of a pair of complex poles, and the height of the peak identifies the damping ratio. More precisely,

$$M_p = \frac{1}{2\zeta\sqrt{1-\zeta^2}} \approx \frac{1}{2\zeta} \quad \text{for } \zeta \text{ small}$$

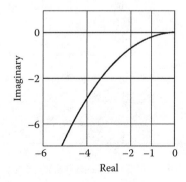

FIGURE 8.21 Nyquist plot of loop gain.

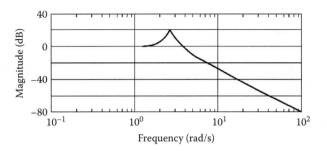

FIGURE 8.22 Magnitude plot of closed-loop frequency response.

and

$$\omega_p = \omega_n\sqrt{1 - 2\zeta^2} \approx \omega_n \text{ for } \zeta \text{ small}$$

Using the approximations for ω_p and M_p that are obtained from the loop gain crossover frequency and phase margin, respectively, the following values are obtained: $\zeta \approx 1/(2M_p) \approx 0.05$ and $\omega_n \approx \omega_p \approx \omega_c \approx 2.5$ rad/s.

If the Nyquist plot of a loop gain does not pass too closely to the -1 point, the closed-loop frequency response does not exhibit a sharp peak. In this case, the dominant poles are well damped or real. The distance of these dominant poles from the origin can be identified by the system's bandwidth, which is given by the frequency at which the closed-loop frequency response begins to decrease. From the $M-$circle concept it can be seen that the frequency at which the closed-loop frequency response starts to decrease is well approximated by the crossover frequency of the loop gain.

Having established the position of the dominant closed-loop poles, it is easy to describe the closed-loop step response. The step response has a percent overshoot given by

$$PO = 100e^{-\left(\frac{\zeta\pi}{\sqrt{1-\zeta^2}}\right)} \approx 85\%$$

The period of the oscillation is given

$$T_d = \frac{2\pi}{\omega_d} = \frac{2\pi}{\omega_n\sqrt{1-\zeta^2}} \approx 2.5\text{s}$$

The first peak in the step response occurs at a time equal to half of the period of oscillations, or about 1.25 s. The envisioned step response is confirmed in the plot of the actual closed-loop step response shown in Figure 8.23.

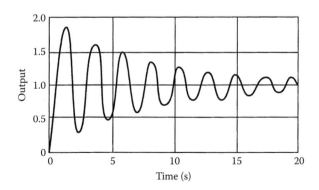

FIGURE 8.23 Closed-loop step response.

The method of the previous example may seem a long way to go in order to get an approximation to the closed-loop step response. Indeed, it is much simpler to calculate the closed-loop transfer function directly from the loop gain transfer function. The importance of the logic in the example is not to create a computational method; the importance lies in the insight that is achieved in predicting problems with the closed-loop transient response by examining the Bode plot of the loop gain. The essence of the insight can be summarized in a few sentences: Assume that the Nyquist plot of the loop gain indicates a stable closed-loop system. *If the Nyquist plot of the loop gain approaches the* −1 *point too closely, the transient response characteristics of the closed-loop system are oscillatory. The speed of the transient response of the closed-loop system is usually indicated by the loop gain crossover frequency.* Detailed information about the loop gain Nyquist plot is available in the loop gain Bode plots. In particular, the crossover frequency and the phase margin can be read from the Bode plots.

Any information that can be wrenched out of the Bode plots of the loop gain is critically important for two reasons. First, the Bode plots are a natural place to judge the properties of the feedback loop. When the magnitude of the loop gain is large, positive feedback properties such as good disturbance rejection and good sensitivity reduction are obtained. When the magnitude of the loop gain is small, these properties are not enhanced. The work of this chapter completes the missing information about transient response that can be read from the loop gain Bode plots. Second, it is the Bode plot that we are able to manipulate directly using series compensation techniques. It is important to be able to establish the qualities of the Bode plots that produce positive qualities in a control system because only then can the Bode plots be manipulated to attain the desired qualities.

References

1. Bode, H.W., *Network Analysis and Feedback Amplifier Design,* Van Nostrand, New York, 1945.
2. Churchill, R.V., Brown, J.W., and Verhey, R.F., *Complex Variables and Applications,* McGraw-Hill, New York, 1976.
3. Horowitz, I.M., *Synthesis of Feedback Systems,* Academic, New York, 1963.
4. Nyquist, H., Regeneration Theory, *Bell System Tech. J.,* 11, 126–147, 1932.
5. Rohrs, C.E., Melsa, J.L., and Schultz, D.G., *Linear Control Systems,* McGraw-Hill, 1993.

8.3 Discrete-Time and Sampled-Data Stability Tests

Mohamed Mansour

8.3.1 Introduction

Discrete-time dynamic systems are described by difference equations. Economic systems are examples of these systems where the information about the system behavior is known only at discrete points of time.

On the other hand, in sampled-data systems some signals are continuous and others are discrete in time. Some of the discrete-time signals come from continuous signals through sampling. An example of a sampled-data system is the control of a continuous process by a digital computer. The digital computer only accepts signals at discrete points of time so that a sampler must transform the continuous time signal to a discrete time signal.

Stability is the major requirement of a control system. For a linear discrete-time system, a necessary and sufficient condition for stability is that all roots of the characteristic polynomial using the z-transform lie inside the unit circle in the complex plane. A solution to this problem was first obtained by Schur [1].

The stability criterion in table and determinant forms was published by Cohn [2]. A symmetrix matrix form was obtained by Fujiwara [3]. Simplifications of the table and the determinant forms were obtained by Jury [4] and Bistritz [5]. A Markov stability test was introduced by Nour Eldin [6]. It is always possible to solve the stability problem of a discrete-time system by reducing it to the stability problem of a continuous system with a bilinear transformation of the unit circle to the left half-plane. For sampled-data systems, if the z-transform is used, then the same criteria apply. If the δ-transform is used, a direct solution of the stability problem, without transformation to the z- or s-plane, is given by Mansour [7] and Premaratne and Jury [8].

8.3.2 Fundamentals

8.3.2.1 Representation of a Discrete-Time System

A linear discrete-time system can be represented by a difference equation, a system of difference equations of first order, or a transfer function in the z-domain.

Difference Equation

$$y(k+n) + a_1 y(k+n-1) + \cdots + a_n y(k) = b_1 u(k+n-1) + \cdots + b_n u(k) \qquad (8.30)$$

If z is the shift operator, i.e., $zy(k) = y(k+1)$ then the difference equation can be written as

$$z^n y(k) + a_1 z^{n-1} y(k) + \cdots + a_n y(k) = b_1 z^{n-1} u(k) + \cdots + b_n u(k) \qquad (8.31)$$

System of Difference Equations of First Order

Equation 8.30 can be decomposed in the following n difference equations using the state variables:

$$x_1(k) \ldots x_n(k)$$
$$x_1(k+1) = x_2(k)$$
$$x_2(k+1) = x_3(k)$$
$$\vdots$$
$$x_{n-1}(k+1) = x_n(k)$$
$$x_n(k+1) = -a_n x_1(k) - \cdots - a_1 x_n(k) + u(k)$$
$$y(k) = b_n x_1(k) + b_{n-1} x_2(k) + \cdots + b_1 x_n(k)$$

This can be written as

$$x(k+1) = Ax(k) + bu(k), \quad y(k) = c^T x(k) \qquad (8.32)$$

where A is in the companion form.

Transfer Function in the z-Domain

The z-transform of Equation 8.30 gives the transfer function

$$G(z) = \frac{Y(z)}{U(z)} = \frac{b_1 z^{n-1} + \cdots + b_n}{z^n + a_1 z^{n-1} + \cdots + a_n} \qquad (8.33)$$

8.3.2.2 Representation of a Sampled-Data System

The digital controller is a discrete-time system represented by a difference equation, a system of difference equations or a transfer function in the z-domain. The continuous process is originally represented by a differential equation, a system of differential equations or a transfer function in the s-domain. However, because the input to the continuous process is normally piecewise constant (constant during a sampling period) then the continuous process can be represented in this special case by difference equations or a transfer function in the z-domain. Figure 8.24 shows a sampled-data control system represented as a discrete-time system.

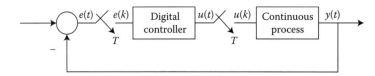

FIGURE 8.24 Sampled-data control system as a discrete time system.

Representation in the δ-Domain

Use the δ-operator which is related to the z-operator by

$$\delta = \frac{z - 1}{\Delta}, \tag{8.34}$$

where Δ is the sampling period.

In this case, δ corresponds to the differentiation operator in continuous systems and tends to it if Δ goes to zero. The continuous system is the limiting case of the sampled-data system when the sampling period becomes very small.

For stability, the characteristic equation with the δ-operator should have all its roots inside the circle of radius $1/\Delta$ in the left half-plane of Figure 8.25.

Representing sampled-data systems with the δ-operator has numerical advantages [9].

8.3.2.3 Definition of Stability

The output of a SISO discrete system is given by

$$y(k) = \sum_{i=0}^{\infty} g(i)u(k - i), \tag{8.35}$$

where $g(i)$ is the impulse response sequence. This system is BIBO-stable if, and only if, a real number $P > 0$ exists so that $\sum_{i=0}^{\infty} |g(i)| \leq P < \infty$.

8.3.2.4 Basics of Stability Criteria for Linear Systems

Stability criteria for linear systems are obtained by a simple idea: an n-degree polynomial is reduced to an (n − 1)-degree polynomial insuring that no root crosses the stability boundary. This can be achieved for discrete-time and sampled-data systems using the Rouché [1] or Hermite-Bieler theorem [10]. This

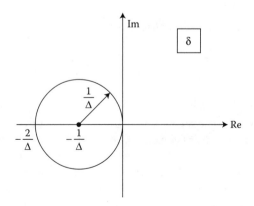

FIGURE 8.25 The stability region in the δ-domain.

reduction operation can be continued, thus obtaining a table form for checking stability, i.e., the impulse response is absolutely summable. This is achieved if all roots of the characteristic equation (or the eigenvalues of the system matrix) lie inside the unit circle. For sampled-data systems represented by the δ-operator, all roots of the characteristic equation must lie inside the circle in Figure 8.25.

8.3.3 Stability Criteria

8.3.3.1 Necessary Conditions for Stability of Discrete-Time Systems [11]

Consider the characteristic polynomial,

$$f(z) = a_0 z^n + a_1 z^{n-1} + \cdots + a_n. \tag{8.36}$$

The following conditions are necessary for the roots of Equation 8.36 to lie inside the unit circle (with $a_0 = 1$):

$$0 < f(1) < 2^n, \quad 0 < (-1)^n f(-1) < 2^n. \tag{8.37}$$

$$|a_n| < 1. \tag{8.38}$$

The ranges of the coefficients $a_1, a_2, \ldots a_n$ are given by the following table.

	a_1	a_2	a_3	a_4	a_5	a_6	a_7	...
$n = 2$	2	1						
	-2	-1						
$n = 3$	3	3	-1					
	-3	-1	1					
$n = 4$	4	6	4	1				
	-4	-2	-4	-1				
$n = 5$	5	10	10	5	1			
	-5	-2	-10	-3	-1			
$n = 6$	6	15	20	15	6	1		
	-6	-3	-20	-5	-6	-1		
$n = 7$	7	21	35	35	21	7	1	
$\vdots$	-7	-3	-35	-5	-21	-5	-1	

Thus for example, a necessary condition for the stability of the characteristic polynomial, $f(z) = z^2 + a_1 z + a_2$, is that $-2 < a_1 < 2$ and $-1 < a_2 < 1$. This table can detect instability without calculations. It is analogous to the positivity of the characteristic polynomial coefficients for a continuous system, but not equivalent to it.

8.3.3.2 Sufficient Conditions (with $a_0 > 0$)

6.1

$$a_0 > \sum |a_k| \tag{8.39}$$

[2].

6.2

$$0 < a_n < a_{n-1} < \cdots < a_1 < a_0 \tag{8.40}$$

[12].

8.3.3.3 Necessary and Sufficient Conditions ($a_0 > 0$)

Frequency Domain Criterion 1

"$f(z)$ has all roots inside the unit circle of the z-plane if, and only if, $f(e^{j\theta})$ has a change of argument of $n\pi$ when θ changes from 0 to π." The proof of this criterion is based on the principle of the argument.

Frequency Domain Criterion 2

"$f(z)$ has all its roots inside the unit circle of the z-plane if, and only if, $f^* = h^* + jg^*$ has a change of argument of $n\pi/2$ when θ changes from 0 to π."

$$f(z) = h(z) + g(z), \tag{8.41}$$

where

$$h(z) = \frac{1}{2}\left[f(z) + z^n f\left(\frac{1}{z}\right)\right] \tag{8.42}$$

and

$$g(z) = \frac{1}{2}\left[f(z) - z^n f\left(\frac{1}{z}\right)\right] \tag{8.43}$$

are the symmetric and antisymmetric parts of $f(z)$ respectively. Also

$$f(e^{j\theta}) = 2e^{jn\theta/2}[h^* + jg^*]. \tag{8.44}$$

For n even, $n = 2m$,

$$h^*(\theta) = \alpha_0 \cos m\theta + \alpha_1 \cos(m-1)\theta + \cdots + \alpha_{m-1}\cos\theta + \frac{\alpha_m}{2} \tag{8.45}$$

$$g*(\theta) = \beta_0 \sin m\theta + \beta_1 \sin(m-1)\theta + \cdots + \beta_{m-1}\sin\theta \tag{8.46}$$

and for n odd, $n = 2m - 1$

$$h^*(\theta) = a_0 \cos\left(m - \frac{1}{2}\right)\theta + a_1 \cos\left(\frac{m-3}{2}\right)\theta + \cdots + \alpha_{m-1}\frac{\cos\theta}{2} \tag{8.47}$$

$$g^*(\theta) = \beta_0 \sin\left(m - \frac{1}{2}\right)\theta + \beta_1 \sin\left(\frac{m-3}{2}\right)\theta + \cdots + \beta_{m-1}\frac{\sin\theta}{2} \tag{8.48}$$

$h^*(x)$ and $g^*(x)$ are the projections of $h(z)$ and $g(z)$ on the real axis with $x = \cos\theta$.

$$\text{For} \quad n = 2m, \quad h^*(x) = \sum_0^{m-1} \alpha_i T_{m-1} + \frac{\alpha_m}{2} \tag{8.49}$$

$$g^*(x) = \sum_0^{m-1} \beta_i U_{m-i} \tag{8.50}$$

where T_k, U_k are Tshebyshef polynomials of the first and second kind, respectively. T_k and U_k can be obtained by the recursions

$$T_{k+1}(x) = 2xT_k(x) - T_{k-1}(x) \tag{8.51}$$
$$U_{k+1}(x) = 2xU_k(x) - U_{k-1}(x) \tag{8.52}$$

where

$$T_0(x) = 1, \quad T_1(x) = x, \quad U_0(x) = 0, U_1(x) = 1$$

Discrete Hermite–Bieler Theorem

"$f(z)$ has all its roots inside the unit circle of the z-plane if, and only if, $h(z)$ and $g(z)$ have simple alternating roots on the unit circle and $\left|\frac{a_n}{a_0}\right| < 1$ [13]. The necessary condition $\left|\frac{a_n}{a_0}\right| < 1$ distinguishes between $f(z)$ and its inverse which has all roots outside the unit circle."

Schur–Cohn Criterion

"$f(z)$ with $|a_n/a_0| < 1$ has all roots inside the unit circle of the z-plane if, and only if, the polynomial $\frac{1}{z}\left[f(z) - \frac{a_n}{a_0}z^n f(1/z)\right]$ has all roots inside the unit circle."

The proof of this theorem is based on the Rouché theorem [1] or on the discrete Hermite–Bieler theorem [10]. This criterion can be translated in table form as follows: where

a_0	$a_1 \quad a_2 \ldots a_n$
a_n	$a_{n-1} \quad a_{n-2} \ldots a_0$
b_0	$b_1 \ldots b_{n-1}$
b_{n-1}	$b_{n-2} \ldots b_0$
c_0	$c_1 \ldots c_{n-2}$
c_{n-2}	$c_{n-3} \ldots c_0$
	$\ddots$
g_0	g_1
g_1	g_0
h_0	

$$b_0 = \begin{vmatrix} a_0 & a_n \\ a_n & a_0 \end{vmatrix}, \quad b_1 = \begin{vmatrix} a_0 & a_{n-1} \\ a_n & a_1 \end{vmatrix}, \quad \ldots,$$

$$b_{n-1} = \begin{vmatrix} a_0 & a_1 \\ a_n & a_{n-1} \end{vmatrix},$$

$$c_0 = \begin{vmatrix} b_0 & b_{n-1} \\ b_{n-1} & b_0 \end{vmatrix}, \quad \ldots$$

The necessary and sufficient condition for stability is that

$$b_0, c_0, \ldots, g_0, \quad h_0 > 0. \tag{8.53}$$

Jury [4] has replaced the last condition by the necessary conditions

$$f(1) > 0, \quad \text{and} \quad (-1)^n f(-1) > 0. \tag{8.54}$$

Example: $f(z) = 6z^2 + z - 1$.

6	1	-1
-1	1	6
35	7	$35 > 0$
7	35	
1176		$1176 > 0$

Hence there is stability.

Bistritz Table [5]

Bistritz used a sequence of symmetric polynomials $T_i(z)$ $i = 0, 1, \ldots, n$,

$$T_n(z) = 2h(z) \tag{8.55}$$
$$T_{n-1}(z) = 2g(z)/(z-1) \tag{8.56}$$
$$T_i(z) = [\delta_{i+2}(z+1)T_{i+1}(z) - T_{i+2}(z)]/z \quad i = n-2, n-3, \ldots, 0 \tag{8.57}$$

where

$$\delta_{i+2} = [T_{i+2}(0)]/[T_{i+1}(0)].$$

"The polynomial $f(z)$ is stable if and only if

    i) $T_i(0) \neq 0, \quad i = n-1, n-2, \ldots, 0,$ and
    ii) $T_n(1), T_{n-1}(1), \ldots, T_0(1)$

have the same sign."
Example: $n = 2$ $a_0 > 0$.

$$T_2(z): \quad (a_0 + a_2)z^2 + 2a_1 z + a_0 + a_2$$
$$T_1(z): \quad (a_0 - a_2)z + (a_0 - a_2)$$
$$T_0(z): \quad 2(a_0 + a_2 - a_1)$$

$$T_2(1) = 2(a_0 + a_1 + a_2) \;=\; 2f(1)$$
$$T_1(1) = 2(a_0 - a_2)$$
$$T_0(1) = 2(a_0 + a_2 - a_1) \;=\; 2f(-1)$$

A necessary and sufficient condition for stability of a second-order system of characteristic equation $f(z)$ is

$$f(1) > 0$$
$$f(-1) > 0$$
$$a_0 - a_2 > 0$$

Determinant Criterion [2]

$$f(z) = a_0 z^n + a_1 z^{n-1} + \ldots + a_n$$

has all roots inside the unit circle if, and only if,

$$\Delta_k < 0 \quad k \text{ odd}$$
$$\Delta_k > 0 \quad k \text{ even} \quad k = 1, 2, \ldots, n \tag{8.58}$$

where

$$\Delta_k = \begin{bmatrix} A_k & B_k^T \\ B_k & A_k^T \end{bmatrix} \tag{8.59}$$

$$A_k = \begin{bmatrix} a_n & & & 0 \\ a_{n-1} & a_n & & \\ \vdots & & \ddots & \ddots \\ a_{n+k-1} & \cdots & a_{n-1} & a_n \end{bmatrix}$$

$$B_k = \begin{bmatrix} a_0 & & & 0 \\ a_1 & a_0 & & \\ \vdots & & \ddots & \ddots \\ a_{k-1} & \cdots & a_1 & a_0 \end{bmatrix}$$

Jury simplified this criterion so that only determinants of dimension $n-1$ are computed [4]. The necessary conditions $f(1) > 0$ and $(-1)f(-1) > 0$ replace the determinants of dimension n. Example: $f(z) = 6z^2 + z - 1$.

$$\Delta_1 = \begin{vmatrix} -1 & 6 \\ 6 & -1 \end{vmatrix} = -35 < 0$$

$$\Delta_2 = \begin{vmatrix} -1 & 0 & 6 & 1 \\ 1 & -1 & 0 & 6 \\ 6 & 0 & -1 & 1 \\ 1 & 6 & 0 & -1 \end{vmatrix} = 161 > 0.$$

Hence there is stability.

Positive Definite Symmetric Matrix

[3] "$f(z)$ has all its roots inside the unit circle, if and only if, the symmetric matrix $C = [c_{ij}]$ is positive definite."

C is given by

$$c_{ij} = \sum_{p=1}^{\min(i,j)} (a_{i-p} - a_{j-p}a_{n-i+p}a_{n-j+p}). \tag{8.60}$$

For $n = 2$,

$$C = \begin{bmatrix} a_0^2 - a_2^2 & a_0a_1 - a_1a_2 \\ a_0a_1 - a_1a_2 & a_0^2 - a_2^2 \end{bmatrix}. \tag{8.61}$$

For $n = 3$,

$$C = \begin{bmatrix} a_0^2 - a_3^2 & a_0a_1 - a_2a_3 & a_0a_2 - a_1a_3 \\ a_0a_1 - a_2a_3 & a_0^2 + a_1^2 - a_2^2 - a_3^2 & a_0a_1 - a_2a_3 \\ a_0a_2 - a_1a_3 & a_0a_1 - a_2a_3 & a_0^2 - a_3^2 \end{bmatrix}. \tag{8.62}$$

For $n = 4$,

$$C = \begin{bmatrix} a_0^2 - a_4^2 & a_0a_1 - a_3a_4 & a_0a_2 - a_2a_4 & a_0a_3 - a_1a_4 \\ a_0a_1 - a_3a_4 & a_0^2 + a_1^2 - a_3^2 - a_4^2 & a_0a_1 + a_1a_2 - a_2a_3 - a_3a_4 & a_0a_2 - a_2a_4 \\ a_0a_2 - a_2a_4 & a_0a_1 + a_1a_2 - a_2a_3 - a_3a_4 & a_0^2 + a_1^2 - a_3^2 - a_4^2 & a_0a_1 - a_3a_4 \\ a_0a_3 - a_1a_4 & a_0a_2 - a_2a_4 & a_0a_1 - a_3a_4 & a_0^2 - a_4^2 \end{bmatrix} \tag{8.63}$$

Jury simplified this criterion. See, for example, [14] and [15].

Example: $f(z) = 6z^2 + z - 1$.

$$C = \begin{bmatrix} 35 & 7 \\ 7 & 35 \end{bmatrix} > 0.$$

Markov Stability Criterion [6]

If

$$\frac{g^*(x)}{h^*(x)} = \frac{s_0}{x} + \frac{s_1}{x^2} + \frac{s_2}{x^3} + \cdots, \tag{8.64}$$

then, for $n = 2m$, the polynomial $f(z)$ has all roots inside the unit circle if, and only if, the Hankel matrices

$$S_a = \begin{bmatrix} s_0 + s_1 & s_1 + s_2 & \cdots & s_{m-1} + s_m \\ s_1 + s_2 & s_2 + s_3 & \cdots & s_m + s_{m+1} \\ \vdots & \vdots & & \vdots \\ s_{m-1} + s_m & s_m + s_{m+1} & \cdots & s_{2m-2} + s_{2m-1} \end{bmatrix} \tag{8.65}$$

and

$$S_b = \begin{bmatrix} s_0 - s_1 & s_1 - s_2 & \cdots & s_{m-1} - s_m \\ s_1 - s_2 & s_2 - s_3 & \cdots & s_m - s_{m+1} \\ \vdots & \vdots & & \vdots \\ s_{m-1} - s_m & s_m - s_{m+1} & \cdots & s_{2m-2} - s_{2m-1} \end{bmatrix} \tag{8.66}$$

are positive definite.

For n odd, $zf(z)$ is considered instead of $f(z)$. Simplifications of the above critrion can be found in [16] and [17].

Stability of Discrete-Time Systems Using Lyapunov Theory

Given the discrete system,

$$x(k+1) = Ax(k) \tag{8.67}$$

and using a quadratic form as a Lyapunov function

$$V(k) = x(k)^T Px(k), \tag{8.68}$$

the change in $V(K)$

$$\Delta V(k) = x(k)^T [A^T PA - P]x(k). \tag{8.69}$$

For stability, $V(k) > 0$, and $\Delta V(k) < 0$.

This is achieved by solving the matrix equation

$$A^T PA - P = -Q \tag{8.70}$$

P and Q are symmetric matrices.

Q is chosen positive definite, e.g., the unity matrix, and P is determined by solving a set of algebraic equations.

Necessary and sufficient conditions for the asymptotic stability of Equation 8.67 are that P is positive definite.

Stability Conditions of Low-Order Polynomials [14]

$n = 2$:
$$f(z) = a_0 z^2 + a_1 z + a_2, \quad a_0 > 0$$
a) $f(1) > 0, \quad f(-1) > 0$
b) $a_0 - a_2 > 0$

$n = 3$:
$$f(z) = a_0 z^3 + a_1 z^2 + a_2 z + a_3, \quad a_0 > 0$$
a) $f(1) > 0, \quad f(-1) < 0$
b) $|a_3| < a_0$
c) $a_3^2 - a_0^2 < a_3 a_1 - a_0 a_2$

$n = 4$:
$$f(z) = a_0 z^4 + a_1 z^3 + a_2 z^2 + a_3 z + a_4, \quad a_0 > 0$$
a) $f(1) > 0, \quad f(-1) > 0$
b) $a_4^2 - a_0^2 - a_4 a_1 + a_3 a_0 < 0$
c) $a_4^2 - a_0^2 + a_4 a_1 - a_3 a_0 < 0$
d) $a_4^3 + 2a_4 a_2 a_0 + a_3 a_1 a_0 - a_4 a_0^2$
$\quad - a_2 a_0^2 - a_4 a_1^2 - a_4^2 a_0 - a_4^2 a_2$
$\quad - a_3^2 a_0 + a_0^3 + a_4 a_3 a_1 > 0$

$n = 5$:
$$f(z) = a_0 z^2 + a_1 z^4 + a_2 z^3 + a_3 z^2 + a_4 z + a_5, \quad a_0 > 0$$
a) $f(1) > 0, \quad f(-1) < 0$
b) $|a_5| < a_0$
c) $a_4 a_1 a_0 - a_5 a_0^2 - a_3 a_0^2 + a_5^3 + a_5 a_2 a_0$
$\quad - a_5 a_1^2 - (a_4^2 a_0 - a_5 a_3 a_0 - a_0^3 + a_5^2 a_2$
$\quad + a_5^2 a_0 - a_5 a_4 a_1) > 0$
d) $a_4 a_1 a_0 - a_5 a_0^2 - a_3 a_0^2 + a_5^3 + a_5 a_2 a_0$
$\quad - a_5 a_1^2 + (a_4^2 a_0 - a_5 a_3 a_0 - a_0^3 + a_5^2 a_2$
$\quad + a_5^2 a_0 - a_5 a_4 a_1) < 0$
e) $(a_5^2 - a_0^2)^2 - (a_5 a_1 - a_4 a_0)^2 + (a_5 a_1$
$\quad - a_4 a_0)(a_1^2 + a_4 a_2 - a_4^2 - a_3 a_1 - a_5^2$
$\quad + a_0^2) + (a_5 a_2 - a_3 a_0)(a_5 a_4 - a_1 a_0$
$\quad - a_5 a_2 + a_3 a_0) - (a_5 a_3 - a_2 a_0)$
$\quad [(a_5^2 - a_0^2) - 2(a_5 a_1 - a_4 a_0)] > 0$

The critical stability constraints that determine the boundary of the stability region in the coefficient space are given by the first condition

a) $f(1) > 0$ and $(-1)^n f(-1) > 0$

and the last condition of the above conditions, i.e., condition b) for $n = 2$, condition c) for $n = 3$ and so on.

Stability Criteria for Delta-Operator Polynomials

The characteristic equation of a sampled-data system, whose characteristic equation is in the δ-domain, is given by
$$f(\delta) = a_0 \delta^n + a_1 \delta^{n-1} + \cdots + a_n, \tag{8.71}$$

where Δ is the sampling period. For stability, the roots of $f(\delta)$ must lie inside the circle in Figure 8.25.

Necessary conditions for stability: [10]

$$\text{i)} \quad a_1, a_2, \ldots a_n > 0, \tag{8.72}$$

$$\text{ii)} \quad (-1)^n f\left(\frac{-2}{\Delta}\right) > 0, \quad \text{and} \tag{8.73}$$

$$\text{iii)} \quad a_i < \binom{n}{i}\left(\frac{2}{\Delta}\right)^i \quad i = 1, 2, \ldots, n. \tag{8.74}$$

Necessary and sufficient conditions: The stability of Equation 8.71 can be checked by one of the following methods:

1. Transforming Equation 8.71 to the *s*-domain by the transformation

$$\delta = \frac{2s}{2 - \Delta s} \tag{8.75}$$

   and applying the Routh–Hurwitz criterion
2. Transforming Equation 8.71 to the *z*-domain by the transformation

$$\delta = \frac{z - 1}{\Delta} \tag{8.76}$$

   and applying Schur-Cohn criterion
3. Using a direct approach such as the one given in [8]

This direct approach is as follows: Let

$$f^*(\delta) = (1 + \Delta\delta)^n f\left(\frac{\delta}{1 + \Delta\delta}\right). \tag{8.77}$$

Consider the sequence of polynomials,

$$T_n(\delta) = f(\delta) + f^*(\delta) \tag{8.78}$$

$$T_{n-1}(\delta) = \frac{1}{\delta}[f(\delta) - f^*(\delta)] \tag{8.79}$$

$$T_j(\delta) = \frac{1}{1 + \Delta\delta}\left[\delta_{j+2}(2 + \Delta\delta)T_{j+1}(\delta) - T_{j+2}(\delta)\right] \tag{8.80}$$

with

$$\delta_{j+2} = \frac{T_{j+2}\left(-\frac{1}{\Delta}\right)}{T_{j+1}\left(-\frac{1}{\Delta}\right)}, \quad j = n - 2, n - 3, \ldots, 0, \tag{8.81}$$

where

$$T_{j-1}(-1/\Delta) \neq 0, \quad j = 1, 2, \ldots, n,$$

Stability is concluded if

$$\text{var}\left\{T_j(0)\right\}_{j=0}^n = 0 \tag{8.82}$$

(no change of sign).

References

1. Schur, I., Ueber Potenzreihen die in Innern des Einheitskreises beschränkt sind. *S. Fuer Math.*, *147*, 205–232, 1917.
2. Cohn, A., Ueber die Anzahl der Wurzeln einer algebraischen Gleichung in einem Kreise. *Math. Z.*, *14*, 110–148, 1922.
3. Fujiwara, M., Ueber die algebraischen Gleichung deren Wurzeln in einem Kreise oder in einer Halbebene liegen. *Math. Z. 24*, 160–169, 1962.
4. Jury, E.I., A simplified stability criterion for linear discrete systems. *IRE Proc. 50(6)*, 1493–1500, 1962.
5. Bistritz, Y., Zero location with respect to the unit circle of discrete-time linear system polynomials. *Proc. IEEE, 72*, 1131–1142, 1984.
6. Nour Eldin, H.A., Ein neues Stabilitaets kriterium fuer abgetastete Regelsysteme. *Regelungstechnik u. Prezess-Daten verabeitung, 7*, 301–307, 1971.
7. Mansour, M., Stability and robust stability of discrete-time systems in the δ-transform. In *Fundamentals of Discrete Time Systems*, M. Jamshidi et al., Eds., TSI Press, Albuquerque, NM, 1993, 133–140.
8. Premaratne, K. and Jury, E.I., Tabular method for determining root distribution of delta-operator formulated real polynomials, *IEEE Trans. AC., 1994, 39(2)*, 352–355, 1994.
9. Middleton, R.H. and Goodwin, G.C., *Digital Control and Estimation. A Unified Approach*. Prentice Hall, Englewood Cliffs, NJ, 1990.
10. Mansour, M., Robust stability in systems described by rational functions. In *Control and Dynamic Systems*, Leondes, Ed., 79–128, Academic Press, 1992, Vol. 51.
11. Mansour, M., Instability criteria of linear discrete systems. *Automatica, 2*, 1985, 167–178, 1965.
12. Ackerman, J., *Sampled-Data Control Systems*. Springer, Berlin, 1985.
13. Scuessler, H.W., A stability theorem for discrete systems. *IEEE Trans, ASSP. 24*, 87–89, 1976.
14. Jury, E.I., *Theory and Applications of the z-Transform Method*. Robert E. Krieger, Huntingdon, NY, 1964.
15. Jury, E.I., *Inners and Stability of Dynamic Systems*. John Wiley & Sons, New York, 1974.
16. Anderson, B.D.O., Jury, E.I. and Chaparro, L.F., On the root distribution of a real polynomial with respect to the unit circle. *Regelungstechnik, 1976, 3*, 101–102, 1976.
17. Mansour, M. and Anderson, B.D.O, On the markov stability criterion for discrete systems. *IEEE Trans. CAS, 37(12)*, 1576–1578, 1990.
18. Astroem, K.J. and Wittenmark, B., *Computer Controlled Systems*, Prentice Hall, Englewood Cliffs, NJ, 1984.

Further Reading

For comprehensive discussions of the stability of linear discrete-time and sampled-data systems, see [14] and [15]. The Delta-operator approach and its advantages is dealt with in [9]. The application of the Nyquist stability criterion, using the discrete frequency response, is given in [12] and [18].

8.4 Gain Margin and Phase Margin

Raymond T. Stefani

8.4.1 Introduction

According* to Nyquist plot stability evaluation methods, a system with no open-loop right half-plane (RHP) poles should have no clockwise (CW) encirclements of the −1 point for stability, and a system with open-loop RHP poles should have as many counterclockwise (CCW) encirclements as there are open-loop RHP poles. It is often possible to determine whether the −1 point is encircled by looking at only that part of the Nyquist plot (or Bode plot) that identifies the presence of an encirclement, that is,

---

* From Stefani, R.T., et al. *Design of Feedback Control Systems*, 4th ed., Oxford University Press, 2002. With permission.

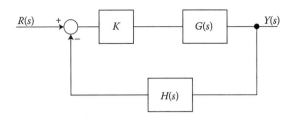

FIGURE 8.26 Closed-loop system. (From Stefani, R.T., et al. *Design of Feedback Control Systems*, 4th ed., Oxford University Press, 2002. With permission.)

the part of the Nyquist plot near the −1 point (the part of the Bode plot near 0 dB for magnitude and −180° for phase).

Similarly, it is possible to examine part of the Nyquist plot (or Bode plot) to determine the factor by which the system magnitude can be changed to make the system marginally stable. That factor is called the *gain margin*. It is also possible to examine part of the Nyquist plot (or Bode plot) to determine the amount of phase shift required to make the system marginally stable. The negative of that phase shift is called the *phase margin*. Both margins are discussed in this chapter.

A polynomial is called *minimum phase* when all the roots are in the left half-plane (LHP) and *non-minimum* phase when there are RHP roots. This means that stability is relatively easy to determine when $G(s)H(s)$ has minimum-phase poles, since there should be no encirclements of −1 for stability (a requirement that is easy to verify), but special care must be taken for the nonminimum-phase RHP pole case where stability demands CCW encirclements.

8.4.2 Gain Margin

In general, a Nyquist plot establishes the stability of a system of the form of Figure 8.26 with $K = 1$ by mapping $G(s)H(s)$ for s along the RHP boundary. One measure of stability arises from use of the *phase crossover frequency*, denoted ω_{PC} and defined as the frequency at which the phase of $G(s)H(s)$ is −180°, that is,

$$\text{phase } G(j\omega_{PC})H(j\omega_{PC}) = -180° = \Phi(\omega)$$

The magnitude of $G(s)H(s)$ at the phase crossover frequency is denoted $A(\omega_{PC})$. The gain margin, GM, is defined to be $1/A(\omega_{PC})$. Suppose that the gain K in Figure 8.26 is not selected to be one; rather K is selected to be $K = GM$. Then

$$KG(j\omega pc)H(j\omega pc) = KA(\omega pc)\angle - 180°$$
$$= 1\angle - 180°,$$

and the system becomes marginally stable.

For example, a system with $G(s)H(s) = 4/s(s+1)(s+2)$ has the Nyquist plot of Figure 8.27. Table 8.1 contains magnitude and phase data spanning the frequency range from zero to infinity.

Figure 8.28a shows the root locus with variable K for the same system as Figure 8.27. Figures 8.28b and 8.28c contain part of the Nyquist plot of $G(s)H(s)$ and Figures 8.28d and 8.28e show the Bode plot of $G(s)H(s)$. The gain margin, GM, for the system of Figure 8.26 with K nominally equals 1 is the K for marginal stability on the root locus of Figure 8.28a. More generally, GM is the gain at which the system

TABLE 8.1 Evaluation for $G(s)H(s)$ for $s = j\omega$

ω	0	0.5	1	1.141	1.414	2	10	∞
$A(\omega)$	∞	3.47	1.26	1.00	0.67	0.31	0.004	0
$\Phi(\omega)$	−90°	−131°	−162°	−169°	−180°	−198°	−253°	−270°

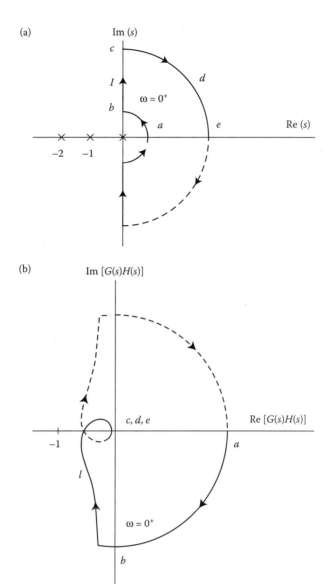

FIGURE 8.27 Nyquist plot for $G(s)H(s) = 4/s(s+1)(s+2)$. (a) RHP boundary; (b) Nyquist plot (not to scale). (From Stefani, R.T., et al. *Design of Feedback Control Systems*, 4th ed., Oxford University Press, 2002. With permission.)

becomes marginally stable divided by the nominal gain. That ratio can be expressed as a base 10 number or in dB.

Viewed from the perspective of a Nyquist plot, $A(\omega_{PC})$ is the distance from the origin to where the Nyquist plot crosses the negative real axis in Figure 8.28b, which occurs for $A(\omega_{PC}) = 0.67$. *GM* is thus measured at $\omega_{PC} = 1.414$ rad/s. *GM* is $1/0.67 = 1.5$ as a base 10 number while

$$\mathrm{dB}(GM) = -\mathrm{dB}[A(\omega_{PC})] = 20\log_{10}(1.5) = 3.5\,\mathrm{dB}$$

The *GM* in dB is the distance on the Bode magnitude plot from the amplitude at the phase crossover frequency up to the 0-dB point (see Figure 8.28d).

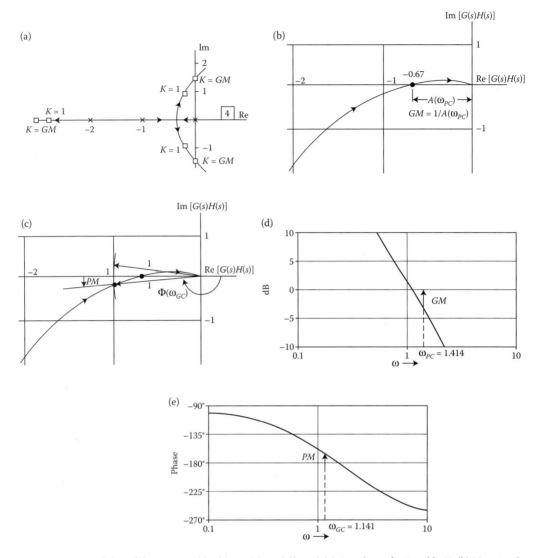

FIGURE 8.28 Stability of the system $G(s)H(s) = 4/s(s+1)/(s+2)$ (a) Root locus for variable K; (b) Nyquist plot of $G(s)H(s)$ showing GM; (c) Nyquist plot of $G(s)H(s)$ showing PM. (d) Bode magnitude plot of $G(s)H(s)$; (e) Bode phase plot of $G(s)H(s)$. (From Stefani, R.T., et al. *Design of Feedback Control Systems*, 4th ed., Oxford University Press, 2002. With permission.)

When a system is stable for all positive K, the phase crossover frequency is generally infinite: $A(\omega_{PC})$ is zero; and GM is infinite. Conversely, when a system is unstable for all positive K, the phase crossover frequency is generally at 0 rad/s; $A(\omega_{PC})$ is infinite; and the GM is zero.

GM can be interpreted in two ways. First, the designer can purposely vary K to some value other than one, and $K = GM$ represents an upper bound on the value of K for which the closed-loop system remains stable. Second, the actual system open-loop transmittance may not actually be $G(s)H(s)$. When the uncertainty in $G(s)H(s)$ is only in the magnitude, the GM is a measure of the allowable margin of error in knowing $|G(s)H(s)|$ before the system moves to marginal stability.

As an open-loop RHP pole example, consider the Nyquist and Bode plots of Figure 8.29. The Nyquist plot of Figure 8.29a has one CCW encirclement of -1 so the number of closed-loop RHP poles is -1 (due

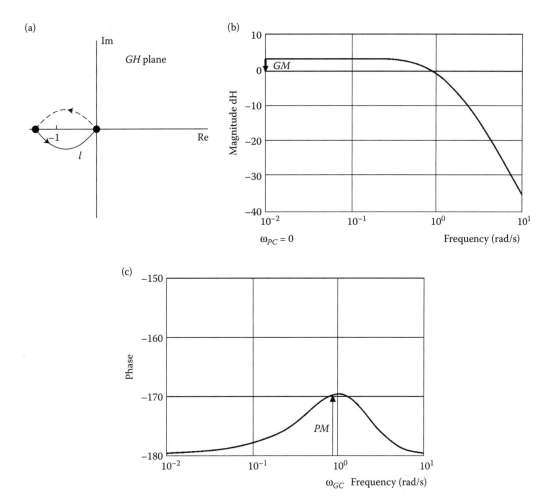

FIGURE 8.29 Stability of the system $G(s)H(s) = 2(s+3)/(s+2)^2(s-1)$. (a) Nyquist plot; (b) Bode magnitude plot of $G(s)H(s)$ showing GM; (c) Bode phase plot of $G(s)H(s)$ showing PM. (From Stefani, R.T., et al. *Design of Feedback Control Systems*, 4th ed., Oxford University Press, 2002. With permission.)

to the CCW encirclement) $+1$ (due to the open-loop RHP pole), which equals zero indicating stability. Here the phase is $-180°$ at $\omega_{PC} = 0$ rad/s and $G(0)H(0)$ is -1.5, so that $A(\omega_{PC})$ is 1.5 and GM is 0.67 or -3.5 dB. In this case, the system is stable with a dB(GM) that is negative.

Figure 8.30a shows part of the Nyquist plot for $G(s)H(s) = 0.75(s+2)^2/s^2(s+0.5)$. From the complete Nyquist plot it is easy to show that there are no CW encirclements of the -1 point; so the system is stable with $K = 1$. The phase crossover frequency is 1.41 rad/s with a GM of $1/1.5 = 2/3 = 0.67(-3.53$ dB). The system is stable for a negative value of dB(GM). If the $G(s)H(s)$ of Figure 8.30a is divided by 3, as in Figure 8.30b, the complete Nyquist plot indicates that the system is unstable. Predictably, the GM of Figure 8.30b is three times that of Figure 8.30a; thus, the GM of Figure 8.30b is 2.0 (6 dB) with a phase crossover frequency of 1.41 rad/s. Here the system is unstable with a positive dB(GM) value.

To be sure of stability, it is good practice to examine the Nyquist plot and the root locus plot. When there is more than one GM value (due to more than one phase crossover frequency) for a stable system, it is good practice to select the smallest GM value to ensure stability.

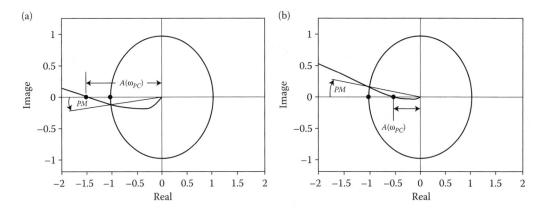

FIGURE 8.30 Partial Nyquist plots. (a) $G(s)H(s) = 0.75(s+2)^2/s^2(s+0.5)$, $GM = 2/3(-3.53\,\text{dB})$, $PM = 7.3°$, stable system; (b) $G(s)H(s) = 0.25(s+2)^2/s^2(s+0.5)$, $GM = 2(6\,\text{dB})$, $PM = -9.2°$, unstable system.

8.4.3 Phase Margin

In contrast to GM, when only the magnitude of $KG(s)H(s)$ is changed compared to that of $G(s)H(s)$, suppose instead that the gain K has unit magnitude and only the phase of $KG(s)H(s)$ is changed compared to that of $G(s)H(s)$.

It is useful to define the *gain crossover frequency* ω_{GC} as the frequency at which the magnitude of $G(s)H(s)$ is one (0 dB). Thus, $A(\omega_{GC}) = 1$. The phase of $G(s)H(s)$ at the gain crossover frequency is denoted by $\Phi(\omega_{GC})$. The phase margin, PM, is defined by

$$PM = 180° + \Phi(\omega_{GC})$$

Suppose the gain K in Figure 8.26 is selected to be $K = 1\angle - PM$. Then at the gain crossover frequency

$$
\begin{aligned}
KG(j\omega_{GC})H(j\omega_{GC}) &= [1\angle - PM][1\angle\Phi(\omega_{GC})], \\
|KG(j\omega_{GC})H(j\omega_{GC})| &= 1, \\
phase\ KG(j\omega_{GC})H(j\omega_{GC}) &= -PM + \Phi(\omega_{GC}) = -180° - \Phi(\omega_{GC}) \\
&\quad + \Phi(\omega_{GC}) = -180°,
\end{aligned}
$$

and the system is marginally stable.

For example, consider again the system with $G(s)H(s) = 4/s(s+1)(s+2)$. From Table 8.1, the gain crossover frequency is at 1.141 rad/s so that $A(\omega_{GC}) = 1$ whereas $\Phi(\omega_{GC}) = \Phi(1.141) = -169°$. Therefore, PM is $180° - 169° = 11°$. The phase margin, PM, is the angle in the Nyquist plot of Figure 8.28c drawn from the negative real axis to the point at which the Nyquist plot penetrates a circle of unit radius (called the unit circle). On the Bode phase plot of Figure 8.28e, the PM is the distance from $-180°$ to the phase at the gain crossover frequency. *The phase margin is therefore the negative of the phase through which the Nyquist plot can be rotated, and similarly the Bode plot can be shifted, so that the closed-loop system becomes marginally stable.*

In order to properly calculate PM, it is generally best to define $\Phi(\omega_{GC})$ as $-270° \leq \Phi(\omega_{GC}) \leq 90°$. For example, a third quadrant $\Phi(\omega_{GC})$ would be written as $-160°$; so PM would be $+20°$.

For a nonminimum-phase example, consider again the system of Figure 8.29. The Nyquist plot indicates stability. Here the gain crossover frequency is at 0.86 rad/s, $\Phi(\omega_{GC})$ is $-170°$ and the PM is $180° - 170°$ or $10°$; hence, the upward-directed arrow on the phase plot of Figure 8.29c.

For the example of Figure 8.30a, which is a stable system, ω_{GC} is 1.71 rad/s and the PM is 7.3°. For the unstable system of Figure 8.30b, ω_{GC} is 1.71 rad/s and the PM is $-9.2°$. It should be noted that there is a positive PM for all the stable systems just examined and a negative PM for all the unstable systems. This sign–stability relationship holds for the PM of most systems, while no such relationship holds for the sign of dB(GM) in the examples just examined.

If there is more than one gain crossover frequency, there is more than one PM. For a stable system, the smallest candidate PM should be chosen.

As noted earlier, when $K = 1\angle - PM$, the system becomes marginally stable. That fact can be interpreted in two ways. First, the designer can purposely vary K away from one and then $1\angle - PM$ represents one extreme of selection of K. Second, the actual system open-loop transmittance may not actually be $G(s)H(s)$. When the uncertainty in $G(s)H(s)$ affects only the phase, the PM is the allowable margin of error in knowing phase $G(s)H(s)$ before the system moves to marginal stability. In most systems, there is uncertainty in both the magnitude and phase of $G(s)H(s)$ so that substantial gain and PMs are required to assure the designer that imprecise knowledge of $G(s)H(s)$ does not necessarily cause instability. In fact, there are examples of systems that have large GM and PM, but small variations in gain *and* phase cause instability. It is of course important to check the complete Nyquist plot when there is any question about stability.

Suppose in Figure 8.26 that $K = 1, H(s) = 1$ and $G(s) = \omega_n^2/s(s + 2\zeta\omega_n)$. The closed-loop transfer function is $T(s) = \omega_n^2/(s^2 + 2\zeta\omega_n s + \omega_n^2)$, the standard form for a second-order system with damping ratio ζ and undamped natural frequency ω_n. For this system, the gain crossover frequency can be found in closed form and the PM follows from a trigonometric identity.

$$\omega_{GC} = k\omega_n,$$
$$k = ((4\zeta^4 + 1)^{0.5} - 2\zeta^2)^{0.5},$$
$$PM = \tan^{-1}(2\zeta/k).$$

Table 8.2 shows values of gain crossover frequency and PM for this standard-form second-order system.

For other systems with two dominant closed-loop underdamped poles, Table 8.2 is often a good approximation. Thus, the PM is approximately 100ζ degrees for damping ratios from zero to about 0.7.

8.4.4 Using MATLAB to Get *GM* and *PM*

GM and PM may be found automatically by the Control System Toolbox option of MATLAB®, a product of The Math Works Inc. These margins are provided by the *margin* command for a stable system. Prior to using the *margin* command, the open-loop system must be defined in either transfer function or state-space form.

TABLE 8.2 Phase Margin for a Standard-Form Second-Order System

Damping Ratio ζ	k ω_{GC}/ω_n	Phase Margin Degrees
0.0	1	0
0.1	0.99	11
0.2	0.96	23
0.3	0.91	33
0.4	0.85	43
0.5	0.79	52
0.6	0.72	59
0.7	0.65	65
0.8	0.59	70
0.9	0.53	74
1.0	0.49	76

TABLE 8.3 Gain and Phase Margins of the Example Systems

System of	Stability	Gain Margin		Phase Margin
		Base 10	dB	
Figure 8.28	Stable	1.5	3.5	11°
Figure 8.29	Stable	0.67	−3.5	10°
Figure 8.30a	Stable	0.67	−3.5	7.3°
Figure 8.30b	Unstable	2.0	6.0	−9.2°

The system in Figure 8.28 has the transfer function $G(s)H(s) = 4/s\,(s+1)(s+2)$. That transfer function may be input to MATLAB via the *zpk* or zero-pole-*k*(gain) command. That command requires that the zeros and poles each be set off within a square bracket followed by the multiplying gain constant. Since there is no zero in this case, the first bracket is empty. Suppose this $G(s)H(s)$ is called *gh1* in MATLAB. The MATLAB command would be

$$gh1 = zpk([\ ],\ [0\ 1\ -2],\ 4).$$

If *gm* represents gain margin, *pm* represents phase margin, *wpc* represents phase crossover frequency, and *wgc* represents gain crossover frequency, the following command causes each of those defined values to be calculated and displayed.

$$[gm,\ pm,\ wpc,\ wgc] = margin(\ gh1).$$

If the command is written as *margin(gh1)*, without the argument list, the margins and crossover frequencies are identified on Bode magnitude and phase plots. The results of Figure 8.28 are obtainable by the above commands.

For the system in Figure 8.29, the following commands will generate those margins.

$$gh2 = zpk([-3],\ [-2\ 2\ 1],\ 2),\quad [gm,\ pm,\ wpc,\ wgc] = margin(gh2).$$

Similarly, for Figure 8.30a, the following commands may be used:

$$gh3 = zpk([-2\ -2],\ [0\ 0\ -0.5],\ 0.75),\quad [gm,\ pm,\ wpc,\ wgc] = margin(gh3).$$

The system of Figure 8.30b may be defined by $gh4 = gh3/3$. For that system it is necessary to obtain either the Bode or Nyquist plot and then find gain and PM directly from the plot, since that system is unstable and margins are calculated by the MATLAB Control System Toolbox only for a stable system.

The GM and PM results are summarized in Table 8.3.

Defining Terms

Gain crossover frequency (rad/s): Frequency at which the magnitude of *GH* is one (zero dB).

Gain margin: Negative of the dB of *GH* measured at the phase crossover frequency (inverse of the base 10 magnitude). When $K = GM$, the system becomes marginally stable.

MATLAB: MAThematical LABoratory produced by The Mathworks, Inc. The software offers a wide range of mathematical commands plus additional toolboxes including control systems and signal processing applications.

Phase crossover frequency (rad/s): Frequency at which the phase of *GH* is $-180°$.

Phase margin: $180° +$ phase of *GH* measured at the gain cross-over frequency. When $K = 1\angle - PM$, the system becomes marginally stable.

Reference

1. Stefani, R.T., Shahian, B.S, Clement, J., and Hostetter, G.H., *Design of Feedback Control Systems*, 4th ed., Oxford University Press, 2002.

Further Reading

Additional information may be found in *IEEE Control Systems Mag.*; *IEEE Trans. Autom. Control*; and *IEEE Trans. Systems, Man, and Cybern.*

9

Design Methods

Jiann-Shiou Yang
University of Minnesota

William S. Levine
University of Maryland

Richard C. Dorf
University of California, Davis

Robert H. Bishop
The University of Texas at Austin

John J. D'Azzo
Air Force Institute of Technology

Constantine H. Houpis
Air Force Institute of Technology

Karl J. Åström
Lund Institute of Technology

Tore Hägglund
Lund Institute of Technology

Katsuhiko Ogata
University of Minnesota

Masako Kishida
University of Illinois at Urbana–Champaign

Richard D. Braatz
University of Illinois at Urbana–Champaign

Z. J. Palmor
Technion–Israel Institute of Technology

Mario E. Salgado
Federico Santa María Technical University

Graham C. Goodwin
The University of Newcastle

9.1 Specification of Control Systems

Jiann-Shiou Yang and William S. Levine

9.1.1 Introduction

Generally, control system specifications can be divided into two categories, performance specifications and robustness specifications. Although the boundaries between the two can be fuzzy, the performance specifications describe the desired response of the nominal system to command inputs. The robustness specifications limit the degradation in performance due to variations in the system and disturbances. Section 9.1.2 of this chapter describes the classical performance specifications for single-input single-output (SISO) linear time-invariant (LTI) systems. This is followed by a discussion of the classical robustness specifications for SISO LTI systems. The fourth section gives some miscellaneous classical specifications. The fifth section describes performance specifications that are unique to multi-input multi-output (MIMO) systems. This is followed by a section on robustness specifications for MIMO systems. The final section contains conclusions.

9.1.2 Performance Specifications for SISO LTI Systems

9.1.2.1 Transient Response Specifications

In many practical cases, the desired performance characteristics of control systems are specified in terms of time-domain quantities, and frequently, in terms of the transient and steady-state response to a unit step input. The unit step signal, one of the three most commonly used test signals (the other two are ramp and parabolic signals), is often used because there is a close correlation between a system response to a unit step input and the system's ability to perform under normal operating conditions. And many control systems experience input signals very similar to the standard test signals. Note that if the response to a unit step input is known, then it is mathematically possible to compute the response to any input. We discuss the transient and steady-state response specifications separately in this section and Section 9.1.2.2. We emphasize that both the transient and steady-state specifications require that the closed-loop system is stable.

The transient response of a controlled system often exhibits damped oscillations before reaching steady state. In specifying the transient response characteristics, it is common to specify the following quantities:

1. Rise time (t_r)
2. Percent overshoot (PO)
3. Peak time (t_p)
4. Settling time (t_s)
5. Delay time (t_d)

The rise time is the time required for the response to rise from x% to y% of its final value. For overdamped second-order systems, the 0% to 100% rise time is normally used, and for underdamped systems (see Figure 9.1), the 10–90% rise time is commonly used.

The peak time is the time required for the response to reach the first (or maximum) peak. The settling time is defined as the time required for the response to settle to within a certain percent of its final value. Typical percentage values used are 2% and 5%. The settling time is related to the largest time constant of the controlled system. The delay time is the time required for the response to reach half of its final value for the very first time. The percent overshoot represents the amount that the response overshoots its steady-state (or final) value at the peak time, expressed as a percentage of the steady-state value. Figure 9.1 shows a typical unit step response of a second-order system

$$G(s) = \frac{\omega_n^2}{s^2 + 2\zeta\omega_n s + \omega_n^2}$$

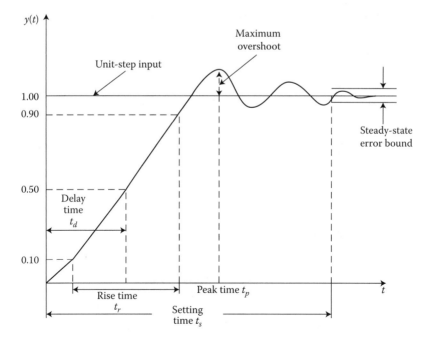

FIGURE 9.1 Typical underdamped unit-step response of a control system. An overdamped unit-step response would not have a peak.

where ζ is the damping ratio and ω_n is the undamped natural frequency. For this second-order system with $0 \leq \zeta < 1$ (an underdamped system), we have the following properties:

$$PO = e^{-\zeta\pi/\sqrt{1-\zeta^2}}$$

$$t_p = \frac{\pi}{\omega_n\sqrt{1-\zeta^2}}$$

$$t_s = \frac{4}{\zeta\omega_n}$$

where the 2% criterion is used for the settling time t_s. If 5% is used, then t_s can often be approximated by $t_s = 3/\zeta\omega_n$. A precise formula for rise time t_r and delay time t_d in terms of damping ratio ζ and undamped natural frequency ω_n cannot be found. But useful approximations are

$$t_d \cong \frac{1.1 + 0.125\zeta + 0.469\zeta^2}{\omega_n}$$

$$t_r \cong \frac{1 - 0.4167\zeta + 2.917\zeta^2}{\omega_n}$$

Note that the above expressions are only accurate for a second-order system. Many systems are more complicated than the pure second-order system. Thus, when using these expressions, the designer should be aware that they are only rough approximations. The time-domain specifications are quite important because most control systems must exhibit acceptable time responses. If the values of t_s, t_d, t_r, t_p, and PO are specified, then the shape of the response curve is virtually determined. However, not all these specifications necessarily apply to any given case. For example, for an overdamped system ($\zeta > 1$) or a critically damped system ($\zeta = 1$), t_p and PO are not useful specifications. Note that the time-domain specifications, such as PO, t_r, ζ, etc., can be applied to discrete-time systems with minor modifications.

Quite often the transient response requirements are described in terms of pole–zero specifications instead of step response specifications. For example, a system may be required to have its poles lying to the left of some constraint boundary in the s-plane, as shown in Figure 9.2.

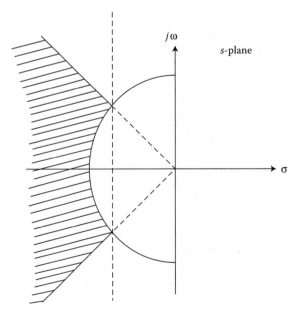

FIGURE 9.2 The shaded region is the allowable region in the s-plane.

The parameters given above can be related to the pole locations of the transfer function. For example, certain transient requirements on the rise time, settling time, and percent overshoot (e.g., t_r, t_s, and PO less than some particular values) may restrict poles to a region of the complex plane. The poles of a second-order system with $\zeta < 1$ are given in terms of ζ and ω_n by

$$p_1 = -\zeta\omega_n + j\omega_n\sqrt{1-\zeta^2}$$
$$p_2 = -\zeta\omega_n - j\omega_n\sqrt{1-\zeta^2}$$

Notice that $|p_1| = |p_2| = \omega_n$ and $\angle p_1 = -\angle p_2 = 180^\circ - tan^{-1}\sqrt{1-\zeta^2}/\zeta$. Thus, contours of constant ω_n are circles in the complex plane, while contours of constant ζ are straight lines ($0 \le \zeta \le 1$), as shown in Figure 9.3.

Because high-order systems can always be decomposed into a parallel combination of first- and second-order subsystems, the parameters related to the time response of these high-order systems can be estimated using the expressions given above. For instance, t_s can be approximated by four (or three) times the slowest time constant (i.e., the slowest subsystem normally determines the system settling time).

It is known that the location of the poles of a transfer function in the s-plane has great effects on the transient response of the system. The poles that are close to the $j\omega$-axis in the left half s-plane give transient responses that decay relatively slowly, whereas those poles far away from the $j\omega$-axis (relative to the dominant poles) correspond to more rapidly decaying time responses. The relative dominance of poles is determined by the ratio of the real parts of the poles, and also by the relative magnitudes of the residues evaluated at these poles (the magnitudes of the residues depend on both the poles and zeros). It has been recognized in practice that if the ratios of the real parts exceed five, then the poles nearest the $j\omega$-axis will dominate in the transient response behavior. The poles that have dominant effects on the transient response behavior are called dominant poles. And those poles with the magnitudes of their real parts at least five times greater than the dominant poles may be regarded as insignificant (as far as the transient response is concerned). Quite often, the dominant closed-loop poles occur in the form of a complex-conjugate pair. It is not uncommon that some high-order systems can be approximated by low-order systems. In other words, they contain insignificant poles that have little effect on the transient response, and may be approximated by dominant poles only. If this is the case, then the parameters

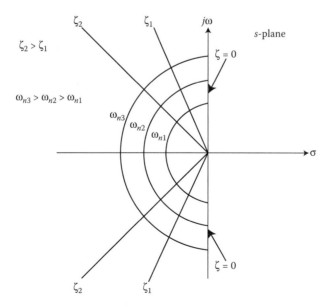

FIGURE 9.3 Contours of constant ζ and ω_n for a second-order system.

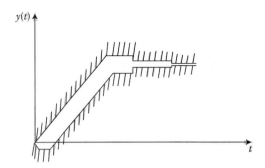

FIGURE 9.4   General envelope specification on a step response. The step response is required to be inside the region indicated.

described above can still be used in specifying the system dynamic behavior, and we can use dominant poles to control the dynamic performance, whereas the insignificant poles are used for ensuring the controller designed can be physically realized.

Although, in general, high-order systems may not have dominant poles (and thus, we can no longer use ζ, ω_n, etc., to specify the design requirements), the time domain requirements on transient and steady-state performance may be specified as bounds on the command step response, such as that shown in Figure 9.4 (i.e., the system has a step response inside some constraint boundaries).

9.1.2.2 Steady-State Accuracy

If the output of a system at steady state does not exactly agree with the input, the system is said to have steady-state error. This error is one measure of the accuracy of the system. Since actual system inputs can frequently be considered combinations of step, ramp, and parabolic types of signals, control systems may be classified according to their ability to follow step, ramp, parabolic inputs, etc. In general, the steady-state error depends not only on the inputs but also on the "type" of control system. Let $G_o(s)$ represent the open-loop transfer function of a stable unity feedback control system (see Figure 9.5), then $G_o(s)$ can generally be expressed as

$$G_o(s) = \frac{k(s - z_1)(s - z_2) \cdots (s - z_m)}{s^N (s - p_1)(s - p_2) \cdots (s - p_n)}$$

where z_i ($\neq 0$, $i = 1, 2, \ldots, m$) are zeros, p_j ($\neq 0$, $j = 1, 2, \ldots, n$) and 0 (a pole at the origin with multiplicity N) are poles, and $m < n + N$.

The type of feedback system refers to the order of the pole of the open-loop transfer function $G_o(s)$ at $s = 0$ (i.e., the value of the exponent N of s in $G_o(s)$). In other words, the classification is based on the number of pure integrators in $G_o(s)$. A system is called type 0, type 1, type 2, ... if $N = 0, 1, 2, \ldots$, respectively. Note that a nonunity feedback system can be mathematically converted to an equivalent unity feedback system from which its "effective" system type and static error constants (to be defined

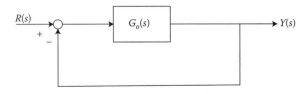

FIGURE 9.5   A simple unity feedback system with open-loop transfer function $G_o(s)$ and closed-loop transfer function $Y(s)/R(s) = G_o(s)/[1 + G_o(s)]$.

later) can be determined. For instance, for a control system with forward path transfer function $G(s)$ and feedback transfer function $H(s)$, the equivalent unity feedback system has the forward path transfer function $G_o(s) = G(s)/(1 + G(s)[H(s) - 1])$.

Static error constants describe the ability of a system to reduce or eliminate steady-state errors. Therefore, they can be used to specify the steady-state performance of control systems. For a stable unity feedback system with open-loop transfer function $G_o(s)$, the position error constant K_p, velocity error constant K_v, and acceleration error constant K_a are defined, respectively, as

$$K_p = \lim_{s \to 0} G_o(s)$$

$$K_v = \lim_{s \to 0} sG_o(s)$$

$$K_a = \lim_{s \to 0} s^2 G_o(s)$$

In terms of K_p, K_v, and K_a, the system's steady-state error for the three commonly used test signals, i.e., a unit step input ($u(t)$), a ramp input ($tu(t)$), and a parabolic input ($\frac{1}{2}t^2 u(t)$), can be expressed, respectively, as

$$e(\infty) = \frac{1}{1 + K_p}$$

$$e(\infty) = \frac{1}{K_v}$$

$$e(\infty) = \frac{1}{K_a}$$

where the error $e(t)$ is the difference between the input and output, and $e(\infty) = \lim_{t \to \infty} e(t)$. Therefore, the value of the steady-state error decreases as the error constants increase. Just as damping ratio (ζ), settling time (t_s), rise time (t_r), delay time (t_d), peak time (t_p), and percent overshoot (PO) are used as specifications for a control system's transient response, so K_p, K_v, and K_a can be used as specifications for a control system's steady-state errors.

To increase the static error constants (and hence, improve the steady-state performance), we can increase the type of the system by adding integrator(s) to the forward path. For example, provided the system is stable, a type-1 system has no steady-state error for a constant input, a type-2 system has no steady-state error for a constant or ramp input, a type-3 system has no steady-state error for a constant, a ramp, or a parabolic input, and so forth. Clearly, as the type number is increased, accuracy is improved. However, increasing the type number aggravates the stability problem. It is desirable to increase the error constants, while maintaining the transient response within an acceptable range. A compromise between steady-state accuracy and relative stability is always necessary.

Step response envelope specifications, similar to that of Figure 9.4, are often used as the time-domain specifications for control system design. These specifications cover both the transient and steady-state performance requirements. For MIMO systems, the diagonal entries of the transfer function matrix represent the functions from the commanded inputs to their associated commanded variables, and the off-diagonal entries are the transfer functions from the commands to other commanded variables. It is generally desirable to require that the diagonal elements lie within bounds such as those shown in Figure 9.4 , while the off-diagonal elements are reasonably small.

9.1.2.3 Frequency-Domain Performance Specifications

A stable LTI system $G(s)$ subjected to a sinusoidal input will, at steady state, have a sinusoidal output of the same frequency as the input. The amplitude and phase of the system output will, in general, be different from those of the system input. In fact, the amplitude of the output is given by the product of the amplitude of the input and $|G(j\omega)|$, while the phase angle differs from the phase angle of the input by

$\angle G(j\omega)$. In other words, if the input is $r(t) = A\,sin\omega t$, the output steady-state response $y(t)$ will be of the form $AR(\omega)\,sin[\omega t + \phi(\omega)]$, where $R(\omega) = |G(s)|_{s=j\omega} = |G(j\omega)|$ and $\phi(\omega) = \angle G(s)|_{s=j\omega} = \angle G(j\omega)$ vary as the input frequency ω is varied. Therefore, the magnitude and phase of the output for a sinusoidal input may be found by simply determining the magnitude and phase of $G(j\omega)$. The complex function $G(j\omega) = G(s)|_{s\to j\omega}$ is referred to as the frequency response of the system $G(s)$. It is understood that the system is required to be stable. In control system design by means of frequency-domain methods, the following specifications are often used in practice.

1. Resonant peak (M_p)
2. Bandwidth (ω_b)
3. Cutoff rate

The resonant peak M_p is defined as the maximum magnitude of the closed-loop frequency response, and the frequency at which M_p occurs is called the resonant frequency (ω_p). More precisely, if we consider a LTI open-loop system described by its transfer function $G_o(s)$ and the unity feedback closed-loop system pictured in Figure 9.5, then the closed-loop transfer function $G_{cl}(s)$ will have transfer function

$$G_{cl}(s) = \frac{G_o(s)}{1 + G_o(s)}$$

and

$$M_p = max_{\omega \geq 0}\,|G_{cl}(j\omega)|$$
$$\omega_p = arg\left\{max_{\omega \geq 0}\,|G_{cl}(j\omega)|\right\}$$

In general, the magnitude of M_p gives an indication of the relative stability of a stable system. Normally, a large M_p corresponds to a large maximum overshoot of the step response in the time domain. For most control systems, it is generally accepted in practice that the desirable M_p should lie between 1.1 and 1.5. The bandwidth, ω_b, is defined as the frequency at which the magnitude of the closed-loop frequency response drops to 0.707 of its zero-frequency value. In general, the bandwidth of a controlled system gives a measure of the transient response properties, in that a large bandwidth corresponds to a faster response. Conversely, if the bandwidth is small, only signals of relatively low frequencies are passed, and the time response will generally be slow and sluggish. Bandwidth also indicates the noise-filtering characteristics and the robustness of the system. Often, bandwidth alone is not adequate as an indication of the characteristics of the system in distinguishing signals from noise. Sometimes it may be necessary to specify the cutoff rate of the frequency response, which is the slope of the closed-loop frequency response at high frequencies. The performance criteria defined above are illustrated in Figure 9.6.

The closed-loop time response is related to the closed-loop frequency response. For example, overshoot in the transient response is related to resonance in the closed-loop frequency response. However, except for first- and second-order systems, the exact relationship is complex and is generally not used. For the standard second-order system, the resonant peak M_p, the resonant frequency ω_p, and the bandwidth ω_b are uniquely related to the damping ratio ζ and undamped natural frequency ω_n. The relations are given by the following equations:

$$\omega_p = \omega_n\sqrt{1 - 2\zeta^2}\quad \text{for } \zeta \leq 0.707$$
$$M_p = \frac{1}{2\zeta\sqrt{1 - \zeta^2}}\quad \text{for } \zeta \leq 0.707$$
$$\omega_b = \omega_n[\,(1 - 2\zeta^2) + \sqrt{4\zeta^4 - 4\zeta^2 + 2}\,]^{1/2}$$

Like the general envelope specifications on a step response (e.g., Figure 9.4), the frequency-domain requirements may also be given as constraint boundaries similar to those shown in Figure 9.7. That is, the closed-loop frequency response of the designed system should lie within the specified bounds. This kind of specification can be applied to many different situations.

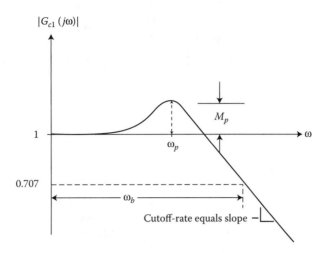

FIGURE 9.6 Frequency response specification.

9.1.3 Robustness Specifications for SISO LTI Systems

9.1.3.1 Relative Stability—Gain and Phase Margins

In control system design, in general, we require the designed system to be not only stable, but to have a certain guarantee of stability. In the time domain, relative stability is measured by parameters such as the maximum overshoot and the damping ratio. In the frequency domain, the resonant peak M_p can be used to indicate relative stability. Gain margin (GM) and phase margin (PM) are two design criteria commonly used to measure the system's relative stability. They provide an approximate indication of the closeness of the Nyquist plot of the system's open-loop frequency response $L(j\omega)$. ($L(j\omega)$ is also often called the loop transfer function) to the critical point, -1, in the complex plane. The open-loop frequency response, $L(j\omega)$, is obtained by connecting all of the elements in the loop in series and not closing the loop. For example, $L(j\omega) = G_o(j\omega)$ for the unity feedback system of Figure 9.5, but $L(j\omega) = H(j\omega)G_p(j\omega)G_c(j\omega)$ for the more complex closed-loop system of Figure 9.8.

The decibel (abbreviated dB) is a commonly used unit for the frequency response magnitude. The magnitude of $L(j\omega)$ is then $20 \log_{10} |L(j\omega)|$ dB. For example, if $|L(j\omega_o)| = 5$, $|L(j\omega_o)| = 20 \log_{10} 5$ dB. The

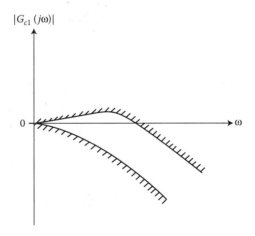

FIGURE 9.7 General envelope specification on the closed-loop frequency response.

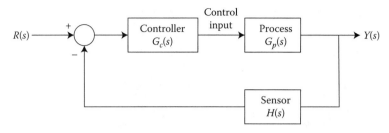

FIGURE 9.8   A typical closed-loop control system.

gain margin is the amount of gain in dB that can be inserted in the loop before the closed-loop system reaches instability. The phase margin is the change in open-loop phase shift required at unity gain to make the closed-loop system unstable, or the magnitude of the minimum angle by which the Nyquist plot of the open-loop transfer function must be rotated in order to intersect the -1 point (for a stable closed-loop system). The Nyquist plot showing definitions of the gain and phase margins is given in Figure 9.9.

Although GM and PM can be obtained directly from a Nyquist plot, they are more often determined from a Bode plot of the open-loop transfer function. The Bode plot (which includes Bode magnitude plot and phase plot) and Nyquist plot (i.e., $L(s)|_{s=j\omega}$ drawn in polar coordinates) differ only in the coordinates and either one can be obtained from the other. The gain crossover frequency ω_c is the frequency at which $|L(j\omega_c)| = 1$ (i.e., the frequency at which the loop gain crosses the 0 dB line). Therefore, comparing both Bode magnitude and phase plots, the distance in degrees between the $-180°$ line and $\angle L(j\omega_c)$ is the PM. If $\angle L(j\omega_c)$ lies above the $-180°$ line, PM is positive, and if it lies below, PM is negative. The phase crossover frequency ω_p is the frequency at which $\angle L(j\omega_p) = -180°$ (i.e., the frequency at which $\angle L(j\omega_p)$ crosses the $-180°$ line). Thus, the distance in decibels between the 0 dB line and $|L(j\omega_p)|$ is the GM (i.e., $-20\log_{10} |L(j\omega_p)|$ dB). If $|L(j\omega_p)|$ lies below the 0 dB line, GM is positive. Otherwise, it is negative. A proper transfer function is called minimum-phase if all its poles and zeros lie in the open left half s-plane. In general, for a minimum-phase $L(s)$, the system is stable if GM (in dB) > 0 and unstable if GM (in dB) < 0. And generally, a minimum-phase system has a positive PM, and it becomes unstable if PM < 0. For nonminimum phase systems, care must be taken in interpreting stability based on the sign of GM (dB) and PM. In this case, the complete Nyquist plot or root locus must be examined for relative stability. For a system in which multiple $-180°$ crossovers occur, the simplest approach is to convert the Bode plot to the corresponding Nyquist plot to determine stability and the frequencies at which the stability

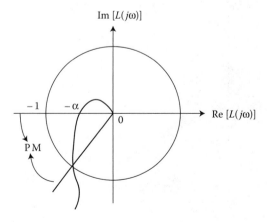

FIGURE 9.9   Definitions of the gain and phase margins. The gain margin is $20\log \frac{1}{\alpha}$.

margins occur. Obviously, systems with greater gain and phase margins can withstand greater changes in system parameter variations before becoming unstable.

It should be noted that neither the GM alone nor the PM alone gives a sufficient indication of the relative stability. For instance, the GM does not provide complete information about the system response; a small GM indicates the Nyquist plot $|L(j\omega_p)|$ is very close to the -1 point. Such a system will have a large M_p (which means an oscillatory closed-loop time response) independent of the PM. A Nyquist plot approaching the -1 point too closely also indicates possible instability in the presence of modeling error and uncertainty. Therefore, both GM and PM should be given in the determination of relative stability. These two values bind the behavior of the closed-loop system near the resonant frequency. Since for most systems there is uncertainty in both the magnitude and phase of the open-loop transfer function $L(s)$, a substantial amount of GM and PM is required in the control design to assure that the possible variations of $L(s)$ will not cause instability of the closed-loop system. For satisfactory performance, the PM should lie between $30°$ and $60°$, and the GM should be greater than $6\,\text{dB}$. For an underdamped second-order system, PM and ζ are related by

$$PM = \tan^{-1} \frac{2\zeta}{\sqrt{\sqrt{4\zeta^4 + 1} - 2\zeta^2}}$$

This gives a close connection between performance and robustness, and allows one to fully specify such a control system by means of phase margin and bandwidth alone. Note that for first- and second-order systems, the Bode phase plot never crosses the $-180°$ line. Thus, GM $= \infty$. In some cases, the gain and phase margins are not helpful indicators of stability. For example, a high-order system may have large GM and PM (both positive); however, its Nyquist plot may still get close enough to the -1 point to incur a large M_p. Such a phenomenon can only occur in high-order systems. Therefore, the designer should check any result obtained by using GM and PM as indicators of relative stability.

9.1.3.2 Sensitivity to Parameters

During the design process, the engineer may want to consider the extent to which changes in system parameters affect the behavior of a system. One of the main advantages of feedback is that it can be used to make the response of a system relatively independent of certain types of changes or inaccuracies in the plant model. Ideally, parameter changes due to heat, humidity, age, or other causes should not appreciably affect a system's performance. The degree to which changes in system parameters affect system transfer functions, and hence performance, is called sensitivity. The greater the sensitivity, the worse is the effect of a parameter change.

A typical closed-loop control system may be modeled as shown in Figure 9.8, where $G_p(s)$ represents the plant or process to be controlled, $G_c(s)$ is the controller, and $H(s)$ may represent the feedback sensor dynamics. The model $G_p(s)$ is usually an approximation to the actual plant dynamic behavior, with parameters at nominal values and high-frequency dynamics neglected. The parameter values in the model are often not precisely known and may also vary widely with operating conditions. For the system given in Figure 9.8, the closed-loop transfer function $G_{cl}(s)$ is

$$G_{cl}(s) = \frac{C(s)}{R(s)} = \frac{G_c(s)G_p(s)}{1 + G_c(s)G_p(s)H(s)}$$

If the loop gain $|L| = |G_c G_p H| \gg 1$, C/R depends almost entirely on the feedback H alone and is virtually independent of the plant and other elements in the forward path and their parameter variations. This is because $|1 + G_c G_p H| \approx |G_c G_p H|$ and $|G_c G_p / G_c G_p H| \approx 1/|H|$. Therefore, the sensitivity of the closed-loop performance to the elements in the forward path reduces as the loop gain increases. This is a major reason for using feedback. With open-loop control (i.e., $H = 0$), $C/R = G_p G_c$. Choice of G_c on the basis of an approximate plant model or a model for which the parameters are subjected to variations will cause errors in C proportional to those in G_p. With feedback, the effects due to approximations and parameter variations in G_p can be greatly reduced. Note that, unlike G_p, the feedback element H is usually under the control of the designer.

The sensitivity of the closed-loop transfer function G_{cl} to changes in the forward path transfer function, especially the plant transfer function G_p, can be defined by

$$S = \frac{\partial G_{cl}}{\partial G_p} \frac{G_p}{G_{cl}} = \frac{1}{1+L}$$

We can plot $S(j\omega)$ as a function of ω, and such a plot shows how sensitivity changes with the frequency of a sinusoidal input R. Obviously, for the sensitivity to be small over a given frequency band, the loop gain L over that band should be large. Generally, for good sensitivity in the forward path of the control system, the loop gain (by definition, the loop gain is $|L(j\omega)|$) is made large over as wide a band of frequencies as possible.

The extent to which the plant model is unknown will be called uncertainty. Uncertainty is another important issue, which designers might need to face. It is known that some uncertainty is always present, both in the environment of the system and in the system itself. We do not know in advance exactly what disturbance and noise signals the system will be subjected to. In the system itself, we know that no mathematical expressions can exactly model a physical system. The uncertainty may be caused by parameter changes, neglected dynamics (especially, high-frequency unmodeled dynamics), or other unspecified effects, which might adversely affect the performance of a control system. Figure 9.10 shows a possible effect of high-frequency plant uncertainty due to dynamics in the high-frequency range being neglected in the nominal plant.

In Figure 9.10, instability can result if an unknown high-frequency resonance causes the magnitude to rise above 1. The likelihood of an unknown resonance in the plant G_p rising above 1 can be reduced if we can keep the loop gain small in the high-frequency range.

In summary, to reduce the sensitivity we need to increase the loop gain. But, in general, increasing the loop gain degrades the stability margins. Hence, we usually have a trade-off between low sensitivity and adequate stability margins.

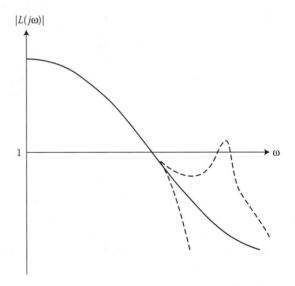

FIGURE 9.10   Effect of high-frequency unmodeled plant uncertainty.

9.1.3.3 Disturbance Rejection and Noise Suppression

All physical systems are subjected to some types of extraneous signals or noise during operation. External disturbances, such as a wind gust acting on an aircraft, are quite common in controlled systems. Therefore, in the design of a control system, consideration should be given so that the system is insensitive to noise and disturbance. The effect of feedback on noise and disturbance depends greatly on where these extraneous signals occur in the system. But in many situations, feedback can reduce the effect of noise and disturbance on system performance. To explain these effects, let us consider a closed-loop unity feedback system as shown in Figure 9.11, where a disturbance $d(t)$ and a sensor noise $n(t)$ have been added to the system.

For simplicity, we assume that the effect of external disturbances is collected and presented at the plant output. The sensor noise, $n(t)$, is introduced into the system via sensors. Both disturbances and sensor noise usually include random high-frequency signals. Let $D(s)$, $N(s)$, $R(s)$, and $Y(s)$ be, respectively, the Laplace transform of the disturbance $d(t)$, sensor noise $n(t)$, system input $r(t)$, and system output $y(t)$. It is easy to find that, by superposition, the total output $Y(s)$ is

$$Y(s) = \frac{G_c(s)G_p(s)}{1 + G_c(s)G_p(s)}R(s) + \frac{1}{1 + G_c(s)G_p(s)}D(s) - \frac{G_c(s)G_p(s)}{1 + G_c(s)G_p(s)}N(s)$$

and the tracking error $e(t)$, defined as $e(t) = r(t) - y(t)$ with its corresponding Laplace transform $E(s)$, becomes

$$E(s) = \frac{1}{1 + G_c(s)G_p(s)}R(s) - \frac{1}{1 + G_c(s)G_p(s)}D(s) + \frac{G_c(s)G_p(s)}{1 + G_c(s)G_p(s)}N(s)$$

In terms of the sensitivity function S and closed-loop transfer function G_{cl} defined in Section 9.1.3.2, the output $Y(s)$ and tracking error $E(s)$ become

$$Y(s) = G_{cl}(s)R(s) + S(s)D(s) - G_{cl}(s)N(s)$$
$$E(s) = S(s)R(s) - S(s)D(s) + G_{cl}(s)N(s)$$

Note that the transfer function G_{cl} is also called the complementary sensitivity function because S and G_{cl} are related by $S(s) + G_{cl}(s) = 1$ for all frequencies. It is clear that $S(s)$ must be kept small to minimize the effects of disturbances. From the definition of S, this can be achieved if the loop gain (i.e., $|L(j\omega)| = |G_c(j\omega)G_p(j\omega)|$) is large. $|G_{cl}(j\omega)|$ must be kept small to reduce the effects of sensor noise on the system's output, and this can be achieved if the loop gain is small. For good tracking, $S(s)$ must be small, which implies that the loop gain should be large over the frequency band of the input signal $r(t)$. Tracking and disturbance rejection require small S, while noise suppression requires small G_{cl}. From the relation between S and G_{cl}, clearly, we cannot reduce both functions simultaneously. However, in practice, disturbances and commands are often low-frequency signals, whereas sensor noises are often high-frequency signals. Therefore, we can still meet both objectives by keeping S small in the low-frequency range and G_{cl} small in the high-frequency range.

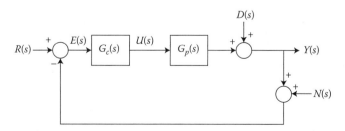

FIGURE 9.11 A unity feedback control system showing sources of noise, $N(s)$, and disturbance, $D(s)$.

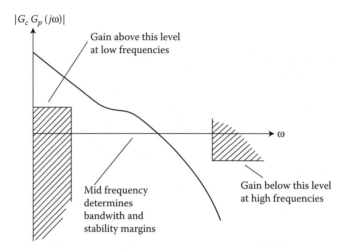

FIGURE 9.12 Desirable shape for the open-loop frequency response.

Putting together the requirements of reducing the sensitivity to parameters, disturbance rejection, and noise suppression, we arrive at a general desired shape for the open-loop transfer function, which is as shown in Figure 9.12.

The general features of the open-loop transfer function are that the gain in the low-frequency region should be large enough, and in the high-frequency region, the gain should be attenuated as much as possible. The gain at intermediate frequencies typically controls the gain and phase margins. Near the gain crossover frequency ω_c, the slope of the log-magnitude curve in the Bode plot should be close to -20 dB/decade (i.e., the transition from the low- to high-frequency range must be smooth). Note that the PM of the feedback system is $180° + \phi_c$ with $\phi_c = \angle\ G_c(j\omega_c)G_p(j\omega_c)$. If the loop transfer function $L(j\omega) = G_c(j\omega)G_p(j\omega)$ is stable, proper, and minimum phase, then ϕ_c is uniquely determined from the gain plot of G_cG_p (i.e., $|G_c(j\omega)G_p(j\omega)|$). Bode actually showed that ϕ_c is given, in terms of the weighted average attenuation rate of $|G_cG_p|$, by

$$\phi_c = \frac{1}{\pi} \int_{-\infty}^{\infty} \frac{d \ln |G_c(j\omega(\mu))G_p(j\omega(\mu))|}{d\mu} (\ln \coth \frac{|\mu|}{2})\, d\mu$$

where $\mu = \ln\ (\omega/\omega_c)$. And ϕ_c is large if $|G_cG_p|$ attenuates slowly and small if it attenuates rapidly. Therefore, a rapid attenuation of $|G_cG_p|$ at or near crossover frequency will decrease the PM and a more than -20 dB/decade slope near ω_c indicates that PM is inadequate. Controlling ϕ_c is important because it is related to the system stability and performance measure.

Based on the loop gain shown in Figure 9.12, therefore, the desirable shapes for the sensitivity and complementary sensitivity functions of a closed-loop system should be similar to those shown in Figure 9.13. That is, the sensitivity function S must be small at low frequencies and roll off to 1 (i.e., 0 dB) at high frequencies, whereas G_{cl} must be 1 (0 dB) at low frequencies and get small at high frequencies.

Notice that Figures 9.12 and 9.13 can be viewed as specifications for a control system. Such graphical specifications of the "loop shape" are suitable for today's computer-aided design packages, which have extensive graphical capabilities. As will be seen shortly, these specifications on the loop shape are particularly easy to adapt to MIMO systems.

In order to achieve the desired performance shown in Figures 9.12 and 9.13, a "loop shaping" method, which presents a graphical technique for designing a controller to achieve robust performance, may be considered. The idea of loop shaping is to design the Bode magnitude plot of the loop transfer function $L(j\omega) = G_c(j\omega)G_p(j\omega)$ to achieve (or at least approximate) the requirements shown in Figure 9.12, and then to back-solve for the controller from the loop transfer function. In other words, we first convert

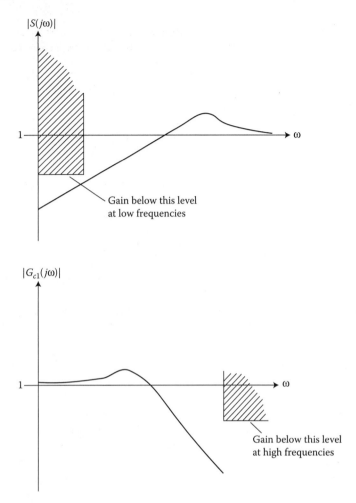

FIGURE 9.13 Desirable shape for the sensitivity and complementary sensitivity functions.

performance specifications on $|S(j\omega)|$ and $|G_{cl}(j\omega)|$ (as given in Figure 9.13) into specifications on $|L(j\omega)|$ (as shown in Figure 9.12). We then shape $|L(j\omega)|$ to make it lie above the first constraint curve ($|L(j\omega)| \gg 1$) at low frequencies, lie below the second constraint curve ($|L(j\omega)| \ll 1$), and roll off as fast as possible in the high-frequency range, and make a smooth transition from low to high frequency, i.e., keep the slope as gentle as possible (about -20 dB/decade) near the crossover frequency ω_c. In loop shaping, the resulting controller has to be checked in the closed-loop system to see whether a satisfactory trade-off between $|S(j\omega)|$ and $|G_{cl}(j\omega)|$ has been reached. Note that it is also possible to directly shape $|S(j\omega)|$ and/or $|G_{cl}(j\omega)|$.

9.1.3.4 Control Effort

It is important to be aware of the limits on actuator signals in specifying controllers. Most actuators have upper limits on their magnitudes and on their rates of change. These limits can severely constrain the performance and robustness of the closed-loop system. For example, actuator saturation can cause instability in conditionally stable systems and very poor transient response because of integral windup (see Chapter 19.1).

The problem should be addressed in two places. The actuator specifications should state the allowable limits on the magnitudes and rates of actuator signals. Proper specification of the actuators greatly facilitates the rest of the design. Secondly, the response of the closed-loop system to "large" inputs should be specified. Such specifications ensure that, if it is necessary to saturate the actuators, the performance and robustness of the closed-loop system remain satisfactory.

9.1.4 Miscellaneous Specifications

There are many other aspects of a control system that are often specified. There are usually constraints on the allowable cost of the controller. In some applications, especially spacecraft, the size, weight, and power required for the controller's operation are restricted.

Control system reliability is also often specified. The simplest such specification is the life expectancy of the controller. This is usually given as the mean time before failure (MTBF). The allowable ways in which a controller may fail are also often specified, especially in applications involving humans. For example, a control system may be required to "fail safe." A stronger requirement is "fail soft."

A good example of a control system that is designed to "fail soft" is the controller that regulates the height of the water in a toilet tank. Sooner or later, the valve that stops the flow of water when the tank is full gets stuck open. This would cause the tank to overflow, creating a mess. The MTBF is a few years, not very long in comparison to the life of the whole system. This is acceptable because the control system includes an overflow tube that causes the overflow to go into the toilet and, from there, into the sewer. This is a soft failure. The controller does not work properly. Water is wasted. But, the failure does not create a serious problem.

9.1.5 Performance Specifications for MIMO LTI Systems

From the discussion given for SISO LTI systems, it is clear that the open-loop transfer function $L(j\omega)$ plays an essential role in determining various performance and robustness properties of the closed-loop system. For MIMO systems, the inputs and outputs are generally interacting. Due to such interactions, it can be difficult to control a MIMO system. However, the classical Bode gain/phase plots can be generalized for MIMO systems. In the following, we describe performance specifications for MIMO systems.

The responses of a MIMO system are generally coupled. That is, every input affects more than one output, and every output is influenced by more than one input. If a controller can be found such that every input affects one and only one output, then we say the MIMO system is decoupled. Exact decoupling can be difficult, if not impossible, to achieve in practice. There are various ways to specify approximate decoupling. Because decoupling is most obvious for square transfer functions, let $G(j\omega)$ be a strictly proper $m \times m$ transfer function and let $g_{ij}(j\omega)$ denote the ijth element of the matrix $G(j\omega)$. Then the requirement that

$$|g_{ij}(j\omega)| < \delta \quad \text{for all } 0 \le \omega < \infty; \quad i,j = 1, 2, \ldots, m; \quad i \ne j$$

would force $G(j\omega)$ to be approximately decoupled provided δ were small enough. Such a specification might be defective because it allows the diagonal transfer functions, the $g_{ii}(j\omega)$, $i = 1, 2, \ldots, m$, to be arbitrary. Typically, one wants $|g_{ii}(j\omega)|$, $i = 1, 2, \ldots, m$ close to one, at least in some range $\omega_1 \le \omega \le \omega_2$. The requirement that

$$\frac{|g_{ij}(j\omega)|}{|g_{ii}(j\omega)|} < \delta \quad \text{for all } 0 \le \omega_1 \le \omega \le \omega_2; \quad i,j = 1, 2, \ldots, m; \quad i \ne j$$

forces the $g_{ij}(j\omega)$ to be small relative to the $g_{ii}(j\omega)$. Of course, nothing prevents adding SISO specifications on the $g_{ii}(j\omega)$ to the decoupling specifications.

Another useful decoupling specification is diagonal dominance. A strictly proper square $m \times m$ transfer function $G(j\omega)$ is said to be row diagonal dominant if

$$\sum_{j=1, j \neq i}^{m} |g_{ij}(j\omega)| < |g_{ii}(j\omega)| \quad \text{for all } i = 1, 2, \ldots, m \quad \text{and} \quad \text{all } 0 \leq \omega < \infty$$

Column diagonal dominance is defined in the obvious way. Decoupling specifications can also be written in the time domain as limits on the impulse or step responses of the closed-loop system.

Decoupling is not always necessary or desirable. Thus, it is necessary to have other ways to specify the performance of MIMO controlled systems. One effective way to do this is by means of the singular value decomposition (SVD). It is known that the SVD is a useful tool in linear algebra, and it has found many applications in control during the past decade. Let G be an $m \times n$ complex (constant) matrix. Then the positive square roots of the eigenvalues of G^*G (where G^* means the complex conjugate transpose of G) are called the singular values of G. These square roots are always real numbers because the eigenvalues of G^*G are always real and ≥ 0. The maximum and minimum singular values of G are denoted by $\bar{\sigma}(G)$ and $\underline{\sigma}(G)$, respectively; and they can also be expressed by

$$\bar{\sigma}(G) = max_{||u||=1} ||Gu||$$

$$\underline{\sigma}(G) = min_{||u||=1} ||Gu||$$

where $u \in C^n$ and the vector norm $|| \cdot ||$ is the Euclidean norm. That is, $\bar{\sigma}(G)$ and $\underline{\sigma}(G)$ are the maximum and minimum gains of the matrix G. For a square G, $\underline{\sigma}(G)$ is a measure of how far G is from singularity, and $\bar{\sigma}(G)/\underline{\sigma}(G)$ is the condition number which is a measure of the difficulty of inverting G. The best way to compute the SVD is by means of an algorithm also known as the singular value decomposition (SVD) (see *Control System Fundamentals*, Chapter 3 and *Control System Advanced Methods*, Chapter 1).

For MIMO systems, the transfer function matrices evaluated at $s = j\omega$ have proven useful in resolving the complexities of MIMO design. The idea is to reduce the transfer function matrices to two critical gains versus frequency, that is, the maximum and minimum singular values of the transfer function matrix. Consider a MIMO system represented by a transfer matrix $G(s)$. Similar to the constant matrix case discussed, if we let $s = j\omega$ $(0 \leq \omega < \infty)$, then the singular values $\sigma_i(G(j\omega))$ will be functions of ω, and a plot of $\sigma_i(G(j\omega))$ is called a singular value plot (or σ-plot) which is analogous to a Bode magnitude plot of a SISO transfer function. The maximum and minimum singular values of $G(j\omega)$ are defined, respectively, as

$$\bar{\sigma}(G(j\omega)) = \sqrt{\lambda_{max}[G(j\omega)^*G(j\omega)]}$$

$$\underline{\sigma}(G(j\omega)) = \sqrt{\lambda_{min}[G(j\omega)^*G(j\omega)]}$$

where $\lambda[\cdot]$ denotes eigenvalues. Note that the H_∞ norm of a stable transfer matrix G is the maximum of $\bar{\sigma}(G)$ over all frequencies, i.e., $||G||_\infty = sup_{\omega \geq 0} \bar{\sigma}(G(j\omega))$. For the performance and robustness measures of MIMO systems, we can examine $\bar{\sigma}$ and $\underline{\sigma}$ in a manner similar to that used to examine the frequency–response magnitude of a SISO transfer function. If $\bar{\sigma}(G)$ and $\underline{\sigma}(G)$ are very close to each other, then we can simply treat the system like a SISO system. In general, they are not close and $\bar{\sigma}$ is important to bound the performance requirements.

Without loss of generality, consider the MIMO unity feedback system shown in Figure 9.11. Like the SISO case, define the sensitivity and complementary sensitivity matrices as $S(s) = [I + G_p(s)G_c(s)]^{-1}$ (where $(\cdot)^{-1}$ means the matrix inverse) and $G_{cl}(s) = I - S(s)$, respectively. Note that $G_{cl}(s) = [I + G_p(s)G_c(s)]^{-1}G_p(s)G_c(s)$ is the closed-loop transfer function matrix which describes the system's input-output relationship. From Figure 9.11, we have

$$Y(s) = G_{cl}(s)R(s) + S(s)D(s) - G_{cl}(s)N(s)$$

$$E(s) = S(s)R(s) - S(s)D(s) + G_{cl}(s)N(s)$$

Similar to the arguments stated for SISO systems, for disturbance rejection, tracking error reduction, and insensitivity to plant parameter variations, we need to make $S(j\omega)$ "small" over the frequency range (say, $0 \leq \omega \leq \omega_0$) where the commands, disturbances, and parameter changes of G_p are significant. That is, to keep $\bar{\sigma}[(I + G_pG_c)^{-1}]$ as small as possible, or $\underline{\sigma}[I + G_pG_c]$ as large as possible $\forall\ \omega \leq \omega_0$ since $\bar{\sigma}[(I + G_pG_c)^{-1}] = 1/\underline{\sigma}[I + G_pG_c]$. The inequalities $\max(0, \underline{\sigma}[G_pG_c] - 1) \leq \underline{\sigma}[I + G_pG_c] \leq \underline{\sigma}[G_pG_c] + 1$ further imply that the loop gain $\underline{\sigma}[G_pG_c]$ should be made as large as possible $\forall\ \omega \leq \omega_0$.

From the equations shown, it is also clear that for sensor noise reduction, the transfer matrix $G_{cl}(j\omega)$ should be "small" over the frequency range (say, $\omega \geq \omega_1$) where the noise is significant. Using the equalities $(I + X)^{-1}X = (I + X^{-1})^{-1}$ and $\bar{\sigma}(X)\underline{\sigma}(X^{-1}) = 1$ (assume X^{-1} exists), this implies that $\bar{\sigma}[(I + G_pG_c)^{-1}G_pG_c] = \bar{\sigma}[(I + (G_pG_c)^{-1})^{-1}] = 1/\underline{\sigma}[I + (G_pG_c)^{-1}]$ should be small, or $\underline{\sigma}[I + (G_pG_c)^{-1}]$ should be large $\forall\ \omega \geq \omega_1$. Since $\underline{\sigma}[I + (G_pG_c)^{-1}] \geq \underline{\sigma}[(G_pG_c)^{-1}]$, we should make $\underline{\sigma}[(G_pG_c)^{-1}]$ large, which further means that $\bar{\sigma}[G_pG_c]$ should be kept as small as possible $\forall\ \omega \geq \omega_1$.

In summary, disturbance rejection, tracking error, and sensitivity (to plant parameter variations) reduction require large $\underline{\sigma}[G_pG_c]$, while sensor noise reduction requires small $\bar{\sigma}[G_pG_c]$. Since $\bar{\sigma}[G_pG_c] \geq \underline{\sigma}[G_pG_c]$, a conflict between the requirements of $S(j\omega)$ and $G_{cl}(j\omega)$ exists, which is exactly the same as that discussed for SISO systems. Therefore, a performance trade-off for command tracking and disturbance reduction versus sensor noise reduction is unavoidable. Typical specifications for MIMO system loop gain requirements will be similar to those of Figure 9.12, with the "low-frequency boundary (or constraint)" being the lower boundary for $\underline{\sigma}[G_pG_c]$ when ω is small, and the "high-frequency boundary" representing the upper boundary for $\bar{\sigma}[G_pG_c]$ when ω becomes large. In other words, we need "loop shaping" of the singular values of the plant transfer function matrix G_p by using G_c (i.e., the design of controller G_c) so that the nominal closed-loop system is stable, $\underline{\sigma}[G_pG_c]$ (thus, $\bar{\sigma}[G_pG_c]$) at low frequencies lies above the low-frequency boundary, and $\bar{\sigma}[G_pG_c]$ (thus, $\underline{\sigma}[G_pG_c]$) at high frequencies lies below the high-frequency boundary. That is, we want to increase the low-frequency value of $\underline{\sigma}$ (thus, $\bar{\sigma}$) of G_pG_c to ensure adequate attenuation of (low-frequency) disturbances and better command tracking, and roll-off $\bar{\sigma}$ and $\underline{\sigma}$ in the high-frequency range to ensure robust stability. Note that the "forbidden" areas shown in Figure 9.12 are problem dependent, and they may be constructed from the design specifications.

It is generally desirable to require that the gap between $\bar{\sigma}[G_pG_c]$ and $\underline{\sigma}[G_pG_c]$ be fairly small and that their slope be close to -20 dB/decade near the gain crossover frequencies. Here, there are a range of gain crossover frequencies from the frequency $\underline{\omega}_c$ at which $\underline{\sigma}[G_p(j\omega)G_c(j\omega)] = 1$ to the frequency $\bar{\omega}_c$ at which $\bar{\sigma}[G_p(j\omega)G_c(j\omega)] = 1$. As for SISO systems, the requirements near the crossover frequencies primarily address robustness.

9.1.6 Robustness Specifications for MIMO LTI Systems

It is known that, in control system design, the plant model used is only an approximate representation of the physical system. The discrepancies between a system and its mathematical representation (model) may lead to a violation of some performance specification, or even to closed-loop instability. We say the system is robust if the design performs satisfactorily under variations in the dynamics of the plant (including parameter variations and various possible uncertainties). Stability and performance robustness are two important issues that should be considered in control design. Generally, the form of the plant uncertainty can be parametric, nonparametric, or both. Typical sources of uncertainty include unmodeled high-frequency dynamics, neglected nonlinearities, plant parameter variations (due to changes of environmental factors), etc. The parametric uncertainty in the plant model can be expressed as $G_p(s, \gamma)$, a parameterized model of the nominal plant $G_p(s)$ with the uncertain parameter γ. The three most commonly used models to represent unstructured uncertainty are additive, input multiplicative, and output multiplicative types; and can be represented, respectively, as

$$\tilde{G}_p(s) = G_p(s) + \Delta_a(s)$$
$$\tilde{G}_p(s) = G_p(s)[I + \Delta_i(s)]$$

$$\tilde{G}_p(s) = [I + \Delta_o(s)]G_p(s)$$

where $\tilde{G}_p(s)$ is the plant transfer matrix as perturbed from its nominal model $G_p(s)$ due to the uncertainty Δ (Δ_a or Δ_i or Δ_o) with $\bar{\sigma}[\Delta(j\omega)]$ bounded above. That is, we have $\bar{\sigma}[\Delta(j\omega)] \leq l(\omega)$, where $l(\omega)$ is a known positive real scalar function. In general, $l(\omega)$ is small at low frequencies because we can model the plant more accurately in the low-frequency range. The plant high-frequency dynamics are less known, resulting in large $l(\omega)$ at high frequencies. Note that Δ_i (Δ_o) assumes all the uncertainty occurred at the plant input (output). Both Δ_i and Δ_o represent relative deviation while Δ_a represents absolute deviation from $G_p(s)$. Of course, there are many possible ways to represent uncertainty, for example, $\tilde{G}_p(s) = [I + \Delta_o(s)]G_p(s)[I + \Delta_i(s)]$, $\tilde{G}_p(s) = [N(s) + \Delta_N(s)][D(s) + \Delta_D(s)]^{-1}$ (a matrix fractional representation with $G_p(s) = N(s)D^{-1}(s)$), etc. We may even model the plant by combining both parametric and unstructured uncertainties in the form of $\tilde{G}_p(s) = [I + \Delta_o(s)]G_p(s, \gamma)[I + \Delta_i(s)]$.

In the following, we examine the robustness of performance together with robust stability under output multiplicative uncertainty Δ_o. Similar results can be derived if other types of unstructured uncertainty are used. Via a multivariable version of the standard Nyquist criterion, the closed-loop system will remain stable under the uncertainty Δ_o if

$$\bar{\sigma}[G_p G_c (I + G_p G_c)^{-1}(j\omega)] < \frac{1}{\bar{\sigma}(\Delta_o)} \quad \forall \omega \geq 0$$

where we assume that the nominal closed-loop system is stable, and $G_p(s)$, $\tilde{G}_p(s)$ have the same number of unstable poles. The above condition is actually necessary and sufficient for robust stability. Note that the expression inside $\bar{\sigma}[\cdot]$ in the left-hand side of the inequality is the complementary sensitivity matrix $G_{cl}(j\omega)$. Thus, the model uncertainty imposes an upper bound on the singular values of $G_{cl}(s)$ for robust stability. Rewriting the inequality, we get

$$\bar{\sigma}(\Delta_o) < \frac{1}{\bar{\sigma}[G_p G_c (I + G_p G_c)^{-1}(j\omega)]} = \underline{\sigma}[I + (G_p G_c)^{-1}(j\omega)] \quad \forall \omega \geq 0$$

where we assume that $G_p G_c$ is invertible. Obviously, $\underline{\sigma}[I + (G_p G_c)^{-1}]$ can be used as a measure of the degree of stability of the feedback system. This is a multivariable version of SISO stability margins (i.e., gain and phase margins) because it allows gain and phase changes in each individual output channel and/or simultaneous gain and phase changes in several channels. The extent to which these changes are allowed is determined by the inequality shown above. Therefore, we can use the singular values to define GM for MIMO systems.

In the high-frequency range (where the loop gain is small), from the inequality given above, we have

$$\bar{\sigma}[G_p(j\omega)G_c(j\omega)] < \frac{1}{\bar{\sigma}(\Delta_o)}$$

This is a constraint on the loop gain $G_p G_c$ for robust stability. This constraint implies that at high frequencies the loop gain on $G_p G_c$ (i.e., $\bar{\sigma}[G_p G_c]$) should lie below a certain limit for robust stability (the same argument described in the previous section). Obviously, satisfaction of the high-frequency boundary shown in Figure 9.12 is mandatory for robust stability (not just desired for sensor noise reduction !).

For robust performance (i.e., good command tracking, disturbance reduction, and insensitivity to plant parameter variations (i.e., structured uncertainty) under all possible $\tilde{G}_p(s)$), we should keep $\bar{\sigma}[(I + \tilde{G}_p G_c)^{-1}]$ of the "perturbed" sensitivity matrix $\tilde{S}(s)$ as small as possible in the low-frequency range. Following the same argument given in the previous section, this further implies that $\underline{\sigma}[\tilde{G}_p G_c]$ should be large at low frequencies. Making $\underline{\sigma}[G_p G_c]$ large enough ensures that $\underline{\sigma}[\tilde{G}_p G_c]$ is large. Clearly, high loop gain can compensate for model uncertainty. Therefore, for robust performance the low-frequency boundary given in Figure 9.12 should be satisfied, although this is not mandatory.

For MIMO systems, as for the requirements on the loop gain in SISO control design, any violation of the low-frequency boundary in Figure 9.12 constitutes a violation of the robust performance specifications, while a violation of the upper boundary (in the high-frequency range) leads to a violation of the robust stability specifications. The main distinction between MIMO and SISO design is the use of singular values of the transfer function matrix to express the "size" of functions.

9.1.7 Conclusions

Writing specifications for control systems is not easy. The different aspects of controller performance and robustness are interrelated and, in many cases, competing. While a good controller can compensate for some deficiencies in the plant (system to be controlled), the plant implies significant limits on controller performance. It is all too easy to impose unachievable specifications on a controller.

For these reasons, it is important to have the plant designer, the person responsible for the controller specifications, and the control designer work together from the beginning to the end of any project with demanding control requirements.

Further Reading

Most undergraduate texts on control systems contain useful descriptions of performance and robustness specifications for SISO LTI systems. Three examples are

1. Franklin, G. F., Powell, J. D., and Emami-Naeni, A., *Feedback Control of Dynamic Systems*, 4th ed., Prentice-Hall, Englewood Cliffs, NJ, 2002.
2. Nise, N. S., *Control Systems Engineering*, 4th ed., John Wiley & Sons Inc., Hoboken, NJ, 2004.
3. Dorf, R. C. and Bishop, R. H., *Modern Control Systems*, 11th ed., Pearson Prentice Hall, Upper Saddle River, NJ, 2008.

Several textbooks have been published recently that include a more loop shaping-oriented discussion of control specifications. These include

4. Boyd, S. P. and Barratt, C. H., *Linear Controller Design—Limits of Performance*, Prentice-Hall, Englewood Cliffs, NJ, 1991.
5. Doyle, J. C., Francis, B. A., and Tannenbaum, A. R., *Feedback Control Theory*, Macmillan, New York, NY, 1992.
6. Belanger, P. R., *Control Engineering—A Modern Approach*, Saunders College Publishing, New York, 1995.
7. Wolovich, W. A., *Automatic Control Systems—Basic Analysis and Design*, Saunders College Publishing, New York, 1994.

Specifications for MIMO LTI systems are covered in

8. Maciejowski, J. M., *Multivariable Feedback Design*, Addison-Wesley, Reading, MA, 1989.
9. Green, M. and Limebeer, D. J. N., *Linear Robust Control*, Prentice-Hall, Englewood Cliffs, NJ, 1995.

The limitations on control systems, including Bode's results on the relation between gain and phase, are described in

10. Freudenberg, J. S. and Looze, D. P., *Frequency Domain Properties of Scalar and Multivariable Feedback Systems*, Springer-Verlag, New York, NY, 1988.

A good, detailed example of the specifications for a complex MIMO control system is

11. Hoh, R. H., Mitchell, D. G., and Aponso, B. L., Handling Qualities Requirements for Military Rotorcraft, Aeronautical Design Standard, ADS-33C, US Army ASC, Aug. 1989.
12. Hoh, R. H., Mitchell, D. G., Aponso, B. L., Key, D. L., and Blanken, C. L., Background Information and User's Guide for Handling Quantities Requirements for Military Rotorcraft, USAAVSCOM TR89-A-8, US Army ASC, Dec. 1989.

Finally, all three publications of the IEEE Control Systems Society, the Transactions on Automatic Control, the Transactions on Control Systems Technology, and the Control Systems Magazine regularly contain articles on the design of control systems.

9.2 Design Using Performance Indices

Richard C. Dorf and Robert H. Bishop

9.2.1 Introduction

Modern control theory assumes that the systems engineer can specify quantitatively the required system performance. Then a *performance index** can be calculated or measured and used to evaluate the system's performance. A quantitative measure of the performance of a system is necessary for automatic parameter optimization of a control system and for the design of optimum systems.

We consider a feedback system as shown in Figure 9.14 where the closed-loop transfer function is

$$\frac{C(s)}{R(s)} = T(s) = \frac{G_c(s)G(s)}{1 + G_c(s)G(s)}. \tag{9.1}$$

Whether the aim is to improve the design of a system or to design a control system, a performance index may be chosen [1]:

The system is considered an *optimum control system* when the system parameters are adjusted so that the index reaches an extremum value, commonly a minimum value. A performance index, to be useful, must be a number that is always positive or zero. Then the best system is defined as the system that minimizes this index.

It is also possible to design an optimum system to achieve a *deadbeat response*, which is characterized by a fast response with minimal overshoot. The desired closed-loop system characteristic equation coefficients are selected to minimize settling time and rise time.

9.2.2 The ISE Index

One performance index is the integral of the square of the error (ISE), which is defined as

$$ISE = \int_0^T e^2(t)\, dt. \tag{9.2}$$

The upper limit T is a finite time chosen somewhat arbitrarily so that the integral approaches a steady-state value. It is usually convenient to choose T as the settling time, T_s. This criterion will discriminate between excessively overdamped and excessively underdamped systems. The minimum value of the integral occurs for a compromise value of the damping. The performance index of Equation 9.2 is easily adapted for practical measurements because a squaring circuit is readily obtained. Furthermore, the squared error is mathematically convenient for analytical and computational purposes.

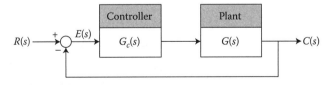

FIGURE 9.14 Feedback control system.

---

* A performance index is a quantitative measure of the performance of a system and is chosen so that emphasis is given to the important system specifications.

9.2.3 The ITAE Index

To reduce the contribution of the relatively large initial error to the value of the performance integral, as well as to emphasize errors occurring later in the response, the following index has been proposed [2]:

$$ITAE = \int_0^T t|e(t)| \, dt, \tag{9.3}$$

where ITAE is the integral of time multiplied by the absolute magnitude of the error.

The coefficients that will minimize the ITAE performance criterion for a step input have been determined for the general closed-loop transfer function [2]

$$T(s) = \frac{C(s)}{R(s)} = \frac{b_0}{s^n + b_{n-1}s^{n-1} + \cdots + b_1 s + b_0}. \tag{9.4}$$

This transfer function has a steady-state error equal to zero for a step input. Note that the transfer function has n poles and no zeros. The optimum coefficients for the ITAE criterion are given in Table 9.1 for a step input. The transfer function, Equation 9.4, implies the plant and controller $G_c(s)G(s)$ have one or more pure integrations to provide zero steady-state error. The responses using optimum coefficients for a step input are given in Figure 9.15 for ITAE. The responses are provided for normalized time, $\omega_n t$. Other standard forms based on different performance indices are available and can be useful in aiding the designer to determine the range of coefficients for a specific problem.

For a ramp input, the coefficients have been determined that minimize the ITAE criterion for the general closed-loop transfer function [2]:

$$T(s) = \frac{b_1 s + b_0}{s^n + b_{n-1}s^{n-1} + \cdots + b_1 s + b_0}. \tag{9.5}$$

This transfer function has a steady-state error equal to zero for a ramp input. The optimum coefficients for this transfer function are given in Table 9.2. The transfer function, Equation 9.5, implies that the plant and controller $G_c(s)G(s)$ have two or more pure integrations, as required to provide zero steady-state error.

9.2.4 Normalized Time

We consider the transfer function of a closed-loop system, $T(s)$. To determine the coefficients that yield the optimal *deadbeat response*, the transfer function is first normalized. An example of this for a third-order system is

$$T(s) = \frac{\omega_n^3}{s^3 + \alpha\omega_n s^2 + \beta\omega_n^2 s + \omega_n^3}. \tag{9.6}$$

Dividing the numerator and denominator by ω_n^3 yields

$$T(s) = \frac{1}{\frac{s^3}{\omega_n^3} + \alpha\frac{s^2}{\omega_n^2} + \beta\frac{s}{\omega_n} + 1}. \tag{9.7}$$

TABLE 9.1 The Optimum Coefficients of $T(s)$ Based on the ITAE Criterion for a Step Input

$$s + \omega_n$$
$$s^2 + 1.4\omega_n s + \omega_n^2$$
$$s^3 + 1.75\omega_n s^2 + 2.15\omega_n^2 s + \omega_n^3$$
$$s^4 + 2.1\omega_n s^3 + 3.4\omega_n^2 s^2 + 2.7\omega_n^3 s + \omega_n^4$$
$$s^5 + 2.8\omega_n s^4 + 5.0\omega_n^2 s^3 + 5.5\omega_n^3 s^2 + 3.4\omega_n^4 s + \omega_n^5$$
$$s^6 + 3.25\omega_n s^5 + 6.6\omega_n^2 s^4 + 8.6\omega_n^3 s^3 + 7.45\omega_n^4 s^2 + 3.95\omega_n^5 s + \omega_n^6$$

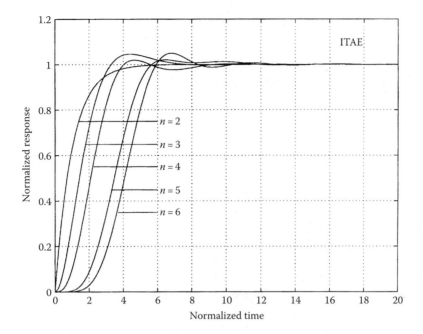

FIGURE 9.15 The step response of a system with a transfer function satisfying the ITAE criterion.

Let $\bar{s} = s/\omega_n$ to obtain

$$T(s) = \frac{1}{\bar{s}^3 + \alpha \bar{s}^2 + \beta \bar{s} + 1}. \tag{9.8}$$

Equation 9.8 is the normalized third-order closed-loop transfer function. For a higher-order system, the same method is used to derive the normalized equation. When we let $\bar{s} = s/\omega_n$, this has the effect in the time-domain of normalizing time, $\omega_n t$. The step response for a normalized system is as shown in Figure 9.15.

9.2.5 Deadbeat Response

Often the goal for a control system is to achieve a fast response to a step command with minimal overshoot. We define a *deadbeat response* as a response that proceeds rapidly to the desired level and holds at that level with minimal overshoot. We use the $\pm 2\%$ band at the desired level as the acceptable range of variation from the desired response. Then, if the response enters the band at time T_s, it has satisfied the settling time T_s upon entry to the band, as illustrated in Figure 9.16. A deadbeat response has the following characteristics:

1. Zero steady-state error
2. Fast response $\rightarrow$ minimum rise time and settling time

TABLE 9.2 The Optimum Coefficients of $T(s)$ Based on the ITAE Criterion for a Ramp Input

$$s^2 + 3.2\omega_n s + \omega_n^2$$
$$s^3 + 1.75\omega_n s^2 + 3.25\omega_n^2 s + \omega_n^3$$
$$s^4 + 2.41\omega_n s^3 + 4.39\omega_n^2 s^2 + 5.14\omega_n^3 s + \omega_n^4$$
$$s^5 + 2.19\omega_n s^4 + 6.5\omega_n^2 s^3 + 6.3\omega_n^3 s^2 + 5.24\omega_n^4 s + \omega_n^5$$

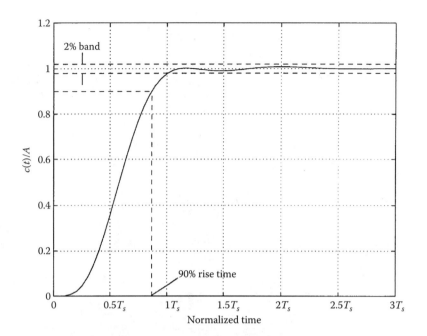

FIGURE 9.16 The deadbeat response, where A is the magnitude of the step input.

3. $0.1\% \leq$ percent overshoot $<2\%$
4. Percent undershoot $<2\%$

Characteristics 3 and 4 require that the response remain within the $\pm2\%$ band so that the entry to the band occurs at the settling time.

A more general normalized transfer function of a closed-loop system may be written as

$$T(s) = \frac{1}{\bar{s}^n + \alpha\bar{s}^{n-1} + \beta\bar{s}^{n-2} + \gamma\bar{s}^{n-3} + \cdots + 1}. \tag{9.9}$$

The coefficients of the denominator equation (α, β, γ, and so on) are then assigned the values necessary to meet the requirement of deadbeat response. The coefficients recorded in Table 9.3 were selected to achieve deadbeat response and to minimize settling time and rise time to 100% of the desired command. The form of Equation 9.9 is normalized since $\bar{s} = s/\omega_n$. Thus, we choose ω_n based on the desired settling time or rise time. For example, if we have a third-order system with a required settling time of 1.2 seconds, we note from Table 9.3 that the normalized settling time is

$$\omega_n T_s = 4.04.$$

Therefore, we require

$$\omega_n = \frac{4.04}{T_s} = \frac{4.04}{1.2} = 3.37.$$

Once ω_n is chosen, the complete third-order closed-loop transfer function is known, having the form of Equation 9.6, where $\alpha = 1.9$ and $\beta = 2.2$. When designing a system to obtain a deadbeat response, the compensator is chosen and the closed-loop transfer function is found. This compensated transfer function is then set equal to Equation 9.9 and the required compensator parameters can be determined.

TABLE 9.3 Coefficients and Response Measures of a Deadbeat System

System Order	Coefficients					Percent Over-shoot P.O.	Percent Under-shoot P.U.	90% Rise Time T_{r90}	100% Rise Time T_r	Settling Time T_s
	α	β	γ	δ	ϵ					
2nd	1.82					0.10%	0.00%	3.47	6.58	4.82
3rd	1.90	2.20				1.65%	1.36%	3.48	4.32	4.04
4th	2.20	3.50	2.80			0.89%	0.95%	4.16	5.29	4.81
5th	2.70	4.90	5.40	3.40		1.29%	0.37%	4.84	5.73	5.43
6th	3.15	6.50	8.70	7.55	4.05	1.63%	0.94%	5.49	6.31	6.04

Note: All time is normalized.

Example 9.1:

Consider a sample plant

$$G(s) = \frac{1}{s(s+p)} \qquad (9.10)$$

and a controller $G_c(s) = K$. The goal is to select the parameters p and K to yield (1) an ITAE response, and alternatively (2) a deadbeat response. In addition, the settling time for a step response is specified as less than 1 second. The closed-loop transfer function is

$$T(s) = \frac{K}{s^2 + ps + K}. \qquad (9.11)$$

For the ITAE system we examine Figure 9.15 and note that for $n = 2$ we have $\omega_n T_s = 8$. Then, for $T_s = 0.8$, we use $\omega_n = 10$. From Table 9.1, we require

$$s^2 + 1.4\omega_n s + \omega_n^2 = s^2 + ps + K.$$

Therefore, we have $K = \omega_n^2 = 100$ and $p = 1.4\omega_n = 14$.

If we seek a deadbeat response, we use Table 9.3 and note that the normalized settling time is

$$\omega_n T_s = 4.82.$$

In order to obtain T_s less than 1 second, we use $\omega_n = 6$ and $T_s = 0.8$. Then we use Table 9.3 to obtain $\alpha = 1.82$. The closed-loop transfer function is

$$T(s) = \frac{\omega_n^2}{s^2 + \alpha\omega_n s + \omega_n^2} = \frac{36}{s^2 + 10.9s + 36}. \qquad (9.12)$$

Then we require $K = 36$ and $p = 10.9$. The deadbeat and the ITAE step response are shown in Figure 9.17.

9.2.6 Conclusions

The design of a feedback system using performance indices or deadbeat control leads to predictable system responses. This permits the designer to select system parameters to achieve desired performance.

9.2.7 Defining Terms

Performance index: A quantitative measure of the performance of a system.

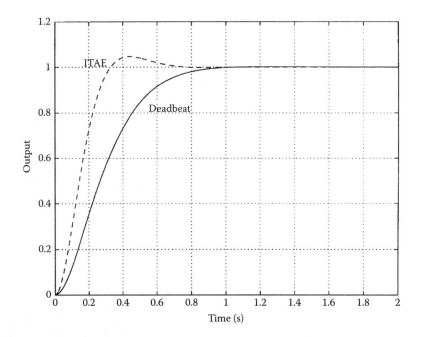

FIGURE 9.17 ITAE ($p = 14$ and $K = 100$) and deadbeat ($p = 10.9$ and $K = 36$) system response.

ISE: Integral of the square of the error.

ITAE: Integral of time multiplied by the absolute error.

Deadbeat response: A system with a rapid response, minimal overshoot, and zero steady-state error for a step input.

References

1. Dorf, R.C. and Bishop, R.H., *Modern Control Systems*, 7th ed., Addison-Wesley, Reading, MA, 1995.
2. Graham, D. and Lathrop, R.C., The synthesis of optimum response: criteria and standard forms. II, *Trans. AIEE 72*, 273–288, November 1953.

9.3 Nyquist, Bode, and Nichols Plots

John J. D'Azzo and Constantine H. Houpis

9.3.1 Introduction

The frequency-response* method of analysis and design is a powerful technique for the comprehensive study of a system by conventional methods. Performance requirements can be readily expressed in terms of the frequency response. Since noise, which is always present in any system, can result in poor overall performance, the frequency-response characteristics of a system permit evaluation of the effect of noise. The design of a passband for the system response may result in excluding the noise and therefore improving the system performance as long as the dynamic tracking performance specifications are met. The frequency response is also useful in situations for which the transfer functions of some or all of

---

* The material contained in this chapter is based on Chapters 8 and 9 from the text *Linear Control System Analysis & Design—Conventional and Modern*, 4th ed., McGraw-Hill, New York, 1995.

the components in a system are unknown. The frequency response can be determined experimentally for these situations, and an approximate expression for the transfer function can be obtained from the graphical plot of the experimental data. The frequency-response method is also a very powerful method for analyzing and designing a robust multi-input/multi-output (MIMO) system with structured uncertain plant parameters. In this chapter, two graphical representations of transfer functions are presented: the logarithmic plot and the polar plot. These plots are used to develop Nyquist's stability criterion [1–4] and closed-loop design procedures. The plots are also readily obtained by use of computer-aided design (CAD) packages like MATLAB® or TOTAL-PC (see [5]). The closed-loop feedback response $M(j\omega)$ is obtained as a function of the open-loop transfer function $G(j\omega)$. Design methods for adjusting the open-loop gain are developed and demonstrated. They are based on the polar plot of $G(j\omega)$ and the Nichols plot. Both methods achieve a peak value M_m and a resonant frequency ω_m of the closed-loop frequency response. A correlation between these frequency-response characteristics and the time response is developed.

9.3.2 Correlation of the Sinusoidal and Time Responses

Once the frequency response [2] of a system has been determined, the time response can be determined by inverting the corresponding Fourier transform. The behavior in the frequency domain for a given driving function $r(t)$ can be determined by the Fourier transform as

$$R(j\omega) = \int_{-\infty}^{\infty} r(t)e^{-j\omega t}\,dt \tag{9.13}$$

For a given control system, the frequency response of the controlled variable is

$$C(j\omega) = \frac{G(j\omega)}{1 + G(j\omega)H(j\omega)}R(j\omega) \tag{9.14}$$

By use of the inverse Fourier transform, the controlled variable as a function of time is

$$c(t) = \frac{1}{2\pi} \int_{-\infty}^{\infty} C(j\omega)e^{j\omega t}\,d\omega \tag{9.15}$$

Equation 9.15 can be evaluated by numerical or graphical integration or by reference to a table of definite integrals. This is necessary if $C(j\omega)$ is available only as a curve and cannot be simply expressed in analytical form, as is often the case. The procedure is described in several books [6]. In addition, methods have been developed based on the Fourier transform and a step input signal, relating $C(j\omega)$ qualitatively to the time solution without actually taking the inverse Fourier transform. These methods permit the engineer to make an approximate determination of the system response through the interpretation of graphical plots in the frequency domain.

Elsewhere [5, sect. 4.12] it is shown that the frequency response is a function of the pole–zero pattern in the s-plane. It is therefore related to the time response of the system. Two features of the frequency response are the maximum value M_m and the resonant frequency ω_m. Section 9.3.8 describes the qualitative relationship between the time response and the values M_m and ω_m. Since the location of the poles can be determined from the root locus, there is a direct relationship between the root-locus and frequency-response methods.

9.3.2.1 Frequency-Response Curves

The frequency domain plots belong to two categories. The first category is the plot of the magnitude of the output–input ratio vs. frequency in rectangular coordinates, as illustrated in [5, sect. 4.12]. In logarithmic coordinates these are known as *Bode plots*. Associated with this plot is a second plot of the corresponding phase angle vs. frequency. In the second category the output–input ratio may be plotted in polar coordinates with frequency as a parameter. Direct polar plots are generally used only for the open-loop response and are commonly referred to as *Nyquist plots*. [7] The plots can be obtained experimentally or by CAD packages. [6,8] When a CAD program is not available, the Bode plots are easily obtained by a graphical procedure. The other plots can then be obtained from the Bode plots.

For a given sinusoidal input signal, the input and steady-state output are of the following forms:

$$r(t) = R \sin \omega t \tag{9.16}$$

$$c(t) = C \sin(\omega t + \alpha) \tag{9.17}$$

The closed-loop frequency response is given by

$$\frac{C(j\omega)}{R(j\omega)} = \frac{G(j\omega)}{1 + G(j\omega)H(j\omega)} = M(\omega)\angle\,[\alpha(\omega)] \tag{9.18}$$

For each value of frequency, Equation 9.18 yields a phasor quantity whose magnitude is M and whose phase angle α is the angle between $C(j\omega)$ and $R(j\omega)$. An ideal system may be defined as one where $\alpha = 0°$ and $R(j\omega) = C(j\omega)$ for $0 < \omega < \infty$ (see curves 1 in Figures 9.18a and 9.18b). However, this definition implies an instantaneous transfer of energy from the input to the output. Such a transfer cannot be achieved in practice since any physical system has some energy dissipation and some energy-storage elements. Curves 2 and 3 in Figures 9.18a and 9.18b represent the frequency responses of practical control systems. The passband, or bandwidth, of the frequency response is defined as the range of frequencies from 0 to the frequency ω_b, where $M = 0.707$ of the value at $\omega = 0$. However, the frequency ω_m is more easily obtained than ω_b. The values M_m and ω_m are often used as figures of merit (F.O.M.).

In any system, the input signal may contain spurious noise signals in addition to the true signal input, or there may be sources of noise within the closed-loop system. This noise may be in a band of frequencies above the dominant frequency band of the true signal. In that case, in order to reproduce the true signal and attenuate the noise, feedback control systems are designed to have a definite passband. In certain cases, the noise frequency may exist in the same frequency band as the true signal. However, when this occurs, the problem of estimating the desired signal is more complicated. Therefore, even if the ideal system were possible, it would not be desirable.

9.3.2.2 Bode Plots (Logarithmic Plots)

The use of semilog paper eliminates the need to take logarithms of very many numbers and expands the low-frequency range, which is of primary importance. The basic factors of the transfer function fall into

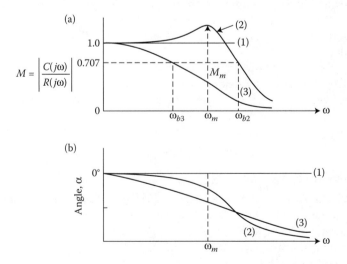

FIGURE 9.18 Frequency-response characteristics of $C(j\omega)/R(j\omega)$ in rectangular coordinates.

three categories, and these can easily be plotted by means of straight-line asymptotic approximations. The straight-line approximations are used to obtain approximate performance characteristics very quickly or to check values obtained from the computer. As the design becomes more firmly specified, the straight-line curves can be corrected for greater accuracy.

Some basic definitions of logarithmic terms follow.

Logarithm The logarithm of a complex number is itself a complex number. The abbreviation log is used to indicate the logarithm to the base 10:

$$\log \left| G(j\omega) \right| e^{j\phi(\omega)} = \log \left| G(j\omega) \right| + \log e^{j\phi(\omega)}$$
$$= \log \left| G(j\omega) \right| + j0.434\phi(\omega) \tag{9.19}$$

The real part is equal to the logarithm of the magnitude, $\log |G(j\omega)|$, and the imaginary part is proportional to the angle, $0.434\phi(\omega)$. In the rest of this chapter, the factor 0.434 is omitted and only the angle $\phi(\omega)$ is used.

Decibels The unit commonly used for the logarithm of the magnitude is the *decibel* (dB). When logarithms of transfer functions of physical systems are used, the input and output variables are not necessarily in the same units; e.g., the output may be speed in radians per second (rad/s), and the input may be voltage in volts (V).

Log magnitude The logarithm of the magnitude of a transfer function $G(j\omega)$ expressed in decibels is

$$20 \log \left| G(j\omega) \right| \text{ dB} \tag{9.20}$$

This quantity is called the *log magnitude,* abbreviated Lm. Thus,

$$\text{Lm} G(j\omega) = 20 \log \left| G(j\omega) \right| \text{ dB} \tag{9.21}$$

Since the transfer function is a function of frequency, the Lm is also a function of frequency.

Octave and decade Two units used to express frequency bands or frequency ratios are the octave and the decade. An octave is a frequency band from f_1 to f_2, where $f_2/f_1 = 2$. Thus, the frequency band from 1 to 2 Hz is 1 octave in width, and the frequency band from 17.4 to 34.8 Hz is also 1 octave in width. Note that 1 octave is not a fixed frequency bandwidth but depends on the frequency range being considered. The number of octaves in the frequency range from f_1 to f_2 is

$$\frac{\log(f_2/f_1)}{\log 2} = 3.32 \log \frac{f_2}{f_1} \text{ octaves} \tag{9.22}$$

There is an increase of 1 decade from f_1 to f_2 when $f_2/f_1 = 10$. The frequency band from 1 to 10 Hz or from 2.5 to 25 Hz is 1 decade in width. The number of decades from f_1 to f_2 is given by

$$\log \frac{f_2}{f_1} \text{ decades} \tag{9.23}$$

The dB values of some common numbers are given in Table 9.4. Note that the reciprocals of numbers differ only in sign. Thus, the dB value of 2 is +6 dB and the dB value of 1/2 is −6 dB. The following two properties are illustrated in Table 9.4:

Property 1 As a number doubles, the decibel value increases by 6 dB. The number 2.70 is twice as large as 1.35, and its decibel value is 6 dB greater. The number 200 is twice as large as 100, and its decibel value is 6 dB greater.

Property 2 As a number increases by a factor of 10, the decibel value increases by 20 dB. The number 100 is 10 times as large as the number 10, and its decibel value is 20 dB greater. The number 200 is 100 times as large as the number 2, and its decibel value is 40 dB greater.

TABLE 9.4 Decibel Values of
Some Common Numbers

Number	Decibels
0.01	−40
0.1	−20
0.5	−6
1.0	0
2.0	6
10.0	20
100.0	40
200.0	46

9.3.2.3 General Frequency Transfer Function Relationships

The frequency transfer function can be written in generalized form as the ratio of polynomials

$$G(j\omega) = \frac{K_m(1+j\omega T_1)(1+j\omega T_2)^r \cdots}{(j\omega)^m(1+j\omega T_a)\left[1+(2\zeta/\omega_n)j\omega+(1/\omega^2)(j\omega)^2\right]\cdots} \tag{9.24}$$

where K_m is the gain constant. The logarithm of the transfer function is a complex quantity; the real portion is proportional to the Lm, and the complex portion is proportional to the angle. Two separate equations are written, one for the Lm and one for the angle, respectively:

$$\mathrm{Lm}G(j\omega) = \mathrm{Lm}K_m + \mathrm{Lm}(1+j\omega T_1) + r\mathrm{Lm}(1+j\omega T_2) + \cdots - m\mathrm{Lm}j\omega - \mathrm{Lm}(1+j\omega T_a)$$
$$- \mathrm{Lm}\left[1+\frac{2\zeta}{\omega_n}j\omega+\frac{1}{\omega^2}(j\omega)^2\right] - \cdots \tag{9.25}$$

$$\angle[G(j\omega)] = \angle[K_m] + \angle[1+j\omega T_1] + r\angle[1+j\omega T_2] + \cdots - m\angle[j\omega] - \angle[1+j\omega T_a]$$
$$- \angle\left[1+\frac{2\zeta}{\omega_n}j\omega+\frac{1}{\omega^2}(j\omega)^2\right] - \cdots \tag{9.26}$$

The angle equation may be rewritten as

$$\angle[G(j\omega)] = \angle[K_m] + \tan^{-1}\omega T_1 + r\tan^{-1}\omega T_2 + \cdots - m90° - \tan^{-1}\omega T_a - \tan^{-1}\frac{2\zeta\omega/\omega_n}{1-\omega^2/\omega_n^2} - \cdots \tag{9.27}$$

The gain K_m is a real number but may be positive or negative; therefore, its angle is correspondingly 0° or 180°. Unless otherwise indicated, a positive value of gain is assumed in this chapter. Both the Lm and the angle given by these equations are functions of frequency. When the Lm and the angle are plotted as functions of the log of frequency, the resulting curves are referred to as the Bode plots or the *Lm diagram and the phase diagram*. Equations 9.25 and 9.26 show that the resultant curves are obtained by the addition and subtraction of the corresponding individual terms in the transfer function equation. The two curves can be combined into a single curve of Lm vs. angle, with frequency ω as a parameter. This curve is called the Nichols or the *log magnitude–angle diagram*.

9.3.2.4 Drawing the Bode Plots

The properties of frequency-response plots are presented in this section, but the data for these plots usually are obtained from a CAD program. The generalized form of the transfer function as given by

Equation 9.24 shows that the numerator and denominator have four basic types of factors:

$$K_m \tag{9.28}$$

$$(j\omega)^{\pm m} \tag{9.29}$$

$$(1 + j\omega T)^{\pm r} \tag{9.30}$$

$$\left[1 + \frac{2\zeta}{\omega_n}j\omega + \frac{1}{\omega_n^2}(j\omega)^2\right]^{\pm p} \tag{9.31}$$

Each of these terms except for K_m may appear raised to an integral power other than 1. The curves of Lm and angle vs. the log of frequency can easily be drawn for each factor. Then these curves for each factor can be added together graphically to get the curves for the complete transfer function. The procedure can be further simplified by using asymptotic approximations to these curves, as shown in the following pages.

Constants Since the constant K_m is frequency invariant, the plot of

$$\text{Lm}K_m = 20 \log K_m \ \text{dB}$$

is a horizontal straight line. The constant raises or lowers the Lm curve of the complete transfer function by a fixed amount. The angle, of course, is zero as long as K_m is positive.

$j\omega$ Factors The factor $j\omega$ appearing in the denominator has an Lm

$$\text{Lm}(j\omega)^{-1} = 20 \log \left|(j\omega)^{-1}\right| = -20 \log \omega \tag{9.32}$$

When plotted against *log* ω, this curve is a straight line with a negative slope of 6 dB/octave or 20 dB/decade. Values of this function can be obtained from Table 9.4 for several values of ω. The angle is constant and equal to $-90°$. When the factor $j\omega$ appears in the numerator, the Lm is

$$\text{Lm}(j\omega) = 20 \log |\omega| = 20 \log \omega \tag{9.33}$$

This curve is a straight line with a positive slope of 6 dB/octave or 20 dB/decade. The angle is constant and equal to $+90°$. Notice that the only difference between the curves for $j\omega$ and for $1/j\omega$ is a change in the sign of the slope of the Lm and a change in the sign of the angle. Both Lm curves go through the point 0 dB at $\omega = 1$. For the factor $(j\omega)^{\pm m}$ the Lm curve has a slope of $\pm 6m$ dB/octave or $\pm 20m$ dB/decade, and the angle is constant and equal to $\pm m90°$.

$1 + j\omega T$ Factors The factor $1 + j\omega T$ appearing in the denominator has an Lm

$$\text{Lm}(1 + j\omega T)^{-1} = 20 \log \left|1 + j\omega T\right|^{-1} = -20 \log \sqrt{\left[1 + \omega^2 T^2\right]} \tag{9.34}$$

For very small values of ω, that is, $\omega T \ll 1$,

$$\text{Lm}(1 + j\omega T)^{-1} \approx \log 1 = 0 \, \text{dB} \tag{9.35}$$

Thus, the plot of the Lm at small frequencies is the $0 - \text{dB}$ line. For very large values of ω, that is, $\omega T \gg 1$,

$$\text{Lm}(1 + j\omega T)^{-1} \approx 20 \log |j\omega T|^{-1} = -20 \log \omega T \tag{9.36}$$

The value of Equation 9.36 at $\omega = 1/T$ is 0. For values of $\omega > 1/T$ this function is a straight line with a negative slope of 6 dB/octave. Therefore, the asymptotes of the plot of $\text{Lm}(1 + j\omega T)^{-1}$ are two straight lines, one of zero slope below $\omega = 1/T$ and one of -6 dB/octave slope above $\omega = 1/T$. These asymptotes are drawn in Figure 9.19.

The frequency at which the asymptotes to the Lm curve intersect is defined as the corner frequency ω_{cf}. The value $\omega_{cf} = 1/T$ is the corner frequency for the function

$$(1 + j\omega T)^{\pm r} = (1 + j\omega/\omega_{cf})^{\pm r}$$

The exact values of Lm $(1 + j\omega T)^{-1}$ are given in Table 9.5 for several frequencies in the range a decade above and below the corner frequency. The exact curve is also drawn in Figure 9.19 The error, in dB,

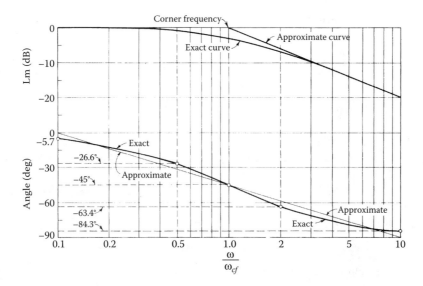

FIGURE 9.19   Log magnitude and phase diagram for $(1 + j\omega T)^{-1} = [1 + j(\omega/\omega_{cf})]^{-1}$.

between the exact curve and the asymptotes is approximately as follows:

- At the corner frequency: 3 dB
- One octave above and below the corner frequency: 1 dB
- Two octaves from the corner frequency: 0.26 dB

Preliminary design studies are often made by using the asymptotes only. The correction to the straight-line approximation to yield the true Lm curve is shown in Figure 9.20. The phase curve for this function is also plotted in Figure 9.19. At zero frequency the angle is $0°$; at the corner frequency $\omega = \omega_{cf}$ the angle is $-45°$; and at infinite frequency the angle is $-90°$. The angle curve is symmetrical about the corner frequency value when plotted against $\log(\omega/\omega_{cf})$ or $\log \omega$. Since the abscissa of the curves in Figure 9.19 is ω/ω_{cf}, the shapes of the angle and Lm curves are independent of the time constant T. Thus, when the curves are plotted with the abscissa in terms of ω, changing T just "slides" the Lm and the angle curves left or right so that the -3 dB and the $-45°$ points occur at the frequency $\omega = \omega_{cf}$. The approximation of the phase curve is a straight line drawn through the following three points:

ω/ω_{cf}	0.1	1.0	10
Angle	$0°$	$-45°$	$-90°$

TABLE 9.5   Values of $\mathrm{Lm}(1 + j\omega T)^{-1}$ for Several Frequencies

$\dfrac{\omega}{\omega_{cf}}$	Exact value, dB	Value of the asymptote, dB	Error, dB
0.1	-0.04	0.	-0.04
0.25	-0.26	0.	-0.26
0.5	-0.97	0.	-0.97
0.76	-2.00	0.	-2.00
1	-3.01	0.	-3.01
1.31	-4.35	-2.35	-2.00
2	-6.99	-6.02	-0.97
4	-12.30	-12.04	-0.26
10	-20.04	-20.0	-0.04

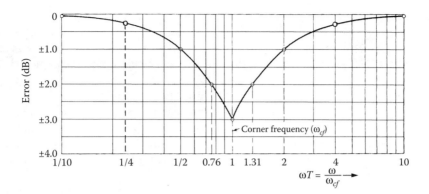

FIGURE 9.20 Log magnitude correction for $(1 + j\omega T)^{\pm 1}$.

The maximum error resulting from this approximation is about $\pm 6°$. For greater accuracy, a smooth curve is drawn through the points given in Table 9.6.

The factor $1 + j\omega T$ appearing in the numerator has the Lm

$$\mathrm{Lm}(1 + j\omega T) = 20 \log \sqrt{(1 + \omega^2 T^2)}$$

This is the same function as its inverse $\mathrm{Lm}\,(1 + j\omega T)^{-1}$ except that it is positive. The corner frequency is the same, and the angle varies from 0 to $90°$ as the frequency increases from zero to infinity. The Lm and angle curves for the function $(1 + j\omega T)$ are symmetrical about the abscissa to the curves for $(1 + j\omega T)^{-1}$.

Quadratic Factors　Quadratic factors in the denominator of the transfer function have the form

$$\left[1 + \frac{2\zeta}{\omega_n} j\omega + \frac{1}{\omega_n^2}(j\omega)^2\right]^{-1} \tag{9.37}$$

For $\zeta > 1$, the quadratic can be factored into two first-order factors with real zeros which can be plotted in the manner shown previously. But for $\zeta < 1$ Equation 9.37 contains conjugate-complex factors, and the entire quadratic is plotted without factoring:

$$\mathrm{Lm}\left[1 + \frac{2\zeta}{\omega_n} j\omega + \frac{1}{\omega_n^2}(j\omega)^2\right]^{-1} = -20 \log \left[\left(1 - \frac{\omega^2}{\omega_n^2}\right)^2 + \left(\frac{2\zeta\omega}{\omega_n}\right)^2\right]^{1/2} \tag{9.38}$$

$$\angle \left[1 + \frac{2\zeta}{\omega_n} j\omega + \frac{1}{\omega_n^2}(j\omega)^2\right]^{-1} = -\tan^{-1} \frac{2\zeta\omega/\omega_n}{1 - \omega^2/\omega_n^2} \tag{9.39}$$

From Equation 9.38 it is seen that for very small values of ω, the low-frequency asymptote is represented by $\mathrm{Lm} = 0\,\mathrm{dB}$. For very high values of frequency, the high-frequency asymptote has a slope of -40 dB/decade. The asymptotes cross at the corner frequency $\omega_{cf} = \omega_n$.

TABLE 9.6　Angles of $(1 + j\omega/\omega_{cf})^{-1}$ for Key Frequency Points

$\frac{\omega}{\omega_{cf}}$	Angle, deg
0.1	-5.7
0.5	-26.6
1.0	-45.0
2.0	-63.4
10.0	-84.3

From Equation 9.38, it is seen that a resonant condition exists in the vicinity of $\omega = \omega_n$, where the peak value of the Lm > 0 dB. Therefore, there may be a substantial deviation of the Lm curve from the straight-line asymptotes, depending on the value of ζ. A family of Lm curves of several values of $\zeta < 1$ is plotted in Figure 9.21. For the appropriate ζ, the Lm curve can be drawn by selecting sufficient points from Figure 9.21 or computed from Equation 9.38.

The phase-angle curve for this function also varies with ζ. At zero frequency the angle is $0°$; at the corner frequency the angle is $-90°$; and at infinite frequency the angle is $-180°$. A family of phase-angle curves for various values of $\zeta < 1$ is plotted in Figure 9.21. Enough values to draw the appropriate phase-angle curve can be taken from Figure 9.21 or computed from Equation 9.39. When the quadratic factor appears in the numerator, the magnitudes of the Lm and phase angle are the same as those in Figure 9.21, except that they are changed in sign.

The Lm$[1 + j2\zeta/\omega_n + (j\omega/\omega_n)^2]^{-1}$ with $\zeta < 0.707$ has a peak value. The magnitude of this peak value and the frequency at which it occurs are important terms. These values, derived in Section 9.3.8 [see 5,

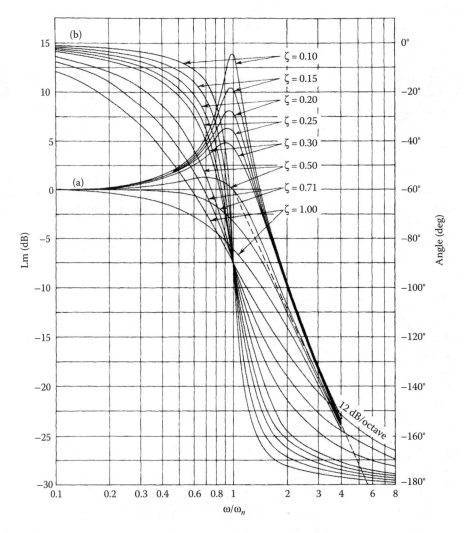

FIGURE 9.21 Log magnitude and phase diagram for $[1 + j2\zeta\omega/\omega_n + (j\omega/\omega_n)^2]^{-1}$.

Sect. 9.3], are repeated here:

$$M_m = \frac{1}{2\zeta\sqrt{(1-\zeta^2)}} \tag{9.40}$$

$$\omega_m = \omega_n\sqrt{(1-2\zeta^2)} \tag{9.41}$$

Note that the peak value M_m depends only on the damping ratio ζ. Since Equation 9.41 is meaningful only for real values of ω_m, the curve of M vs. ω has a peak value greater than unity only for $\zeta < 0.707$. The frequency at which the peak value occurs depends on both the damping ratio ζ and the undamped natural frequency ω_n. This information is used when adjusting a control system for good response characteristics. These characteristics are discussed in Sections 9.3.8 and 9.3.9.

The Lm curves for poles and zeros lying in the right-half (RH) s-plane are the same as those for poles and zeros located in the left-half (LH) s-plane. However, the angle curves are different. For example, the angle for the factor $(1 - j\omega T)$ varies from 0 to $-90°$ as ω varies from zero to infinity. Also, if ζ is negative, the quadratic factor of Equation 9.37 contains RH s-plane poles or zeros. Its angle varies from $-360°$ at $\omega = 0$ to $-180°$ at $\omega = \infty$. This information can be obtained from the pole–zero diagram [5, Sect. 4.12] with all angles measured in a *counter-clockwise* (CCW) direction. Some CAD packages do not consistently use a CCW measurement direction, thus resulting in inaccurate angle values.

9.3.3 System Type and Gain as Related to Lm Curves

The steady-state error of a closed-loop system depends on the system type and the gain. The system error coefficients are determined by these two characteristics [5, Chap. 6]. For any given Lm curve, the system type and gain can be determined. Also, with the transfer function given so that the system type and gain are known, they can expedite drawing the Lm curve. This is described for Type 0, 1, and 2 systems.

9.3.3.1 Type 0 System

A first-order Type 0 system has a transfer function of the form

$$G(j\omega) = \frac{K_0}{1 + j\omega T_a}$$

At low frequencies, $\omega < 1/Ta$, $LmG(j\omega) \approx 20 \log K_0$, which is a constant. The slope of the Lm curve is zero below the corner frequency $\omega_1 = 1/T_a$ and $-20\,dB/decade$ above the corner frequency. The Lm curve is shown in Figure 9.22.

For a Type 0 system the characteristics are as follows:

1. The slope at low frequencies is zero.

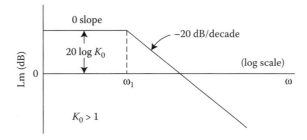

FIGURE 9.22 Log magnitude plot for $G(j\omega) = K_0/(1 + j\omega T_a)$.

2. The magnitude at low frequencies is $20 \log K_0$.
3. The gain K_0 is the steady-state step error coefficient.

9.3.3.2 Type-1 System

A second-order Type 1 system has a transfer function of the form

$$G(j\omega) = \frac{K_1}{j\omega(1+j\omega T_a)}$$

At low frequencies, $\omega < 1/T_a$, $\mathrm{Lm}[G(j\omega)] \approx \mathrm{Lm}(K_1/j\omega) = \mathrm{Lm}\ K_1 - \mathrm{Lm}\ j\omega$, which has a slope of -20 dB/decade. At $\omega = K_1$, $\mathrm{Lm}(K1/j\omega) = 0$. If the corner frequency $\omega_1 = 1/T_a$ is greater than K_1, the low-frequency portion of the curve of slope -20 dB/decade crosses the 0–dB axis at a value of $\omega_x = K_1$, as shown in Figure 9.23a. If the corner frequency is less than K_1, the low-frequency portion of the curve of slope -20 dB/decade may be extended until it does cross the 0–dB axis. The value of the frequency at which the extension crosses the 0–dB axis is $\omega_x = K_1$. In other words, the plot $\mathrm{Lm}\ (K_1/j\omega)$ crosses the 0–dB value at $\omega_x = K_1$, as illustrated in Figure 9.23b.

At $\omega = 1$, $\mathrm{Lm}\ j\omega = 0$; therefore, $\mathrm{Lm}\ (K_1/j\omega)_{\omega=1} = 20 \log(K_1)$. For $T_a < 1$ this value is a point on the slope of -20 dB/decade. For $T_a > 1$ this value is a point on the extension of the initial slope, as shown in Figure 9.23b. The frequency ω_x is smaller or larger than unity according as K_1 is smaller or larger than unity. For a Type-1 system the characteristics are as follows:

1. The slope at low frequencies is -20 dB/decade.
2. The intercept of the low-frequency slope of -20 dB/decade (or its extension) with the 0–dB axis occurs at the frequency ω_x, where $\omega_x = K_1$.
3. The value of the low-frequency slope of -20 dB/decade (or its extension) at the frequency $\omega = 1$ is equal to $20 \log(K_1)$.
4. The gain K_1 is the steady-state ramp error coefficient.

9.3.3.3 Type-2 System

A Type-2 system has a transfer function of the form

$$G(j\omega) = \frac{K_2}{(j\omega)^2(1+j\omega T_a)}$$

At low frequencies, $\omega < 1/T_a$, $\mathrm{Lm}\ [G(j\omega)] = \mathrm{Lm}\ [K_2/(j\omega)^2] = \mathrm{Lm}\ [K_2] - \mathrm{Lm}[j\omega]^2$, for which the slope is -40 dB/decade. At $\omega^2 = K_2$, $\mathrm{Lm}\ [K_2/(j\omega)^2] = 0$; therefore, the intercept of the initial slope

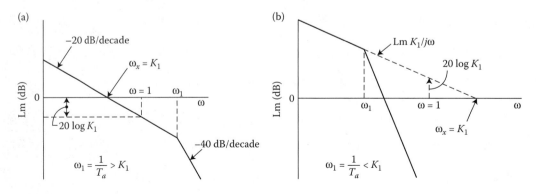

FIGURE 9.23   Log magnitude plot for $G(j\omega) = K_1/j\omega(1+j\omega T_a)$.

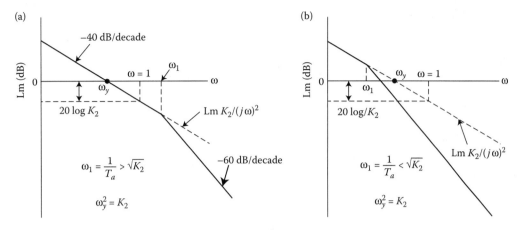

FIGURE 9.24 Log magnitude plot for $G(j\omega) = K_2/(j\omega)^2(1 + j\omega T_a)$.

of -40 dB/decade (or its extension, if necessary) with the $0-$dB axis occurs at the frequency ω_y, where $\omega_y^2 = K_2$.

At $\omega = 1$, Lm $[j\omega]^2 = 0$; therefore, Lm $[K_2/(\omega)^2]_{\omega=1} = 20\log[K_2]$. This point occurs on the initial slope or on its extension, according as $\omega_1 = 1/T_a$ is larger or smaller than $\sqrt{K_2}$. If $K_2 > 1$, the quantity $20\log[K_2]$ is positive, and if $K_2 < 1$, the quantity $20\log[K_2]$ is negative.

The Lm curve for a Type-2 transfer function is shown in Figure 9.24. The determination of gain K_2 from the graph is shown. For a Type-2 system the characteristics are as follows:

1. The slope at low frequencies is -40 dB/decade.
2. The intercept of the low-frequency slope of -40 dB/decade (or its extension, if necessary) with the $0-$dB axis occurs at a frequency ω_y, where $\omega_y^2 = K_2$.
3. The value on the low-frequency slope of -40 dB/decade (or its extension) at the frequency $\omega = 1$ is equal to $20\log[K_2]$.
4. The gain K_2 is the steady-state parabolic error coefficient.

9.3.4 Experimental Determination of Transfer Functions

The magnitude and angle of the ratio of the output to the input can [4,6] be obtained experimentally for a steady-state sinusoidal input signal at a number of frequencies. For stable plants, the Bode data for the plant is used to obtain the exact Lm and angle diagram. Asymptotes are drawn on the exact Lm curve, using the fact that their slopes must be multiples of ± 20 dB/decade. From these asymptotes and their intersections, the system type and the approximate time constants are determined. In this manner, the transfer function of the system can be synthesized. Care must be exercised in determining whether any zeros of the transfer function are in the RH s-plane. A system that has no open-loop zeros in the RH s-plane is defined as a *minimum-phase* system, [5,9,10] and all factors have the form $(1 + Ts)$ and/or $(1 + As + Bs^2)$. A system that has open-loop zeros in the RH s-plane is a *nonminimum-phase* system. The stability is determined by the location of the poles and does not affect the designation of minimum or nonminimum phase. The angular variation for poles or zeros in the RH s-plane is different from those in the LH plane [9]. For this situation, one or more terms in the transfer function have the form $(1 - Ts)$ and/or $(1 \pm As \pm Bs^2)$. Care must be exercised in interpreting the angle plot to determine whether any factors of the transfer function lie in the RH s-plane. Many practical systems are minimum phase. Unstable plants must be handled with care. That is, first a stabilizing compensator must be added to form a stable closed-loop system. From the experimental data for the stable closed-loop system, the plant transfer function is determined by using the known compensator transfer function.

9.3.5 Direct Polar Plots

The magnitude and angle of $G(j\omega)$, for sufficient frequency points, are readily obtainable from the $\text{Lm}[G(j\omega)]$ and $\angle[G(j\omega)]$ vs. $\log[\omega]$ curves or by the use of a CAD program. It is also possible to visualize the complete shape of the frequency-response curve from the pole–zero diagram, because the angular contribution of each pole and zero is readily apparent. The polar plot of $G(j\omega)$ is called the *direct polar plot*. The polar plot of $[G(j\omega)]^{-1}$ is called the *inverse polar plot* [11].

9.3.5.1 Lag–Lead Compensator [5]

The compensator transfer function is

$$G_c(s) = \frac{1 + (T_1 + T_2)s + T_1 T_2 s^2}{1 + (T_1 + T_2 + T_{12})s + T_1 T_2 s^2} \tag{9.42}$$

As a function of frequency, the transfer function is

$$G_c(j\omega) = \frac{E_0(j\omega)}{E_i(j\omega)} = \frac{(1 - \omega^2 T_1 T_2) + j\omega(T_1 + T_2)}{(1 - \omega^2 T_1 T_2) + j\omega(T_1 + T_2 + T_{12})} \tag{9.43}$$

By the proper choice of the time constants, the compensator acts as a lag network [i.e., the output signal $E_0(j\omega)$ lags the input signal $E_i(j\omega)$] in the lower-frequency range of 0 to ω_x and as a lead network [i.e., the output signal leads the input signal] in the higher-frequency range of ω_x to ∞. The polar plot of this transfer function is a circle with its center on the real axis and lying in the first and fourth quadrants. Its properties are

1. $\displaystyle\lim_{\omega \to 0} [G(j\omega T_1)] \to 1\angle 0°$

2. $\displaystyle\lim_{\omega \to \infty} [G(j\omega T_1)] \to 1\angle 0°$

3. At $\omega = \omega_x$, for which $\omega_x^2 T_1 T_2 = 1$,

Equation 9.43 yields the value

$$G(j\omega_x T_1) = \frac{T_1 + T_2}{T_1 + T_2 + T_{12}} = |G(j\omega_x T_1)| \angle 0° \tag{9.44}$$

Note that Equation 9.44 represents the minimum value of the transfer function in the whole frequency spectrum. For frequencies below ω_x, the transfer function has a negative or lag angle. For frequencies above ω_x, it has a positive or lead angle.

9.3.5.2 Type 0 Feedback Control System

The field-controlled servomotor [11] illustrates a typical Type 0 device. It has the transfer function

$$G(j\omega) = \frac{C(j\omega)}{E(j\omega)} = \frac{K_0}{(1 + j\omega T_f)(1 + j\omega T_m)} \tag{9.45}$$

$$\text{Note: } G(j\omega) \to \begin{pmatrix} K_0\angle 0° & \text{as} & \omega \to 0^+ \\ 0\angle -180° & \text{as} & \omega \to \infty \end{pmatrix} \tag{9.46}$$

Also, for each term in the denominator the angular contribution to $G(j\omega)$, as ω goes from 0 to ∞, goes from 0 to $-90°$. Thus, the polar plot of this transfer function must start at $G(j\omega) = K_0\angle 0°$ for $\omega = 0$ and proceed first through the fourth and then through the third quadrants to $\lim_{\omega \to \infty} G(j\omega) = 0\angle -180°$ as the frequency approaches infinity. In other words, the angular variation of $G(j\omega)$ is continuously decreasing, going in a clockwise (CW) direction from $0°$ to $-180°$. The exact shape of this plot is determined by the particular values of the time constants T_f and T_m.

Consider the transfer function

$$G(j\omega) = \frac{K_0}{(1+j\omega T_f)(1+j\omega T_m)(1+j\omega T)} \tag{9.47}$$

In this case, when $\omega \to \infty$, $G(j\omega) \to 0\angle -270°$. Thus, the curve crosses the negative real axis at a frequency ω_x for which the imaginary part of the transfer function is zero. When a term of the form $(1+j\omega T)$ appears in the numerator, the transfer function experiences an angular variation of 0 to $90°$ (a CCW rotation) as the frequency is varied from 0 to ∞. Thus, the angle of $G(j\omega)$ may not change continuously in one direction. Also, the resultant polar plot may not be as smooth as the one for Equations 9.45 and 9.47.

In the same manner, a quadratic in either the numerator or the denominator of a transfer function results in an angular contribution of 0 to $\pm180°$, respectively, and the polar plot of $G(j\omega)$ is affected accordingly. It can be seen from the examples that the polar plot of a Type 0 system always starts at a value K_0 (step error coefficient) on the positive real axis for $\omega = 0$ and ends at zero magnitude (for $n > \omega$) and tangent to one of the major axes at $\omega = \infty$. The final angle is $-90°$ times the order n of the denominator minus the order w of the numerator of $G(j\omega)$.

9.3.5.3 Type-1 Feedback Control System

A typical Type-1 system containing only poles is

$$G(j\omega) = \frac{C(j\omega)}{E(j\omega)} = \frac{K_1}{j\omega(1+J\omega T_m)(1+j\omega T_c)(1+j\omega T_q)} \tag{9.48}$$

$$\text{Note: } G(j\omega) \to \begin{pmatrix} \infty\angle -90° & \text{as} & \omega \to 0^+ \\ 0\angle -360° & \text{as} & \omega \to \infty \end{pmatrix} \tag{9.49}$$

Note that the $j\omega$ term in the denominator contributes the angle $-90°$ to the total angle of $G(j\omega)$ for all frequencies. Thus, the basic difference between Equations 9.47 and 9.48 is the presence of the term $j\omega$ in the denominator of the latter equation. Since all the $(1+j\omega T)$ terms of Equation 9.48 appear in the denominator, the angle of the polar plot of $G(j\omega)$ decreases continuously (CW) in the same direction from -90 to $-360°$ as ω increases from 0 to ∞. The frequency of the crossing point on the negative real axis of the $G(j\omega)$ function is that value of frequency ω_x for which the imaginary part of $G(j\omega)$ is equal to zero. The real-axis crossing point is very important, because it determines closed-loop stability, as described in later sections dealing with system stability.

9.3.5.4 Type-2 Feedback Control System

The transfer function of a Type-2 system is

$$G(j\omega) = \frac{C(j\omega)}{E(j\omega)} = \frac{K_2}{(j\omega)^2(1+j\omega T_f)(1+j\omega T_m)} \tag{9.50}$$

Its properties are

$$G(j\omega) \to \begin{pmatrix} \infty\angle -180° & \text{as} & \omega \to 0^+ \\ 0\angle -360° & \text{as} & \omega \to +\infty \end{pmatrix} \tag{9.51}$$

The presence of the $(j\omega)^2$ term in the denominator contributes $-180°$ to the total angle of $G(j\omega)$ for all frequencies. The polar plot for the transfer function of Equation 9.50 is a smooth curve whose angle $\phi(\omega)$ decreases continuously from -180 to $-360°$. The introduction of an additional pole and a zero can alter the shape of the polar plot. It can be shown that as $\omega \to 0^+$, the polar plot of a Type-2 system is below the real axis if

$$\sum T_{numerator} - \sum T_{denominator} \tag{9.52}$$

is a positive value, and above the real axis if it is a negative value.

9.3.5.5 Summary: Direct Polar Plots

To obtain the direct polar plot of a system's forward transfer function, the following steps are used to determine the key parts of the curve. Figure 9.25 shows the typical polar plot shapes for different system types.

Step 1

The forward transfer function has the general form

$$G(j\omega) = \frac{K_m(1+j\omega T_a)(1+j\omega T_b)\cdots(1+j\omega T_w)}{(j\omega)^m(1+j\omega T_1)(1+j\omega T_2)\cdots(1+j\omega T_u)} \tag{9.53}$$

For this transfer function, the system type is equal to the value of m and determines the portion of the polar plot representing the $\lim_{\omega\to 0}[G(j\omega)]$. The low-frequency polar-plot characteristics (as $\omega \to 0$) of the different system types are determined by the angle at $\omega = 0$, i.e., $\angle G(j0) = m(-90°)$.

Step 2

The high-frequency end of the polar plot can be determined as follows:

$$\lim_{\omega\to +\infty}[G(j\omega)] = 0\angle[(w - m - u)90°] \tag{9.54}$$

Note that since the degree of the denominator of Equation 9.53 is always greater than the degree of the numerator, the angular condition of the high-frequency point ($\omega = \infty$) is approached in the CW sense. The plot ends at the origin and is tangent to the axis determined by Equation 9.54. Tangency may occur on either side of the axis.

Step 3

The asymptote that the low-frequency end approaches, for a Type-1 system, is determined by taking the limit as $\omega \to 0$ of the real part of the transfer function.

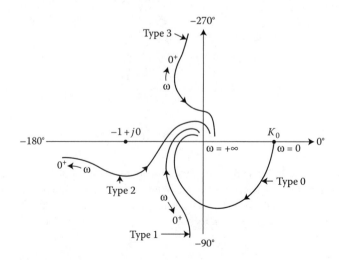

FIGURE 9.25 A summary of direct polar plots of different types of systems.

Step 4

The frequencies at the points of intersection of the polar plot with the negative real axis and the imaginary axis are determined, respectively, by

$$Im[G(j\omega)] = 0 \tag{9.55}$$

$$Re[G(j\omega)] = 0 \tag{9.56}$$

Step 5

If there are no frequency-dependent terms in the numerator of the transfer function, the curve is a smooth one in which the angle of $G(j\omega)$ continuously decreases as ω goes from 0 to ∞. With time constants in the numerator, and depending upon their values, the angle may not change continuously in the same direction, thus creating "dents" in the polar plot.

Step 6

As is seen later, it is important to know the exact shape of the polar plot of $G(j\omega)$ in the vicinity of the $-1 \pm j0$ point and the crossing point on the negative real axis.

9.3.6 Nyquist's Stability Criterion

The Nyquist stability criterion [1–4] provides a simple graphical procedure for determining closed-loop stability from the frequency-response curves of the open-loop transfer function $G(j\omega)H(j\omega)$ (for the case of no poles or zeros on the imaginary axis, etc.). The application of this method in terms of the polar plot is covered in this section; application in terms of the log magnitude–angle (Nichols) diagram is covered in a later section.

For a stable closed-loop system, the roots of the characteristic equation

$$B(s) = 1 + G(s)H(s) = 0 \tag{9.57}$$

cannot be permitted to lie in the RH s-plane or on the $j\omega$ axis. In terms of $G = N_1/D_1$ and $H = N_2/D_2$, Equation 9.57 becomes

$$B(s) = 1 + \frac{N_1 N_2}{D_1 D_2} = \frac{D_1 D_2 + N_1 N_2}{D_1 D_2} = \frac{(s - Z_1)(s - Z_2) \cdots (s - Z_n)}{(s - p_1)(s - p_2) \cdots (s - p_n)} \tag{9.58}$$

Note that the numerator and denominator of $B(s)$ have the same degree and the *poles of the open-loop transfer function $G(s)H(s)$ are the poles of $B(s)$*. The closed-loop transfer function of the system is

$$\frac{C(s)}{R(s)} = \frac{G(s)}{1 + G(s)H(s)} = \frac{N_1 D_2}{D_1 D_2 + N_1 N_2} \tag{9.59}$$

The denominator of $C(s)/R(s)$ is the same as the numerator of $B(s)$. The condition for stability may therefore be restated as: For a stable system none of the zeros of $B(s)$ can lie in the RH s-plane or on the imaginary axis. Nyquist's stability criterion relates the number of zeros and poles of $B(s)$ that lie in the RH s-plane to the polar plot of $G(s)H(s)$.

9.3.6.1 Limitations

In this analysis, it is assumed that all the control systems are inherently linear or that their limits of operation are confined to give a linear operation. This yields a set of linear differential equations that describe the dynamic performance of the systems. Because of the physical nature of feedback control systems, the degree of the denominator $D_1 D_2$ is equal to or greater than the degree of the numerator $N_1 N_2$ of the open-loop transfer function $G(s)H(s)$. Mathematically, this means that $\lim_{s \to \infty}[G(s)H(s)] \to 0$ or a constant. These two factors satisfy the necessary limitations to the generalized Nyquist stability criterion.

9.3.6.2 Generalized Nyquist's Stability Criterion

Consider a closed contour Q such that the whole RH s-plane is encircled (see Figure 9.26a with $\epsilon \to 0$), thus enclosing all zeros and poles of $B(s)$ that have positive real parts. The theory of complex variables used in the rigorous derivation requires that the contour Q must not pass *through any poles or zeros of $B(s)$*. When these results are applied to the contour Q, the following properties are noted:

1. The total number of CW rotations of $B(s)$ due to its zeros is equal to the total number of zeros Z_R in the RH s-plane.
2. The total number of CCW rotations of $B(s)$ due to its poles is equal to the total number of poles P_R in the RH s-plane.
3. The *net* number of rotations N of $B(s) = 1 + G(s)H(s)$ about the origin is equal to the total number of poles P_R minus the total number of zeros Z_R in the RH s-plane. N may be positive (CCW), negative (CW), or zero.

The essence of these three conclusions can be represented by

$$N = \frac{\text{phase change of } [1 + G(s)H(s)]}{2\pi} = P_R - Z_R \tag{9.60}$$

where CCW rotation is defined as being positive and CW rotation is negative. In order for $B(s)$ to realize a net rotation N, the directed line segment representing $B(s)$ (see Figure 9.27a) must rotate about the origin $360N°$, or N complete revolutions. Solving for Z_R in Equation 9.60 yields $Z_R = P_R - N$. Since $B(s)$ can have no zeros Z_R in the RH s-plane for a stable system, and it is therefore concluded that, *for a stable system, the net number of rotations of $B(s)$ about the origin must be CCW and equal to the number of poles P_R that lie in the RH plane*. In other words, if $B(s)$ experiences a net CW rotation, this indicates that $Z_R > P_R$, where $P_R \geq 0$, and thus the closed-loop system is unstable. If there are zero net rotations, then $Z_R = P_R$ and the system may or may not be stable, according as $P_R = 0$ or $P_R > 0$.

9.3.6.3 Obtaining a Plot of $B(s)$

Figures 9.27a and 9.27b show a plot of $B(s)$ and a plot of $G(s)H(s)$, respectively. By moving the origin of Figure 9.27b to the $-1 + j0$ point, the curve is now equal to $1 + G(s)H(s)$, which is $B(s)$. Since $G(s)H(s)$ is known, this function is plotted and then the origin is moved to the -1 point to obtain $B(s)$. In general, the open-loop transfer functions of many physical systems do not have any poles P_R in the RH s-plane.

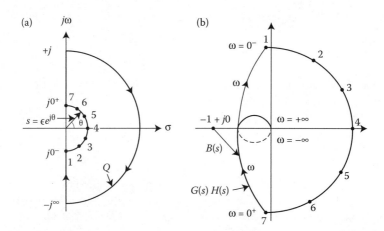

FIGURE 9.26 (a) The contour Q, which encircles the right-half s-plane; (b) complete plot for $G(s)H(s) = K_1/s(1 + T_1s)(1 + T_2s)$.

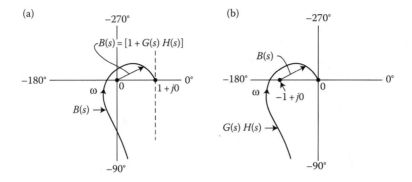

FIGURE 9.27 A change of reference for $B(s)$.

In this case, $Z_R = N$. *Thus, for a stable system the net number of rotations about the $-1 + j0$ point must be zero when there are no poles of $G(s)H(s)$ in the RH s-plane.*

9.3.6.4 Analysis of Path Q

In applying Nyquist's criterion, the whole RH s-plane must be encircled to ensure the inclusion of all poles or zeros in this portion of the plane. In Figure 9.26, the entire RH s-plane is enclosed by the closed path Q which is composed of the following four segments:

1. One segment is the imaginary axis from $-j\infty$ to $j0^-$.
2. The second segment is the semicircle of radius $\epsilon \to 0$.
3. The third segment is the imaginary axis from $j0^+$ to $+j\infty$.
4. The fourth segment is a semicircle of infinite radius that encircles the entire RH s-plane.

The portion of the path along the imaginary axis is represented mathematically by $s = j\omega$. Thus, replacing s by $j\omega$ in Equation 9.58 and letting ω take on all values from $-\infty$ to $+\infty$ gives the portion of the $B(s)$ plot corresponding to that portion of the closed contour Q that lies on the imaginary axis.

One of the requirements of the Nyquist criterion is that $\lim_{s\to\infty}[G(s)H(s)] \to 0$ or a constant. Thus, $\lim_{s\to\infty}[B(s)] = \lim_{s\to\infty}[1 + G(s)H(s)] \to 1$ or 1 plus the constant. As a consequence, the segment of the closed contour represented by the semicircle of infinite radius, the corresponding portion of the $B(s)$ plot is a fixed point. As a result, the movement along only the imaginary axis from $-j\infty$ to $+j\infty$ results in the same net rotation of $B(s)$ as if the whole contour Q were considered. *In other words, all the rotation of $B(s)$ occurs while the point O, in Figure 9.26a, goes from $-j\infty$ to $+j\infty$ along the imaginary axis.* More generally, this statement applies only to those transfer functions $G(s)H(s)$ that conform to the limitations stated earlier in this section [3].

9.3.6.5 Effect of Poles at the Origin on the Rotation of B(s)

The manner in which the $\omega = 0^-$ and $\omega = 0^+$ portions of the plot in Figure 9.26a are joined is now investigated for those transfer functions $G(s)H(s)$ that have s^m in the denominator. Consider the transfer function with positive values of T_1 and T_2:

$$G(s)H(s) = \frac{K_1}{s(1 + T_1 s)(1 + T_2 s)} \tag{9.61}$$

The direct polar plot of $G(j\omega)H(j\omega)$ of this function is obtained by substituting $s = j\omega$ into Equation 9.61, as shown in Figure 9.26b. The plot is drawn for both positive and negative frequency values. *The polar plot drawn for negative frequencies $(0^- > \omega > -\infty)$ is the conjugate of the plot drawn for positive frequencies.*

This means that the curve for negative frequencies is symmetrical to the curve for positive frequencies, with the real axis as the axis of symmetry.

The closed contour Q of Figure 9.26a, in the vicinity of $s = 0$, has been modified as shown. In other words, the point O is moved along the negative imaginary axis from $s = -j\infty$ to a point where $s = -j\epsilon = 0^- \angle -\pi/2$ becomes very small. Then the point O moves along a semicircular path of radius $s = \epsilon e^{j\theta}$ in the RH s-plane with a very small radius ϵ until it reaches the positive imaginary axis at $s = +j\epsilon = j0^+ = 0^+ \angle \pi/2$. From here the point O proceeds along the positive imaginary axis to $s = +j\infty$. Then, letting the radius approach zero, $\epsilon \to 0$, for the semicircle around the origin ensures the inclusion of all poles and zeros in the RH s-plane. To complete the plot of $B(s)$ in Figure 9.27, the effect of moving point O on this semicircle around the origin must be investigated. For the semicircular portion of the path Q represented by $s = \epsilon e^{j\theta}$, where $\epsilon \to 0$ and $-\pi/2 \le \theta \le \pi/2$, Equation 9.61 becomes

$$G(s)H(s) = \frac{K_1}{s} = \frac{K_1}{\epsilon e^{j\theta}} = \frac{K_1}{\epsilon} e^{-j\theta} = \frac{K_1}{\epsilon} e^{j\psi} \tag{9.62}$$

where $K_1/\epsilon \to \infty$ as $\epsilon \to 0$, and $\psi = -\theta$ goes from $\pi/2$ to $-\pi/2$ as the directed segment s goes CCW from $\epsilon \angle -\pi/2$ to $\epsilon \angle +\pi/2$. Thus, in Figure 9.26b, the end points from $\omega \to 0^-$ and $\omega \to 0^+$ are joined by a semicircle of infinite radius in the first and fourth quadrants. Figure 9.26b shows the completed contour of $G(s)H(s)$ as the point O moves along the contour Q in the s-plane in the CW direction. When the origin is moved to the $-1 + j0$ point, the curve becomes $B(s)$. The plot of $B(s)$ in Figure 9.26b does not encircle the $-1 + j0$ point; therefore, the encirclement N is zero. From Equation 9.61, there are no poles within Q; that is, $P_R = 0$. Thus, when Equation 9.60 is applied, $Z_R = 0$ and the closed-loop system is stable.

Transfer functions that have the term s^m in the denominator have the general form, with $s = \epsilon e^{j\theta}$ as $\epsilon \to 0$,

$$G(s)H(s) = \frac{K_m}{s^m} = \frac{K_m}{(\epsilon^m)e^{jm\theta}} = \frac{K_m}{\epsilon^m} e^{-jm\theta} = \frac{K_m}{\epsilon^m} e^{jm\psi} \tag{9.63}$$

where $m = 1, 2, 3, 4, \ldots$. It is seen from Equation 9.63 that, as s moves from 0^- to 0^+, the plot of $G(s)H(s)$ traces m CW semicircles of infinite radius about the origin. Since the polar plots are symmetrical about the real axis, all that is needed is to determine the shape of the plot of $G(s)H(s)$ for a range of values of $0 < \omega < +\infty$. The net rotation of the plot for the range of $-\infty < \omega < +\infty$ is twice that of the plot for the range of $0 < \omega < +\infty$.

9.3.6.6 When $G(j\omega)H(j\omega)$ Passes through the Point $-1 + j0$

When the curve of $G(j\omega)H(j\omega)$ passes through the $-1 + j0$ point, the number of encirclements N is indeterminate. This corresponds to the condition where $B(s)$ has zeros on the imaginary axis. A necessary condition for applying the Nyquist criterion is that the path encircling the specified area must not pass through any poles or zeros of $B(s)$. When this condition is violated, the value for N becomes indeterminate and the Nyquist stability criterion cannot be applied. Simple imaginary zeros of $B(s)$ mean that the closed-loop system will have a continuous steady-state sinusoidal component in its output that is independent of the form of the input. Unless otherwise stated, this condition is considered unstable.

9.3.6.7 Nyquist's Stability Criterion Applied to Systems Having Dead Time

The transfer function representing transport lag (dead time) is

$$G_\tau(s) = e^{-\tau s} \to G_\tau(j\omega) = e^{-j\omega\tau} = 1\angle -\omega\tau \tag{9.64}$$

It has a magnitude of unity and a negative angle whose magnitude increases directly in proportion to frequency. The polar plot of Equation 9.64 is a unit circle that is traced indefinitely, as shown in Figure 9.28a. The corresponding Lm and phase-angle diagram shows a constant value of 0 dB and a phase angle that decreases with frequency. When the contour Q is traversed and the polar-plot characteristic of dead time, shown in Figure 9.28, is included, the effects on the complete polar plot are as follows:

1. In traversing the imaginary axis of the contour Q between $0^+ < \omega < +\infty$, the portion of the polar plot of $G(j\omega)H(j\omega)$ in the third quadrant is shifted CW, closer to the $-1 + j0$ point (see Figure 10.28c). Thus, if the dead time is increased sufficiently, the $-1 + j0$ point is enclosed by the polar plot and the system becomes unstable.
2. As $\omega \to +\infty$, the magnitude of the angle contributed by the transport lag increases indefinitely. This yields a spiraling curve as $|G(j\omega)H(j\omega)| \to 0$.

A transport lag therefore tends to make a system less stable. This is illustrated for the transfer function

$$G(s)H(s) = \frac{K_1 e^{-\tau s}}{s(1 + T_1 s)(1 + T_2 s)} \tag{9.64a}$$

Figure 9.28b shows the polar plot without transport lag; Figure 9.28c shows the destabilizing effect of transport lag.

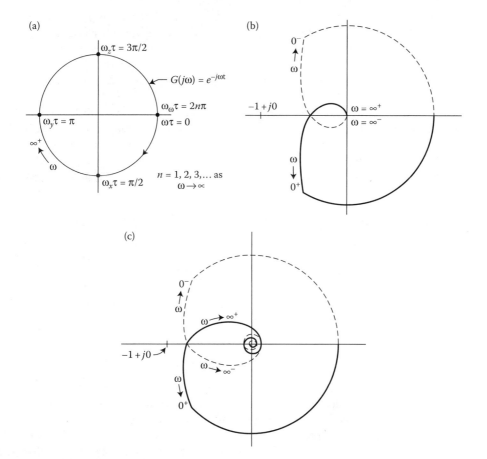

FIGURE 9.28 (a) Polar-plot characteristic for transport lag, Equation 9.64; (b) polar plot for Equation 9.64a without transport lag ($\tau = 0$); and (c) destabilizing effect of transport lag in Equation 9.64a.

9.3.7 Definitions of Phase Margin and Gain Margin and Their Relation to Stability

The stability and approximate degree of stability [5] can be determined from the Lm and phase diagram. The stability characteristic is specified in terms of the following quantities:

- *Gain crossover* This is the point on the plot of the transfer function at which the magnitude of $G(j\omega)$ is unity [$\text{Lm}G(j\omega) = 0$ dB]. The frequency at gain crossover is called the *phase-margin frequency* ω_ϕ.
- *Phase-margin angle* γ This angle is 180° plus the negative trigonometrically considered angle of the transfer function at the gain crossover point. It is designated as the angle γ, which can be expressed as $\gamma = 180° + \phi$, where $\angle[G(j\omega_\phi)] = \phi$ is negative.
- *Phase crossover* This is the point on the plot of the transfer function at which the phase angle is $-180°$. The frequency at which phase crossover occurs is called the *gain-margin frequency* ω_c.
- *Gain margin* The gain margin is the factor a by which the gain must be changed in order to produce instability., i.e.,

$$|G(j\omega_c)|a = 1 \rightarrow |G(j\omega)| = \frac{1}{a}$$
$$\rightarrow \text{Lm } a = -\text{Lm}G(j\omega_c) \tag{9.65}$$

These quantities are illustrated in Figure 9.29 on both the Lm and the polar curves. Note the algebraic sign associated with these two quantities as marked on the curves. Figures 9.29a and 9.29b represent a stable system, and Figures 9.29c and 9.29d represent an unstable system. The phase-margin angle is the amount of phase shift at the frequency ω_ϕ that would just produce instability. The γ for minimum-phase systems must be positive for a stable system, whereas a negative γ means that the system is unstable.

It is shown later that γ is related to the effective damping ratio ζ of the system. Satisfactory response is usually obtained in the range of $40° \leq \gamma \leq 60°$. As an individual acquires experience, the value of γ to be used for a particular system becomes more evident. This guideline for system performance applies only to those systems where behavior is that of an equivalent second-order system. The gain margin must be positive when expressed in decibels (greater than unity as a numeric) for a stable system. A negative gain margin means that the system is unstable. The damping ratio ζ of the system is also related to the gain margin. However, γ gives a better estimate of damping ratio, and therefore of the transient overshoot of the system, than does the gain margin.

The values of ω_ϕ, γ, ω_c, and Lm a are also readily identified on the Nichols plot as shown in Figure 9.30 and described in the next section. Further information about the speed of response of the system can be obtained from the maximum value of the control ratio and the frequency at which this maximum occurs. The relationship of stability and gain margin is modified for a conditionally stable system. [3] Instability can occur with both an increase or a decrease in gain. Therefore, both "upper" and "lower" gain margins must be identified, corresponding to the upper crossover frequency ω_{cu} and the lower crossover frequency ω_{cl}.

9.3.7.1 Stability Characteristics of the Lm and Phase Diagram

The total phase angle of the transfer function at any frequency is closely related to the slope of the Lm curve at that frequency. A slope of -20 dB/decade is related to an angle of $-90°$; a slope of -40 dB/decade is related to an angle of $-180°$; a slope of -60 dB/decade is related to an angle of $-270°$; etc. Changes of slope at higher and lower corner frequencies, around the particular frequency being considered, contribute to the total angle at that frequency. The farther away the changes of slope are from the particular frequency, the less they contribute to the total angle at that frequency. The stability of a minimum-phase system requires that $\gamma > 0$. For this to be true, the angle at the gain crossover

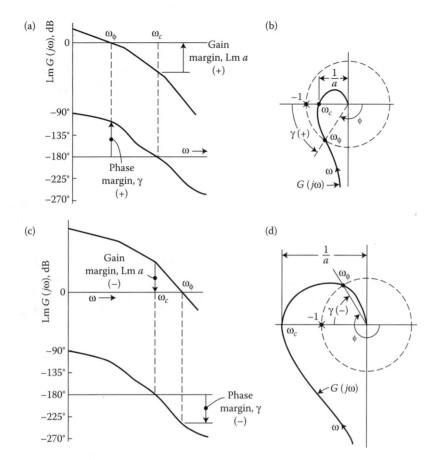

FIGURE 9.29 Log magnitude and phase diagram and polar plots of $G(j\omega)$, showing gain margin and phase margin: (a) and (b) stable; (c) and (d) unstable.

[Lm $G(j\omega) = 0$ dB] must be greater than $-180°$. This places a limit on the slope of the Lm curve at the gain crossover. *The slope at the gain crossover should be more positive than -40 dB/decade if the adjacent corner frequencies are not close.* A slope of -20 dB/decade is preferable. This is derived from the consideration of a theorem by Bode. Thus, the Lm and phase diagram reveals some pertinent information. For example, the gain can be adjusted (this raises or lowers the Lm curve) to achieve the desirable range of $45° \le \gamma \le 60°$. The shape of the low-frequency portion of the curve determines system type and therefore the degree of steady-state accuracy. The system type and the gain determine the error coefficients and therefore the steady-state error. The value of ω_ϕ gives a qualitative indication of the speed of response of a system.

9.3.7.2 Stability from the Nichols Plot (Lm–Angle Diagram)

The Lm–angle diagram is drawn by picking for each frequency the values of Lm and angle from the Lm and phase diagrams vs. ω (Bode plot). The resultant curve has frequency as a parameter. The curve for the example shown in Figure 9.30, shows a positive gain margin and phase margin angle; therefore, this represents a stable system. Changing the gain raises or lowers the curve without changing the angle characteristics. Increasing the gain raises the curve, thereby decreasing the gain margin and phase-margin angle, with the result that the stability is decreased. Increasing the gain so that the curve has a positive Lm at $-180°$ results in negative gain margin and phase-margin angle; therefore, an unstable system results.

Decreasing the gain lowers the curve and increases stability. However, a large gain is desired to reduce steady-state errors [5].

The Lm–angle diagram for $G(s)H(s)$ can be drawn for all values of s on the contour Q of Figure 9.26a. The resultant curve for minimum-phase systems is a closed contour. Nyquist's criterion can be applied to this contour by determining the number of points (having the values 0 dB and odd multiples of 180°) enclosed by the curve of $G(s)H(s)$. This number is the value of N that is used in the equation $Z_R = N - P_R$ to determine the value of Z_R. An example for $G(s) = K_1/[s(1 + Ts)]$ is shown in Figure 9.31. The Lm–angle contour for a nonminimum-phase system does not close [5]; thus, it is more difficult to determine the value of N. For these cases, the polar plot is easier to use to determine stability.

It is not necessary to obtain the complete Lm–angle contour to determine stability for minimum-phase systems. Only that portion of the contour is drawn representing $G(j\omega)$ for the range of values $0^+ < \omega < \infty$. The stability is then determined from the position of the curve of $G(j\omega)$ relative to the (0 dB, −180°) point. In other words, the curve is traced in the direction of increasing frequency, i.e., walking along the curve in the direction of increasing frequency. The system is stable if the (0 dB, −180°) point is to the right of the curve. This is a simplified rule of thumb, which is based on Nyquist's stability criterion for a minimum-phase system.

A conditionally stable system is one in which the curve crosses the −180° axis at more than one point. The gain determines whether the system is stable or unstable. That is, instability (or stability) can occur with both an increase (or a decrease) in gain.

9.3.8 Closed-Loop Tracking Performance Based on the Frequency Response

A correlation between the frequency and time responses of a system, leading to a method of gain setting in order to achieve a specified closed-loop frequency response, is now developed [12]. The closed-loop frequency response is obtained as a function of the open-loop frequency response. Although the design is performed in the frequency domain, the closed-loop responses in the time domain are also obtained. Then a "best" design is selected by considering both the frequency and the time responses. Both the polar plot and the Lm–angle diagram (Nichols plot) are used.

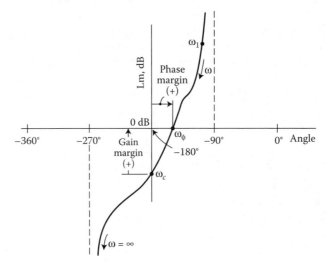

FIGURE 9.30　Typical log magnitude–angle diagram for $G(j\omega) = \dfrac{4(1+j0.5\omega)}{j\omega(1+j2\omega)[1+j0.5\omega+(j0.125\omega)^2]}$.

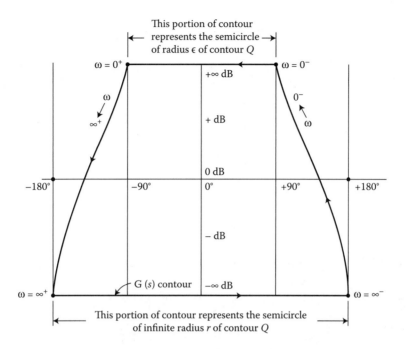

FIGURE 9.31 The log magnitude–angle diagram for $G(s) = K_1/s(1 + Ts)$.

9.3.8.1 Direct Polar Plot

The frequency control ratio $C(j\omega)/R(j\omega)$ for a unity feedback system is given by

$$\frac{C(j\omega)}{R(j\omega)} = \frac{A(j\omega)}{B(j\omega)} = \frac{|A(j\omega)| \, e^{j\phi(\omega)}}{|B(j\omega)| \, e^{j\lambda(\omega)}} = \frac{G(j\omega)}{1 + G(j\omega)}$$

$$\frac{C(j\omega)}{R(j\omega)} = \left| \frac{A(j\omega)}{B(j\omega)} \right| e^{j(\phi - \lambda)} = M(\omega) e^{j\alpha(\omega)} \tag{9.66}$$

where $A(j\omega) = G(j\omega)$ and $B(j\omega) = 1 + G(j\omega)$. Since the magnitude of the angle $\phi(\omega)$, as shown in Figure 9.32, is greater than the magnitude of the angle $\lambda(\omega)$, the value of the angle $\alpha(\omega)$ is negative. Remember that CCW rotation is taken as positive. The error control ratio $E(j\omega)/R(j\omega)$ is given by

$$\frac{E(j\omega)}{R(j\omega)} = \frac{1}{1 + G(j\omega)} = \frac{1}{|B(j\omega)| \, e^{j\lambda}} \tag{9.67}$$

From Equation 9.67 and Figure 9.32, it is seen that the greater the distance from the $-1 + j0$ point to a point on the $G(j\omega)$ locus, for a given frequency, the smaller the steady-state sinusoidal error for a stated sinusoidal input. Thus, the usefulness and importance of the polar plot of $G(j\omega)$ have been enhanced.

9.3.8.2 Determination of M_m and ω_m for a Simple Second-Order System

The frequency at which the maximum value of $|C(j\omega)/R(j\omega)|$ occurs (see Figure 9.33) is referred to as the *resonant frequency* ω_m. The maximum value is labeled M_m. These two quantities are figures of merit (F.O.M.) of a system. Compensation to improve system performance is based upon a knowledge of ω_m and M_m. For a *simple second-order system* a direct and simple relationship can be obtained for M_m and

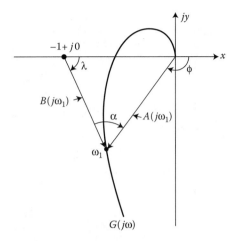

FIGURE 9.32 Polar plot of $G(j\omega)$ for a unity-feedback system.

ω_m in terms of the system parameters [5, Sect. 9.3]. These relationships are

$$\omega_m = \omega_n\sqrt{1 - 2\zeta^2} \quad M_m = \frac{1}{2\zeta\sqrt{1 - \zeta^2}} \tag{9.68}$$

From these equations, it is seen that the curve of M vs. ω has a peak value, other than at $\omega = 0$, for only $\zeta < 0.707$. Figure 9.34 shows a plot of M_m vs. ζ for a simple second-order system. It is seen for values of $\zeta < 0.4$ that M_m increases very rapidly in magnitude; the transient oscillatory response is therefore excessively large and might damage the physical system. The correlation between the frequency and time responses is shown qualitatively in Figure 9.35.

The corresponding time domain F.O.M [5] are

- The damped natural frequency ω_d
- The peak value M_p
- The peak time T_p
- The settling time $t_s(\pm2\%)$

For a unit-step forcing function, these F.O.M. for the transient of a simple second-order system are

$$M_p = 1 + e^{-\zeta\pi\sqrt{1-\zeta^2}} \tag{9.69}$$

$$T_p = \pi/\omega_d \tag{9.70}$$

$$\omega_d = \omega_n\sqrt{1 - \zeta^2} \tag{9.71}$$

$$t_s = 4/\zeta\omega_n \tag{9.72}$$

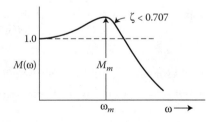

FIGURE 9.33 A closed-loop frequency-response curve indicating M_m and ω_m.

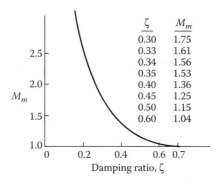

FIGURE 9.34   A plot of M_m vs. ζ for a simple second-order system.

Therefore, for a simple second-order system the following conclusions are obtained in correlating the frequency and time responses:

1. Inspection of Equation 9.68 reveals that ω_m is a function of both ω_n and ζ. Thus, for a given ζ, the larger the value of ω_m, the larger ω_n, and the faster the transient time of response for this system given by Equation 9.72.
2. Inspection of Equations 9.68 and 9.69 shows that both M_m and M_p are functions of ζ. The smaller ζ becomes, the larger in value M_m and M_p become. Thus, it is concluded that the larger the value of M_m, the larger the value of M_p. For values of $\zeta < 0.4$, the correspondence between M_m and M_p is only qualitative for this simple case. In other words, for $\zeta = 0$ the time domain yields $M_p = 2$, whereas the frequency domain yields $M_m = \infty$. When $\zeta > 0.4$, there is a close correspondence between M_m and M_p.
3. Note that the shorter the distance between the $-1 + j0$ point and a particular $G(j\omega)$ plot (see Figure 9.36), the smaller the damping ratio. Thus, M_m is larger and consequently M_p is also larger.

From these characteristics, a designer can obtain a good approximation of the time response of a simple second-order system by knowing only the M_m and ω_m of its frequency response. A corresponding correlation for ω_m and M_m becomes tedious for more complex systems. Therefore, a graphic procedure is generally used, as shown in the following sections [13,14].

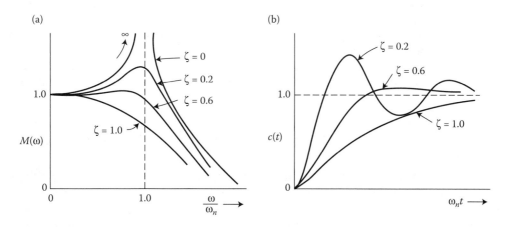

FIGURE 9.35   (a) Plots of M vs. ω/ω_n for a simple second-order system and (b) corresponding time plots for a step input.

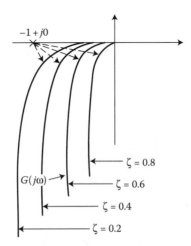

FIGURE 9.36 Polar plots of $G(j\omega) = K_1/[j\omega(1+j\omega T)]$ for different values of K_1 and the resulting closed-loop damping ratios.

9.3.8.3 Correlation of Sinusoidal and Time Response

It has been found by experience [5] that M_m is also a function of the *effective* ζ and ω_n for higher-order systems. The effective ζ and ω_n of a higher-order system is dependent upon the ζ and ω_n of each second-order term, the zeros of $C(s)/R(s)$, and the values of the real roots in the characteristic equation of $C(s)/R(s)$. Thus, in order to alter the M_m, the location of some of the roots must be changed. Which ones should be altered depends on which are dominant in the time domain. From the analysis for a simple second-order system, whenever the frequency response has the shape shown in Figure 9.33, the following correlation exists between the frequency and time responses for systems of any order:

1. The larger ω_m is made, the faster the time of response for the system.
2. The value of M_m gives a qualitative measure of M_p within the acceptable range of the *effective* damping ratio $0.4 < \zeta < 0.707$. In terms of M_m, the acceptable range is $1 < M_m < 1.4$.
3. The closer the $G(j\omega)$ curve comes to the $-1 + j0$ point, the larger the value of M_m.

The larger K_p, K_v, or K_a is made, the greater the steady-state accuracy for a step, a ramp, and a parabolic input, respectively. In terms of the polar plot, the farther the point $G(j\omega)\big|_{\omega=0} = K_0$ is from the origin, the more accurate is the steady-state time response for a step input. For a Type-1 system, the farther the low-frequency asymptote (as $\omega \to 0$) is from the imaginary axis, the more accurate is the steady-state time response for a ramp input. All the factors mentioned above are merely *guideposts* in the *frequency domain* to assist the designer in obtaining an *approximate* idea of the time response of a system. They serve as "stop-and-go signals" to indicate if one is headed in the right direction in achieving the desired time response. If the desired performance specifications are not satisfactorily met, compensation techniques must be used.

9.3.8.4 Constant $M(\omega)$ and $\alpha(\omega)$ Contours of $C(j\omega)/R(j\omega)$ on the Complex Plane (Direct Plot)

The contours of constant values of M drawn in the complex plane yield a rapid means of determining the values of M_m and ω_m and the value of gain required to achieve a desired value of M_m. In conjunction with the contours of constant values of $\alpha(\omega)$, also drawn in the complex plane, the plot of $C(j\omega)/R(j\omega)$ can be obtained rapidly. The M and α contours are developed only for unity-feedback systems by inserting $G(j\omega) = x + jy$ into $M(j\omega)$ [5]. The derivation of the M and α contours yields the equation of a circle with its center at the point (a, b) and having radius r. The location of the center and the radius for a specified value of M are given by

$$x_0 = -\frac{M^2}{M^2 - 1} \tag{9.73}$$

$$y_0 = 0 \tag{9.74}$$

$$r_0 = \left| \frac{M}{M^2 - 1} \right| \tag{9.75}$$

This circle is called a *constant M contour* for $M = M_a$. Figure 9.37 shows a family of circles in the complex plane for different values of M. Note that the larger the value M, the smaller its corresponding M circle. A further inspection of Figure 9.37 and Equation 9.73 reveals the following:

1. For $M \to \infty$, which represents a condition of oscillation ($\zeta \to 0$), the center of the M circle $x_0 \to -1 + j0$ and the radius $r_0 \to 0$. Thus, as the $G(j\omega)$ plot comes closer to the $-1 + j0$ point, the effective ζ becomes smaller and the degree of stability decreases.
2. For $M(\omega) = 1$, which represents the condition where $C(j\omega) = R(j\omega)$, $r_0 \to \infty$ and the M contour becomes a straight line perpendicular to the real axis at $x = -1/2$.
3. For $M \to 0$, the center of the M circle $x_0 \to 0$ and the radius $r_0 \to 0$.
4. For $M > 1$, the centers x_0 of the circles lie to the left of $x = -1 + j0$; and for $M < 1$, x_0 of the circles lie to the right of $x = 0$. All centers are on the real axis.

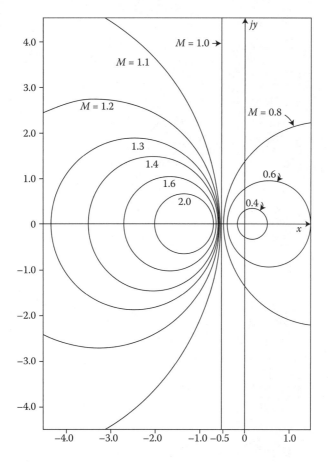

FIGURE 9.37 Constant M contours.

$\alpha(\omega)$ *Contours*　　The $\alpha(\omega)$ contours, representing constant values of phase angle $\alpha(\omega)$ for $C(j\omega)/R(j\omega)$, can also be determined in the same manner as for the M contours [5]. The derivation results in the equation of a circle, with $N = \tan\alpha$ as a parameter, given by

$$\left(x+\frac{1}{2}\right)^2 + \left(y-\frac{1}{2N}\right)^2 = \frac{1}{4}\frac{N^2+1}{N^2} \tag{9.76a}$$

whose center is located at $x_q = -\frac{1}{2}$, $y_q = \frac{1}{2N}$ with a radius

$$r_q = \frac{1}{2}\left(\frac{N^2+1}{N^2}\right)^{1/2} \tag{9.76b}$$

Different values of α result in a family of circles in the complex plane with centers on the line represented by $(-1/2, y)$, as illustrated in Figure 9.38.

Tangents to the M Circles [5]　　The line drawn through the origin of the complex plane and tangent to a given M circle plays an important part in setting the gain of $G(j\omega)$. Referring to Figure 9.39 and recognizing that $bc = r_0$ is the radius and $ob = x_0$ is the distance to the center of the particular M circle yields $\sin\psi = 1/M$ and $oa = 1$.

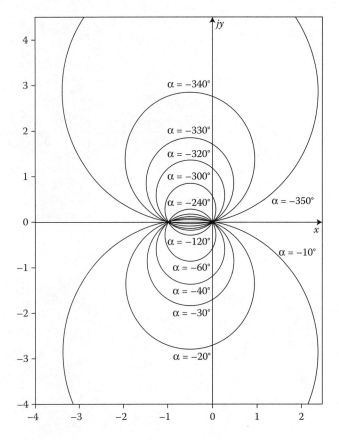

FIGURE 9.38　Constant α contours.

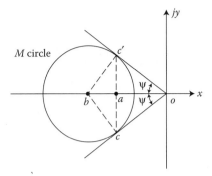

FIGURE 9.39 Determination of sin ψ.

9.3.9 Gain Adjustment for a Desired M_m of a Unity-Feedback System (Direct Polar Plot)

Gain adjustment is the first step in adjusting the system for the desired performance. The procedure for adjusting the gain is outlined in this section. Figure 9.40a shows $G_x(j\omega)$ with its respective M_m circle in the complex plane. Since

$$G_x(j\omega) = x + jy = K_x G'_x(j\omega) = K_x(x' + jy') \tag{9.77}$$

then

$$x' + jy' = \frac{x}{K_x} + j\frac{y}{K_x}$$

where $G'_x(j\omega) = G_x(j\omega)/K_x$ is defined as the frequency-sensitive portion of $G_x(j\omega)$ with unity gain. Note that changing the gain merely changes the amplitude and not the angle of the locus of points of $G_x(j\omega)$. Thus, if in Figure 9.40a a change of scale is made by dividing the x, y coordinates by K_x so that the new coordinates are x', y', the following are true:

1. The $G_x(j\omega)$ plot becomes the $G'_x(j\omega)$ plot.
2. The M_m circle becomes a circle that is simultaneously tangent to $G'_x(j\omega)$ and the line representing $\sin \psi = 1/M_m$.

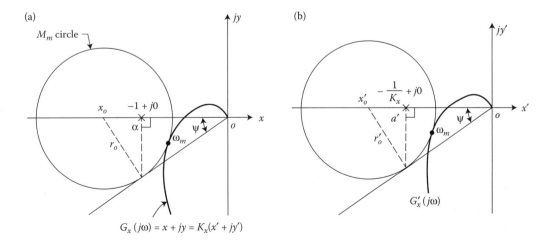

FIGURE 9.40 (a) Plot of $G_x(j\omega)$ with respective M_m circle and (b) circle drawn tangent to both the plot of $G'_x(j\omega)$ and the line representing the angle $\psi = \sin^{-1}(1/M_m)$.

3. The $-1 + j0$ point becomes the $-1/K_x + j0$ point.
4. The radius r_0 becomes $r'_0 = r_0/K_x$.

It is possible to determine the required gain to achieve a desired M_m for a given system by using the following graphical procedure:

Step 1

If the original system has a transfer function

$$G_x(j\omega) = K_x G'_x(j\omega) = \frac{K_x(1 + j\omega T_1)(1 + j\omega T_2) \cdots}{(j\omega)^m (1 + j\omega T_a)(1 + j\omega T_b)(1 + j\omega T_c) \cdots} \qquad (9.78)$$

with an original gain K_x, only the frequency-sensitive portion $G'_x(j\omega)$ is plotted.

Step 2

Draw a straight line through the origin at the angle $\psi = \sin^{-1}(1/M_m)$, measured from the negative real axis.

Step 3

By trial and error, find a circle whose center lies on the negative real axis and is simultaneously tangent to both the $G'_x(j\omega)$ plot and the line drawn at the angle ψ.

Step 4

Having found this circle, locate the point of tangency on the ψ−angle line. Draw a vertical line from this point of tangency perpendicular to the real axis. Label the point where this line intersects the real axis as a'.

Step 5

For this circle to be an M circle representing M_m, the point a' must be the $-1 + j0$ point. Thus, the x', y' coordinates must be multiplied by a gain factor K_m in order to convert this plot into a plot of $G(j\omega)$. From the graphical construction the gain value is $K_m = 1/oa'$.

Step 6

The original gain must be changed by a factor $A = K_m/K_x$.

Note that if $G_x(j\omega)$, which includes a gain K_x, is already plotted, it is possible to work directly with the plot of the function $G_x(j\omega)$. Following the procedure just outlined results in the determination of the *additional* gain required to produce the specified M_m; that is, the additional gain is

$$A = \frac{K_m}{K_x} = \frac{1}{oa'} \qquad (9.79)$$

9.3.10 Constant M and α Curves on the Lm–Angle Diagram (Nichols Chart)

The transformation of the constant M curves (circles) [5,14] on the polar plot to the Lm-angle diagram is done more easily by starting from the inverse polar plot since all the M^{-1} circles have the same center at $-1 + j0$. Also, the constant α contours are radial lines drawn through this point [5]. There is a change of sign of the Lm and angle obtained, since the transformation is from the inverse transfer function on the inverse polar plot to the direct transfer function on the Lm vs. ϕ plot. This transformation results in constant M and α curves that have symmetry at every 180° interval. An expanded 300° section of the constant M and α graph is shown in Figure 9.41. This graph is commonly referred to as the *Nichols chart*. Note that the $M = 1$ (0 dB) curve is asymptotic to $\phi = -90°$ and $\phi = -270°$ and the curves for

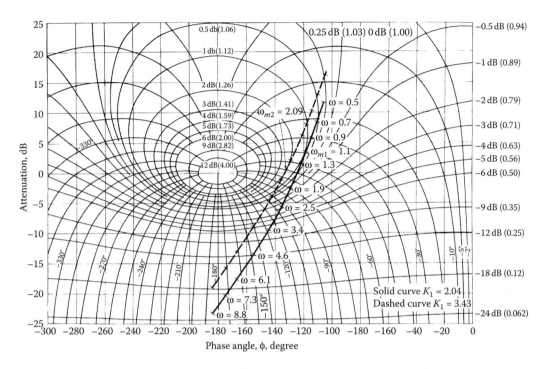

FIGURE 9.41 Use of the log magnitude–angle diagram (Nichols Chart) for $G_x(j\omega) = K_x G_x'(j\omega)$.

$M < 1/2(-6 \text{ dB})$ are always negative. The curve for $M = \infty$ is the point at 0 dB, $-180°$, and the curves for $M > 1$ are closed curves inside the limits $\phi = -90°$ and $\phi = -270°$. These loci for constant M and α on the Nichols Chart apply only for stable unity-feedback systems.

The Nichols Chart has the Cartesian coordinates of dB vs. phase angle ϕ. Standard graph paper with loci of constant M and α for the closed-loop transfer function is available. The open-loop frequency response $G(j\omega)$, with the frequencies noted along the plot, is superimposed on the Nichols Chart as shown in Figure 9.41. The intersections of the $G(j\omega)$ plot with the M and α contours yield the closed-loop frequency response $M\angle\alpha$.

By plotting Lm $G_x'(j\omega)$ vs. $\angle G_x'(j\omega)$ on the Nichols Chart, the value of K_x required to achieve a desired value of M_m can be determined. The amount Δ dB required to raise or lower this plot of $G_x(j\omega)$ vs. ϕ in order to make it just tangent to the desired $M = M_m$ contour yields Lm$K_x = \Delta$. The frequency value at the point of tangency, i.e., Lm$G_x(j\omega_m)$, yields the value of the resonant frequency $\omega = \omega_m$.

9.3.11 Correlation of Pole–Zero Diagram with Frequency and Time Responses

Whenever the closed-loop control ratio $M(j\omega)$ has the characteristic [5] form shown in Figure 9.33, the system may be approximated as a simple second-order system. This usually implies that the poles, other than the dominant complex pair, are either far to the left of the dominant complex poles or are close to zeros. When these conditions are not satisfied, the frequency response may have other shapes. This can be illustrated by considering the following three control ratios:

$$\frac{C(s)}{R(s)} = \frac{1}{s^2 + s + 1} \tag{9.80}$$

$$\frac{C(s)}{R(s)} = \frac{0.313(s + 0.8)}{(s + 0.25)(s^2 + 0.3s + 1)} \tag{9.81}$$

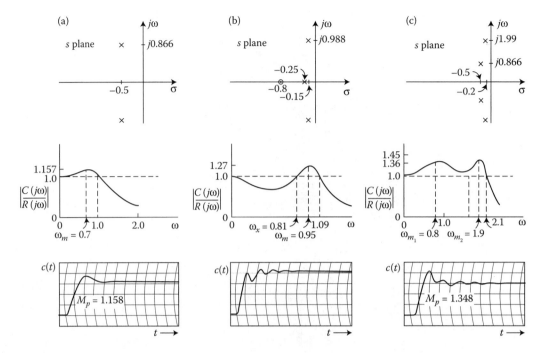

FIGURE 9.42 Comparison of frequency and time responses for three pole–zero patterns.

$$\frac{C(s)}{R(s)} = \frac{4}{(s^2 + s + 1)(s^2 + 0.4s + 4)} \tag{9.82}$$

The pole–zero diagram, the frequency response, and the time response to a step input for each of these equations are shown in Figure 9.42. For Equation 9.80 the following characteristics are noted from Figure 9.42a:

1. The control ratio has only two complex dominant poles and no zeros.
2. The frequency-response curve has the following characteristics:
   a. A single peak $M_m = 1.157$ at $\omega_m = 0.7$.
   b. $1 < M < M_m$ in the frequency range $0 < \omega < 1$.
3. The time response has the typical waveform for a simple underdamped second-order system. That is, the first maximum of $c(t)$ due to the oscillatory term is greater than $c(t)_{ss}$, and the $c(t)$ response after this maximum oscillates around the value of $c(t)_{ss}$.

For Equation 9.81 the following characteristics are noted from Figure 9.42b:

1. The control ratio has two complex poles and one real pole, all dominant, and one real zero.
2. The frequency-response curve has the following characteristics:
   a. A single peak, $M_m = 1.27$ at $\omega_m = 0.95$.
   b. $M < 1$ in the frequency range $0 < \omega < \omega_x$.
   c. The peak M_m occurs at $\omega_m = 0.95 > \omega_x$.
3. The time response does not have the conventional waveform. That is, the first maximum of $c(t)$ due to the oscillatory term is less than $c(t)_{ss}$ because of the transient term $A_3 e^{-0.25t}$.

For Equation 9.82 the following characteristics are noted from Figure 9.42c:

1. The control ratio has four complex poles, all dominant, and no zeros.
2. The frequency-response curve has the following characteristics:

a. There are two peaks, $M_{m1} = 1.36$ at $\omega_{m1} = 0.81$ and $M_{m2} = 1.45$ at $\omega_{m2} = 1.9$.
b. $1 < M < 1.45$ in the frequency range $0 < \omega < 2.1$.
c. The time response does not have the simple second-order waveform. That is, the first maximum of $c(t)$ in the oscillation is greater than $c(t)_{ss}$, and the oscillatory portion of $c(t)$ does not oscillate about a value of $c(t)_{ss}$. This time response can be predicted from the pole locations in the s-plane and from the two peaks in the plot of M vs. ω.

9.3.12 Summary

The different types of frequency-response plots are presented in this chapter. All of these plots indicate the type of system under consideration. Both the polar plot and the Nichols plot can be used to determine the necessary gain adjustment that must be made to improve its response. The methods presented for obtaining the Lm frequency-response plots stress graphical techniques. For greater accuracy, a CAD program should be used to calculate this data. This chapter shows that the polar plot of the transfer function $G(s)$, in conjunction with Nyquist's stability criterion, gives a rapid means of determining whether a system is stable or unstable. The phase-margin angle and gain margin are also used as a means of measuring stability. This is followed by the correlation between the frequency and time responses. The F.O.M. M_m and ω_m are established as guideposts for evaluating the tracking performance of a system. The addition of a pole to an open-loop transfer function produces a CW shift of the direct polar plot, which results in a larger value of M_m. The time response also suffers because ω_m becomes smaller. The reverse is true if a zero is added to the open-loop transfer function. This agrees with the analysis using the root locus, which shows that the addition of a pole or zero results in a less stable or more stable system, respectively. Thus, the qualitative correlation between the root locus and the frequency response is enhanced. The M and α contours are an aid in adjusting the gain to obtain a desired M_m. The methods described for adjusting the gain for a desired M_m are based on the fact that generally the desired values of M_m are slightly greater than 1. This yields a time response having an underdamped response, with a small amplitude of oscillation that reaches steady state rapidly. When the gain adjustment does not yield a satisfactory value of ω_m, the system must be compensated in order to increase ω_m without changing the value of M_m.

References

1. Maccoll, L.A., *Fundamental Theory of Servomechanisms,* Van Nostrand, Princeton, NJ, 1945.
2. James, H.M., Nichols, N.B., and Phillips, R.S., *Theory of Servomechanisms,* McGraw-Hill, New York, 1947.
3. Bode, H.W., *Network Analysis and Feedback Amplifier Design,* Van Nostrand, Princeton, NJ, 1945, chap. 8.
4. Sanathanan, C.K. and Tsukui, H., Synthesis of transfer function from frequency response data, *Int. J. Syst. Sci.,* 5(1), 41–54, 1974.
5. D'Azzo, J.J. and Houpis, C.H., *Linear Control System Analysis and Design: Conventional and Modern,* 4th ed., McGraw-Hill, New York, 1995.
6. Bruns, R.A. and Saunders, R.M., *Analysis of Feedback Control Systems,* McGraw-Hill, New York, 1955, chap. 14.
7. Chestnut, H. and Mayer, R.W., *Servomechanisms and Regulating System Design,* Vol. 1, 2nd ed., Wiley, New York, 1959.
8. Nyquist, H., Regeneration theory, *Bell Syst. Tech. J.,* 11, 126–147, 1932.
9. Balabanian, N. and LePage, W.R., What is a minimum-phase network?, *Trans. AIEE,* 74, pt. II, 785–788, 1956.
10. Freudenberg, J.S. and Looze, D.P., Right half-plane poles and zeros and design trade-offs in feedback systems, *IEEE Trans. Autom. Control,* AC-30, 555–565, 1985.
11. D'Azzo, J.J. and Houpis, C.H., *Feedback Control System Analysis and Synthesis,* 2nd ed., McGraw-Hill, New York, 1966.
12. Brown, G.S. and Campbell, D.P., *Principles of Servomechanisms,* Wiley, New York, 1948.

13. Chu, Y., Correlation between frequency and transient responses of feedback control systems, *Trans. AIEE*, 72, pt. II, 81–92, 1953.

14. James, H.M., Nichols, N.B., and Phillips, R.S., *Theory of Servomechanisms*, McGraw-Hill, New York, 1947, chap. 4.

15. Higgins, T.J., and Siegel, C.M., Determination of the maximum modulus, or the specified gain of a servomechanism by complex variable differentiation, *Trans. AIEE*, 72, pt. II, 467, 1954.

9.4 The Root Locus Plot

William S. Levine

9.4.1 Introduction

The root locus plot was invented by W. R. Evans around 1948 [1,2]. This is somewhat surprising because the essential ideas behind the root locus were available many years earlier. All that is really needed is the Laplace transform, the idea that the poles of a linear time-invariant system are important in control design, and the geometry of the complex plane. One could argue that the essentials were known by 1868 when Maxwell published his paper "On Governors" [3]. It is interesting to speculate on why it took so long to discover such a natural and useful tool.

It has become much easier to produce root locus plots over the last few years. Evans's graphical construction has been superseded by computer software. Today it takes just a few minutes to input the necessary data to the computer. An accurate root locus plot is available seconds later [4]. In fact, the computer makes it possible to extend the basic idea of the root locus to study graphically almost any property of a system that can be parameterized by a single real number.

The detailed discussion of root locus plots and their uses begins with an example and a definition. This is followed by a description of the original rules and procedures for constructing root locus plots. Using the computer introduces different questions. These are addressed in Section 9.4.4. The use of root locus plots in the design of control systems is described in Section 9.4.5. In particular, the design of lead, lag, and lead/lag compensators, as well as the design of notch filters, is described. This is followed by a brief introduction to other uses of the basic idea of the root locus. The final section summarizes and mentions some limitations.

9.4.2 Definition

The situation of interest is illustrated in Figure 9.43, where $G(s)$ is the transfer function of a single-input single-output (SISO) linear time-invariant system and k is a real number. The closed-loop system has the transfer function

$$G_{cl}(s) = \frac{y(s)}{r(s)} = \frac{kG(s)}{1 + kG(s)} \tag{9.83}$$

The standard root locus only applies to the case where $G(s)$ is a rational function of s. That is,

$$G(s) = n(s)/d(s) \tag{9.84}$$

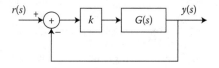

FIGURE 9.43　Block diagram for a simple unity feedback control system.

and $n(s)$ and $d(s)$ are polynomials in s with real coefficients. If this is true then it is easy to show that

$$G_{cl}(s) = \frac{kn(s)}{d(s) + kn(s)} \qquad (9.85)$$

Note that the numerators of $G(s)$ and of $G_{cl}(s)$ are identical. We have just proved that, except possibly for pole–zero cancellations, the open-loop system, $G(s)$, and the closed-loop system, $G_{cl}(s)$, have exactly the same zeros regardless of the value of k.

What happens to the poles of $G_{cl}(s)$ as k varies? This is precisely the question that is answered by the root locus plot. By definition, the root locus plot is a plot of the poles of $G_{cl}(s)$ in the complex plane as the parameter k varies. It is very easy to generate such plots for simple systems. For example, if

$$G(s) = \frac{3}{s^2 + 4s + 3}$$

then

$$G_{cl}(s) = \frac{3k}{s^2 + 4s + (3 + 3k)}$$

The poles of $G_{cl}(s)$, denoted by p_1 and p_2, are given by

$$p_1 = -2 + \sqrt{1 - 3k}$$
$$p_2 = -2 - \sqrt{1 - 3k} \qquad (9.86)$$

It is straightforward to plot p_1 and p_2 in the complex plane as k varies. This is done for $k \geq 0$ in Figure 9.44. Note that, strictly speaking, $G_{cl}(s) = 0$ when $k = 0$. However, the denominators of Equations 9.84 and 9.85 both give the same values for the closed-loop poles when $k = 0$, namely -1 and -3. Those values are the same as the open-loop poles. By convention, all root locus plots use the open-loop poles as the closed-loop poles when $k = 0$.

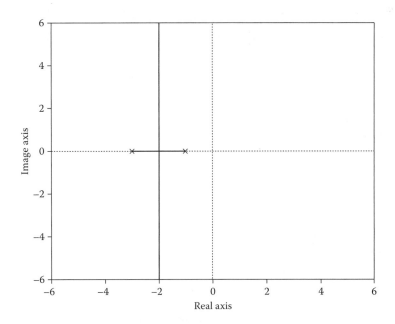

FIGURE 9.44 Root locus plot for $G(s) = 3/(s^2 + 4s + 3)$ and $k \geq 0$.

The plot provides a great deal of useful information. First, it gives the pole locations for every possible closed-loop system that can be created from the open-loop plant and any positive gain k. Second, if there are points on the root locus for which the closed-loop system would meet the design specifications, then simply applying the corresponding value of k completes the design. For example, if a closed-loop system with damping ratio $\zeta = 0.707$ is desired for the system whose root locus is plotted in Figure 9.44, then simply choose the value of k that puts the poles of $G_{cl}(s)$ at $-2 \pm j2$. That is, from Equation 9.86, choose $k = 5/3$.

The standard root locus can be easily applied to nonunity feedback control systems by using block diagram manipulations to put the system in an equivalent unity feedback form (see *Control System Fundamentals*, Chapter 6). Because the standard root locus depends only on properties of polynomials it applies equally well to discrete-time systems. The only change is that $G(s)$ is replaced by $G(z)$, the z-transform transfer function.

9.4.3 Some Construction Rules

Evans's procedure for plotting the Root Locus consists of a collection of rules for determining if a test point, s_t, in the complex plane is a pole of $G_{cl}(s)$ for some value of k. The first such rule has already been explained.

Rule 1

The open-loop poles, i.e., the roots of $d(s) = 0$, are all points in the root locus plot corresponding to $k = 0$.
The second rule is also elementary. Suppose that

$$d(s) = s^n + a_{n-1}s^{n-1} + a_{n-2}s^{n-2} + \cdots + a_0$$
$$n(s) = b_m s^m + b_{m-1}s^{m-1} + \cdots + b_0 \tag{9.87}$$

For physical systems it is always true that $n > m$. Although it is possible to have reasonable mathematical models that violate this condition, it will be assumed that $n > m$. The denominator of Equation 9.85 is then

$$d_{cl}(s) = s^n + a_{n-1}s^{n-1} + \cdots + a_{n-m-1}s_{n-m-1} + (a_{n-m} + kb_m)s^m + \cdots + (a_0 + kb_0) \tag{9.88}$$

Rule 2 is an obvious consequence of Equation 9.88, the assumption that $n > m$, and the fact that a polynomial of degree n has exactly n roots.

Rule 2

The root locus consists of exactly n branches.
The remaining rules are derived from a different form of the denominator of $G_{cl}(s)$. Equation 9.83 shows that the denominator of $G_{cl}(s)$ can be written as $1 + kG(s)$. Even though $1 + kG(s)$ is not a polynomial it is still true that the poles of $G_{cl}(s)$ must satisfy the equation

$$1 + kG(s) = 0 \tag{9.89}$$

Because s is a complex number, $G(s)$ is generally complex and Equation 9.89 is equivalent to two independent equations. These could be, for instance, that the real and imaginary parts of Equation 9.89 must separately and independently equal zero. It is more convenient, and equivalent, to use the magnitude and angle of Equation 9.89. That is, Equation 9.89 is equivalent to the two equations

$$|kG(s)| = 1 \tag{9.90}$$
$$\angle(kG(s)) = \pm(2h + 1)180°, \quad \text{where } h = 0, 1, 2, \ldots$$

The first equation explicitly states that, for Equation 9.89 to hold, the magnitude of $kG(s)$ must be one. The second equation shows that the phase angle of $kG(s)$ must be $\pm180°$, or $\pm540°$, etc. It is possible to simplify Equations 9.90 somewhat because k is a real number. Thus, for $k \geq 0$ Equations 9.90 become

$$|G(s)| = 1/k \tag{9.91}$$
$$\angle(G(s)) = \pm(2h+1)180°, \quad \text{where } h = 0, 1, 2, \ldots$$

The form for $k \leq 0$ is the same for the magnitude of $G(s)$, except for a minus sign ($|G(s)| = -\frac{1}{k}$), but the angle condition becomes integer multiples of 360°.

Equations 9.91 are the basis for plotting the root locus. The first step in producing the plot is to mark the locations of the open-loop poles and zeros on a graph of the complex plane. The poles are denoted by the symbol, ×, as in Figure 9.44, while the zeros are denoted by the symbol, o. If the poles and zeros are accurately plotted it is then possible to measure $|G(s_t)|$ and $\angle(G(s_t))$ for any given test point s_t.

For example, suppose $G(s) = 10(s+4)/(s+3+j4)(s+3-j4)$. The poles and zeros of this transfer function are plotted in Figure 9.45. Notice that the plot does not depend, in any way, on the gain 10. It is generally true that pole–zero plots are ambiguous with respect to pure gain. Figure 9.45 contains a plot of the complex number $(s+4)$ for the specific value $s_t = -1+j3$. It is exactly the same length as, and parallel to, the vector drawn from the zero to the point $s = -1+j3$, also shown. The same is true of the vectors corresponding to the two poles. To save effort, only the vectors from the poles to the test point $s_t = -1+j3$ are drawn. Once the figure is drawn, simple measurements with a ruler and a protractor provide

$$|G(-1+j3)| = \frac{l_1}{l_2 l_3}$$
$$\angle(G(-1+j3)) = \phi_1 - \phi_2 - \phi_3$$

One can then check the angle condition in Equations 9.91 to see if $s_t = -1+j3$ is a point on the root locus for this $G(s)$.

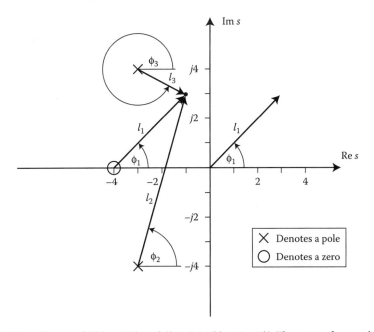

FIGURE 9.45 Poles and zeros of $G(s) = [10(s+4)/(s+3+j4)(s+3-j4)]$. The vectors from each of the singularities to the test point $s_t = -1+j3$ are also shown, as is the vector $s+4|_{s_t=-1+j3}$.

Of course, it would be tedious to check every point in the complex plane. This is not necessary. There is a collection of rules for finding points on the root locus plot. A few of these are developed below. The others can be found in most undergraduate textbooks on control, such as [5,6].

Rule 3

For $k \geq 0$, any point on the real axis that lies to the left of an odd number of singularities (poles plus zeros) on the real axis is a point on the root locus. Any other point on the real axis is not. (Change "odd" to "even" for negative k.)

A proof follows immediately from applying the angle condition in Equations 9.91 to test points on the real axis. The angular contributions due to poles and zeros that are not on the real axis cancel as a result of symmetry. Poles and zeros to the left of the test point have angles equal to zero. Poles and zeros to the right of the test point contribute angles of $-180°$ and $+180°$, respectively. In fact, a fourth rule follows easily from the symmetry.

Rule 4

The root locus is symmetric about the real axis.

We already know that all branches start at the open-loop poles. Where do they end?

Rule 5

If $G(s)$ has n poles and m finite zeros ($m \leq n$) then exactly m branches terminate, as $k \to \infty$, on the finite zeros. The remaining $n - m$ branches go to infinity as $k \to \infty$.

The validity of the rule can be proved by taking limits as $k \to \infty$ in the magnitude part of Equations 9.91. Doing so gives

$$\lim_{k \to \infty} |G(s)| = \lim_{k \to \infty} \frac{1}{k} = 0$$

Thus, as $k \to \infty$, it must be true that $|G(s)| \to 0$. This is true when s coincides with any finite zero of $G(s)$. From Equations 9.84 and 9.87,

$$\lim_{s \to \infty} G(s) = \lim_{s \to \infty} \frac{b_m s^m + b_{m-1} s^{m-1} + \cdots + b_0}{s^n + a_{n-1} s^{n-1} + \cdots + a_0}$$
$$= \lim_{s \to \infty} \frac{s^{m-n}(b_m + b_{m-1} s^{-1} + \cdots + b_0 s^{-m})}{(1 + a_{n-1} s^{-1} + \cdots + a_0 s^{-n})}$$

Finally,

$$\lim_{s \to \infty} G(s) = \lim_{s \to \infty} b_m s^{m-n} = 0 \tag{9.92}$$

The b_m factors out and the fact that $|G(s)| \to 0$ as $s \to \infty$ with multiplicity $n - m$ is apparent. One can think of this as a demonstration that $G(s)$ has $n - m$ zeros at infinity. Equation 9.92 plays an important role in the proof of the next rule as well.

Rule 6

If $G(s)$ has n poles and m finite zeros ($n \geq m$) and $k \geq 0$ then the $n - m$ branches that end at infinity asymptotically approach lines that intersect the real axis at a point σ_0 and that make an angle γ with the real axis, where

$$\gamma = \pm \frac{(1+2h)180°}{n-m} \quad \text{where } h = 0,1,2,\ldots$$

and

$$\sigma_0 = \frac{\sum\limits_{i=1}^{n} Re(p_i) - \sum\limits_{l=1}^{m} Re(z_l)}{n-m}$$

A proof of the formula for γ follows from applying the angle condition of Equations 9.91 to Equation 9.92. That is,

$$\angle(s^{m-n}) = \pm(1+2h)180°$$

so

$$\gamma = \angle(s) = \frac{\mp(1+2h)180°}{n-m}$$

A proof of the equation for σ_0 can be found in [5, pp. 239–249]. Most textbooks include around a dozen rules for plotting the root locus; see, for example, [5–8]. These are much less important today than they were just a few years ago because good, inexpensive software for plotting root loci is now widely available.

9.4.4 Use of the Computer to Plot Root Loci

There are many different software packages that can be used to plot the root locus. A particularly well-known example is MATLAB. To some extent, the software is foolproof. If the data are input correctly, the resulting root locus is calculated correctly. However, there are several possible pitfalls. For example, the software automatically scales the plot. The scaling can obscure important aspects of the root locus, as is described in the next section. This, and other possible problems associated with computer-generated root locus plots, is discussed in more detail in [4].

9.4.5 Uses

The root locus plot can be an excellent tool for designing single-input single-output control systems. It is particularly effective when the open-loop transfer function is accurately known and is, at least approximately, reasonably low order. This is often the case in the design of servomechanisms. The root locus is also very useful as an aid to understanding the effect of feedback and compensation on the closed-loop system poles. Some ways to use the root locus are illustrated below.

9.4.5.1 Design of a Proportional Feedback Gain

Consider the open-loop plant $G(s) = 1/s(s+1)(0.1s+1)$. This is a typical textbook example. The plant is third order and given in factored form. The plant has been normalized so that $\lim_{s \to 0} sG(s) = 1$. This is particularly helpful in comparing different candidate designs. A servo motor driving an inertial load would typically have such a description. The motor plus load would correspond to the poles at 0 and -1. The electrical characteristics of the motor add a pole that is normally fairly far to the left, such as the pole at -10 in this example.

The simplest controller is a pure gain, as in Figure 9.43. Suppose, for illustration, that the specifications on the closed-loop system are that ζ, the damping ratio, must be exactly 0.707 and the natural frequency, ω_n, must be as large as possible. Because the closed-loop system is actually third order, both damping ratio and natural frequency are not well defined. However, the open-loop system has a pair of dominant poles, those at 0 and −1. The closed-loop system will also have a dominant pair of poles because the third pole will be more than a factor of 10 to the left of the complex-conjugate pair of poles. The damping ratio and natural frequency will then be defined by the dominant poles.

A root locus plot for this system, generated by MATLAB, is shown in Figure 9.46. (Using the default scaling it was difficult to see the exact intersection of the locus with the $\zeta = 0.707$ line because the automatic scaling includes all the poles in the visible plot. A version of the same plot with better scaling was created and is the one shown.) The diagonal lines on the plot correspond to $\zeta = 0.707$. The four curved lines are lines of constant ω_n, with values 1, 2, 3, 4. The value of gain k corresponding to $\zeta = 0.707$ is approximately 5, and the closed-loop poles resulting from this gain are roughly $−10, −1/2 \pm j/2$. These values are easily obtained from MATLAB [4]. The bandwidth of the closed-loop system is fairly small because, as you can see from Figure 9.46, the dominant poles have $\omega_n < 1$. This indicates that the transient response will be slow. If a closed-loop system that responds faster is needed, two approaches could be used. The plant could be changed. For instance, if the plant is a motor then a more powerful motor could be obtained. This would move the pole at −1 to the left, say to −3. A faster closed-loop system could then be designed using only a feedback gain. The other option is to introduce a lead compensator.

9.4.5.2 Design of Lead Compensation

A lead compensator is a device that can be added in series with the plant and has a transfer function $G_c(s) = (s − z)/(s − p)$. Both z and p are real and negative. The zero, z, lies to the right of the pole, p ($z > p$). Because the magnitude of the transfer function of a lead compensator will increase with frequency between the zero and the pole, some combination of actuator limits and noise usually forces $p/z < 10$. Lead compensators are always used in conjunction with a gain, k.

The purpose of a lead compensator is to speed up the transient response. The example we have been working on is one for which lead compensation is easy and effective. For a system with three real poles and no zeros one normally puts the zero of the lead compensator close to the middle pole of the plant. The compensator pole is placed as far to the left as possible. The result of doing this for our example is shown

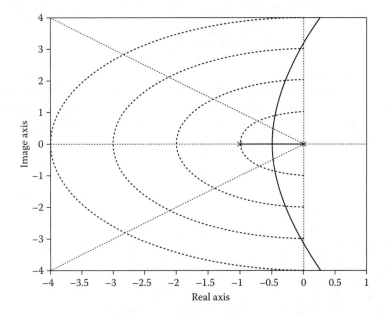

FIGURE 9.46 Root locus plot of $1/s(s + 1)(s + 10)$. The dotted diagonal lines correspond to $\zeta = 0.707$. The dotted elliptical lines correspond to $\omega_n = 1, 2, 3, 4$, respectively.

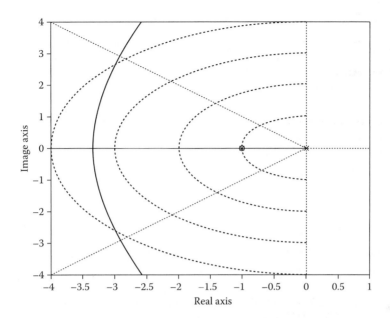

FIGURE 9.47 Root locus plot of $G_c(s)G(s) = (s+1)/(s+10)\,[1/s(s+1)(s+10)]$. The dotted diagonal lines correspond to $\zeta = 0.707$. The dotted elliptical curves correspond to $w_n = 1, 2, 3, 4$.

in Figure 9.47. Comparison of the root loci in Figures 9.46 and 9.47 shows that the lead compensator has made it possible to find a gain, k, for which the closed-loop system has $\zeta = 0.707$ and $w_n > 4$. This approach to designing lead compensators basically maximizes the bandwidth of the closed-loop system for a given damping ratio. If the desired transient response of the closed-loop system is specified in more detail, there are other procedures for placing the compensator poles and zeros [7, pp. 514–530].

9.4.5.3 Design of Lag Compensation

A lag compensator is a device that can be added in series with the plant and has a transfer function $G_c(s) = (s-z)/(s-p)$. Both z and p are real and negative. The zero, z, lies to the left of the pole, p ($z < p$). Lag compensators are always used in series with a gain, k. Again, it is usually not feasible to have the pole and the zero of the compensator differ by more than a factor of 10. One important reason for this is explained below.

The purpose of a lag compensator is to improve the steady-state response of the closed-loop system. Again, the example we have been working on illustrates the issues very well. Because our example already has an open-loop pole at the origin it is a type-1 system. The closed-loop system will have zero steady-state error in response to a unit step input. Adding another open-loop pole close to the origin would make the steady-state error of the closed-loop system in response to a unit ramp smaller. In fact, if we put the extra pole at the origin we would reduce this error to zero. Unfortunately, addition of only an open-loop pole close to the origin will severely damage the transient response. No choice of gain will produce a closed-loop system with a fast transient response.

The solution is to add a zero to the left of the lag compensator's pole. Then, if the gain, k, is chosen large enough, the compensator's pole will be close to the compensator's zero so they will approximately cancel each other. The result is that the closed-loop transient response will be nearly unaffected by the lag compensator while the steady-state error of the closed-loop system is reduced. Note that this will not be true if the gain is too low. In that case, the closed-loop transient response will be slowed by the compensator pole.

9.4.5.4 Design of Lead/Lag Compensation

Conceptually, the lead and lag compensators are independent. One can design a lead compensator so as to produce a closed-loop system that satisfies the specifications of the transient response while ignoring the steady-state specifications. One can then design a lag compensator to meet the steady-state requirements knowing that the effect of this compensator on the transient response will be negligible. As mentioned above, this does require that the gain is large enough to move the pole of the lag compensator close to its zero. Otherwise, the lag compensator will slow down the transient response, perhaps greatly.

There are a number of different ways to implement lead/lag compensators. One relatively inexpensive implementation is shown in [8, p. 700]. It has the disadvantage that the ratio of the lead compensator zero to the lead compensator pole must be identical to the ratio of the lag compensator pole to the lag compensator zero. This introduces some coupling between the two compensators which complicates the design process. See [7, pp. 537–547], for a discussion.

9.4.5.5 Design of a Notch Filter

The notch filter gets its name from the appearance of a notch in the plot of the magnitude of its transfer function versus frequency. Nonetheless, there are aspects of the design of notch filters that are best understood by means of the root locus plot. It is easiest to begin with an example where a notch filter would be appropriate. Such an example would have open-loop transfer function $G(s) = 1/s(s + 1)$ $(s + 0.1 + j5)(s + 0.1 - j5)$. This transfer function might correspond to a motor driving an inertial load at the end of a long and fairly flexible shaft. The flexure of the shaft introduces the pair of lightly damped poles and greatly complicates the design of a good feedback controller for this plant. While it is fairly rare that a motor can only be connected to its load by a flexible shaft, problems where the open-loop system includes a pair of lightly damped poles are reasonably common.

The obvious thing to do is to add a pair of zeros to cancel the offending poles. Because the poles are stable, although lightly damped, it is feasible to do this. The only important complication is that you cannot implement a compensator that has two zeros and no poles. In practice, one adds a compensator consisting of the desired pair of zeros and a pair of poles. The poles are usually placed as far to the left as feasible and close to the real axis. Such a compensator is called a notch filter.

One rarely knows the exact location of the lightly damped poles. Thus, the notch filter has to work well, even when the poles to be canceled are not exactly where they were expected to be. Simply plotting the root locus corresponding to a plant plus notch filter for which the zeros are above the poles, as is done in Figure 9.48, shows that such a design is relatively safe. The root locus lies to the left of the lightly damped poles and zeros. The root locus plot corresponding to the situation where the compensator zeros are below the poles curves the opposite way, showing that this is a dangerous situation, in the sense that such a system can easily become unstable as a result of small variations in the plant gain.

9.4.5.6 Other Uses of the Root Locus Plot

Any time the controller can be characterized by a single parameter it is possible to plot the locus of the closed-loop poles as a function of that parameter. This creates a kind of root locus that, although often very useful and easy enough to compute, does not necessarily satisfy the plotting rules given previously. An excellent example is provided by a special case of the optimal linear quadratic regulator. Given a SISO linear time-invariant system described in state-space form by

$$\dot{x}(t) = Ax(t) + bu(t)$$
$$y(t) = cx(t)$$

find the control that minimizes

$$J = \int_0^\infty (y^2(t) + ru^2(t)) \, dt$$

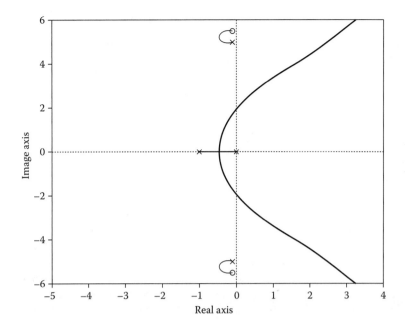

FIGURE 9.48 Root locus plot for the system $G_c(s)G(s) = [(s + 0.11 + j5.5)(s + 0.11 - j5.5)]/(s + 8)^2 \, 1/[s(s + 1)(s + 0.1 + j5)(s + 0.1 - j5)]$.

The solution, assuming that the state vector $x(t)$ is available for feedback, is $u(t) = kx(t)$, where k is a row vector containing n elements. The vector k is a function of the real number r so it is possible to plot the locus of the closed-loop poles as a function of r. Under some mild additional assumptions this locus demonstrates that these poles approach a Butterworth pattern as r goes to zero. The details can be found in [9, pp. 218–233].

9.4.6 Conclusions

The root locus has been one of the most useful items in the control engineer's toolbox since its invention. Modern computer software for plotting the root locus has only increased its utility. Of course, there are situations where it is difficult or impossible to use it. Specifically, when the system to be controlled is not accurately known or cannot be well approximated by a rational transfer function, then it is better to use other tools.

References

1. Evans, W.R., Control system synthesis by root locus method, *AIEE Trans.*, 69, 66–69, 1950.
2. Evans, W.R., Graphical analysis of control systems, *AIEE Trans.*, 67, 547–551, 1948.
3. Maxwell, J.C., On governors, *Proc. R. Soc. London*, 16, 270–283, 1868. (Reprinted in *Selected Papers on Mathematical Trends in Control Theory*, R. Bellman and R. Kalaba, Eds., Dover Publishing, New York, NY, 1964.)
4. Leonard, N.E. and Levine, W.S., *Using MATLAB to Analyze and Design Control Systems*, 2nd ed., Benjamin/Cummings, Menlo Park, CA, 1995.
5. Franklin, G.F., Powell, J.D., and Emami-Naeni, A., *Feedback Control of Dynamic Systems*, 5th ed., Pearson Prentice Hall, Upper Saddle River, NJ, 2006.
6. D'Azzo, J.J. and Houpis, C.H., *Linear Control System Analysis and Design*, 2nd ed., McGraw-Hill, New York, NY, 1981.
7. Nise, N.S., *Control Systems Engineering*, 4th ed., John Wiley and Sons Inc., Hoboken, NJ, 2004.

8. Dorf, R.C. and Bishop, R.H., *Modern Control Systems*, 11th ed., Pearson Prentice Hall, Upper Saddle River, NJ, 2008.

9. Kailath, T., *Linear Systems*, Prentice-Hall, Englewood Cliffs, NJ, 1980.

10. Kuo, B.C., *Automatic Control Systems*, 7th ed., Prentice-Hall, Englewood Cliffs, NJ, 1995.

9.5 PID Control

Karl J. Åström and Tore Hägglund

9.5.1 Introduction

The proportional–integral–derivative (PID) controller is by far the most commonly used controller. About 90–95% of all control problems are solved by this controller which comes in many forms. It is packaged in standard boxes for process control and in simpler versions for temperature control. It is a key component of all distributed systems for process control. Specialized controllers for many different applications are also based on PID control. The PID controller can thus be regarded as the "bread and butter" of control engineering. The PID controller has gone through many changes in technology. The early controllers were based on relays and synchronous electric motors or pneumatic or hydraulic systems. These systems were then replaced by electronics and, lately, microprocessors [2–6,10–12].

Much interest was devoted to PID control in the early development of automatic control. For a long time researchers paid very little attention to the PID controller. Lately, there has been a resurgence of interest in PID control because of the possibility of making PID controllers with automatic tuning, automatic generation of gain schedules and continuous adaptation. See the chapter "Automatic Tuning of PID Controllers" in this handbook.

Even if PID controllers are very common, they are not always used in the best way. The controllers are often poorly tuned. It is quite common that derivative action is not used. The reason is that it is difficult to tune three parameters by trial and error.

In this chapter, we will first present the basic PID controller in Section 9.5.2. When using PID control, it is important to be aware of the fact that PID controllers are parameterized in several different ways. This means for example that "integral time" does not mean the same thing for different controllers. PID controllers cannot be understood from linear theory. Amplitude and rate limitations in the actuators are key elements that lead to the windup phenomena. This is discussed in Section 9.5.4 where different ways to avoid windup are also discussed. Mode switches also are discussed in the same section.

Most PID controllers are implemented as digital controllers. In Section 9.5.5, we discuss digital implementation. In Section 9.5.6, we discuss uses of PID control, and in Section 9.5.7 we describe how complex control systems are obtained in a "bottom up" fashion by combining PID controllers with other simple systems.

We also refer to the companion chapter "Automatic Tuning of PID Controllers" in this handbook, which treats design and tuning of PID controllers. Examples of industrial products are also given in that chapter.

9.5.2 The Control Law

In a PID controller the control action is generated as a sum of three terms. The control law is thus described as

$$u(t) = u_P(t) + u_I(t) + u_D(t) \tag{9.93}$$

where u_P is the proportional part, u_I the integral part and u_D the derivative part.

9.5.2.1 Proportional Control

The proportional part is a simple feedback

$$u_P(t) = Ke(t) \tag{9.94}$$

where e is the control error, and K is the controller gain. The error is defined as the difference between the set point y_{sp} and the process output y, i.e.,

$$e(t) = y_{sp}(t) - y(t) \tag{9.95}$$

The modified form,

$$u_P(t) = K(by_{sp}(t) - y(t)) \tag{9.96}$$

where b is called *set point weighting,* admits independent adjustment of set point response and load disturbance response. *The setpoint response can also be influenced by a prefilter.*

9.5.2.2 Integral Control

Proportional control normally gives a system that has a steady-state error. Integral action is introduced to remove this. Integral action has the form

$$u_I(t) = k_i \int^t e(s)\, ds = \frac{K}{T_i} \int^t e(s)\, ds \tag{9.97}$$

The idea is simply that control action is taken even if the error is very small provided that the average of the error has the same sign over a long period.

Automatic Reset

A proportional controller often gives a steady-state error. A manually adjustable reset term may be added to the control signal to eliminate the steady-state error. The proportional controller given by Equation 9.94 then becomes

$$u(t) = Ke(t) + u_b(t) \tag{9.98}$$

where u_b is the reset term. Historically, integral action was the result of an attempt to obtain automatic adjustment of the reset term. One way to do this is shown in Figure 9.49.

The idea is simply to filter out the low frequency part of the error signal and add it to the proportional part. Notice that the closed loop has positive feedback. Analyzing the system in the figure we find that

$$U(s) = K(1 + \frac{1}{sT_i})E(s)$$

which is the input–output relation of a proportional–integral (PI) controller. Furthermore, we have

$$u_b(t) = \frac{K}{T_i} \int^t e(s)\, ds = u_I(t)$$

The automatic reset is thus the same as integral action.

Notice, however, that set point weighting is not obtained when integral action is obtained as automatic reset.

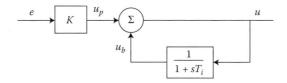

FIGURE 9.49   Controller with integral action implemented as automatic reset.

9.5.2.3 Derivative Control

Derivative control is used to provide anticipative action. A simple form is

$$u_D(t) = k_d \frac{de(t)}{dt} = KT_d \frac{de(t)}{dt} \tag{9.99}$$

The combination of proportional and derivative action is then

$$u_P(t) + u_D(t) = K \left[e(t) + T_d \frac{de(t)}{dt} \right]$$

This means that control action is based on linear extrapolation of the error T_d time units ahead. See Figure 9.50. Parameter T_d, which is called derivative time, thus has a good intuitive interpretation.

The main difference between a PID controller and a more complex controller is that a dynamic model admits better prediction than straight-line extrapolation.

In many practical applications, the set point is piecewise constant. This means that the derivative of the set point is zero except for those time instances when the set point is changed. At these time instances, the derivative becomes infinitely large. Linear extrapolation is not useful for predicting such signals. Also, linear extrapolation is inaccurate when the measurement signal changes rapidly compared to the prediction horizon T_d.

A better realization of derivative action is, therefore,

$$U_D(s) = \frac{KT_d s}{1 + sT_d/N}(cY_{sp}(s) - Y(s)) \tag{9.100}$$

The signals pass through a low-pass filter with time constant T_d/N. Parameter c is a set point weighting, which is often set to zero.

Filtering of Process Variable

Instead of filtering the the derivative as discusssed above one can also filter the measured signal by a first or second-order filter

$$G_f(s) = \frac{1}{1 + sT_f} \quad G_f(s) = \frac{1}{1 + sT_f + (sT_f)^2/2} \tag{9.101}$$

An ideal PID controller can then be applied to the filtered measurement. For PID controllers that are implemented digitally, the filter can be combined with the antialiasing filter as discussed in Section 9.5.5.

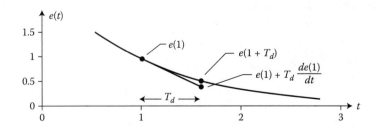

FIGURE 9.50   Interpretation of derivative action as prediction.

Set Point Weighting

The PID controller introduces extra zeros in the transmission from set point to output. From Equations 9.96, 9.97, and 9.99, the zeros of the PID controller can be determined as the roots of the equation

$$cT_iT_ds^2 + bT_is + 1 = 0 \tag{9.102}$$

There are no extra zeros if $b = 0$ and $c = 0$. If only $c = 0$, then there is one extra zero at

$$s = -\frac{1}{bT_i} \tag{9.103}$$

This zero can have a significant influence on the set point response. The overshoot is often too large with $b = 1$. It can be reduced substantially by using a smaller value of b. This is a much better solution than the traditional way of detuning the controller [2,3].

This is illustrated in Figure 9.51, which shows PI control of a system with the transfer function

$$G_p(s) = \frac{1}{s+a} \tag{9.104}$$

9.5.3 Different Representations

The PID controller discussed in the previous section can be described by

$$U(s) = G_{sp}(s)Y_{sp}(s) - G_c(s)Y(s) \tag{9.105}$$

where

$$G_{sp}(s) = K\left(b + \frac{1}{sT_i} + c\frac{sT_d}{1 + sT_d/N}\right)$$
$$G_c(s) = K\left(1 + \frac{1}{sT_i} + \frac{sT_d}{1 + sT_d/N}\right) \tag{9.106}$$

The linear behavior of the controller is thus characterized by two transfer functions: $G_{sp}(s)$, which gives the signal transmission from the set point to the control variable, and $G_c(s)$, which describes the signal transmission from the process output to the control variable.

Notice that the signal transmission from the process output to the control signal is different from the signal transmission from the set point to the control signal if either set point weighting parameter $b \neq 1$ or $c \neq 1$. The PID controller then has two degrees of freedom.

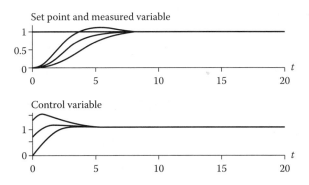

FIGURE 9.51 The usefulness of set point weighting. The values of the set point weighting parameter are $0, 0.5,$ and 1.

Another way to express this is that the set point parameters make it possible to modify the zeros in the signal transmission from set point to control signal.

The PID controller is thus a simple control algorithm that has seven parameters: Controller gain K, integral time T_i, derivative time T_d, maximum derivative gain N, set point weightings b and c, and filter time constant T_f. Parameters K, T_i and T_d are the primary parameters that are normally discussed. Parameter N is a constant, whose value typically is between 5 and 20. The set point weighting parameter b is often 0 or 1, although it is quite useful to use different values. Parameter c is mostly zero in commercial controllers.

9.5.3.1 The Standard Form

The controller given by Equations 9.105 and 9.106 is called the *standard form*, or the ISA (Instrument Society of America) form. The standard form admits complex zeros, which is useful when controlling systems with oscillatory poles. The parameterization given in Equation 9.106 is the normal one. There are, however, also other parameterizations.

9.5.3.2 The Parallel Form

A slight variation of the standard form is the *parallel form*, which is described by

$$U(s) = k[bY_{sp}(s) - Y(s)] + \frac{k_i}{s}[Y_{sp}(s) - Y(s)] + \frac{k_d s}{1 + sT_{df}}[cY_{sp}(s) - Y(s)] \tag{9.107}$$

This form has the advantage that it is easy to obtain pure proportional, integral or derivative control simply by setting appropriate parameters to zero. The interpretation of T_i and T_d as integration time and prediction horizon is, however, lost in this representation. The parameters of the controllers given by Equations 9.105 and 9.107 are related by

$$k = K$$
$$k_i = \frac{K}{T_i}$$
$$k_d = KT_d \tag{9.108}$$

Use of the different forms causes considerable confusion, particularly when parameter $1/k_i$ is called integral time and k_d derivative time.

The form given by Equation 9.107 is often useful in analytical calculations, because the parameters appear linearly. However, the parameters do not have nice physical interpretations.

The Parallel Form with Signal Filtering

If the measured signal is filtered [7] the controller transfer function is

$$C(s) = k_i \frac{1 + sT_i + s^2 T_i T_d}{s(1 + sT_f)} \tag{9.109}$$

$$C(s) = k_i \frac{1 + sT_i + s^2 T_i T_d}{s(1 + sT_f + s^2 T_i T_f)} \tag{9.110}$$

for systems with first- or second-order filtering. The controller with a first-order filter is sometimes called a complex proportional–integral lead, [1]. The input–output relation for a controller with setpoint weighting becomes

$$U(s) = k[bY_{sp}(s) - Y_f(s)] + \frac{k_i}{s}[Y_{sp}(s) - Y_f(s)] + k_d s[cY_{sp}(s) - Y_f(s)]$$

$$Y_f(s) = \frac{1}{1 + sT_f + (sT_f)^2/2} \tag{9.111}$$

Since the above forms are a cascade of and ideal PID controller and a filter, the filter can be combined with the process and the design can be carried out for an ideal PID controller.

9.5.3.3 The Series Forms

If $N = 0$ and if $T_i > 4T_d$ the transfer function $G_c(s)$ can be written as

$$G'_c(s) = K'\left(1 + \frac{1}{sT'_i}\right)(1 + sT'_d) \tag{9.112}$$

This form is called the series form. The parameters are related to the parameters of the parallel form in the following way:

$$K = K'\frac{T'_i + T'_d}{T'_i}$$

$$T_i = T'_i + T'_d$$

$$T_d = \frac{T'_i T'_d}{T'_i + T'_d} \tag{9.113}$$

The inverse relation is

$$K' = \frac{K}{2}\left(1 + \sqrt{1 - 4T_d/T_i}\right)$$

$$T'_i = \frac{T_i}{2}\left(1 + \sqrt{1 - 4T_d/T_i}\right)$$

$$T'_d = \frac{T_i}{2}\left(1 - \sqrt{1 - 4T_d/T_i}\right) \tag{9.114}$$

Similar, but more complicated, formulas are obtained for $N \neq 0$. Notice that the parallel form admits complex zeros while the series form has real zeros.

The parallel form given by Equations 9.105 and 9.106 is more general. The series form is also called the classical form because it is obtained naturally when a controller is implemented as automatic reset. The series form has an attractive interpretation in the frequency domain because the zeros of the feedback transfer function are the inverse values of T'_i and T'_d. Because of tradition, the form of the controller remained unchanged when technology changed from pneumatic via electric to digital.

It is important to keep in mind that different controllers may have different structures. This means that if a controller in a certain control loop is replaced by another type of controller, the controller parameters may have to be changed. Note, however, that the series and parallel forms differ only when both the integral and the derivative parts of the controller are used.

The parallel form is the most general form, because pure proportional or integral action can be obtained with finite parameters. The controller can also have complex zeros. In this way, it is the most flexible form. However, it is also the form where the parameters have little physical interpretation. The series form is the least general, because it does not allow complex zeros in the feedback path.

9.5.3.4 Velocity Algorithms

The PID controllers given by Equations 9.105, 9.107, and 9.112 are called positional algorithms, because the output of the algorithms is the control variable. In some cases, it is more natural to let the control algorithm generate the rate of change of the control signal. Such a control law is called a velocity algorithm. In digital implementations, velocity algorithms are also called incremental algorithms.

Many early controllers that were built around motors used velocity algorithms. Algorithms and structure were often retained by the manufacturers when technology was changed in order to have products that were compatible with older equipment. Another reason is that many practical issues, like windup protection and bumpless parameter changes, are easy to implement using the velocity algorithm.

A velocity algorithm cannot be used directly for a controller without integral action, because such a controller cannot keep the stationary value. The system will have an unstable mode, an integrator, that is canceled. Special care must therefore be exercised for velocity algorithms that allow the integral action to be switched off.

9.5.4 Nonlinear Issues

So far we have discussed only the linear behavior of the PID controller. There are several nonlinear issues that also must be considered. These include effects of actuator saturation, mode switches, and parameter changes.

9.5.4.1 Actuator Saturation and Windup

All actuators have physical limitations, a control valve cannot be more than fully open or fully closed, a motor has limited velocity, etc. This has severe consequences for control [2,8]. Integral action in a PID controller is an unstable mode. This does not cause any difficulty when the loop is closed. The feedback loop will, however, be broken when the actuator saturates, because the output of the saturating element is then not influenced by its input. The unstable mode in the controller may then drift to very large values. When the actuator desaturates it may then take a long time for the system to recover. It may also happen that the actuator bounces several times between high and low values before the system recovers.

Integrator windup is illustrated in Figure 9.52, which shows simulation of a system where the process dynamics is a saturation at a level of ± 0.1 followed by a linear system with the transfer function

$$G(s) = \frac{1}{s(s+1)}$$

The controller is a PI controller with gain $K = 0.27$ and $T_i = 7.5$. The set point is a unit step. Because of the saturation in the actuator, the control signal saturates immediately when the step is applied. The control signal then remains at the saturation level and the feedback is broken. The integral part continues to increase, because the error is positive. The integral part starts to decrease when the output equals the set point, but the output remains saturated because of the large integral part. The output finally decreases around time $t = 14$ when the integral part has decreased sufficiently. The system then settles. The net effect is that there is a large overshoot. This phenomenon, which was observed experimentally very early, is called "integrator windup." Many so-called antiwindup schemes for avoiding windup have been developed; conditional integration and tracking are two common methods.

9.5.4.2 Conditional Integration

Integrator windup can be avoided by using integral action only when certain conditions are fulfilled. Integral action is thus switched off when the actuator saturates, and it is switched on again when it desaturates. This scheme is easy to implement, but it leads to controllers with discontinuities. Care must also be exercised when formulating the switching logic so that the system does not come to a state where integral action is never used.

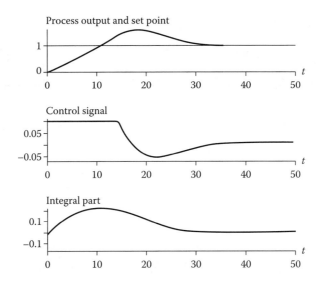

FIGURE 9.52   Simulation that illustrates integrator windup.

9.5.4.3 Tracking

Tracking or back calculation is another way to avoid windup. The idea is to make sure that the integral is kept at a proper value when the actuator saturates so that the controller is ready to resume action as soon as the control error changes. This can be done as shown in Figure 9.53. The actuator output is measured and the signal e_t, which is the difference between the input v and the output u of the actuator, is formed. The signal e_t is different from zero when the actuator saturates. The signal e_t is then fed back to the integrator. The feedback does not have any effect when the actuator does not saturate because the signal e_t is then zero. When the actuator saturates, the feedback drives the integrator output to a value such that the error e_t is zero.

Figure 9.54 illustrates the effect of using the antiwindup scheme. The simulation is identical to the one in Figure 9.52, and the curves from that figure are copied to illustrate the properties of the system. Notice the drastic difference in the behavior of the system. The control signal starts to decrease before the output reaches the set point. The integral part of the controller is also initially driven towards negative values.

The signal y_t may be regarded as an external signal to the controller. The PID controller can then be represented as a block with three inputs, y_{sp}, y and y_t, and one output v, and the antiwindup scheme can then be shown as in Figure 9.55. Notice that tracking is disabled when the signals y_t and v are the same.

The signal y_t is called the tracking signal because the output of the controller tracks this signal. The time constant T_t is called the tracking time constant.

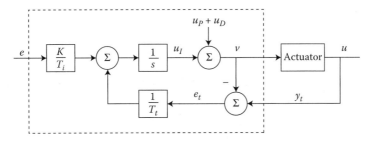

FIGURE 9.53   PID controller that avoids windup by tracking.

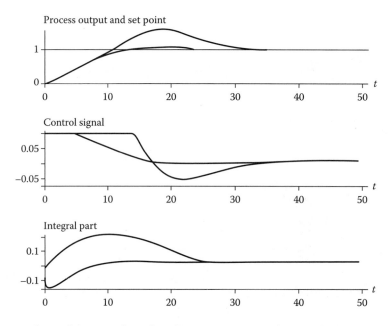

FIGURE 9.54 Simulation of PID controller with tracking. For comparison, the response for a system without windup protection is also shown. Compare with Figure 9.52.

The configuration with a tracking input is very useful in many contexts. Manual control signals can be introduced at the tracking input. Several controllers can also be combined to build complex systems. One example is when controllers are coupled in parallel or when selectors are used as discussed in Section 9.5.8.

The tracking time constant influences the behavior of the system as shown in Figure 9.56. The values of the tracking constant are 1, 5, 20, and 100. The system recovers faster with smaller tracking constants. It is, however, not useful to make the time constant too small, because tracking may then be introduced accidentally by noise. It is reasonable to choose $T_t < T_i$ for a PI controller and $T_d < T_t < T_i$ for a PID controller [2].

9.5.4.4 The Proportional Band

Let u_{max} and u_{min} denote the limits of the control variable. The proportional band K_p of the controller is then

$$K_p = \frac{u_{max} - u_{min}}{K}$$

This is sometimes used instead of the gain of the controller; the value is often expressed in percent (%).

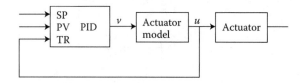

FIGURE 9.55 Antiwindup in PID controller with tracking input.

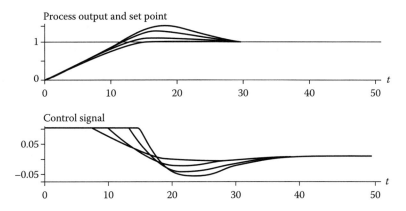

FIGURE 9.56 Effect of the tracking time constant on the antiwindup. The values of the tracking time constant are 1, 5, 20, and 100.

For a PI controller, the values of the process output that correspond to the limits of the control signal are given by

$$y_{\max} = by_{sp} + \frac{u_I - u_{\max}}{K}$$
$$y_{\min} = by_{sp} + \frac{u_I - u_{\min}}{K}$$

The controller operates linearly only if the process output is in the range $(y_{\min}, y_{\max})$. The controller output saturates when the predicted output is outside this band. Notice that the proportional band is strongly influenced by the integral term. A good insight into the windup problem and antiwindup schemes is obtained by investigating the proportional band. To illustrate this, Figure 9.57 shows the same simulation as Figure 9.52, but the proportional band is now also shown. The figure shows that the output is outside the proportional band initially. The control signal is thus saturated immediately. The signal desaturates as soon as the output leaves the proportional band. The large overshoot is caused by windup, which increases the integral when the output saturates.

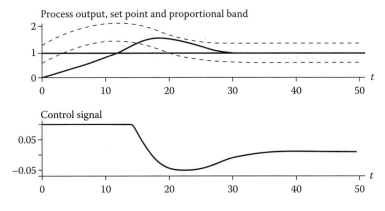

FIGURE 9.57 Proportional band for simulation in Figure 9.52.

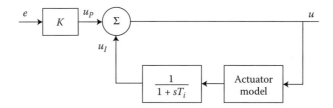

FIGURE 9.58 A scheme for avoiding windup in a controller with a series implementation.

9.5.4.5 Antiwindup in Controller on Series Form

A special method is used to avoid windup in controllers with a series implementation. Figure 9.58 shows a block diagram of the system. The idea is to make sure that the integral term that represents the automatic reset is always inside the saturation limits. The proportional and derivative parts do, however, change the output directly. It is also possible to treat the input to the saturation as an external tracking signal.

Notice that the tracking time constant in the controller in Figure 9.58 is equal to the integration time. Better performance can often be obtained with smaller values. This is a limitation of the scheme in Figure 9.58.

9.5.4.6 Antiwindup in Velocity Algorithms

In a controller that uses a velocity algorithm we can avoid windup simply by limiting the input to the integrator. The behavior of the system is then similar to a controller with conditional integration.

9.5.4.7 Mode Switches

Most PID controllers can be operated in one of two modes, manual or automatic [2,3]. So far we have discussed the automatic mode. In the manual mode the controller output is manipulated directly. This is often done by two buttons labeled "increase" and "decrease." The output is changed with a given rate when a button is pushed. To obtain this function the buttons are connected to the output via an integrator. *The integrator used for integral action can also be used if the manual input is introduced as a tracking signal.*

It is important that the system be implemented in such a way that there are no transients when the modes are switched. This is very easy to arrange in a controller based on a velocity algorithm, where the same integrator is used in both modes.

It is more complicated to obtain bumpless parameter changes in the other implementations. It is often handled via the tracking mode.

9.5.4.8 Parameter Changes

Switching transients may also occur when parameters are changed. Some transients cannot be avoided, but others are implementation dependent. In a proportional controller it is unavoidable to have transients if the gain is changed when the control error is different from zero.

For controllers with integral action, it is possible to avoid switching transients even if the parameters are changed when the error is not zero, provided that the controller is implemented properly.

If integral action is implemented as

$$\frac{dx}{dt} = e$$

$$I = \frac{K}{T_i} x$$

there will be a transient whenever K or T_i is changed when $x \neq 0$.

If the integral part is realized as

$$\frac{dx}{dt} = \frac{K}{T_i} e$$
$$I = x$$

we find that the transient is avoided. This is a manifestation that linear time-varying systems do not commute.

9.5.5 Digital Implementation

Most controllers are implemented using digital controllers. In this handbook several chapters deal with these issues. Here we will summarize some issues of particular relevance to PID control. The following operations are performed when a controller is implemented digitally:

- Step 1. Wait for clock interrupt.
- Step 2. Read analog input.
- Step 3. Compute control signal.
- Step 4. Set analog output.
- Step 5. Update controller variables.
- Step 6. Go to 1.

To avoid unnecessary delay, it is useful to arrange the computations so that as many as possible of the calculations are performed in Step 5. In Step 3, it is then sufficient to do two multiplications and one addition.

When computations are based on sampled data, it is good practice to introduce a prefilter that effectively eliminates all frequencies above the Nyquist frequency, $f_N = \pi/h$, where h is the sampling period. If this is not done, high-frequency disturbances may be aliased so that they appear as low-frequency disturbances. In commercial PID controllers this is often done by a first-order system.

9.5.5.1 Discretization

So far we have characterized the PID controller as a continuous time system. To obtain a computer implementation, we have to find a discrete time approximation. There are many ways to do this. Refer to the section on digital control for a general discussion; here we do approximations specifically for the PID controller. We will discuss discretization of the different terms separately. The sampling instants are denoted as t_k where $k = 0, 1, 2, \ldots$. It is assumed that the sampling instants are equally spaced. The sampling period is denoted by h. The proportional action, which is described by

$$u_p = K(by_{sp} - y)$$

is easily discretized by replacing the continuous variables with their sampled versions. This gives

$$u_p(t_k) = K\left(by_{sp}(t_k) - y(t_k)\right) \tag{9.115}$$

The integral term is given by

$$u_I(t) = \frac{K}{T_i} \int_0^t e(\tau)\, d\tau$$

Differentiation with respect to time gives

$$\frac{du_I}{dt} = \frac{K}{T_i} e$$

There are several ways to discretize this equation. Approximating the derivative by a forward difference gives

$$u_I(t_{k+1}) = u_I(t_k) + \frac{Kh}{T_i} e(t_k) \tag{9.116}$$

If the derivative is approximated by a backward difference we get instead

$$u_I(t_k) = u_I(t_{k-1}) + \frac{Kh}{T_i} e(t_k) \tag{9.117}$$

Another possibility is to approximate the integral by the trapezoidal rule, which gives

$$u_I(t_{k+1}) = u_I(t_k) + \frac{Kh}{T_i} \frac{e(t_{k+1}) + e(t_k)}{2} \tag{9.118}$$

Yet, another method is called ramp equivalence. This method gives exact outputs at the sampling instants if the input signal is continuous and piecewise linear between the sampling instants. In this particular case, the ramp equivalence method gives the same approximation of the integral term as the Tustin approximation. The derivative term is given by

$$\frac{T_d}{N} \frac{du_D}{dt} + u_D = -KT_d \frac{dy}{dt}$$

This equation can be approximated in the same way as the integral term.

The forward difference approximation is

$$u_D(t_{k+1}) = \left(1 - \frac{Nh}{T_d}\right) u_D(t_k) - KN(y(t_{k+1}) - y(t_k)) \tag{9.119}$$

The backward difference approximation is

$$u_D(t_k) = \frac{T_d}{T_d + Nh} u_D(t_{k-1}) - \frac{KT_dN}{T_d + Nh}(y(t_k) - y(t_{k-1})) \tag{9.120}$$

Tustin's approximation gives

$$u_D(t_k) = \frac{2T_d - Nh}{2T_d + Nh} u_D(t_{k-1}) - \frac{2KT_dN}{2T_d + Nh}(y(t_k) - y(t_{k-1})) \tag{9.121}$$

The ramp equivalence approximation gives

$$u_D(t_k) = e^{-Nh/T_d} u_D(t_{k-1}) - \frac{KT_d(1 - e^{-Nh/T_d})}{h}(y(t_k) - y(t_{k-1})) \tag{9.122}$$

9.5.5.2 Unification

The approximations of the integral and derivative terms have the same form, namely

$$u_I(t_k) = u_I(t_{k-1}) + b_{i1}e(t_k) + b_{i2}e(t_{k-1})$$
$$u_D(t_k) = a_d u_D(t_{k-1}) - b_d(y(t_k) - y(t_{k-1})) \tag{9.123}$$

The parameters for the different approximations are given in Table 9.7.

The controllers obtained can be written as

$$u(t_k) = t_0 y_{sp}(t_k) + t_1 y_{sp}(t_{k-1}) + t_2 y_{sp}(t_{k-2}) - s_0 y(t_k) - s_1 y(t_{k-1}) - s_2 y(t_{k-2})$$
$$+ (1 + a_d)u(t_{k-1}) - a_d u(t_{k-2}) \tag{9.124}$$

where $s_0 = K + b_{i1} + b_d$; $s_1 = -K(1 + a_d) - b_{i1}a_d + b_{i2} - 2b_d$; $s_2 = Ka_d - b_{i2}a_d + b_d$; $t_0 = Kb + b_{i1}$; $t_1 = -Kb(1 + a_d) - b_{i1}a_d + b_{i2}$; and $t_2 = Kba_d - b_{i2}a_d$.

Equation 9.124 gives the linear behavior of the controller. To obtain the complete controller, we have to add the antiwindup feature and facilities for changing modes and parameters.

TABLE 9.7   Parameters for the Different Approximations

	Forward	Backward	Tustin	Ramp Equivalence
b_{i1}	0	$\dfrac{Kh}{T_i}$	$\dfrac{Kh}{2T_i}$	$\dfrac{Kh}{2T_i}$
b_{i2}	$\dfrac{Kh}{T_i}$	0	$\dfrac{Kh}{2T_i}$	$\dfrac{Kh}{2T_i}$
a_d	$1-\dfrac{Nh}{T_d}$	$\dfrac{T_d}{T_d+Nh}$	$\dfrac{2T_d-Nh}{2T_d+Nh}$	e^{-Nh/T_d}
b_d	KN	$\dfrac{KT_dN}{T_d+Nh}$	$\dfrac{2KT_dN}{2T_d+Nh}$	$\dfrac{KT_d(1-e^{-Nh/T_d})}{h}$

9.5.5.3 Discussion

There is no significant difference between the different approximations of the integral term. The approximations of the derivative term have, however, quite different properties.

The approximations are stable when $|a_d| < 1$. For the forward difference approximation, this implies that $T_d > Nh/2$. The approximation is thus unstable for small values of T_d. The other approximations are stable for all values of T_d. Tustin's approximation and the forward difference method give negative values of a_d if T_d is small. This is undesirable, because the approximation then exhibits ringing. The backward difference approximation gives good results for all values of T_d.

Tustin's approximation and the ramp equivalence approximation give the best agreement with the continuous time case; the backward approximation gives less phase advance; and the forward approximation gives more phase advance. The forward approximation is seldom used because of the problems with instability for small values of derivative time T_d. Tustin's algorithm has the ringing problem for small T_d. Ramp equivalence requires evaluation of an exponential function. The backward difference approximation is used most commonly. The backward difference is well behaved.

9.5.5.4 Computer Code

As an illustration we give the computer code for a reasonably complete PID controller that has set point weighting, limitation of derivative gain, bumpless parameter changes and antiwindup protection by tracking.

Code
```
Compute controller coefficients
bi=K*h/Ti
ad=(2*Td-N*h)/(2*Td+N*h)
bd=2*K*N*Td/(2*Td+N*h)
a0=h/Tt
Bumpless parameter changes
uI=uI+Kold*(bold*ysp-y)-Knew*(bnew*ysp-y)
Read set point and process output from AD converter
ysp=adin(ch1)
y=adin(ch2)
Compute proportional part
uP=K*(b*ysp-y)
Update derivative part
uD=ad*uD-bd*(y-yold)
Compute control variable
```

v=uP+uI+uD
u=sat(v,ulow,uhigh)
Command analog output
daout(ch1)
Update integral part with windup protection
uI=uI+bi*(ysp-y)+ao*(u-v)
yold=y

Precomputation of the controller coefficients *ad, ao, bd*, and *bi* in Equation 9.124 saves computer time in the main loop. These computations are made only when the controller parameters are changed. The main program is called once every sampling period. The program has three states: *yold, uI*, and *uD*. One state variable can be eliminated at the cost of a less readable code.

PID controllers are implemented in many different computers, standard processors as well as dedicated machines. Word length is usually not a problem if general-purpose machines are used. For special-purpose systems, it may be possible to choose word length. It is necessary to have sufficiently long word length to properly represent the integral part.

9.5.5.5 Velocity Algorithms

The velocity algorithm is obtained simply by taking the difference of the position algorithm

$$\Delta u(t_k) = u(t_k) - u(t_{k-1}) = \Delta u_P(t_k) + \Delta I(t_k) + \Delta D(t_k)$$

The differences are then added to obtain the actual value of the control signal. Sometimes the integration is done externally. The differences of the proportional, derivative, and integral terms are obtained from Equations 9.115 and 9.123.

$$\begin{aligned}
\Delta u_P(t_k) &= u_P(t_k) - u_P(t_{k-1}) \\
&= K\left(by_{sp}(t_k) - y(t_k) - by_{sp}(t_{k-1}) + y(t_{k-1})\right) \\
\Delta u_I(t_k) &= u_I(t_k) - u_I(t_{k-1}) \\
&= b_{i1}\,e(t_k) + b_{i2}\,e(t_{k-1}) \\
\Delta u_D(t_k) &= u_D(t_k) - u_D(t_{k-1}) \\
&= a_d\Delta u_D(t_{k-1}) - b_d\left(y(t_k) - 2y(t_{k-1}) + y(t_{k-2})\right)
\end{aligned}$$

One advantage with the incremental algorithm is that most of the computations are done using increments only. Short word-length calculations can often be used. It is only in the final stage where the increments are added that precision is needed. Another advantage is that the controller output is driven directly from an integrator. This makes it very easy to deal with windup and mode switches. A problem with the incremental algorithm is that it cannot be used for controllers with P or proportional–derivative (PD) action only. Therefore, Δu_P has to be calculated in the following way when integral action is not used:

$$\Delta u_P(t_k) = K\left(by_{sp}(t_k) - y(t_k)\right) + u_b - u(t_{k-1})$$

where u_b is the bias term. When there is no integral action, it is necessary to adjust this term to obtain zero steady-state error.

9.5.6 Uses of PID Control

The PID controller is by far the control algorithm that is most commonly used. It is interesting to observe that in order to obtain a functional controller it is necessary to consider linear and nonlinear behavior of the controller as well as operational issues such as mode switches and tuning. For a discussion of tuning, we refer to the chapter "Automatic Tuning of PID Controllers" in this handbook. These questions have been worked out quite well for PID controllers, and the issues involved are quite well understood.

The PID controller in many cases gives satisfactory performance. It can often be used on processes that are difficult to control provided that extreme performance is not required. There are, however, situations when it is possible to obtain better performance by other types of controllers. Typical examples are processes with long relative dead times and oscillatory systems.

There are also cases where PID controllers are clearly inadequate. If we consider the fact that a PI controller always has phase lag and that a PID controller can provide a phase lead of at most 90°, it is clear that neither will work for systems that require more phase advance. A typical example is the stabilization of unstable systems with time delays.

A few examples are given as illustrations.

9.5.6.1 Systems with Long Time Delays

Processes with long time delays are difficult to control [2,9–11]. The loop gain with proportional control is very small so integral action is necessary to get good control. Such processes can be controlled by PI controllers, but the performance can be increased by more sophisticated controllers. The reason derivative action is not so useful for processes of this type is that prediction by linear extrapolation of the output is not very effective. To make a proper prediction, it is necessary to take account of the past control signals that have not yet shown up in the output. To illustrate this, we consider a process with the transfer function

$$G(s) = \frac{e^{-10s}}{(s+1)^3}$$

The dynamics of this process is dominated by the time delay. A good PI controller that gives a step response without overshoot has a gain $K = 0.27$ and $T_i = 4.8$. The response to set point changes and load disturbances of the system is shown in Figure 9.59. This figure shows the response to a step in the set point at time $t = 0$ and a step at the process input at time $t = 50$.

One way to obtain improved control is to use a controller with a Smith predictor. This controller requires a model of the process. If a model in the form of a first-order system with gain K_p, time constant

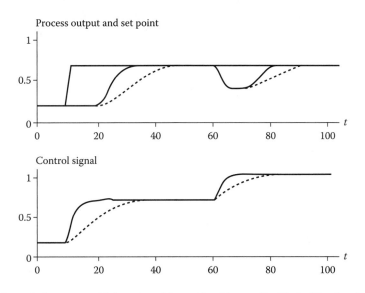

FIGURE 9.59 Control of a process with long time delays with a PI controller (dashed lines) and a Smith predictor (solid lines).

T, and a time delay L is used, the controller becomes

$$U(s) = K\left(1 + \frac{1}{sT_i}\right)\left(E(s) - \frac{K_p}{1+sT}(1 - e^{-sL})U(s)\right) \tag{9.125}$$

The controller can predict the output better than a PID controller because of the internal process model. The last term in the right-hand side of Equation 9.125 can be interpreted as the effect on the output of control signals that have been applied in the time interval $(t - T, t)$. Because of the time delay the effect of these signals has not appeared in the output at time t. The improved performance is seen in the simulation in Figure 9.59.

If load disturbance response is evaluated with the integrated absolute error (IAE), we find that the Smith predictor is about 30% better than the PI controller. There are situations when the increased complexity is worth while.

9.5.6.2 Systems with Oscillatory Modes

Systems with poorly damped oscillatory modes are another case where more complex controllers can outperform PID control. The reason for this is that it pays to have a more complex model in the controller. To illustrate this, we consider a system with the transfer function

$$G(s) = \frac{25}{(s+1)(s^2+25)}$$

This system has two complex undamped poles.

The system cannot be stabilized with a PI controller with positive coefficients. To stabilize the undamped poles with a PI controller, it is necessary to have controllers with a zero in the right half-plane. Some damping of the unstable poles can be provided in this way. It is advisable to choose set point weighting $b = 0$ in order to avoid unnecessary excitation of the modes. The response obtained with such a PID controller is shown in Figure 9.60. In this figure, a step change in the set point has been introduced at time $t = 0$, and a step change in the load disturbance has been applied at time $t = 20$. The set point weighting b is zero. Because of this we avoid a right half-plane zero in the transfer function from set point to output, and the oscillatory modes are not excited much by changes in the set point. The oscillatory modes are, however, excited by the load disturbance.

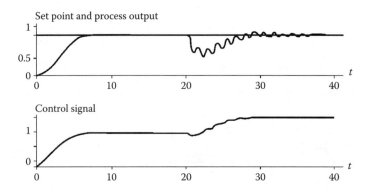

FIGURE 9.60 Control of an oscillatory system with PI control. The controller parameters are $K = -0.25$, $T_i = -1$ and $b = 0$.

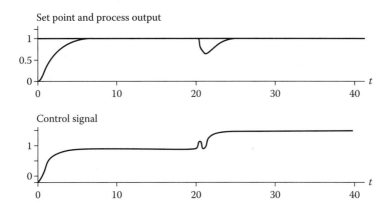

FIGURE 9.61 Control of the system in Figure 9.60 with a third-order controller.

By using a controller that is more complex than a PID controller it is possible to introduce damping in the system. This is illustrated by the simulation in Figure 9.61. The controller has the transfer functions

$$G_c(s) = \frac{21s^3 - 14s^2 + 65s + 100}{s(s^2 + 16s + 165)}$$

$$G_{sp}(s) = \frac{100}{s(s^2 + 16s + 165)}$$

The transfer function $G_c(s)$ has poles at 0 and $-8 \pm 10.05i$ and zeros at -1 and $0.833 \pm 2.02i$. Notice that the controller has two complex zeros in the right half-plane. This is typical for controllers of oscillatory systems. The controller transfer function can be written as

$$G_c(s) = 0.6061 \left(1 + \frac{1}{s}\right) \frac{1 - 0.35s + 0.21s^2}{1 + 0.0970s + 0.00606s^2}$$

$$G_{sp} = \frac{0.6061}{s} \frac{1}{1 + 0.0970s + 0.00606s^2}$$

The controller can thus be interpreted as a PI controller with an additional compensation. Notice that the gain of the controller is 2.4 times larger than the gain of the PI controller used in the simulation in Figure 9.60. This gives faster set point response and a better rejection of load disturbances.

9.5.7 Bottom-Up Design of Complex Systems

Control problems are seldom solved by a single controller. Many control systems are designed using a "bottom-up" approach where PID controllers are combined with other components, such as filters, selectors and others [2,3,11].

9.5.7.1 Cascade Control

Cascade control is used when there are several measured signals and one control variable. It is particularly useful when there are significant dynamics (e.g., long dead times or long time constants) between the control variable and the process variable. Tighter control can then be achieved by using an intermediate measured signal that responds faster to the control signal. Cascade control is built up by nesting the control loops, as shown in Figure 9.62. The system in this figure has two loops. The inner loop is called *the secondary loop*; the outer loop is called *the primary loop*. The reason for this terminology is that the outer loop controls the signal we are primarily interested in. It is also possible to have a cascade control with

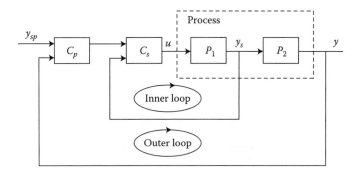

FIGURE 9.62   Block diagram of a system with cascade control.

more nested loops. The performance of a system can be improved with a number of measured signals, up to a certain limit. If all state variables are measured, it is often not worthwhile to introduce other measured variables. In such a case, the cascade control is the same as state feedback.

9.5.7.2 Feedforward Control

Disturbances can be eliminated by feedback. With a feedback system it is, however, necessary that there be an error before the controller can take actions to eliminate disturbances. In some situations, it is possible to measure disturbances before they have influenced the processes. It is then natural to try to eliminate the effects of the disturbances before they have created control errors. This control paradigm is called *feedforward*. The principle is illustrated simply in Figure 9.63. Feedforward can be used for both linear and nonlinear systems. It requires a mathematical model of the process.

As an illustration we consider a linear system that has two inputs, the control variable u and the disturbance v, and one output y. The transfer function from disturbance to output is G_v, and the transfer function from the control variable to the output is G_u. The process can be described by

$$Y(s) = G_u(s)U(s) + G_v(s)V(s)$$

where the Laplace transformed variables are denoted by capital letters. The feedforward control law

$$U(s) = -\frac{G_v(s)}{G_u(s)}\,V(s)$$

makes the output zero for all disturbances v. The feedforward transfer function should thus be chosen as

$$G_{ff}(s) = -\frac{G_v(s)}{G_u(s)}$$

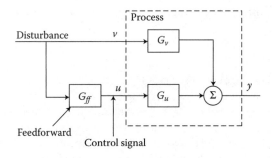

FIGURE 9.63   Block diagram of a system with feedforward control from a measurable disturbance.

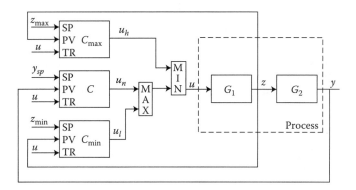

FIGURE 9.64 Control system with selector control.

9.5.8 Selector Control

Selector control can be viewed as the inverse of split range control. In split range, there is one measured signal and several actuators. In selector control, there are many measured signals and only one actuator. A selector is a static device with many inputs and one output. There are two types of selectors: *maximum* and *minimum*. For a maximum selector, the output is the largest of the input signals.

There are situations where several controlled process variables must be taken into account. One variable is the primary controlled variable, but it is also required that other process variables remain within given ranges. Selector control can be used to achieve this. The idea is to use several controllers and to have a selector that chooses the controller that is most appropriate. For example, selector control is used when the primary controlled variable is temperature and we must ensure that pressure does not exceed a certain range for safety reasons.

The principle of selector control is illustrated in Figure 9.64. The primary controlled variable is the process output y. There is an auxiliary measured variable z that should be kept within the limits z_{min} and z_{max}. The primary controller C has process variable y, setpoint y_{sp}, and output u_n. There are also secondary controllers with measured process variables that are the auxiliary variable z and with set points that are bounds of the variable z. The outputs of these controllers are u_h and u_l. The controller C is an ordinary PI or PID controller that gives good control under normal circumstances. The output of the minimum selector is the smallest of the input signals; the output of the maximum selector is the largest of the inputs.

Under normal circumstances, the auxiliary variable is larger than the minimum value z_{min} and smaller than the maximum value z_{max}. This means that the output u_h is large and the output u_l is small. The maximum selector, therefore, selects u_n and the minimum selector also selects u_n. The system acts as if the maximum and minimum controller were not present. If the variable z reaches its upper limit, the variable u_h becomes small and is selected by the minimum selector. This means that the control system now attempts to control the variable z and drive it toward its limit. A similar situation occurs if the variable z becomes smaller than z_{min}. To avoid windup, the finally selected control u is used as a tracking signal for all controllers.

References

1. Messner, W.C., Bedillion, M.D., and Karns, D.C. Lead and lag compensators with complex poles and zeros. *IEEE Control Systems Magazine* (27) February, 45–54, 2007.
2. Åström, K.J. and Hägglund, T., *Advanced PID Control,* ISA—The Instrumentation, Systems, and Automation Society, Research Triangle Park, NC, 2005.

3. Åström, K.J. and Hägglund, T., *PID Control—Theory, Design and Tuning*, 2nd ed., Instrument Society of America, Research Triangle Park, NC, 1995.
4. O'Dwyer, A. *Handbook of Pi and Pid Controller Tuning Rules.* Imperial College Press, London, 2009.
5. Åström, K.J., Hägglund, T., Hang, C.C., and Ho, W.K., Automatic tuning and adaptation for PID controllers—A survey, *Control Eng. Pract.*, 1(4), 699–714, 1993.
6. Åström, K.J., Hang, C.C., Persson, P., and Ho, W.K., Towards intelligent PID control, *Automatica*, 28(1), 1–9, 1992.
7. Fertik, H.A. Tuning controllers for noisy processes, *ISA Trans.*, 14, 292–304, 1975.
8. Fertik, H.A. and Ross, C.W., Direct digital control algorithms with anti-windup feature, *ISA Trans.*, 6(4), 317–328, 1967.
9. Ross, C.W., Evaluation of controllers for deadtime processes, *ISA Trans.*, 16(3), 25–34, 1977.
10. Seborg, D.E., Edgar, T.F., and Mellichamp, D.A., *Process Dynamics and Control*, Wiley, New York, 1989.
11. Shinskey, F.G. *Process-Control Systems. Application, Design, and Tuning*, 3rd ed., McGraw-Hill, New York, 1988.
12. Smith, C.L. and Murrill, P.W., A more precise method for tuning controllers, *ISA Journal*, May, 50–58, 1966.

9.6 State-Space–Pole Placement

Katsuhiko Ogata

9.6.1 Introduction

In this chapter*, we present a design method commonly called the pole placement or pole assignment technique. We assume that all state variables are measurable and are available for feedback. It will be shown that if the system considered is completely state controllable, then poles of the closed-loop system may be placed at any desired locations by means of state feedback through an appropriate state feedback gain matrix.

The present design technique begins with a determination of the desired closed-loop poles based on the transient-response and/or frequency-response requirements, such as speed, damping ratio, or bandwidth, as well as steady-state requirements.

Let us assume that we decide that the desired closed-loop poles are to be at $s = \mu_1, s = \mu_2, \ldots, s = \mu_n$. By choosing an appropriate gain matrix for state feedback, it is possible to force the system to have closed-loop poles at the desired locations, provided that the original system is completely state controllable.

In what follows, we treat the case where the control signal is a scalar and prove that a necessary and sufficient condition that the closed-loop poles can be placed at any arbitrary locations in the s plane is that the system be completely state controllable. Then we discuss three methods for determining the required state feedback gain matrix.

It is noted that when the control signal is a vector quantity, the state feedback gain matrix is not unique. It is possible to choose freely more than n parameters; that is, in addition to being able to place n closed-loop poles properly, we have the freedom to satisfy some of the other requirements, if any, of the closed-loop system. This chapter, however, discusses only the case where the control signal is a scalar quantity. (For the case where the control signal is a vector quantity, refer to MIMO LTI systems in this handbook.)

9.6.2 Design via Pole Placement

In the conventional approach to the design of a single-input, single-output control system, we design a controller (compensator) such that the dominant closed-loop poles have a desired damping ratio ζ and undamped natural frequency ω_n. In this approach, the order of the system may be raised by 1 or 2 unless

---

* Most of the material presented here is from [1].

pole–zero cancellation takes place. Note that in this approach we assume the effects on the responses of nondominant closed-loop poles to be negligible.

Different from specifying only dominant closed-loop poles (conventional design approach), the present pole placement approach specifies all closed-loop poles. (There is a cost associated with placing all closed-loop poles, however, because placing all closed-loop poles requires successful measurements of all state variables or else requires the inclusion of a state observer in the system.) There is also a requirement on the part of the system for the closed-loop poles to be placed at arbitrarily chosen locations. The requirement is that the system be completely state controllable.

Consider a control system

$$\dot{x} = Ax + Bu \tag{9.126}$$

where x = state vector (n-vector); u = control signal (scalar); $A = n \times n$ constant matrix; $B = n \times 1$ constant matrix.

We shall choose the control signal to be

$$u = -Kx \tag{9.127}$$

This means that the control signal is determined by instantaneous state. Such a scheme is called *state feedback*. The $1 \times n$ matrix K is called the state feedback gain matrix. In the following analysis, we assume that u is unconstrained.

Substituting Equation 9.127 into Equation 9.126 gives

$$\dot{x}(t) = (A - BK)x(t)$$

The solution of this equation is given by

$$x(t) = e^{(A-BK)t}x(0) \tag{9.128}$$

where $x(0)$ is the initial state caused by external disturbances. The stability and transient response characteristics are determined by the eigenvalues of matrix $A - BK$. If matrix K is chosen properly, then matrix $A - BK$ can be made an asymptotically stable matrix, and for all $x(0) \neq 0$ it is possible to make $x(t)$ approach 0 as t approaches infinity. The eigenvalues of matrix $A - BK$ are called the regulator poles. If these regulator poles are located in the left half of the s plane, then $x(t)$ approaches 0 as t approaches infinity. The problem of placing the closed-loop poles at the desired location is called a pole placement problem.

Figure 9.65a shows the system defined by Equation 9.126. It is an open-loop control system, because the state x is not fed back to the control signal u. Figure 9.65b shows the system with state feedback. This is a closed-loop control system because the state x is fed back to the control signal u.

In what follows, we prove that arbitrary pole placement for a given system is possible if and only if the system is completely state controllable.

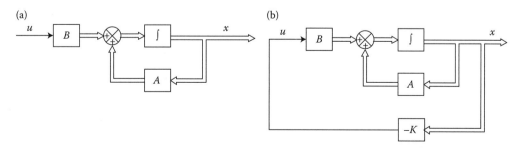

FIGURE 9.65 (a) Open-loop control system and (b) closed-loop control system with $u = -Kx$. (From Ogata, K., *Modern Control Engineering*, 2nd ed., Prentice Hall, Inc., Englewood Cliffs, NJ, 1990, 777. With permission.)

9.6.3 Necessary and Sufficient Condition for Arbitrary Pole Placement

Consider the control system defined by Equation 9.126. We assume that the magnitude of the control signal u is unbounded. If the control signal u is chosen as

$$u = -Kx$$

where K is the state feedback gain matrix ($1 \times n$ matrix), then the system becomes a closed-loop control system as shown in Figure 9.65b and the solution to Equation 9.126 becomes as given by Equation 9.128, or

$$x(t) = e^{(A-BK)t}x(0)$$

Note that the eigenvalues of matrix $A - BK$ (which we denote $\mu_1, \mu_2, \ldots, \mu_n$) are the desired closed-loop poles.

We now prove that a necessary and sufficient condition for arbitrary pole placement is that the system be completely state controllable. We first derive the necessary condition. We begin by proving that if the system is not completely state controllable, then there are eigenvalues of matrix $A - BK$ that cannot be controlled by state feedback.

Suppose the system of Equation 9.126 is not completely state controllable. Then the rank of the controllability matrix is less than n, or

$$\text{rank}[B|AB|\cdots|A^{n-1}B] = q < n$$

This means that there are q linearly independent column vectors in the controllability matrix. Let us define such q linearly independent column vectors as $f_1, f_2, \ldots, f_q$. Also, let us choose $n - q$ additional n-vectors $v_{q+1}, v_{q+2}, \ldots, v_n$ such that

$$P = [f_1|f_2|\cdots|f_q|v_{q+1}|v_{q+2}|\cdots|v_n]$$

is of rank n. Then it can be shown that

$$\hat{A} = P^{-1}AP = \left[\begin{array}{c|c} A_{11} & A_{12} \\ \hline 0 & A_{22} \end{array}\right], \quad \hat{B} = P^{-1}B = \left[\begin{array}{c} B_{11} \\ \hline 0 \end{array}\right]$$

Define

$$\hat{K} = KP = [k_1 \mid k_2]$$

Then we have

$$
\begin{aligned}
|sI - A + BK| &= |P^{-1}(sI - A + BK)P| \\
&= |sI - P^{-1}AP + P^{-1}BKP| \\
&= |sI - \hat{A} + \hat{B}\hat{K}| \\
&= \left| sI - \left[\begin{array}{c|c} A_{11} & A_{12} \\ \hline 0 & A_{22} \end{array}\right] + \left[\begin{array}{c} B_{11} \\ \hline 0 \end{array}\right][k_1 \mid k_2] \right| \\
&= \left| \begin{array}{cc} sI_q - A_{11} + B_{11}k_1 & -A_{12} + B_{11}k_2 \\ 0 & sI_{n-q} - A_{22} \end{array} \right| \\
&= |sI_q - A_{11} + B_{11}k_1| \cdot |sI_{n-q} - A_{22}| \\
&= 0
\end{aligned}
$$

where I_q is a q-dimensional identity matrix and I_{n-q} is an $(n - q)$-dimensional identity matrix.

Notice that the eigenvalues of A_{22} do not depend on K. Thus, if the system is not completely state controllable, then there are eigenvalues of matrix A that cannot be arbitrarily placed. Therefore, to place the eigenvalues of matrix $A - BK$ arbitrarily, the system must be completely state controllable (necessary condition).

Next we prove a sufficient condition: that is, if the system is completely state controllable (meaning that matrix M given by Equation 9.130 has an inverse), then all eigenvalues of matrix A can be arbitrarily placed.

In proving a sufficient condition, it is convenient to transform the state equation given by Equation 9.126 into the controllable canonical form.

Define a transformation matrix T by

$$T = MW \tag{9.129}$$

where M is the controllability matrix

$$M = [B|AB| \cdots |A^{n-1}B] \tag{9.130}$$

and

$$W = \begin{bmatrix} a_{n-1} & a_{n-2} & \cdots & a_1 & 1 \\ a_{n-2} & a_{n-3} & \cdots & 1 & 0 \\ \vdots & \vdots & & \vdots & \vdots \\ a_1 & 1 & \cdots & 0 & 0 \\ 1 & 0 & \cdots & 0 & 0 \end{bmatrix} \tag{9.131}$$

where each a_i is a coefficient of the characteristic polynomial

$$|sI - A| = s^n + a_1 s^{n-1} + \cdots + a_{n-1}s + a_n$$

Define a new state vector $\hat{x}$ by

$$x = T\hat{x}$$

If the rank of the controllability matrix M is n (meaning that the system is completely state controllable), then the inverse of matrix T exists and Equation 9.126 can be modified to

$$\dot{\hat{x}} = T^{-1}AT\hat{x} + T^{-1}Bu \tag{9.132}$$

where

$$T^{-1}AT = \begin{bmatrix} 0 & 1 & 0 & \cdots & 0 \\ 0 & 0 & 1 & \cdots & 0 \\ \vdots & \vdots & \vdots & & \vdots \\ 0 & 0 & 0 & \cdots & 1 \\ -a_n & -a_{n-1} & -a_{n-2} & \cdots & -a_1 \end{bmatrix} \tag{9.133}$$

$$T^{-1}B = \begin{bmatrix} 0 \\ 0 \\ \vdots \\ 0 \\ 1 \end{bmatrix} \tag{9.134}$$

Equation 9.132 is in the controllable canonical form. Thus, given a state equation, Equation 9.126, it can be transformed into the controllable canonical form if the system is completely state controllable and if we transform the state vector x into state vector $\hat{x}$ by use of the transformation matrix T given by Equation 9.129.

Let us choose a set of the desired eigenvalues as $\mu_1, \mu_2, \ldots, \mu_n$. Then the desired characteristic equation becomes

$$(s - \mu_1)(s - \mu_2) \cdots (s - \mu_n) = s^n + \alpha_1 s^{n-1} + \cdots + \alpha_{n-1} s + \alpha_n = 0 \qquad (9.135)$$

Let us write

$$\hat{K} = KT = [\delta_n \ \delta_{n-1} \ \cdots \ \delta_1] \qquad (9.136)$$

When $u = -\hat{K}\hat{x} = -KT\hat{x}$ is used to control the system given by Equation 9.132, the system equation becomes

$$\dot{\hat{x}} = T^{-1}AT\hat{x} - T^{-1}BKT\hat{x}$$

The characteristic equation is

$$|sI - T^{-1}AT + T^{-1}BKT| = 0$$

This characteristic equation is the same as the characteristic equation for the system, defined by Equation 9.126, when $u = -Kx$ is used as the control signal. This can be seen as follows: Since

$$\dot{x} = Ax + Bu = (A - BK)x$$

the characteristic equation for this system is

$$|sI - A + BK| = |T^{-1}(sI - A + BK)T|$$
$$= |sI - T^{-1}AT + T^{-1}BKT| = 0$$

Now let us simplify the characteristic equation of the system in the controllable canonical form. Referring to Equations 9.133, 9.134, and 9.136, we have

$$|sI - T^{-1}AT + T^{-1}BKT|$$

$$= \left| sI - \begin{bmatrix} 0 & 1 & \cdots & 0 \\ \vdots & \vdots & & \vdots \\ 0 & 0 & \cdots & 1 \\ -a_n & -a_{n-1} & \cdots & -a_1 \end{bmatrix} + \begin{bmatrix} 0 \\ \vdots \\ 0 \\ 1 \end{bmatrix} [\delta_n \ \delta_{n-1} \ \cdots \ \delta_1] \right|$$

$$= \begin{vmatrix} s & -1 & \cdots & 0 \\ 0 & s & \cdots & 0 \\ \vdots & \vdots & & \vdots \\ a_n + \delta_n & a_{n-1} + \delta_{n-1} & \cdots & s + a_1 + \delta_1 \end{vmatrix}$$

$$= s^n + (a_1 + \delta_1)s^{n-1} + \cdots + (a_{n-1} + \delta_{n-1})s + (a_n + \delta_n) = 0 \qquad (9.137)$$

This is the characteristic equation for the system with state feedback. Therefore, it must be equal to Equation 9.135, the desired characteristic equation. By equating the coefficients of like powers of s, we get

$$a_1 + \delta_1 = \alpha_1$$
$$a_2 + \delta_2 = \alpha_2$$
$$\vdots$$
$$a_n + \delta_n = \alpha_n$$

Solving the preceding equations for each δ_i and substituting them into Equation 9.136, we obtain

$$K = \hat{K}T^{-1} = [\delta_n \, \delta_{n-1} \cdots \delta_1] \, T^{-1}$$

$$= [\alpha_n - a_n | \alpha_{n-1} - a_{n-1} | \cdots | \alpha_2 - a_2 | \alpha_1 - a_1] \, T^{-1} \qquad (9.138)$$

Thus, if the system is completely state controllable, all eigenvalues can be arbitrarily placed by choosing matrix K according to Equation 9.138 (sufficient condition).

We have thus proved that the necessary and sufficient condition for arbitrary pole placement is that the system be completely state controllable.

9.6.4 Design Steps for Pole Placement

Suppose that the system is defined by

$$\dot{x} = Ax + Bu$$

and the control signal is given by

$$u = -Kx$$

The feedback gain matrix K that forces the eigenvalues of $A - BK$ to be $\mu_1, \mu_2, \ldots, \mu_n$ (desired values) can be determined by the following steps. (If μ_i is a complex eigenvalue, then its conjugate must also be an eigenvalue of $A - BK$.)

Step 1

Check the controllability condition for the system. If the system is completely state controllable, then use the following steps.

Step 2

From the characteristic polynomial for matrix A:

$$|sI - A| = s^n + a_1 s^{n-1} + \cdots + a_{n-1}s + a_n$$

determine the values of $a_1, a_2, \ldots, a_n$.

Step 3

Determine the transformation matrix T that transforms the system state equation into the controllable canonical form. (If the given system equation is already in the controllable canonical form, then $T = I$.) It is not necessary to write the state equation in the controllable canonical form. All we need here is to find the matrix T. The transformation matrix T is given by Equation 9.129, or

$$T = MW$$

where M is given by Equation 9.130 and W is given by Equation 9.131.

Step 4

Using the desired eigenvalues (desired closed-loop poles), write the desired characteristic polynomial

$$(s - \mu_1)(s - \mu_2) \cdots (s - \mu_n) = s^n + \alpha_1 s^{n-1} + \cdots + \alpha_{n-1}s + \alpha_n$$

and determine the values of $\alpha_1, \alpha_2, \ldots, \alpha_n$.

Step 5

The required state feedback gain matrix K can be determined from Equation 9.138, rewritten thus:

$$K = [\alpha_n - a_n | \alpha_{n-1} - a_{n-1} | \cdots | \alpha_2 - a_2 | \alpha_1 - a_1] T^{-1}$$

9.6.5 Comments

Note that if the system is of lower order ($n \leq 3$), then direct substitution of matrix K into the desired characteristic polynomial may be simpler. For example, if $n = 3$, then write the state feedback gain matrix K as

$$K = [k_1 \; k_2 \; k_3]$$

Substitute this K matrix into the desired characteristic polynomial $|sI - A + BK|$ and equate it to $(s - \mu_1)(s - \mu_2)(s - \mu_3)$, or

$$|sI - A + BK| = (s - \mu_1)(s - \mu_2)(s - \mu_3)$$

Since both sides of this characteristic equation are polynomials in s, by equating the coefficients of the like powers of s on both sides it is possible to determine the values of k_1, k_2, and k_3. This approach is convenient if $n = 2$ or 3. (For $n = 4, 5, 6, \ldots$, this approach may become very tedious.)

There are other approaches for the determination of the state feedback gain matrix K. In what follows, we present a well-known formula, known as Ackermann's formula, for the determination of the state feedback gain matrix K.

9.6.6 Ackermann's Formula

Consider the system given by Equation 9.126, rewritten thus:

$$\dot{x} = Ax + Bu$$

We assume that the system is completely state controllable. We also assume that the desired closed-loop poles are at $s = \mu_1, s = \mu_2, \ldots, s = \mu_n$.

Use of the state feedback control

$$u = -Kx$$

modifies the system equation to

$$\dot{x} = (A - BK)x \tag{9.139}$$

Let us define

$$\tilde{A} = A - BK$$

The desired characteristic equation is

$$
\begin{aligned}
|sI - A + BK| &= |sI - \tilde{A}| \\
&= (s - \mu_1)(s - \mu_2) \cdots (s - \mu_n) \\
&= s^n + \alpha_1 s^{n-1} + \cdots + \alpha_{n-1} s + \alpha_n = 0
\end{aligned}
$$

Since the Cayley–Hamilton theorem states that $\tilde{A}$ satisfies its own characteristic equation, we have

$$\phi(\tilde{A}) = \tilde{A}^n + \alpha_1 \tilde{A}^{n-1} + \cdots + \alpha_{n-1} \tilde{A} + \alpha_n I = 0 \tag{9.140}$$

We utilize Equation 9.140 to derive Ackermann's formula. To simplify the derivation, we consider the case where $n = 3$. (For any other positive integer n, the following derivation can be easily extended.)

Consider the following identities:

$$I = I$$
$$\tilde{A} = A - BK$$
$$\tilde{A}^2 = (A - BK)^2 = A^2 - ABK - BK\tilde{A}$$
$$\tilde{A}^3 = (A - BK)^3 = A^3 - A^2BK - ABK\tilde{A} - BK\tilde{A}^2$$

Multiplying the preceding equations in order by $\alpha_3, \alpha_2, \alpha_1, \alpha_0$ (where $\alpha_0 = 1$), respectively, and adding the results, we obtain

$$\alpha_3 I + \alpha_2 \tilde{A} + \alpha_1 \tilde{A}^2 + \tilde{A}^3$$
$$= \alpha_3 I + \alpha_2(A - BK) + \alpha_1(A^2 - ABK - BK\tilde{A}) + A^3 - A^2BK - ABK\tilde{A} - BK\tilde{A}^2 \qquad (9.141)$$
$$= \alpha_3 I + \alpha_2 A + \alpha_1 A^2 + A^3 - \alpha_2 BK - \alpha_1 ABK - \alpha_1 BK\tilde{A} - A^2BK - ABK\tilde{A} - BK\tilde{A}^2$$

Referring to Equation 9.140, we have

$$\alpha_3 I + \alpha_2 \tilde{A} + \alpha_1 \tilde{A}^2 + \tilde{A}^3 = \phi(\tilde{A}) = 0$$

Also, we have

$$\alpha_3 I + \alpha_2 A + \alpha_1 A^2 + A^3 = \phi(A) \neq 0$$

Substituting the last two equations into Equation 9.141, we have

$$\phi(\tilde{A}) = \phi(A) - \alpha_2 BK - \alpha_1 BK\tilde{A} - BK\tilde{A}^2 - \alpha_1 ABK - ABK\tilde{A} - A^2BK$$

Since $\phi(\tilde{A}) = 0$, we obtain

$$\phi(A) = B(\alpha_2 K + \alpha_1 K\tilde{A} + K\tilde{A}^2) + AB(\alpha_1 K + K\tilde{A}) + A^2 BK$$

$$= [B \mid AB \mid A^2 B] \begin{bmatrix} \alpha_2 K + \alpha_1 K\tilde{A} + K\tilde{A}^2 \\ \alpha_1 K + K\tilde{A} \\ K \end{bmatrix} \qquad (9.142)$$

Since the system is completely state controllable, the inverse of the controllability matrix

$$[B \mid AB \mid A^2 B]$$

exists. Premultiplying the inverse of the controllability matrix to both sides of Equation 9.142, we obtain

$$[B \mid AB \mid A^2 B]^{-1} \phi(A) = \begin{bmatrix} \alpha_2 K + \alpha_1 K\tilde{A} + K\tilde{A}^2 \\ \alpha_1 K + K\tilde{A} \\ K \end{bmatrix}$$

Premultiplying both sides of this last equation by $[0\ 0\ 1]$, we obtain

$$[0\ 0\ 1][B \mid AB \mid A^2 B]^{-1} \phi(A) = [0\ 0\ 1] \begin{bmatrix} \alpha_2 K + \alpha_1 K\tilde{A} + K\tilde{A}^2 \\ \alpha_1 K + K\tilde{A} \\ K \end{bmatrix} = K$$

which can be rewritten as

$$K = [0\ 0\ 1][B \mid AB \mid A^2 B]^{-1} \phi(A)$$

This last equation gives the required state feedback gain matrix K.

For an arbitrary positive integer n, we have

$$K = [0\,0\,\cdots\,0\,1][B\,|\,AB\,|\,\cdots\,|\,A^{n-1}B]^{-1}\phi(A) \tag{9.143}$$

Equations 9.139 to 9.143 collectively are known as Ackermann's formula for the determination of the state feedback gain matrix K.

Example 9.2:

Consider the system defined by

$$\dot{x} = Ax + Bu$$

where

$$A = \begin{bmatrix} 0 & 1 & 0 \\ 0 & 0 & 1 \\ -1 & -5 & -6 \end{bmatrix}, \quad B = \begin{bmatrix} 0 \\ 0 \\ 1 \end{bmatrix}$$

By using the state feedback control $u = -Kx$, it is desired to have the closed-loop poles at $s = -2 \pm j4$, and $s = -10$. Determine the state feedback gain matrix K.

First, we need to check the controllability of the system. Since the controllability matrix M is given by

$$M = [B\,|\,AB\,|\,A^2B] = \begin{bmatrix} 0 & 0 & 1 \\ 0 & 1 & -6 \\ 1 & -6 & 31 \end{bmatrix}$$

we find that $\det M = -1$ and therefore rank $M = 3$. Thus, the system is completely state controllable and arbitrary pole placement is possible.

Next, we solve this problem. We demonstrate each of the three methods presented in this chapter.

Method 1

The first method is to use Equation 9.128. The characteristic equation for the system is

$$|sI - A| = \begin{vmatrix} s & -1 & 0 \\ 0 & s & -1 \\ 1 & 5 & s+6 \end{vmatrix}$$

$$= s^3 + 6s^2 + 5s + 1$$

$$= s^3 + a_1 s^2 + a_2 s + a_3 = 0$$

Hence,

$$a_1 = 6, \quad a_2 = 5, \quad a_3 = 1$$

The desired characteristic equation is

$$(s+2-j4)(s+2+j4)(s+10) = s^3 + 14s^2 + 60s + 200$$

$$= s^3 + \alpha_1 s^2 + \alpha_2 s + \alpha_3$$

$$= 0$$

Hence,

$$\alpha_1 = 14, \quad \alpha_2 = 60, \quad \alpha_3 = 200$$

Referring to Equation 9.138 we have

$$K = [\alpha_3 - a_3\,|\,\alpha_2 - a_2\,|\,\alpha_1 - a_1]T^{-1}$$

where $T = I$ for this problem because the given state equation is in the controllable canonical form. Then we have

$$K = [200 - 1 \mid 60 - 5 \mid 14 - 6]$$
$$= [199 \quad 55 \quad 8]$$

Method 2

By defining the desired state feedback gain matrix K as

$$K = [k_1 \, k_2 \, k_3]$$

and equating $|sI - A + BK|$ with the desired characteristic equation, we obtain

$$|sI - A + BK| = \left| \begin{bmatrix} s & 0 & 0 \\ 0 & s & 0 \\ 0 & 0 & s \end{bmatrix} - \begin{bmatrix} 0 & 1 & 0 \\ 0 & 0 & 1 \\ -1 & -5 & -6 \end{bmatrix} + \begin{bmatrix} 0 \\ 0 \\ 1 \end{bmatrix} [k_1 \, k_2 \, k_3] \right|$$

$$= \begin{vmatrix} s & -1 & 0 \\ 0 & s & -1 \\ 1+k_1 & 5+k_2 & s+6+k_3 \end{vmatrix}$$

$$= s^3 + (6+k_3)s^2 + (5+k_2)s + 1 + k_1$$

$$= s^3 + 14s^2 + 60s + 200$$

Thus,

$$6 + k_3 = 14, \quad 5 + k_2 = 60, \quad 1 + k_1 = 200$$

from which we obtain

$$k_1 = 199, \quad k_2 = 55, \quad k_3 = 8$$

or

$$K = [199 \quad 55 \quad 8]$$

Method 3

The third method is to use Ackermann's formula. Referring to Equation 9.143, we have

$$K = [0 \, 0 \, 1][B \mid AB \mid A^2B]^{-1}\phi(A)$$

Since

$$\phi(A) = A^3 + 14A^2 + 60A + 200I$$

$$= \begin{bmatrix} 0 & 1 & 0 \\ 0 & 0 & 1 \\ -1 & -5 & -6 \end{bmatrix}^3 + 14 \begin{bmatrix} 0 & 1 & 0 \\ 0 & 0 & 1 \\ -1 & -5 & -6 \end{bmatrix}^2 + 60 \begin{bmatrix} 0 & 1 & 0 \\ 0 & 0 & 1 \\ -1 & -5 & -6 \end{bmatrix} + \begin{bmatrix} 200 & 0 & 0 \\ 0 & 200 & 0 \\ 0 & 0 & 200 \end{bmatrix}$$

$$= \begin{bmatrix} 199 & 55 & 8 \\ -8 & 159 & 7 \\ -7 & -43 & 117 \end{bmatrix}$$

and

$$[B \mid AB \mid A^2B] = \begin{bmatrix} 0 & 0 & 1 \\ 0 & 1 & -6 \\ 1 & -6 & 31 \end{bmatrix}$$

we obtain

$$K = [0\ 0\ 1] \begin{bmatrix} 0 & 0 & 1 \\ 0 & 1 & -6 \\ 1 & -6 & 31 \end{bmatrix}^{-1} \begin{bmatrix} 199 & 55 & 8 \\ -8 & 159 & 7 \\ -7 & -43 & 117 \end{bmatrix}$$

$$= [0\ 0\ 1] \begin{bmatrix} 5 & 6 & 1 \\ 6 & 1 & 0 \\ 1 & 0 & 0 \end{bmatrix} \begin{bmatrix} 199 & 55 & 8 \\ -8 & 159 & 7 \\ -7 & -43 & 117 \end{bmatrix}$$

$$= [199\ \ 55\ \ 8]$$

As a matter of course, the feedback gain matrix K obtained by the three methods are the same. With this state feedback, the closed-loop poles are located at $s = -2 \pm j4$ and $s = -10$, as desired.

It is noted that if the order n of the system is 4 or higher, methods 1 and 3 are recommended, since all matrix computations can be carried by a computer. If method 2 is used, hand computations become necessary because a computer may not handle the characteristic equation with unknown parameters $k_1, k_2, \ldots, k_n$.

9.6.7 Comments

It is important to note that matrix K is not unique for a given system, but depends on the desired closed-loop pole locations (which determine the speed and damping of the response) selected. Note that the selection of the desired closed-loop poles or the desired characteristic equation is a compromise between the rapidity of the response of the error vector and the sensitivity to disturbances and measurement noises. That is, if we increase the speed of error response, then the adverse effects of disturbances and measurement noises generally increase. If the system is of second order, then the system dynamics (response characteristics) can be precisely correlated to the location of the desired closed-loop poles and the zero(s) of the plant. For higher-order systems, the location of the closed-loop poles and the system dynamics (response characteristics) are not easily correlated. Hence, in determining the state feedback gain matrix K for a given system, it is desirable to examine by computer simulations the response characteristics of the system for several different matrices K (based on several different desired characteristic equations) and to choose the one that gives the best overall system performance.

References

1. Ogata, K., *Modern Control Engineering,* 2nd ed., Prentice Hall, Inc., Englewood Cliffs, NJ, 1990.
2. Ogata, K., *Designing Linear Control Systems with MATLAB,* Prentice-Hall, Inc., Englewood Cliffs, NJ, 1994.
3. Ogata, K., *Discrete-Time Control Systems,* 2nd ed., Prentice-Hall, Inc., Englewood Cliffs, NJ, 1995.
4. Willems, J. C. and Mitter, S. K., Controllability, observability, pole allocation, and state reconstruction, *IEEE Trans. Autom. Control,* 16, 582–595, 1971.
5. Wonham, W. M., On pole assignment in multi-input controllable linear systems, *IEEE Trans. Autom. Control,* 12, 660–665, 1967.

9.7 Internal Model Control

Masako Kishida and Richard D. Braatz

9.7.1 Introduction

The field of process control experienced a surge of interest during the 1960s, as engineers worked to apply the newly developed state-space optimal control theory to industrial processes. Although these methods had been applied successfully to the control of many mechanical and electrical systems, applications to most industrial processes were not so forthcoming. By the 1970s, both industrialists and academicians began to understand that certain characteristics of most industrial processes make it very difficult to directly apply the optimal control design procedures available at that time in a consistent and reproducible manner.

Unknown disturbances, inaccurate values for the physical parameters of the process, and lack of complete understanding of the underlying physical phenomena make it impossible to generate a highly accurate model for most industrial processes, either phenomenologically or via input–output identification. Another consideration for process control problems was the overwhelming importance of constraints on the manipulated variables (e.g., valve positions, pump and compressor throughput) and the controlled variables (e.g., pressure, temperature, or capacity limits). Both model uncertainty and process constraints were not satisfactorily addressed by the state-space optimal control theory of the 1960s, and this to a large part explained the difficulties in applying this theory to process control problems.

An approach for explicitly addressing industrial process control problems began to coalesce in the late 1970s that came to be known as *Internal Model Control* (IMC). IMC became widely used in the process industries, mostly in the form of Proportional–Integral–Derivative (PID) tuning rules, in which a single parameter provides a clear tradeoff between closed-loop performance and robustness to model uncertainty. IMC also provided a convenient theoretical framework for understanding Smith predictors, multiple-degree-of-freedom problems, and the performance limitations due to nonminimum phase behavior and model uncertainty. The main focus of this chapter is IMC for stable processes, as the greatest strengths of the IMC framework occur in this case. The results will be developed initially for continuous-time linear time-invariant systems, followed by extensions to nonlinear and spatially distributed processes and comments on more complicated control structures and more advanced methods for handling constraints.

9.7.2 Fundamentals

Here the Internal Model and classical control structures are compared, which will illustrate the advantages of IMC in terms of addressing model uncertainty and actuator constraints in the control design. Then the IMC design procedure is presented.

9.7.2.1 Classical Control Structure

Before describing the IMC structure, let us consider the classical control structure used for the feedback control of *single-input single-input* (SISO) linear time-invariant (LTI) processes (shown in Figure 9.66). Here *p* refers to the transfer function of the process; *d* and *l* refer to the output and load disturbances, respectively; *y* refers to the controlled variable; *n* refers to measurement noise; *r* refers to the setpoint; and *u* refers to the manipulated variable specified by the controller *k*. The controlled variable is related to the setpoint, measurement noise, and unmeasured disturbances by

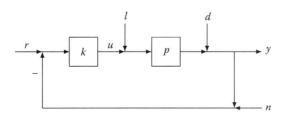

FIGURE 9.66 Classical control structure.

$$y = \frac{pk}{1+pk}(r-n) + \frac{1}{1+pk}(d+pl) = T(r-n) + S(d+pl), \tag{9.144}$$

where

$$S = \frac{1}{1+pk}, \quad T = 1 - S = \frac{pk}{1+pk} \tag{9.145}$$

are the *sensitivity* and *complementary sensitivity* functions, respectively.

A well-known requirement of any closed-loop system is internal stability, that is, that bounded signals injected at any point in the control system generate bounded signals at any other point. From the viewpoint of internal stability, only the boundedness of outputs y and u needs to be considered, since all other signals in the system are bounded provided u, y, and all inputs are bounded. Similarly, in terms of internal stability only the inputs r and l need to be considered. Thus, the classic control structure is internally stable if and only if all elements in the following 2×2 transfer matrix are stable (i.e., have all their poles in the open left-half plane)

$$\begin{pmatrix} y \\ u \end{pmatrix} = \begin{pmatrix} pS & T \\ -T & kS \end{pmatrix} \begin{pmatrix} l \\ r \end{pmatrix}. \tag{9.146}$$

The closed-loop system is internally stable if and only if the transfer functions pS, T, and kS are stable. For a stable process p the stability of these three transfer functions is implied by the stability of only one transfer function, kS (the stability of p and kS implies the stability of $T = pkS$, the stability of 1 and T implies the stability of $S = 1 - T$, and the stability of p and S implies the stability of pS). Thus, for a stable process p, the closed-loop system is internally stable if and only if

$$kS = \frac{k}{1+pk} \tag{9.147}$$

is stable. For good setpoint tracking, it is desirable in Equation 9.144 to have $T(j\omega) \approx 1$, and for good disturbance rejection it is desirable to have $S(j\omega) \approx 0$ for all frequencies. These performance requirements are commensurate, since $S + T = 1$. On the other hand, to avoid magnifying measurement noise at high frequencies, it is desirable to have $|T(j\omega)|$ roll off there. This is a fundamental tradeoff between system performance (that corresponds to $S \approx 0$) and insensitivity of the closed-loop system to measurement noise (that corresponds to $T \approx 0$).

To explicitly account for model uncertainty, it is necessary to quantify the accuracy of the process model $\tilde{p}$ used in the control design procedure. A natural and convenient method for quantifying model uncertainty is as a frequency-dependent bound on the difference between the process model $\tilde{p}$ and the true process p

$$\left| \frac{p(j\omega) - \tilde{p}(j\omega)}{\tilde{p}(j\omega)} \right| \leq |w_u(j\omega)|, \quad \forall \omega. \tag{9.148}$$

It should be expected that the inaccuracy of the model described by the uncertainty weight w_u would have a magnitude that increases with frequency and eventually exceeds 1, as it would be difficult to ascertain whether the true process has unmodeled zeros on the imaginary axis at sufficiently high frequencies (which would happen, for example, if there were any uncertainty in a process time delay), and these zeros would give $|(p(j\omega) - \tilde{p}(j\omega))/\tilde{p}(j\omega)| = |0 - \tilde{p}(j\omega)/\tilde{p}(j\omega)| = 1$.

The Nyquist Theorem can be used to show that the closed-loop system is internally stable for all processes that satisfy Equation 9.148 if and only if the nominal closed-loop system is internally stable and

$$|\tilde{T}(j\omega)| = \left| \frac{\tilde{p}(j\omega)k(j\omega)}{1 + \tilde{p}(j\omega)k(j\omega)} \right| < \frac{1}{|w_u(j\omega)|}, \quad \forall \omega. \tag{9.149}$$

As the magnitude of the uncertainty weight $|w_u(j\omega)|$ is expected to be greater than one at high frequencies, it is necessary for the nominal complementary sensitivity $\tilde{T}(j\omega)$ to be detuned at high frequencies to prevent the control system from being sensitive to model uncertainty. This is a fundamental trade-off between nominal performance (that corresponds to $\tilde{S} \approx 0$) and insensitivity of the closed-loop system to model uncertainty (that corresponds to $\tilde{T} \approx 0$). For industrial processes, closed-loop performance is usually limited more by model uncertainty than measurement noise.

A disadvantage of the classical control structure is that the controller k enters the stability and performance specifications (Equations 9.147 and 9.149) in an inconvenient manner. Also, in the presence of actuator constraints it is not simple to design the classical feedback controller k to ensure internal stability and performance for the closed-loop system. It is well known that a controller implemented using the classical control structure can give arbitrarily poor performance or even instability when the control action becomes limited.

9.7.2.2 Internal Model Control Structure

The IMC structure is shown in Figure 9.67, where $\tilde{p}$ refers to a model of the true process p and q refers to the IMC controller. Simple block diagram manipulations show that the IMC structure is equivalent to the classical control structure provided

$$k = \frac{q}{1 - \tilde{p}q}, \text{ or, equivalently, } q = \frac{k}{1 + \tilde{p}k}. \tag{9.150}$$

This control structure is referred to as *Internal Model Control*, because the process *model* $\tilde{p}$ is explicitly an *internal* part of the controller k.

In terms of the IMC controller q, the transfer functions between the controlled variable and the setpoint, measurement noise, and unmeasured disturbances are given by

$$y = T(r - n) + S(d + pl) = \frac{pq}{1 + q(p - \tilde{p})}(r - n) + \frac{1 - \tilde{p}q}{1 + q(p - \tilde{p})}(d + pl). \tag{9.151}$$

When the process model $\tilde{p}$ is not equal to the true process p, then the closed-loop transfer functions S and T in Equation 9.151 do not appear to be any simpler for the IMC control structure than for the classical control structure (Equation 9.144). However, when the process model $\tilde{p}$ is equal to the true process p, then Equation 9.151 simplifies to

$$y = \tilde{T}(r - n) + \tilde{S}(d + pl) = \tilde{p}q(r - n) + (1 - \tilde{p}q)(d + \tilde{p}l), \tag{9.152}$$

and the IMC controller is related to the classical controller by Equation 9.150

$$q = \frac{k}{1 + \tilde{p}k} = k\tilde{S}. \tag{9.153}$$

For stable processes, the stability of q in Equation 9.153 is exactly the condition for internal stability derived for the classical control structure. This replaces the somewhat inconvenient task of selecting a

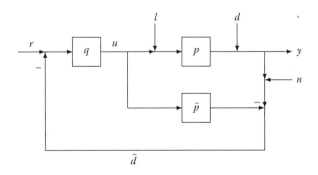

FIGURE 9.67 Internal model control structure.

controller k to stabilize $k/(1 + pk)$ with the simpler task of selecting any stable transfer function q. Also, the IMC controller q enters the closed-loop transfer functions $\tilde{T}$ and $\tilde{S}$ in Equation 9.152 in an affine manner, that is,

$$\tilde{T} = \tilde{p}q, \quad \tilde{S} = 1 - \tilde{p}q. \tag{9.154}$$

This makes the trade-off between nominal performance and model uncertainty very simple, which will be exploited in the IMC design procedure described later.

Another advantage of the IMC structure over the classical control structure is that the explicit consideration of the process model provides a convenient means for understanding the role of model uncertainty in the control system design. To see this, let us interpret the feedback signal $\tilde{d}$ in Figure 9.67 for the case where the process model is not an exact representation of the true process ($\tilde{p} \neq p$):

$$\tilde{d} = (p - \tilde{p})u + n + d + pl. \tag{9.155}$$

When there are no unknown disturbances or measurement noise ($n = d = l = 0$), and no model uncertainty ($\tilde{p} = p$), then the feedback signal $\tilde{d}$ is zero and the control system is an open loop, that is, no feedback is necessary. If there are disturbances, measurement noise, or model uncertainty, then the feedback signal $\tilde{d}$ is not equal to zero. All that is unknown about the system is expressed in $\tilde{d}$. This motivates the idea of placing a filter on $\tilde{d}$ to reduce the effects of the deleterious signals on the system; this is an important step in the IMC design procedure discussed in the next section.

In addition, for the case of no model uncertainty ($p = \tilde{p}$), actuator constraints cannot destabilize the closed-loop system if the constrained process input is sent to the model $\tilde{p}$ (see Figure 9.68). In this case, inspection of Figure 9.68 indicates that

$$\tilde{d} = n + d + pl, \tag{9.156}$$

which is independent of the controller q. When the process model is equal to the true process then the control system is open loop, and the system is internally stable if and only if all the blocks in series are stable. In this case internal stability is implied by the stability of the IMC controller q, the process p, and the actuator nonlinearity. When model uncertainty is taken into account, then

$$\tilde{d} = (p\alpha - \tilde{p}\alpha)u + n + d + pl = (p - \tilde{p})\alpha u + n + d + pl, \tag{9.157}$$

where α represents a stable nonlinear operator, in this case, the static actuator limitation nonlinearity. Observe that $\tilde{d}$ still represents all that is unknown about the process.

9.7.2.3 IMC Design Procedure

The objectives of setpoint tracking and disturbance rejection are to minimize the error $e = y - r$. When the process model $\tilde{p}$ is equal to the true process p, then the error derived from Equation 9.152 is

$$e = r - y = \tilde{p}qn + (1 - \tilde{p}q)(r - d - pl) = \tilde{T}n + \tilde{S}(r - d - pl). \tag{9.158}$$

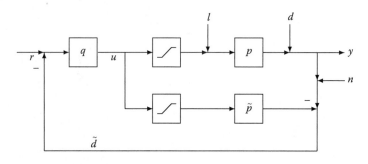

FIGURE 9.68   IMC implementation with actuator constraints.

The error e is an affine function of the IMC controller q. The previous section discussed how introducing a filter in the IMC feedback path (in $\tilde{d}$) reduces the effects of model uncertainty and measurement noise on the system. This filter can just as easily be introduced elsewhere in the feedback path, such as directly into q. This motivates the IMC design procedure, which consists of designing the IMC controller q in two steps:

- *Step 1: Nominal Performance:* A nominal IMC controller $\tilde{q}$ is designed to yield optimal setpoint tracking and disturbance rejection while ignoring measurement noise, model uncertainty, and constraints on the manipulated variable.
- *Step 2: Robust Stability and Performance:* An IMC filter f is used to detune the controller $q = \tilde{q}f$, to trade off performance with smoothing the control action and reducing the sensitivity to measurement noise and model uncertainty.

These steps are described below.

Performance Measures

Almost any reasonable performance measure can be used in the design of the nominal IMC controller. For fixed inputs (i.e., disturbances and/or setpoint), two of the most popular performance measures are the *integral absolute error* (IAE) and the *integral squared error* (ISE):

$$\text{IAE}\{e\} \equiv \int_0^\infty |e(t)| \, dt, \tag{9.159}$$

$$\text{ISE}\{e\} \equiv \int_0^\infty e^2(t) \, dt. \tag{9.160}$$

When the inputs are best described as being a set of stochastic signals (e.g., filtered white noise with zero mean and specified variance), a popular performance measure is

$$\text{Expected Value} \left\{ \int_0^\infty e^2(t) \, dt \right\}. \tag{9.161}$$

Another popular performance measure defined for a set of inputs v is the worst-case integral squared error,

$$\sup_{\int_0^\infty v^2(t)dt \leq 1} \int_0^\infty e^2(t) \, dt, \tag{9.162}$$

which is commonly used in the design of closed-loop transfer functions to have desired frequency-domain properties. Since the ISE performance measure for fixed inputs is the most popular in IMC, it will be used in what follows.

Irrespective of the closed-loop performance measure that a control engineer may prefer, it is usually important that the closed-loop system satisfies certain steady-state properties. For example, a common control system requirement is that the error signal resulting from step inputs approaches zero at steady state. The final value theorem applied to Equation 9.151 implies that this is equivalent to

$$q(0) = \tilde{p}^{-1}(0). \tag{9.163}$$

Another typical requirement is that the error signal resulting from ramp inputs approaches zero at steady state. The final value theorem implies that this requirement is equivalent to having both of the following conditions satisfied

$$q(0) = \tilde{p}^{-1}(0); \quad \frac{d}{ds}\left(\tilde{p}(0)q(0)\right) = 0. \tag{9.164}$$

Such conditions are used when selecting the IMC filter.

ISE-Optimal Performance

The ISE-optimal controller can be solved via a simple analytical procedure when the process is stable and has no zeros on the imaginary axis. The first step in this procedure is a factorization of the process model $\tilde{p}$ into an allpass portion $\tilde{p}_A$ and a minimum phase portion $\tilde{p}_M$

$$\tilde{p} = \tilde{p}_A \tilde{p}_M, \tag{9.165}$$

where $\tilde{p}_A$ includes all the open right-half plane zeros and delays of $\tilde{p}$ and has the form

$$\tilde{p}_A = e^{-s\theta} \prod_i \frac{-s + z_i}{s + \bar{z}_i}, \quad \mathrm{Re}\{z_i\}, \theta > 0, \tag{9.166}$$

θ is the time delay and z_i is a right-half plane zero in the process model, and $\bar{z}_i$ is the complex conjugate of z_i. The unmeasured input v in Equation 9.158 is given by $v = d + pl - r$. The controller that minimizes the ISE for these inputs is given in the following theorem.

Theorem 9.1:

Assume that the process model $\tilde{p}$ is stable. Factor $\tilde{p}$ and the input v into allpass and minimum phase portions

$$\tilde{p} = \tilde{p}_A \tilde{p}_M, \quad v = v_A v_M. \tag{9.167}$$

The controller that minimizes the ISE is given by

$$\tilde{q} = \left(\tilde{p}_M v_M\right)^{-1} \left\{\tilde{p}_A^{-1} v_M\right\}_*, \tag{9.168}$$

where the operator $\{\cdot\}_*$ denotes that, after a partial fraction expansion of the operand, all terms involving the poles of $\tilde{p}_A^{-1}$ are omitted.

   Provided that the input v has been chosen to be of the appropriate type (e.g., step or ramp), the ISE-optimal controller $\tilde{q}$ will satisfy the appropriate asymptotic steady-state performance requirements (Equation 9.163 or 9.164). In general, the $\tilde{q}$ given by Theorem 9.1 will not be proper, and the complementary sensitivity $\tilde{T} = \tilde{p}\tilde{q}$ will have undesirable high-frequency behavior. The $\tilde{q}$ is augmented by a low-pass filter f (that is, $q = \tilde{q}f$) to provide desirable high-frequency behavior, to prevent sensitivity to model uncertainty and measurement noise, and to avoid overly rapid or large control actions. This filter f provides the compromise between performance and robustness, and its selection is described next.

IMC Filter Forms

The IMC filter f should be selected so that the closed-loop system retains its asymptotic properties as $\tilde{q}$ is detuned for robustness. In particular, for the error signal resulting from step inputs to approach zero at steady state, the filter f must satisfy

$$f(0) = 1. \tag{9.169}$$

   Filters that satisfy this form include

$$f(s) = \frac{1}{(\lambda s + 1)^n}, \tag{9.170}$$

and

$$f(s) = \frac{\beta s + 1}{(\lambda s + 1)^n}, \tag{9.171}$$

where λ is an adjustable filter parameter that provides the tradeoff between performance and robustness, n is selected large enough to make q proper, and β is another free parameter that can be useful for some applications (its use is described in Example 9.8). For the error signal resulting from ramp inputs to approach zero at steady state, the filter f must satisfy

$$f(0) = 1 \quad \text{and} \quad \frac{df}{ds}(0) = 0. \tag{9.172}$$

A filter form that satisfies these conditions is

$$f(s) = \frac{n\lambda s + 1}{(\lambda s + 1)^n}. \tag{9.173}$$

The parameter β that is free in Equation 9.171 becomes fixed in Equation 9.173 to satisfy the additional condition in Equation 9.172.

The IMC controller q is calculated from $q = \tilde{q}f$, and the adjustable parameters tuned to arrive at the appropriate tradeoff between performance and robustness. The corresponding classical controller, if desired, can be calculated by substituting q into Equation 9.150.

Rapid changes in the control action are generally undesirable, as they waste energy and may cause the actuators to wear out prematurely. The IMC filter allows the control engineer to directly detune the control action, as can be seen from (ignoring model uncertainty in Figure 9.67)

$$u = \tilde{q}f(r - n - d - pl). \tag{9.174}$$

9.7.3 Applications

The IMC design procedure is applied to models for common industrial processes.

Example 9.3: A Nonminimum Phase Integrating Process

Processes with inverse response are common in industry. In the SISO case, these correspond to processes with nonminimum phase (right-half plane) zeros. For example, a model of the level in a reboiler (located at the base of a distillation column) to a change in steam duty is

$$\tilde{p} = \frac{-3s + 1}{s(s + 1)}. \tag{9.175}$$

Developing an accurate model for the level in a reboiler is difficult because the level depends on frothing, which does not respond in completely reproducible manner. The uncertainty in this model of the process can be described by a frequency-dependent bound on the difference between the process model $\tilde{p}$ and the true process p

$$\left| \frac{p(j\omega) - \tilde{p}(j\omega)}{\tilde{p}(j\omega)} \right| \le |w_u(j\omega)|, \quad \forall \omega, \tag{9.176}$$

where the Bode magnitude plot of the uncertainty weight $w_u(s) = (2s + 0.2)/(s + 1)$ is shown in Figure 9.69. This uncertainty covers up to 20% error in the steady-state gain, and up to 200% error at high frequencies.

The performance specification is to minimize the integral squared error in rejecting ramp output disturbances, $d = 1/s^2$. Because the process is not stable, a control system implemented using the Internal Model structure (Figure 9.67) would not be internally stable, as bounded load disturbances would lead to unbounded process outputs. On the other hand, Theorem 9.1 can still be used to design the ISE-optimal controller, as long as the controller is implemented using the classical control structure (Figure 9.66), and the integrators in the process also appear in the input v [1].

The first step in calculating $\tilde{q}$ is to factor the process model $\tilde{p}$ and the input $v = d$ into allpass and minimum phase portions (Equation 9.167)

$$\tilde{p}_A = \frac{-3s+1}{3s+1}, \quad \tilde{p}_M = \frac{3s+1}{s(s+1)}, \quad v_A = 1, \quad v_M = \frac{1}{s^2}. \tag{9.177}$$

The ISE-optimal controller (Equation 9.168) is

$$\tilde{q} = \left(\tilde{p}_M v_M\right)^{-1} \left\{\tilde{p}_A^{-1} v_M\right\}_* \tag{9.178}$$

$$\tilde{q} = \left(\frac{1}{s}\frac{3s+1}{s+1}\frac{1}{s^2}\right)^{-1} \left\{\left(\frac{-3s+1}{3s+1}\right)^{-1}\frac{1}{s^2}\right\}_* \tag{9.179}$$

$$\tilde{q} = \frac{s^3(s+1)}{3s+1} \left\{\frac{6}{s} + \frac{1}{s^2} + \frac{18}{-3s+1}\right\}_* \tag{9.180}$$

$$\tilde{q} = \frac{s^3(s+1)}{3s+1} \left(\frac{6}{s} + \frac{1}{s^2}\right) \tag{9.181}$$

$$\tilde{q} = \frac{s(s+1)(6s+1)}{3s+1}. \tag{9.182}$$

This is augmented with a filter form appropriate for ramp inputs (Equation 9.173), where the order of the denominator is chosen so that the IMC controller

$$q = \tilde{q}f = \frac{s(s+1)(6s+1)}{3s+1}\frac{3\lambda s+1}{(\lambda s+1)^3} = \frac{s(s+1)(6s+1)(3\lambda s+1)}{(3s+1)(\lambda s+1)^3} \tag{9.183}$$

is proper. The value of λ is selected just large enough that the inequality described by Equation 9.149,

$$|\tilde{T}(j\omega)| = |\tilde{p}(j\omega)q(j\omega)| = \left|\frac{(6j\omega+1)(-3j\omega+1)(3\lambda j\omega+1)}{(3j\omega+1)(\lambda j\omega+1)^3}\right| < \left|\frac{j\omega+1}{2j\omega+0.2}\right|, \tag{9.184}$$

is satisfied for all frequencies. A value of $\lambda = 5.4$ is adequate (Figure 9.69) for the closed-loop system to be robust to model uncertainty (any larger would result in overly sluggish performance). As stated

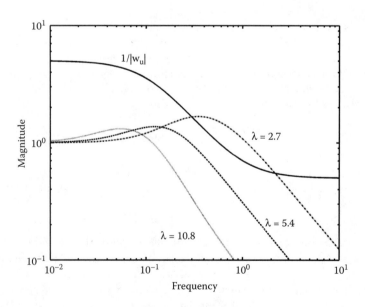

FIGURE 9.69 Bode magnitude plots for designing an IMC controller that achieves robust stability: $1/|w_u|$ and $|\tilde{T}|$ for $\lambda = 2.7, 5.4,$ and $10.8.$

earlier, this controller cannot be implemented using the IMC structure, but must be implemented using the classical control structure with Equation 9.150

$$k = \frac{q}{1 - \tilde{p}q} = \frac{(s+1)(6s+1)(3\lambda s+1)}{s(3\lambda^3 s^2 + (\lambda^2 + 9\lambda + 54)\lambda s + 3\lambda^2 + 18)}. \tag{9.185}$$

Closed-loop responses to ramp output disturbances for several processes included in the uncertainty description (two have the same dynamics but different steady-state error; the other two were chosen so that the inequality in Equation 9.176 is satisfied as an equality) are shown in Figure 9.70. The closed-loop responses are bounded as desired.

Example 9.4: Minimum Phase Process Models

Although most processes have some nonminimum phase character, a large number of processes can be approximated by a minimum phase model. In this case, $\tilde{p}_A = 1$, and the ISE-optimal controller (Equation 9.168) is

$$\tilde{q} = (\tilde{p}_M v_M)^{-1} \left\{ \tilde{p}_A^{-1} v_M \right\}_* \tag{9.186}$$

$$\tilde{q} = (\tilde{p}_M v_M)^{-1} \{v_M\}_* \tag{9.187}$$

$$\tilde{q} = \tilde{p}_M^{-1}. \tag{9.188}$$

The value for $\tilde{q}$ is the inverse of the process model (for this reason, IMC is often referred to as being *model inverse-based control*). In this case, the integral squared error for the unfiltered nominal system is zero irrespective of the disturbances and setpoint, because the resulting error (Equation 9.158),

$$e = y - r = (1 - \tilde{p}\tilde{q})(d + pl - r) = (1 - \tilde{p}_M \tilde{p}_M^{-1})(d + pl - r) = 0, \tag{9.189}$$

is zero. For the nominal process model, augmenting the ISE-optimal controller with an IMC filter gives the controlled output (Equation 9.152)

$$y = \tilde{T}(r - n) + \tilde{S}(d + pl) = \tilde{p}q(r - n) + (1 - \tilde{p}q)(d + pl) = fr + (1 - f)(d + pl). \tag{9.190}$$

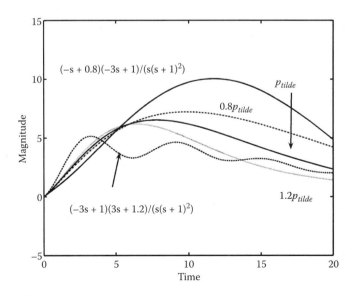

FIGURE 9.70 Closed-loop output responses y for a ramp output disturbance for Example 1 for $p = \tilde{p}$, $1.2\tilde{p}$, $0.8\tilde{p}$, $(-3s + 1)(3s + 1.2)/s(s + 1)^2$, and $(-3s + 1)(-s + .8)/s(s + 1)^2$ ($\tilde{p}$ is marked as p_{tilde} in the figure).

If the process has a relative degree of one, then the IMC filter can be chosen to have a relative degree of one to give

$$y = \frac{1}{\lambda s + 1}(r - n) + \frac{\lambda s}{\lambda s + 1}(d + pl).$$

(9.191)

The bandwidth of $\tilde{T}$ is very nearly equal to the bandwidth of $\tilde{S}$, with λ specifying the exact location of these bandwidths. When the process model is not minimum phase, then the bandwidths of $\tilde{T}$ and $\tilde{S}$ can be far apart; the IMC filter parameter will still provide the trade-off between nominal performance ($\tilde{S} \approx 0$) and insensitivity to measurement noise ($\tilde{T} \approx 0$).

Example 9.5: Processes with Common Stochastic Disturbances

Alhough the IMC design procedure was presented in terms of minimizing the integral squared error for a fixed input, industrial disturbances are often more naturally modeled as stochastic inputs. Fortunately, Parseval's Lemma informs us that Theorem 9.1 provides the minimum variance controller for stochastic inputs, if v is chosen correctly. For example, the minimum variance controller for integrated white-noise inputs is equal to the ISE-optimal controller designed for a step input v. This and other equivalences are provided in Table 9.8.

Since most industrial process disturbances are represented well by one of these stochastic descriptions, for convenience the simplified expressions for the minimum variance (or ISE-optimal) controller are given in the third column of Table 9.8. These expressions follow by analytically performing the partial fraction expansion and applying the $\{\cdot\}_*$ operator in Theorem 9.1. For example, for $v = 1/s$, the ISE-optimal controller (Equation 9.168) is

$$\tilde{q} = \left(\tilde{p}_M v_M\right)^{-1} \left\{\tilde{p}_A^{-1} v_M\right\}_*$$

(9.192)

$$\tilde{q} = \left(\tilde{p}_M \frac{1}{s}\right)^{-1} \left\{\tilde{p}_A^{-1} \frac{1}{s}\right\}_*$$

(9.193)

$$\tilde{q} = s\tilde{p}_M^{-1} \left\{\frac{\tilde{p}_A^{-1}(0)}{s} + \cdots\right\}_*$$

(9.194)

$$\tilde{q} = \tilde{p}_M^{-1}.$$

(9.195)

Example 9.6: Processes with Time Delay

It is common for processes to include time delays. This may be due to transport delays in reactors or process piping, or to approximating high-order dynamics. Consider the design of an IMC controller

TABLE 9.8 Minimum Variance Controllers for Common Stochastic Inputs

v	Stochastic Inputs	Minimum Variance $\tilde{q}$
$\dfrac{1}{s}$	Integrated white noise	$\tilde{p}_M^{-1}$
$\dfrac{1}{\tau s + 1}$	Filtered white noise	$p_M^{-1} p_A^{-1}(-1/\tau)$
$\dfrac{1}{s(\tau s + 1)}$	Filtered integrated white noise	$p_M^{-1}\left(1 + \left(1 - p_A^{-1}(-1/\tau)\right)\tau s\right)$
$\dfrac{1}{s^2}$	Double integrated white noise	$p_M^{-1}\left(1 - s\dfrac{dp_A}{ds}(0)\right)$

for the process

$$\tilde{p} = \bar{p}e^{-s\theta}, \tag{9.196}$$

where θ is the time delay and $\bar{p}$ is the delay-free part of the process and can include nonminimum phase zeros. Irrespective of the assumptions on the nature of the inputs, the IMC controller will have the form

$$q = \tilde{q}f. \tag{9.197}$$

From Equation 9.150, the corresponding classical controller has the form

$$k = \frac{\tilde{q}f}{1 - \tilde{q}f\bar{p}e^{-\theta s}}. \tag{9.198}$$

This controller is closely related to the well-known Smith predictor shown in Figure 9.71 and given by [2]:

$$k = \frac{c}{1 + c\bar{p}(1 - e^{-\theta s})}. \tag{9.199}$$

Actually, setting Equation 9.198 equal to Equation 9.199 and rearranging gives the Smith predictor controller in terms of the IMC controller and vice versa

$$c = \frac{\tilde{q}f}{1 - \bar{p}\tilde{q}f}, \quad \tilde{q}f = \frac{c}{1 + \bar{p}c}. \tag{9.200}$$

This implies that the Smith predictor and IMC structures are equivalent.

Smith states in his original manuscript that the Smith predictor control structure in Figure 9.71 allows the controller c to be designed via any optimal controller design method applied to the delay-free process. This seems to be confirmed by Equation 9.200, where c would be the form of the classical controller designed via IMC applied to the delay-free process. Although c *could* be designed by ignoring the delay in the process, the nominal closed-loop performance depends on the sensitivity

$$\tilde{S} = 1 - \tilde{p}\tilde{q} = 1 - \bar{p}e^{-\theta s}\tilde{q}f, \tag{9.201}$$

which is a function of the time delay and thus its effect should be considered in the controller design. An appropriate method of designing the controller c would be to design $\tilde{q}$ based on the process model with delay, and tune the IMC filter based on $\tilde{S}$ and $\tilde{T}$ taking performance and robustness into account. Thus IMC provides a transparent method for designing Smith predictor controllers. Alternatively, the controller could be implemented in the IMC control structure in Figure 9.68, which would have the advantage of ensuring closed-loop stability in the presence of actuator constraints.

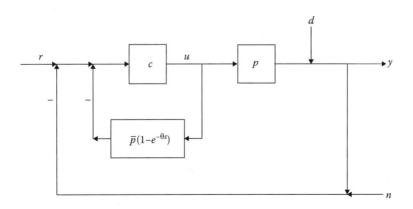

FIGURE 9.71 Smith predictor control structure.

Example 9.7: PID Tuning Rules for Low-Order Processes

In IMC, the resulting controller is of an order roughly equivalent to that of the process model. Many models for SISO industrial processes are low order, so that an IMC controller based on such a model is of low order and can be exactly or approximately described as a *Proportional–Integral–Derivative* (PID) controller:

$$k = k_c \left(1 + \tau_D s + \frac{1}{\tau_I s} \right), \tag{9.202}$$

where k_c is the *gain*, τ_I is the *integral time constant*, and τ_D is the *derivative time constant*. PID controllers are the most popular and reliable controllers in the process industries. To a large part, this explains why the largest number of industrial applications of IMC to SISO processes is for the tuning of PID controllers.

To provide an example of the derivation of *IMC PID tuning rules*, consider a first-order process model

$$\tilde{p} = \tilde{p}_M = \frac{k_p}{\tau s + 1}, \tag{9.203}$$

where k_p is its steady-state gain and τ is its time constant. Table 9.8 gives the IMC controller (with a first-order filter) for step inputs as

$$q = \tilde{p}_M^{-1} = \frac{\tau s + 1}{k_p(\lambda s + 1)}. \tag{9.204}$$

The corresponding classical controller is given by Equation 9.150

$$k = \frac{\tau s + 1}{k_p \lambda s}. \tag{9.205}$$

This can be rearranged to be in the form of an ideal *Proportional-Integral* (PI) controller

$$k = k_c \left(1 + \frac{1}{\tau_I s} \right) \tag{9.206}$$

with

$$k_c = \frac{\tau}{k_p \lambda}, \quad \tau_I = \tau. \tag{9.207}$$

An advantage of designing PID controllers via IMC is that only one parameter is required to provide a clear tradeoff between robustness and performance; whereas PID has three parameters that do not provide this clear tradeoff. IMC PID tuning rules for low-order process models and the most common disturbance and setpoint model (step) are listed in Table 9.9.

TABLE 9.9　IMC PID Tuning Rules

Process Model $\tilde{p}$	k_c	τ_I	τ_D
$\dfrac{k_p}{\tau s + 1}$	$\dfrac{\tau}{\lambda k_p}$	τ	-
$\dfrac{k_p}{\tau^2 s^2 + 2\zeta \tau s + 1}$	$\dfrac{2\zeta \tau}{\lambda k_p}$	$2\zeta \tau$	$\dfrac{\tau}{2\zeta}$
$\dfrac{k_p}{s}$	$\dfrac{1}{\lambda k_p}$	-	-
$\dfrac{k_p}{s(\tau s + 1)}$	$\dfrac{1}{\lambda k_p}$	-	τ

Source:　Data from D. E. Rivera, S. Skogestad, and M. Morari. *Ind. Eng. Chem. Proc. Des. Dev.*, 25:252–265, 1986.

Example 9.8: Processes with a Single Dominant Lag

Many processes have a time lag that is substantially slower than the other time lags and the time delay. Several researchers over the last 25 years have claimed that IMC gives poor rejection of load disturbances for these processes. To aid in the understanding of this claim, consider a process modeled by a dominant lag

$$\tilde{p} = \frac{1}{100s + 1}. \tag{9.208}$$

An IMC controller designed for this process model is (see Example 9.4)

$$q = \frac{100s + 1}{\lambda s + 1}. \tag{9.209}$$

For simplicity of presentation only, let $r = n = 0$ and ignore model uncertainty in what follows. The controlled output is related to the output disturbance d and load disturbance l by Equation 9.191

$$y = \tilde{S}(d + pl) = (1 - \tilde{p}q)d + (1 - \tilde{p}q)pl, \tag{9.210}$$

$$y = \frac{\lambda s}{\lambda s + 1}d + \frac{\lambda s}{(\lambda s + 1)(100s + 1)}l. \tag{9.211}$$

The value for λ is selected as 20 to provide nominal performance approximately five times faster than open loop. The closed-loop responses to unit step load and output disturbances are shown in Figure 9.72. As expected, the control system rejects the unit step output disturbance d with a time constant of approximately 20 time units. On the other hand, the control system rejects the load disturbance l very slowly. This difference in behavior is easily understood from Equation 9.211, since the slow process time lag appears in the transfer function between the load disturbance and the controlled output, irrespective of the magnitude of the filter parameter λ (as long as $\lambda \neq 0$). The open-loop dynamics appear in the closed-loop dynamics, resulting in the long tail in Figure 9.72.

Several researchers have proposed *ad hoc* fixes for this problem. The simplest solution is presented here. The reason that the closed-loop dynamics are poor is because the commonly used IMC filter forms are designed for output disturbances, not load disturbances. Thus, a simple fix is to design the correct filter for the load disturbance. Consider the IMC filter (Equation 9.171) that provides an extra degree of freedom (β) over the other filter forms (Equations 9.170 and 9.173). The order n of the filter is chosen equal to 2 so that the IMC controller will be proper. Then the controlled output is related to the output disturbance d and load disturbance l by Equation 9.210

$$y = \tilde{S}d + \tilde{S}pl = (1 - \tilde{p}q)d + (1 - \tilde{p}q)pl, \tag{9.212}$$

$$y = \frac{\lambda^2 s^2 + (2\lambda - \beta)s}{(\lambda s + 1)^2}d + \frac{\lambda^2 s^2 + (2\lambda - \beta)s}{(\lambda s + 1)^2(100s + 1)}l. \tag{9.213}$$

Since the sluggish response to load disturbances is due to the open-loop pole at $s = -1/100$, select the extra degree of freedom β to cancel this pole

$$\beta = 2\lambda - \frac{\lambda^2}{100}. \tag{9.214}$$

Then the controlled output (Equation 9.213) is

$$y = \frac{\lambda^2 s^2 + \lambda^2 s/100}{(\lambda s + 1)^2}d + \frac{\lambda^2 s^2 + \lambda^2 s/100}{(\lambda s + 1)^2(100s + 1)}l, \tag{9.215}$$

$$y = \frac{s\lambda^2(s + 1/100)}{(\lambda s + 1)^2}d + \frac{s\lambda^2/100}{(\lambda s + 1)^2}l. \tag{9.216}$$

The closed-loop responses to unit step load and output disturbances are shown in Figure 9.72 (with $\lambda = 20$). This time the undesirable open-loop time constant does not appear in the controlled variable.

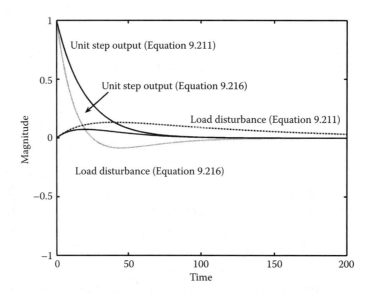

FIGURE 9.72 Closed-loop output responses y for unit step output and load disturbances with common filter design (Equation 9.211) and correct filter design (Equation 9.216).

9.7.4 Extension to Nonlinear Processes

This section describes an extension of IMC to open-loop stable nonlinear processes with stable inverses. To simplify the presentation, the focus will be on controller design for setpoint tracking for SISO processes, with similar ideas applying for disturbance rejection.

The IMC controller is a nonlinear inverse of the process model augmented by a nonlinear filter to produce a physically realizable controller with a single tuning parameter. Assume that the nominal process model $\tilde{p}$ available for controller design has the form

$$\frac{\mathrm{d}}{\mathrm{d}t}\tilde{x} = \tilde{f}(\tilde{x}) + \tilde{g}(\tilde{x})u, \tag{9.217}$$

$$\tilde{y} = \tilde{h}(\tilde{x}), \tag{9.218}$$

where $\tilde{x}$ is an $\tilde{n}$-dimensional state vector, u is the manipulated variable, $\tilde{y}$ is the model output, $\tilde{f}(\tilde{x})$ and $\tilde{g}(\tilde{x})$ are $\tilde{n}$-dimensional vectors of nonlinear functions, and $\tilde{h}(\tilde{x})$ is a scalar nonlinear function. The true process p is assumed to have a similar form

$$\frac{\mathrm{d}}{\mathrm{d}t}x = f(x) + g(x)u, \tag{9.219}$$

$$y = h(x), \tag{9.220}$$

where x is an n-dimensional state vector, y is the true process output, $f(x)$ and $g(x)$ are n-dimensional vectors of nonlinear functions, and $h(x)$ is a scalar nonlinear function.

The nonlinear IMC controller has the form

$$q = \tilde{q}f, \tag{9.221}$$

where f is a filter and $\tilde{q}$ minimizes the tracking error

$$\|r(t) - y(t)\| = \|(1 - \tilde{p}\tilde{q})r\|. \tag{9.222}$$

When the nominal process model is stable with a stable inverse then

$$\tilde{q} = \tilde{p}_r^{-1} \tag{9.223}$$

where $\tilde{p}_r^{-1}$ is the right inverse of the nominal process model, which is the operator that satisfies

$$\tilde{p}\tilde{p}_r^{-1}y = y. \tag{9.224}$$

For determining this right inverse, assume that the vector fields $\tilde{f}(\tilde{x})$ and $\tilde{g}(\tilde{x})$ and scalar field $\tilde{h}(\tilde{x})$ have continuous derivatives of all order. Also assume that

$$L_{\tilde{g}}L_{\tilde{f}}^{r-1}\tilde{h}(\tilde{x}) \neq 0, \quad \forall \tilde{x} \in R^{\tilde{n}}, \tag{9.225}$$

where $L_{\tilde{f}}\tilde{h}(\tilde{x})$ is a *Lie derivative* of $\tilde{h}(\tilde{x})$ with respect to the function $\tilde{f}(\tilde{x})$ and r is the *relative degree*. Then, the first r time derivatives of the nominal process output are:

$$\tilde{y}^{(k)} := \frac{d^k}{dt^k}\tilde{y}(t) = L_{\tilde{f}}^k\tilde{h}(\tilde{x}), \quad 1 \leq k \leq r - 1, \tag{9.226}$$

$$\tilde{y}^{(r)} = L_{\tilde{f}}^r\tilde{h}(\tilde{x}) + L_{\tilde{g}}L_{\tilde{f}}^{r-1}\tilde{h}(\tilde{x})u. \tag{9.227}$$

Solving Equation 9.227 for the manipulated variable u and substituting the result into Equation 9.217 produces the *Hirschorn inverse*:

$$\frac{d}{dt}\tilde{x} = \tilde{f}(\tilde{x}) - \frac{L_{\tilde{f}}^r\tilde{h}(\tilde{x})}{L_{\tilde{g}}L_{\tilde{f}}^{r-1}\tilde{h}(\tilde{x})}\tilde{g}(\tilde{x}) + \frac{\tilde{g}(\tilde{x})}{L_{\tilde{g}}L_{\tilde{f}}^{r-1}\tilde{h}(\tilde{x})}\tilde{y}^{(r)}, \tag{9.228}$$

$$u = -\frac{L_{\tilde{f}}^r\tilde{h}(\tilde{x})}{L_{\tilde{g}}L_{\tilde{f}}^{r-1}\tilde{h}(\tilde{x})} + \frac{1}{L_{\tilde{g}}L_{\tilde{f}}^{r-1}\tilde{h}(\tilde{x})}\tilde{y}^{(r)}. \tag{9.229}$$

These two equations define an inverse of the nominal process model, in which the manipulated variable $u(t)$ is reconstructed from the rth derivative of the process output. To determine the manipulated variable to track a setpoint r, replace $\tilde{y}$ with the filtered error e:

$$v^{(r)} = \alpha_1 e - \alpha_r L_{\tilde{f}}^{(r-1)}\tilde{h}(\tilde{x}) - \alpha_{r-1}L_{\tilde{f}}^{(r-2)}\tilde{h}(\tilde{x}) - \cdots - \alpha_1\tilde{h}(\tilde{x}), \tag{9.230}$$

where $\{\alpha_i\}$ are tuning parameters. The IMC controller $q = \tilde{q}f$ is obtained by combining Equations 9.229 and 9.230 to give

$$u = \frac{\alpha_1 e - \sum_{k=1}^{r+1}\alpha_k L_{\tilde{f}}^{k-1}\tilde{h}(\tilde{x})}{L_{\tilde{g}}L_{\tilde{f}}^{r-1}\tilde{h}(\tilde{x})} \tag{9.231}$$

where $\alpha_{r+1} \equiv 1$. The above equations can be combined to show that, if

$$\tilde{y}^{(k)}(0) = e^{(k)}(0), \quad 0 \leq k \leq r - 1, \tag{9.232}$$

then

$$\frac{\tilde{y}(s)}{e(s)} = \frac{\alpha_1}{s^r + \alpha_r s^{r-1} + \cdots + \alpha_2 s + \alpha_1}. \tag{9.233}$$

The tuning parameters $\{\alpha_i\}$ can be reduced to a single IMC tuning parameter λ by setting this transfer function equal to

$$\frac{\tilde{y}(s)}{e(s)} = \frac{1}{(\lambda s + 1)^r}. \tag{9.234}$$

For a perfect model ($\tilde{p} = p$) with no disturbances or noise, the closed-loop relationship between the controlled variable and setpoint is

$$\frac{y(s)}{r(s)} = \frac{1}{(\lambda s + 1)^r}. \tag{9.235}$$

Example 9.9: Chemical Reactor

Consider a continuous-flow stirred-tank reactor with first-order kinetics that are irreversible and exothermic:

$$\frac{dx_1}{dt} = -x_1 + D_a(1 - x_1)e^{x_2/(1+x_2/\gamma)}, \tag{9.236}$$

$$\frac{dx_2}{dt} = -(1 + \beta)x_2 + HD_a(1 - x_1)e^{x_2/(1+x_2/\gamma)} + \beta u, \tag{9.237}$$

where x_1 is the concentration, x_2 is the temperature, D_a is the Damköhler number, γ is the activation energy, β is the cooling rate, and H is the heat of reaction (all parameters have been nondimensionalized). The manipulated variable u acts by varying the temperature of water in a cooling jacket. The control objective is to regulate the concentration subject to uncertainties in the model parameters D_a, H, β, and γ.

Writing Equations 9.236 and 9.237 in the form of Equations 9.217 and 9.218 for the nominal process model results in

$$\tilde{f}_1(\tilde{x}) = -\tilde{x}_1 + \tilde{D}_a(1 - \tilde{x}_1)e^{\tilde{x}_2/(1+\tilde{x}_2/\tilde{\gamma})}, \tag{9.238}$$

$$\tilde{f}_2(\tilde{x}) = -(1 + \tilde{\beta})\tilde{x}_2 + \tilde{H}\tilde{D}_a(1 - \tilde{x}_1)e^{\tilde{x}_2/(1+\tilde{x}_2/\tilde{\gamma})}, \tag{9.239}$$

$$\tilde{g}_1(\tilde{x}) = 0, \quad \tilde{g}_2(\tilde{x}) = \tilde{\beta}, \quad \tilde{h}(\tilde{x}) = \tilde{x}_1. \tag{9.240}$$

This nominal process model has a relative degree $r = 2$. Taking the Lie derivatives gives

$$L_{\tilde{g}}L_{\tilde{f}}^{r-1}\tilde{h}(\tilde{x}) = L_{\tilde{g}}L_{\tilde{f}}\tilde{h}(\tilde{x}) = \tilde{\beta}\frac{\partial\tilde{f}_1}{\partial\tilde{x}_2}, \tag{9.241}$$

and

$$L_{\tilde{f}}^2\tilde{h}(\tilde{x}) = \tilde{f}_1\frac{\partial\tilde{f}_1}{\partial\tilde{x}_1} + \tilde{f}_2\frac{\partial\tilde{f}_1}{\partial\tilde{x}_2}. \tag{9.242}$$

For the IMC tuning parameter $\lambda = 1/10$, the appropriate filter parameters

$$\alpha_1 = \frac{1}{\lambda^2} = 100, \quad \alpha_2 = \frac{2}{\lambda} = 20, \quad \alpha_3 = 1, \tag{9.243}$$

can be determined by matching term-by-term the powers of s in Equations 9.233 and 9.234.

The IMC controller $q = \tilde{q}f$ is

$$u = \frac{\alpha_1 e - \alpha_1\tilde{h} - \alpha_2\tilde{f}_1 - \alpha_3\left(\tilde{f}_1\dfrac{\partial\tilde{f}_1}{\partial\tilde{x}_1} + \tilde{f}_2\dfrac{\partial\tilde{f}_1}{\partial\tilde{x}_2}\right)}{\tilde{\beta}\dfrac{\partial\tilde{f}_1}{\partial\tilde{x}_2}}, \tag{9.244}$$

where $e = r - (y - \tilde{y})$ in the case of model uncertainties. The closed-loop process outputs track the setpoint trajectory for $\pm 30\%$ perturbations in each of the parameters (see Figure 9.73). The spikes obtained for some values of the parameters could be reduced by increasing the value of the tuning parameter λ, which would make the setpoint tracking more sluggish.

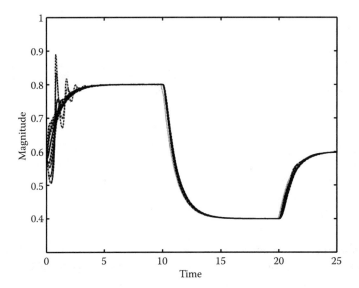

FIGURE 9.73 The setpoint ($\cdots$) and the closed-loop output responses when there is no model uncertainty (—) and when there is model uncertainty (- · -). The nominal model parameters are $(\tilde{D}_a, \tilde{H}, \tilde{\beta}, \tilde{\gamma}) = (0.072, 1, 0.3, 20)$ and the "real" parameters are $\pm 30\%$ deviations in each parameter (hence 2^4 plots).

9.7.5 Extension to Infinite-Dimensional Systems

This section describes the extension of IMC design to linear infinite-dimensional systems. An infinite-dimensional filter is coupled with the inverse of the process model to produce a physically realizable controller with a single tuning parameter. Two design approaches are described: (1) inversion of process model following by augmentation with an infinite-dimensional filter, or (2) augmentation of a process model to be semiproper, followed by inversion.

9.7.5.1 Method 1

The IMC controller consists of the optimal controller for the process model combined with filtering to reduce sensitivity to noise, provide robustness to model uncertainties, and smooth the control action.

Nominal performance

The first step is to determine the operator $\tilde{q}$ that optimizes the nominal performance:

$$\min_{\tilde{q}} \left\| w_p \tilde{S} \right\|_\infty, \tag{9.245}$$

where w_p is a stable minimum-phase performance weight and

$$\left\| w_p \tilde{S} \right\|_\infty := \sup_\omega \left| w_p(j\omega)\tilde{S}(j\omega) \right|. \tag{9.246}$$

This performance objective, which is known as the H_∞-norm, is equivalent to the worst-case integral squared error (Equation 9.162). Several algorithms have been developed for solving the minimization Equation 9.245 for infinite-dimensional processes [4]. As in the finite-dimensional case, the solution for minimum-phase processes is

$$\tilde{q} = \tilde{p}_M^{-1}. \tag{9.247}$$

This $\tilde{q}$ is usually improper since the nominal process model is usually strictly proper.

Robust Stability and Performance

The IMC controller

$$q(s) = \tilde{q}(s)f(s, \lambda) \qquad (9.248)$$

augments $\tilde{q}$ from Equation 9.245 with a filter f to detune the optimal controller. As for finite-dimensional processes, the corresponding classical feedback controller is given by

$$k = \frac{q}{1 - \tilde{p}q}. \qquad (9.249)$$

Usually an infinite-dimensional filter is needed to make the IMC controller q proper and the feedback controller k physically realizable. The filter f should be selected so that the closed-loop system retains desired asymptotic properties as q is detuned for robustness. In particular, for the error signal resulting from a step input to approach zero at steady state, the filter should satisfy

$$\lim_{s \to 0} f(s, \lambda) = 1. \qquad (9.250)$$

To provide a one-to-one correspondence to the IMC design method for finite-dimensional systems, the tuning parameter λ in the infinite-dimensional filter f should be defined so that the optimal nominal performance is achieved as $\lambda \to 0$:

$$\lim_{\lambda \to 0} f(s, \lambda) = 1, \qquad (9.251)$$

so that increasing λ slows the closed-loop dynamics to increase robustness to model uncertainties.

The specific finite value for the tuning parameter λ can be selected in a number of ways, corresponding to the same criteria used to tune IMC controllers for finite-dimensional systems. For example, λ can be selected as small as possible while satisfying the robust stability condition

$$\left\| w_u \tilde{T} \right\|_\infty < 1, \qquad (9.252)$$

which equivalent to condition (Equation 9.149). Another criterion commonly used to tune λ that is applicable to both finite- and infinite-dimensional systems is that the controller achieve *robust performance*, that is,

$$\left\| w_p S \right\|_\infty < 1, \qquad (9.253)$$

for all processes that satisfy the uncertainty description (Equation 9.148). It can be shown that this condition is equivalent to

$$\left\| |w_p \tilde{S}| + |w_u \tilde{T}| \right\|_\infty < 1. \qquad (9.254)$$

Another approach for selecting λ is as the solution to the one-parameter optimization

$$\min_{\lambda > 0} \left\| |w_p \tilde{S}| + |w_u \tilde{T}| \right\|_\infty, \qquad (9.255)$$

with a controller achieving robust performance if the attained objective is less than one.

9.7.5.2 Method 2

This approach first defines a stably invertible super-set of the process model. For a minimum-phase process model, construct $\tilde{p}_s(s, \lambda) \supset \tilde{p}(s)$ for $\lambda > 0$ such that $\tilde{p}_s(s, \lambda)$ is minimum phase and biproper and satisfies

$$\lim_{\lambda \to 0} \tilde{p}_s(s, \lambda) = \tilde{p}(s). \qquad (9.256)$$

Then the IMC controller is

$$q(s, \lambda) = \tilde{p}_s^{-1}(s, \lambda), \qquad (9.257)$$

where the IMC tuning parameter λ is selected as described in Method 1. If the nominal process model is nonminimum phase, then $\tilde{p}_s(s, \lambda)$ should be constructed so that the IMC controller q optimizes the nominal performance (Equation 9.245) as $\lambda \to 0$.

Implementation

Method 1 is closest in character to the IMC method for finite-dimensional systems, whereas Method 2 is more convenient when inspecting Laplace transform tables to identify suitable forms for the IMC controller. As in IMC for finite-dimensional process models, the form of the filter is up to the designer. The transfer function of the classical controller is determined by Equation 9.249 with a time-domain description for the controller constructed by analytical or numerical solution of the inverse Laplace transform. When the processes in the model uncertainty set are all stable, then the control system can be implemented using either the IMC or classical feedback structure.

Example 9.10: Diffusion Equation

Consider the diffusion equation

$$\frac{\partial C}{\partial t} = D\frac{\partial^2 C}{\partial x^2}, \quad \forall x \in (0, a), \quad \forall t > 0, \tag{9.258}$$

where $C(x, t)$ is the concentration at spatial location x and time t, D is the diffusion coefficient with nominal value $\tilde{D} = 10^{-10}$ m^2/s, and the distance across the domain $a = 10^{-5}$ m. The Dirichlet and Neumann boundary conditions

$$C(0, t) = u(t), \tag{9.259}$$

$$\left.\frac{\partial C}{\partial x}\right|_{x=a} = 0, \tag{9.260}$$

are assumed, in which the manipulated variable is the concentration at $x = 0$ and the controlled variable is the concentration at $x = a$.

The minimum-phase process model

$$\tilde{p}(s) = \frac{1}{\cosh\sqrt{s}} \tag{9.261}$$

between the manipulated variable $u(t)$ and the process output $C(a, t)$ is obtained by taking Laplace transforms of the partial differential equation 9.258 and boundary conditions (Equations 9.259 and 9.260) with respect to t, and solving the resulting ordinary differential equation in x.

The performance weight

$$w_p(s) = 0.5\frac{0.06s + 1}{0.06s}, \tag{9.262}$$

is selected to specify zero steady-state error for a step input (i.e., integral action), a peak sensitivity less than 2, and a closed-loop time constant of 0.06 s.

The model uncertainty is described by the frequency-dependent bound

$$\left|\frac{p(j\omega) - \tilde{p}(j\omega)}{\tilde{p}(j\omega)}\right| \le |w_u(j\omega)|, \quad \forall \omega, \tag{9.263}$$

with

$$w_u(s) = \frac{\cosh\sqrt{s}}{\cosh\sqrt{s/1.2}} - 0.8, \tag{9.264}$$

which includes variations in the diffusion coefficient, $D \in [0.72, 1.2] \times 10^{-10}$ m^2/s. The model uncertainty set (Equation 9.264) also includes processes that have the same dynamics as the nominal process model but have up to 20% uncertainty in the steady-state gain.

An invertible semiproper super-set of the nominal process model

$$\tilde{p}_s(s, \lambda) = \frac{\cosh \lambda \sqrt{s}}{\cosh \sqrt{s}} \qquad (9.265)$$

follows naturally from Equation 9.261. The corresponding IMC controller is

$$q(s, \lambda) = \frac{\cosh \sqrt{s}}{\cosh \lambda \sqrt{s}}, \qquad (9.266)$$

which is the optimal solution of Equation 9.245 for any fixed $\lambda \geq 0$ with $\tilde{p}_s(s, \lambda)$ in place of the nominal process model.

The nominal sensitivity and complementary sensitivity for the above q and $\tilde{p}$ (Equation 9.261) are

$$\tilde{S} = 1 - \tilde{p}q = 1 - \frac{1}{\cosh \lambda \sqrt{s}}, \qquad (9.267)$$

$$\tilde{T} = \tilde{p}q = \frac{1}{\cosh \lambda \sqrt{s}}. \qquad (9.268)$$

Figure 9.74 shows that $\lambda = 0.3$ satisfies the robust stability condition (Equation 9.252) and robust performance condition (Equation 9.254), and nearly minimizes Equation 9.255. The insensitivity of the closed-loop response to +20% uncertainty in the diffusion coefficient is seen in Figure 9.75.

9.7.6 Defining Terms

Complementary Sensitivity: The transfer function T between the setpoint r and the controlled variable y. For the classical control structure, this transfer function is given by

$$T = pk/(1 + pk),$$

where p and k are the process and controller transfer functions, respectively.

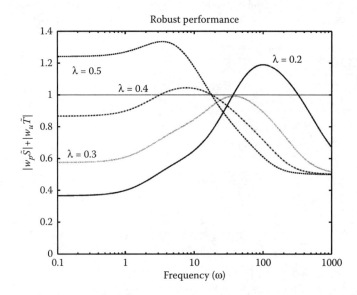

FIGURE 9.74 Bode magnitude plots for the evaluation of robust performance for the infinite-dimensional IMC controller in Example 9.10 for $\lambda = 0.5$ (---), 0.4 (---), 0.3 ($\cdots$), and 0.2 (—).

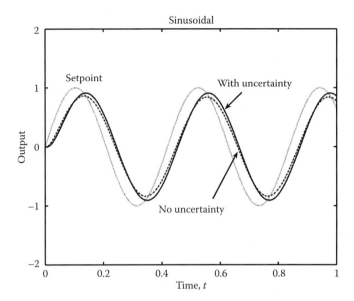

FIGURE 9.75 Setpoint trajectory and closed-loop output responses for Example 9.10 with no uncertainty and with uncertainty in the diffusion coefficient, $D = 1.2 \times 10^{-10} \, \text{m}^2/\text{s}$. The "real" process was simulated by the finite-difference method with 50 grid points.

Control Structure: The placement of controllers and their interconnections with the process.

H_∞-norm: The largest value of the ISE-norm of the output that can be obtained using inputs with ISE-norm less than one.

Hirschorn Inverse: The right inverse of a nonlinear process defined by Equations 9.228 and 9.229.

Stochastic Performance Measure: A closed-loop performance measure appropriate for stochastic inputs that consists of the expected variable of the integral squared closed-loop error e over the stochastic inputs:

$$\text{Expected Value} \left\{ \int_0^\infty e^2(t) \, dt \right\}. \tag{9.269}$$

Internal Model Control (IMC): A method of implementing and designing controllers, in which the process model is explicitly an internal part of the controller.

Integral Squared Error (ISE): A closed-loop performance measure appropriate for fixed inputs:

$$\text{ISE}\{e\} = \int_0^\infty e^2(t) \, dt, \tag{9.270}$$

where e is the closed-loop error. The ISE (Equation 9.270) is commonly referred in the modern control literature as the L_2-norm.

Internal Stability: The condition where bounded signals injected at any point in the control system generates bounded signals at any other point.

Inverse-Based Control: Any control design method in which an explicit inverse of the model is used in the design procedure.

Left Inverse: A left inverse (Figure 9.76) produces the manipulated variable given the process output.

Lie Derivative: The Lie derivative of a scalar function $\tilde{h}(\tilde{x})$ with respect to a vector function $\tilde{f}(\tilde{x})$ is defined as

$$L_{\tilde{f}} \tilde{h}(\tilde{x}) = \frac{\partial \tilde{h}}{\partial \tilde{x}} \tilde{f}(\tilde{x}). \tag{9.271}$$

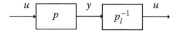

FIGURE 9.76   Left inverse.

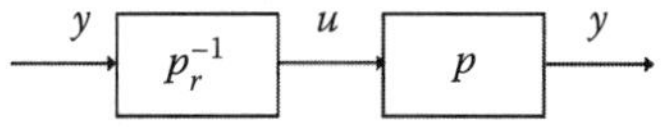

$$\xrightarrow{y}\boxed{p_r^{-1}}\xrightarrow{u}\boxed{p}\xrightarrow{y}$$

FIGURE 9.77   Right inverse.

Higher-order Lie derivatives can be defined recursively as:

$$L_{\tilde{f}}^{k}\tilde{h}(\tilde{x}) = \sum_{j=1}^{n}\frac{\partial}{\partial \tilde{x}_{j}}\{L_{\tilde{f}}^{k-1}\tilde{h}(\tilde{x})\}\tilde{f}_{j}(\tilde{x}), \tag{9.272}$$

where

$$L_{\tilde{f}}^{0}\tilde{h}(\tilde{x}) = \tilde{h}(\tilde{x}). \tag{9.273}$$

Load Disturbance: A disturbance that enters the input of the process.

Model Inverse-Based Control: Same as inverse-based control.

Proportional-Integral-Derivative (PID) controller: The most common controller in the process industries. The ideal form of this controller is given by

$$k = k_{c}\left(1 + \tau_{D}s + \frac{1}{\tau_{I}s}\right). \tag{9.274}$$

Relative Degree: The input in Equation 9.217 is said to have relative degree r at a point $\tilde{x}_0$ if

1. $L_{\tilde{g}}L_{\tilde{f}}^{k}\tilde{h}(\tilde{x}_0) = 0$, $\forall x$ in a neighborhood of $\tilde{x}_0$ and $\forall k < r - 1$, and
2. $L_{\tilde{g}}L_{\tilde{f}}^{r-1}\tilde{h}(\tilde{x}_0) \neq 0$.

Right Inverse: A right inverse (Figure 9.77) produces the manipulated variable required to obtain a given process output, which makes the right inverse an ideal feedforward controller.

Robust Stability: A closed-loop system that is internally stable for all processes within a well-defined set.

Robust Performance: A closed-loop system that is internally stable and satisfies some performance criterion for all processes within a well-defined set.

Sensitivity: The transfer function S between disturbances d at the process output and the controlled variable y. For the classical control structure, this transfer function is given by $S = 1/(1 + pk)$, where p and k are the process and controller transfer functions, respectively.

Single-input Single-output (SISO): A process with one manipulated variable and one controlled variable that is measured.

Smith Predictor: A strategy for designing controllers for processes with significant time delay, in which a predictor in the control structure seems to allow the controller to be designed ignoring the time delay.

References

1. Morari, M. and Zafiriou, E., *Robust Process Control*. Prentice-Hall, Englewood Cliffs, NJ, 1989.
2. Smith, O. J. M., Closer control of loops with dead time. *Chem. Eng. Prog.*, 25:217–219, 1957.
3. Rivera, D. E., Skogestad, S., and Morari, M., Internal model control 4: PID controller design. *Ind. Eng. Chem. Proc. Des. Dev.*, 25:252–265, 1986.

4. Curtain, R. F. and Zwart, H., *An Introduction to Infinite-Dimensional Linear Systems Theory*. Springer-Verlag, New York, 1995.
5. Frank, P. M., *Entwurf von Regelkreisen mit vorgeschriebenem Verhalten*. G. Braun, Karlsruhe, 1974.
6. Brosilow, C. B., The structure and design of Smith predictors from the viewpoint of inferential control. In *Proceedings of the Joint Automatic Control Conference*, pp. 288–288, 1979.
7. Bequette, B. W., *Process Control: Modeling, Design and Simulation*. Prentice-Hall, Upper Saddle River, NJ, 2003.
8. Horn, I. G., Arulandu, J. R., Gombas, C. J., VanAntwerp, J. G., and Braatz, R. D., Improved filter design in internal model control. *Ind. Eng. Chem. Res.*, 35:3437–3441, 1996.
9. Henson, M. A. and Seborg, D. E., An internal model control strategy for nonlinear systems. *AIChE J.*, 37:1065–1081, 1991.
10. Kishida, M. and Braatz, R. D., Internal model control of infinite dimensional systems. In *Proceedings of the IEEE Conference on Decision and Control*, pp. 1434–1441, IEEE Press, Piscataway, NJ, 2008.
11. Braatz, R. D., Ogunnaike, B. A., and Schwaber, J. S., Failure tolerant globally optimal linear control via parallel design. In *AIChE Annual Meeting*, San Francisco, CA, 1994. Paper 232b.
12. Garcia, C. E., Prett, D. M., and Morari, M., Model predictive control: Theory and practice—A survey. *Automatica*, 25:335–348, 1989.
13. Alvarez, J. and Zazueta, S., An internal-model controller for a class of single-input single-output nonlinear systems: Stability and robustness. *Dynamics and Control*, 8:123–144, 1998.

For Further Information

A review of the origins of the IMC control structure is provided in the PhD thesis of P. M. Frank [5]. Coleman B. Brosilow and Manfred Morari popularized the structure in a series of conference presentations and journal articles published in the late 1970s and 1980s. Brosilow [6] showed that the Smith predictor is equivalent to the IMC structure for processes with time delays. More thorough descriptions of IMC that include many of the equations in this chapter are provided in a research monograph [1] and a textbook [7].

The IMC control structure cannot be implemented for unstable processes. The IMC approach can be used to design feedback controllers for unstable processes, provided that the controllers are implemented in the classical feedback structure, and any unstable poles in the nominal process model are cancelled by unstable zeros in q [1].

Alternative techniques have been proposed for designing IMC controllers that provide reasonable rejection of load disturbances for processes with a dominant lag. The approach described in Example 9.8 is the simplest to apply. A table of PID controller parameters for other process models is available [8].

A more detailed description of the nonlinear IMC method is available [9], which also surveys alternative generalizations of IMC to nonlinear processes.

A much more detailed description of IMC for infinite-dimensional processes that includes all of the proofs and Example 9.10 is available [10]. For most performance objectives, the solution of the optimal control problem for $\tilde{q}$ is nontrivial when the nominal process model is nonminimum phase.

A control structure is said to have *multiple degrees of freedom* if it has more than one controller, with each controller having different input signals. Examples common in the process industries include combined cascade control and feedforward–feedback control. Strategies for placing and designing the controllers in an IMC setting for the simple cases were derived in the 1970s to 1980s [1,5]. A general method is available for constructing the optimal and most general multiple degree of freedom control structures, in which each controller is designed independently [11].

When there are multiple process inputs and/or outputs, IMC is usually treated in discrete time, and the performance objective is optimized on-line subject to the constraints. A linear or quadratic optimization is typically solved at each sampling instance, with off-the-shelf software available for performing these calculations. This method of control is referred to by many names, including *Model Predictive Control*, *Generalized Predictive Control*, and *Receding Horizon Control*, and is the most popular multivariable control method applied in industry today. A survey of these methods is available [12].

Alternative approaches for the control of the chemical reactor in Example 9.9 have been reported [13].

9.8 Time-Delay Compensation: Smith Predictor and Its Modifications

Z. J. Palmor

Despite the many articles and several book chapters that were published on this topic since the appearance of the first edition of this chapter and in spite of the fact that the Smith predictor continues to be a very active subject even today, we did not modify the presentation of this chapter as we believe that the original material is still relevant and useful. Hence, we added at the end of the chapter (mainly in the "Further Reading" section) brief comments pointing to newer references containing new related results.

9.8.1 Introduction

Time delays or dead times (DTs) between inputs and outputs are common phenomena in industrial processes, engineering systems, and economical and biological systems. Transportation and measurement lags, analysis times, and computation and communication lags, all introduce DTs into control loops. DTs are also inherent in distributed parameter systems and frequently are used to compensate for model reduction where high-order systems are represented by low-order models with delays. The presence of DTs in the control loops has two major consequences: (1) it greatly complicates the analysis and the design of feedback controllers for such systems and (2) it makes satisfactory control more difficult to achieve.

In 1957, O.J.M. Smith [1] presented a control scheme for single-input single-output (SISO) systems, which has the potential of improving the control of loops with DTs. This scheme became known as the Smith predictor (SP) or Smith dead-time compensator (DTC). It can be traced back to optimal control [2]. Early attempts to apply the SP demonstrated that classical design methods were not suitable for the SP or similar schemes. Theoretical investigations performed in the late 1970s and early 1980s clarified the special properties of the SP and provided tools for understanding and designing such algorithms. Over the years, numerous studies on the properties of the SP have been performed, both in academia and in industry. Many modifications have been suggested, and the SP was extended to multi-input and multi-output (MIMO) cases with multiple DTs.

The SP contains a model of the process with a DT. Its implementation on analog equipment was therefore difficult and inconvenient. When digital process controllers began to appear in the marketplace at the beginning of the 1980s, it became relatively easy to implement the DTC algorithms. Indeed, in the early 1980s some microprocessor-based industrial controllers offered the DTC as a standard algorithm like the proportional–integral-derivative (PID).

It is impossible to include all the available results on the topic and the many modifications and extensions in a single chapter. Hence, in this chapter, the SISO continuous case is treated. This case is a key to understanding the sampled data and the multivariable cases. Attention is paid to both theoretical and practical aspects. To make the reading more transparent, proofs are omitted but are referenced.

9.8.2 Control Difficulties due to Time Delays

A linear time-invariant (LTI) SISO plant with an input delay is represented in the state space as follows:

$$\dot{x}(t) = Ax(t) + Bu(t - \theta),$$
$$y(t) = Cx(t). \tag{9.275}$$

where $x \in R^n$ is the state vector, $u \in R$ is the input, and $y \in R$ is the output. A, B, and C are matrices of appropriate dimensions and θ is the time delay (or DT). Similarly, an LTI SISO plant with an output delay

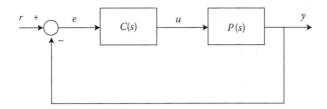

FIGURE 9.78 A feedback control system with plant P and controller C.

is given by

$$\dot{x}(t) = Ax(t) + Bu(t),$$
$$y(t) = Cx(t - \theta). \tag{9.276}$$

The transfer function of both Equations 9.275 and 9.276 is

$$y(s)/u(s) = P(s) = P_o(s)e^{-\theta s}, \tag{9.277}$$

where

$$P_o(s) = C(sI - A)^{-1}B. \tag{9.278}$$

$P_o(s)$ is seen to be a rational transfer function of order n. The presence of a DT in the control loop complicates the stability analysis and the control design of such systems. Furthermore, it degrades the quality of control due to unavoidable reduction in control gains as is demonstrated by the following simple example.

Assume that in the feedback control loop shown in Figure 9.78 the controller, $C(s)$, is a proportional (P) controller [i.e., $C(s) = K$,] and that $P_o(s)$ is a first-order filter [i.e., $P_o(s)=1/(\tau s+1)$]. $P(s)$ is thus given by

$$P(s) = e^{-\theta s}/(\tau s + 1). \tag{9.279}$$

The transfer function of the closed-loop system relating the output, $y(s)$, to the setpoint (or reference), $r(s)$, is

$$\frac{y(s)}{r(s)} = \frac{Ke^{-\theta s}}{\tau s + 1 + Ke^{-\theta s}}. \tag{9.280}$$

First, note that the characteristic equation contains $e^{-\theta s}$. Hence, it is a transcendental equation in s, which is more difficult to analyze than a polynomial equation. Second, the larger the ratio between the DT, θ, and the time constant, τ, the smaller the maximum gain, K_{max}, for which stability of the closed loop holds. When $\theta/\tau = 0$ (i.e., the process is DT free), then $K_{max} \to \infty$, at least theoretically. When $\theta/\tau = 1$ (i.e., the DT equals the time constant), the maximum gain reduces drastically, from ∞ to about 2.26, and when $\theta/\tau \to \infty$, $K_{max} \to 1$.

The preceding example demonstrates clearly that when DTs are present in the control loop, controller gains have to be reduced to maintain stability. The larger the DT is relative to the timescale of the dynamics of the process, the larger the reduction required. Under most circumstances, this results in poor performance and sluggish responses. One of the first control schemes aimed at improving the closed-loop performance for systems with DTs was that proposed by Smith [1]. This scheme is discussed next.

9.8.3 Smith Predictor (DTC)

9.8.3.1 Structure and Basic Properties

The classical configuration of a system containing an SP is depicted in Figure 9.79. $P(s)$ is the transfer function of the process, which consists of a stable rational transfer function $P_o(s)$ and a DT as in Equation 9.277. $\hat{P}_o(s)$ and $\hat{P}(s)$ are models, or nominal values, of $P_o(s)$ and $P(s)$, respectively. The shaded area in

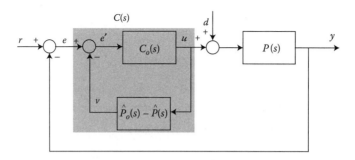

FIGURE 9.79   Classical configuration of a system incorporating SP.

Figure 9.79 is the SP, or the DTC. It consists of a *primary controller*, $C_o(s)$, which in industrial controllers is usually the conventional proportional–integral (PI) controller or PID controller, and a minor feedback loop, which contains the model of the process with and without the DT. The overall transfer function of the DTC is given by

$$C(s) = C_o(s)/[1 + C_o(s)(\hat{P}_o(s) - \hat{P}(s))]. \tag{9.281}$$

The underlying idea of the SP is clear, if one notes that the signal $v(t)$ (see Figure 9.79), contains a prediction of $y(t)$ DT units of time into the future. For that reason, the minor feedback around the primary controller is called a "predictor." It is noted that $e' = r - P_o u$, whereas $e = r - Pu$. Therefore, the "adjusted" error, $e'(t)$, which is fed into the primary controller, carries that part of the error that is "directly" caused by the primary controller. This eliminates the overcorrections associated with conventional controllers that require significant reductions in gains as was discussed earlier. Thus, it is seen that the SP should permit higher gains to be used.

The above qualitative arguments can be supported analytically. Assuming perfect model matching (which is called in the sequel the *ideal case*), that is, $\hat{P}(s) = P(s)$, the transfer function of the closed loop in Figure 9.79 from the setpoint to the output is

$$G_{r(s)} = \frac{y(s)}{r(s)} = \frac{C_o P}{1 + C_o P_o}, \tag{9.282}$$

where the arguments have been dropped for convenience. It is observed that the DT has been removed from the denominator of Equation 9.282. This is a direct consequence of using the predictor. In fact, the denominator of Equation 9.282 is the same as the one of a feedback system with the DT-free process, P_o, and the controller C_o, without a predictor. Furthermore, Equation 9.282 is also the transfer function of the system shown in Figure 9.80, which contains neither DT nor DTC inside the closed loop.

The input–output equivalence of the two systems in Figures 9.79 and 9.80 may lead to the conclusion that one can design the primary controller in the SP by considering the system in Figure 9.80 as if DT did not exist. While the elimination of the DT from the characteristic equation is the main source for the potential improvement of the SP, its design cannot be based on Equation 9.282 or on the system in Figure 9.80. The reason is that the two input–output equivalent systems possess completely different

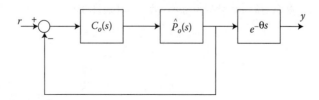

FIGURE 9.80   An input–output equivalent system.

sensitivity and robustness properties. It turns out that, under certain circumstances (discussed in the next section), an asymptotically stable design, which is based on Equation 9.282, with seemingly large stability margins may in fact be *practically unstable*. That is, the overall system with an SP may lose stability under infinitesimal model mismatchings.

An alternative way from which it can be concluded that the design and tuning of the SP should not rely on the equivalent system in Figure 9.80 is to write down G_r, the closed-loop transfer function from r to y, for the more practical situation, in which mismatchings, or uncertainties, are taken into account. This transfer function is denoted G'_r. Thus, when $\hat{P} \neq P$, G'_r takes the following form:

$$G'_r(s) = \frac{y(s)}{r(s)} = \frac{C_o P}{1 + C_o \hat{P}_o - C_o \hat{P} + C_o P}. \tag{9.283}$$

It is evident from Equation 9.283 that when mismatching exists, the DT is not removed in its totality from the denominator and therefore affects the stability. It is therefore more appropriate to state that the SP minimizes the effect of the DT on stability, thereby allowing tighter control to be used. Also, note that the DT has not been removed from the numerator of G_r (and G'_r). Consequently, the SP tracks reference variations with a time delay.

The transfer function from the input disturbance d (see Figure 9.79) to the output y, in the ideal case, is denoted G_d and is given by

$$G_d(s) = \frac{y(s)}{d(s)} = P \left[1 - \frac{C_o P}{1 + C_o P_o} \right]. \tag{9.284}$$

It is seen that the closed-loop poles consist of the zeros of $1 + C_o P_o$ and the poles of P, the open-loop plant. Consequently, the "classical" SP, shown in Figure 9.79, can be used for stable plants only. In Section 9.8.5, modified SP schemes for unstable plants are presented. In addition, the presence of the open-loop poles in G_d strongly influences the regulatory capabilities of the SP. This is discussed further in Section 9.8.3.3.

An equivalent configuration of the SP is shown in Figure 9.81. Since the scheme in Figure 9.81 results from a simple rearrangement of the block diagram in Figure 9.79, it is apparent that it leaves all input–output relationships unaffected. Although known long ago, this scheme is commonly known as the IMC (internal model control) form of the SP [4,5]. The dashed area in Figure 9.81 is the IMC controller, $q(s)$, which is related to the primary controller $C_o(s)$ via the following relationship:

$$q(s) = C_o(s) / [1 + C_o(s) \hat{P}_o(s)]. \tag{9.285}$$

The controller $q(s)$ is usually cascaded with a filter $f(s)$. The filter parameters are adjusted to comply with robustness requirements. Thus, the overall IMC controller, $\bar{q}(s)$, is

$$\bar{q}(s) = f(s) q(s). \tag{9.286}$$

The IMC parameterization is referred to in Section 9.8.3.4.

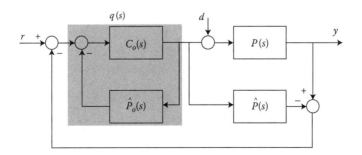

FIGURE 9.81 SP in IMC form.

9.8.3.2 Practical, Robust, and Relative Stability

Several stability results that are fundamental to understanding the special stability properties of the SP are presented. Among other things, they clarify why the design of the SP cannot be based on the ideal case. To motivate the development to follow, let us examine the following simple example.

Example 9.11:

Let the process in Figure 9.79 be given by $P(s) = e^{-s}/(s+1)$ and the primary controller by $C_0(s) = 4(0.5s + 1)$, an ideal proportional derivative (PD) controller. In the ideal case (i.e., perfect matching), the overall closed loop including the SP has, according to Equation 9.282, a single pole at $s = -5/3$. The system not only is asymptotically stable, but possesses a gain margin and a phase margin of approximately 2 and 80°, respectively, as indicated by the Nyquist plot (the solid line) in Figure 9.82.

However, for a slight mismatch in the DT, the overall system goes unstable, as is clearly observed from the dashed line in Figure 9.82, which shows the Nyquist plot for the nonideal case with 5% mismatch in the DT. In other words, the system is practically unstable.

For methodological reasons, the next definition of *practical instability* is presented in rather nonrigorous fashion.

Definition

A system that is asymptotically stable in the ideal case but becomes unstable for infinitesimal modeling mismatches is called a practically unstable system.

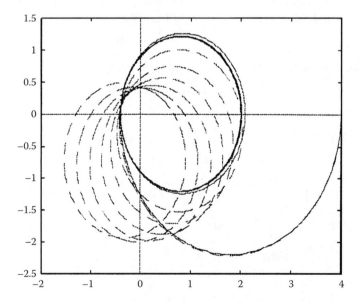

FIGURE 9.82 Nyquist plots of the system in Example 9.11. Solid line: ideal case; dashed line: nonideal case (5% mismatch in DT).

A necessary condition for practical stability of systems with SP is developed next. To this end, the following quantity is defined:

$$Q(s) = C_o \hat{P}_o / (1 + C_o \hat{P}_o). \tag{9.287}$$

It is noted that $Q(s)$ is $G_r(s)/e^{-\theta s}$, where G_r, defined in Equation 9.282, is the transfer function of the closed loop in the ideal case. Hence, it is assumed in the sequel that $Q(s)$ is *stable*. Denoting $Im(s)$ by ω, we have the following theorem:

Theorem 9.2: [3]

For the system with an SP to be closed-loop practically stable, it is necessary that

$$\lim_{\omega \to \infty} |Q(j\omega)| < 1/2. \tag{9.288}$$

Remark 9.1

If only mismatches in DT are considered, then it can be shown that the condition in Equation 9.288 is sufficient as well.

Remark 9.2

For $Q(s)$ to satisfy Equation 9.288, it must be at least proper. If $Q(s)$ is strictly proper, then the system is practically stable.

Example 9.12:

Equation 9.288 is applied to Example 9.11. It is easily verified that $Q(s)$ in that case is $Q(s) = (2s + 4)/(3s + 5)$. Hence, $\lim_{s \to \infty} |Q(s)| = 2/3 > 1/2$ and the system is practically unstable as was confirmed in Example 9.11.

Unless stated otherwise, it is assumed in all subsequent results that $Q(s)$ satisfies Equation 9.288. When the design complies with the condition in Theorem 9.2, one still may distinguish between two possible cases: one in which the design is, stability wise, completely insensitive to mismatches in the DT; and the second, where there is a finite maximum mismatch in the DT below which the system remains stable. $\Delta\theta$ denotes the mismatch in DT and is given by

$$\Delta\theta = \theta - \hat{\theta}, \tag{9.289}$$

where $\hat{\theta}$ is the estimated DT used in the SP. In the following theorem, it is assumed that $P_o = \hat{P}_o$, that is, mismatches may exist only in the DTs.

Theorem 9.3: [3]

a. The closed-loop system is asymptotically stable for any $\Delta\theta$ if

$$|Q(j\omega)| < 1/2 \forall \omega \geq 0. \tag{9.290}$$

b. If

$$|Q(j\omega)| \leq 1 \forall \omega \geq 0 \quad and \quad \lim_{\omega \to \infty} |Q(j\omega)| < 1/2, \tag{9.291}$$

then there exists a finite positive $(\Delta\theta)_m$ such that the closed loop is asymptotically stable for all $|\Delta\theta| < (\Delta\theta)_m$.

Remark 9.3

In [3], it is shown that a rough (and frequently conservative) estimate of $(\Delta\theta)_m$ is given by

$$(\Delta\theta)_m = \frac{\pi}{(3\omega_o)}, \tag{9.292}$$

where ω_o is the frequency above which $|Q(j\omega)| < 1/2$.

The next example demonstrates the application of Theorem 9.3a.

Example 9.13:

If in Example 9.11 the gain of the primary controller is reduced such that $C_0=0.9(0.5s+1)$, then the corresponding $Q(s)$ satisfies the condition in Equation 9.290. Consequently, the system with the above primary controller not only is practically stable but also maintains stability for any mismatch in DT.

The conditions for robust stability presented so far were associated with uncertainties just in the DT. While the SP is largely sensitive to mismatches in DTs (particularly when the DTs are large as compared to the time constants of the process), conditions for robust stability of the SP for simultaneous uncertainties in all parameters, or even for structural differences between the model used in SP and the plant, may be derived. When modeling error is represented by uncertainty in several parameters, it is often mathematically convenient to approximate the uncertainty with a single multiplicative perturbation. Multiplicative perturbations on a nominal plant are commonly represented by

$$P(s) = \hat{P}(s)\,[1 + \ell_m(s)], \tag{9.293}$$

where, as before, $\hat{P}(s)$, is the model used in the SP. Hence, $\ell_m(s)$, the multiplicative perturbation, is given by

$$\ell_m(s) = (P(s) - \hat{P}(s))/\hat{P}(s) = \frac{P_o(s)}{\hat{P}_o(s)}e^{-\Delta\theta s} - 1. \tag{9.294}$$

A straightforward application of the well-known robust stability theorem (see, e.g., [4],) leads to the following result:

Theorem 9.4: [3]

Assume that $Q(s)$ is stable. Then the closed-loop system will be asymptotically stable for any multiplicative perturbation satisfying the following condition:

$$\left|Q(j\omega)\ell_m(j\omega)\right| < 1 \quad \forall\omega \geq 0. \tag{9.295}$$

Remark 9.4

It is common, where possible, to norm bound the multiplicative error $\ell_m(j\omega)$. If that bound is denoted by $\ell(\omega)$, then, the condition in Equation 9.295 can be restated as

$$\left|Q(j\omega)\right|\ell(\omega) < 1 \quad \forall\omega \geq 0. \tag{9.296}$$

In [5], for example, the smallest possible $\ell(\omega)$ for the popular first order with DT model

$$P(s) = k_p e^{-\theta s}/(\tau s + 1) \tag{9.297}$$

was found for simultaneous uncertainties in gain, k_p, time-constant, τ, and DT, θ. When $\ell(\omega)$ is available or can be determined, it is quite easy to check whether the SP design complies with Equation 9.296. This can be done by plotting the amplitude Bode diagram of $|Q(j\omega)|$ and verifying that it stays below $1/\ell(\omega)$ for all frequencies.

Considering further the properties of the rational function $Q(s)$, conditions under which the closed-loop system containing the SP possesses some attractive relative stability properties may be derived. The following result is due to [3]:

Theorem 9.5: [3]

Let $Q(s)$ be stable. If

$$|Q(j\omega)| \leq 1 \quad \forall \omega \geq 0, \tag{9.298}$$

then the closed-loop system has

   a. A minimum gain margin of 2.
   b. A minimum phase margin of $60°$.

Remark 9.5

It should be emphasized that unless the design is practically stable, the phase-margin property may be misleading. That is to say that a design may satisfy Equation 9.298 but not Equation 9.288. Under such circumstances, the system will go unstable for an infinitesimal mismatch in DT despite the $60°$ phase margin.

Remark 9.6

Note that the gain margin property relates to the overall gain of the loop and not to the gain of the primary controller.

9.8.3.3 Performance

Several aspects related to the performance of the SP are briefly discussed. The typical improvements in performance, due to the SP, that can be expected are demonstrated. Both reference tracking and disturbance attenuation are considered. It is shown that while the potential improvement in reference tracking is significant, the SP is less effective in attenuating disturbances. The reasons for that are clarified and the feedforward SP is presented.

First, the steady-state errors to step changes in the reference (r) and in the input disturbance (d) (see Figure 9.79) are examined. The following assumptions are made:

1. $P(s)$, the plant, is asymptotically stable.
2. $P(s)$ does not contain a pure differentiator.
3. $C_o(s)$, the primary controller, contains an integrator.
4. $Q(s)$ is asymptotically stable and practically stable.

With the above assumptions, it is quite straightforward to prove the following theorem by applying the final-value theorem to G_r in Equation 9.282 and to G_d in Equation 9.284.

Theorem 9.6:

Under Assumptions 1–4, the SP system will display zero steady-state errors to step reference inputs and to step disturbances.

Remark 9.7

Theorem 9.5 remains valid for all uncertainties (mismatchings) for which the closed-loop stability is maintained.

The next example is intended to demonstrate the typical improved performance to be expected by employing the SP and to motivate further discussion on one of the structural properties of the SP that directly affects its regulation capabilities.

Example 9.14:

The SP is applied to the following second order with DT process:

$$P(s) = e^{-.5s}/(s+1)^2. \qquad (9.299)$$

The primary controller, $C_o(s)$, is chosen to be the following ideal PID controller:

$$C_o(s) = K(s+1)^2/s; \quad K = 6. \qquad (9.300)$$

This choice of $C_o(s)$ is discussed in more detailed form in Section 9.8.3.4. It should be emphasized that the value of the gain $K = 6$ is quite conservative. This can be concluded from the inspection of $Q(s)$ in Equation 9.287 (or equivalently, G_r in Equation 9.282), which in this case is

$$Q(s) = K/(s+K). \qquad (9.301)$$

It is seen that one may employ K as large as desired in the ideal case, without impairing stability. K will have, however, to be reduced to accommodate mismatching. For reasons to be elaborated in Section 9.8.3.4, the value of 6 was selected for the gain. Note that without the SP, the maximum gain allowed would be 4.7 approximately.

In Figure 9.83, the time responses, in the ideal case, of the SP and a conventional PID controller to a step change in reference are compared. The PID controller settings were determined via the well-known Ziegler–Nichols rules, which can be found in *Control System Advanced Methods*, Chapter 32. It is evident

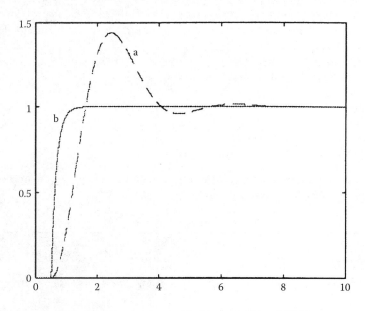

FIGURE 9.83 Responses to step change in setpoint, the ideal case: (a) PID and (b) SP.

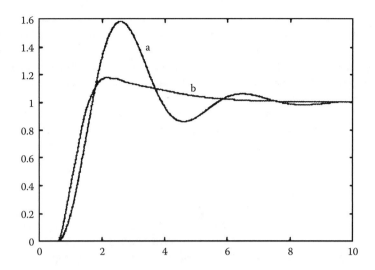

FIGURE 9.84 Setpoint step responses, the nonideal case: (a) PID and (b) SP.

that despite the relatively conservative tuning of the SP it outperformed the PID in all respects: better rise time, better settling time, no overshoot, etc. One may try to optimize the PID settings, but the response of the SP is hard to beat.

In Figure 9.84, the same comparison of responses is made, but with a mismatch of 20% in the DT, namely, $\hat{\theta} = 0.5$, but the DT has been changed to 0.6. The effect of the mismatch on the responses of both the SP and the PID is noticeable. However, the response of the SP is still considerably better.

In Figure 9.85, the corresponding responses to a unit input disturbance are depicted. While the response of the SP has a smaller overshoot, its settling time is inferior to that of the PID. This point is elaborated upon next.

Example 9.14 demonstrates that, on the one hand, the SP provides significant improvements in tracking properties over conventional controllers, but on the other hand, its potential enhancements in regulatory

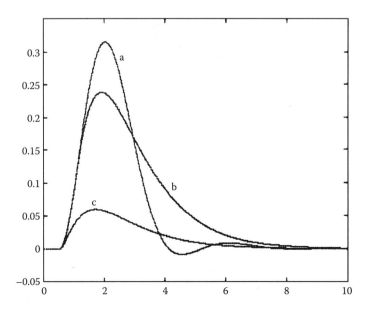

FIGURE 9.85 Responses to a unit-step input disturbance: (a) PID; (b) SP; and (c) feed forward SP.

capabilities are not as apparent. The reason for this has been pointed out in Section 9.8.3.1, where it was shown that the open-loop poles are present in the transfer function G_d. These poles are excited by input disturbances but not by the reference. Depending on their locations relative to the closed-loop poles, these poles may dominate the response. The slower the open-loop poles, the more sluggish the response to input disturbances. This is exactly the situation in Example 9.14: the closed-loop pole (the zero of $1 + C_o P_o$) is $s = -6$, while the two open-loop poles are located at $s = -1$. The presence of the open-loop poles in G_d is a direct consequence of the structure of the SP, and many modifications aimed at improving that shortcoming of the SP were proposed. Several modifications are presented in Section 9.8.4. It is worth noting, however, that the influence of the open-loop poles on the response to disturbances is less pronounced in cases with large DTs. In such a circumstance, the closed-loop poles cannot usually be shifted much to the left, mainly due to the effect of model uncertainties. Hence, in such situations the closed-loop poles do not differ significantly from the open-loop ones, and their influence on the response to disturbances is less prominent.

For the reasons mentioned above, the SP is usually designed for tracking, and if necessary, a modification aimed at improving the disturbance rejection properties is added. When other than step inputs are considered, C_o should and can be designed to accommodate such inputs. For a given plant and inputs, an H_2-optimal design in the framework of IMC (see Section 9.8.3.1) was proposed. The interested reader is referred to [4,5] for details. A DTC of a special structure that can accommodate various disturbances is discussed in Section 9.8.4.3.

Another way to improve on the regulation capabilities of the SP is to add a standard feedforward controller, which requires on-line measurements of the disturbances. However, if disturbances are measurable, the SP can provide significant improvements in disturbance attenuation in a direct fashion. This may be achieved by transmitting the measured disturbance into the predictor. The general idea, in the most simple form, is shown in Figure 9.86.

In this fashion, the plant model in the predictor is further exploited to predict the effect of the disturbance on the output. The advantage of this scheme may be better appreciated by comparing the closed-loop relations between the control signal, u, and the disturbance, d, in both the "conventional" SP, Figure 9.79, and the one in Figure 9.86. In the conventional SP scheme (Figure 9.79), that relation is given by

$$u(s)/d(s) = -C_o P_o e^{-\theta s}/(1 + C_o P_o). \qquad (9.302)$$

The corresponding relation in the scheme in Figure 9.86 is

$$u(s)/d(s) = -C_o P_o/(1 + C_o P_o) \qquad (9.303)$$

and it is evident that the DT by which the control action is delayed in the conventional SP is effectively canceled in the scheme in Figure 9.86. By counteracting the effect of the disturbance before it can appreciably change the output, the scheme in Figure 9.86 behaves in a manner similar to that of a

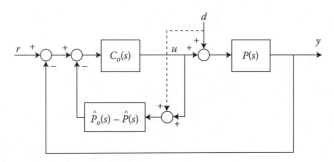

FIGURE 9.86 A simple form of the feedforward SP for measurable disturbances.

conventional feedforward controller. For this reason, the scheme in Figure 9.86 is called *feedforward SP*. In fact, the feedforward SP eliminates the need for a separate feedforward controller in many circumstances under which it would be employed. The advantage of the scheme is demonstrated in Figure 9.85. It should be noted that the feedforward SP in Figure 9.86 is presented in its most simplistic form. More realistic forms and other related issues can be found in [6].

Finally, it is worth noting that it has been found from practical experience that a properly tuned SP performs better than a PID in many loops typical to the process industries, even when the model used in the SP is of lower order than the true behavior of the loop. This applies in many circumstances even to DT-free loops. In those cases, the DT in the model is used to compensate for the order reduction.

9.8.3.4 Tuning Considerations

Since the SP is composed of a primary controller and a model of the plant, its tuning, in practice, involves the determination of the parameters of the model and the settings of the primary controller. In this chapter, however, it is assumed that the model is available, and the problem with which we are concerned is the design and setting of the primary controller. From the preceding sections, it is clear that the tuning should be related to the stability and robustness properties of the SP.

A simple tuning rule for simple primary controllers, $C_o(s)$, is presented. For low-order plant models with poles and zeros in the left half-plane (LHP), a simple structure for $C_o(s)$, which can be traced back to optimal control [2], is given by

$$C_o(s) = \frac{K}{s} \hat{P}_o(s)^{-1}. \tag{9.304}$$

When $\hat{P}_o(s)$ is a first-order filter or a second-order one, then the resulting $C_o(s)$ is the classical PI or PID controller, respectively. More specifically, in the first-order and the second-order cases, the corresponding $\hat{P}_o$ is given as in Equations 9.305 and 9.306, respectively:

$$\hat{P}_o(s) = k_p/(\tau s + 1), \tag{9.305}$$

$$\hat{P}_o(s) = k_p/(\tau^2 s^2 + 2\tau \xi s + 1). \tag{9.306}$$

The "textbook" transfer functions of the PI and PID controllers are as follows:

$$K_c \left(1 + \frac{1}{\tau_i s} \right), \tag{9.307}$$

$$K_c \left(1 + \frac{1}{\tau_i s} + \tau_D s \right). \tag{9.308}$$

If Equation 9.305 is substituted into Equation 9.304, the resulting primary controller, C_o, is equivalent to the PI controller in Equation 9.307 if

$$K_c = K\tau/k_p; \quad \tau_i = \tau. \tag{9.309}$$

Similarly, for the second-order case (Equation 9.306), C_o will be equivalent to Equation 9.308 if:

$$K_c = K(2\tau \xi)/k_p; \quad \tau_i = 2\tau \xi; \quad \tau_D = \tau/2\xi. \tag{9.310}$$

Commercially available SPs offer PI and PID controllers for C_o. Thus, the structure in Equation 9.304 is well suited to SP with simple plant models. It should be noted that if the pole excess in $\hat{P}_o$ is larger than 2, then C_o in Equation 9.304 must be supplemented by an appropriate filter to make it proper.

The particular choice in Equation 9.304 leads to a $Q(s)$ with the following simple form:

$$Q(s) = K/(s + K), \tag{9.311}$$

and it is evident that stability of the closed loop, in the ideal case, is maintained for any positive K. However, model uncertainty imposes an upper limit on K. Since the SP is mostly sensitive to mismatches

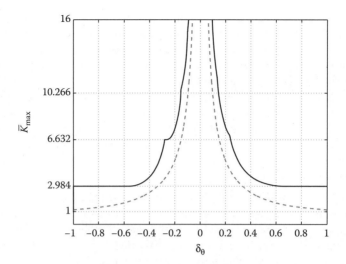

FIGURE 9.87 Maximum $\bar{K}$ for stability versus δ_θ (solid line) and the rule of thumb (dashed line) suggested in [4].

in the DT, a simple rule for the setting of K, the single tuning parameter of C_o in Equation 9.304, can be obtained by calculating $K_{\max}$ [7] for stability as a function of δ_θ, the relative error in DT.

The solid line in Figure 9.87 depicts $\bar{K}_{\max}$ as a function of δ_θ. $\bar{K}$ and δ_θ are defined as follows:

$$\bar{K}\Delta = K\hat{\theta}; \quad \delta_\theta\Delta = \Delta\theta/\hat{\theta}. \tag{9.312}$$

Figure 9.87 reveals an interesting property, namely that the choice of $\bar{K} < 2.984$ (for all practical purposes and for convenience 2.984 may be replaced by 3) assures stability for $\pm100\%$ mismatches in DT. Thus, a simple tuning rule is to set K as follows:

$$K = 3/\hat{\theta}. \tag{9.313}$$

This rule was applied in Example 9.14 in Section 9.8.3.3. While the solid line in Figure 9.87 displays the exact $\bar{K}_{\max}$ for all cases for which C_o is given by Equation 9.304, some caution must be exercised in using it, or Equation 9.313, since it takes into account mismatches in DT only. The dashed line in Figure 9.87 displays a rule of thumb suggested in [4] where $\bar{K}$ is chosen as $1/\delta_\theta$. This latter rule was developed for first-order systems (see Equation 9.297) and mismatches in the DT only.

A method for tuning the primary controller of the SP for robust performance in the presence of simultaneous uncertainties in parameters was developed in [5] in the framework of IMC. For the three parameters first order with DT model given in Equation 9.297, the overall H_2-optimal IMC controller (see Equation 9.286) for step references and equipped with a filter, is given by

$$\bar{q}(s) = (\tau s + 1)/\left[k_p(\lambda s + 1)\right], \tag{9.314}$$

where λ is the time constant of the filter $f(s)$. In Equation 9.314, $\bar{q}(s)$ is equivalent to C_o in Equation 9.304, for this case, with $\lambda=1/K$. Using various robust performance criteria, tuning tables for λ for various simultaneous uncertainties in the three parameters have been developed and can be found in [5].

9.8.4 Modified Predictors

In the previous section, it was seen that the improvements in disturbance attenuation offered by the SP are not as good as for tracking. The reasons for that were discussed and a "remedy" in the form of the feedforward SP was presented. The feedforward SP is applicable, however, only if disturbances can be measured online. This may not be possible in many cases. Quite a number of researchers have recognized this fact and proposed modifications aimed at improving on the regulatory capabilities of the SP. Three such designs are briefly described in this section. In the first two, the structure of the SP is kept, but a new component is added or an existing one is modified. In the third one, a new structure is proposed.

9.8.4.1 Internal Cancellation of Model Poles

The scheme suggested in [8] has the same structure as the SP in Figure 9.79, but $\hat{P}_1$ replaces $\hat{P}_o$ in the minor feedback around the primary controller. Hence, the minor feedback consists of $(\hat{P}_1 - \hat{P})$ instead of $(\hat{P}_o - \hat{P})$. $\hat{P}_1$ may be considered to be a modified plant model without a DT. $\hat{P}_1$ is the nominal model of P_1, which is given by:

$$P_1(s) = Ce^{-A\theta}(sI - A)^{-1}B - \int_{\theta}^{0} Ce^{-A\tau}B\,d\tau, \tag{9.315}$$

where A, B, and C are the "true" matrices in the state-space representation of the plant given in Equation 9.275. The role of P_1 will be clarified later. Note that $Q(s)$ may be defined in a similar fashion to the one defined in Equation 9.287 for the conventional SP. For the scheme under consideration, it is given by

$$Q(s) = C_o\hat{P}_o/(1 + C_o\hat{P}_1) \tag{9.316}$$

and all the previous results on practical and robust stability of Section 9.8.3.2 apply to this case as well.

In [8], some general results, applicable to the scheme considered here with any stable $\hat{P}_1$ were stated and proven. Under Assumptions 1–4 of Section 9.8.3.3, and for the particular $\hat{P}_1$ in Equation 9.315, the application of the general results yields the following theorem:

Theorem 9.7:

Under assumptions 1 to 4, the modified SP, with $\hat{P}_o$ replaced by $\hat{P}_1$ in the minor feedback, has the following properties:

   a. *A zero steady-state error to step reference.*
   b. *A zero steady-state error to step disturbance.*
   c. *The poles of $(\hat{P}_1 - \hat{P})$, the minor feedback are canceled with its zeros.*

The following remarks explain and provide some insight into the properties stated in Theorem 9.7.

Remark 9.8

Property (a) holds for any P_1 satisfying $\lim_{s \to 0} P_o(s)/P_1(s) = 1$. It can be verified that the P_1 in Equation 9.315 satisfies the latter condition.

Remark 9.9

Property (b) holds for any stable P_1 satisfying $\lim_{s \to 0}(P_1(s) - P(s)) = 0$. It is easy to show that the P_1 in Equation 9.315 satisfies that condition.

Remark 9.10

Property (c) represents the major advantage of the scheme discussed here and is the source for its potential improvement in regulatory capabilities. Due to the pole–zero cancellation in the minor feedback, G_d (see Equation 9.284), the transfer function relating y to d, no longer contains the open-loop poles. The response to disturbances, under these circumstances, is governed by the zeros of $(1 + C_oP_1)$, after θ units of time from the application time of the disturbances. That is, the error decay can be made as fast as desired (in the ideal case) after the DT has elapsed. See the discussion following Example 9.15.

Remark 9.11

An equivalent way to express properties (b) and (c) is to say that the states of the minor feedback, as well as the state of the integrator of the primary controller, are unobservable in v (see Figure 9.79).

Remark 9.12

Note that for given A, B, C, and θ, the integral in P_1 (Equation 9.315) is a constant scalar. Its sole purpose is to nullify the steady-state gain of the minor feedback thus assuring an integral action in the DTC if the primary controller includes one.

Remark 9.13

If the minor feedback, $\hat{P}_1 - \hat{P}$, is realized as a dynamical system (i.e., by Equations 9.315 and 9.278), then this scheme is not applicable to unstable plants. However, with a different realization, to be discussed in Section 9.8.5, it can be applied to unstable plants.

Example 9.15:

Two cases are considered. In both, the plant is a first–order with DT (see Equation 9.297). In the first case, the plant parameters are $k_p = 1$, $\tau = 1$, $\theta = 0.2$. In the second case θ, the DT, is increased to $\theta = 1$. Primary controllers, C_o, are designed for the SP and for the modified predictor, and the responses to input disturbances are compared. For a fair comparison, both designs are required to guarantee stability for $\pm 60\%$ mismatching in the DT. C_o for the SP is taken to be as in Equation 9.304 with K according to Equation 9.313. Thus, the C_o for the SP is a PI controller, which clearly satisfies the above stability requirement. A PI controller is also selected for the modified predictor. K_c and τ_i, the parameters of the PI controller (Equation 9.307), for the modified predictor are determined such that the stability requirement is satisfied and the response optimized. The resulting parameters of the PI controllers for both schemes and for the two cases are as follows:

1. SP
   a. Case 1 –$K_c = 15$, $\tau_i = 1/15$
   b. Case 2 –$K_c = 3$, $\tau_i = 0.33$.
2. Modified predictor
   a. Case 1 –$K_c = 3$, $\tau_i = 0.14$
   b. Case 2 –$K_c = 0.46$, $\tau_i = 1.92$.

Note the substantial reduction in the gains of the modified predictor, relative to the SP, required to guarantee stability for the same range of mismatches in the DT. The responses to a unit-step input disturbance of the SP [curve (1)] and the modified predictor [curve (2)] are compared in Figure 9.88a for Case 1 and in Figure 9.88b for Case 2.

While the improvement achieved by the modified predictor is evident in Case 1, it is insignificant in Case 2. This point is elaborated on next.

Example 9.15 demonstrates the potential improvement in the regulatory performance of the modified scheme. However, practical considerations reduce the effectiveness of this scheme in certain cases. First, note that P_1 is a proper transfer function and frequently is nonminimum phase. Hence, the design of C_o is usually more involved. Simple primary controllers, like those in Equation 9.304, are not applicable, and in many cases the conventional PI and PID controllers may not stabilize the system even in the ideal case. Second, when it is designed for robust stability or robust performance, the resulting gains of the primary controllers are considerably lower than those allowed in the conventional SP for the same range of uncertainties. Therefore, the improvements in disturbance attenuation are usually less prominent. It turns out, as is also evident in Example 9.15, that the modified scheme is advantageous in cases where

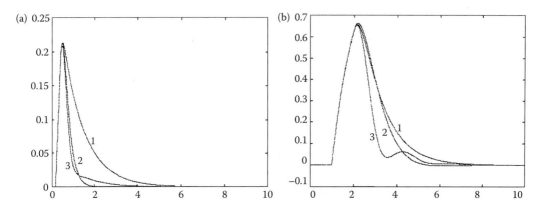

FIGURE 9.88 Responses to a unit-step input disturbance: (a) Case 1 ($\theta/\tau = 0.2$); (b) Case 2 ($\theta/\tau = 1$). (1) SP. (2) Modified predictor with internal cancellations. (3) SP with approximate inverse of DT.

DTs are small relative to the time constants of the plant, or when uncertainties are small. The chief reason for this is that it is possible to improve on the conventional SP only if the closed-loop poles (i.e., zeros of $1 + C_o P_1$) can be made considerably faster than the open-loop ones. This can be achieved if relatively high gains are allowed.

It was pointed out in the previous paragraph that the design of C_o in the modified scheme is considerably more involved than in the conventional SP, more so when plant models are of order higher than two. In those cases, C_o may be determined by input–output pole placement. A realizable C_o and of low order, if possible, is sought such that the closed-loop poles in the ideal case (i.e., the zeros of $1 + C_o \hat{P}_1$) are placed in predetermined locations. Then the design is checked for robust stability and robust performance. If the design fails the robustness tests, then C_o is modified in a trial-and-error fashion until it complies with the robustness requirements.

9.8.4.2 An Approximate Inverse of DT

A simple modification was suggested in [9]. It consists of a simple predictor, $M(s)$, which is placed in the major feedback of the SP as shown in Figure 9.89.

It is desired to have $M(s)$ equal to the inverse of the DT, that is, $M(s) = e^{\hat{\theta}s}$. In this fashion, the output, y, is forecast one DT into the future. This in turn eliminates the DT between the disturbance, d, and the control, u, in a similar fashion to the feedforward SP (see Figure 9.86). Indeed, with $M(s)$ as above, the transfer function $u(s)/d(s)$ is exactly the one in Equation 9.303. It is clear, however, that it is impossible

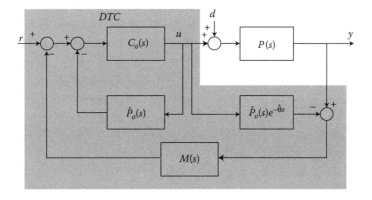

FIGURE 9.89 Modified SP with an approximate inverse of DT.

to realize an inverse of the DT. Hence, a realizable approximation of the inverse of the DT is employed. In [9], the following $M(s)$ is suggested:

$$M(s) = (1 + B(s))/[1 + B(s)e^{-Ls}]. \tag{9.317}$$

If $B(s)$ is a high-gain low-pass filter given by

$$B(s) = K_m/(\tau_m s + 1), \tag{9.318}$$

then $M(s)$ in Equation 9.317 approximates e^{Ls} at low frequencies.

A method for the design of $M(s)$ is suggested in [9]. It consists of two steps. First, an SP is designed based on methods like the one in [5], or the one described in Section 9.8.3.4. In the second step, the $M(s)$ in Equations 9.317 and 9.318 is designed to cope with uncertainties. With $M(s)$ in the major feedback, the condition for stability under multiplicative error, which corresponds to the one in Equation 9.296, is easily shown to be

$$M(j\omega)Q(j\omega)|\ell(\omega) < 1 \quad \forall \omega \geq 0$$

or equivalently,

$$|M(j\omega)| < [|Q(j\omega)|\ell(\omega)]^{-1} \quad \forall \omega \geq 0. \tag{9.319}$$

For good performance, it is desired to have $|M(j\omega)Q(j\omega)|$ close to one at frequencies below the bandwidth of the closed loop. Thus, the design of $M(s)$ is to choose the three parameters, k_m, τ_m, and L, such that the magnitude curve of $|M(j\omega)|$ lies as close as possible to $|Q(j\omega)|^{-1}$ and beneath the curve $[|Q(j\omega)|\ell(\omega)]^{-1}$.

The three parameters of $M(s)$ optimizing the regulatory properties of the SP with the primary controller in Equation 9.304 tuned according to Equation 9.313 were determined experimentally in [6] for the first-order case (Equation 9.297) and are summarized in Table 9.10.

It was found that for $\hat{\theta}/\hat{\tau}$ up to 2, the inclusion of the simple filter $M(s)$ enhances the disturbance attenuation properties of the SP. However, for $\hat{\theta}/\hat{\tau}$ above 2, the improvement via $M(s)$ is minor and the use of $M(s)$ is not justified in those cases. In Figure 9.88, the responses to a step input disturbance for the two cases considered in Example 9.15, with $M(s)$ in place, are shown [curve (3)], and compared to those of the two predictors discussed in Example 9.15. The primary controllers are the same as those used in the conventional SP, and the parameters of $M(s)$ were determined from Table 9.10. The improvement achieved by $M(s)$ in both cases shown in Figure 9.88 is evident.

9.8.4.3 Observer–Predictor

The structure of the observer–predictor (OP) is depicted in Figure 9.90. It consists of

- An asymptotic observer that estimates the states of both the plant and the disturbance.
- A predictor that uses the estimated state to forecast the output one DT into the future.

TABLE 9.10 Parameters of $M(s)$ for the First Order with DT Case with the C_o in Equation 9.304 and Tuning Rule in Equation 9.313

$\hat{\theta}/\hat{\tau}$	k_m	$L/\hat{\theta}$	L/τ_m
0.3	10	0.75	0.05
0.6	8	0.6	0.1
1.0	4	0.45	0.2
2.0	2	0.27	0.3

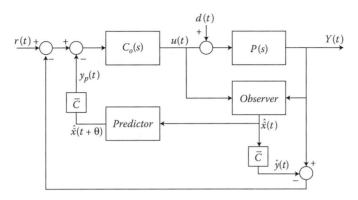

FIGURE 9.90   The structure of the observer–predictor (OP).

- A primary controller, C_o.

The basic structure of the OP is not a new one. It was suggested in [10], where a static-state feedback was employed and where the main concern was the stability properties of the scheme. No attention was paid to tracking and regulation capabilities. Indeed, the performance of the OP in [10] is poor. The OP outlined in this section was developed in [11] and contains several modifications aimed at improving its regulatory properties. First, it contains a dynamical primary controller that operates on the forecast output. Second, a model of the dynamics of the disturbance is incorporated in the observer [12]. It enables the online estimation of the disturbance. Third, an additional feedback, similar to the one used in the SP, which carries the difference between the measured and the estimated outputs, is introduced. The main objective of that feedback is to compensate for uncertainties and disturbances. The equations of the OP in Figure 9.90, are given next, followed by brief comments on several properties of the OP. An example demonstrating the effectiveness of the OP in attenuating disturbances concludes this section.

As in Equation 9.275, the plant is given by

$$\dot{x}(t) = Ax(t) + B_1 u(t - \theta) + B_2 d(t - \theta),$$
$$y(t) = Cx(t). \tag{9.320}$$

The plant model in Equation 9.320 is slightly more general than before, as the control, u, and the disturbance, d, go through different input matrices. The disturbance is assumed to obey the following model:

$$\dot{z}(t) = Dz(t),$$
$$d(t) = Hz(t), \tag{9.321}$$

where $z \in R^m$ is the state vector of the disturbance model. It is further assumed that the pairs (A, C) and (D, H) are observable. Substitution of d from Equation 9.321 into Equation 9.320 yields:

$$\dot{x}(t) = Ax(t) + B_1 u(t - \theta) + B_2 H e^{-D\theta} z(t). \tag{9.322}$$

Next, an augmented state vector is defined:

$$\bar{x}(t)^T = (x(t)^T, z(t)^T). \tag{9.323}$$

By means of Equation 9.323, the plant and the disturbance models are combined:

$$\dot{\bar{x}}(t) = \bar{A}\bar{x}(t) + \bar{B}u(t - \theta),$$
$$y(t) = \bar{C}\bar{x}(t) \tag{9.324}$$

where

$$\bar{A} = \begin{pmatrix} A & B_2 H e^{-D\theta} \\ 0 & D \end{pmatrix}; \quad \bar{B} = \begin{pmatrix} B_1 \\ 0 \end{pmatrix};$$

$$\bar{C} = (C \quad 0).$$

(9.325)

The observer is given by

$$\dot{\hat{\bar{x}}}(t) = \bar{A}\hat{\bar{x}}(t) + \bar{B}u(t-\theta) - L(y(t) - \bar{C}\hat{\bar{x}}(t)),$$

(9.326)

where $\hat{\bar{x}}$ is the estimate of $\bar{x}$, and L is the vector of gains of the observer. The predictor is given by

$$y_p(t) \triangleq \hat{y}(t+\theta) = Ce^{A\theta}\hat{x}(t) + C\int_{-\theta}^{0} e^{-Ah} \left[B_1 u(t+h) + B_2 \hat{d}(t+h) \right] dh,$$

(9.327)

where $y_p(t)$ is the forecast of the estimated output θ units of time ahead, and $\hat{x}(t)$ and $\hat{d}(t)$ are the estimates of the state and of the disturbance, respectively, both generated by the observer. Equation 9.327 presents the *integral form* of the predictor. If Equation 9.327 is Laplace transformed, then the *dynamical form* of the predictor is obtained:

$$y_p(s) \triangleq Ce^{A\theta}\hat{x}(s) + C(I - e^{-(sI-A)\theta})(sI - A)^{-1} \left[B_1 u(s) + B_2 \hat{d}(s) \right].$$

(9.328)

Finally, the control signal $u(s)$ is

$$u(s) = C_o(s) \left[r(s) - y_p(s) - (y(s) - \hat{y}(s)) \right].$$

(9.329)

The design of the OP consists of the selection of L, the observer's gain vector, and of $C_o(s)$. L may be determined by the well-known pole-placement techniques. If, in addition to the observability assumptions stated above, it is further assumed that no pole–zero cancellations between the models of the plant and the disturbances occur, then the pair $(\bar{A}, \bar{C})$ is observable and the observer poles can be placed at will. The design of C_o is referred to in the subsequent remarks.

The main properties of the OP are summarized in the following remarks:

Remark 9.14

In the ideal case, the DT is eliminated from the characteristic equation of the closed loop.

Remark 9.15

The overall closed-loop transfer function relating r to y is identical, in the ideal case, to that of the SP given in Equation 9.282.

Remark 9.16

If the predictor is realized in the integral form of Equation 9.327, then the closed-loop poles consist of the observer poles (i.e., the zeros of $\det[sI - \bar{A} - L\bar{C}]$) and of the zeros of $1 + C_o P_o$.

Remark 9.17

According to the *internal model principle* (see, e.g., [12]), C_o should contain the disturbance poles in order to reject disturbances. Note that the poles of the disturbance in Equation 9.321 are the zeros of $\det(sI - D)$. With the latter requirement the design of C_o may be carried out by appropriate placement of the zeros of $1 + C_o P_o$.

Remark 9.18

While the OP has far better disturbance rejection properties than the SP, as demonstrated in Example 9.16, it is, in general, considerably more sensitive to uncertainties.

Example 9.16 demonstrates the improved capability of the OP to reject disturbances.

Example 9.16:

The plant is the one used in Example 9.14, Equation 9.299. The disturbance, however, in this case is given by

$$d(t) = \sin 2t. \tag{9.330}$$

The *D* in Equation 9.321 is therefore:

$$D = \begin{pmatrix} 0 & 1 \\ -4 & 0 \end{pmatrix}$$

and $\det(sI - D) = s^2 + 4$. The latter factor is included (according to Remark 9.17) in the denominator of C_o.

The observer gains were selected such that all the observer poles are placed at $s = -2$. In addition to containing the disturbance poles, C_o was required to have an integrator. The rest of the design of C_o was based on placing the zeros of $1 + C_o P_o$ at $s = -2$. No claim is made that this is an optimal design. For comparison purposes, the same C_o was used in the SP. The responses of the OP and the SP to the sinusoidal disturbance in Equation 9.330 are shown in Figure 9.91.

It is apparent that while the SP cannot handle such a disturbance, the OP does a remarkable job.

9.8.5 Time-Delay Compensation for Unstable Plants

In this section, we briefly discuss which of the schemes presented in the preceding sections is applicable to unstable plants with DTs and under what conditions.

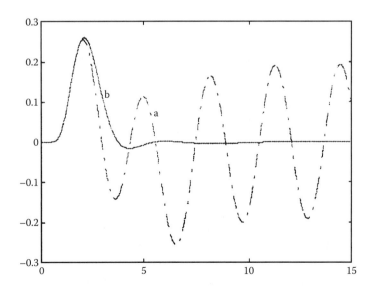

FIGURE 9.91 Response of sinusoidal input disturbance: (a) SP and (b) OP.

In Section 9.8.3.1, it was pointed out that the SP cannot be applied to unstable plants. It has been shown that the plant models in the minor feedback of the SP are the cause for the appearance of the poles of the plant as poles of the closed loop. This fact was evident in the transfer function G_d in Equation 9.284. A straightforward calculation of the closed-loop poles of a system with an SP shows that they consist of the zeros of $1 + C_o P_o$ and the poles of the open-loop plant. Hence, it is concluded that the SP is internally unstable if the plant is unstable [13]. An alternative way to arrive at the same conclusion is to look at the IMC form of the SP in Figure 9.81. It is seen that the control signal, u, is fed in parallel to the plant and to the model. Such a structure is clearly uncontrollable. The above conclusion applies to the modified SP with the approximate inverse of the DT as well.

The modified predictor with internal pole–zero cancellations in Section 9.8.4.1 is in the same situation. Although the poles of the minor feedback are canceled with its zeros and therefore do not show up in G_d, for example, these poles are clearly internal modes of the closed loop. Upon noting that the poles of the minor feedback are those of the plant, we may conclude that the modified predictor with the internal cancelation will be internally unstable for unstable plants. It is possible, however, to make the modified predictor cope with unstable plants. Fortunately enough, the minor feedback in the modified predictor can be realized as a finite impulse response (FIR) system which does not possess poles. The derivation of the FIR form of the minor feedback is outlined next. Recall that the minor feedback is

$$P_1 - P. \tag{9.331}$$

Upon substitution of the P_1 in Equation 9.315 and P in Equations 9.277 and 9.278, Equation 9.331 can be written explicitly as follows:

$$P_1 - P = Ce^{-A\theta}(sI - A)^{-1}B - \int_0^\theta Ce^{-A\tau}B\,d\tau - C(sI - A)^{-1}Be^{-\theta s}. \tag{9.332}$$

With the aid of the following, easily verified identity:

$$Ce^{-A\theta}(sI - A)^{-1}B - C(sI - A)^{-1}Be^{-\theta s} = Ce^{-A\theta}\int_{-\theta}^0 e^{\tau(sI-A)}B\,d\tau. \tag{9.333}$$

Equation 9.332 becomes:

$$P_1 - P = Ce^{-A\theta}\int_{-\theta}^0 e^{\tau(sI-A)}B\,d\tau - \int_0^\theta Ce^{-A\tau}B\,d\tau. \tag{9.334}$$

Finally, inverse Laplace transformation of Equation 9.332 yields

$$v(t) = Ce^{-A\theta}\int_{-\theta}^0 e^{-A\tau}Bu(t+\tau)\,d\tau - \left[\int_0^\theta Ce^{-A\tau}B\,d\tau\right]u(t), \tag{9.335}$$

where $v(t)$ is the output variable of the minor feedback (see Figure 9.79). Due to the finite limits of the integral, the right-hand side of Equation 9.335 is an entire function that does not possess singularities. Consequently, if the minor feedback is realized via Equation 9.335, the modified predictor with the internal cancelations is applicable to unstable plants. By applying exactly the same arguments to the case of the OP of Section 9.8.4.3 it may be concluded, at once, that the OP can be applied to unstable plants if the predictor is realized in the integral form given in Equation 9.327.

9.8.5.1 Concluding Remarks

In this chapter, we have presented the basic features and the special properties of the SP. The advantages and drawbacks of the SP were discussed. For the benefit of potential users, attention was paid to both theoretical and practical aspects. Several modifications and alternative schemes that were developed over the years to improve, in certain cases, on some of the properties of the SP were presented also. Due to the lack of space, however, we confined our attention to the continuous SISO case. As mentioned in the introduction, vast material exists on this topic and it was impossible to cover many additional contributions and extensions in a single chapter.

References

1. Smith, O.J.M., *Chem. Eng. Prog.,* 53, 217, 1959.
2. Palmor, Z.J., *Automatica,* 18(1), 107–116, 1982.
3. Palmor, Z.J., *Int. J. Control,* 32(6), 937–949, 1980.
4. Morari, M. and Zafiriou, E., *Robust Process Control,* Prentice-Hall, Englewood Cliffs, NJ, 1989.
5. Laughlin, D.L., Rivera, D.E., and Morari, M., *Int. J. Control,* 46(2), 477–504, 1987.
6. Palmor, Z.J. and Powers, D.V., *AIChE J.,* 31(2), 215–221, 1985.
7. Palmor, Z.J. and Blau, M., *Int. J. Control,* 60(1), 117–135, 1994.
8. Watanabe, K. and Ito, M., *IEEE Trans. Autom. Control,* 26(6), 1261–1269, 1981.
9. Huang, H.P., Chen, C.L., Chao, Y.C., and Chen, P.L., *AIChE J.,* 36(7), 1025–1031, 1990.
10. Furakawa, T. and Shimemura, E., *Int. J. Control,* 37(2), 399–412, 1983.
11. Stein, A., Robust dead-time compensator for disturbance absorption, M.Sc. Thesis, Faculty of Mech. Eng., Technion, Haifa, Israel, 1994.
12. Johnson, C.D., *Control and Dynamic Systems,* Vol. 12, Leondes, C.T., Ed. Academic Press, New York, 1976, 389–489.
13. Palmor, Z.J. and Halevi, Y., *Automatica,* 26(3), 637–640, 1990.
14. Normey-Rico, J.E. and Camacho, E.F., *Control of Dead-Time Processes,* Springer-Verlag, London, 2007.
15. Mirkin, L. and Palmor, Z.J., Control issues in systems with loop delays, in *The Handbook of Networked and Embedded Systems,* Hristu-Varsakelis, D. and Levine, W.S. Eds., pp. 628–648, Birkhauser, Boston, 2005.
16. Mirkin, L., *Systems Control Lett.,* 51, 331–342, 2004.
17. Mirkin, L. and Raskin, N., *Automatica,* 39, 1747–1754, 2003.
18. Ogunnaike, B.A. and Ray, W.H., *AIChE J.,* 25(6), 1043, 1979.
19. Holt, B.R., Morari, M., *Chem. Eng. Sci.,* 40(7), 1229–1237, 1985.
20. Jerome, N.F. and Ray, W.H., *AIChE J.,* 32(6), 914–931, 1986.
21. Shneideman, D., Palmor Z.J., and Mirkin L., In *Proceedings of the 8th IFAC Workshop on Time-Delay Systems,* Sinaia, Romania, Sept. 1–3, 2009, Paper no. 63.
22. Mirkin, L., Palmor, Z.J., and Shneideman, D., In *Proceedings of the 6th IFAC Symposium on Robust Control Design, ROCOND09,* Haifa, Israel, June 2009, pp. 307–312.
23. Mirkin, L., Palmor, Z.J., and Shneideman, D., In *Proceedings of the Joint 48th IEEE Conference on Decision and Control and 28th Chinese Control Conference,* Shanghai, P.R. China, December 16–18, pp. 257–262, 2009.
24. Palmor, Z.J. and Halevi, Y., *Automatica,* 19(3), 255–264, 1983.
25. Astrom, K.J., Hang, C.C., and Lim, B.C., *IEEE Trans. Autom. Control,* 39(2), 343–345, 1994.
26. Matausek, M.R. and Micic, A.D., *IEEE Trans. Autom. Control,* 41(8), 1199–1203, 1996.

Further Reading

For the interested reader, we mention a few references that contain additional results and extensions. This section has been extended relative to the original article by mentioning significant new results on the topic that appeared in the literature in later years since the publication of the original article.

Results on the sampled-data version of SP can be found in [13], where it was shown that while some of the properties of the continuous SP carry over to its discrete counterpart, there are properties unique to the sampled-data case.

A simple automatic tuner for SP with simple models and that simultaneously identifies the model and tunes the primary controller can be found in [6].

In [14] a quite similar modification to the one described in Section 9.8.4.2, called filtered Smith predictor, where a low-pass filter is placed in the main feedback (similarly to M(s) in Figure 9.89) was suggested. Its design is based upon robustness issues.

The modified scheme discussed in Sections 9.8.4.1 and 9.8.5 was generalized and received over the years the name MSP (modified Smith predictor), see [15] where the state-space version of the MSP is discussed as well.

Issues related to possible discretized approximations of the integral in Equation 9.335 were discussed and analyzed in [16]. It was shown there that some discretization schemes may lead to practical instabilities even when discretization interval approaches zero and methods for safe implementations of that integral were proposed.

Parametrizations of all stabilizing DTCs for systems with a single DT have been derived in [17] showing that every stabilizing DTC can be recast in an observer-predictor-based form. Furthermore it has been shown that if the nominal plant is stable then independent of the DT the DTC guarantees the same robustness level for uncertainties in the rational part of the plant as its primary controller for the delay free plant.

Surprisingly, very few attempts to extend the SP to MIMO plants with multiple delays were reported in the literature. A straightforward extension in which all DTs are eliminated from the closed-loop characteristic polynomial was given in [18]. More general extensions which were shown to extend various properties of the SISO SP and which were based on the dynamic resilience theory [19] were developed in [20], see also [14]. Just recently [21–23] novel DTCs for MIMO plants with multiple input and output DTs were obtained via new H_2 solutions. Those novel DTCs contain interchannel feedforward compensators alongside the conventional feedback predictor and are shown to be the source for potential improvements in resilience and performance.

The stability properties of the multivariable SP were analyzed in [24].

An SP-like schemes, specific for plants with an integral mode, one which decouples tracking from regulation, has been presented in [25] and another one with fewer number of tuning parameters was suggested in [26]. See also [14].

9.9 Architectural Issues in Control System Design

Mario E. Salgado and Graham C. Goodwin

9.9.1 Introduction

Feedback is a rich and powerful tool for solving control problems. However, it is also well known that feedback is not a panacea but comes with a set of associated limitations. Perhaps the best known of those limitations is the Bode Sensitivity Integral, which, inter alia, states that for a stable loop

$$\int_{-\infty}^{\infty} \log |S_o(j\omega)| \, d\omega = 0 \tag{9.336}$$

where $S_o(s)$ is the nominal sensitivity function. The implication of this result is that attempts to reduce the sensitivity below 1 (0 dB) over a range of frequencies, must be accompanied by an increase in sensitivity above 1 (>0 dB) at other frequencies.

This result has wide spread implications. For example, if one places feedback around a quantizer, then one can shift the frequency content of the quantization error (or quantization noise) to frequency bands where they have less impact, but the errors cannot be eliminated. This principle is used in CD mastering, where quantization noise arising from the A/D converter is placed at high frequencies, beyond the range of human hearing.

There are other well-known fundamental limitations in feedback loops [1,2]. For example, real non-minimum phase zeros always lead to undershoot in the step response. Moreover, the magnitude of the undershoot depends on the distance of the zero from the $j\omega$ axis.

In this chapter, we wish to place a caveat on the above facts, namely they apply to a *given feedback architecture*. Frequently, one is stuck with the given architecture but, on other occasions, it is possible to enrich the architecture to mitigate one or more of the fundamental limitations.

To be more specific, the fundamentals of linear feedback control theory are usually developed, explained, and evaluated using the basic feedback structure shown in Figure 9.92. This architecture is usually known as a one degree of freedom (ODOF) control loop.

In Figure 9.92, $D_g(s)$ is the Laplace transform of the generalized disturbance signal $d_g(t)$. The plant nominal model is denoted by $G_o(s)$, where $G_o(s) = G_{o1}(s)G_{o2}(s)$. If $G_{o1}(s) = 1$, then $d_g(t)$ is an input disturbance; on the other hand, if $G_{o2}(s) = 1$, then $d_g(t)$ is an output disturbance. Also, $D_m(s)$ is the Laplace transform of $d_m(t)$, which models measurement noise; apart from this noisy characteristic, the measurement is assumed to be ideal, that is, linear, precise, and highly responsive to changes in the measured variable $y(t)$.

Key issues in the ODOF loop are stability, closed-loop dynamics, reference tracking, disturbance rejection, noise immunity, and robustness to plant modeling errors, among others. Furthermore, these issues are closely connected to other topics such as fundamental constraints and performance limitations. Indeed, the design of the feedback controller with transfer function $C(s)$ needs to consider a complex web of conflicting requirements. In fact, when using the ODOF architecture, this set of compromises creates severe, sometimes unacceptable, performance constraints. We will briefly review how and why these constraints appear. To do that, we first recall the definitions of loop sensitivity functions, and we will then write the formulae quantifying the main performance indicators. During the discussion we will frequently refer to the *closed-loop bandwidth*, we will use this expression as a synonym for the bandwidth of the complementary sensitivity, $T_o(s)$.

9.9.1.1 Loop Nominal Sensitivities and Design Objectives

The nominal sensitivities are defined as [1,2]

$$S_o(s) = \frac{1}{1 + G_o(s)C(s)} = 1 - T_o(s) \qquad \text{Sensitivity}$$

$$T_o(s) = \frac{G_o(s)C(s)}{1 + G_o(s)C(s)} = 1 - S_o(s) \qquad \text{Complementary sensitivity}$$

$$S_{uo}(s) = \frac{C(s)}{1 + G_o(s)C(s)} = \frac{T_o(s)}{G_o(s)} \qquad \text{Control sensitivity}$$

$$S_{io}(s) = \frac{G_o(s)}{1 + G_o(s)C(s)} = G_o(s)S_o(s) \qquad \text{Input sensitivity}$$

$$S_{do}(s) = \frac{G_{o2}(s)}{1 + G_o(s)C(s)} = G_{o2}(s)S_o(s) \qquad \text{Disturbance sensitivity}$$

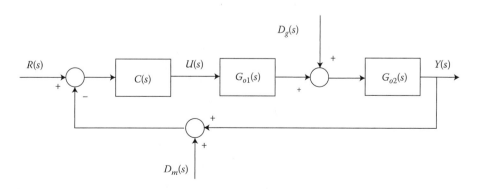

FIGURE 9.92 Basic control loop with noisy measurements and a generalized disturbance.

9.9.1.1.1 Reference Tracking

Reference tracking performance is closely related to the expression

$$Y(s) = T_o(s)R(s) \tag{9.337}$$

If we assume that the reference is a signal with significant energy only in the frequency band $[0; \omega_r]$, then good tracking requires that $T_o(j\omega) \approx 1$ in that band, or equivalently that $|S_o(j\omega)| \approx 0$ in that band. It can then be concluded that the larger the bandwidth of T_o then, the broader range of signals can be tracked with acceptable errors.

9.9.1.1.2 Disturbance Rejection

Disturbance rejection in the loop can be quantified using the expression

$$Y(s) = S_d(s)D_g(s) = S_o(s)G_{o2}(s)D_g(s) \tag{9.338}$$

Consider the case when the disturbance filtered through $G_{o2}(s)$ is a signal with significant energy only in the frequency band $[0; \omega_d]$, then good disturbance rejection is achieved if and only if $|S_o(j\omega)| \approx 0$ in that band, or equivalently if and only if $T_o(j\omega) \approx 1$ in that band. The most demanding case is when $G_{o2}(s) = 1$; otherwise, the fact that $G_{o2}(s)$ typically has lowpass characteristics, helps to reduce the equivalent disturbance bandwidth. As in the tracking case, a broader range of disturbances can be rejected by having a large closed-loop bandwidth.

9.9.1.1.3 Noise Immunity

The ability of the control loop to reject measurement noise is quantified by the expression

$$Y(s) = -T_o(s)D_m(s) \tag{9.339}$$

Therefore, to obtain good immunity to noise with significant energy only in the frequency band $[\omega_{n1}, \omega_{n2}]$, it is necessary and sufficient that $|T_o(j\omega)| \approx 0$ in that band. Given that the noise is typically a high-frequency signal, this requirement is equivalent to setting an upper bound on the loop bandwidth. This requirement defines the first conflict, since we have already seen that disturbance compensation encourages us to use a wide bandwidth. One of the most frequent cases is when the measurement signal contains a d.c. component; this bias usually originates in hardware offsets. This offset should be compensated independently, on a case by case basis; otherwise an error will always appear, since we cannot make $T_o(0) = 0$, because then we could not achieve zero tracking error for constant reference and disturbance signals. The above reasoning assumes that the compensation of the bias is complete.

9.9.1.1.4 Control Limitations

The control signal is related to the loop inputs (reference, disturbance and noise) through the expression

$$U(s) = S_{uo}(s)\left(R(s) - G_{o2}(s)D_g(s) - D_m(s)\right) \tag{9.340}$$

Ideally, $u(t)$ should exhibit no significant peaks and have low energy in the high-frequency band. These desirable features have to do with the need to avoid exceeding amplitude constraints, and also with the need to avoid unnecessary actuator stress. Thus, ideally S_{uo} should be lowpass. However, this desirable feature will usually conflict with the reference tracking and disturbance rejection design objectives. This conflict arises from the consequent need to have the bandwidth of T_o larger than that of $G_o(s)$. To illustrate this phenomenon, consider the following simple example:

$$G_o(s) = \frac{2}{(s+1)(s+2)} \tag{9.341}$$

We also assume that we require to track a reference in the frequency band $[0, 8]$ (rad/s), and to reject a disturbance in the frequency band $[0, 6]$ (rad/s). In this framework, a sensible choice for the bandwidth of T_o is approximately 20 (rad/s); for example, we could choose

$$T_o(s) = \frac{400}{s^2 + 28s + 400} \tag{9.342}$$

This choice leads to

$$S_{uo}(s) = \frac{200(s + 1)(s + 2)}{s^2 + 28s + 400} \tag{9.343}$$

This sensitivity function exhibits a frequency response, which emphasizes the high-frequency component of the input. Note that the sensitivity gain at d.c. is 1; however, at very high frequencies (much larger than 20 (rad/s)), the gain approaches 200.

It is thus natural to conclude that the bandwidth of T_o should be bounded above.

A key observation regarding this issue is the following. Assume that we specify that the output $y(t)$ must follow a particular trajectory for $t \geq 0$; then, for a given disturbance $d_g(t)$ there is one, and only one input $u(t)$, which allows one to achieve the specific trajectory for $y(t)$.

9.9.1.1.5 Robustness to Modeling Errors

A basic idea in control system design is to use a nominal plant model, $G_o(s)$, to synthesize a controller, and then to test that controller in conjunction with the real plant, or with a more realistic model (sometimes called the calibration model), $G(s)$. A basic requirement is that the controller $C(s)$, designed for the nominal model, also stabilizes the calibration model. This property is known as robust stability. However, we also usually aim to have robust performance. This property means that the performance of the control loop defined by the pair $\{G(s), C(s)\}$ should be close to that of the nominal loop defined by the pair $\{G_o(s), C(s)\}$. We will next examine how we can translate these robustness requirements into sensitivity constraints.

We first assume that the nominal model and the calibration model are connected by the relation

$$G(s) = G_o(s)(1 + G_\Delta(s)) \tag{9.344}$$

where $G_\Delta(s)$ is known as the multiplicative modeling error (MME).

This connection can be appreciated by examining some common situations, which are illustrated in Table 9.11, where only stable models are considered.

The examples in Table 9.11 illustrate typical features of modeling errors: the multiplicative error is small at low frequencies, and becomes more significant as the frequency increases. This characteristic has implications on design, as described below.

TABLE 9.11 Examples of Multiplicative Modeling Errors

$G_o(s)$	$G(s)$	$G_\Delta(s)$
$W(s)$	$W(s)\dfrac{1}{Ts+1}$	$\dfrac{-Ts}{Ts+1}$
$W(s)$	$W(s)e^{-\tau s}$	$e^{-\tau s} - 1$
$W(s)$	$W(s)\dfrac{b^2}{s^2 + 2abs + b^2}$	$\dfrac{-s(s + 2ab)}{s^2 + 2abs + b^2}$
$W(s)\left(\dfrac{-\tau s + 2k}{\tau s + 2k}\right)^k$	$W(s)e^{-\tau s}$	$e^{-\tau s}\left(\dfrac{-\tau s + 2k}{\tau s + 2k}\right)^{-k} - 1$

The achieved sensitivity $S(s)$ and the nominal sensitivity are related by

$$S(s) = \frac{1}{1 + G(s)C(s)} = \frac{S_o(s)}{1 + T_o(s)G_\Delta(s)} = S_\Delta(s)S_o(s) \qquad (9.345)$$

This result has been obtained by using Equation 9.344. The expression (Equation 9.345) leads to the conclusion that the nominal design will achieve robust performance if $||S_\Delta(j\omega)| - 1| \ll 1$, where

$$S_\Delta(s) = \frac{1}{1 + T_o(s)G_\Delta(s)} \qquad (9.346)$$

A sufficient condition is $|T_o(j\omega)G_\Delta(j\omega)| \ll 1$. This condition can be met if $|T_o(j\omega)|$ is small over the range of frequencies where $|G_\Delta(j\omega)|$ is significant. Given that the MME is usually significant only at high frequencies, we have that, to achieve robustness, the loop bandwidth must be bounded from above.

9.9.1.1.6 Summary

The design objectives described above conform to a complex web of conflicting constraints on the control loop bandwidth. These are summarized in Table 9.12.

The problem is that, at this stage, we have only one degree of freedom to satisfy those conflicting requirements. We next examine how these conflicts can be more readily dealt with by enriching the control architecture with additional degrees of freedom.

9.9.2 Reference Feedforward

One possible strategy is to deal with reference tracking as a separate problem. We first design the feedback controller $C(s)$ taking into account only objectives O.2 to O.5 in Table 9.12. This process leads to a specific $T_o(s)$. Then a new block with transfer function $H(s)$ is introduced [2,3] as shown in Figure 9.93.

From Figure 9.93 we see that reference tracking satisfies

$$Y(s) = T_o(s)R_H(s) = T_o(s)H(s)R(s) \qquad (9.347)$$

Thus, given $T_o(s)$ and the reference spectrum, by choosing an appropriate $H(s)$ we can shape the tracking performance, without modifying the ability of the loop to reject disturbance, the noise immunity, or the performance robustness. The selected $H(s)$ must be, at least, proper and stable.

Since the tracking properties are defined by the product $T_o(s)H(s)$, we should choose an $H(s)$ that satisfies the above constraints and compensates the limited bandwidth of $T_o(s)$. This compensation is basically achieved by making $H(j\omega)T(j\omega) \approx 1$ over the desired closed-loop bandwidth. In other words, $H(s)$ must be chosen to be a good inverse of $T_o(s)$ in the desired bandwidth.

TABLE 9.12   Conflicting Requirements on the Closed-Loop Bandwidth

Objective	Description	Key Sensitivity	Loop Bandwidth
O.1	Reference tracking	$T_o(s)$	Large
O.2	Disturbance compensation	$S_{do}(s)$	Large
O.3	Control limitations	$S_{uo}(s)$	Reduced
O.4	Noise immunity	$T_o(s)$	Reduced
O.5	Robustness	$S_o(s)$	Reduced

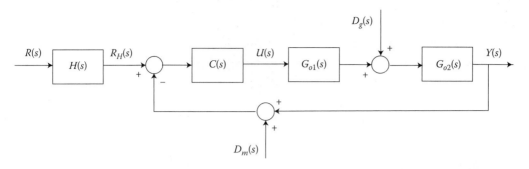

FIGURE 9.93 Alternative 1—Control loop with reference feedforward.

To illustrate the basic ideas in this approach, we will consider an example. Assume that the nominal plant model is given by

$$G_o(s) = \frac{2}{(s+1)(s+2)} \tag{9.348}$$

Further, assume that, due to noise and robustness constraints, the maximum closed loop bandwidth is limited to 3 rad/s. However, the desired reference signal has significant energy in the frequency band $[0; 8]$ rad/s. It is then requested to choose a convenient $H(s)$.

We first choose $T_o(s)$, with the maximum possible bandwidth, as

$$T_o(s) = \frac{9}{s^2 + 4.5s + 9} \tag{9.349}$$

Note that with this choice of $T_o(s)$, the resulting controller is biproper, and the loop is internally stable.

Given that we need a bandwidth of $Y(s)/R(s)$ which is, at least, equal to 8 rad/s, we choose

$$T_o(s)H(s) = \frac{400}{s^2 + 28s + 400} \implies H(s) = \frac{400(s^2 + 4.5s + 9)}{9(s^2 + 28s + 400)} \tag{9.350}$$

By inspection, we conclude that this choice of $H(s)$ satisfies the given design requirements. As a further comment, we note that the underlying philosophy (enhancing reference tracking in a desired frequency range, using a high-pass filter) is similar to that used in the Dolby technique of audio processing.

A second possible architecture aimed at dealing with the reference tracking objective is shown in Figure 9.94. The design block in this case, is $F(s)$; this function must be, at least, stable and proper.

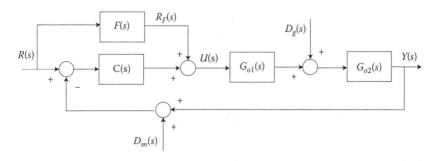

FIGURE 9.94 Alternative 2—Control loop with reference feedforward.

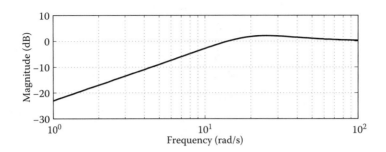

FIGURE 9.95 Bode plot (magnitude) of the term $G_o(s)F(s) - 1$.

From Figure 9.94, we conclude that the tracking performance is quantified by the expressions

$$Y(s) = T_o(s)R(s) + S_{io}(s)F(s)R(s) = R(s) + (G_o(s)F(s) - 1)S_o(s)R(s) \qquad (9.351)$$

$$E(s) = R(s) - Y(s) = (1 - G_o(s)F(s))S_o(s)R(s) \qquad (9.352)$$

Thus, the ideal choice of $F(s)$ would be $G_o(s)^{-1}$; however, this selection will typically yield an improper transfer function. Hence, the basic idea is to choose $F(s)$ so as to make $|G_o(j\omega)F(j\omega) - 1|$ small at the frequencies where the spectrum $|S_o(j\omega)R(j\omega)|$ is significant. Given that $G_o(s)$ is typically lowpass, then $F(s)$ will typically be high pass. If we now consider the same example developed for the first proposed reference feedforward architecture, a sensible choice would be

$$F(s) = \frac{200(s+1)(s+2)}{s^2 + 28s + 400} \qquad (9.353)$$

To examine the effect of this choice, we plot the (magnitude) Bode diagram of the term $G_o(s)F(s) - 1$. This is shown in Figure 9.95, where we can verify that the feedforward block contributes with a gain smaller than 1 in the tracking error expression (Equation 9.352), in the frequency band $[0; 8]$ (rad/s).

Faster reference tracking can be achieved by extending the frequency band in which $H(s)$ is a good inverse of $T_o(s)$ (first scheme) or where $F(s)$ is a good inverse of $G_o(s)$ (second scheme). This will not affect the ability of the controller $C(s)$ with respect to disturbance compensation, noise immunity, or performance robustness. However, as anticipated, it does affect the requirement on the plant input $u(t)$. To gain insight into this problem, we compute the plant input for both alternatives:

First scheme $U(s) = S_{uo}(s)R_H(s) = S_{uo}(s)H(s)R(s) \qquad (9.354)$

Second scheme $U(s) = S_{uo}(s)R(s) + S_o(s)F(s)R(s) \qquad (9.355)$

Recall that $C(s)$ should be designed to achieve a limited bandwidth $T_o(s)$; this design most likely yields a modest plant input. However, in the above expressions, this situation changes due to the intervention of the high-pass transfer functions $H(s)$ and $F(s)$. Then, in both cases, the demand on the plant input would increase as the desired bandwidth for reference tracking increases. ("There is no such a thing as a free lunch"). The main reflection on this issue is that there is no way in which we can improve the reference tracking without negatively impacting on the size of plant input.

9.9.3 Disturbance Rejection—Disturbance Feedforward

A second strategy aimed at resolving the conflicting design requirements is to add an open-loop mechanism, which preempts, to a large extent, the impact of the disturbance on the process output [2,4,5]. The basic idea is shown in Figure 9.96.

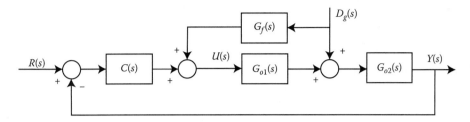

FIGURE 9.96 Control loop with disturbance feedforward.

We can see that the effect of the disturbance on the output can now be expressed as

$$Y(s) = (1 + G_{o1}(s)G_f(s))S_{do}(s)D_g(s) = (G_{o1}(s)^{-1} + G_f(s))S_{io}(s)D_g(s) \quad (9.356)$$

It is thus clear that the ideal choice is

$$G_f(s) = -G_{o1}(s)^{-1} \quad (9.357)$$

That is, $G_f(s)$ should be equal to the negative inverse of $G_{o1}(s)$. However, this would normally yield an improper, noncausal, and unstable transfer function. We consider, the following example to illustrate this point

$$G_{o1}(s) = \frac{-s+2}{s^2+1.2s+0.5} \implies G_f^{ideal}(s) = -\frac{s^2+1.2s+0.5}{-s+2} \quad (9.358)$$

Here, the ideal $G_f(s)$ is both improper and unstable.

Careful analysis of Equation 9.356 suggests that a more modest choice of $G_f(s)$ will suffice to provide significant disturbance pre-compensation. The appropriate design approach is to choose $G_f(s)$, so that $|1 + G_{o1}(j\omega)G_f(j\omega)| \ll 1$ in the frequency band where $|S_o(j\omega)D_g(j\omega)|$ is significant. If this criterion is satisfied, then we achieve a net gain over the case when no disturbance feedforward is used, that is, when $G_f(s) = 0$.

When using this architecture, the cost/benefit ratio must be carefully evaluated. There are two issues to consider: the financial cost (initial and maintenance) of a new measurement subsystem, including sensor, signal adaptation and signal transmission, and the introduction of additional measurement noise. The preceding analysis was made under the assumption of a perfect disturbance measurement. If measurement noise is considered on the disturbance, then a better description of the situation is the one shown in Figure 9.97.

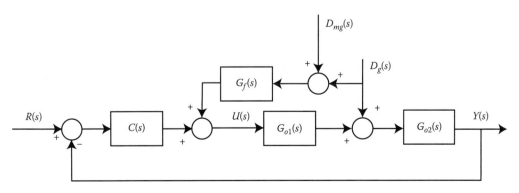

FIGURE 9.97 Control loop with noisy disturbance feedforward.

When disturbance measurement noise is added, the output is given by

$$Y(s) = (G_{o1}(s)^{-1} + G_f(s))S_{io}(s)D_g(s) + S_{io}(s)G_f(s)D_{mg}(s) \tag{9.359}$$

The impact of the additional term may be significant, given that the measurement noise in the loop forward path is typically a wide-band disturbance. The situation is made worse by the fact that $G_f(s)$ is typically high pass or, at least, band pass with a pass band in the high frequency region. On the other hand, $S_{io}(s)$ is typically band pass, with a pass band in the low-frequency region. Therefore, $S_{io}(s)$ provides measurement noise attenuation.

We will next consider the impact of the disturbance feedforward mechanism on the plant input $u(t)$. The input, assuming $D_{mg}(s) = 0$, is now described by

$$U(s) = -S_{uo}(s)G_{o2}(s)D_g(s) + \underbrace{S_o(s)G_f(s)D_g(s)}_{\text{feedforward contribution}} \tag{9.360}$$

If we assume that $G_f(s)$ is chosen close to the ideal given in Equation 9.357, we have

$$U(s) \approx - \left(S_{uo}(s)G_{o2}(s) + \frac{S_o(s)}{G_{o1}(s)} \right) D_g(s) = -(S_o(s) + T_o(s)) \frac{D_g(s)}{G_{o1}(s)} = \frac{D_g(s)}{G_{o1}(s)} \tag{9.361}$$

It is then clear from Equation 9.361 (and on noting that $G_{o1}(s)$ is typically lowpass) that perfect disturbance rejection will demand enhanced control effort, which is a negative byproduct of this architecture.

In summary, disturbance feedforward is an open-loop technique, which pre-compensates a disturbance at the point it enters the loop. This strategy will not adversely affect other design objectives, such as tracking, noise immunity, and robustness. Morever, it can be superimposed on a pre-existing ODOF control loop.

9.9.4 Disturbance Rejection: Cascade Architecture

The disturbance feedforward architectures presented in the previous section require that the disturbance be measured. This architecture is also based on an open-loop philosophy, which means that any inaccuracy in the model for $G_{o1}(s)$ will translate into erroneous feedforward action. If, however, another plant internal variable is available to be measured, another architecture can be devised to achieve a similar goal, that is, to compensate the disturbance to a significant extent, before it produces a deleterious effect at the plant output.

Consider the architecture shown in Figure 9.98, where $V(s)$ is the Laplace transform of a plant inner variable, $v(t)$. Inspection of that architecture reveals two main differences with the disturbance feedforward scheme shown in Figure 9.96, namely

- Cascade control is based upon a closed-loop strategy

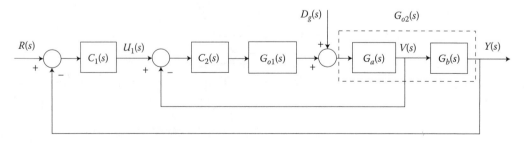

FIGURE 9.98 Cascade control loop.

- Cascade control design aimed at improving disturbance rejection needs to be carried out in conjunction with other design objectives. It is clear that the primary controller $C_1(s)$ cannot be designed if the secondary controller $C_2(s)$ is not known in advance.

To build a step-by-step design strategy, we first derive the key relationships in the cascade architecture shown in Figure 9.98.

$$V(s) = \underbrace{\frac{G_{o1}(s)G_a(s)C_2(s)}{1 + G_{o1}(s)G_a(s)C_2(s)}}_{T_{o2}(s)} U_1(s) + \underbrace{\frac{1}{1 + G_{o1}(s)G_a(s)C_2(s)}}_{S_{o2}(s)} G_a(s)D_g(s) \qquad (9.362)$$

The above expression gives rise to an equivalent control loop, which is shown in Figure 9.99.

The benefits of the cascade architecture are now evident from Figure 9.99, namely it can be seen that the disturbance can be attenuated by the impact of $C_2(s)$ in the inner sensitivity function $S_{o2}(s)$. The design goal is then to choose the bandwidth of the inner loop so that it covers the frequency band where $|D_g(j\omega)|$ is significant.

We can now sketch the full design procedure for the cascade architecture

Step 1 Design $C_2(s)$ to achieve the best possible disturbance compensation
Step 2 Compute the blocks appearing in Figure 9.99
Step 3 Design $C_1(s)$ by considering the following equivalent plant:

$$G_{eq}(s) = T_{o2}(s)G_b(s) \qquad (9.363)$$

The design should focus on reference tracking, noise immunity, and robustness.
Step 4 Evaluate the performance of the complete control system design. If not satisfactory, then go back to Step 1 or Step 3. If no significant improvement is achieved, then go to Step 5.
Step 5 Add reference feedforward and/or disturbance feedforward, as necessary, and concentrate on the design of $C_1(s)$ to achieve the remaining design objectives.
Step 6 If the resulting performance is still unsatisfactory, then rethink the problem, and possibly consider a process redesign.

We next write the expressions for $Y(s)$ and $U_1(s)$, as

$$Y(s) = T_{o1}(s)R(s) + S_{d1}(s)S_{o2}(s)D_g(s) \qquad (9.364)$$
$$U_1(s) = S_{uo1}(s)R(s) - S_{uo1}G_{o2}(s)S_{o2}(s)D_g(s) \qquad (9.365)$$

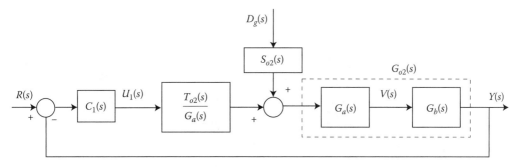

FIGURE 9.99 Equivalent cascade control loop.

where

$$T_{o1}(s) = \frac{C_1(s)T_{o2}(s)G_b(s)}{1 + C_1(s)T_{o2}(s)G_b(s)} \tag{9.366}$$

$$S_{d1}(s) = S_{o1}(s)G_{o2}(s) \tag{9.367}$$

$$S_{uo1}(s) = \frac{C_1(s)}{1 + C_1(s)T_{o2}(s)G_b(s)} \tag{9.368}$$

The benefits of using the cascade architecture must be weighed against the cost of an additional measurement, and the deleterious effect on performance of the additional measurement noise.

9.9.5 Case Studies

In this section, we will present three case studies, which illustrate the benefits that the architectures described above can provide. One of the key issues regarding feedforward and cascade control is that they are especially effective when their action is significantly faster than the feedback loop (in the feedforward strategy) or than the outer or main loop (in the cascade strategy). Otherwise, their contribution may possibly be overshadowed by the associated implementation costs.

The first case is simple, but it is a widespread application of cascade control. The second case is a more complex case where several of the ideas presented above can be used to the designer's advantage.

9.9.5.1 Cascade Control in Valves

We assume that we want to control a variable $y(t)$ by manipulating the flow rate $q(t)$. The simplest architecture for achieving this is the one shown in part (a) of Figure 9.100. Note that the controller output commands the valve opening; however, in this case, changes in the supply pressure $p_s(t)$ will yield different flow rates for the same value of $u(t)$ and thus affect the control goal.

An alternate cascade architecture is shown in part (b) of Figure 9.100. A second loop has been introduced to control the flow rate $q(t)$. This loop requires that one also measure $q(t)$, denoted by $q_m(t)$ in the figure. Note that the first controller output provides the reference for the second loop.

If we assume that the flow $q(t)$ feeds a process with slow dynamics, such as a large reactor tank or a large furnace, then the benefits of the cascade architecture are evident. When the ODOF control loop is used, the disturbance can be compensated only when it impacts the output, after going through the reactor or the furnace, and that will introduce a significant lag. If we build the cascade architecture, then there will be a preemptive action to compensate the disturbance before it enters the reactor or the furnace.

In this particular case, it is advisable to use cascade, since the inner loop is a flow control loop. The latter loop is well understood, easy to design and, in general, not very expensive. The above strategy will be part of the proposed control architecture in the next case.

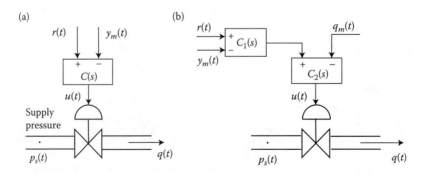

FIGURE 9.100   Example of application of cascade control.

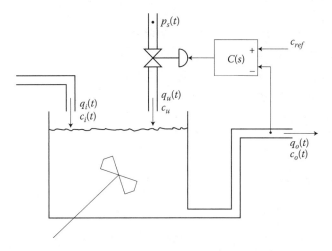

FIGURE 9.101 Mixing stirred tank.

9.9.5.2 Concentration Control in a Mixing Stirred Tank

We will next examine the case of a mixing stirred tank, as shown in Figure 9.101.

In the process shown in the figure, there is an unknown inflow $q_i(t)$ having an unknown concentration $c_i(t)$ of a component X. The system aims to obtain a specified concentration c_{ref} of X in the outflow $q_o(t)$, that is, it is desirable that $c_o(t) = c_{ref}$ $\forall t \geq 0$. To achieve this goal, a controlled flow $q_u(t)$ with known constant concentration c_u is added. It is assumed that $c_i(t) \leq c_{ref} \leq c_u$. Figure 9.101 shows a ODOF control architecture, aimed at achieving the specified goal. In this setting, $q_i(t)$, $c_i(t)$ and the supply pressure $p_s(t)$ are disturbances. The main difficulty is that big changes in those variables will be detected only once their effect appears on the output concentration $c_o(t)$.

To improve the rejection of the disturbances, we can introduce feedforward and cascade structures as shown in Figure 9.102.

The way in which these structures work is as follows:

Cascade To compensate the effect of changes in the supply pressure $p_s(t)$, an inner loop with a controller $C_2(s)$ is introduced, to ensure that the command issued by $C_1(s)$ has direct authority on the flow $q_u(t)$.

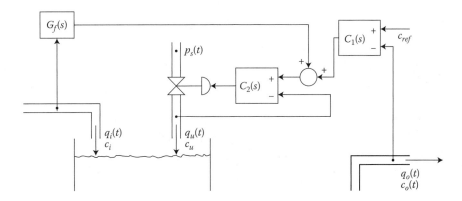

FIGURE 9.102 Feedforward and cascade control for the mixing stirred tank.

Feedforward Changes in $q_i(t)$ and in $c_i(t)$ are fed to a block $G_f(s)$, which additively adjusts the value of the reference for the inner control loop for $q_u(t)$. In this case, two additional measurements are needed, one to measure the incoming flow and the other to measure the concentration of X in that flow.

9.9.5.3 Centerline Gauge Control in Steel Reversing Mills

A typical reversing mill is shown schematically in Figure 9.103.

The basic function of the mill is to change the thickness of the steel strip as well as to make adjustments to the metallurgical properties.

There are many industrial control issues in these systems. A key point is that the control system architecture plays a fundamental role in achieving high performance. We briefly describe two such issues below.

1. Measurement delay

   The centerline strip thickness is typically measured by a radiation gauge placed downstream from the mill, say displaced from the roll gap by about one meter. If the strip is traveling at 60 (km/h), then this displacement amounts to a 60 (ms) delay. Due to robustness considerations, this delay effectively limits the closed-loop bandwidth to about 100 (rad/s). This constraint severely limits achievable performance. A well-known architectural change is to add an extra measurement of the roll force via a strain gauge. We next show how we can use this additional measurement to improve the performance of the controller.

   The mill can be approximately modeled by the spring equation

   $$f(t) = M[h(t) - \sigma(t)] \tag{9.369}$$

   where $f(t)$: force; $\sigma(t)$: unloaded roll gap; $h(t)$: output thickness; and M: mill modulus (constant). Then, an estimate of the exit thickness is

   $$\hat{h}(t) = \frac{f(t)}{M} + \sigma(t) \tag{9.370}$$

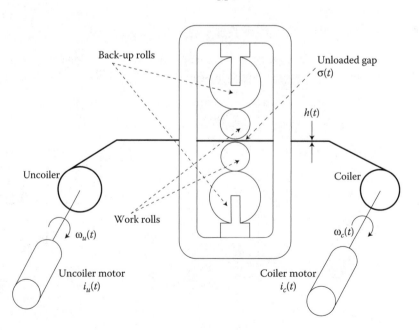

FIGURE 9.103 Reversing mill schematic representation.

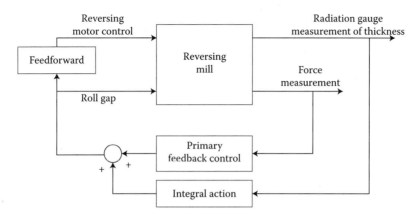

FIGURE 9.104 Block diagram of the control architecture for the reversing mill.

Typically, the estimate provided by Equation 9.370 is used for fast disturbance rejection whilst the the delayed thickness provided by the radiation gauge is used for steady-state error compensation.

2. Strip stretch compensation

One might believe that improved measurement architectures, such as the one described in (1) might be sufficient to achieve very wide bandwidth control. However, when this is tried, then one observes a limit to the achievable bandwidth. This limit has a mathematical description (see [2]). The phenomenon also has a nice physical explanation, as follows

- We first note that strip tension plays a key role in the thickness reduction process; the higher the tension, then the greater the reduction.
- Now if we notice that the exit thickness is too high, then we would close the roll gap.
- Then less metal exits the mill at constant exit strip velocity.
- Then the velocity of the strip entering the mill must fall.
- Due to inertia of the uncoiler, its radial velocity remains substantially unchanged.
- Hence, the tension in the strip will drop and the exit thickness goes up (contrary what was trying to be achieved).

It turns out that the above limitation (known as the *hold-up effect*) is fundamental and no ODOF architecture can overcome the problem. However, this does not mean that an alternative architecture will not help.

Indeed, a little thought indicates that one should try to slow down the uncoiler whenever the roll gap is reduced. This can be easily achieved by using feedforward into the uncoiler motor current (so that when the roll gap diminishes, then the uncoiler current goes up). Indeed, one can adjust the feedforward gain to completely remove the hold-up effect. The final control architecture is then as in Figure 9.104.

References

1. Åström, K. and Murray, R., *Feedback Systems: An Introduction for Scientists and Engineers*. Princeton University Press, Princeton, NJ, 2008.
2. Goodwin, G. et al. *Control System Design*. Prentice-Hall, Englewood Cliffs, NJ, 2001.
3. Kuo, B., *Automatic Control Systems*, 7th ed., Prentice-Hall, Englewood Cliffs, NJ, 1995.
4. Shinskey, F., *Process Control Systems: Application, Design and Adjustment*. 3rd ed., McGraw–Hill, New York, 1998.
5. Stephanopoulos, G., *Chemical Process Control: An Introduction to Theory and Practice*. Prentice–Hall, Englewood Cliffs, NJ, 1984.

IV

Digital Control

10

Discrete-Time Systems

Michael Santina
The Boeing Company

Allen R. Stubberud
University of California, Irvine

10.1 Discrete-Time Systems

10.1.1 Introduction to Digital Control

Rapid advances in digital system technology have radically altered the control system design options. It has become routinely practicable to design very complicated digital controllers and to carry out the extensive calculations required for their design. These advances in implementation and design capability can be obtained at low cost because of the widespread availability of inexpensive and powerful digital computers and their related devices.

A *digital control system* uses digital hardware, usually in the form of a programmed digital computer, as the heart of the controller. A typical digital controller has analog components at its periphery to interface with the plant. It is the processing of the controller equations that distinguishes analog from digital control.

In general, digital control systems have many advantages over analog control systems. Some of the advantages are

1. Low cost, low weight, and low power consumption
2. Zero drift of system parameters despite wide variations in temperature, humidity, and component aging
3. High accuracy
4. High reliability and ease of making software and design changes

The signals used in the description of digital control systems are termed *discrete-time signals*. Discrete-time signals are defined only for discrete instants of time, usually at evenly spaced time steps. Discrete-time computer-generated signals have discrete (or *quantized*) amplitudes and thus attain only discrete values. Figure 10.1 shows a continuous amplitude signal that is represented by a 3-bit binary code at evenly spaced time instants. In general, an n-bit binary code can represent only 2^n different values. Because of the complexity of dealing with quantized signals, digital control system design proceeds as if the signals involved are not of discrete amplitude. Further analysis usually must be performed to determine whether the proposed level of quantization is acceptable.

A discrete-time system is said to be *linear* if it satisfies the principle of *superposition*. Any linear combination of inputs produces the same linear combination of corresponding output components. If a system is not linear, then it is termed *nonlinear*. A discrete-time system is *step invariant* if its properties do

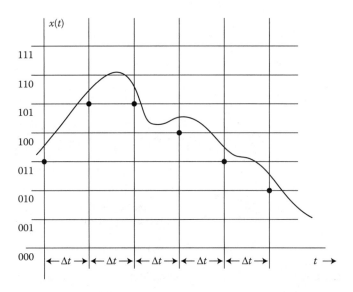

FIGURE 10.1   An example of a 3-bit quantized signal.

not change with time step. Any time shift of the inputs produces an equal time shift of every corresponding output signal.

Figure 10.2 shows a block diagram of a typical digital control system for a continuous-time plant. The system has two reference inputs and five outputs, two of which are measured directly by analog sensors. The *analog-to-digital converters* (A/D) *sample* the analog sensor signals and produce equivalent binary representations of these signals. The sampled sensor signals are then modified by the digital controller algorithms, which are designed to produce the necessary digital control inputs $u_1(k)$ and $u_2(k)$. Consequently, the control inputs $u_1(k)$ and $u_2(k)$ are converted to analog signals $u_1(t)$ and $u_2(t)$ using *digital-to-analog converters* (D/A). The D/A transforms the digital codes to signal *samples* and then produces *step reconstruction* from the signal samples by transforming the binary-coded digital input to voltages. These voltages are held constant during the *sampling period T* until the next sample arrives. This process of holding each of the samples is termed *sample and hold*. Then the analog signals $u_1(t)$ and

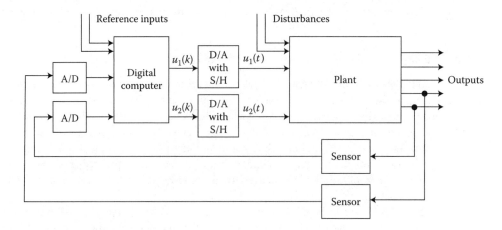

FIGURE 10.2   A digital control system controlling a continuous-time plant.

$u_2(t)$ are applied to control the behavior of the plant. Not shown in Figure 10.2 is a real-time clock that synchronizes the actions of the A/D, D/A, and shift registers.

Of course, there are many variations on this basic theme, including situations where the signals of the analog sensors are sampled at different sampling periods and where the system has many controllers with different sampling periods. Other examples include circumstances where (1) the A/D and D/A are not synchronized; (2) the sampling rate is not fixed; (3) the sensors produce digital signals directly; (4) the A/D conversion is different from sample and hold; and (5) the actuators accept digital commands.

10.1.2 Discrete-Time Signals and Systems

A *discrete-time signal* $f(k)$ is a sequence of numbers called samples. It is a function of the discrete variable k, termed the *step index*. Figure 10.3 shows some fundamental sequences, all having samples that are zero prior $k = 0$. In this figure, the step and the ramp sequences consist of samples that are values of the corresponding continuous-time functions at evenly spaced points in time. But the unit pulse sequence and the unit impulse function are not related this way, because the pulse has a unit sample at $k = 0$ while the impulse is infinite at $t = 0$.

10.1.2.1 z-Transformation

The one-sided z-transform of a sequence $f(k)$ is defined by the equation

$$Z[f(k)] = F(z) = \sum_{k=0}^{\infty} f(k)z^{-k}$$

It is termed one-sided, because samples before step zero are not included in the transform. The z-transform plays much the same role in the analysis of discrete-time systems as the Laplace transform does with

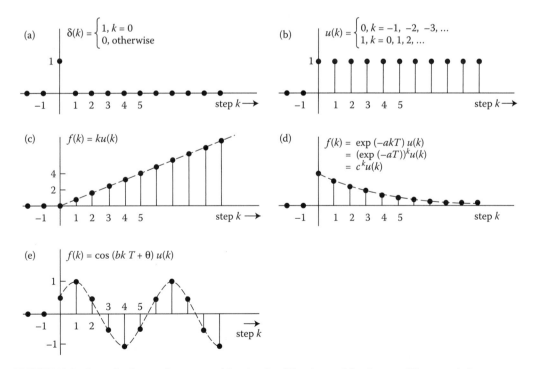

FIGURE 10.3 Some fundamental sequences. (a) unit pulse; (b) unit step; (c) unit ramp; (d) geometric (or exponential); and (e) sinusoidal.

TABLE 10.1   *z*-Transform Pairs

$f(k)$	$F(z)$
$\delta(k)$, unit pulse	1
$u(k)$, unit step	$\dfrac{z}{z-1}$
$ku(k)$	$\dfrac{z}{(z-1)^2}$
$c^k u(k)$	$\dfrac{z}{z-c}$
$kc^k u(k)$	$\dfrac{cz}{(z-c)^2}$
$u(k)\sin\Omega k$	$\dfrac{z\sin\Omega}{z^2 - 2z\cos\Omega + 1}$
$u(k)\cos\Omega k$	$\dfrac{z(z-\cos\Omega)}{z^2 - 2z\cos\Omega + 1}$
$u(k)c^k \sin\Omega k$	$\dfrac{z(c\sin\Omega)}{z^2 - (2c\cos\Omega)z + c^2}$
$u(k)c^k \cos\Omega k$	$\dfrac{z(z-c\cos\Omega)}{z^2 - (2c\cos\Omega)z + c^2}$

continuous-time systems. Important sequences and their *z*-transforms are listed in Table 10.1, and properties of the *z*-transform are summarized in Table 10.2.

A sequence that is zero prior to $k = 0$ is recovered from its *z*-transform via the inverse *z*-transform

$$f(k) = \frac{1}{2\pi j} \oint F(z)z^{k-1}dz$$

in which the integration is performed in a counterclockwise direction along a closed contour on the complex plane. In practice, the integrals involved are often difficult; so other methods of inversion have

TABLE 10.2   *z*-Transform Properties

$$\mathcal{Z}[f(k)] = \sum_{k=0}^{\infty} f(k)z^{-k} = F(z)$$

$\mathcal{Z}[cf(k)] = cF(z)$ c a constant

$\mathcal{Z}[f(k) + g(k)] = F(z) + G(z)$

$\mathcal{Z}[kf(k)] = -z\dfrac{dF(z)}{dz}$

$\mathcal{Z}[c^k f(k)] = F(z/c)$ c a constant

$\mathcal{Z}[f(k-1)] = f(-1) + z^{-1}F(z)$

$\mathcal{Z}[f(k-2)] = f(-2) + z^{-1}f(-1) + z^{-2}F(z)$

$\mathcal{Z}[f(k-n)] = f(-n) + z^{-1}f(1-n) + z^{-2}f(2-n) + \cdots + z^{1-n}f(-1) + z^{-n}F(z)$

$\mathcal{Z}[f(k+1)] = zF(z) - zf(0)$

$\mathcal{Z}[f(k+2)] = z^2 F(z) - z^2 f(0) - zf(1)$

$\mathcal{Z}[f(k+n)] = z^n F(z) - z^n f(0) - z^{n-1}f(1) - \cdots - z^2 f(n-2) - zf(n-1)$

$f(0) = \lim\limits_{z\to\infty} F(z)$

If $\lim\limits_{k\to\infty} f(k)$ exists and is finite, $\lim\limits_{k\to\infty} f(k) = \lim\limits_{z\to 1}\left[\dfrac{z-1}{z}F(z)\right]$

$\mathcal{Z}\left[\sum\limits_{i=0}^{k} f_1(k-i)f_2(i)\right] = F_1(z)F_2(z)$

been developed to replace the inverse transform calculations. For rational $F(z)$, sequence samples can be obtained from $F(z)$ by long division. Another method of recovering the sequence from its z-transform is to expand $F(z)/z$ in a partial fraction expansion and use appropriate transform pairs from Table 10.1. Rather than expanding a z-transform $F(z)$ directly in a partial fraction, the function $F(z)/z$ is expanded so that terms with a z in the numerator result. Yet another method of determining the inverse z-transform that is well suited to digital computation is to construct a difference equation from the rational function and then solve the difference equation recursively.

10.1.2.2 Difference Equations

Analogous to the differential equations that describe continuous-time systems, the input–output behavior of a discrete-time system can be described by difference equations. Linear discrete-time systems are described by linear difference equations. If the coefficients of a difference equation are constant, the system is step invariant and the difference equation has the form

$$y(k+n) + a_{n-1}y(k+n-1) + a_{n-2}y(k+n-2) + \cdots + a_1 y(k+1) + a_0 y(k)$$
$$= b_m r(k+m) + b_{m-1}r(k+m-1) + \cdots + b_1 r(k+1) + b_0 r(k), \tag{10.1}$$

where r is the input and y is the output. The order of the difference equation is n, which is the number of past output steps that are involved in calculating the present output:

$$y(k+n) = \underbrace{-a_{n-1}y(k+n-1) - a_{n-2}y(k+n-2) - \cdots - a_1 y(k+1) - a_0 y(k)}_{n \text{ terms}}$$
$$\underbrace{+ b_m r(k+m) + b_{m-1}r(k+m-1) + \cdots + b_1 r(k+1) + b_0 r(k)}_{m+1 \text{ terms}}$$

Returning to Equation 10.1, a discrete-time system is said to be *causal* if $m \le n$ so that only past and present inputs, not future ones, are involved in the calculation of the present output. An alternative equivalent form of the difference equation is obtained by replacing k by $k - n$ in Equation 10.1.

Difference equations can be solved recursively using the equation and solutions in the previous steps to calculate the solution in the next step. For example, consider the difference equation

$$y(k) - y(k-1) = 2u(k) \tag{10.2}$$

with the initial condition $y(-1) = 0$ and $u(k) = 1$ for all k. Letting $k = 0$ and substituting into the difference equation gives

$$y(0) - y(-1) = 2u(0),$$
$$y(0) = 2.$$

Letting $k = 1$ and substituting gives

$$y(1) - y(0) = 2u(1),$$
$$y(1) = 2 + 2 = 4.$$

At Step 2,

$$y(2) - y(1) = 2,$$
$$y(2) = 6,$$

and so on.

A difference equation can be constructed using a computer by programming its recursive solution. A digital hardware realization of the difference equation can also be constructed by coding the signals as binary words, storing present and past values of the input and output in registers, and using binary arithmetic devices to multiply the signals by the equation coefficients and adding them to form the output.

10.1.3 z-Transfer Function Methods

Solutions of linear step-invariant difference equations can be found using z-. For example, consider the single-input, single-output system described by Equation 10.2. Using z-transformation,

$$Y(z) - z^{-1}Y(z) = \frac{2z}{z-1},$$

$$\frac{Y(z)}{z} = \frac{2z}{(z-1)^2} = \frac{k_1}{(z-1)} + \frac{k_2}{(z-1)^2}$$

$$Y(z) = \frac{2z}{(z-1)} + \frac{2z}{(z-1)^2},$$

and

$$y(k) = 2u(k) + 2ku(k).$$

Checking,

$$y(0) = 2,$$
$$y(1) = 2 + 2 = 4,$$
$$y(2) = 2 + 4 = 6,$$

which agrees with the recursive solution in the previous example.

In general, an nth-order linear discrete-time system is modeled by a difference equation of the form

$$y(k+n) + a_{n-1}y(k+n-1) + \cdots + a_1 y(k+1) + a_0 y(k) = b_m r(k+m) + \cdots + b_1 r(k+1) + b_0 r(k)$$

or

$$y(k) + a_{n-1}y(k-1) + \cdots + a_1 y(k-n+1) + a_0 y(k-n)$$
$$= b_m r(k+m-n) + \cdots + b_1 r(k-n+1) + b_0 r(k-n),$$

which has a z-transform given by

$$Y(z) + a_{n-1}[z^{-1}Y(z) + y(-1)] + \cdots + a_1[z^{-n+1}Y(z) + z^{-n+2}y(-1) + \cdots + y(-n+1)]$$
$$+ a_0[z^{-n}Y(z) + z^{-n+1}y(-1) + \cdots + y(-n)]$$
$$= b_m[z^{-n+m}R(z) + z^{-n+m-1}r(-1) + \cdots + r(-n+m-1)] + \cdots$$
$$+ b_0[z^{-n}R(z) + z^{-n+1}r(-1) + \cdots + r(-n)],$$

$$Y(z) = \underbrace{\frac{b_m z^m + b_{m-1}z^{m-1} + \cdots + b_1 z + b_0}{z^n + a_{n-1}z^{n-1} + \cdots + a_1 z + a_0}R(z)}_{\text{Zero-state component}}$$

$$+ \underbrace{\frac{\text{(Polynomial in } z \text{ of degree } n \text{ or less with coefficients}}{z^n + a_{n-1}z^{n-1} + \cdots + a_1 z + a_0}}_{\text{Zero-input component}}.$$

If all the initial conditions are zero, the zero-input component of the response is zero. The zero-state component of the response is the product of the system z-transfer function

$$T(z) = \frac{b_m z^m + b_{m-1}z^{m-1} + \cdots + b_1 z + b_0}{z^n + a_{n-1}z^{n-1} + \cdots + a_1 z + a_0}$$

FIGURE 10.4 Unit pulse response of a discrete-time system.

and the z-transform of the system input:

$$Y_{\text{zero state}}(z) = T(z)R(z).$$

Analogous to continuous-time systems, the *transfer function* of a linear step-invariant discrete-time system is the ratio of the z-transform of the output to the z-transform of the input when all initial conditions are zero.

It is also common practice to separate the system response into *natural* (or *transient*) and *forced* (or *steady-state*) components. The natural response is a solution of the *homogeneous* difference equation. This is the solution of the difference equation due to initial conditions only with all independent inputs set to zero. The remainder of the response, which includes a term in a form dependent on the specific input, is the forced response component.

10.1.3.1 Stability and Response Terms

As shown in Figure 10.4, when the input to a linear step-invariant discrete-time system is the unit pulse $\delta(k)$ and all initial conditions are zero, the response of the system is given by

$$Y_{\text{pulse}}(z) = R(z)T(z) = T(z).$$

A linear step-invariant discrete-time system is said to be *input–output stable* if its pulse response decays to zero asymptotically. This occurs if and only if *all* the roots of the denominator polynomial of the transfer function are inside the unit circle in the complex plane.

Figure 10.5 shows pulse responses corresponding to various *pole* (denominator root) locations. Sequences corresponding to pole locations inside the unit circle decay to zero asymptotically; hence they are stable. Systems with poles that are outside the unit circle or repeated on the unit circle have outputs that expand with step and are thus *unstable*. Systems with nonrepeated poles on the unit circle have responses that neither decay nor expand with step and are termed *marginally stable*. Methods for testing the location of the roots of the denominator polynomial of a transfer function are presented in Chapter 8 of this handbook.

A pole–zero plot of a z-transfer function consists of Xs denoting poles and Os denoting zeros in the complex plane. The z-transfer function

$$T(z) = \frac{3z + 3z^3}{4z^4 + 6z^3 - 4z^2 + z + 2}$$

$$= \left(\frac{3}{4}\right) \frac{z(z+j)(z-j)}{(z+2)\left(z+\frac{1}{2}\right)\left(z-\frac{1}{2}+j\frac{1}{2}\right)\left(z-\frac{1}{2}-j\frac{1}{2}\right)}$$

has the pole–zero plot shown in Figure 10.6. It represents an unstable discrete-time system because it has a pole at $z = -2$, which is outside the unit circle.

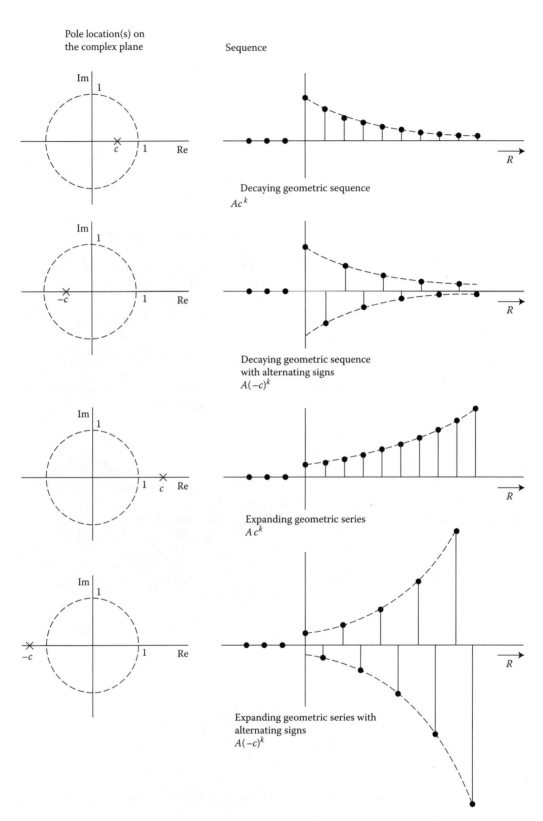

FIGURE 10.5   Sequences corresponding to various *z*-transform pole locations.

Pole location(s) on
the complex plane

Sequence

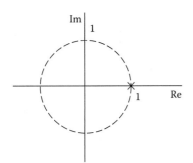

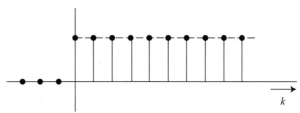

Constant sequence
$A(1)^k = A$

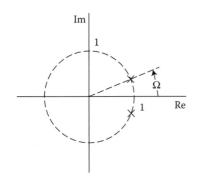

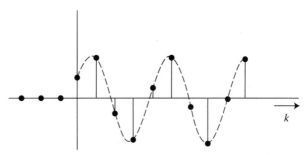

Sinusoidal sequence
$A \cos (\Omega k + \theta)$

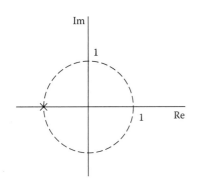

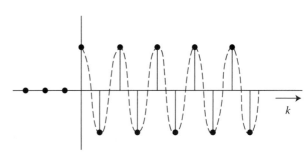

Alternating sequence
$A(-1)^k$

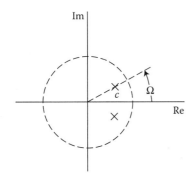

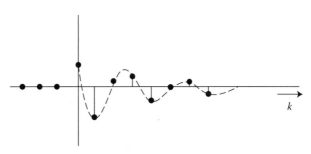

Damped sinusoidal sequence
$Ac^k \cos (\Omega k + \theta)$

FIGURE 10.5 *Continued.*

Pole location(s) on
the complex plane

Sequence

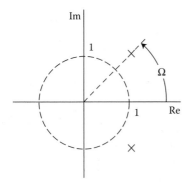

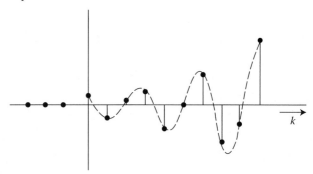

Exponentially expanding sinusoidal
sequence
$A^k \cos(\Omega k + \theta)$

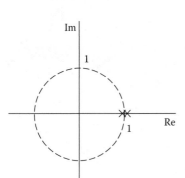

Ramp sequence
$Ak(1)^k = Ak$

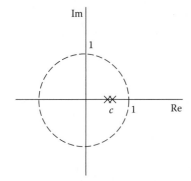

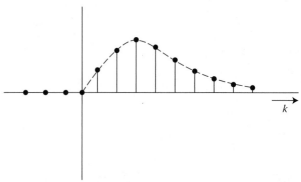

Ramp-weighted geometric sequence
Akc^k

FIGURE 10.5 *Continued.*

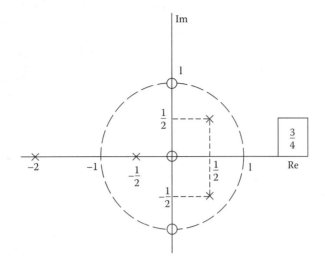

FIGURE 10.6 An example.

10.1.3.2 Block Diagram Algebra

The rules of block diagram algebra for linear time-invariant continuous-time systems apply to linear step-invariant discrete-time systems as well. Combining blocks in cascade or in tandem or moving a pick-off point in front of or behind a block, etc. with discrete-time systems is done the same way as with continuous-time systems. However, as we see in Chapter 11, these rules do not necessarily apply for sampled data systems containing discrete-time as well as continuous-time components.

Similar to a continuous-time system, when a discrete-time system has several inputs and/or outputs, there is a z-transfer function relating each one of the inputs to each one of the outputs, with all other inputs set to zero:

$$T_{ij}(z) = \frac{Y_i(z)}{R_j(z)}\bigg|, \quad \substack{\text{when all initial conditions} \\ \text{are zero and when all inputs} \\ \text{except } R_j \text{ are zero}}$$

In general, when all the initial conditions of a system are zero, the outputs of the system are given by

$$Y_1(z) = T_{11}(z)R_1(z) + T_{12}(z)R_2(z) + T_{13}(z)R_3(z) + \cdots$$
$$Y_2(z) = T_{21}(z)R_1(z) + T_{22}(z)R_2(z) + T_{23}(z)R_3(z) + \cdots$$
$$Y_3(z) = T_{31}(z)R_1(z) + T_{32}(z)R_2(z) + T_{33}(z)R_3(z) + \cdots$$
$$\vdots$$

For example, the four transfer functions of the two-input, two-output system shown in Figure 10.7 are as follows:

$$T_{11}(z) = \frac{Y_1(z)}{R_1(z)} = \frac{z^2}{z^2 + 0.5z - 0.5},$$

$$T_{21}(z) = \frac{Y_2(z)}{R_1(z)} = \frac{z^2(z + 0.5)}{z^2 + 0.5z - 0.5},$$

$$T_{12}(z) = \frac{Y_1(z)}{R_2(z)} = \frac{z(z - 1)}{z^2 + 0.5z - 0.5},$$

$$T_{22}(z) = \frac{Y_2(z)}{R_2(z)} = \frac{(z - 1)(z + 0.5)}{z^2 + 0.5z - 0.5}.$$

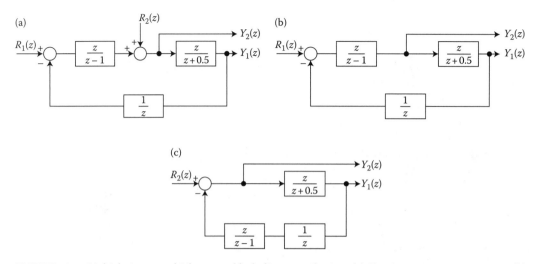

FIGURE 10.7 Multiple-input, multiple-output block diagram reduction. (a) Two-input, two-output system; (b) block diagram reduction to determine $T_{11}(z)$ and $T_{12}(z)$; and (c) block diagram reduction to determine $T_{21}(z)$ and $T_{22}(z)$.

A linear step-invariant discrete-time multiple-input, multiple-output system is input–output stable if, and only if, *all* the poles of *all* its z-transfer functions are inside the unit circle on the complex plane.

10.1.3.3 Discrete-Frequency Response

As shown in Figure 10.8, when the input to a linear step-invariant discrete-time system is a sinusoidal sequence of the form

$$r(k) = A\cos(\Omega k + \theta),$$

the forced output $y(k)$ of the system includes another sinusoidal sequence with the same frequency, but generally with different amplitude B and different phase ϕ.

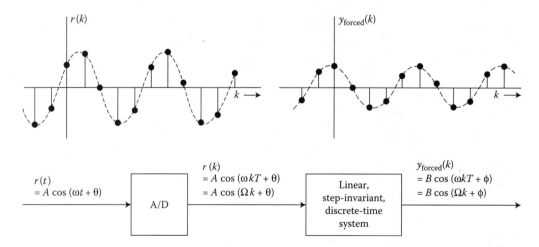

FIGURE 10.8 Discrete-time system with a sinusoidal input sequence and the sinusoidal forced output sequence.

If a discrete-time system is described by the difference equation

$$y(k+n) + a_{n-1}y(k+n-1) + \cdots + a_1y(k+1) + a_0y(k)$$
$$= b_m r(k+m) + b_{m-1}r(k+m-1) + \cdots + b_0 r(k),$$

its transfer function is given by

$$T(z) = \frac{b_m z^m + b_{m-1}z^{m-1} + \cdots + b_1 z + b_0}{z^n + a_{n-1}z^{n-1} + \cdots + a_1 z + a_0}.$$

The *magnitude* of the z-transfer function, evaluated at $z = \exp(j\Omega)$, is the ratio of the amplitude of the forced output to the amplitude of the input:

$$\frac{B}{A} = |T(z = e^{j\Omega})|.$$

The angle of the z-transfer function, evaluated at $z = \exp(j\Omega)$, is the phase difference between the input and output:

$$\phi - \theta = \underline{/T(z = e^{j\Omega})}.$$

These results are similar to the counterpart for continuous-time systems, in which the transfer function is evaluated at $s = j\omega$.

Frequency response plots for a linear step-invariant discrete-time system are plots of the magnitude and angle of the z-transfer function, evaluated at $z = \exp(j\Omega)$, vs. Ω. The plots are periodic, because $\exp(j\Omega)$ is periodic in Ω with period 2π. This is illustrated in Figure 10.9, which gives the frequency response for the z-transfer function in the accompanying pole–zero plot. In general, the frequency response plots for discrete-time systems are symmetric about $\Omega = \pi$, as shown. The amplitude ratio is even symmetric while the phase shift is odd symmetric about $\Omega = \pi$. Therefore, the frequency range of Ω from 0 to π is adequate to completely specify the frequency response of a discrete-time system. Logarithmic frequency response plots for the system given in Figure 10.9 are shown in Figure 10.10.

10.1.4 Discrete-Time State Equations and System Response

We now make the transition from classical system description and analysis methods to state variable methods. System response is expressed in terms of discrete convolution and in terms of z-transforms. z-Transfer function matrices of multiple-input, multiple-output systems are found in terms of the state equations. The state equations and response of step-varying systems are also discussed.

10.1.4.1 State Variable Models of Linear Step-Invariant Discrete-Time Systems

An nth-order linear discrete-time system can be modeled by a state equation of the form

$$x(k+1) = Ax(k) + Bu(k) \tag{10.3a}$$

where x is the n-vector state of the system, u is an r-vector of input signals, the state coupling matrix A is $n \times n$, and the input coupling matrix B is $n \times r$. The m-vector of system measurement outputs y is related to the state and inputs by a measurement equation of the form

$$y(k) = Cx(k) + Du(k), \tag{10.3b}$$

where the output coupling matrix C is $m \times n$, and the input-to-output coupling matrix D is $m \times r$. A block diagram showing how the various quantities of the state and output equations are related is shown in Figure 10.11. In the diagram, wide arrows represent vectors of signals. A system given by Equations 10.3a, b is termed *step invariant* if the matrices A, B, C, and D do not change with step.

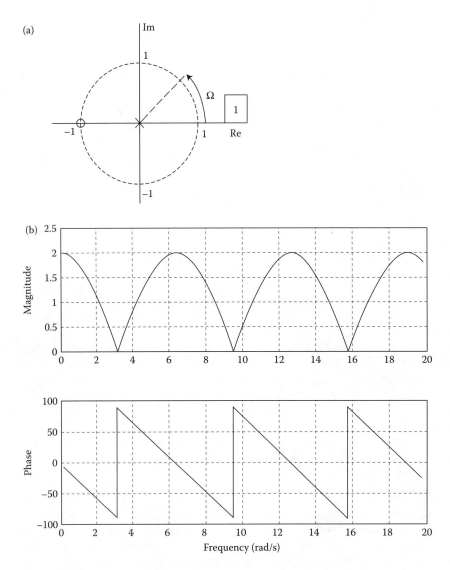

FIGURE 10.9 Periodicity of the frequency response of a discrete-time system. (a) Pole–zero plot of a transfer function $T(z)$ and (b) frequency response plots for $T(z)$.

10.1.4.2 Response in Terms of Discrete Convolution

In terms of the initial state $x(0)$ and the inputs $u(k)$ at step zero and beyond, the solution for the state after step zero can be calculated recursively. From $x(0)$ and $u(0)$, $x(1)$ can be calculated:

$$x(1) = Ax(0) + Bu(0).$$

Then, using $x(1)$ and $u(1)$,

$$x(2) = Ax(1) + Bu(1) = A^2 x(0) + ABu(0) + Bu(1).$$

From $x(2)$ and $u(2)$,

$$x(3) = Ax(2) + Bu(2) = A^3 x(0) + A^2 Bu(0) + ABu(1) + Bu(2),$$

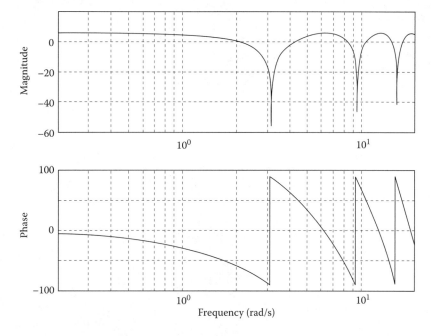

FIGURE 10.10 Logarithmic frequency response plots for the system shown in Figure 10.9.

and in general

$$x(k) = \underbrace{A^k x(0)}_{\text{Zero-input component}} + \underbrace{\sum_{i=0}^{k-1} A^{k-1-j} Bu(i)}_{\text{Zero-state component}}, \qquad (10.4)$$

and the solution for the output $y(k)$ in Equation 10.3b is

$$y(k) = CA^k x(0) + \left\{ \sum_{i=0}^{k-1} CA^{k-i-1} Bu(i) \right\} + Du(k).$$

The system output when all initial conditions are zero is termed the *zero-state* response of the system. When the system initial conditions are not all zero, the additional components in the outputs are termed the *zero-input* response components.

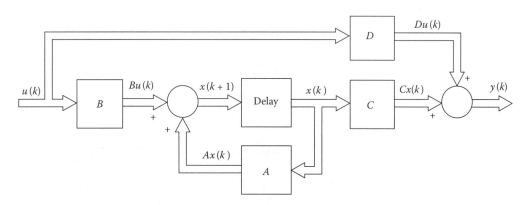

FIGURE 10.11 Block diagram showing the relations between signal vectors in a discrete-time state-variable model.

10.1.4.3 Response in Terms of z-Transform

The response of a discrete-time system described by the state Equations 10.3a, b can be found by z-transforming the state equations

$$x(k+1) = Ax(k) + Bu(k).$$

That is,

$$zX(z) - zx(0) = AX(z) + BU(z)$$

or

$$(zI - A)X(z) = zx(0) + BU(z).$$

Hence,

$$X(z) = z(zI - A)^{-1}x(0) + (zI - A)^{-1}BU(z)$$

and

$$Y(z) = Cz(zI - A)^{-1}x(0) + \{C(zI - A)^{-1}B + D\}U(z). \tag{10.5}$$

The solution for the state is then

$$x(k) = \mathcal{Z}^{-1}[z(zI - A)^{-1}]x(0) + \mathcal{Z}^{-1}[(zI - A)^{-1}BU(z)].$$

Comparing this result with Equation 10.4 shows that

$$A^k = \mathcal{Z}^{-1}[z(zI - A)^{-1}],$$

which is analogous to the continuous-state transition matrix. Setting the initial conditions in Equation 10.5 to zero gives the $m \times r$ transfer function matrix

$$T(z) = C(zI - A)^{-1}B + D,$$

where m is the number of outputs in $y(k)$ and r is the number of inputs in $u(k)$. The elements of $T(z)$ are functions of the variable z, and the element in the ith row and jth column of $T(z)$ is the transfer function relating the ith output to the jth input:

$$T_{ij}(z) = \left. \frac{Y_i(z)}{U_j(z)} \right|_{\substack{\text{Zero initial conditions} \\ \text{and all other inputs zero}}}.$$

For an $n \times n$ matrix A,

$$(zI - A)^{-1} = \frac{\text{adj}(zI - A)}{|zI - A|},$$

where $|zI - A|$ is the determinant of $zI - A$, and hence each element of $T(z)$ is a ratio of polynomials in z that shares the denominator polynomial

$$q(z) = |zI - A|$$

which is the characteristic polynomial of the matrix A. Each transfer function of $T(z)$ then has the same poles, although there may be pole-zero cancelation. Stability requires that all the system poles (or *eigenvalues*) be within the unit circle on the complex plane.

As an example, consider the second-order three-input, two-output system described by

$$
\begin{bmatrix} x_1(k+1) \\ x_2(k+1) \end{bmatrix} = \begin{bmatrix} 2 & -5 \\ \frac{1}{2} & -1 \end{bmatrix} \begin{bmatrix} x_1(k) \\ x_2(k) \end{bmatrix} + \begin{bmatrix} 1 & -2 & 0 \\ 0 & 1 & 3 \end{bmatrix} \begin{bmatrix} u_1(k) \\ u_2(k) \\ u_3(k) \end{bmatrix}
$$

$$
= Ax(k) + Bu(k),
$$

$$
\begin{bmatrix} y_1(k) \\ y_2(k) \end{bmatrix} = \begin{bmatrix} 2 & 0 \\ 1 & -1 \end{bmatrix} \begin{bmatrix} x_1(k) \\ x_2(k) \end{bmatrix} + \begin{bmatrix} 0 & 4 & 0 \\ 0 & 0 & -2 \end{bmatrix} \begin{bmatrix} u_1(k) \\ u_2(k) \\ u_3(k) \end{bmatrix}
$$

$$
= Cx(k) + Du(k).
$$

This system has the characteristic equation

$$
|zI - A| = \begin{vmatrix} (z-2) & 5 \\ -\frac{1}{2} & (z+1) \end{vmatrix} = z^2 - z + \frac{1}{2}
$$

$$
= \left(z - \frac{1}{2} - j\frac{1}{2} \right) \left(z - \frac{1}{2} + j\frac{1}{2} \right) = 0.
$$

All its six transfer functions share the poles

$$
z_1 = \frac{1}{2} + j\frac{1}{2} \qquad z_2 = \frac{1}{2} - j\frac{1}{2}.
$$

The transfer function matrix for this system, which is stable, is given by

$$
T(z) = C(zI - A)^{-1}B + D = \begin{bmatrix} 2 & 0 \\ 1 & -1 \end{bmatrix} \begin{bmatrix} (z-2) & 5 \\ -\frac{1}{2} & (z+1) \end{bmatrix}^{-1} \begin{bmatrix} 1 & -2 & 0 \\ 0 & 1 & 3 \end{bmatrix} + \begin{bmatrix} 0 & 4 & 0 \\ 0 & 0 & -2 \end{bmatrix}
$$

$$
= \begin{bmatrix} \dfrac{2z+2}{z^2 - z + \frac{1}{2}} & 4 + \dfrac{-4z - 14}{z^2 - z + \frac{1}{2}} & \dfrac{-30}{z^2 - z + \frac{1}{2}} \\[4mm] \dfrac{z + \frac{1}{2}}{z^2 - z + \frac{1}{2}} & \dfrac{-3z - 4}{z^2 - z + \frac{1}{2}} & -2 + \dfrac{-3z - 9}{z^2 - z + \frac{1}{2}} \end{bmatrix}.
$$

Linear step-invariant discrete-time causal systems must have transfer functions with numerator polynomials of an order less than or equal to that of the denominator polynomials. Only causal systems can be represented by the standard state-variable models.

10.1.4.4 State Equations and Response of Step-Varying Systems

A linear step-varying discrete-time system has a state equation of the form

$$
x(k+1) = A(k)x(k) + B(k)u(k),
$$
$$
y(k) = C(k)x(k) + D(k)u(k).
$$

In terms of the initial state $x(0)$ and the inputs $u(k)$ at step zero and beyond, the solution for the state after step zero is

$$
x(k) = \Phi(k, 0)x(0) + \sum_{i=0}^{k-1} \Phi(k, i+1)B(i)u(i),
$$

where the state transition matrices, $\Phi(\cdot, \cdot)$, are the $n \times n$ products of state coupling matrices

$$\Phi(k, j) = A(k-1)A(k-2)\cdots A(j-)A(j) \; k > j,$$
$$\Phi(i, i) = I,$$

where I is the $n \times n$ identity matrix.

A linear step-varying discrete-time system of the form

$$x(k+1) = A(k)x(k) + B(k)u(k)$$

is said to be *zero-input stable* if, and only if, for every set of initial conditions $x_{\text{zero-input}}(0)$, the zero-input component of the state, governed by

$$x_{\text{zero-input}}(k+1) = A(k)x_{\text{zero-input}}(k)$$

approaches zero with step. That is,

$$\lim_{k \to \infty} ||x_{\text{zero-input}}(k)|| = 0,$$

where the symbol $||.||$ denotes the Euclidean norm of the quantity.

The system is *zero-state stable* if, and only if, for zero initial conditions and every bounded input

$$||u(k)|| < \delta \quad k = 0, 1, 2, \ldots$$

the zero-state component of the state, governed by

$$\begin{cases} x_{\text{zero-state}}(k+1) = A(k)x_{\text{zero-state}}(k) + B(k)u(k) \\ x_{\text{zero-state}}(0) = 0 \end{cases}$$

is bounded

$$||x_{\text{zero-state}}(k)|| < \sigma \quad k = 0, 1, 2, \ldots$$

10.1.4.5 Change of Variables

A nonsingular change of state variables

$$x(k) = P\bar{x}(k) \quad \bar{x}(k) = P^{-1}x(k)$$

in discrete-time state-variable equations

$$x(k+1) = Ax(k) + Bu(k),$$
$$y(k) = Cx(k) + Du(k)$$

gives new equations of the same form

$$\bar{x}(k+1) = (P^{-1}AP)\bar{x}(k) + (P^{-1}B)u(k) = \bar{A}\bar{x}(k) + \bar{B}u(k),$$
$$y(k) = (CP)\bar{x}(k) + Du(k) = \bar{C}\bar{x}(k) + Du(k).$$

The system transfer function matrix is unchanged by a nonsingular change of state variables

$$\begin{aligned} \bar{T}(z) &= \bar{C}(zI - \bar{A})^{-1}\bar{B} + D \\ &= CP(zP^{-1}P - P^{-1}AP)^{-1}P^{-1}B + D \\ &= CP[P^{-1}(zI - A)P]^{-1}P^{-1}B + D \\ &= CPP^{-1}(zI - A)^{-1}PP^{-1}B + D \\ &= C(zI - A)^{-1}B + D = T(z). \end{aligned}$$

Each different set of state-variable equations having the same z-transfer function matrix is termed a *realization* of the z-transfer functions. The transformation

$$\bar{A} = P^{-1}AP$$

is called a *similarity transformation*. The transformation matrix P can be selected to take the system to a convenient realization such as the *controllable form, observable form, diagonal form, block Jordan form*, etc. Using these forms, it is especially easy to synthesize systems having desired transfer functions. For example, if the eigenvalues of the A matrix are distinct, there exists a nonsingular matrix P such that the state coupling matrix A of the new system

$$\bar{x}(k+1) = \bar{A}\bar{x}(k) + \bar{B}u(k),$$
$$y(k) = \bar{C}\bar{x}(k) + \bar{D}u(k)$$

is diagonal with the eigenvalues of A as the diagonal elements. The new state equations are decoupled from one another, and each equation involves only one state variable. In this example, the columns of the P matrix are the eigenvectors of the A matrix. It should be noted that taking a system from one realization to another may not always be possible. This depends on the characteristics of the system.

10.1.4.6 Controllability and Observability

A discrete-time system is said to be *completely controllable* if, by knowing the system model and its state $x(k)$ at any specific step k, a control input sequence $u(k), u(k+1), \ldots, u(k+i-1)$ can be determined that it will take the system to any desired later state x in a finite number of steps. For a step-invariant system, if it is possible to move the state at any step, say $x(0)$, to an arbitrary state at a later step, then it is possible to move it to an arbitrary desired state starting with any beginning step.

For an nth-order step-invariant system with r inputs

$$x(k+1) = Ax(k) + Bu(k),$$
$$y(k) = Cx(k) + Du(k)$$

and a desired state δ, the system state at step n, in terms of the initial state $x(0)$ and the inputs, is

$$\delta = x(n) = A^n x(0) + \sum_{i=0}^{n-1} A^{n-1-i} Bu(i)$$

or

$$Bu(n-1) + ABu(n-2) + \cdots + A^{n-2}Bu(1)$$
$$+ A^{n-1}Bu(0) = \delta - A^n x(0),$$

where the terms on the right-hand side are known. These equations have a solution for the inputs $u(0), u(1), \ldots, u(n-1)$ if, and only if, the $n \times (rn)$ array of coefficients

$$M_c = [B|AB| \cdots |A^{n-2}B|A^{n-1}B],$$

called the *controllability matrix* of the system, is of full rank. Additional steps, giving additional equations with coefficients $A^n B, A^{n+1} B$, and so on, do not affect this result because, by the Cayley–Hamilton theorem, these equations are linearly dependent on the others.

For a multiple-input system, the smallest possible integer η for which the matrix

$$M_c(\eta) = [B|AB|A^2 B| \cdots |A^{\eta-1}B]$$

has full rank is called the *controllability index* of the system. It is the minimum number of steps needed to control the system state.

A discrete-time system is said to be *completely observable* if its state $x(k)$ at any specific step k can be determined from the system model and its inputs and measurement outputs for a finite number of steps. For a step-invariant system, if it is possible to determine the state at any step, $x(0)$, then with a shift of step, the state at any other step can be determined in the same way.

For an nth-order step-invariant system with m outputs,

$$x(k+1) = Ax(k) + Bu(k),$$
$$y(k) = Cx(k) + Du(k)$$

the initial state $x(0)$, in terms of the outputs and inputs, is given by

$$\begin{cases} y(0) = Cx(0) + Du(0) \\ y(1) = Cx(1) + Du(1) = CAx(0) \\ \qquad + CBu(0) + Du(1) \\ y(2) = Cx(2) + Du(2) \\ \qquad = CA^2x(0) + CABu(0) + CBu(1) + Du(2) \\ \vdots \\ y(n-1) = Cx(n-1) + Du(n-1) \\ \qquad = CA^{n-1}x(0) + CA^{n-2}Bu(0) + CA^{n-3}Bu(1) + \cdots \\ \qquad + CBu(n-2) + Du(n-1) \end{cases}$$

Collecting the $x(0)$ terms on the left

$$\begin{cases} Cx(0) = y(0) - Du(0) \\ CAx(0) = y(1) - CBu(0) - Du(1) \\ CA^2x(0) = y(2) - CABu(0) - CBu(1) - Du(2) \\ \vdots \\ CA^{n-1}x(0) = y(n-1) - CA^{n-2}Bu(0) - \cdots \\ \qquad - CBu(n-2) - Du(n-1), \end{cases}$$

where the terms on the right-hand side are known and $x(0)$ is unknown. This set of linear algebraic equations can be solved for $x(0)$ only if the array of coefficients

$$M_0 = \begin{bmatrix} C \\ \hline CA \\ \hline CA^2 \\ \vdots \\ \hline CA^{n-1} \end{bmatrix}$$

is of full rank. Additional outputs are of no help, because they give additional equations with coefficients $CA^n, CA^{n+1}, \ldots$, which are linearly dependent on the others.

For a multiple-output system, the smallest possible integer ν for which

$$M_0 = \begin{bmatrix} C \\ \hline CA \\ \hline CA^2 \\ \vdots \\ \hline CA^{\nu-1} \end{bmatrix}$$

has full rank is called the *observability index* of the system. It is the minimum number of steps needed to determine the system state.

The replacements

$$\begin{cases} A \to A^\dagger \\ B \to C^\dagger \\ C \to B^\dagger, \end{cases}$$

where † denotes matrix transposition, creates a system with a controllability matrix that is the observability matrix of the original system and an observability matrix that is the controllability matrix of the original system. Every controllability result has a corresponding observability result and vice versa, a concept termed *duality*.

Bibliography

1. Santina, M.S., Stubberud, A.R., and Hostetter, G.H., *Digital Control System Design*, 2nd ed. Oxford University Press Inc., New York, NY, 1994.
2. DiStefano, J.J., III, Stubberud, A.R., and Williams, I.J., *Feedback and Control Systems (Schaum's Outline)*, 2nd ed., McGraw-Hill, New York, NY, 1994.
3. Jury, E.I., *Theory and Application of the z-Transform Method,* John Wiley & Sons, New York, NY, 1964.
4. Oppenheim, A.V. et al, *Signals and Systems*, 2nd ed., Prentice-Hall, Englewood Cliffs, NJ, 1996.
5. Proakis, J.G., et al. *Digital Signal Processing*, 4th ed., Prentice-Hall, Englewood Cliffs, NJ, 2006.
6. Papoulis, A., *Circuits and Systems: A Modern Approach,* Saunders College Publishing, Philadelphia, PA, 1980.
7. Chen, C.T., *System and Signal Analysis*, 3rd ed., Oxford University Press Inc., New York, NY, 2004.
8. Lathi, B.P., *Linear Systems and Signals*, Oxford University Press Inc., New York, NY, 2005.
9. Brogan, W.L. *Modern Control Theory*, 3rd ed., Prentice-Hall, Englewood Cliffs, NJ, 1991.
10. Kailath, T., *Linear Systems,* Prentice-Hall, Englewood Cliffs, NJ, 1980.
11. DeRusso, P.M., et al., *State Variables for Engineers,* Wiley, New York, NY, 1998.
12. Shenoi, B.A., *Introduction to Digital Signal Processing and Filter Design*, Wiley-Interscience, Hoboken, NJ, 2005.
13. Mitra, S., *Digital Signal Processing Laboratory Using MATLAB*, McGraw-Hill, New York, NY, 2005.

11

Sampled-Data Systems

A. Feuer
Technion–Israel Institute of Technology

Graham C. Goodwin
The University of Newcastle

11.1 Introduction and Mathematical Preliminaries

The advances in digital computer technology have led to its application in a very wide variety of areas. In particular, it has been used to replace the analog controller in many control systems. However, to use the digital computer as a controller one has to overcome the following problem: The input and output signals of the physical plant are analog, namely, continuous-time signals, and the digital computer can only accept and generate sequences of numbers, namely, discrete-time signals (we do not discuss here the quantization problem).

This problem was solved by developing two types of interfacing units: a sampler, which transforms the analog signal to a discrete one, and a hold unit, which transforms the discrete signal to an analog one. A typical configuration of the resulting system, Figure 11.1, shows that the control system contains continuous-time signals and discrete-time signals (drawn in full and broken lines, respectively).

Because mathematical tools were available for analyzing systems with either continuous-time signals or discrete-time signals, the approach to controller design evolved accordingly. One approach is to design a continuous-time controller and then approximate it by a discrete-time system. Another approach is to look at the system, from the input to the hold unit to the output of the sampler, as a discrete-time system. Then, use discrete-time control design methods (the development of which was prompted by that approach) to design the controller.

Both approaches are acceptable when the sampling is fast enough. However, in many applications the sampling rate may be constrained and as a result the approaches above may prove inappropriate. This realization prompted many researchers to develop tools to analyze systems containing both continuous- and discrete-time signals, referred to as sampled-data systems.

The purpose of this chapter is to introduce one such tool which is based on frequency-domain considerations and seems to be a very natural approach. Because we will heavily rely on Fourier transforms (FT), it will be helpful to review some of the definitions and properties.

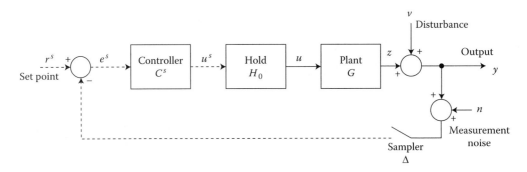

FIGURE 11.1 Configuration of a typical sampled data control system.

11.1.1 Fourier Transform (FT)

$$X(\omega) = \mathcal{F}\{x(t)\}$$
$$= \int_{-\infty}^{\infty} x(t)e^{-j\omega t}\, dt, \tag{11.1}$$

$x(t)$, a continuous-time signal, ω, angular frequency (in rad/s).

11.1.2 Inverse Fourier Transform (IFT)

$$x(t) = \frac{1}{2\pi} \int_{-\infty}^{\infty} X(\omega)e^{-j\omega t}\, d\omega. \tag{11.2}$$

11.1.3 Discrete-Time Fourier Transform (DTFT)

$$X^s(\omega) = \Delta \sum_{k=-\infty}^{\infty} x[k]e^{-j\omega \Delta k}, \tag{11.3}$$

$x[k]$, a discrete-time signal obtained by sampling $x(t)$ at the instants $t = k\Delta \quad k = 0, 1, 2, \ldots$, Δ, the associated sampling time interval.

11.1.4 Inverse Discrete-Time Fourier Transform (IDTFT)

$$x[k] = \frac{1}{2\pi} \int_{-\pi/\Delta}^{\pi/\Delta} X^s(\omega)e^{j-\omega \Delta k}\, d\omega. \tag{11.4}$$

Given that $x[k]$ are samples of $x(t)$, namely,

$$x[k] = x(k\Delta), \tag{11.5}$$

$$X^s(\omega) = \sum_{k=-\infty}^{\infty} X\left(\omega - k\frac{2\pi}{\Delta}\right) \tag{11.6}$$

(see Chapter 15 in this Handbook). $X^s(\omega)$ results from "folding" $X(\omega)$ every $2\pi/\Delta$ and repeating it periodically (this is sometimes referred to as "aliasing").

We adopt the following notation

$$[X]^s \triangleq \sum_{k=-\infty}^{\infty} X\left(\omega - k\frac{2\pi}{\Delta}\right) \tag{11.7}$$

and readily observe the following properties:

$$[X_1 + X_2]^s = [X_1]^s + [X_2]^s \tag{11.8}$$

$$[C^s X]^s = C^s [X]^s \tag{11.9}$$

where C^s is the frequency response of a discrete-time system.

We should also point out that the most commonly used hold unit is the zero order hold (ZOH) given by

$$H_o(\omega) = \frac{1 - e^{-j\omega\Delta}}{j\omega\Delta} \tag{11.10}$$

11.2 "Sensitivity Functions" in Sampled-Data Control Systems

It is well known that sensitivity functions play a key role in control design, be it a continuous-time controller for a continuous-time system or a discrete-time controller for a discrete-time system. Let us start our discussion with a brief review of commonly known facts about sensitivity functions. Consider the system in Figure 11.2.

Denoting by capital letters the Fourier transforms of their lower-case counterparts in the time domain, from Figure 11.2:

$$Y = TR + SV - TN, \tag{11.11}$$

where

$$S = \frac{1}{1 + GC} \tag{11.12}$$

is the sensitivity function, and

$$T = \frac{GC}{1 + GC} \tag{11.13}$$

is the complementary sensitivity function as well as the closed-loop transfer function. Here are some facts regarding these functions:

1. Let ΔG denote the change in the open-loop transfer function G, and let ΔT denote the corresponding change in the closed-loop transfer function. Then

$$\frac{\Delta T}{T} \simeq S\frac{\Delta G}{G}. \tag{11.14}$$

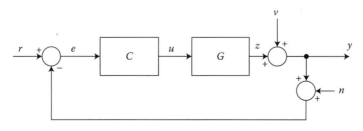

FIGURE 11.2   Control system with uniform type of signals (either all continuous-time or all discrete-time).

2. Clearly,

$$S + T = 1. \tag{11.15}$$

3. The zeros of T are the open-loop zeros. Hence, if z_o is an open-loop zero, $S(z_o) = 1$.
4. The zeros of S are the open-loop poles. Hence if p_o is an open-loop pole, $T(p_o) = 1$.
5. $|T|$ is usually made to approach 1 at low frequencies to give zero steady-state errors at d.c.
6. $|T|$ is usually made small at high frequencies to give insensitivity to high-frequency noise n (this implies $|S| \approx 1$ at high frequencies).
7. Because of (3) and (5), to avoid peaks in $|T|$ it is desirable that, moving from low to high frequencies, we meet a closed-loop pole before we meet each open-loop zero.
8. Because of (4) and (6), to avoid large peaks in $|S|$ it is desirable that, moving from high to low frequencies, we meet a closed-loop pole before we meet each open-loop pole.
9. For stable, well-damped open-loop poles and zeros, (7) and (8) can easily be achieved by cancellation. However, open-loop unstable poles and zeros place fundamental limitation on the desirable closed-loop bandwidth.

   In particular, the following bandwidth limitations are necessary to avoid peaks in either S or T:

$$\left. \begin{array}{l} \text{bandwidth} < \text{open-loop unstable zeros} \\ \text{bandwidth} > \text{open-loop unstable poles} \end{array} \right\} \tag{11.16}$$

With the above in mind, one may adopt the approach mentioned in Section 11.1. View the system of Figure 11.1 as a discrete-time system by looking at the discrete-time equivalent of the ZOH, the plant, and the sampler. Then, using the above, one gets a discrete-time system for which a discrete-time controller can be designed. The problem is that this approach guarantees desired behavior only *at the sampled outputs*. However, there is no *a priori* reason to presume that the response between samples would not deviate significantly from what is observed at the sample points. Indeed we shall see later that it is quite possible for the intersample response to be markedly different from the sampled response. It is then clear that the sensitivity functions calculated for the discrete equivalents are unsatisfactory tools for the sampled data system. In the following, we will develop equivalent functions for the sampled-data system.

Let us again consider the system in Figure 11.1. We have the following key result describing the *continuous output* response under the digital control law:

Theorem 11.1:

Subject to closed-loop stability, the Fourier transform of the continuous-time output of the single-input, single-output system in Figure 11.1 is given by

$$Y(\omega) = P(\omega)(R^s(\omega) - N^s(\omega)) + D(\omega)V(\omega) - P(\omega) \sum_{\substack{k=-\infty \\ k \neq 0}}^{\infty} V\left(\omega - k\frac{2\pi}{\Delta}\right),$$

$$= P(\omega)(R^s(\omega) - N^s(\omega)) + V(\omega) - P(\omega)[V]^s. \tag{11.17}$$

$P(\omega)$ *and* $D(\omega)$ *are frequency response functions given, respectively, by*

$$P(\omega) \triangleq C^s(\omega)G(\omega)H_o(\omega)S^s(\omega) \tag{11.18}$$

and

$$D(\omega) = 1 - P(\omega) \tag{11.19}$$

where $S^s(\omega)$ is the usual discrete sensitivity calculated for the discrete equivalent system given by

$$S^s(\omega) \triangleq \frac{1}{1 + C^s(\omega)[GH_o]^s}. \tag{11.20}$$

Note that $[GH_o]^s$ is the frequency response of the discrete equivalent of GH_o.

Proof. Observing Figure 11.1 we have (using Equation 11.6)

$$
\begin{aligned}
Y^s(\omega) &= \sum_{k=-\infty}^{\infty} Y\left(\omega - k\frac{2\pi}{\Delta}\right), \\
&= \sum_{k=-\infty}^{\infty} \left(Z\left(\omega - k\frac{2\pi}{\Delta}\right) + V\left(\omega - k\frac{2\pi}{\Delta}\right)\right), \\
&= \sum_{k=-\infty}^{\infty} G\left(\omega - k\frac{2\pi}{\Delta}\right) H_o\left(\omega - k\frac{2\pi}{\Delta}\right) C^s\left(\omega - k\frac{2\pi}{\Delta}\right)\left(R^s\left(\omega - k\frac{2\pi}{\Delta}\right)\right. \\
&\quad \left. - Y^s\left(\omega - k\frac{2\pi}{\Delta}\right) - N^s\left(\omega - k\frac{2\pi}{\Delta}\right)\right) + V\left(\omega - k\frac{2\pi}{\Delta}\right).
\end{aligned}
$$

Because, C^s, R^s, Y^s, and N^s are periodic functions of ω,

$$
\begin{aligned}
Y^s(\omega) &= C^s(\omega) \sum_{k=-\infty}^{\infty} G\left(\omega - k\frac{2\pi}{\Delta}\right) H_o\left(\omega - k\frac{2\pi}{\Delta}\right)\left[R^s(\omega) - Y^s(\omega) - N^s(\omega)\right] + V^s(\omega) \\
&= C^s(\omega)\left[GH_o\right]^s\left[R^s(\omega) - Y^s(\omega) - N^s(\omega)\right] + V^s(\omega).
\end{aligned}
$$

Hence,

$$Y^s(\omega) = [1 - S^s(\omega)][R^s(\omega) - N^s(\omega)] + S^s(\omega)V^s(\omega). \tag{11.21}$$

Now, from Figure 11.1,

$$Y(\omega) = G(\omega)H_o(\omega)C^s(\omega)(R^s(\omega) - Y^s(\omega) - N^s(\omega)) + V(\omega)$$

and substituting Equation 11.21 results in

$$Y(\omega) = G(\omega)H_o(\omega)C^s(\omega)S^s(\omega)[R^s(\omega) - N^s(\omega)] - G(\omega)H_o(\omega)C^s(\omega)S^s(\omega)V^s(\omega) + V(\omega)$$

which, by substituting Equations 11.18 and 11.19, leads to Equation 11.17. ☐

Comparing Equations 11.17 and 11.11, the roles that D and P play in a sampled-data system are very similar to the roles that sensitivity and complementary sensitivity play in Figure 11.2.

We also note that the functions $P(\omega)$ and $D(\omega)$ allow computing the continuous output in the frequency domain using the input $R^s(\omega)$, the disturbance $V(\omega)$, and the noise $N^s(\omega)$. We will thus refer to $P(\omega)$ and $D(\omega)$ as the *reference* and *disturbance gain functions*, respectively. Observe that the infinite sum defining $[GH_o]^s$ is convergent provided the transfer function GH_o is strictly proper.

The result in Theorem 11.1 holds for general reference, noise, and disturbance inputs. However, it is insightful to consider the special case of sinusoidal signals. In this case, the result simplifies as follows. Let

$r[k]$ be a sampled sinewave

$$r[k] = \cos(\omega_o k \Delta)$$

Then

$$R^s(\omega) = \pi \sum_{k=-\infty}^{\infty} \left[\delta \left(\omega - \omega_o - k\frac{2\pi}{\Delta} \right) + \delta \left(\omega + \omega_o - k\frac{2\pi}{\Delta} \right) \right] \quad (11.22)$$

Hence, using Equation 11.17 (assuming $V(\omega) = N^s(\omega) = 0$),

$$Y(\omega) = \pi \sum_{k=-\infty}^{\infty} P(\omega) \left[\delta \left(\omega - \omega_o - k\frac{2\pi}{\Delta} \right) + \delta \left(\omega + \omega_o - k\frac{2\pi}{\Delta} \right) \right] \quad (11.23)$$

Thus, the continuous-time output in this case is *multifrequency* with corresponding magnitudes and phases determined by the *reference gain function* $P(\omega)$. In particular, for a sinusoidal reference signal as above, where $0 < \omega_o < \pi/\Delta$, the first two components in the output are at frequencies ω_o and $2\pi/\Delta - \omega_o$ and have amplitudes $|P(\omega)|$ and $|P(2\pi/\Delta - \omega_o)|$, respectively. Similar observations can be made for $N^s(\omega)$ and $V(\omega)$.

In the next section we will show that the connections of P and D with T^s and S^s go beyond the apparent similarity in roles.

11.3 Sensitivity Consideration

In the previous section we found that the reference gain function $P(\omega)$ and the disturbance gain function $D(\omega)$ allow computing the *continuous-time* output response in a sampled-data system, namely, a *digital* controller in a closed loop, with a *continuous-time* plant. We recall the definitions for convenience

$$P(\omega) = C^s(\omega)G(\omega)H_o(\omega)S^s(\omega) \quad (11.24)$$

$$D(\omega) = 1 - P(\omega) \quad (11.25)$$

$$S^s(\omega) = \frac{1}{1 + C^s(\omega)[GH_o]^s} \quad (11.26)$$

$$T^s(\omega) = \frac{C^s(\omega)[GH_o]^s}{1 + C^s(\omega)[GH_o]^s} \quad (11.27)$$

where C^s, G, and H_o are the frequency responses of the controller, plant, and ZOH, respectively.

First we note that, as in Equation 11.15 for S^s and T^s,

$$P + D = 1 \quad (11.28)$$

Next it is interesting to note that the *open-loop continuous-time* zeros of the plant appear as zeros of $P(\omega)$. Thus, *irrespective* of any *discrete-time consideration*, the locations of the continuous-time plant zeros are of concern because they affect the continuous-time output responses. Specifically, the existence of a nonminimum phase zero in the plant results in performance constraints for any type of controller, continuous time or discrete time.

The magnitude of the ratio between the P and T^s at frequency ω_o is

$$\left| \frac{P(\omega_o)}{T^s(\omega_o)} \right| = \left| \frac{G(\omega_o)H_o(\omega_o)}{[GH_o]^s} \right| \quad (11.29)$$

Hence, to avoid large peaks in $|P(\omega)|$, $|T^s(\omega)|$ must not be near 1 at any frequency where the gain of the composite continuous-time transfer function GH_o is significantly greater than the gain of its discrete-time equivalent $[GH_o]^s$. Otherwise, as Equation 11.28 indicates, a large $|P(\omega)|$ and a large $|D(\omega)|$ result, showing large sensitivity to disturbances. There are several reasons why the gain of GH_o might be significantly greater than that of $[GH_o]^s$. Two common reasons are

1. For continuous plants having a relative degree exceeding one, there is usually a discrete zero near the point $z = -1$. Thus, the gain of $[GH_o]^s$ typically falls near $\omega = \pi/\Delta$ (i.e., the folding frequency). Hence, it is rarely desirable to have a discrete closed-loop bandwidth approaching the folding frequency as will be demonstrated in later examples.
2. Sometimes high-frequency resonances can perturb the discrete transfer function away from the continuous-time transfer function by folding effects leading to differences between GH_o and $[GH_o]^s$.

One needs to be careful about the effect these factors have on the differences between P and T^s. In particular, the bandwidth must be kept well below any frequency where folding effects reduce $[GH_o]^s$ relative to the continuous plant transfer function. This will also be illustrated in the examples presented later.

Finally, we look at the sensitivity of the closed-loop system to changes in the open-loop plant transfer function.

Recall, using Equations 11.15 and 11.26 that

$$T^s(\omega) = \frac{C^s(\omega)[GH_o]^s}{1 + C^s(\omega)[GH_o]^s} \tag{11.30}$$

and

$$P(\omega) = C^s(\omega)G(\omega)H_o(\omega)S^s(\omega). \tag{11.31}$$

Clearly,

$$T^s(\omega) = [P]^s. \tag{11.32}$$

Furthermore, we have the following result which extends Equation 11.14 to the case of mixed continuous and discrete signals:

Lemma 11.1:

The relative changes in the reference gain function and the closed-loop discrete-time transfer function are

$$\frac{\Delta P}{P} = D\frac{\Delta G}{G} - \sum_{\substack{k=-\infty \\ k \neq 0}}^{\infty} P\left(\omega - k\frac{2\pi}{\Delta}\right) \frac{\Delta G(\omega - k\frac{2\pi}{\Delta})}{G(\omega - k\frac{2\pi}{\Delta})} \tag{11.33}$$

and

$$\frac{\Delta T^s}{T^s} = \frac{S^s(\omega)}{[GH_o]^s} \sum_{k=-\infty}^{\infty} G\left(\omega - k\frac{2\pi}{\Delta}\right) \cdot H\left(\omega - k\frac{2\pi}{\Delta}\right) \frac{\Delta G(\omega - k\frac{2\pi}{\Delta})}{G(\omega - k\frac{2\pi}{\Delta})}. \tag{11.34}$$

Proof. By differentiating Equations 11.30 and 11.31 with respect to G, we see that, up to first order,

$$\Delta P \simeq \frac{C^s(\omega)H_o(\omega)}{\left(1 + C^s(\omega)[GH_o]^s\right)^2} \left[(1 + C^s(\omega)[H_oG]^s)\Delta G(\omega) - G(\omega)C^s(\omega)[H_o\Delta G]^s\right]$$

$$\Delta T^s \simeq \frac{C^s(\omega)}{\left(1 + C^s(\omega)[GH_o]^s\right)^2} [H_o\Delta G]^s.$$

Dividing by P and T^s, respectively, and recalling Equation 11.7 leads to Equations 11.33 and 11.34. □

Again note the similarity of the roles of P and D in Equation 11.33 to those of T and S in Equation 11.14. Typically, the magnitude of $(\Delta G/G)$ approaches unity at high frequencies. Hence, Equations 11.33 and 11.34 show that a sensitivity problem will exist even for the discrete-time transfer function, if $[GH_o]^s$ is small at a frequency where GH_o is large, unless, of course, S^s is small at the same frequency. This further reinforces the claim that the bandwidth must be kept well below any frequencies where folding effects reduce $[GH_o]^s$ relative to the continuous plant transfer function GH_o.

11.4 Examples

In this section, we present some simple examples which illustrate the application of Theorem 11.1 in computing the intersample behavior of continuous-time systems under the action of sampled data control. The examples are not intended to illustrate good sampled data control design but have been chosen to show the utility of the functions $P(\omega)$ and $D(\omega)$ in giving correct qualitative and quantitative understanding of the continuous time response in a difficult situation (when it is significantly different from that indicated by the sampled response).

For each example we give G, Δ and the desired T^s. Then we show the resulting functions, P, D, and S^s and make comparisons enabling us to predict the effects of various test signals, the results of which will also be shown.

Example 11.1:

$$G(\omega) = \frac{1}{j\omega(j\omega + 1)} \quad \Delta = 0.4 \text{ s}$$

C^s is chosen so that

$$T^s(\omega) = e^{-0.4j\omega}.$$

Figure 11.3 shows $|G(\omega)H_0(\omega)|$ and $|[GH_0]^s|$. Over the range $(0, \pi/\Delta)$ the two are very nearly equal. Only near $\omega = \pi/\Delta = 7.85$ is there "some" discrepancy due to the sampling zero of $[GH_0]^s$. Figure 11.4 shows $|T^s|$ and $|P|$. Although T^s seems ideal, the graph for P indicates "trouble" around $\omega = \pi/\Delta$ rad/s. The peak in P results from the discrepancy in Figure 11.3 around the same frequency. Similar *trouble* is, naturally, observed in Figure 11.5 from the graph of D. These peaks indicate that a large continuous-time response is to be expected whenever reference input, noise, or disturbance have frequency content around that frequency. This is demonstrated in Figure 11.6 for a step reference input and in Figure 11.7 for a sinusoidal disturbance of frequency $3/\Delta$ rad/s and unit amplitude. The exact expressions for both responses can be derived from Equation 11.17. In particular, as marked in Figures 11.4 and 11.5 the disturbance response of the two dominant frequencies will be $3/\Delta$ with amplitude $|D(3/\Delta)| \approx 4$ and $(2\pi - 3)/\Delta$ with amplitude $\left| P\left(\frac{2\pi-3}{\Delta}\right) \right| \approx 2.4$. Adding the two, with the appropriate phase shift, will give the signal in Figure 11.7. Note that the sampled responses for both cases (marked in both figures) are very misleading.

Example 11.2:

The plant and Δ are the same as in Example 11.1. However, C^s is chosen so that

$$T^s(\omega) = \frac{B(\omega)}{B(0)} e^{-0.8j\omega}$$

where $B(\omega)$ is the numerator of $[GH_0]^s$ (this is a deadbeat control). In this case, both P and T^s and D and S^s are very close in the range $(0, \pi/\Delta)$ as in Figures 11.8 and 11.9. Predictably, the same test signal

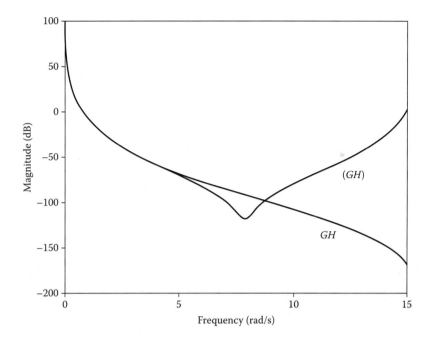

FIGURE 11.3   Continuous and discrete frequency response.

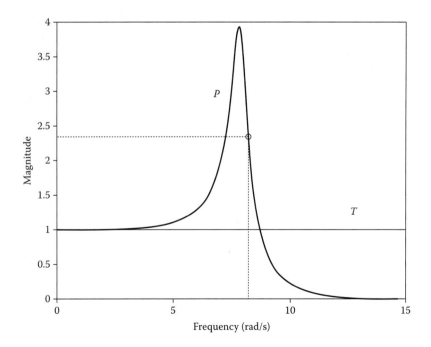

FIGURE 11.4   Complementary sensitivity and reference gain functions.

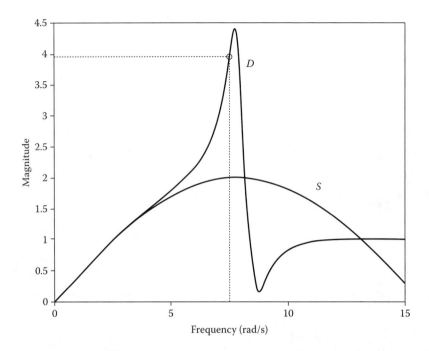

FIGURE 11.5   Sensitivity and disturbance gain functions.

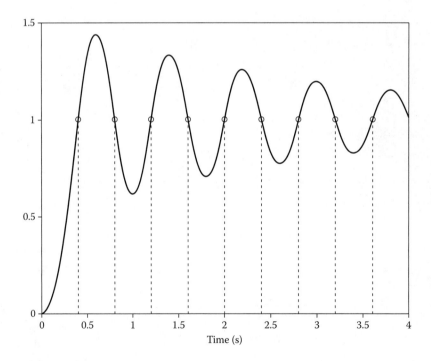

FIGURE 11.6   Step response.

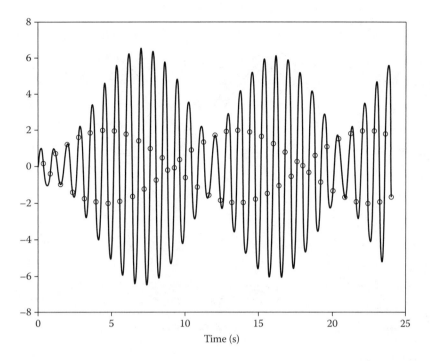

FIGURE 11.7 Disturbance response.

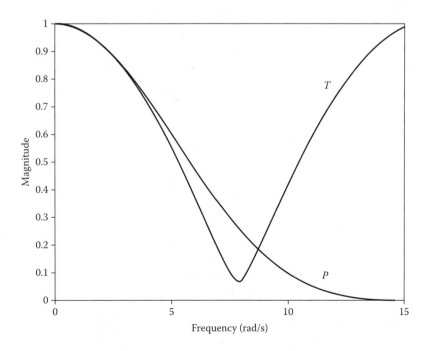

FIGURE 11.8 Complementary sensitivity and reference gain functions.

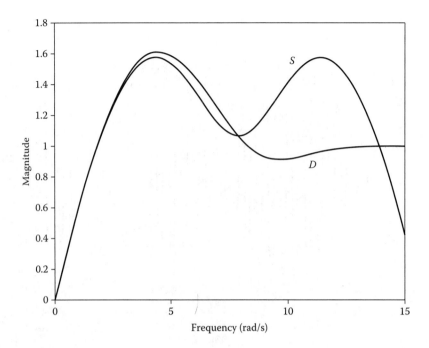

FIGURE 11.9 Sensitivity and disturbance gain functions.

as in Example 11.1 produces a sampled response more indicative of the continuous-time response. This is clearly noted in Figures 11.10 and 11.11.

Examples 11.1 and 11.2 represent two controller designs; the design in Example 11.2 is more along the recommendations in the previous section with clearly superior *continuous-time* performance.

Example 11.3:

$$G(\omega) = \frac{100}{(j\omega)^2 + 2j\omega + 100} \quad \Delta = 0.5 \text{ s}$$

C^s is chosen so that

$$T^s(\omega) = \frac{0.5}{e^{0.5j\omega} - 0.5}.$$

Figure 11.12 compares the $|[GH_o]^s|$ and $|GH_o|$. The resonant peak has been folded into the low-frequency range in the discrete-frequency response. Figure 11.13 shows $|P|$ and $|T^s|$. $P(\omega)$ has a significant peak at $\omega = \pi/\Delta$, reflected in the intersample behavior of the step response in Figure 11.15. However, the sampled response in Figure 11.15 is a simple exponential as could be predicted from T^s in Figure 11.13.

Significant differences in S^s and D can also be observed in Figure 11.14. When a sinusoidal disturbance of frequency $1/\Delta$ rad/s was applied, we observe a multifrequency continuous-time response, but the sampled response is a single sinusoid of frequency $1/\Delta$ shown in Figure 11.16. This result can again be predicted from Equation 11.17 and Figures 11.13 and 11.14.

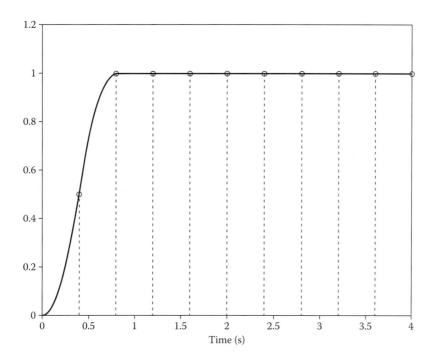

FIGURE 11.10 Step response.

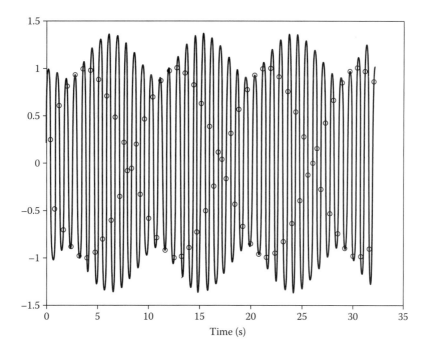

FIGURE 11.11 Disturbance response.

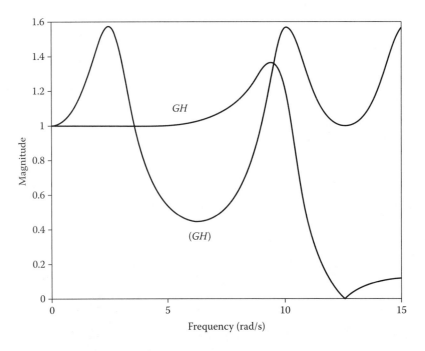

FIGURE 11.12 Continuous and discrete-frequency responses.

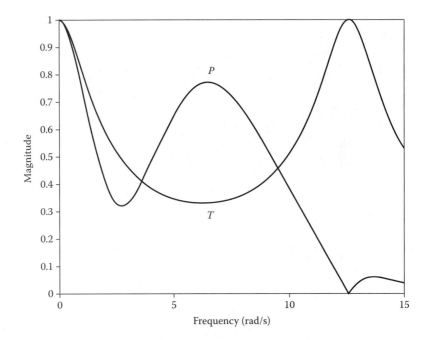

FIGURE 11.13 Complementary sensitivity and reference gain functions.

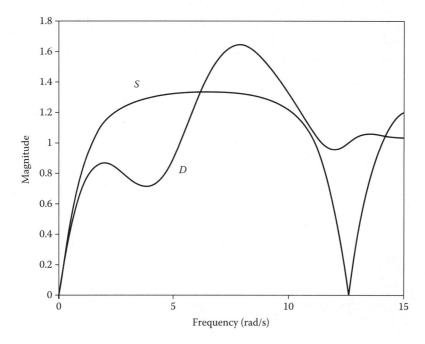

FIGURE 11.14 Sensitivity and disturbance gain functions.

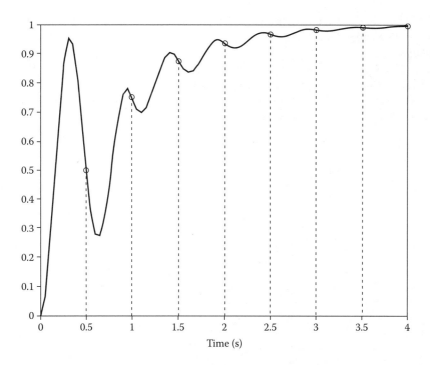

FIGURE 11.15 Step response.

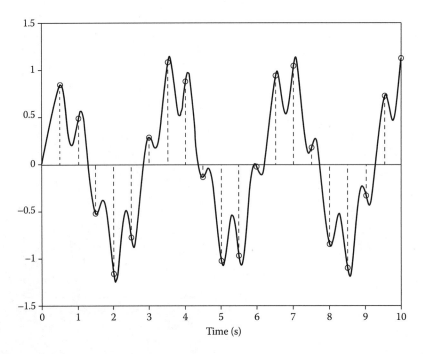

FIGURE 11.16   Disturbance response.

11.4.1 Observations and Comments from Examples

1. The reference and disturbance gain functions, $P(\omega)$ and $D(\omega)$, give qualitative and quantitative information about the true continuous-time response resulting from reference, noise, or disturbance input.
2. In many cases, the first two components in the multifrequency output response suffice for an accurate qualitative description of the response to a sinewave disturbance input.
3. For a sinewave disturbance of frequency $\omega_o \in [0, \pi/\Delta)$, the continuous-time response will (if we consider only the first two frequency components).

12

Discrete-Time Equivalents of Continuous-Time Systems

Michael Santina
The Boeing Company

Allen R. Stubberud
University of California, Irvine

12.1 Introduction

Similar to continuous-time methods, digital controllers are usually designed using frequency-domain or time-domain methods so that the overall closed-loop system meets its transient and steady-state requirements. When a digital controller is designed to control a continuous-time plant it is important to have a good understanding of the plant to be controlled as well as of the controller and its interfaces with the plant. This chapter deals with digitizing continuous-time plants and continuous-time controllers. In the first approach, called the *direct method*, the plant is discretized first and then a discrete-time controller is designed directly to control the discretized plant. The second approach, called the *indirect method*, deals with discretizing the existing continuous-time controllers to be implemented digitally. Several discretization methods, using transfer function and state-space techniques, are presented to approximate the behavior of continuous-time plants and continuous-time controllers. In general, as the sampling period is reduced, the discrete-time response using these discretization methods will be nearly indistinguishable from that of the continuous-time counterpart. As we shall see later, reducing the sampling period may not be practical and may cause numerical problems. Simulation of the closed-loop system including the plant and the controller is essential to verify the properties of a preliminary design and to test its performance under conditions (e.g., noise, disturbances, delays, quantization, parameter variations, and nonlinearities) that might be difficult or cumbersome to study analytically. Through simulation, difficulties with between-sample plant response are also discovered and solved.

12.2 Design of Discrete-Time Control Systems for Continuous-Time plants

There are two fundamental approaches to designing discrete-time control systems for continuous-time plants. The first approach is to derive a discrete-time equivalent of the plant and then design a discrete-time controller directly to control the discretized plant. This approach is discussed in Section 12.2.3. The other and more traditional approach to designing discrete-time control systems for continuous-time plants is to first design a continuous-time controller for the plant, then derive a discrete-time equivalent that closely approximates the behavior of the original analog controller. This approach is especially useful when an existing continuous-time controller or a part of the controller is to be replaced with a discrete-time controller. Usually, however, even for small sampling periods, the discrete-time approximation performs less well than the continuous-time controller from which it was derived. The approach to deriving a discrete-time controller that closely approximates the behavior of the original analog controller is discussed in Section 12.3.

Before we discuss discrete-time equivalents of continuous-time systems, it is instructive to briefly discuss sampling and reconstruction in order to gain greater insight into the process of discretizing continuous-time systems.

12.2.1 Sampling and Analog-to-Digital Conversion

Sampling is the process of deriving a discrete-time sequence from a continuous-time function. As shown in Figure 12.1, an incoming continuous-time signal $f(t)$ is sampled by an analog-to-digital (A/D) converter to produce the discrete-time sequence $f(k)$. Usually, but not always, the samples are evenly spaced in time. The sampling interval T is generally known and is indicated on the diagram or elsewhere.

The A/D converter produces a binary representation, using a finite number of bits, of the applied input signal at each sample time. Using a finite number of bits to represent a signal sample generally results in quantization errors in the A/D process. For example, the maximum quantization error in 16-bit A/D conversion is $2^{-16} = 0.0015\%$, which is very low compared with typical errors in analog sensors. This

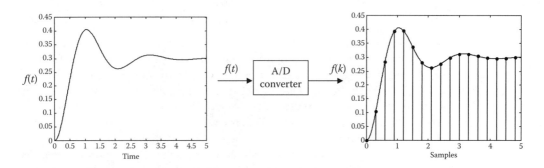

FIGURE 12.1 Sampling of a continuous-time signal using an A/D converter.

TABLE 12.1   Laplace and z-Transform Pairs

$f(t)$	$F(s)$	$f(k)$	$F(z)$
$u(t)$, unit step	$\dfrac{1}{s}$	$u(k)$, unit step	$\dfrac{z}{z-1}$
$tu(t)$	$\dfrac{1}{s^2}$	$kTu(k)$	$\dfrac{Tz}{(z-1)^2}$
$e^{-at}u(t)$	$\dfrac{1}{s+a}$	$(e^{-aT})^k u(k)$	$\dfrac{z}{z-e^{-aT}}$
$te^{-at}u(t)$	$\dfrac{1}{(s+a)^2}$	$kT(e^{-aT})^k u(k)$	$\dfrac{Tze^{-aT}}{(z-e^{-aT})^2}$
$\sin(\omega t)u(t)$	$\dfrac{\omega}{s^2+\omega^2}$	$\sin(k\omega T)u(k)$	$\dfrac{z\sin\omega T}{z^2-2z\cos\omega T+1}$
$\cos(\omega t)u(t)$	$\dfrac{s}{s^2+\omega^2}$	$\cos(k\omega T)u(k)$	$\dfrac{z(z-\cos\omega T)}{z^2-2z\cos\omega T+1}$

error, if taken to be "noise," gives a signal-to-noise (SNR) of $20\log_{10}(2^{-16}) = -96.3$ db which is much better than that of most control systems. The control system designer must ensure that enough bits are used to give the desired system accuracy. That is, it is important to use adequate word lengths in fixed or floating point computations. Years ago, digital hardware was very expensive, so minimizing word length was much more important than it is today. Study of the effects of roundoff or truncation errors in digital computation is presented in Chapter 14.

When a continuous-time signal $f(t)$ is sampled to form the sequence $f(k)$, there exists a relationship between the Laplace transform of $f(t)$ and the z-transform of $f(k)$. If a rational Laplace transform is expanded into partial fraction terms, the corresponding continuous-time signal components in the time domain are powers of time, exponentials, sinusoids, and so on. Uniform samples of these elementary signal components have, in turn, simple z-transforms that can be summed to give the z-transform of the entire sampled signal. Table 12.1 lists some Laplace transform terms and the resulting z-transforms when the corresponding time functions are sampled uniformly.

As an example, consider the continuous-time function with Laplace transform

$$F(s) = \frac{2}{s(s+2)} = \frac{1}{s} + \frac{-1}{s+2}$$

The z-transform of the sampled signal with a sampling interval $T = 0.1\,\text{s}$ is

$$F(z) = \frac{z}{z-1} - \frac{z}{z-e^{-0.2}} = \frac{0.18z}{(z-1)(z-0.82)}$$

12.2.2 Reconstruction and Digital-to-Analog Conversion

Reconstruction is the formation of a continuous-time function from a sequence of samples. Many different continuous-time functions can have the same set of samples; so a reconstruction is not unique. Reconstruction is performed using digital-to-analog (D/A) converters. Electronic D/A converters typically produce a step reconstruction from incoming signal samples by converting the binary-coded digital input to a voltage, transferring the voltage to the output, and holding the output voltage constant until the next sample is available. The symbol for a D/A converter that generates the step reconstruction $f^0(t)$ from signal samples $f(k)$ is shown in Figure 12.2a. Sample and hold (S/H) is the operation of holding each of these samples for a sampling interval T to form the step reconstruction. The step reconstruction of a continuous-time signal from samples can be represented as the conversion of the sequence $f(k)$ to its

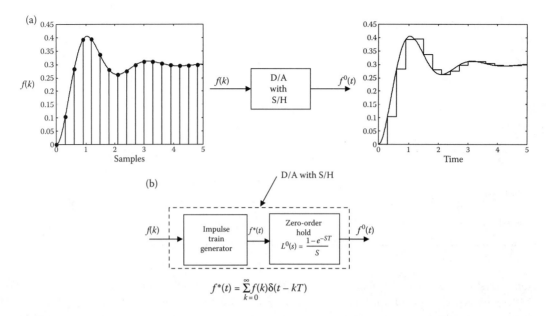

FIGURE 12.2 D/A conversion with S/H (a) Symbol for D/A conversion with S/H. (b) Representation of a D/A converter with S/H.

corresponding impulse train $f^*(t)$, where

$$f^*(t) = \sum_{k=0}^{\infty} f(k)\delta(t - kT), \tag{12.1}$$

and then conversion of the impulse train to the step reconstruction as shown in Figure 12.2b. This viewpoint neatly separates conversion of the discrete sequence to a continuous-time waveform and the details of the shape of the reconstructed waveform. The continuous-time transfer function that converts the impulse train with sampling interval T to a step reconstruction is termed zero-order-hold (ZOH). Each incoming impulse in Equation 12.1 to the ZOH produces a rectangular pulse of duration T. Therefore, the transfer function of the ZOH is given by

$$\text{Lo}(s) = \frac{1}{s}(1e^{-sT}).$$

One way to improve the accuracy of the reconstruction is to employ higher-order holds than the ZOH. An nth-order hold produces a piecewise nth-degree polynomial that passes through the most recent $n + 1$ input samples. It can be shown that, as the order of the hold is increased, a well-behaved signal is reconstructed with increased accuracy. For example, a first-order hold (FOH) uses the previous two samples to construct a straight-line approximation during each interval. The transfer function of the FOH is

$$L_1(s) = \frac{(Ts + 1)(1 - e^{-sT})^2}{Ts^2}.$$

A model of the FOH is shown in Figure 12.3a. If the hardware of the FOH is not available, one can implement an FOH as shown in Figure 12.3b.

12.2.3 Discrete-Time Equivalents of Continuous-Time Plants

The first approach to designing discrete-time control systems for continuous-time plants is to derive a discrete-time equivalent of the plant and then design a discrete-time controller directly to control the

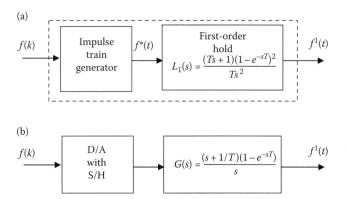

FIGURE 12.3 First-order hold reconstruction. (a) A model of first-order hold; (b) implementation of first order hold using ZOH.

discretized plant. Consider the general configuration shown in Figure 12.4a where it is desired to design a discrete-time controller transfer function $G_c(z)$ to control the continuous-time plant described by the transfer function $G_p(s)$. The first step is to derive a discrete-time equivalent of the plant described by $G_p(z)$ as shown in Figure 12.4b. To do so, the dashed portion of Figure 12.4b has been redrawn in Figure 12.5a to emphasize the relationship between the discrete-time signals, $f(k)$ and $y(k)$. It is desired now to find the discrete-time transfer function $G_p(z)$ of the arrangement, and this can be done by finding its pulse response.

For a unit pulse input of

$$f(k) = \delta(k),$$

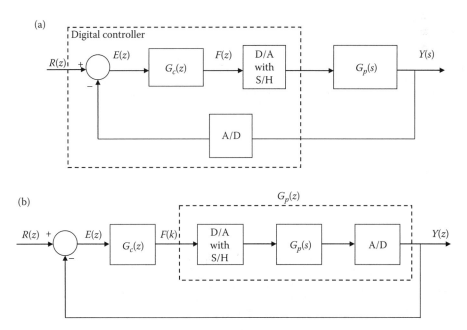

FIGURE 12.4 A discrete-time equivalent of a continuous-time plant. (a) General configuration of a digital control system; (b) rearranging the system in (a) to discretize the plant.

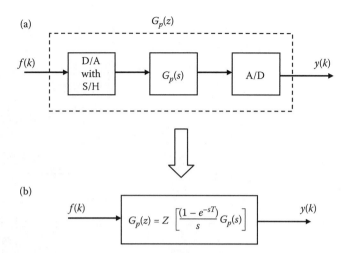

FIGURE 12.5 Discretizing a continuous-time plant. (a) Continuous-time plant with discrete inputs and outputs; (b) discretized plant.

the sampled-and-held continuous-time signal that is the input to $G_p(s)$ is given by

$$f^0(t) = u(t) - u(t - T)$$

or

$$F^0(s) = \frac{1 - e^{-sT}}{s},$$

where T is the sampling interval. Then

$$Y(s) = F^0(s)G_p(s) = \frac{1 - e^{-sT}}{s} G_p(s), \qquad (12.2)$$

and therefore

$$G_p(z) = Z\left[\frac{1 - e^{-sT}}{s} G_p(s)\right], \qquad (12.3)$$

where Z is the z-transform as given in Table 12.1. This equivalence is shown in Figure 12.5b.

As a numerical example, suppose that the continuous-time transfer function of the plant is given by

$$G_p(s) = \frac{4}{s(s + 2)},$$

and the sampling interval is $T = 0.2$ s. According to Equation 12.2,

$$Y(s) = F^0(s)G_p(s) = \frac{1 - e^{-0.2s}}{s}\left[\frac{4}{s(s + 2)}\right] = (1 - e^{-0.2s})\left[\frac{4}{s^2(s + 2)}\right]$$

$$= (1 - e^{-0.2s})\left[\frac{-1}{s} + \frac{2}{s^2} + \frac{1}{(s + 2)}\right].$$

Using Table 12.1, the discrete-time plant transfer function, for $T = 0.2$, is determined using Equation 12.3 as

$$G_p(z) = (1 - z^{-1})\left[\frac{-z}{z - 1} + \frac{2(0.2)z}{(z - 1)^2} + \frac{z}{z - e^{-0.4}}\right] = \frac{0.0703z + 0.0616}{(z - 1)(z - 0.6703)}.$$

Knowing $G_p(z)$, and returning to Figure 12.4b, the control system designer can now proceed to specify the digital controller $G_c(z)$ using classical design techniques to meet the control system requirements.

The classical approach to designing a digital controller directly, which has many variations, parallels the classical approach to analog controller design. We begin with simple discrete-time controllers, increasing their complexity until the performance requirements can be met. Classical discrete-time control system design is beyond our scope in this chapter and therefore will not be discussed.

In the following section, we present several methods for discretizing continuous-time controllers. In the latter section, the relationship between continuous-time state variable plant models and their discrete counterparts are derived with the results being useful for designing digital controllers for discrete-time systems.

12.3 Digitizing Analog Controllers

The second approach to designing discrete-time control systems for continuous-time plants is to derive a continuous-time (or analog) controller and then approximate the behavior of the analog controller with a digital one.

Consider the situation shown in Figure 12.6 where it is desired to derive a discrete-time controller $G_c(z)$ that approximates the behavior of the continuous-time controller described by $G_c(s)$. For the sake

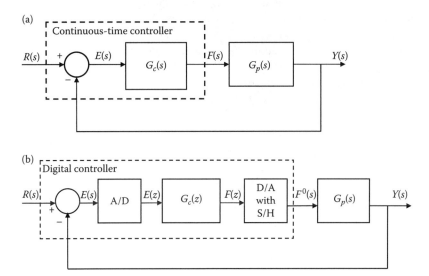

FIGURE 12.6 Discrete-time equivalents of continuous-time controllers. (a) Continuous-time controller; (b) discrete-time equivalent of analog controllers.

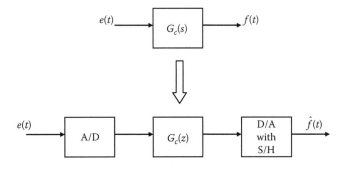

FIGURE 12.7 Digitizing a continuous-time controller.

of clarity, the dashed portion of Figure 12.6 has been redrawn in Figure 12.7. As shown in the figure, the digital controller consists of an A/D converter driving the discrete-time controller described by the z-transfer function $G_c(z)$ followed by a D/A converter with the sample and the hold. If the sample rate is sufficiently high and the approximation sufficiently good, the behavior of the digital controller will be nearly indistinguishable from that of the analog controller. The digital controller will have such advantages as high reliability, low drift with temperature, power supply and age, and the ability to make changes in software.

We now discuss several methods for discretizing continuous-time controllers. These methods are

1. Numerical approximation of differential equations.
2. Matching step and other responses.
3. Pole–zero matching.

In the material to follow, the accent ˆ over a symbol denotes an approximation of the quantity.

12.3.1 Numerical Approximation of Differential Equations

One way to approximate an analog controller with a digital one is to convert the analog controller transfer function $G_c(s)$ to a differential equation and then obtain a numerical approximation to the solution of the differential equation. There are two basic methods of numerical approximation of the solution of differential equations. They are (a) numerical integration and (b) numerical differentiation. We first discuss numerical integration and then summarize numerical differentiation.

Numerical integration is an important computational problem in its own right and is fundamental to the numerical solution of differential equation. The most common approach of performing numerical integration is to divide the interval of integration into many T subintervals and approximate the contribution to the integral in each T strip by the integral of a polynomial approximation to the integrand in that strip.

Consider the transfer function

$$G_c(s) = \frac{F(s)}{E(s)} = \frac{1}{s},$$ (12.4)

which has the corresponding differential equation

$$\frac{df}{dt} = e(t).$$ (12.5)

Integrating both sides of Equation 12.5 from t_0 to t,

$$f(t) = f(t_0) + \int_{t_0}^{t} e(t)\, dt, \quad t \geq t_0.$$

For evenly spaced sample times $t = kT, k = 0, 1, 2, \ldots$ and during one sampling interval $t_0 = kT$ to $t = kT + T$, the solution is

$$f(kT + T) = f(kT) + \int_{kT}^{kT+T} e(t)\, dt.$$ (12.6)

12.3.1.1 Euler's Forward Method (One Sample)

The simplest approximation of the integral in Equation 12.6 is simply to approximate the integrand by a constant equal to the value of the integrand at the *left* endpoint of each T subinterval and multiply by the sampling interval T as in Figure 12.8a. Thus,

$$\hat{f}(kT + T) = \hat{f}(kT) + Te(kT).\tag{12.7}$$

Z-transforming both sides of Equation 12.7,

$$z\hat{F}(z) - \hat{F}(z) = TE(z),$$

and, therefore,

$$G_c(z) = \frac{\hat{F}(z)}{E(z)} = \frac{T}{z-1}.\tag{12.8}$$

Comparing Equation 12.8 with the analog controller transfer function Equation 12.4 implies that a discrete-time equivalence of an analog controller can be determined with Euler's forward method by simply replacing each s in the analog controller transfer function with $(z-1)/T$, that is,

$$G_c(z) = G_c(s)|_{s=z-1/T}.$$

12.3.1.2 Euler's Backward Method (One Sample)

Instead of approximating the integrand in Equation 12.6 during one sampling interval by its value at the left endpoint, *Euler's backward* method approximates the integrand by its value at the *right* endpoint

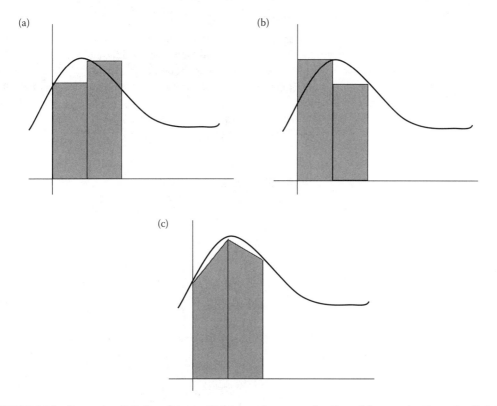

FIGURE 12.8 Comparing Euler's and trapezoidal integration approximations. (a) approximation using Euler's forward method; (b) approximation using Euler's backward method; and (c) approximation using the trapezoidal method.

of each T subinterval and multiplies by the sampling interval, as in Figure 12.8b. Then, Equation 12.6 becomes

$$\hat{f}(kT + T) = \hat{f}(kT) + Te(kT + T) \tag{12.9}$$

or, using the z-transformation,

$$G_c(z) = \frac{\hat{F}(z)}{E(z)} = \frac{Tz}{z - 1} \tag{12.10}$$

Comparing Equation 12.10 with Equation 12.4 shows that the equivalent discrete-time transfer function of the analog controller can be obtained by replacing each s in $G_c(s)$ with $(z - 1)/Tz$, that is,

$$G_c(z) = G_c(s)|_{s=z-1/Tz}.$$

For the analog controller,

$$G_c(s) = \frac{a}{s + a},$$

for example, the discrete-time equivalent controller using Euler's backward method is

$$G_c(z) = \frac{a}{(z - 1)/Tz + a} = \frac{aTz}{(1 + aT)z - 1}$$

and the discrete-time equivalent controller using Euler's forward method is

$$G_c(z) = \frac{aT}{z - 1 + aT}.$$

12.3.1.3 Trapezoidal Method (Two Samples)

Euler's forward and backward methods are sometimes called *rectangular* methods because, during the sampling interval, the area under the curve is approximated with a rectangle. Additionally, Euler's methods are also called *first order* because they use one sample during each sampling interval.

The performance of the digital controller can be improved over the simpler approximation by either Euler's forward or backward methods if more than one sample is used to update the approximation of the analog controller transfer function during a sampling interval. As in Figure 12.8c, the trapezoidal approximation approximates the integrand with a straight line. Applying the trapezoidal rule to the integral in Equation 12.6 gives

$$\hat{f}(kT + T) = \hat{f}(kT) + \frac{T}{2}\{e(kT) + e(kT + T)\},$$

which has a corresponding z-transfer function,

$$(z - 1)\hat{F}(z) = \frac{T}{2}(z + 1)E(z),$$

or

$$G_c(z) = \frac{\hat{F}(z)}{E(z)} = \frac{T}{2}\left(\frac{z + 1}{z - 1}\right). \tag{12.11}$$

Comparing Equation 12.11 with Equation 12.4, the digital controller transfer function can be obtained by simply replacing each s in the analog controller transfer function with $\frac{2}{T}(z - 1)/(z + 1)$.
That is,

$$G_c(z) = G_c(s)|_{s=(2/T)(z-1/z+1)}.$$

The trapezoidal method is also called *Tustin's* method, or the *bilinear* transformation. Higher-order polynomial integrals can be approximated in the same way, but for a recursive numerical solution, an integral approximation should involve only present and past values of the integrand, not future ones.

A summary of some common approximations for integrals, along with the corresponding z-transfer function of each integral approximation, is shown in Table 12.2. Higher-order approximations result in digital controllers of progressively higher order. The higher the order of the approximation, the better the approximation to the analog integrations and the more accurately the digital controller output tends to track samples of the analog controller output for any input. The digital controller, however, probably has a sample-and-held output between samples, so that accurate tracking of samples is of less concern to the designer than the sample rate.

12.3.1.4 An Example

As an example of deriving a digital controller for a continuous-time plant, consider the system shown in Figure 12.9a, where it is assumed that the continuous-time controller

$$G_c(s) = \frac{0.5(s+5)}{s(s+2)}$$

has been designed to control the continuous-time plant with the transfer function

$$G_p(s) = \frac{5}{s^2+4s+8}$$

Using Euler's forward method, the discrete-time controller transfer function is

$$G_c(z) = \frac{0.5T(z-1+5T)}{(z-1)(z-1+2T)},$$

as shown in Figure 12.9b. The plant and the discrete-time controller are shown in Figure 12.9c. The step response of the overall feedback system for various sampling intervals T is shown in Figure 12.10. When $T = 0.4$ s, which is relatively large as compared to the continuous-time closed-loop time constants and/or frequencies, the discrete-time approximation of the overall feedback response deviates significantly from the continuous-time response. As the sampling interval is decreased so that there are several steps during each time constant of the closed-loop poles, the step responses of the analog and digital system are nearly the same. Similar conclusions can be made when comparing the step responses of the analog and

TABLE 12.2   Some Integral Approximations Using Present and Past Integrand Samples

Approximation to the Integral Over One Step	Difference Equation for the Approximate Integral	z-Transmittance of the Approximate Integral
One-Sample		
$\int_{kT}^{kT+T} e(t)dt \cong Te(kT)$	$\hat{f}[(k+1)T] = \hat{f}(kT) + Te(kT)$	$\dfrac{T}{z-1}$
$\int_{kT}^{kT+T} e(t)dt \cong Te(kT+T)$	$\hat{f}[(k+1)T] = \hat{f}(kT) + Te(kT+T)$	$\dfrac{Tz}{z-1}$
Two-Sample (Tustin approximation)		
$\int_{kT}^{kT+T} e(t)dt \cong T\left\{\dfrac{1}{2}e[(k+1)T] + \dfrac{1}{2}e(kT)\right\}$	$\hat{f}[(k+1)T] = \hat{f}(kT) + \dfrac{T}{2}e[(k+1)T]$ $+ \dfrac{T}{2}e(kT)$	$\dfrac{T(z+1)}{2(z-1)}$
Three sample		
$\int_{kT}^{kT+T} e(t)dt \cong T\left\{\dfrac{5}{12}e[(k+1)T] + \dfrac{8}{12}e(kT)\right.$ $\left. - \dfrac{1}{12}e[(k-1)T]\right\}$	$\hat{f}[(k+1)T] = \hat{f}(kT) + \dfrac{5T}{12}e[(k+1)T]$ $+ \dfrac{8T}{12}e(kT) - \dfrac{T}{12}e[(k-1)T]$	$\dfrac{T[(5/12)z^2 + (8/12)z - (1/12)]}{z(z-1)}$

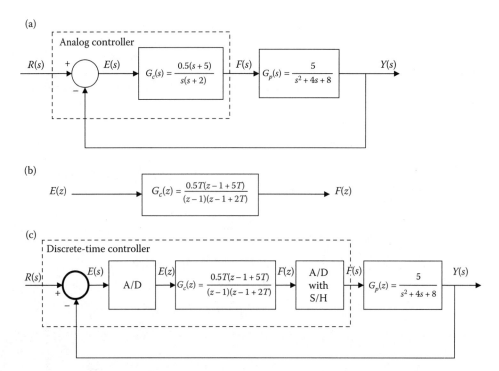

FIGURE 12.9 An example of digitizing an analog control system. (a) Continuous-time control system, (b) discrete-time equivalent of analog controller using Euler's forward method, and (c) discrete-time control system.

discrete-time controllers. When $T = 0.4$ s, which is relatively large as compared to the continuous-time controller's fastest mode e^{-2t}, the discrete-time controller approximation deviates significantly from the continuous-time controller. As the sampling interval is decreased, the step responses of the analog and discrete-time controllers are the same.

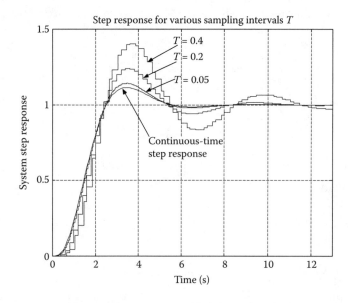

FIGURE 12.10 Step response of the example system using Euler's forward method.

If, on the other hand, we use the trapezoidal method, the transfer function of the digital controller becomes

$$G_c(z) = \frac{0.5[(5T^2 + 2T)z^2 + 10T^2 z + (5T^2 - 2T)]}{(4T + 4)z^2 - 8z - 4T + 4}.$$

For the same sampling interval, the step response of the overall feedback system using Tustin's method tends to track the output of the continuous-time system more accurately at the sample times than Euler's forward method because the approximations to the analog integration are better. The step responses of the overall feedback system using Tustin's and Euler's forward methods are shown in Figure 12.11a–c for the sampling intervals $T = 0.4, 0.2$, and 0.1 s, respectively.

Because Euler's and Tustin's methods result in controllers of the same order, the designer usually opts for Tustin's approximation. In general, if the sampling interval is sufficiently small and the approximation is sufficiently good, the behavior of the discrete-time controller, designed using any one of the approximation methods, will be similar to the behavior of the continuous-time controller. One should not be too hasty in abandoning the simple Euler approximation for higher-order approximation. In modeling physical systems, poor accuracy of the Euler approximation with very small sampling interval is often indicative of an underlying lack of physical robustness that probably ought to be carefully examined.

Warning   The approximation methods summarized in Table 12.2 apply by replacing each s in the analog controller transfer function with the corresponding z-transmittance. Every z-transmittance is a mapping from the s-plane to the z-plane. As shown in Figure 12.12b, Euler's forward method has the potential of mapping poles in the left-half of the s-plane, as shown in Figure 12.12a, to poles outside the unit circle on the complex plane. Then, some stable analog controllers may produce unstable digital controllers. Euler's backward rule maps the left hand of the s-plane to a region inside the unit circle, as shown in Figure 12.12c. The trapezoidal rule, however, maps the left-half of the s-plane to the interior of the unit circle on the z-plane, the right-half of the s-plane to the exterior of the unit circle, and the imaginary axis of the s-plane to the boundaries of the unit circle, as shown in Figure 12.12d.

12.3.1.5 Frequency Response Approximations

Consider the second-order, continuous-time, low-pass Butterworth filter described by the transfer function

$$G(s) = \frac{\omega_c^2}{s^2 + \sqrt{2}\omega_c s + \omega_c^2},$$

where $\omega_c = 2\pi(10)$ rad/s. The discrete-time transfer function of this filter can be derived using any one of the methods summarized in Table 12.2 above. For a sampling rate of 100 Hz, which is 10 times higher than the filter corner frequency, the frequency responses of the filter using Euler's and Tustin's methods are shown in Figures 12.13a and b. As shown in this figure, for low frequencies, the analog and discrete-time magnitude and phase responses are nearly the same. As the frequency increases beyond the filter corner frequency, the magnitude and phase responses using Euler's methods deviate significantly from the analog frequency response. On the other hand, the frequency response using Tustin's method tend to track the analog filter response more accurately than Euler's. As the sampling rate increases, the behavior of the discrete-time filter, derived using any one of the approximation methods, will be nearly indistinguishable from that of the analog filter.

12.3.1.6 Bilinear Transformation with Frequency Prewarping

In many digital control and digital signal processing applications, it is desired to design a digital filter $G(z)$ that closely approximates the frequency response of a continuous-time filter $G(s)$ within the

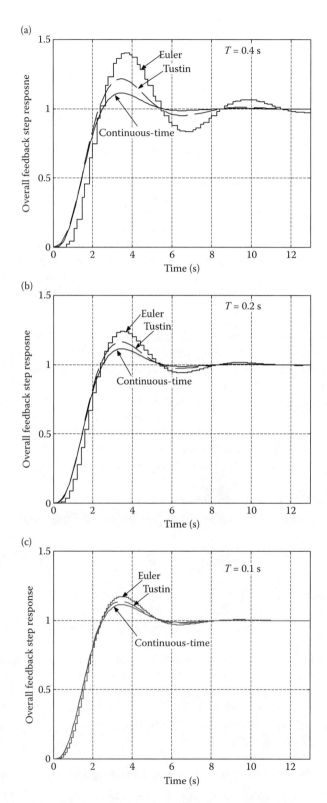

FIGURE 12.11 Step responses of the example system using Tustin's and Euler's approximations. (a) Step response using Tustin's and Euler's forward methods ($T = 0.4$ s); (b) step response using Tustin's and Euler's forward methods ($T = 0.2$ s); and (c) step response using Tustin's and Euler's forward methods ($T = 0.1$ s).

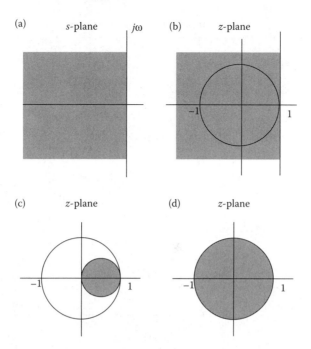

FIGURE 12.12 Mapping between the *s*-plane and the *z*-plane. (a) Stability region in the *s*-plane; (b) corresponding region in the *z*-plane using Euler's forward approximation; (c) corresponding region in the *z*-plane using Euler's backward approximation; and (d) corresponding region in the *z*-plane using Tustin approximation.

band-limited range

$$G(z = e^{j\omega T}) \cong G(s = j\omega) \quad 0 \le \omega < \omega_0 = \frac{\pi}{T}.$$

The bilinear (trapezoidal) method applies but with minor modifications. If the frequency response of the digital filter $G(z)$ is to approximate the frequency response of the analog controller $G(s)$, then

$$G(z) = G\left(s = \frac{2}{T}\frac{z-1}{z+1}\right),$$

$$G(z = e^{j\omega_d T}) = G\left(j\omega_c = \frac{2}{T}\frac{e^{j\omega_d T}-1}{e^{j\omega_d T}+1}\right)$$

$$= G\left[j\omega_c = \frac{2}{T}\left(\frac{e^{(j\omega_d T/2)} - e^{(-j\omega_d T/2)}}{e^{(j\omega_d T/2)} + e^{(-j\omega_d T/2)}}\right)\right]$$

$$= G\left[j\omega_c = j\frac{2}{T}\frac{\sin\omega_d T/2}{\cos\omega_d T/2}\right]$$

$$= G\left[j\omega_c = j\frac{2}{T}\tan\omega_d T\right],$$

and, therefore,

$$\omega_c = \frac{2}{T}\tan\frac{\omega_d T}{2}, \tag{12.12}$$

where ω_c is the continuous frequency and ω_d is the discrete frequency. This nonlinear relationship arises because the *entire j\omega*-axis of the *s*-plane is mapped into one complete revolution of the unit circle in the *z*-plane.

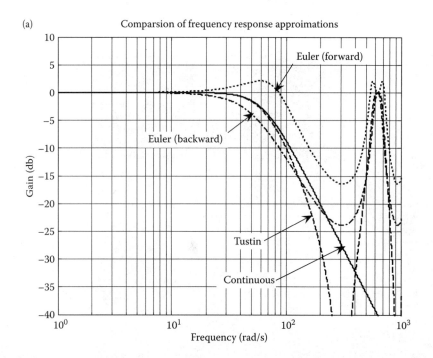

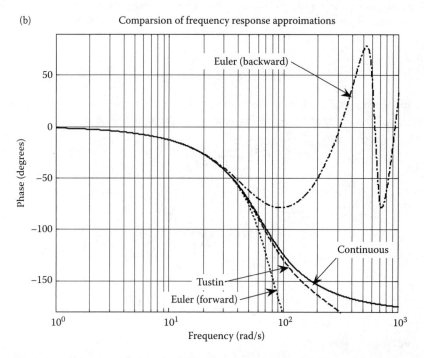

FIGURE 12.13 Frequency response of Butterworth filter using Tustin's and Euler's Methods. (a) Magnitude response of Butterworth filter using Tustin's and Euler's methods and (b) phase response of Butterworth filter using Tustin's and Euler's methods.

For relatively small values of ω_d as compared to the *folding frequency* π/T, then,

$$j\omega_c \cong j\frac{2}{T}\frac{\omega_d T}{2} = j\omega_d,$$

and the behavior of the discrete-time filter closely approximates the frequency response of the corresponding continuous-time filter. When ω_d approaches the folding frequency π/T,

$$j\omega_c = j\frac{2}{T}\tan\frac{\omega_d T}{2} \to j\infty,$$

the continuous frequency approaches infinity, and distortion becomes evident. However, if the bilinear transformation is applied together with Equation 12.12 near the frequencies of interest, the frequency distortion can be reduced considerably.

The general design procedure for discretizing a continuous-time filter using the bilinear transformation with frequency *prewarping* is as follows:

1. Beginning with the continuous-time filter $G(s)$, obtain a new continuous-time filter with transfer function $G'(s)$ whose poles and zeros with critical frequencies $(s+\alpha')$ are related to those of the original $G(s)$ by

$$(s+\alpha) \to (s+\alpha')|_{\alpha'=2/T\tan\alpha T/2}$$

   in the case of real roots, and by

$$s^2 + 2\zeta\omega_n s + \omega_n^2 \to s^2 + 2\zeta\omega_n's + \omega_n'^2|_{\omega_n'=2/T\tan\omega_n T/2}$$

   in the case of complex roots.

2. Apply the bilinear transformation to $G'(s)$ by replacing each s in $G'(s)$ with

$$s = \frac{2}{T}\frac{z-1}{z+1}.$$

3. Scale the multiplying constant of $G(z)$ to match the multiplying constant of the continuous-time filter $G(z)$ at a specific frequency.

   To illustrate the above steps, consider the second-order low-pass filter described by the transfer function

$$G(s) = \frac{\omega_n^2}{s^2 + 0.4\omega_n s + \omega_n^2}, \tag{12.13}$$

   where $\omega_n = 2\pi(10)$ rad/s. This filter has a unity DC gain, undamped natural frequency $f = 10$ Hz, and a damping ratio $\zeta = 0.2$. For a sampling interval $T = 0.02$ s, the folding frequency of $f_0 = 25$ Hz, which is above the 10 Hz cutoff of the filter. At $\omega_n = 2\pi(10)$,

$$\omega_n' = \frac{2}{T}\tan\frac{\omega_n T}{2} = 100\tan\frac{20\pi}{100} = 72.65 \text{ rad/s},$$

   and hence the warped transfer function is

$$G'(s) = \frac{K}{s^2 + 29.06s + (72.65)^2}.$$

Therefore,

$$G'(z) = \frac{K}{[100(z-1)/(z+1)]^2 + 2906(z-1)/(z+1) + (72.65)^2}$$

$$= \frac{K(z+1)^2}{18184z^2 - 9444z + 12372}.$$

For unity DC gain of the continuous filter,

$$G'(1) = \frac{4K}{21112} = 1$$

then

$$K = 5278$$

and hence

$$G'(z) = \frac{0.2903(z+1)^2}{z^2 - 0.5193z + 0.6804} \qquad (12.14)$$

which is the required filter.

Figure 12.14a and b show the frequency responses of the continuous-time filter and the digital filters given by Equations 12.13 and 12.14. For the sake of comparison, the frequency response using Tustin's approximation is also shown. As expected, the digital filter obtained using the bilinear transformation with frequency prewarping approximates the frequency response more accurately within the bandlimited range than the filter obtained using Tustin's method. Also, as the sampling interval T is reduced to 0.005 s, the frequency response of the digital controllers using either Tustin or the bilinear transformation with frequency prewarping will be nearly indistinguishable from that of the continuous-time filter for frequencies below 100 rad/s. Frequency response plots of digital equivalents are periodic in ω with period $2\pi/T$. The magnitude plots are even-symmetric about the Nyquist frequency π/T while the phase plots are odd-symmetric about the Nyquist frequency π/T. In Figure 12.14, the Nyquist frequency is 157.08 rad/s. More details on Nyquist frequency and the sampling rate selection will be discussed in Chapter 15.

The other main approach to approximating the solution of a differential equation is *numerical differentiation*. The approximate solution of the differential equation is obtained by replacing the derivative terms in the differential equation with finite difference approximations. The resulting difference equation can then be solved numerically. Three methods of first- and second-order derivative approximations listed in Table 12.3 are called finite-difference approximations of derivatives. The corresponding z-transmittance of each of these finite difference approximations is also listed in Table 12.3.

As an example, for the analog controller

$$G_c(s) = \frac{K(s+a)}{(s+b)}$$

the forward-difference approximation gives the digital controller,

$$G_c(z) = \frac{K[(z-1)/T + a]}{(z-1)/T + b} = \frac{K(z-1+aT)}{z-1+bT}$$

and the backward-difference approximation yields the digital controller

$$G_c(z) = \frac{K[(z-1)/Tz + a]}{(z-1)/Tz + b} = \frac{K(1+aT)}{1+bT}\left[\frac{z - \frac{1}{1+aT}}{z - \frac{1}{1+bT}}\right]$$

Similar to integral approximations, higher-order derivative approximations can be generated in the same way, but for a recursive numerical solution, a derivative approximation should involve only present and past values of the input, not future ones.

12.3.2 Matching Step and Other Responses

Another way of approximating the behavior of the analog controller with a digital controller is to require that, at the sampling times, the step response of the digital controller *matches* the analog controller step

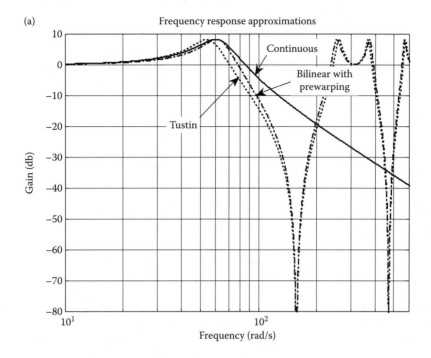

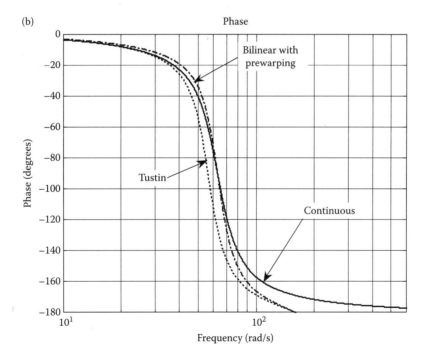

FIGURE 12.14 Frequency response of a second-order filter using bilinear transformation. (a) Magnitude response approximation of a second order filter ($T = 0.02$ s) and (b) phase approximation of a second order filter ($T = 0.02$ s).

TABLE 12.3 Finite Difference Approximations of Derivatives

	Derivative Approximation	z-Transmittance of the Approximate Differentiator
First-Order Derivative		
Forward difference	$\dfrac{f[(k+1)T]-f(kT)}{T}$	$\dfrac{z-1}{T}$
Backward difference	$\dfrac{f(kT)-f[(k-1)T]}{T}$	$\dfrac{z-1}{Tz}$
Central difference	$\dfrac{f[(k+1)T]-f[(k-1)T]}{2T}$	$\dfrac{z^2-1}{2Tz}$
Second-Order Derivative		
Forward difference	$\dfrac{f[(k+2)T]-2f[(k+1)T]+f(kT)}{T^2}$	$\dfrac{z^2-2z+1}{T^2}$
Backward difference	$\dfrac{f(kT)-2f[(k-1)T]+f[(k-2)T]}{T^2}$	$\dfrac{z^2-2z+1}{T^2z^2}$
Central difference	$\dfrac{f[(k+1)T]-2f(kT)+f[(k-1)T]}{T^2}$	$\dfrac{z^2-2z+1}{T^2z}$

response. Consider the unit step response $f_{\text{step}}(t)$ of the analog controller with transfer function $G_c(s)$ shown in Figure 12.15a. Our objective is to design a discrete-portion $G_c(z)$ of the digital controller, as in Figure 12.15b, such that its step response $f_{\text{step}}(k)$ to a unit step input consists of samples of $f_{\text{step}}(t)$. Then, as in Figure 12.15c, the digital controller has a step response that equals the step response of the analog controller at the sample times. This method is termed a *step-invariant approximation* of the analog system by a digital system.

As an example, consider the continuous-time controller that has the transfer function

$$G_c(s) = \frac{20}{s^2 + 4s + 20},$$

which can be written as

$$G_c(s) = 5\frac{b}{(s+a)^2 + b^2},$$

where $a = 2$ and $b = 4$.

This controller has the unit step response

$$F_{\text{step}}(s) = \frac{1}{s}G_c(s) = 5\frac{b}{a^2+b^2}\left[\frac{1}{s} - \frac{(s+a)}{(s+a)^2+b^2} - \frac{a}{(s+a)^2+b^2}\right].$$

Using the transform pairs in Table 12.1, the samples of $f_{\text{step}}(t)$ have the z-transform

$$F_{\text{step}}(z) = 5\frac{b}{a^2+b^2}\left[\frac{z}{z-1} - \frac{z(z-c\cos\Omega)}{z^2-2c\cos\Omega z+c^2} - \frac{a}{b}\frac{zc\sin\Omega}{z^2-2c\cos\Omega z+c^2}\right], \qquad (12.15)$$

where

$$\Omega = bT \quad \text{and} \quad c = e^{-aT}.$$

Taking these samples to be the output of a discrete-time system $G_c(z)$ driven by a unit step sequence

$$F_{\text{step}}(z) = \left(\frac{z}{z-1}\right)G_c(z),$$

the step-invariant approximation is

$$G_c(z) = \left(\frac{z-1}{z}\right)F_{\text{step}}(z).$$

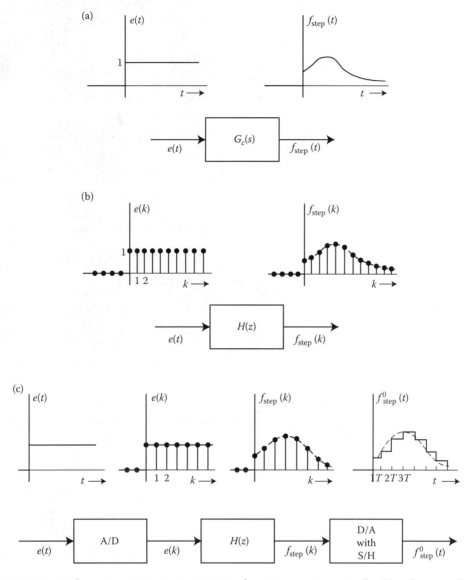

FIGURE 12.15 Finding a step-invariant approximation of a continuous-time controller. (a) analog controller step response; (b) discrete step response consisting of samples of the analog controller step response; and (c) step-invariant digital controller.

Using Equation 12.15, and for a sampling interval $T = 0.25$ s, then

$$G_c(z) = \frac{0.4171z + 0.2954}{z^2 - 0.6554z + 0.3679}.$$

Figure 12.16a shows the step response of the continuous-time controller and the step response of the step-invariant approximation. As shown in the figure, the step response of the digital controller equals the step response of the continuous-time controller at the sample times. Reducing the sampling interval to $T = 0.1$ s, the step-invariant approximation becomes

$$G_c(z) = \frac{0.0865z + 0.0756}{z^2 - 1.508z + 0.6703}.$$

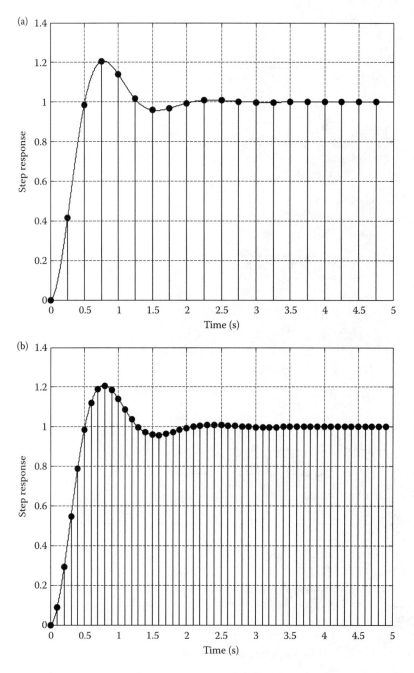

FIGURE 12.16   Matching step response of a continuous-time controller. (a) Matching step response of a continuous-time controller ($T = 0.25$ s). (b) Matching step response of a continuous-time controller ($T = 0.1$ s).

The step response of this controller and the continuous-time one are shown in Figure 12.16b.

It is occasionally desirable to design digital controllers so that their response to some input other than a step consists of a sampled-and-held version of an analog controller's response to that input. The *impulse-invariant approximation*, for example, has a discrete-time response to a unit pulse sequence that consists of samples of the unit impulse response of the continuous-time system. The *ramp-invariant approximation* is another possibility.

12.3.3 Pole–Zero Matching

Yet another method of approximating an analog controller by a digital one is to map the poles and zeros of the analog controller transfer function $G_c(s)$ to those of the corresponding digital controller $G_c(z)$ as follows:

$$(s+a) \to z - e^{-aT},$$

for real roots, and

$$(s+a)^2 + b^2 \to z^2 - 2(e^{-aT} \cos bT)z + e^{-2aT},$$

for complex conjugate pairs.

Usually, an analog controller has more finite poles than zeros. In this case, its high-frequency response tends to zero as ω_c approaches infinity. Because the entire $j\omega$ axis of the s-plane is mapped into one complete revolution of the unit circle in the z-plane, the highest possible frequency on the $j\omega$-axis is at $\omega_c = \pi/T$. Hence,

$$z = e^{sT} = e^{j(\pi/T)T} = -1,$$

and, therefore, infinite zeros of the analog controller map into finite zeros located at $z = -1$ in the corresponding digital equivalence. The resulting transfer function of the digital controller will always have the number of poles equal to the number of zeros.

For example, the analog controller,

$$G_c(s) = \frac{6s + 10}{s^2 + 2s + 5} = \frac{6(s + \frac{5}{3})}{(s+1)^2 + 2^2},$$

has two finite poles and one finite zero. For a sampling interval $T = 0.1$ s,

$$G_c(z) = \frac{K(z+1)(z - e^{-(5/30)})}{z^2 - 2(e^{-0.1}\cos 2T)z + e^{-0.2}} = \frac{K(z+1)(z - 0.85)}{z^2 - 1.773z + 0.818}.$$

The DC gain of the analog controller is

$$G_c(s = j0) = 2.$$

For the identical DC gain of the digital controller

$$G_c(z = 1) = \frac{K(2)(0.15)}{1 - 1.773 + 0.818} = 2,$$

then $K = 0.3$. Hence,

$$G_c(z) = \frac{0.3(z+1)(z - 0.85)}{z^2 - 1.773z + 0.818}.$$

If the analog controller has poles or zeros at the origin of the s-plane, the multiplying constant of the digital controller is selected to match the gain of the analog controller at a specified frequency.

At low frequency,

$$G_c(z)|_{z=1} = G_c(s)|_{s=0},$$

and at high frequency,

$$G_c(z)|_{z=-1} = G_c(s)|_{s=\infty}.$$

All of the approximation methods presented thus far in this section for digitizing analog controllers tend to perform well for sufficiently short sampling intervals. For longer sampling intervals, dictated by design and cost constraints, one approximation method or another might perform the best in a given situation.

Digitizing an analog controller is not a good general design technique, although it is very useful when replacing all or part of an existing analog controller with a digital one. The technique requires beginning with a good analog design, which is probably as difficult as creating a good digital design. The digital design usually performs less well than the analog counterpart from which it was originally derived. Furthermore, the step-invariant and other approximations are not easily extended to systems with multiple inputs and outputs. When the resulting feedback system performance is inadequate, the designer may have few options besides raising the sampling rate.

12.4 Discretization of Continuous-Time State Variable Models

We now discuss the relationship between continuous-time state variable plant models and discrete-time models of plant signal samples.

12.4.1 Discrete-Time Models of Continuous-Time Systems

Consider an nth-order continuous-time system described by the state variable equation

$$
\begin{aligned}
\dot{x}(t) &= \mathcal{A}x(t) + \mathcal{B}u(t), \\
y(t) &= \mathcal{C}x(t) + \mathcal{D}u(t).
\end{aligned}
\tag{12.16}
$$

Script symbols are now used for state and input coupling matrices to distinguish between them and the corresponding discrete-time models. The time t is a continuous variable, x is the n-vector state of the system, u is an r-vector of system inputs, and y is an m-vector of system outputs. The remaining matrices in Equation 12.16 are of appropriate dimensions.

The solution for the state for $t \geq 0$ is given by the convolution,

$$
x(t) = e^{\mathcal{A}t}x(0) + \int_0^t e^{\mathcal{A}(t-\tau)}\mathcal{B}u(\tau)\,d\tau.
$$

At the sample times $kT, k = 0, 1, 2, \ldots$, the state is

$$
x(kT) = e^{\mathcal{A}kT}x(0) + \int_0^{kT} e^{\mathcal{A}(kT-\tau)}\mathcal{B}u(\tau)\,d\tau.
$$

The state at the $(k+1)$th step can be expressed in terms of the state at the kth step as follows:

$$
\begin{aligned}
x(kT+T) &= e^{\mathcal{A}(kT+T)}x(0) + \int_0^{kT+T} e^{\mathcal{A}(kT+T-\tau)}\mathcal{B}u(\tau)\,d\tau \\
&= e^{\mathcal{A}T}e^{\mathcal{A}kT}x(0) + \int_0^{kT} e^{\mathcal{A}(kT+T-\tau)}\mathcal{B}u(\tau)\,d\tau + \int_{kT}^{kT+T} e^{\mathcal{A}(kT+T-\tau)}\mathcal{B}u(\tau)\,d\tau \\
&= e^{\mathcal{A}T}\left[e^{\mathcal{A}kT}x(0) + \int_0^{kT} e^{\mathcal{A}(kT-\tau)}\mathcal{B}u(\tau)\,d\tau\right] + \int_{kT}^{kT+T} e^{\mathcal{A}(kT+T-\tau)}\mathcal{B}u(\tau)\,d\tau \\
&= e^{\mathcal{A}T}x(kT) + \text{(input term)}.
\end{aligned}
$$

When the input $u(t)$ is constant during each sampling interval, as it is when driven by sample-and-hold, then the input term becomes

$$
\int_{kT}^{kT+T} e^{\mathcal{A}(kT+T-\tau)}\mathcal{B}u(\tau)\,d\tau = \left[\int_{kT}^{kT+T} e^{\mathcal{A}(kT+T-\tau)}\,d\tau\right]\mathcal{B}u(kT).
$$

The discrete-time input coupling matrix is

$$B = \left[\int_{kT}^{kT+T} e^{A(kT+T-\tau)} d\tau \right] \mathcal{B}.$$

Let

$$\gamma = kT + T - \tau, \quad d\gamma = -d\tau;$$

then

$$B = \left[\int_0^T e^{A\gamma} d\gamma \right] \mathcal{B}. \tag{12.17}$$

Expanding the integrand into a power series,

$$e^{A\gamma} = I + \frac{A\gamma}{1!} + \frac{A^2\gamma^2}{2!} + \cdots + \frac{A^i\gamma^i}{i!} + \cdots,$$

and integrating term by term results in

$$\begin{aligned}
B &= \left\{ \int_0^T \left[I + \frac{A\gamma}{1!} + \frac{A^2\gamma^2}{2!} + \cdots + \frac{A^i\gamma^i}{i!} + \cdots \right] d\gamma \right\} \mathcal{B} \\
&= \left[IT + \frac{AT^2}{2!} + \frac{A^2T^3}{3!} + \cdots + \frac{A^i T^{i+1}}{(i+1)!} + \cdots \right] \mathcal{B}
\end{aligned}$$

or

$$B = \left[T \sum_{k=0}^{\infty} \frac{(AT)^k}{(k+1)!} \right] \mathcal{B} = \Psi \mathcal{B}. \tag{12.18}$$

Because

$$A B = \left[\frac{AT}{1!} + \frac{A^2T^2}{2!} + \cdots + \frac{A^{i+1}T^{i+1}}{(i+1)!} + \cdots \right] \mathcal{B} = (e^{AT} - I)\mathcal{B}$$

and, if A is nonsingular, then

$$B = A^{-1}(e^{AT} - I)\mathcal{B} = (e^{AT} - I)A^{-1}\mathcal{B}.$$

The discrete-time model of Equation 12.16 is then

$$\begin{aligned}
x[(k+1)T] &= Ax(kT) + Bu(kT), \\
y(kT) &= Cx(kT) + Du(kT)
\end{aligned} \tag{12.19}$$

or

$$\begin{aligned}
x(k+1) &= Ax(k) + Bu(k), \\
y(k) &= Cx(k) + Du(k),
\end{aligned}$$

where

$$A = e^{AT} = \left[\sum_{k=0}^{\infty} \frac{(AT)^k}{k!} \right] = I + A\Psi = I + \Psi A \tag{12.20}$$

$$B = \left[IT + \frac{AT^2}{2!} + \frac{A^2T^3}{3!} + \cdots + \frac{A^i T^{i+1}}{(i+1)!} + \cdots \right] \mathcal{B} = \Psi \mathcal{B} \tag{12.21}$$

and where

$$B = A^{-1}[e^{AT} - I]\mathcal{B} = [e^{AT} - I]A^{-1}\mathcal{B}$$

when A is nonsingular.

As a numerical example, consider the continuous-time system,

$$\begin{bmatrix} \dot{x}_1(t) \\ \dot{x}_2(t) \end{bmatrix} = \begin{bmatrix} -2 & 2 \\ 1 & -3 \end{bmatrix} \begin{bmatrix} x_1(t) \\ x_2(t) \end{bmatrix} + \begin{bmatrix} -1 \\ 5 \end{bmatrix} u(t),$$

$$y(t) = [2 \quad -4] \begin{bmatrix} x_1(t) \\ x_2(t) \end{bmatrix} + 6u(t),$$

with a sampling interval $T = 0.2$ s. The matrix exponential is

$$e^{AT} = e^{0.2A} = \begin{bmatrix} 0.696 & 0.246 \\ 0.123 & 0.572 \end{bmatrix},$$

which can be calculated by truncating the power series

$$e^{AT} \cong I + AT + \frac{(AT)^2}{2!} + \frac{(AT)^3}{3!} + \cdots + \frac{(AT)^i}{i!}.$$

By examining the finite series as more and more terms are added, it can be decided when to truncate the series. However, there are pathologic matrices for which the series converges slowly, for which the series seems to converge first to one matrix then to another, and for which numerical rounding can give misleading results.

Continuing with the example, if the input $u(t)$ is constant in each interval from kT to $kT + T$, then

$$B = [e^{AT} - I]A^{-1}B$$

$$= \begin{bmatrix} -0.304 & 0.246 \\ 0.123 & -0.428 \end{bmatrix} \begin{bmatrix} -\dfrac{3}{4} & -\dfrac{1}{2} \\ -\dfrac{1}{4} & -\dfrac{1}{2} \end{bmatrix} \begin{bmatrix} -1 \\ 5 \end{bmatrix} = \begin{bmatrix} -0.021 \\ 0.747 \end{bmatrix},$$

which could also have been found using Equation 12.21. The discrete-time model of the continuous-time system is then

$$\begin{bmatrix} x_1(kT + T) \\ x_2(kT + T) \end{bmatrix} = \begin{bmatrix} 0.696 & 0.246 \\ 0.123 & 0.572 \end{bmatrix} \begin{bmatrix} x_1(kT) \\ x_2(kT) \end{bmatrix} + \begin{bmatrix} -0.021 \\ 0.747 \end{bmatrix} u(kT),$$

$$y(kT) = [2 \quad -4] \begin{bmatrix} x_1(kT) \\ x_2(kT) \end{bmatrix} + 6u(kT). \tag{12.22}$$

12.4.2 Approximation Methods

Another method for finding discrete-time equivalents of continuous-time systems described by state variable equations is to integrate Equation 12.16 as follows:

$$x(t) = x(t_0) + \int_{t_0}^{t} [Ax(t) + Bu(t)] \, dt. \tag{12.23}$$

For evenly spaced samples at $t = kT, k = 0, 1, 2, \ldots$

$$x(kT + T) = x(kT) + \int_{kT}^{kT+T} [Ax(t) + Bu(t)] \, dt. \tag{12.24}$$

Applying Euler's forward rectangular approximation of the integral,

$$x(kT + T) \cong x(kT) + [Ax(kT) + Bu(kT)]T$$

or

$$x(kT + T) \cong [I + AT]x(kT) + BTu(kT).$$

For the previous example, Euler's forward rectangular rule gives

$$\begin{bmatrix} x_1(kT + T) \\ x_2(kT + T) \end{bmatrix} \cong \left\{ \begin{bmatrix} 1 & 0 \\ 0 & 1 \end{bmatrix} + \begin{bmatrix} -2 & 2 \\ 1 & -3 \end{bmatrix} (0.2) \right\} \begin{bmatrix} x_1(kT) \\ x_2(kT) \end{bmatrix} + \begin{bmatrix} -1 \\ 5 \end{bmatrix} (0.2)u(kT)$$

or

$$\begin{bmatrix} x_1(kT + T) \\ x_2(kT + T) \end{bmatrix} \cong \begin{bmatrix} 0.6 & 0.4 \\ 0.2 & 0.4 \end{bmatrix} \begin{bmatrix} x_1(kT) \\ x_2(kT) \end{bmatrix} + \begin{bmatrix} -0.2 \\ 1 \end{bmatrix} u(kT)$$

and

$$y(kT) = [2 \quad -4] \begin{bmatrix} x_1(kT) \\ x_2(kT) \end{bmatrix} + 6u(kT),$$

which does not match well with the result given by Equation 12.22.

Reducing the sampling interval to $T = 0.01$ s, Euler's approximation gives

$$\begin{bmatrix} x_1(kT + T) \\ x_2(kT + T) \end{bmatrix} \cong \begin{bmatrix} 0.98 & 0.02 \\ 0.01 & 0.97 \end{bmatrix} \begin{bmatrix} x_1(kT) \\ x_2(kT) \end{bmatrix} + \begin{bmatrix} -0.01 \\ 0.05 \end{bmatrix} u(kT),$$

$$y(kT) = [2 \quad -4] \begin{bmatrix} x_1(kT) \\ x_2(kT) \end{bmatrix} + 6u(kT).$$

Taking the first two terms in the series in Equations 12.20 and 12.21 results in

$$\begin{bmatrix} x_1(kT + T) \\ x_2(kT + T) \end{bmatrix} \cong \begin{bmatrix} 0.9803 & 0.0195 \\ 0.0098 & 0.9706 \end{bmatrix} \begin{bmatrix} x_1(kT) \\ x_2(kT) \end{bmatrix} + \begin{bmatrix} -0.0094 \\ 0.0492 \end{bmatrix} u(kT),$$

which is in close agreement with Euler's result.

Euler's backward approximation of the integral in Equation 12.24 gives

$$x(kT + T) \cong x(kT) + [Ax(kT + T) + Bu(kT + T)]T$$

or

$$[I - AT]x(kT + T) \cong x(kT) + BTu(kT + T).$$

Hence,

$$x(kT + T) \cong [I - AT]^{-1}x(kT) + [I - AT]^{-1}BTu(kT + T).$$

Letting

$$\bar{x}(kT + T) = x(kT)$$

then

$$[I - AT]x(kT + T) \cong \bar{x}(kT + T) + BTu(kT + T)$$

or

$$[I - AT]x(kT) \cong \bar{x}(kT) + BTu(kT).$$

Hence,

$$x(kT) \cong [I - AT]^{-1}\bar{x}(kT) + [I - AT]^{-1}BTu(kT),$$

and, therefore,

$$\bar{x}(kT + T) \cong [I - \mathcal{A}T]^{-1}\bar{x}(kT) + [I - \mathcal{A}T]^{-1}\mathcal{B}Tu(kT).$$

The output equation

$$y(kT) = Cx(kT) + Du(kT)$$

in terms of the new variable $\bar{x}$ becomes

$$y(kT) \cong C[I - \mathcal{A}T]^{-1}\bar{x}(kT) + C[I - \mathcal{A}T]^{-1}\mathcal{B}Tu(kT) + Du(kT)$$

or

$$y(kT) \cong C[I - \mathcal{A}T]^{-1}\bar{x}(kT) + \left\{ C[I - \mathcal{A}T]^{-1}\mathcal{B}T + D \right\} u(kT).$$

Some formulas for discretizing continuous-time state variable equations using numerical integration are listed in Table 12.4. Derivative approximations, such as those listed in Table 12.3, are also possibilities for discretizing continuous-time state equations. Improved accuracy and a reduced sampling interval may also result from using more involved approximations, such as the predictor-corrector or Runge–Kutta methods.

12.4.3 Discrete-Time Equivalents of Pulsed Inputs

In the previous sections, the control input $u(t)$ is held constant during the entire sampling period T. In some applications, however, the control $u(t)$ may be constant for a fraction of the sampling period, T_w, that is, $0 < T_w \leq T$ as shown in Figure 12.17. In this situation, the state-space representation described in Section 1.3.4.1 above should be modified as follows:

$$x(kT + T) = e^{\mathcal{A}T}x(kT) + \int_{kT}^{kT+T} e^{\mathcal{A}(kT+T-\tau)}\mathcal{B}u(\tau)\, d\tau$$

$$= e^{\mathcal{A}T}x(kT) + \int_{kT}^{kT+T_w} e^{\mathcal{A}(kT+T-\tau)}\mathcal{B}u(\tau)\, d\tau + \int_{kT+T_w}^{kT+T} e^{\mathcal{A}(kT+T-\tau)}\mathcal{B}u(\tau)\, d\tau$$

$$= e^{\mathcal{A}T}x(kT) + \int_{kT}^{kT+T_w} e^{\mathcal{A}(kT+T-\tau)}\mathcal{B}u(\tau)\, d\tau,$$

because the value of the control input between $kT + T_w$ and $kT + T$ is zero. Letting

$$\eta = kT + T - \tau, \quad d\eta = -d\tau,$$

then

$$x(kT + T) = e^{\mathcal{A}T}x(kT) + \int_{T-T_w}^{T} e^{\mathcal{A}\eta}\mathcal{B}\, d\eta u(kT)$$

Again, letting

$$\gamma = \eta - (T - T_w), \quad d\gamma = d\eta,$$

then

$$x(kT + T) = e^{\mathcal{A}T}x(kT) + e^{\mathcal{A}(T-T_w)}\left[\int_{0}^{T_w} e^{\mathcal{A}\gamma}\mathcal{B}\, d\gamma \right] u(kT) \tag{12.25}$$

which is in the form given by Equation 12.19. The term $e^{\mathcal{A}(T-T_w)}$ on the right-hand side of Equation 12.25 can be easily obtained using Equation 12.20 by substituting $T - T_w$ instead of T. The term in brackets in Equation 12.25 can be evaluated using Equation 12.17 or 12.18 except T is replaced with T_w.

TABLE 12.4 Some Formulas for Discretizing Continuous-Time State Variable Models

For the nth-order continuous-time plant described by

$$\dot{x}(t) = \mathcal{A}x(t) + \mathcal{B}u(t)$$
$$y(t) = Cx(t) + Du(t)$$

its discrete-time equivalence is given by

1. Zero-order hold method

$$x[(k+1)T] = Ax(kT) + Bu(kT)$$
$$y(kT) = Cx(kT) + Du(kT)$$

where

$$A = e^{\mathcal{A}T} = I + \frac{\mathcal{A}T}{1!} + \frac{\mathcal{A}^2 T^2}{2!} + \cdots + \frac{\mathcal{A}^i T^i}{i!} + \cdots$$

$$B = \left[IT + \frac{\mathcal{A}T^2}{2!} + \frac{\mathcal{A}^2 T^3}{3!} + \cdots + \frac{\mathcal{A}^i T^{i+1}}{i+1!} + \cdots \right] \mathcal{B}$$

or

$$B = \mathcal{A}^{-1}[\exp(\mathcal{A}T) - I]\mathcal{B} = [\exp(\mathcal{A}T) - I]\mathcal{A}^{-1}\mathcal{B}$$

when $\mathcal{A}$ is nonsingular.

2. Euler's forward rectangular method

$$x(kT + T) \cong x(kT) + [\mathcal{A}x(kT) + \mathcal{B}u(kT)]T$$
$$y(kT) = Cx(kT) + Du(kT)$$

3. Euler's backward rectangular method

$$\bar{x}(kT + T) \cong [I - \mathcal{A}T]^{-1}\bar{x}(kT) + [I - \mathcal{A}T]^{-1}\mathcal{B}Tu(kT)$$
$$y(kT) \cong C[I - \mathcal{A}T]^{-1}\bar{x}(kT) + \{C[I - \mathcal{A}T]^{-1}\mathcal{B}T + D\}u(kT)$$

where

$$x(kT) = \bar{x}(kT + T)$$

4. Trapezoidal method

$$\bar{x}(kT + T) = \left(I - \frac{\mathcal{A}T}{2} \right)^{-1} \left(I + \frac{\mathcal{A}T}{2} \right) \bar{x}(kT)$$
$$+ \frac{T}{2} \left(I - \frac{\mathcal{A}T}{2} \right)^{-1} \mathcal{B}u(kT - T) + \frac{T}{2} \left(I - \frac{\mathcal{A}T}{2} \right)^{-1} \mathcal{B}u(kT)$$

$$y(kT) = C \left(I - \frac{\mathcal{A}T}{2} \right)^{-1} \left(I + \frac{\mathcal{A}T}{2} \right) \bar{x}(kT)$$
$$+ \frac{T}{2}C \left(I - \frac{\mathcal{A}T}{2} \right)^{-1} \mathcal{B}u(kT - T) + \left[\frac{T}{2}C \left(I - \frac{\mathcal{A}T}{2} \right)^{-1} \mathcal{B} + D \right] u(kT)$$

where

$$x(kT) = \bar{x}(kT + T)$$

For the example presented in Section 12.4.1, letting $T = 0.2$ and $T_w = 0.08$ gives

$$\begin{bmatrix} x_1(kT + T) \\ x_2(kT + T) \end{bmatrix} = \begin{bmatrix} 0.696 & 0.246 \\ 0.123 & 0.572 \end{bmatrix} \begin{bmatrix} x_1(kT) \\ x_2(kT) \end{bmatrix} + \begin{bmatrix} 0.0265 \\ 0.2463 \end{bmatrix} u(kT),$$

$$y(kT) = \begin{bmatrix} 2 & -4 \end{bmatrix} \begin{bmatrix} x_1(kT) \\ x_2(kT) \end{bmatrix} + 6u(kT).$$

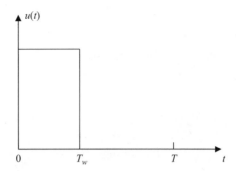

FIGURE 12.17   Control input $u(t)$ is constant for a fraction of the sampling period.

12.5 The Delta Operator

It is well known that when a digital controller, formulated using the z-transform and the shift operator methods discussed in the previous sections, is implemented with a finite-precision device, it may achieve a lower-than-predicted performance, or even become unstable due to the finite word length (FWL) effects. Furthermore, when the sampling period of a z-transform discrete-time system is very small, the response of the system may not converge smoothly to its continuous counterpart and hence may cause significant implementation issues.

On the other hand, the delta operator, given by

$$\delta = \frac{z-1}{T} \qquad (12.26)$$

offers superior numerical performance in FWL implementation over the z-transform and the shift operator. And, as the sampling rate is increased, the discrete-time results and models converge to their continuous counterparts. In this section, we will discuss the mathematical representation of discrete time controllers using the delta operator. The implementation of these digital controllers using FWL registers and finite-precision arithmetic is discussed in detail in Chapter 14.

12.5.1 Transfer Function Representation

Consider the second-order analog filer

$$G(s) = \frac{s^2 + 2\xi_z \omega_n s + \omega_n^2}{s^2 + 2\xi_p \omega_n s + \omega_n^2} \qquad (12.27)$$

Using the pole–zero matching method, the equivalent discrete-time filter $G(z)$ is determined by transforming the numerator and the denominator of $G(s)$ using the pair

$$(s+a)^2 + b^2 \rightarrow z^2 - 2(e^{-aT}\cos bT)z + e^{-2aT}$$

Then

$$G(z) = \frac{K\left[z^2 - 2e^{-\xi_z \omega_n T}\cos\left(\omega_n T\sqrt{1-\xi_z^2}\right)z + e^{-2\xi_z \omega_n T}\right]}{\left[z^2 - 2e^{-\xi_p \omega_n T}\cos\left(\omega_n T\sqrt{1-\xi_p^2}\right)z + e^{-2\xi_p \omega_n T}\right]}$$

or

$$G(z) = \frac{K(z^2 + a_1 z + a_0)}{z^2 + b_1 z + b_0} \qquad (12.28)$$

where $a_1 = -2e^{-\xi_z \omega_n T} \cos \omega_n T \sqrt{1 - \xi_z^2}$; $a_0 = e^{-2\xi_z \omega_n T}$; $b_1 = -2e^{-\xi_p \omega_n T} \cos \omega_n T \sqrt{1 - \xi_p^2}$; and $b_0 = e^{-2\xi_p \omega_n T}$.

The gain K is selected such that the DC gain of the digital filter is identical to the continuous-time filter given in Equation 12.27. Hence,

$$K = \frac{1 + b_1 + b_0}{1 + a_1 + a_0}.$$

In terms of the delta operator δ, substituting Equation 12.26 into Equation 12.28 gives

$$G(\delta) = \frac{K \left[\delta^2 + \dfrac{2 + a_1}{T} \delta + \dfrac{1 + a_1 + a_0}{T^2} \right]}{\delta^2 + \dfrac{2 + b_1}{T} \delta + \dfrac{1 + b_1 + b_0}{T^2}}.$$

As a numerical example, let

$$\xi_z = 0.07, \quad \xi_p = 0.7, \quad \omega_n = 2\pi(0.2), \quad T = 0.02.$$

Then, one can easily show that

$$G(s) = \frac{s^2 + 0.1759s + 1.5791}{s^2 + 1.7593s + 1.5791}, \tag{12.29}$$

$$G(z) = \frac{0.9843z^2 - 1.9646z + 0.9809}{z^2 - 1.9648z + 0.9654}, \tag{12.30}$$

$$G(\delta) = \frac{0.9843\delta^2 + 0.2039\delta + 1.5516}{\delta^2 + 1.7597\delta + 1.5516}. \tag{12.31}$$

The relationship between the Laplace transform, the z transform, and the delta operator is given by

$$z = e^{sT} = 1 + sT + h.o.t,$$

$$\delta = \frac{e^{sT} - 1}{T} = \frac{sT}{T} + h.o.t.$$

As the sampling interval approaches 0, $z \rightarrow sT + 1$, and $\delta \rightarrow s$, which means that the z transform model is sensitive to roundoff while the delta operator model converges to its continuous counterpart and is less sensitive to roundoff. This will be discussed further in Article 15. Also, it is instructive to compare the coefficients of the numerator and the denominator polynomials in Equations 12.29 and 12.31 to those in Equation 12.30.

12.5.2 State-Space Representation

Consider the nth-order continuous-time system described by the state variable Equation 12.16. The corresponding discrete-time system using the delta operator is

$$\delta[x(kT)] = A_\delta x(kT) + B_\delta u(kT),$$
$$y(kT) = Cx(kT) + Du(kT), \tag{12.32}$$

where

$$A_\delta = \Omega A$$
$$B_\delta = \Omega B \tag{12.33}$$

and

$$\Omega = \frac{1}{T} \int_0^T e^{A\gamma} d\gamma. \tag{12.34}$$

Recall that the delta operator is given by

$$\delta(.) = \frac{z-1}{T}.$$

An easy way to derive Equations 12.32 through 12.34 is to subtract $x(kT)$ from both sides of Equation 12.19 as follows:

$$x[(k+1)T] - x(kT) = Ax(kT) - x(kT) + Bu(kT).$$

Dividing both sides of the resulting Equation by T gives

$$\frac{x[(k+1)T] - x(kT)}{T} = \frac{A-I}{T}x(kT) + \frac{B}{T}u(kT).$$

But the left-hand side of this equation is simply the delta operator of the state vector x. Then

$$\delta[x(kT)] = \frac{A-I}{T}x(kT) + \frac{B}{T}u(kT). \tag{12.35}$$

Comparing Equation 12.32 to Equation 12.35 gives

$$A_\delta = \frac{A-I}{T}, \quad B_\delta = \frac{B}{T}, \tag{12.36}$$

where A and B are determined using Equations 12.20 and 12.21 and repeated here for convenience:

$$A = e^{AT} \tag{12.37}$$

and

$$B = \left[IT + \frac{AT^2}{2!} + \frac{A^2 T^3}{3!} + \cdots + \frac{A^i T^{i+1}}{(i+1)!} + \cdots \right] B \tag{12.38}$$

Finally, the equations in Equation 12.36 can be easily verified using Equations 12.33 and 12.34 and employing Equations 12.37 and 12.38.

It is evident from Equations 12.36 through 12.38 that as T approaches zero (neglect higher-order terms),

$$A_\delta = \frac{e^{AT} - I}{T} = \frac{I + AT - I}{T} = A$$

$$B_\delta = \frac{B}{T} = \frac{1}{T}\left[IT + \frac{AT^2}{2!} + \frac{A^2 T^3}{3!} + \cdots + \frac{A^i T^{i+1}}{(i+1)!} + \cdots \right] B = B$$

Hence, the discrete-time system described by Equation 12.35 approaches its continuous-time counterpart described by Equation 12.16.

Table 12.5 shows the Matlab code for simulating the unit step response of the second-order filter described by Equation 12.27. The results are shown in Figure 12.18 where the unit step response of the continuous-time filter and its discrete-time equivalent using the delta operator are indistinguishable.

One of the most important control system design tools is simulation, computer modeling of the plant and controller to verify the properties of a preliminary design and to test its performance under conditions (e.g., noise, disturbances, parameter variations, and nonlinearities) that might be difficult or cumbersome to study analytically. Through simulation, difficulties with between-sample plant response are discovered and solved.

When a continuous-time plant is simulated on a digital computer, its response is computed at closely spaced discrete times. It is plotted by joining the closely spaced calculated response values with straight line segments approximating a continuous curve. Digital computer simulation of discrete-time control of a continuous-time system involves at least two sets of discrete-time calculations. One runs at high rate to simulate the continuous-time plant. The other runs at a lower rate (say once every 10–50 of the former calculations) to generate new control signals at each discrete control step.

TABLE 12.5 Matlab Code to Simulate the Unit Step Response of a Second-Order Filter Using the Delta Operator

```
clear all
% Input continous-time filter parameters
    dz = 0.07;
    dp = 0.7;
    wn = 2*pi*0.2;
    T = 0.02; % sampling period
    numc = [1 2*dz*wn wn^2];
    denc = [1 2*dp*wn wn^2];
    sysc = tf(numc,denc);
% State space and T.F. using Dleta Operator
    sysd=c2d(sysc,T);
    [phi,gamma,cd,dd]=ssdata(sysd);
    Adel = (phi-eye(size(phi)))/T;
    Bdel = gamma/T;
    sys=ss(Adel,Bdel,cd,dd);
    sysdelta=canon(sys,'companion');
    [numdDel,dendDel]=ss2tf(Adel,Bdel,cd,dd);
% Unit step input
    u=1;
% parameters
    tend = 12;
    i=0;
    x=zeros(length(phi),1);
% Simulation loop
    for t=0:T:tend
        i=i+1;
        time(i) = t;
        y=cd*x+dd*u;
        delx = Adel*x+Bdel*u;
        x = x+T*delx;
        y2(:,i) = y;
    end

    figure(1)
    plot(time,y2);
    hold on
    [z,time]=step(sysc,time);
    plot(time,z);
    xlabel('Time (sec)')
    ylabel('filter step response')
    grid
    figure(2)
    plot(time,z'-y2);
    xlabel('Time (sec)')
    ylabel('error')
    grid
```

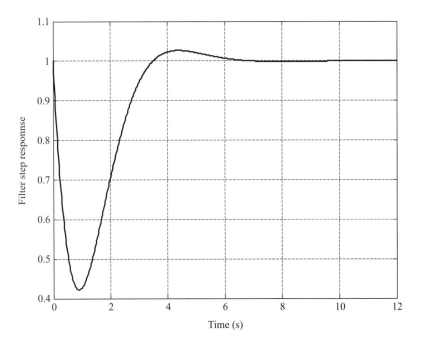

FIGURE 12.18   Analog filter unit step response and discrete-time filter unit step response using delta operator.

Bibliography

1. Santina, M.S., Stubberud, A.R., and Hostetter, G.H., *Digital Control System Design*, 2nd edn, Oxford University Press Inc., New York, NY, 1994.
2. DiStefano III, J.J., Stubberud, A.R., and Williams, I.J., *Feedback and Control Systems (Schaum's Outline)*, 2nd edn, McGraw-Hill, New York, NY, 1994.
3. Kuo, B.C., *Digital Control Systems*, 2nd edn, Oxford University Press Inc., New York, NY, 1995.
4. Franklin, G.F., Powell, J.D., and Workman, M.L., *Digital Control of Dynamic Systems*, 3rd edn, Ellis-Kagle Press, CA, 2006.
5. Phillips, C.L. and Nagle, Jr., H.T., *Digital Control System Analysis and Design*, 3rd edn, Prentice-Hall, Englewood Cliffs, NJ, 1994.
6. Ogata, K., *Discrete-Time Control Systems*, 2nd edn, Prentice-Hall, Englewood Cliffs, NJ, 1994.
7. Åström, K.J. and Wittenmark, B., *Computer Controlled Systems*, 3rd edn, Prentice-Hall, Englewood Cliffs, NJ, 1996.
8. Landau, I., *Digital Control Systems: Design, Identification and Implementation*, Springer, New York, NY, 2006.
9. Santina, M. and Stubberud, A., Discrete-time equivalents to continuous-time systems, *Control Systems, Robotics, and Automation, Encyclopedia of Life Support Systems (EOLSS)*, 2004.
10. Middleton, R. and Goodwin, G., *Digital Control and Estimation: A Unified Approach*, Prentice-Hall, Englewood Cliffs, NJ, 1990.
11. Moudgalya, K., *Digital Control*, Wiley-Interscience, MA, 2007.

13

Design Methods for Discrete-Time, Linear Time-Invariant Systems

Michael Santina
The Boeing Company

Allen R. Stubberud
University of California, Irvine

13.1 An Overview

The starting point of most beginning studies of classical and state-space control is with control of a linear, single-input–single-output, time-invariant plant. The tools of classical discrete-time linear control system design, which parallel the tools for continuous-time systems, are the z-transform, stability testing, root locus, and frequency response methods.

As in the classical approach to designing analog controllers, one begins with simple digital controllers, increasing their complexity until the performance requirements are met. The digital controller parameters are chosen to give feedback system pole locations that result in acceptable zero-input (transient) response. At the same time, requirements are placed on the overall system's zero-state response components for representative discrete-time reference inputs, such as steps or ramps.

Extending classical single-input–single-output control system design methods to the design of complicated feedback structures involving many loops, each including a compensator, is not easy. Put another way, modern control systems require the design of compensators having multiple inputs and multiple outputs. Design *is* iterative, involving considerable trial and error. Therefore with many design variables, it is important to deal efficiently with those design decisions that need not be iterative. The powerful methods of state space offer insights into what is possible and what is not. They also provide an excellent framework for general methods of approaching and accomplishing design objectives.

A *tracking system* is one in which the plant outputs are controlled so that they become and remain nearly equal to externally applied *reference inputs*. A special case of a tracking system in which the desired tracking system output is zero is termed a *regulator*. In general, tracking control system design has two basic concerns:

1. Obtaining acceptable zero-input response.
2. Obtaining acceptable zero-state system response to reference inputs.

In addition, if the plant is continuous-time, and the controller is discrete-time, it is necessary to

3. Obtain acceptable between-sample response of the continuous-time plant.

Through superposition, the zero-input response due to initial conditions and the individual zero-state response contributions of each input can be dealt with separately. The character of a system's zero-input response is determined by its pole locations, so that the first concern of tracking system design is met by selecting a controller that places all of the overall system poles in acceptable locations. Having designed a feedback structure placing all of the overall system poles to achieve the desired character of zero-input response, additional design freedom can then be used to obtain good tracking of reference inputs.

The above three concerns of digital tracking system design are the subject of this section.

13.2 Classical Control System Design Methods

A typical classical control system design problem is to determine and specify the transfer function $G_c(z)$ of a cascade compensator that results in a feedback tracking system with prescribed performance requirements. This is only a part of complete control system design, of course. It applies after a suitable model has been found and the performance requirements are quantified. In describing solution methods for idealized problems such as these, we separate general design principles from the highly specialized details of a particular application.

The basic system configuration for this problem is shown in Figure 13.1. There are, of course, many variations on this basic theme, including situations where the system structure is more involved, where there is a feedback transmittance $H(z)$, and where there are disturbance inputs to be considered. Usually, these disturbances are undesirable inputs that the plant *should not* track.

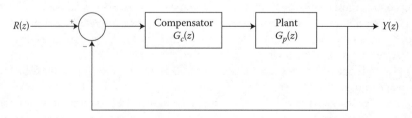

FIGURE 13.1 Cascade compensation of a unity feedback system.

The first concern of tracking system design is met by choosing the compensator $G_c(z)$ that results in acceptable pole locations for the overall transfer function

$$T(z) = \frac{Y(z)}{R(z)} = \frac{G_c(z)G_p(z)}{1 + G_c(z)G_p(z)}$$

Root locus is an important design tool because, with it, the effects on closed-loop system pole location of varying design parameters are quickly and easily visualized.

The second concern of tracking system design is obtaining acceptable closed-loop zero-state response to reference inputs. Zero-state performance is simple to deal with if it can be expressed as a maximum steady-state error to a power-of-time input. For the discrete-time system shown in Figure 13.2, the open-loop transfer function may have the form

$$KG(z)H(z) = \frac{K(z+\alpha_1)(z+\alpha_2)\cdots(z+\alpha_l)}{(z-1)^n(z+\beta_1)(z+\beta_2)\cdots(z+\beta_m)} = \frac{KN(z)}{(z-1)^nD(z)} \tag{13.1}$$

If n is nonnegative, the system is said to be *type n*.

The error between the input and the output of the system is

$$E(z) = R(z) - Y(z)H(z)$$

but

$$Y(z) = KE(z)G(z)$$

Then

$$T_E(z) = \frac{E(z)}{R(z)} = \frac{1}{1 + KG(z)H(z)}$$

The steady-state error to a power-of-time input is given by the final value theorem

$$\lim_{k \to \infty} e(k) = \lim_{z \to 1} (1 - z^{-1})E(z) = \lim_{z \to 1} \frac{(1 - z^{-1})R(z)}{1 + KG(z)H(z)} \tag{13.2}$$

provided that the limit exists. A necessary condition for the limit to exist and be finite is that all the closed-loop poles of the system be inside the unit circle on the z-plane.

Similar to continuous-time systems, there are three reference inputs for which steady-state errors are commonly defined. They are the step (position), ramp (velocity), and parabolic (acceleration) inputs. The

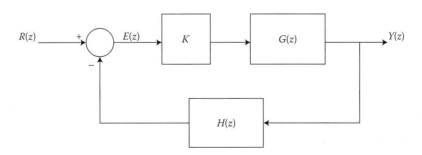

FIGURE 13.2 A discrete-time control system.

TABLE 13.1 Steady State Errors to Power-of-Time Inputs

System Type	Steady-State Error to Step Input $R(z) = \dfrac{Az}{z-1}$	Steady-State Error to Ramp Input $R(z) = \dfrac{ATz}{(z-1)^2}$	Steady-State Error to Parabolic Input $R(z) = \dfrac{T^2}{2}\dfrac{Az(z+1)}{(z-1)^3}$
0	$\dfrac{A}{1+K\dfrac{N(1)}{D(1)}}$	∞	∞
1	0	$\dfrac{AT}{K\dfrac{N(1)}{D(1)}}$	∞
2	0	0	$\dfrac{AT^2}{K\dfrac{N(1)}{D(1)}}$
$\vdots$	$\vdots$	$\vdots$	$\vdots$
n	0	0	0

step input has the form

$$r(kT) = Au(kT)$$

or, in the z-domain,

$$R(z) = \frac{Az}{z-1}$$

and the ramp input is

$$r(kT) = AkTu(kT)$$

or

$$R(z) = \frac{ATz}{(z-1)^2}$$

The parabolic input is

$$r(kT) = \frac{1}{2}A(kT)^2u(kT)$$

or

$$R(z) = \frac{T^2}{2}\frac{Az(z+1)}{(z-1)^3}$$

Table 13.1 summarizes steady-state errors using Equations 13.1 and 13.2 for various system types for power-of-time inputs.

We now present an overview of classical discrete-time control system design using an example.

13.2.1 Root Locus Design Methods

Similar to continuous-time systems, the root locus plot of a discrete-time system consists of the loci of the poles of a transfer function as some parameter is varied. For the configuration shown in Figure 13.2 where the constant gain K is the parameter of interest, the overall transfer function of this system is

$$T(z) = \frac{KG(z)}{1 + KG(z)H(z)}$$

and the poles of the overall system are the roots of

$$1 + KG(z)H(z) = 0$$

which depend on the parameter K. The rules for constructing the root locus of discrete-time systems are identical to the rules for plotting the root locus of continuous-time systems (see Chapter 9.4). The root locus plot, however, must be interpreted relative to the z-plane.

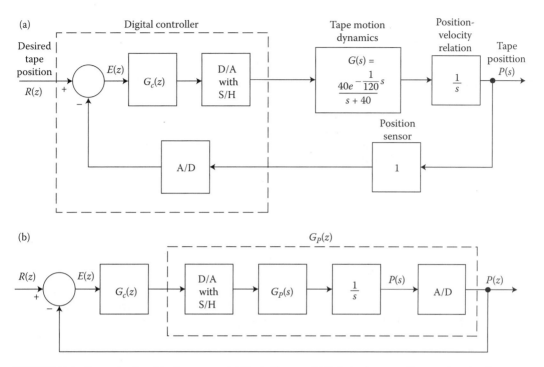

FIGURE 13.3 Example of positioning system. (a) Block diagram. (b) Relation between discrete-time signals.

Consider the block diagram of a positioning system shown in Figure 13.3a. The transfer function $G(s)$ relates the input to a motion actuator and the speed of the element being positioned at two different points. The delay term accounts for the propagation of speed changes over the distance physically separating the two points. The pole term in $G(s)$ represents the dynamics of the actuator and the element being positioned. The position variable is sensed directly.

It is desired to design a digital controller that results in zero steady-state error to any step change in desired position. Also, the system should have a zero-input (or transient) response that decays to no more than 10% of any initial value within a 1/30 s interval. The sampling period of the controller is chosen as $T = 1/120$ s. In Figure 13.3b the diagram of Figure 13.3a has been rearranged to emphasize the discrete-time input $R(z)$ and the discrete-time samples $P(z)$ of the positioning system output.

The open-loop transfer function of the system is

$$
\begin{aligned}
G_p(z) &= Z\left[\left(\frac{1-e^{-(1/120)s}}{s}\right)\left(\frac{40e^{-(1/120)s}}{s+40}\right)\left(\frac{1}{s}\right)\right] \\
&= Z\left\{[1-e^{-(1/120)s}]e^{-(1/120)s}\left[\frac{-(1/140)}{s}+\frac{1}{s^2}+\frac{1/40}{s+40}\right]\right\} \\
&= (1-z^{-1})z^{-1}\left[\frac{-z/40}{z-1}+\frac{z/120}{(z-1)^2}+\frac{z/40}{z-0.72}\right] \\
&= \frac{0.00133(z+0.75)}{z(z-1)(z-0.72)}
\end{aligned}
$$

The position error signal, in terms of the compensator's z-transfer function $G_c(z)$, is

$$
\begin{aligned}
E(z) &= R(z) - Y(z) \\
&= \left[1-\frac{G_c(z)G_p(z)}{1+G_c(z)G_p(z)}\right]R(z) = \frac{1}{1+G_c(z)G_p(z)}R(z)
\end{aligned}
$$

For a unit step input sequence,

$$E(z) = \frac{1}{1 + G_c(z)G_p(z)} \left(\frac{z}{z-1}\right)$$

Assuming that the feedback system is stable,

$$\lim_{k \to \infty} e(k) = \lim_{z \to 1}\left[\frac{z-1}{z}E(z)\right] = \lim_{z \to 1}\left[\frac{1}{1 + G_c(z)G_p(z)}\right]$$

Provided that the compensator does not have a zero at $z = 1$, the system type is 1 and, therefore according to Table 13.1, the steady-state error to a step input is zero. For the feedback system transient response to decay by at least a factor 1/10 within 1/30 s, the desired closed-loop poles must be located so that a decay of at least this amount occurs every 1/120 s step. This implies that the closed-loop poles must lie within a radius c of the origin on the z-plane, where

$$c^4 = 0.1, \quad c = 0.56$$

Similar to continuous-time systems, one usually begins with the simplest compensator consisting of only a gain K. The feedback system is stable for

$$0 < K < 95$$

but, as shown in Figure 13.4, this compensator is inadequate because there are always poles at distances from the origin greater than the required $c = 0.56$ regardless of the value of K. As shown in Figure 13.5a, another compensator with z-transfer function,

$$G_c(z) = \frac{K(z - 0.72)}{z}$$

which cancels the plant pole at $z = 0.72$, is tried. The root locus plot for this system is shown in Figure 13.5b. For $K = 90$, the design is close to meeting the requirements, but it is not quite good

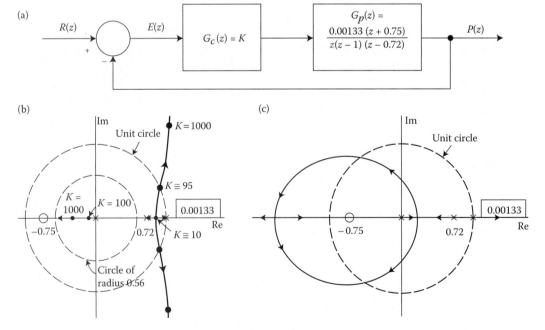

FIGURE 13.4 Constant-gain compensator. (a) Block diagram. (b) Root locus for positive K. (c) Root locus for negative K.

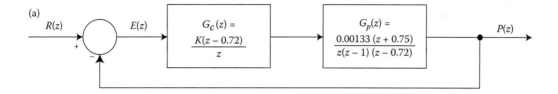

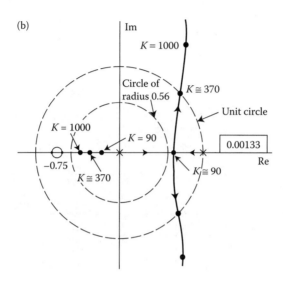

FIGURE 13.5 Compensator with zero at $z = 0.72$ and pole at $z = 0$. (a) Block diagram. (b) Root locus for positive K.

enough. However, if the compensator pole is moved from the origin to the left as shown in Figure 13.6, the root locus is pulled to the left and the performance requirements are met.

For the compensator with z-transfer function

$$G_c(z) = \frac{150(z - 0.72)}{z + 0.4} \tag{13.3}$$

the feedback system z-transfer function is

$$T(z) = \frac{G_c(z)G_p(z)}{1 + G_c(z)G_p(z)} = \frac{0.2(z + 0.75)}{z^3 - 0.6z^2 - 0.2z + 0.15}$$

$$= \frac{0.2(z + 0.75)}{(z - 0.539 - j0.155)(z - 0.539 + j0.155)} \frac{1}{(z + 0.477)}$$

As expected, the steady-state error to a step input is zero,

$$\lim_{z \to 1} \left\{ \left(\frac{z - 1}{z} \right)[1 - T(z)]\left(\frac{z}{z - 1} \right) \right\} = \lim_{z \to 1} \frac{z^3 - 0.6z^2 - 0.4z}{z^3 - 0.6z^2 - 0.2z + 0.15} = 0$$

The steady-state error to a unit ramp input is

$$\lim_{z \to 1} \left\{ \left(\frac{z - 1}{z} \right)[1 - T(z)]\left[\frac{Tz}{(z - 1)^2} \right] \right\}$$

$$= \lim_{z \to 1} \frac{\frac{1}{120}(z^2 + 0.4z)}{z^3 - 0.6z^2 - 0.2z + 0.15} = \frac{1}{30}$$

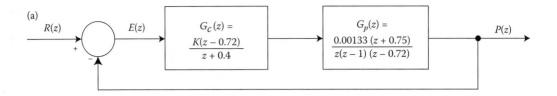

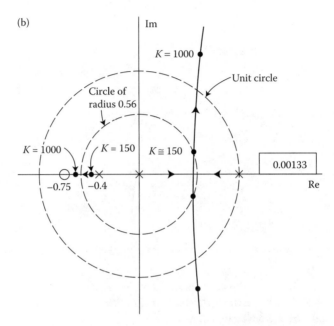

FIGURE 13.6 Compensator with zero at $z = 0.72$ and pole at $z = 0.4$. (a) Block diagram. (b) Root locus for positive K.

For a compensator with a z-transfer function of the form

$$G_c(z) = \frac{150(z - 0.72)}{z + a}$$

the feedback system has the z-transfer function,

$$
\begin{aligned}
T(z) &= \frac{G_c(z)G_p(z)}{1 + G_c(z)G_p(z)} = \frac{0.2(z + 0.75)}{(z + a)(z^2 - z) + 0.2(z + 0.75)} \\[6pt]
&= \frac{0.2(z + 0.75)}{z^3 - z^2 + 0.2z + 0.15 + a(z^2 - z)} = \frac{0.2(z + 0.75)/[z^3 - z^2 + 0.2z + 0.15]}{1 + az(z - 1)/[z^3 - z^2 + 0.2z + 0.15]} \\[6pt]
&= \frac{\text{numerator}}{1 + az(z - 1)/[(z - 0.637 - j0.378)(z - 0.637 + j0.378)(z + 0.274)]}
\end{aligned}
$$

A root locus plot in terms of positive a in Figure 13.7 shows that choices of a between 0.4 and 0.5 give a controller that meets the performance requirements.

Classical discrete-time control system design is an iterative process just like its continuous-time counterpart. Increasingly complicated controllers are tried until both the steady-state error and transient

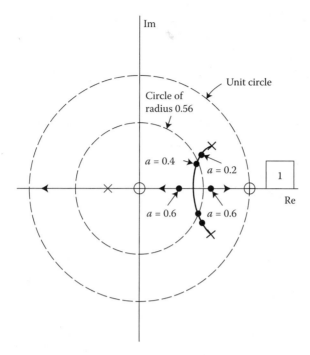

FIGURE 13.7   Root locus plot as a function of the compensator pole location.

performance requirements are met. Root locus is an important tool because it easily indicates qualitative closed-loop system pole locations as a function of a parameter. Once feasible controllers are selected, root locus plots are refined to show quantitative results.

13.2.2 Frequency-Domain Methods

Frequency response characterizations of systems have long been popular because steady-state sinusoidal response methods are easy and practical. Furthermore, frequency response methods do not require explicit knowledge of system transfer function models.

For the positioning system in the previous example, the open-loop z-transfer function that includes the compensator given by Equation 13.3 is

$$G_c(z)G_p(z) = \frac{(150)(0.00133)(z + 0.75)}{z(z + 0.4)(z - 1)}$$

Substituting $z = e^{j\omega T}$, then

$$G_c(e^{j\omega T})G_p(e^{j\omega T}) = \frac{0.1995(e^{j\omega T} + 0.75)}{e^{j\omega T}(e^{j\omega T} + 0.4)(e^{j\omega T} - 1)} \tag{13.4}$$

which has frequency response plots shown in Figure 13.8. At the *phase crossover frequency* (114.2 rad/s), the *gain margin* is 11.48 dB, and, at the *gain crossover frequency* (30 rad/s), the *phase margin* is about 66.5°.

For ease in generating frequency response plots and gaining greater insight into the design process, frequency-domain methods such as Nyquist, Bode, Nichols, etc. for discrete-time systems are best developed with the *w-transform*. In the *w*-plane, the wealth of tools and techniques developed for continuous-time systems are directly applicable to discrete-time systems as well.

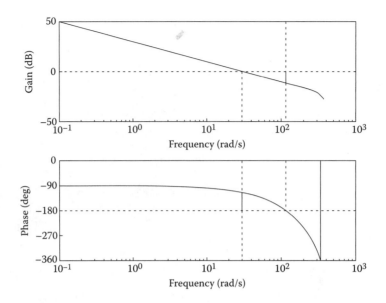

FIGURE 13.8   Frequency response plots of the positioning system.

The *w*-transform is

$$w = \frac{z-1}{z+1}, \quad z = \frac{w+1}{1-w}$$

which is a bilinear transformation between the *w*-plane and the *z*-plane.

The general procedure for analyzing and designing discrete-time systems with the *w*-transform is summarized as follows:

1. Apply the *w*-transform to the open-loop transfer function $G(z)H(z)$ by replacing each *z* in $G(z)H(z)$ with

$$z = \frac{w+1}{1-w}$$

   to obtain $G(w)H(w)$. Note that the functions *G* and *H* actually are different after the substitution.

2. Visualizing the *w*-plane as if it were the *s*-plane, substitute $w = jv$ into $G(w)H(w)$ and generate frequency response plots in terms of the real frequency *v*, such as Nyquist, Bode, Nichols, etc.

3. Determine the gain margin, phase margin, crossover frequencies, bandwidth, closed-loop frequency response, or any other desired frequency response characteristics.

4. If it is necessary, design an appropriate compensator $G_c(w)$ to satisfy the frequency-domain performance requirements.

5. Convert critical frequencies *v* in the *w*-plane to frequencies ω in the *z*-domain according to

$$v = \tan \frac{\omega T}{2}$$

   or

$$\omega = \frac{2}{T} \tan^{-1} v$$

6. Transform the controller $G_c(w)$ to $G_c(z)$ according to

$$w = \frac{z-1}{z+1}$$

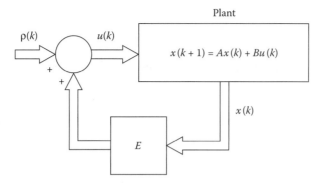

FIGURE 13.9 State feedback.

Control system design for discrete-time systems using Bode, Nyquist, or Nichols methods can be found in [1,2]. Frequency response methods are most useful in developing models from experimental data, in verifying the performance of a system designed by other methods, and in dealing with those systems and situations in which rational transfer function models are not adequate.

13.3 Eigenvalue Placement with State Feedback

All of the results for eigenvalue placement with state feedback for continuous-time systems carry over to discrete-time systems. For a linear, step-invariant nth-order system, described by the state equation

$$x(k+1) = Ax(k) + Bu(k)$$

consider the state-feedback arrangement

$$u(k) = Ex(k) + \rho(k)$$

where $\rho(k)$ is a vector of external inputs, as shown in Figure 13.9. Provided that the plant is completely controllable, and that the state is accessible for feedback,* the feedback gain matrix E can always be chosen so that each of the eigenvalues of the feedback system

$$x(k+1) = (A + BE)x(k) + B\rho(k)$$

is at an arbitrary desired location selected by the designer. This is to say that the designer can freely choose the character of the overall system's transient performance.

13.3.1 Eigenvalue Placement for Single-Input Systems

There are a number of methods for finding the state feedback gain vector of single-input plants, one is summarized below, and additional ones can be found in [3,4] and in *Control System Fundamentals*, Chapter 5 and *Control System Advanced Methods*, Chapter 16.

---

* When the plant state vector is not available for feedback, as is usually the case, an *observer* is designed to estimate the state vector. The observer state estimate is used for feedback in place of the state itself.

One method for calculating the state feedback gain vector is given by *Ackermann's formula*:

$$e^\dagger = -j_n^\dagger M_c^{-1} \Delta_c(A)$$

where $j_n^\dagger$ is the transpose of the *n*th-unit coordinate vector

$$j_n^\dagger = [0 \quad 0 \quad \cdots \quad 0 \quad 1]$$

M_c is the controllability matrix of the system, and $\Delta_c(A)$ is the desired characteristic equation with the matrix A substituted for the variable z.

For example, for the completely controllable system,

$$\begin{bmatrix} x_1(k+1) \\ x_2(k+1) \end{bmatrix} = \begin{bmatrix} 1 & -1 \\ 3 & 0 \end{bmatrix} \begin{bmatrix} x_1(k) \\ x_2(k) \end{bmatrix} + \begin{bmatrix} 1 \\ 2 \end{bmatrix} u(k)$$

it is desired to place the feedback system eigenvalues at $z = 0, -0.5$. Then,

$$\Delta_c(z) = z(z+0.5) = z^2 + 0.5z$$

and

$$\Delta_c(A) = A^2 + 0.5A$$

Using Ackermann's formula, the state-feedback gain vector is

$$e^\dagger = -[0 \quad 1] \begin{bmatrix} 1 & -1 \\ 2 & 3 \end{bmatrix}^{-1} \left\{ \begin{bmatrix} -2 & -1 \\ 3 & -3 \end{bmatrix} + 0.5 \begin{bmatrix} 1 & -1 \\ 3 & 0 \end{bmatrix} \right\} = [-1.5 \quad 0]$$

13.3.2 Eigenvalue Placement with Multiple Inputs

If the plant has multiple inputs and if it is completely controllable from one of the inputs, then that one input alone can be used for feedback. If the plant is not completely controllable from a single input, a single input can usually be distributed to the multiple ones so that the plant is completely controllable from the single input.

For example, for the system

$$\begin{bmatrix} x_1(k+1) \\ x_2(k+1) \\ x_3(k+1) \end{bmatrix} = \begin{bmatrix} -0.5 & 0 & 1 \\ 0 & 0.5 & 2 \\ 1 & -1 & 0 \end{bmatrix} \begin{bmatrix} x_1(k) \\ x_2(k) \\ x_3(k) \end{bmatrix} + \begin{bmatrix} 1 & 0 \\ 0 & -2 \\ -1 & 1 \end{bmatrix} \begin{bmatrix} u_1(k) \\ u_2(k) \end{bmatrix}$$

letting

$$u_1(k) = 3\mu(k)$$

and

$$u_2(k) = \mu(k)$$

then

$$\begin{bmatrix} x_1(k+1) \\ x_2(k+1) \\ x_3(k+1) \end{bmatrix} = \begin{bmatrix} -0.5 & 0 & 1 \\ 0 & 0.5 & 2 \\ 1 & -1 & 0 \end{bmatrix} \begin{bmatrix} x_1(k) \\ x_2(k) \\ x_3(k) \end{bmatrix} + \begin{bmatrix} 3 \\ -2 \\ -2 \end{bmatrix} [\mu(k)]$$

which is a controllable single-input system. If the desired eigenvalues are located at $z_1 = -0.1, z_2 = -0.15$. and $z_3 = 0.1$, Ackermann's formula gives

$$e^\dagger = [0.152 \quad 0.0223 \quad 0.2807]$$

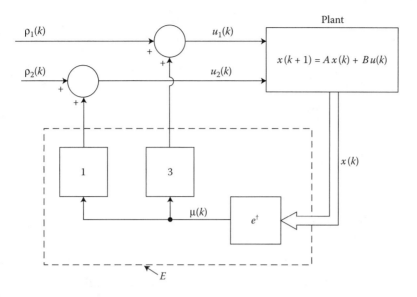

FIGURE 13.10   State feedback to a plant with multiple inputs.

and hence, the feedback gain matrix for the multiple input system is

$$E = \begin{bmatrix} 0.4559 & 0.0669 & 0.8420 \\ 0.1520 & 0.0223 & 0.2807 \end{bmatrix}$$

The structure of this system is shown in Figure 13.10.

13.3.3 Eigenvalue Placement with Output Feedback

It is the measurement vector of a plant, not the state vector, that is available for feedback. For the nth-order plant with state and output equations

$$x(k+1) = Ax(k) + Bu(k)$$
$$y(k) = Cx(k) + Du(k)$$

if the output coupling matrix C has n linearly independent rows, then the plant state can be recovered from the plant inputs and the measurement outputs and the method of the previous section applied:

$$x(k) = C^{-1}\{y(k) - Du(k)\}$$

When the nth-order plant does not have n linearly independent measurement outputs, it still might be possible to select a feedback matrix E in

$$u(k) = E\{y(k) - Du(k)\} + \rho(k) = ECx(k) + \rho(k)$$

to place all of the feedback system eigenvalues, those of $(A + BEC)$, acceptably. Generally, however, measurement feedback alone is insufficient for arbitrary eigenvalue placement.

13.3.4 Pole Placement with Feedback Compensation

Similar to output feedback, pole placement with feedback compensation assumes that the measurement outputs of a plant, not the state vector, are available for feedback.

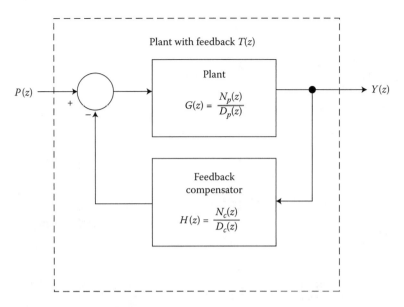

FIGURE 13.11   Pole placement with feedback compensation.

Consider the single-input–single-output, nth-order, linear, step-invariant, discrete-time system described by the transfer function $G(z)$. Arbitrary pole placement of the feedback system can be accomplished with an mth-order feedback compensator as shown in Figure 13.11.

Let the numerator and the denominator polynomials of $G(z)$ be $N_p(z)$ and $D_p(z)$, respectively. Also, let the numerator and the denominator of the compensator transfer function $H(z)$ be $N_c(z)$, and $D_c(z)$, respectively. Then, the overall transfer function of the system is

$$T(z) = \frac{G(z)}{1 + G(z)H(z)} = \frac{N_p(z)/D_p(z)}{1 + [N_p(z)/D_p(z)][N_c(z)/D_c(z)]}$$

$$= \frac{N_p(z)D_c(z)}{D_p(z)D_c(z) + N_p(z)N_c(z)} = \frac{P(z)}{Q(z)}$$

which has closed-loop zeros in $P(z)$ that are those of the plant, in $N_p(z)$, together with zeros that are the poles of the feedback compensator, in $D_c(z)$.

For a desired set of poles of $T(z)$, given with an unknown multiplicative constant by the polynomial $Q(z)$,

$$D_p(z)D_c(z) + N_p(z)N_c(z) = Q(z)$$

The desired polynomial $Q(z)$ has the form

$$Q(z) = \alpha_0(z^{n+m} + \beta_{n+m-1}z^{n+m-1} + \cdots + \beta_1 z + \beta_0)$$

where the βs are known coefficients, but the α_0 is unknown. In general, for a solution to exist, there must be at least as many unknowns as equations

$$n + m + 1 \leq 2m + 2$$

or

$$m \geq n - 1 \tag{13.5}$$

where n is the order of the plant and m is the order of the compensator. Equation 13.5 states that the order of the feedback controller cannot be less than the order of the plant minus one. If the plant transfer function has *coprime* numerator and denominator polynomials (i.e., plant pole–zero cancellations have been made), then a solution is guaranteed to exist.

For example, consider the second-order plant

$$G(z) = \frac{(z+1)(z+0.5)}{z(z-1)} = \frac{N_p(z)}{D_p(z)} \tag{13.6}$$

According to Equation 13.5, a first-order feedback compensator of the form

$$H(z) = \frac{\alpha_1 z + \alpha_2}{z + \alpha_3} = \frac{N_c(z)}{D_c(z)}$$

places the three closed-loop poles of the feedback system at any desired location in the z-plane by appropriate choice of α_1, α_2, and α_3. Let the desired poles of the plant with feedback be at $z = 0.1$. Then,

$$Q(z) = \alpha_0(z - 0.1)^3 = \alpha_0(z^3 - 0.3z^2 + 0.03z - 0.001) \tag{13.7}$$

In terms of the compensator coefficients, the characteristic equation of the feedback system is

$$\begin{aligned}
D_p(z)D_c(z) &+ N_p(z)N_c(z) \\
&= z(z-1)(z+\alpha_3) + (z+1)(z+0.5)(\alpha_1 z + \alpha_2) \\
&= (\alpha_1 + 1)z^3 + (1.5\alpha_1 + \alpha_2 + \alpha_3 - 1)z^2 + (0.5\alpha_1 + 1.5\alpha_2 - \alpha_3)z + 0.5\alpha_2
\end{aligned} \tag{13.8}$$

Equating coefficients in Equations 13.7 and 13.8 and solving for the unknowns,

$$\alpha_0 = 1.325 \quad \alpha_1 = 0.325 \quad \alpha_2 = -0.00265 \quad \alpha_3 = 0.1185$$

Therefore, the compensator

$$H(z) = \frac{0.325z - 0.00265}{z + 0.1185}$$

will place the closed-loop poles where desired.

As far as feedback system pole placement is concerned, a feedback compensator of order $n - 1$, where n is the order of the plant, can always be designed. It is possible, however, that a lower-order feedback controller may give acceptable feedback pole locations even though those locations are constrained and not completely arbitrary. This is the thrust of classical control system design, in which increasingly higher-order controllers are tested until satisfactory results are obtained.

For the plant given by Equation 13.6, for example, a *zeroth*-order feedback controller of the form

$$H(z) = K$$

gives overall closed-loop poles at $z = 0.1428$ and $z = 0.5$ for $K = 1/6$ which might be an adequate pole placement design.

13.4 Step-Invariant Discrete-Time Observer Design

When the plant state vector is not entirely accessible, as is usually the case, the state is *estimated* with an observer, and the estimated state is used in place of the actual state for feedback. See the article entitled "Observers."

13.4.1 Full-Order Observers

A full-order state observer of an nth-order step-invariant discrete-time plant

$$x(k+1) = Ax(k) + Bu(k)$$
$$y(k) = Cx(k) + Du(k)$$

$$(13.9)$$

is another nth-order system of the form

$$\xi(k+1) = F\xi(k) + Gy(k) + Hu(k) \tag{13.10}$$

driven by the inputs and outputs of the plant so that the error between the plant state and the observer state

$$x(k+1) - \xi(k+1) = Ax(k) + Bu(k) - F\xi(k) - Gy(k) - Hu(k)$$
$$= Ax(k) + Bu(k) - F\xi(k) - GCx(k) - GDu(k) - Hu(k)$$
$$= (A - GC)x(k) - F\xi(k) + (B - GD - H)u(k)$$

is governed by an autonomous equation. This requires that

$$F = A - GC \tag{13.11}$$
$$H = B - GD \tag{13.12}$$

so that the error satisfies

$$x(k+1) - \xi(k+1) = (A - GC)[x(k) - \xi(k)]$$

or

$$x(k) - \xi(k) = (A - GC)^k[x(0) - \xi(0)] = F^k[x(0) - \xi(0)]$$

The eigenvalues of $F = A - GC$ can be placed arbitrarily by the choice of G, provided that the system is completely observable. The observer error, then, approaches zero with step regardless of the initial values of $x(0)$ and $\xi(0)$ that is, the observer state $\xi(k)$ will approach the plant state $x(k)$. The full-order observer relations are summarized in Table 13.2. If all n of the observer eigenvalues (eigenvalues of F) are selected to be zero, then the characteristic equation of F is

$$\lambda^n = 0$$

and, because every matrix satisfies its own characteristic equation, then

$$F^n = 0$$

At the nth step, the error between the plant state and the observer state is

$$x(n) - \xi(n) = F^n[x(0) - \xi(0)]$$

so that

$$x(n) = \xi(n)$$

and the observer state equals the plant state. Such an observer is termed *deadbeat*. In subsequent steps, the observer state continues to equal the plant state.

TABLE 13.2 Full-Order State Observer
Relations

Plant model

$x(k+1) = Ax(k) + Bu(k)$

$y(k) = Cx(k) + Du(k)$

Observer

$\xi(k+1) = F\xi(k) + Gy(k) + Hu(k)$

where

$F = A - GC$

$H = B - GD$

Observer error

$x(k+1) - \xi(k+1) = F[x(k) - \xi(k)]$

$x(k) - \xi(k) = F^k[x(0) - \xi(0)]$

There are a number of ways for calculating the observer gain matrix g for single-output plants. Similar to the situation with state feedback, the eigenvalues of $F = A - gc^\dagger$ can be placed arbitrarily by choice of g as given by Ackermann's formula

$$g = \Delta_0(A)M_0^{-1}j_n \qquad (13.13)$$

provided that $(A, c^\dagger)$ is completely observable. In Equation 13.13, $\Delta_0(A)$ is the desired characteristic equation of the observer eigenvalues with the matrix A substituted for the variable z, M_0 is the observability matrix

$$M_0 = \begin{bmatrix} c^\dagger \\ c^\dagger A \\ c^\dagger A^2 \\ \vdots \\ c^\dagger A^{n-1} \end{bmatrix}$$

and j_n is the nth-unit coordinate vector

$$j_n = \begin{bmatrix} 0 \\ 0 \\ 0 \\ \vdots \\ 1 \end{bmatrix}$$

It is enlightening to express the full-order observer equations given by Equations 13.10 through 13.12 in the form

$$\xi(k+1) = (A - GC)\xi(k) + Gy(k) + (B - GD)u(k)$$
$$= A\xi(k) + Bu(k) + G[y(k) - w(k)]$$

where

$$w(k) = C\xi(k) + Du(k)$$

The observer consists of a model of the plant driven by the input $u(k)$ and the error between the plant output $y(k)$ and the plant output that is estimated by the model $w(k)$.

13.4.2 Reduced-Order State Observers

Rather than estimating the entire state vector of a plant, if a completely observable nth-order plant has m linearly independent outputs, a *reduced-order* state observer, of order $n - m$, having an output that observes the plant state can be constructed. See the article entitled "Observers."

When an observer's state

$$\xi(k+1) = F\xi(k) + Gy(k) + Hu(k)$$

estimates a linear combination $Mx(k)$ of the plant state rather than the state itself, the error between the observer state and the plant state transformation is given by

$$Mx(k+1) - \xi(k+1) = MAx(k) + MBu(k) - F\xi(k) - Gy(k) - Hu(k)$$
$$= (MA - GC)x(k) - F\xi(k) + (MB - GD - H)u(k)$$

where M is $(n-m)xn$. For the observer error system to be autonomous,

$$FM = MA - GC \tag{13.14}$$
$$H = MB - GD$$

so that the error is governed by

$$Mx(k+1) - \xi(k+1) = F[Mx(k) - \xi(k)]$$

For a completely observable plant, the observer gain matrix g can always be chosen so that all of the eigenvalues of F are inside the unit circle on the complex plane. Then the observer error

$$Mx(k) - \xi(k) = F^k[Mx(0) - \xi(0)]$$

will approach zero asymptotically with step, and then

$$\xi(k) \rightarrow Mx(k)$$

If the plant outputs, which also involve linear transformation of the plant state, are used in the formulation of a state observer, the dynamic order of the observer can be reduced. For the nth-order plant given by Equation 13.9 with the m rows of C linearly independent, an observer of order $n-m$ with n outputs

$$w'(k) = \begin{bmatrix} 0 \\ I \end{bmatrix} \xi(k) + \begin{bmatrix} I \\ 0 \end{bmatrix} y(k) - \begin{bmatrix} D \\ 0 \end{bmatrix} u(k)$$

observes

$$w'(k) \rightarrow \begin{bmatrix} C \\ M \end{bmatrix} x(k) = Nx(k)$$

Except in special cases, the rows of M and the rows of C are linearly independent. If they are not so, slightly different observer eigenvalues can be chosen to give linear independence. Therefore,

$$w(k) = N^{-1}w'(k)$$

observes $x(k)$.

13.4.3 Eigenvalue Placement with Observer Feedback

When observer feedback is used in place of plant state feedback, the eigenvalues of the feedback system are those the plant would have, if the state feedback were used, and those of the observer. This result is known as the *separation theorem* for observer feedback. For a completely controllable and completely observable plant, an observer of the form

$$\xi(k+1) = F\xi(k) + Gy(k) + Hu(k) \tag{13.15}$$
$$w(k) = L\xi(k) + N[y(k) - Du(k)] \tag{13.16}$$

with feedback to the plant given by

$$u(k) = Ew(k) \tag{13.17}$$

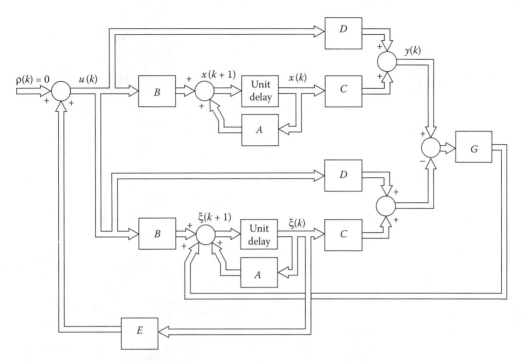

FIGURE 13.12 Eigenvalue placement with full-order observer feedback.

can thus be designed so that the overall feedback system eigenvalues are specified by the designer. The design procedure can proceed in two steps. First, the state feedback is designed to place the n state feedback system eigenvalues at desired locations as if the state vector is accessible. Second, the state feedback is replaced by feedback of an observer estimate of the same linear transformations of the state. As an example of eigenvalue placement with observer feedback, Figure 13.12 shows eigenvalue placement with full-order state observer. The eigenvalues of the overall system are those of the state feedback and those of the full-order observer.

13.5 Tracking System Design

The second concern of tracking system design, obtaining acceptable zero-state system response to reference inputs, is now discussed.

A tracking system is one in which the plant's outputs are controlled so that they become and remain nearly equal to externally applied reference signals $r(k)$ as shown in Figure 13.13a. The outputs $\bar{y}(k)$ are said to *track* or *follow* the reference inputs.

As shown in Figure 13.13b, a linear, step-invariant controller of a multiple-input–multiple-output plant is described by two transfer function matrices: one relating the reference inputs to the plant inputs and the other relating the output feedback vector to the plant inputs. The feedback compensator is used for shaping the plant's zero-input response by placing the feedback system eigenvalues at desired locations as discussed in the previous subsections. The input compensator, on the other hand, is designed to achieve good tracking of the reference inputs by the system outputs.

The output of any linear system can always be decomposed into two parts: the zero-input component due to the initial conditions alone and the zero-state component due to the input alone, that is,

$$\bar{y}(k) = \bar{y}_{\text{zero-input}}(k) + \bar{y}_{\text{zero-state}}(k)$$

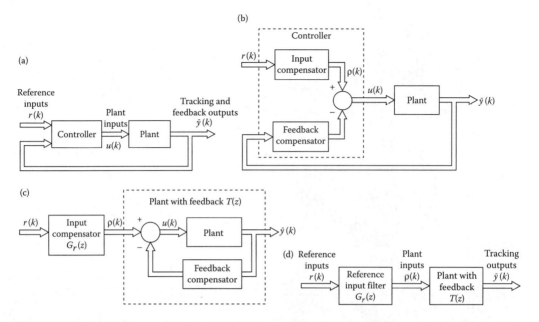

FIGURE 13.13   Controlling a multiple-input–multiple-output plant. The output $\bar{y}(k)$ is to track the reference input $r(k)$. (a) A tracking system using the reference inputs and plant outputs. (b) Representing a controller with a feedback compensator and an input compensator. (c) Feedback compensator combined with plant to produce a plant-with-feedback transfer function matrix $T(z)$. (d) Using a reference input filter for tracking.

Basically, there are three methods for tracking system design:

1. Ideal tracking system design
2. Response model design
3. Reference model design

13.5.1 Ideal Tracking System Design

In this first method, *ideal tracking* is obtained if the measurement output equals the tracking input

$$\bar{y}_{\text{zero-state}}(k) = r(k)$$

The tracking outputs $\bar{y}(k)$ have initial transient error due to any nonzero plant initial conditions, after which they are equal to the reference inputs $r(k)$, no matter what these inputs are.

As shown in Figure 13.13c, if the plant with feedback has the z-transfer function matrix $T(z)$ relating the tracking output to the plant inputs, then

$$\bar{Y}(z) = T(z)\rho(z)$$

An input compensator or a *reference input filter*, as shown in Figure 13.13d, with the transfer function matrix $G_r(z)$, for which

$$\rho(z) = G_r(z)R(z)$$

gives

$$\bar{Y}(z) = T(z)G_r(z)R(z)$$

Ideal tracking is achieved if

$$T(z)G_r(z) = I$$

where I is the identity matrix with dimensions equal to the number of reference inputs and tracking outputs. This is to say that ideal tracking is obtained if the reference input filter is an *inverse filter* of the plant with feedback. Reference input filters do not change the eigenvalues of the plant with feedback which are assumed to have been previously placed with output or observer feedback.

When a solution exists, ideal tracking system design achieves *exact* zero-state tracking of any reference input. Because it involves constructing inverse filters, ideal tracking system design may require unstable or noncausal filters. An ideal tracking solution can also have other undesirable properties, such as unreasonably large gains, highly oscillatory plant control inputs, and the necessity of canceling plant poles and zeros when the plant model is not known accurately.

13.5.2 Response Model Design

When ideal tracking is not possible or desirable, the designer can elect to design *response model* tracking, for which

$$T(z)G_r(z) = \Omega(z)$$

where the response model z-transfer function matrix $\Omega(z)$ characterizes an acceptable relation between the tracking outputs and the reference inputs. Clearly, the price one pays for the added freedom designing a reference model can be degraded tracking performance. However, performance can be improved by increasing the order of the reference input filter. Response model design is a generalization of the classical design technique of imposing requirements for a controller's steady-state response to power-of-time inputs.

The difficulty with the response model design method is in selecting suitable model systems. For example, when two or more reference input signals are to be tracked simultaneously, the response model z-transfer functions selected include those relating plant tracking outputs and the reference inputs they are to track, and those relating unwanted coupling between each tracking output and the other reference inputs.

13.5.3 Reference Model Tracking System Design

The practical response model performance is awkward to design because it is difficult to relate performance criteria to the z-transfer functions of response models. An alternative design method models the reference input signals $r(k)$ instead of the system response. This method, termed reference model tracking system design, allows the designer to specify a class of representative reference inputs that are to be tracked exactly, rather than having to specify acceptable response models for all the possible inputs.

In reference model tracking system design, additional external input signals $r(k)$ to the composite system are applied to the original plant inputs and to the observer state equations so that the feedback system, instead of Equations 13.15 through 13.17, is described by Equation 13.9 and

$$\xi(k+1) = F\xi(k) + Gy(k) + Hu(k) + Jr(k) \tag{13.18}$$

$$w(k) = L\xi(k) + N[y(k) - Du(k)] \tag{13.19}$$

with

$$u(k) = Ew(k) + Pr(k) \tag{13.20}$$

Then, the overall composite system has the state equations

$$\begin{bmatrix} x(k+1) \\ \xi(k+1) \end{bmatrix} = \begin{bmatrix} A + BENC & BEL \\ GC + \bar{H}ENC & F + \bar{H}EL \end{bmatrix} \begin{bmatrix} x(k) \\ \xi(k) \end{bmatrix} + \begin{bmatrix} BP \\ \bar{H}P + J \end{bmatrix} r(k)$$

$$= \bar{A}\bar{x}(k) + \bar{B}r(k),$$

and the output equation

$$\bar{y}(k) = [C + DENC \quad DEL]\bar{x}(k) + D\,Pr(k)$$

$$= \bar{C}\bar{x}(k) + \bar{D}r(k)$$

where

$$\bar{H} = H + GD$$

The composite state coupling matrix $\bar{A}$ above shows that the coupling of external inputs $r(k)$ to the feedback system does not affect its eigenvalues. The input coupling matrix $\bar{B}$ has matrices P and J which are entirely arbitrary and thus can be selected by the designer. Our objective is to select P and J so that the system output $\bar{y}(k)$ tracks the reference input $r(k)$.

Consider the class of reference signals, generated by the autonomous state variable model of the form

$$\sigma(k+1) = \Psi\sigma(k)$$
$$r(k) = \Theta\sigma(k) \tag{13.21}$$

The output of this *reference input model system* may consist of step, ramp, parabolic, exponential, sinusoidal, and other common sequences. For example, the model,

$$\begin{bmatrix} \sigma_1(k+1) \\ \sigma_2(k+1) \end{bmatrix} = \begin{bmatrix} 2 & 1 \\ -1 & 0 \end{bmatrix} \begin{bmatrix} \sigma_1(k) \\ \sigma_2(k) \end{bmatrix}$$
$$r(k) = \begin{bmatrix} 1 & 0 \end{bmatrix} \begin{bmatrix} \sigma_1(k) \\ \sigma_2(k) \end{bmatrix}$$

has an arbitrary constant plus an arbitrary ramp,

$$r(k) = \beta_1 + \beta_2 k$$

In reference model tracking system design, the concept of an observer is used in a new way; it is the plant with feedback that is an observer of the *fictitious* reference input model system in Figure 13.14. When driven by $r(k)$, the state of the composite system observes

$$\bar{x}(k) \to M\sigma(k)$$

where M, according to Equation 13.14, satisfies

$$M\Psi - \bar{A}M = \bar{b}\Theta \tag{13.22}$$

The plant tracking output $\bar{y}(k)$ observes

$$\bar{y}(k) = \bar{C}\bar{x}(k) + \bar{D}r(k) \to \bar{C}M\sigma(k) + \bar{D}r(k)$$

and, for

$$\bar{y}(k) \to r(k)$$

it is necessary that

$$\bar{C}M\sigma(k) + \bar{D}r(k) = r(k) \tag{13.23}$$

Equations 13.22 and 13.23 constitute a set of linear algebraic equations where the elements of M, P, and J are unknowns. If, for an initial problem formulation, there is no solution to the equations, one

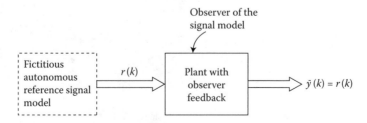

FIGURE 13.14 Observing a reference signal model.

can reduce the order of the reference signal model and/or raise the order of the observer used for plant feedback until an acceptable solution is obtained.

The autonomous reference input model has no physical existence; the actual reference input $r(k)$ likely deviates somewhat from the prediction of the model. The designer deals with representative reference inputs, such as constants and ramps, and, by designing for exact tracking of these, obtains acceptable tracking performance for other reference inputs.

13.6 Designing Between-Sample Response

The first two design problems for a tracking system, obtaining acceptable zero-input response and zero-state response, were discussed and solved in the previous subsections.

When a digital controller is to control a continuous-time plant, a third design problem is achieving good between-sample response of the continuous-time plant. A good discrete-time design will insure that samples of the plant response are well-behaved, but satisfactory response between the discrete-time steps is also necessary. Signals in a continuous-time plant can fluctuate wildly, even though discrete-time samples of those signals are well-behaved. The basic problem is illustrated in Figure 13.15 with the zero-input continuous-time system

$$\begin{bmatrix} \dot{x}_1(t) \\ \dot{x}_2(t) \end{bmatrix} = \begin{bmatrix} -0.2 & 1 \\ -1.01 & 0 \end{bmatrix} \begin{bmatrix} x_1(t) \\ x_2(t) \end{bmatrix} = Ax(t)$$

$$y(t) = \begin{bmatrix} 1 & 0 \end{bmatrix} \begin{bmatrix} x_1(t) \\ x_2(t) \end{bmatrix} = c^\dagger x(t)$$

This system has a response of the form

$$y(t) = Me^{-0.1t} \cos(t + \theta)$$

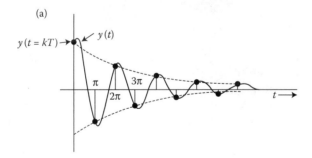

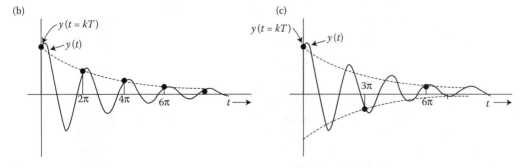

FIGURE 13.15   Hidden oscillations in a sampled continuous-time signal. (a) $T = \pi$; (b) $T = 2\pi$; and (c) $T = 3\pi$.

where M and θ depend on the initial conditions $x(0)$. When the output of this system is sampled with $T = \pi$, the output samples are

$$y(k) = y(t = k\pi) = Me^{-0.1k\pi}\cos(k\pi + \theta)$$
$$= M(e^{-0.1\pi})^k(-1)^k\cos\theta = M\cos\theta(-0.73)^k$$

as shown in Figure 13.15a. But these samples are the response of a first-order discrete-time system with a single geometric series model. The wild fluctuations of $y(t)$ between sampling times, termed *hidden oscillations*, cannot be determined from the samples $y(k)$.

As one might expect, the discrete-time model of this continuous-time system

$$x(k+1) = [\exp(\mathcal{A}T)]x(k)$$
$$y(k) = c^\dagger x(k)$$

or

$$\begin{bmatrix} x_1(k+1) \\ x_2(k+1) \end{bmatrix} = \begin{bmatrix} -0.73 & 0 \\ 0 & -0.73 \end{bmatrix}\begin{bmatrix} x_1(k) \\ x_2(k) \end{bmatrix} = Ax(k)$$

$$y(k) = \begin{bmatrix} 1 & 0 \end{bmatrix}\begin{bmatrix} x_1(k) \\ x_2(k) \end{bmatrix} = c^\dagger x(k)$$

is not completely observable in this circumstance because

$$M_0 = \begin{bmatrix} c^\dagger \\ c^\dagger A \end{bmatrix} = \begin{bmatrix} 1 & 0 \\ -0.73 & 0 \end{bmatrix}$$

This phenomenon is called *loss of observability due to sampling*. The discrete-time system would normally have two modes, those given by its characteristic equation

$$\begin{bmatrix} (z+0.73) & 0 \\ 0 & (z+0.73) \end{bmatrix} = (z+0.73)^2 = 0$$

which are $(-0.73)^k$ and $k(-0.73)^k$. Only the $(-0.73)^k$ mode appears in the output, however.

Hidden oscillations occur at any other integer multiple of the same sampling period $T = \pi$. Figure 13.15b shows sampling with $T = 2\pi$, for which only a $[(-0.73)^2]^k = (0.533)^k$ mode is observable from $y(k)$. In Figure 13.15c, with $T = 3\pi$, only a $[(-0.73)^3]^k = (-0.39)^k$ is observable from $y(k)$. For a slightly different sampling interval, for example, $T = 3$ s, there are no hidden oscillations, and the discrete-time model is completely observable.

Hidden oscillations in a continuous-time system and the accompanying loss of observability of the discrete-time model occur when the sampling interval is half the period of oscillation of an oscillatory mode or an integer multiple of that period. Although it is very unlikely that the sampling interval chosen for a plant control system would be precisely one resulting in hidden oscillations, intervals close to these result in modes in discrete-time models that are "almost unobservable." With limited numerical precision in measurements and in controller computations, these modes can be difficult to detect and control.

Between input changes, the continuous-time plant behaves as if it has a constant input *without* feedback. Often, the abrupt changes in the plant inputs at the sample times are the major cause of poor between-sample plant response. If the between-sample response of a continuous-time plant is not acceptable, there are a number of ways of improving the between-sample response.

The first approach is to *increase the controller sampling rate* so that each step change in the plant input is of smaller amplitude. Higher sample rates are often expensive, however, because the amount of computation that must be performed by the controller is proportional to the sample rate.

The second approach is to *change the plant*, perhaps by adding continuous-time feedback. The continuous-time plant with feedback, rather than the original plant, then becomes the plant to be controlled digitally. This too is often undesirable because of its susceptibility to noise and drift at the expense of routing analog signals.

The third approach is to *change the shape of the plant input signals* from having step changes at the controller sample rate to a shape that gives improved plant response. This third approach is now examined. First, input signal shaping with analog filters is considered, and then input signal shaping with high-speed dedicated digital filters is discussed.

13.6.1 Analog Plant Input Filtering

Analog filters between the D/A converters and the plant inputs are usually acceptable in a controller design. As indicated in Figure 13.16, the idea is to use a filter or filters to smooth or shape the plant inputs so that the undesirable modes of the plant's open-loop response are not excited as much as they would be with abrupt changes in the plant inputs at each step. Figure 13.17 shows simulation results for the continuous-time plant

$$\begin{bmatrix} \dot{x}_1(t) \\ \dot{x}_2(t) \end{bmatrix} = \begin{bmatrix} -0.6 & 1 \\ -9 & 0 \end{bmatrix} \begin{bmatrix} x_1(t) \\ x_2(t) \end{bmatrix} + \begin{bmatrix} -1 \\ 1 \end{bmatrix} u(t) = \mathcal{A}x(t) + \mathcal{B}u(t)$$

$$\bar{y}(t) = \begin{bmatrix} 1 & 0 \end{bmatrix} \begin{bmatrix} x_1(t) \\ x_2(t) \end{bmatrix} = \bar{c}^\dagger x(t) \tag{13.24}$$

driven by a discrete-time control system

$$u(k) = -9r(k)$$

where $r(k)$ is the reference input and $T = 5$. The highly undamped zero-input plant response results in large fluctuations ("ringing") of the tracking output each time there is a step change in the plant input by the controller. The plant response is improved considerably by the insertion of a plant input analog filter with the transfer function

$$G(s) = \frac{1/3}{s + 1/3}$$

as shown in Figure 13.17b. The filter was designed with a 3-s time constant to smooth the plant input waveform during each 5-s sampling interval, resulting in much less ringing of the tracking output.

In general, insertion of a plant input filter with the transfer function matrix $G(s)$ before a plant with transfer function matrix $T(s)$ results in a composite continuous-time *model plant* with the transfer function matrix

$$M(s) = G(s)T(s)$$

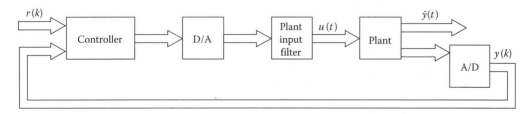

FIGURE 13.16 Use of an analog plant input filter to improve between-sample response.

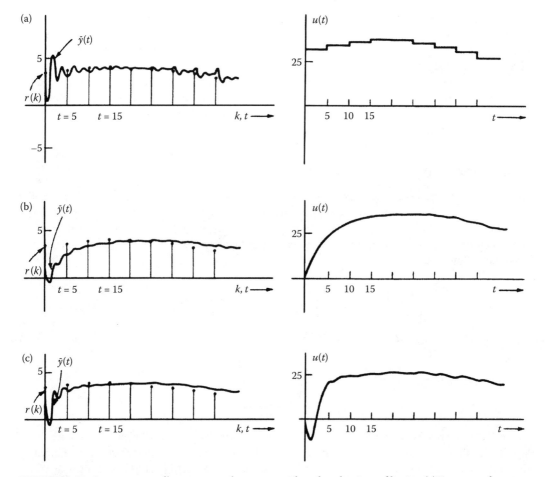

FIGURE 13.17 Improvement of between-sample response with analog plant input filtering. (a) Response of a system without a plant input filter. (b) Response with an added first-order analog plant input filter. (c) Response with a better analog plant input filter.

Designing analog plant input filters is quite similar to designing discrete-time reference input filters. The objective of the design is to improve the between-sample plant response by obtaining a model plant with acceptable step response. As an example, consider again the continuous-time plant Equation 13.26 with the transfer function

$$T(s) = \bar{c}^\dagger (sI - \mathcal{A})^{-1}\mathcal{B} = \frac{-s+1}{s^2 + 0.6s + 9}$$

A plant input filter with the transfer function

$$G(s) = \frac{s^2 + 0.6s + 9}{(s+1)^2} = 1 + \frac{-1.4s + 8}{s^2 + 2s + 1}$$

cancels the plant poles and results in a model plant transfer function

$$M(s) = G(s)T(s) = \frac{-s+1}{(s+1)^2}$$

The resulting response is shown in Figure 13.17c. The original plant's zero-input response, excited by nonzero plant initial conditions, is apparent at first but eventually decays to zero.

13.6.2 Higher-Rate Digital Filtering and Feedback

Another method for performing plant input filtering is to use a digital filter that operates at many times the rate of the controller, as in Figure 13.18. A/D conversion is usually accomplished with repeated approximations, one for each bit. With a given technology, one can perform several D/A conversions as fast as one A/D conversion. The cost of a digital plant input filter is thus relatively low because it requires higher D/A, not A/D, speed.

As an example, consider again the continuous-time plant given by Equation 13.26. The discrete-time controller for this plant operates with a sampling interval $T = 5$ s.

It was found earlier that an analog filter with transfer function

$$G(s) = \frac{1/3}{s + 1/3}$$

improves the plant's between-sample response. The step response of the plant alone and the response of the plant with this input filter are shown in Figure 13.19a.

A state variable realization of the analog filter is

$$\dot{\varepsilon}(t) = -\frac{1}{3}\varepsilon(t) + \frac{1}{3}w(t)$$

$$u(t) = \varepsilon(t)$$

A discrete-time model of this filter with sampling interval $\Delta t = 1$, one-fifth the controller interval, is

$$\varepsilon(k'+1) = 0.717\varepsilon(k') + 0.283w(k')$$

$$u(k') = \varepsilon(k')$$

where k' is the index for steps of size Δt. The step response of the plant with a digital filter that approximates the continuous-time filter is shown in Figure 13.19b. Figure 13.19c is a tracking response plot for the combination of plant, high-speed digital plant input filter, and lower speed digital controller.

Higher-rate digital plant feedback can improve a plant's between-sample response, but this requires high-rate A/D as well as high-rate D/A conversion. Another possibility is to sample the plant measurement outputs at a lower rate, but estimate the plant state at a high rate, feeding the state estimate back at high rate.

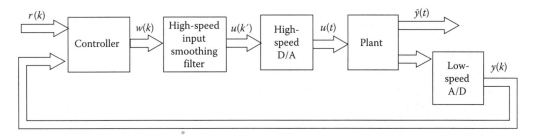

FIGURE 13.18 Use of high-speed digital plant input filtering to improve between-sample response.

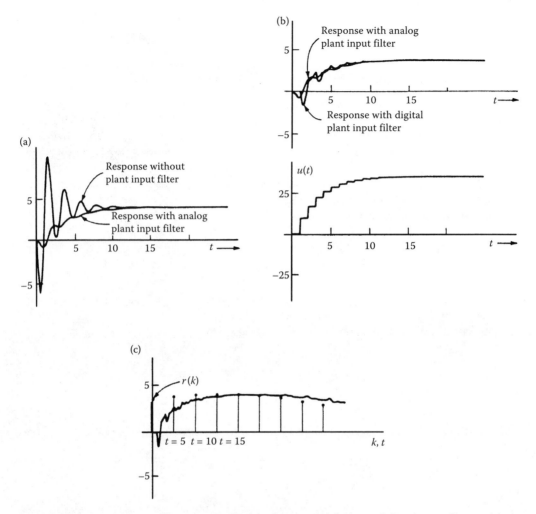

FIGURE 13.19 Improvement of between-sample plant response with high-speed digital input filtering. (a) Step response of the example system, with and without an analog input filter. (b) Step response with an analog input filter and with a digital input filter. (c) Tracking response of the system with a digital input filter.

13.6.3 Higher-Order Holds

A traditional approach to improving reconstruction is to employ higher-order holds than the zero-order ones. An nth-order hold produces a piecewise nth-degree polynomial output that passes through the most recent $n + 1$ input samples. As the order of the hold is increased, a well-behaved signal is reconstructed with increasing accuracy. Several holds and their typical responses are shown in Figure 13.20.

Although higher-order holds do have a smoothing effect on the plant inputs, the resulting improvement of plant between-sample response is generally poor compared with that possible with a conventional filter of comparable complexity. Hardware for holds of higher than zero order (which is the sample-and-hold operation) is not routinely available. One approach is to employ high-speed digital devices and D/A conversion, as in the technique of Figure 13.18, but where the high-speed input-smoothing filter performs the interpolation calculations for a hold.

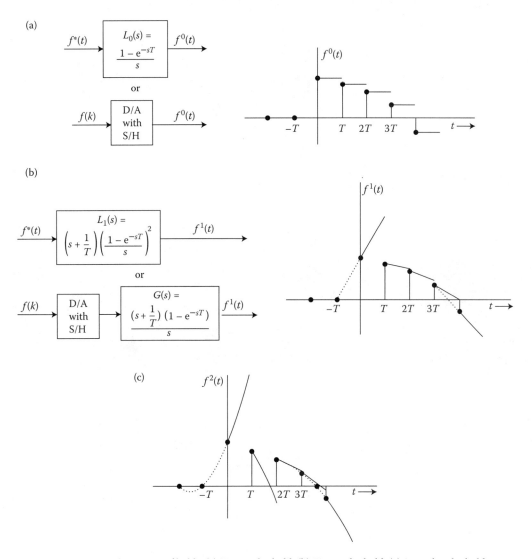

FIGURE 13.20 Typical response of holds. (a) Zero-order hold. (b) First-order hold. (c) Second-order hold.

Bibliography

1. DiStefano III, J.J., Stubberud, A.R., and Williams, I.J., *Feedback and Control Systems (Schaum's Outline)*, 2nd edn, McGraw-Hill, New York, NY, 1990.
2. Kuo, B.C., *Digital Control Systems*, 2nd edn, Oxford University Press Inc., New York, NY, 1995.
3. Santina, M.S., Stubberud, A.R., and Hostetter, G.H., *Digital Control System Design*, 2nd edn, Oxford University Press Inc., New York, NY, 1994.
4. Kailath, T., *Linear Systems*, Prentice-Hall, Englewood Cliffs, NJ, 1980.
5. Franklin, G.F., Powell, J.D., and Workman, M.L., *Digital Control of Dynamic Systems*, 3rd edn, Ellis-Kagle Press, Half Moon Bay, CA, 2006.
6. Chen, C.T., *Linear System Theory and Design*, 3rd edn, Oxford University Press, New York, NY, 1998.
7. Åström, K.J. and Wittenmark, B., *Computer Controlled Systems*, 3rd edn, Prentice-Hall, Englewood Cliffs, NJ, 1996.
8. Ogata, K., *Discrete-Time Control Systems*, 2nd edn, Prentice–Hall, Englewood Cliffs, NJ, 1994.
9. Friedland, B., *Control System Design,* Dover Publications, New York, NY, 2005.
10. Nise, N., *Control Systems Engineering*, 4th edn, Wiley & Sons, New York, NY, 2003.

14

Quantization Effects

Michael Santina
The Boeing Company

Allen R. Stubberud
University of California, Irvine

Peter Stubberud
The University of Nevada, Las Vegas

14.1 Overview

The digital control system analysis and design methods that were presented in the previous chapters proceeded as if the controller signals and coefficients were of continuous amplitude. However, because digital controllers are implemented with finite word length (FWL) registers and finite precision arithmetic, their signals and coefficients can attain only discrete values. Therefore, further analysis is needed to determine if the performance of the resulting digital controller in the presence of signal and coefficient quantization is acceptable.

In this chapter, we discuss three error sources that may occur in digital processing of controllers. These error sources are (1) coefficient quantization, (2) quantization in A/D conversion, and (3) arithmetic operations. Limit cycles and deadbands are also discussed. Before discussing these errors, however, a brief review of fixed- and floating-point number arithmetic is presented.

14.2 Fixed-Point and Floating-Point Number Representations

There are many choices of arithmetic that can be used to implement digital controllers. The two most popular ones are *fixed-point* and *floating-point* binary arithmetic. Other nonstandard arithmetic such as *logarithmic* and *residue* representations are also possibilities.

14.2.1 Fixed-Point Arithmetic

In general, an n-bit fixed-point binary number can be expressed as [1]

$$
N = \sum_{j=-m}^{n-1} b_j 2^j = b_{n-1} 2^{n-1} + b_{n-2} 2^{n-2} + \cdots + b_1 2^1 + b_0 2^0 + b_{-1} 2^{-1}
$$
$$
+ b_{-2} 2^{-2} + \cdots + b_{-m} 2^{-m} \tag{14.1}
$$
$$
= (b_{n-1} \cdots b_0 \bullet b_{-1} b_{-2} \cdots b_{-m})_2
$$

where b_j is either zero or one.

The bit b_{n-1} is termed the *most significant bit* (MSB) and b_{-m} is termed the *least significant bit* (LSB). The *integer* portion of the number, $b_n b_{n-1} b_0$, is separated from the *fractional portion*, $b_{-1} b_{-2} \cdots b_{-m}$, by the *binary point* or *radix point*.

In the binary representation, each bit can be either a zero or a one. For example, the binary number 1101.101 has the decimal value

$$
1101.101 = 1(2^3) + 1(2^2) + 0(2^1) + 1(2^0) + 1(2^{-1}) + 0(2^{-2}) + 1(2^{-3})
$$
$$
= 13.625
$$

In fixed-point arithmetic, numbers are always normalized as binary fractions (i.e., less than one) of the form $b_0 \bullet b_1 b_2 \cdots b_c$ where b_0 is the sign bit. The $C + 1$ bit normalized number is stored in a register as shown in Figure 14.1 where the sign bit is separated from the C-bit number by a *fictitious* binary point. The binary point is fictitious because it does not occupy any bit location in the register. The *word length*, C is defined as the number of bit locations in the register to the right of the binary point.

There are three commonly used methods for representing signed numbers:

1. Signed-magnitude
2. Two's complement
3. One's complement

Consider the $C+1$ bit binary fraction $b_0 \bullet b_1 b_2 \cdots b_c$ where b_0 is the sign bit. In the *signed-magnitude* representation, the fractional number is positive if b_0 is zero, and it is negative if b_0 is one. For example, the decimal number 0.75 equals 0.11 in binary representation, and -0.75 equals 1.11.

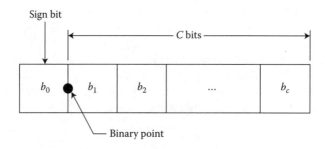

FIGURE 14.1 Normalized fixed-point numbers in a register.

In signed-magnitude representation, binary numbers can be converted to decimal numbers using the relation

$$N = (-1)^{b_0} \sum_{i=1}^{C} b_i 2^{-i} \tag{14.2}$$

The *two's complement* representation of positive numbers is the same as signed-magnitude representation. The two's complement representations of negative numbers, however, are obtained by *complementing* (i.e., replacing every 1 with 0 and every 0 with 1) all of the bits of the positive number and adding one to the LSB of the complemented number. For example, the two's complement representation of the decimal number 0.75 is 0.11 and the two's complement representation of −0.75 is 1.01.

A decimal number can be recovered from its two's complement representation using the relationship

$$N = -b_0 + \sum_{i=1}^{C} b_i 2^{-i} \tag{14.3}$$

The one's complement representation of fractional numbers is the same as the two's complement without the addition of one to the LSB. For example, the one's complement representation of 0.75 is 0.11 and the one's complement representation of −0.75 is 1.00. A decimal number can be recovered from its one's complement representation via the relationship

$$N = b_0(2^{-c} - 1) + \sum_{i=1}^{C} b_i 2^{-i} \tag{14.4}$$

The two's complement representation of binary numbers has several advantages over the signed-magnitude and the one's complement representations and therefore is more popular.

In general, the sum of two normalized C-bit numbers using fixed-point arithmetic is a C-bit number while the product of two C-bit numbers is a $2C$-bit number. Hence, if the register word length is fixed to C bits, a *quantization* error will be introduced in multiplication but not in addition.* The product is quantized either by *rounding* or by *truncation*. For example, rounding the binary number 0.010101 to four bits after the binary point gives 0.0101 but rounding it to three bits yields 0.011. When a number is truncated, all of the bits to the right of its LSB are discarded. For example, truncating the number 0.010101 to three bits after the binary point gives 0.010.

14.2.2 Floating-Point Arithmetic

A major disadvantage of fixed-point arithmetic is the limited range of numbers that can be represented with a given word length. Another type of arithmetic which, for the same number of bits, has a much larger range of numbers is *floating-point arithmetic*. In general, a floating-point number can be expressed as

$$N = M \times 2^E \tag{14.5}$$

where M and E, both expressed in binary form, are termed the *mantissa* and the *exponent* of the number, respectively. In binary floating-point representation, numbers are always normalized by scaling M as a fraction whose decimal value lies in the range $0.5 \le M < 1$.

Figure 14.2 shows a floating-point number stored in a register. The register is divided into the mantissa and the exponent. Both the mantissa and the exponent have fictitious binary points to separate the sign bits from the numbers. In floating-point arithmetic, negative mantissas and negative exponents are coded the same way as in fixed-point arithmetic using two's complement, signed-magnitude, or one's complement.

---

* When normalized signed numbers are added and the result is larger than one, then *overflow* occurs. Overflow does not occur in multiplication, because the product of two normalized numbers is always less than one.

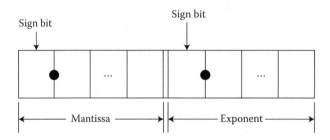

FIGURE 14.2 Normalized floating-point number in a register.

The product of two floating-point numbers is

$$(M_1 \times 2^{E_1})(M_2 \times 2^{E_2}) = (M_1 \times M_2)2^{(E1+E2)}$$

Thus, if the mantissa is limited to C bits, the product $M_1 \times M_2$ must be rounded or truncated to C bits. The sum of two floating-point numbers is performed by shifting the bits of the mantissa of the smaller number to the right and increasing its exponent until the two exponents are equal. Then the two mantissas are added, and, if necessary, normalized to satisfy Equation 14.5. The shifted mantissa may exceed its limited range and thus must be quantized. Hence, in floating-point arithmetic, quantization errors are introduced in both addition and multiplication. Roundoff or truncation errors will be introduced in the mantissa M but not in the exponent E, because the exponent, E, is always a positive or negative integer and integers have exact binary representations. Of course, if the number is too large or too small then over- or underflow can occur.

14.3 Truncation and Roundoff

Because of the FWL of registers in digital computers, errors are always introduced when the numbers to be processed are quantized. These errors depend on (1) the way the numbers are represented (fixed- or floating-point arithmetic, signed-magnitude, two's or one's complement, etc.) and (2) how the numbers are quantized.

Consider the normalized binary number $b_0 \bullet b_1 b_2 \cdots b_c$, where b_0 is the sign bit, and $b_1 b_2 \cdots b_c$ is the binary code of a fixed-point number or the mantissa of a floating-point number. Denoting the number before quantization by x, the error introduced by quantization is

$$e_q = Q[x] - x$$

where $Q[x]$ is the quantized value of x. The range of quantization error depends on the type of arithmetic and the type of quantization used. Figure 14.3 shows the transfer characteristics of truncation and roundoff for signed-magnitude, two's complement, and one's complement representations.

For fixed-point arithmetic, the error caused by truncating a number to C bits is [2]

$$\begin{aligned} -2^{-c} < e_T \leqslant 0, \quad x \geqslant 0 \\ 0 \leqslant e_T < 2^{-c}, \quad x < 0 \end{aligned} \tag{14.6}$$

for the signed-magnitude and one's complement representations. For two's complement, the truncation error is

$$-2^{-c} < e_T \leqslant 0 \tag{14.7}$$

for all x.

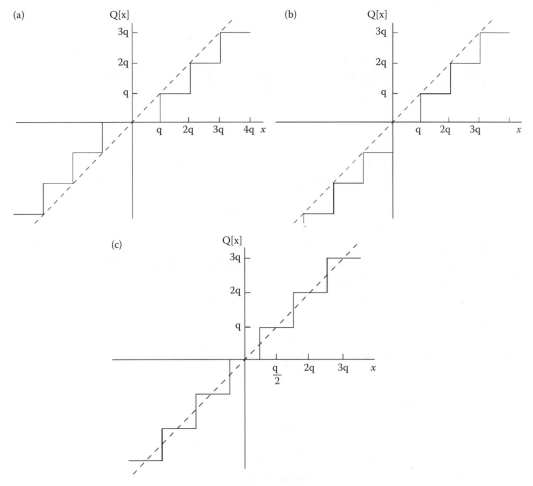

FIGURE 14.3 Transfer characteristics of truncation and rounding. (a) Truncation with signed-magnitude and one's complement; (b) truncation with two's complement; and (c) rounding with signed-magnitude, one's complement, and two's complement.

On the other hand, the error caused by rounding a number to C bits is

$$\frac{-2^{-c}}{2} \leqslant e_R < \frac{2^{-c}}{2} \tag{14.8}$$

for signed-magnitude, one's complement and two's complement representations.

In fixed-point arithmetic, truncation or roundoff errors are independent of the magnitude of the original unquantized numbers. However, in floating-point arithmetic, these errors depend on the magnitude of the unquantized number. In floating-point arithmetic, roundoff and truncation errors occur only in the mantissa. Thus, if the mantissa is truncated to C bits the quantized number is

$$x_q = (1 + \varepsilon)x$$

where ε is the relative error in x. In the case of truncation, for signed-magnitude and one's complement representations, the relative error in the value of the floating-point word is

$$-2.2^{-c} < \varepsilon \leqslant 0 \tag{14.9}$$

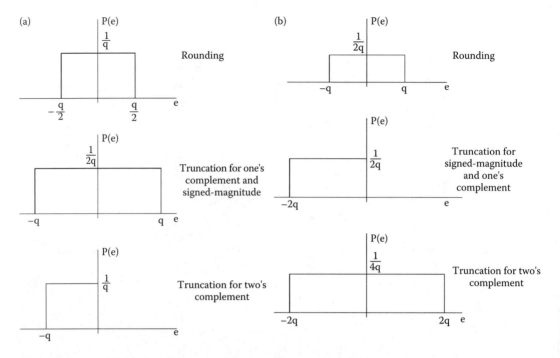

FIGURE 14.4 Probability density functions of fixed- and floating-point errors. (a) Fixed-point errors. (b) Floating-point errors.

and for two's complement truncation, the error is

$$-2.2^{-c} < \varepsilon \leqslant 0, \quad x \geqslant 0 \tag{14.10}$$

$$0 \leqslant \varepsilon < 2.2^{-c}, \quad x < 0 \tag{14.11}$$

On the other hand, the roundoff error in the mantissa is of the form

$$-2^{-c} \leqslant \varepsilon \leqslant 2^{-c} \tag{14.12}$$

for all three types of representations. Figure 14.4 shows the probability density functions of fixed- and floating-point errors. As shown, the probability density function is *uniformly distributed* over the range of quantization. In the remainder of this chapter, and unless otherwise stated, we model fixed- and floating-point errors as stationary, white noise, random processes.

We now discuss the major sources of error caused by FWL and then determine their effects on the behavior of digital controllers. These errors are coefficient quantization, quantization errors in A/D converters and quantization errors in arithmetic operations.

14.4 Effects of Coefficient Quantization

The digital control system design methods that were presented in the previous articles resulted in controllers whose coefficients are of arbitrary precision. However, because the controller is implemented with FWL registers, each of its coefficients must be quantized. For example, consider the digital controller described by the transfer function

$$H(z) = \frac{z^3 + 1.584z^2 + 1.2769z + 0.5642}{z^4 + 2.689z^3 + 3.3774z^2 + 2.3823z + 0.6942} \tag{14.13}$$

which has poles located at $z_1 = 0.999$, $z_2 = 0.697$, and $z_{3,4} = 0.4965 \pm j0.8663$. If the binary form of the coefficients of this controller are truncated to three bits to the right of the binary point, then the quantized controller transfer function becomes

$$H_q(z) = \frac{z^3 + 1.5z^2 + 1.25z + 0.5}{z^4 + 2.625z^3 + 3.3749z^2 + 2.3748z + 0.6249} \tag{14.14}$$

which has two poles outside the unit circle. Another example is quoted in reference [3] in which a stable fifth-order controller can become unstable even if it is realized with 18-bit arithmetic. One can also find examples where stable controllers can become unstable even if they are realized with 18-bit arithmetic.

In general, consider the digital controller described by the transfer function

$$H(z) = \frac{\sum_{k=0}^{m} b_k z^{-k}}{1 - \sum_{k=1}^{n} a_k z^{-k}} \tag{14.15}$$

If the controller coefficients are quantized to C bits, then

$$\hat{a}_k = a_k + \delta_k$$

for fixed-point arithmetic or

$$\hat{a}_k = a_k(1 + \delta_k)$$

for floating-point arithmetic, where δ_k is bounded in absolute value by 2^{-c}. Similarly,

$$\hat{b}_k = b_k + \eta_k$$

for fixed-point arithmetic or

$$\hat{b}_k = b_k(1 + \eta_k)$$

for floating-point arithmetic.

In terms of the quantized coefficients, the controller transfer function becomes

$$H_q(z) = \frac{\sum_{k=0}^{m} \hat{b}_k z^{-k}}{1 - \sum_{k=1}^{n} \hat{a}_k z^{-k}} \tag{14.16}$$

One approach for analyzing the effects of coefficient quantization on system performance is to compare the response of the quantized controller with that of the ideal controller before quantization. One can also apply *sensitivity* analysis to determine the *variations* of the controller response to variations in its numerator and denominator coefficients. If the controller transfer function given by Equation 14.15 is rewritten in the form (assuming $b_o = 1$)

$$H(z) = \frac{\prod_{i=1}^{m} (1 + \beta_i z^{-1})}{\prod_{j=1}^{n} (1 - \alpha_i z^{-1})} = \frac{N(z^{-1})}{D(z^{-1})}$$

where α_i is the location of the ith pole of $H(z)$ and if the product of the quantized poles is expressed as

$$D_q(z^{-1}) = \prod_{j=1}^{n} (1 - \hat{\alpha}_j z^{-1})$$

where

$$\hat{\alpha}_j = \alpha_j + \Delta\alpha_j$$

then

$$\Delta\alpha_j = \sum_{k=1}^{n} \frac{\alpha_j^{n-k}}{\prod_{\substack{i=1 \\ i \neq j}}^{n} (\alpha_j - \alpha_i)} \Delta a_k \tag{14.17}$$

which relates the incremental changes of the jth pole of the controller transfer function to incremental changes in the a_k coefficient of the denominator polynomial of the controller transfer function. Recall

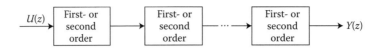

FIGURE 14.5 Cascade first- or second-order subsystems.

that

$$\Delta a_k = \delta_k$$

for fixed-point arithmetic and

$$\Delta a_k = a_k \delta_k$$

for floating-point arithmetic. Similar results can also be obtained for the controller's zeros.

Equation 14.17 shows that, when the controller poles are close to each other, small changes in the coefficients a_k of the denominator polynomial cause large changes in the locations of the controller poles. As the order of the controller increases, the roots of its denominator become more sensitive to changes in the coefficients of the denominator polynomial.

To avoid the coefficient sensitivity problem, higher-order controller transfer functions are decomposed into *cascaded* first- or second-order transfer functions as in Figure 14.5. When all the poles and zeros of the controller transfer function are real, the cascaded transfer functions are all of first order. Complex conjugate pairs of roots should be grouped into second-order subsystems to avoid complex number arithmetic operations.

Another way to avoid the coefficient sensitivity problem is to use the *parallel* form as shown in Figure 14.6. The parallel form is obtained by decomposing the controller transfer function into first- or second-order subsystems using the method of partial fraction expansion. Using either form, that is cascade or parallel, each first-order subsystem can be realized using the structure shown in Figure 14.7a, where

$$H(z) = \frac{1 + \beta_1 z^{-1}}{1 - \alpha_1 z^{-1}}$$

and each second-order subsystem may be realized as shown in Figure 14.7b, where

$$H(z) = \frac{1 + \beta_1 z^{-1} + \beta_2 z^{-2}}{1 - \alpha_1 z^{-1} - \alpha_2 z^{-2}}$$

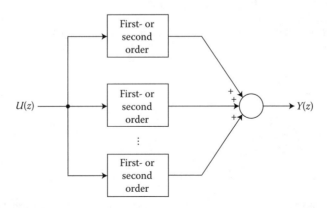

FIGURE 14.6 Parallel first- or second-order subsystems.

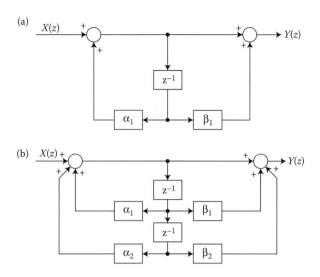

FIGURE 14.7 First- and second-order realizations. (a) First-order realization. (b) Second-order realization.

As a numerical example, consider again the controller transfer function given by Equation 14.13. Rewriting the transfer function in factored form yields

$$H(z) = \left[\frac{z+0.862}{z+0.999}\right]\left[\frac{1}{z+0.697}\right]\left[\frac{z^2+0.722z+0.6545}{z^2+0.993z+0.997}\right]$$

As in the previous example, if the binary representations of the coefficients of each factor are truncated to three bits to the right of the binary point, then the resulting quantized transfer function is

$$H_q(z) = \left[\frac{z+0.75}{z+0.875}\right]\left[\frac{1}{z+0.625}\right]\left[\frac{z^2+0.625z+0.625}{z^2+0.875z+0.875}\right]$$

which is stable and can be realized using first- and second-order subsystems. This controller is significantly different from the quantized controller given by Equation 14.14.

14.5 Digital Filter Design Using the Delta Operator

We mentioned in Chapter 12 that for high sampling applications where the sampling intervals are very small, the dynamic response of the z-transformed discrete-time system does not converge smoothly to the continuous-time counterpart causing significant implementation issues. We also mentioned that the delta operator [4], given by

$$\delta = \frac{z-1}{T} \tag{14.18}$$

offers superior numerical performance in FWL implementation over the z-transform and shift operator. As the sampling interval approaches zero, the filter coefficients using the delta operator have the property that they converge to the equivalent continuous time filter coefficients, and therefore, the response of the filter based on the delta operator converges to it continuous-time counterpart.

Consider the continuous-time notch filter

$$G(s) = \frac{N_c(s)}{D_c(s)} = \frac{s^2 + 2\xi_z\omega_0 s + \omega_0^2}{s^2 + 2\xi_p\omega_0 s + \omega_0^2} \tag{14.19}$$

The discrete-time equivalent of this filter, using the pole–zero matching method presented in Chapter 12, is given by

$$G(z) = \frac{N_d(z)}{D_d(z)} = \frac{K(z^2 + \alpha_1 z + \alpha_0)}{z^2 + \beta_1 z + \beta_0} \tag{14.20}$$

where

$$
\begin{aligned}
\alpha_1 &= -2e^{-\xi_z\omega_0 T}\cos\left(\omega_0 T\sqrt{1 - \xi_z^2}\right) \\
\alpha_0 &= e^{-2\xi_z\omega_0 T} \\
\beta_1 &= -2e^{-\xi_p\omega_0 T}\cos\left(\omega_0 T\sqrt{1 - \xi_p^2}\right) \\
\beta_0 &= e^{-2\xi_p\omega_0 T}
\end{aligned}
\tag{14.21}
$$

and where the gain K is selected to match the DC gain of the continuous-time filter which, in this example, is unity. That is,

$$K = \frac{D_d(1)}{N_d(1)} = \frac{1 + \beta_1 + \beta_0}{1 + \alpha_1 + \alpha_0}$$

Let the filter parameters $\omega_0 = 2\pi(0.2)$ rad/s, $\xi_z = 0.07$, $\xi_p = 0.7$; then

$$G(s) = \frac{s^2 + 0.1759292s + 1.5791367}{s^2 + 1.7592918s + 1.5791367} \tag{14.22}$$

Let the sampling interval be $T = 0.02$ s; then, the digital filter transfer function in Equation 14.20 becomes

$$G(z) = \frac{0.9843413z^2 - 1.9646046z + 0.9808839}{z^2 - 1.9648053z + 0.9654259} \tag{14.23}$$

Assuming a fixed-point arithmetic processor with 12 bits accuracy, the transfer function of the digital filter in Equation 14.23 becomes, after rounding off,

$$G_q(z) = \frac{N_q}{D_q} = \frac{0.9843750z^2 - 1.9645996z + 0.9809570}{z^2 - 1.9648437z + 0.9653320} \tag{14.24}$$

where the numerator and the denominator of $G_q(z)$ are determined using the MATLAB® commands:

$$
\begin{aligned}
N_q &= q * \text{round}(N_d/q) \\
D_q &= q * \text{round}(D_d/q)
\end{aligned}
$$

In this example, the quantum size is $q = 2^{-12}$. As shown in Figures 14.8a and b, the frequency response plots of the infinite precision digital filter, given by Equation 14.23 match very well with the frequency response plots of the continuous-time filter given by Equation 14.22. But, as shown in the same figure, the frequency response plots of the filter with quantized coefficients, given by Equation 14.24, differ significantly from those given by Equation 14.23 resulting in an unacceptable filter performance.

One can observe from Equations 14.21 that as the sampling period approaches zero, the coefficients of the discrete-time filter using the shift operator approach fixed values that are independent of the coefficients of the continuous-time filter.

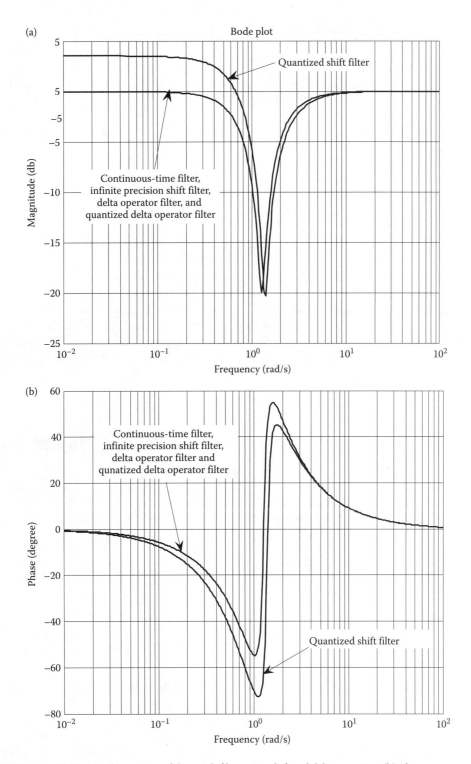

FIGURE 14.8 (a) Magnitude response of the notch filter using shift and delta operators. (b) Phase response of the notch filter using shift and delta operators.

Using the delta operator, let us substitute Equation 14.18 into Equation 14.20. Then,

$$G_d(\delta) = \frac{K\left[\delta^2 + \left(\dfrac{2+\alpha_1}{T}\right)\delta + \left(\dfrac{1+\alpha_1+\alpha_0}{T^2}\right)\right]}{\delta^2 + \left(\dfrac{2+\beta_1}{T}\right)\delta + \left(\dfrac{1+\beta_1+\beta_0}{T^2}\right)} \tag{14.25}$$

In terms of the specified filter parameters and the specified sampling period, the delta operator transfer function using infinite precision is given by

$$G_d(\delta) = \frac{0.9843413\delta^2 + 0.2039019\delta + 1.5515964}{\delta^2 + 1.7597325\delta + 1.5515964} \tag{14.26}$$

Using 12-bit accuracy, the quantized delta operator filter transfer function is

$$G_{dq}(\delta) = \frac{0.9843750\delta^2 + 0.2038574\delta + 1.5515136}{\delta^2 + 1.7597656\delta + 1.5515136} \tag{14.27}$$

The frequency response plots of this quantized delta operator filter are also shown in Figures 14.8a and b. Its performance compares very well with the continuous-time filter. Also, note the resemblance of the coefficients in Equations 14.26 and 14.27 to those of the continuous-time filter given in Equation 14.22.

As another example, consider the 4th-order elliptic analog low-pass filter whose passband-edge frequency is 20 Hz, peak-to-peak ripple is of 1 db, and has a minimum stopband attenuation of 80 db. The discrete-time equivalent of this filter, using Tustin's method with prewarping at 20 Hz, and $T = 0.00166$ s is given by

$$G(z) = 2.8808099 \times 10^{-4}\left[\frac{z^2 + 0.2813048z + 1}{z^2 - 1.8552558z + 0.8670579}\right]\left[\frac{z^2 - 1.24580094z + 1}{z^2 - 1.9030691z + 0.9450667}\right]$$

Using floating-point arithmetic with $p = 13$-bit mantissa, the transfer function of the quantized digital filter becomes

$$G_q(z) = 2.88069248 \times 10^{-4}\left[\frac{z^2 + 0.2811842z + 1}{z^2 - 1.8459314z + 0.86177303}\right]\left[\frac{z^2 - 1.24580248z + 1}{z^2 - 1.9123693z + 0.9508977}\right]$$

where the numerator and the denominator of the quantized transfer function have been determined using the MATLAB implementation of floating-point quantization given in reference [2] as

$$[f, e] = \log 2(N_d)$$
$$dxp = sign(N_d). * pow2(max(e, -1021) + 52 - p)$$
$$N_q = (N_d + dxp) - dxp$$
$$[f, e] = \log 2(D_d)$$
$$dxp = sign(D_d). * pow2(max(e, -1021) + 52 - p)$$
$$D_q = (D_d + dxp) - dxp$$

where p is the number of floating-point quantization bits, N_d and D_d are the numerator and the denominator of the transfer function to be quantized, and N_q and D_q are the numerator and the denominator of the quantized transfer function. As shown in Figure 14.9, the step response of the quantized filter does not match the step response of the continuous-time filter. As a matter of fact, this quantized filter goes unstable for $p = 12$. As shown in this figure, the quantized delta operator filter compares very well with the continuous-time filter. The transfer function of the delta operator filter is

$$G(\delta) = 1 \times 10^{-4}\left[\frac{\delta^2 + 0.5847\delta + 4.5641}{\delta^2 - 1.8556\delta + 0.8673}\right]\left[\frac{\delta^2 - 1.1228\delta + 0.923}{\delta^2 - 1.9027\delta + 0.9447}\right]$$

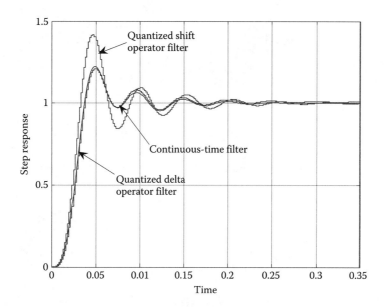

FIGURE 14.9 Step response of the 4th-order elliptic filter using shift and delta operators.

The delta operator described by Equation 14.18 can be modified as follows:

$$\delta = \frac{z-1}{\Delta} \tag{14.28}$$

where Δ is a fraction of the sampling period. This is called the *modified delta operator* and is useful, because Δ can be selected to optimize the numerical performance of the filter and allow lower rounding errors. One can repeat the above examples and demonstrate performance improvements using the modified delta operator.

14.6 Quantization Effects in A/D Conversion

The second source of error that we shall discuss is quantization in analog-to-digital conversion. Conceptually, A/D conversion involves two steps: sampling and quantization. Sampling is the process of converting a continuous-time signal $x(t)$ to a discrete-time sequence $x(k)$, and quantization is the process of approximating each sample of the sequence with a digital code word, that is, each sample is rounded or truncated to fit into the finite length register. Rounding approximates the sample by the nearest *quantization level* and truncating approximates the sample by the highest quantization level that is smaller than or equal to the sample value. Figure 14.10a shows a block diagram of an A/D converter in which the input signal $x(t)$ is sampled and then quantized to produce $x_q(k)$.

Let the word length of the A/D converter be C bits and let the number converted be of the form

$$x_q = b_1 2^{-1} + b_2 2^{-2} + b_3 2^{-3} + \cdots + b_c 2^{-c}$$

where $b_1, b_2, \ldots b_c$ are the binary codes. The sign bit, however coded, will always be present. The largest x_q possible that is produced by the A/D converter is

$$x_q = 2^{-1} + 2^{-2} + 2^{-3} + \cdots + 2^{-c} = \frac{1}{2} \sum_{i=0}^{C-1} \left(\frac{1}{2}\right)^i$$

$$= 1 - 2^{-C}$$

FIGURE 14.10   Block diagram of an A/D converter and its statistical model. (a) Block diagram of an A/D converter. (b) Statistical model of an A/D converter.

and the smallest nonzero x_q is 2^{-C}. The *dynamic range* of the converter is commonly defined as

$$DR = \frac{1 - 2^{-C}}{2^{-C}} = 2^C - 1$$

Thus,

$$C \geqslant \log_2(DR + 1) \tag{14.29}$$

For roundoff, taking the error e as a random variable with a uniform probability density shown in Figure 14.4,

$$E\{e\} = \int_{-q/2}^{q/2} x p(x)\, dx = 0 \tag{14.30}$$

where E denotes expected value, and

$$E\{e^2\} = \int_{-\infty}^{\infty} x^2\, p(x)\, dx = \int_{-q/2}^{q/2} \frac{x^2}{q}\, dx = \frac{q^2}{12}$$

In terms of the number of bits, C, the *variance* is

$$E\{e^2\} = \frac{2^{-2C}}{12} \tag{14.31}$$

For signed-magnitude and one's complement truncation, the error is uniformly distributed between $-q$ and q as shown in Figure 14.4. Then

$$E\{e\} = \int_{-\infty}^{\infty} x p\,(x)\, dx = 0 \tag{14.32}$$

and

$$E\{e^2\} = \int_{-\infty}^{\infty} x^2\, p(x)\, dx = \int_{-q}^{q} \frac{x^2}{2q}\, dx = \frac{q^2}{3}$$

In terms of the number of bits, C,

$$E\{e^2\} = \frac{2^{-2C}}{3} \tag{14.33}$$

Comparing Equation 14.31 with Equation 14.33 for roundoff quantization gives

$$E\{e^2\} = \frac{q^2}{12} = \frac{2^{-2C}}{12} = \frac{2^{-2(C+1)}}{3}$$

For two's complement truncation, the error is uniformly distributed as shown in Figure 14.4. Thus,

$$E\{e\} = \frac{-2^{-C}}{2} \tag{14.34}$$

and

$$E\{e^2\} = \frac{2^{-C}}{12} = \frac{2^{-2(C+1)}}{3} \tag{14.35}$$

As will be discussed in Section 14.7, the quantization error $e(k)$ can be viewed as an additive, stationary, white-noise process, as shown in Figure 14.10b, with mean and variance given by Equations 14.30 through 14.35 depending on whether the quantization is due to roundoff or truncation. Using superposition, the output of the digital controller can be decomposed into two parts: one due to the input $x(k)$ alone, and the other due to $e(k)$ which is assumed to be uncorrelated with $x(k)$.

14.6.1 Signal-to-Noise Ratio of an A/D Converter

Referring to Figure 14.10, the *signal-to-noise ratio* (SNR) of an *A/D* converter is defined as

$$(SNR)_{dB} = 10 \log_{10} \left(\frac{P_s}{P_e} \right)$$

where P_s is the output signal power and P_e is the output noise power. Assuming that the output signal and the output noise are both zero mean signals,

$$P_s = P_x = E\left\{x^2(k)\right\} = \text{var}\left\{x(k)\right\} = \sigma_x^2$$

and

$$P_e = E\left\{e^2(k)\right\} = \text{var}\left\{e(k)\right\} = \sigma_e^2$$

and, therefore, the SNR becomes

$$(SNR)_{dB} = 10 \log_{10} \frac{E\left\{x^2(k)\right\}}{E\left\{e^2(k)\right\}} = 10 \log_{10} \frac{\sigma_x^2}{\sigma_e^2} = 20 \log_{10} \frac{\sigma_x}{\sigma_e} \tag{14.36}$$

The SNR of an A/D converter that quantizes by rounding which generates zero-mean quantization noise can be determined by substituting Equation 14.31 into Equation 14.36 to give

$$(SNR)_{dB} = 10 \log_{10} \frac{\sigma_x^2}{\frac{2^{-2C}}{12}} = 10 \log_{10} \sigma_x^2 + 10 \log_{10}(12) - 10 \log_{10} 2^{-2C}$$
$$= 10 \log_{10} \sigma_x^2 + 10.79 + 6.02C \tag{14.37}$$

A/D converters are typically marketed using all of the A/D converter's bits including its sign bit. Thus, for a B bit A/D converter, where $C = B - 1$, Equation 14.37 can be written as

$$(SNR)_{dB} = 10 \log_{10} \sigma_x^2 + 4.77 + 6.02B \tag{14.38}$$

Equation 14.38 shows that the SNR increases 6.02 dB for every additional bit of resolution.

As an example, consider the input signal $x(k) = \sin(\omega k)$. For this signal,

$$E\{x^2(k)\} = \sigma_x^2 = \frac{1}{2} \tag{14.39}$$

Substituting Equation 14.39 into 14.37,

$$(SNR)_{dB} = -3.01 + 10.79 + 6.02C = 7.78 + 6.02C \tag{14.40}$$

which is equivalent to

$$(SNR)_{dB} = 7.78 + 6.02(B - 1) = 1.76 + 6.02B$$

where $B = C + 1$, so that B includes all the A/D converter's bits including its sign bit.

If the input sequence $x(k)$ is a random signal which has a zero mean Gaussian probability density so that $3\sigma = 1$ (i.e., the unity input level is a 3σ event), then

$$E\{x^2(k)\} = \sigma_x^2 = \left(\frac{1}{3}\right)^2 = \frac{1}{9} \tag{14.41}$$

Substituting Equation 14.41 into Equation 14.37,

$$(SNR)_{dB} = -9.54 + 10.79 + 6.02C = 1.25 + 6.02C$$

which implies that

$$C = 0.166(SNR)_{dB} - 0.208$$

As a simple design procedure, one may choose the number of bits of the A/D converter to be the larger of the two values necessary for required dynamic range and required SNR, that is,

$$C = \max\left\{\log_2(DR + 1), 0.166(S/N)_{dB} - 0.208\right\}$$

The quantization error of an A/D converter is not a serious problem. In an ideal 16-bit A/D converter, that is, $C = 15$, the maximum quantization error is $2^{-C}/2 = 2^{-16} = 0.0015\%$, which is quite low compared with typical errors in analog sensors. This error, if taken to be "noise," gives an SNR of $20\log_{10}(2^{-16}) = -96.3$ dB, which is much better than that of most high-fidelity audio systems. The designer must insure that enough bits are used to give the desired system accuracy.

14.6.2 SNR of an Oversampling A/D Converter

Oversampling A/D converters are A/D converters that sample signals at rates significantly higher than their Nyquist rate which is the minimum sampling frequency from which the analog signal can be reconstructed from its samples. For example, if a real signal of interest has a spectrum which contains frequencies from DC to f_{BW} Hz, which implies that it contains frequencies from $-f_{BW}$ Hz to f_{BW} Hz, then the signal's Nyquist rate would be $2f_{BW}$. If this signal is oversampled by a factor M, then the signal is sampled at a rate of $2Mf_{BW}$. This factor M is often referred to as an *oversampling rate* (OSR). After a signal has been oversampled, it is typically filtered digitally, and then the filtered signal's sampling rate is reduced to a rate near the original analog signal's Nyquist rate. Oversampling A/D converters are typically used to simplify antialiasing filters and improve an A/D converter's SNR, or to implement certain classes of A/D converters such as delta sigma ($\Delta\Sigma$) modulators.

When a signal is sampled at or near its Nyquist rate, an analog antialiasing filter with a very narrow transition band is often required. Such filters are typically difficult and expensive to implement and may also have highly nonlinear phase responses. The implementation of this system can be simplified by increasing the A/D converter's sampling rate by a factor of M which increases the antialiasing filter's transition bandwidth. The oversampled signal can be filtered digitally using a linear phase finite impulse response (FIR) antialiasing filter, and then the filtered signal's sampling rate can be reduced by only retaining every Mth sample of the digitally filtered signal. To illustrate, consider the system shown in Figure 14.11, where $x(t)$ is a real analog signal that has a spectrum which contains frequencies from DC to f_{BW} Hz. If the A/D converter oversamples $x(t)$ by a factor of M, then the A/D converter samples the signal at a rate of $2Mf_{BW}$, and thus, the antialiasing filter's transition band would extend from f_{BW} Hz to $2Mf_{BW} - f_{BW}$ Hz. The oversampled signal, $x_o(k)$, is then filtered by a digital linear phase FIR filter that has a cutoff frequency of f_{BW} Hz, or π/M radians/sample, and a very narrow transition band. The sampling rate of the resulting signal can then be reduced to the Nyquist rate of the signal of interest by only retaining every Mth sample of the digitally filtered signal, $x_f(k)$.

Theoretically, the oversampling process in Figure 14.11 can also improve the SNR of the sampling process. To illustrate, let us model the quantization noise of the A/D converter in Figure 14.11 as an

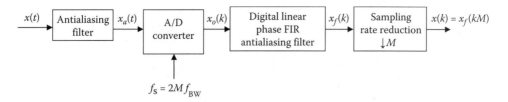

FIGURE 14.11 A system that uses an oversampling A/D converter that simplifies the design and implementation of the analog antialiasing filter. The sampling frequency of the A/D converter is fs and M is the OSR.

additive, stationary, white-noise process, denoted by $e_o(k)$, as shown in Figure 14.12. Because the signal of interest is bandlimited from DC to f_{BW} Hz, the signal of interest will pass through the digital antialiasing filter and the sampling rate reduction block. Therefore, if the signal of interest has a zero mean, then the power P_x of the signal of interest can be expressed as $P_x = E\{x^2(k)\} = \sigma_x^2$. However, because the quantization from the A/D converter is a broadband signal, the digital antialiasing filter reduces its power.

To illustrate, recall that

$$P_e = E\{e^2(k)\} = E\{e(k)e(k+n)\}|_{n=0} = \phi_{ee}(n)|_{n=0} = \phi_{ee}(0)$$

where $\phi_{ee}(n)$ is the autocorrelation of $e(k)$. Therefore,

$$P_e = F^{-1}\{\Phi_{ee}(e^{j\omega})\}|_{n=0} = \frac{1}{2\pi} \int_{\omega=-\pi}^{\pi} \Phi_{ee}(e^{j\omega})e^{j\omega n} d\omega|_{n=0} = \frac{1}{2\pi} \int_{\omega=-\pi}^{\pi} \Phi_{ee}(e^{j\omega}) \, d\omega \qquad (14.42)$$

where F^{-1} is the inverse Fourier transform for discrete signals, $\Phi_{ee}(e^{j\omega})$ is the Fourier transform of $\phi_{ee}(n)$ or the power spectral density (PSD) of $e(k)$, and ω represents frequency in radians/sample. Assuming that the digital antialiasing filter is linear and time invariant, then

$$\Phi_{ee}(e^{j\omega}) = |H(e^{j\omega})|^2 \Phi_{e_o e_o}(e^{j\omega}) \qquad (14.43)$$

where $\Phi_{e_o e_o}(e^{j\omega})$ is the PSD of $e_o(k)$, and $H(e^{j\omega})$ is the frequency response of the digital antialiasing filter which is ideally a low-pass filter with a very narrow transition band and a cutoff frequency of π/M radians/sample. Therefore, substituting Equation 14.43 into Equation 14.42 gives

$$P_e = \frac{1}{2\pi} \int_{\omega=-\pi/M}^{\pi/M} \Phi_{e_o e_o}(e^{j\omega}) \, d\omega = \frac{P_{e_o}}{M} \qquad (14.44)$$

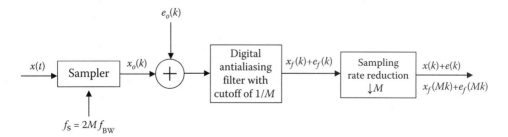

FIGURE 14.12 Simple linear model of the oversampling system shown in Figure 14.11.

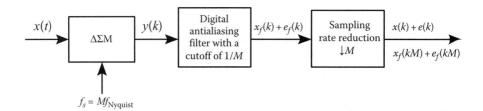

FIGURE 14.13 An oversampling system that uses a $\Delta\Sigma M$. The sampling rate of the $\Delta\Sigma M$'s is f_s, M is the OSR, and $f_{Nyquist}$ is the Nyquist rate of the signal of interest.

Assuming that the A/D converter performs quantization by rounding, $e_o(k)$ will be a zero mean random process which implies that $P_{e_o} = \sigma_{e_o}^2$. Therefore, Equation 14.44 can be written as

$$P_e = \frac{\sigma_{e_o}^2}{M} \tag{14.45}$$

Substituting Equation 14.31 into Equation 14.45 gives

$$P_e = \frac{\sigma_{e_o}^2}{M} = \frac{2^{-2C}/12}{M}$$

Therefore, the oversampling system's SNR can be determined as

$$(SNR)_{dB} = 10\log_{10}\frac{\sigma_x^2}{2^{-2C}/12M} = 10\log_{10}\sigma_x^2 + 10.79 + 6.02C + 10\log_{10}M \tag{14.46}$$

or as

$$(SNR)_{dB} = 10\log_{10}\sigma_x^2 + 4.77 + 6.02B + 10\log_{10}M$$

Equation 14.46 shows that the oversampling system shown in Figure 14.11 improves the SNR from the A/D converter's output by $10\log_{10}M$ dB, which implies that for every quadrupling of the oversampling factor M, the output signal's SNR improves by 6.02 dB which is equivalent to adding a bit of resolution. This increase in resolution is only possible if the A/D converter can generate samples at this higher resolution and if the digital antialiasing filter is designed to process signals at this higher resolution.

14.6.3 SNR of a Delta Sigma Modulator

Delta sigma modulators ($\Delta\Sigma Ms$) are an A/D converter architecture that uses an oversampling A/D converter with a small number of bits and a feedback loop that shapes the A/D converter's quantization noise to achieve high SNRs and large dynamic ranges. The feedback loop can also be designed to act as an antialiasing filter. The output of a delta sigma modulator ($\Delta\Sigma M$) is processed in a manner similar to the oversampling A/D converters discussed in the previous section as shown in Figure 14.13.

Figure 14.14 shows a block diagram of a typical $\Delta\Sigma M$. Figure 14.15 shows a linear model of the $\Delta\Sigma M$ given in Figure 14.14 where the A/D converter in Figure 14.14 has been modeled as an additive, stationary, random, white noise source, $E(s)$, that has a uniform probability density function, and the digital-to-analog (D/A) converter in the feedback loop has been modeled by the system function $H(s)$. The transfer function that describes the relationship between the $\Delta\Sigma M$'s quantization noise, $E_o(s)$, and the $\Delta\Sigma M$'s output, $Y(s)$, is referred to as the $\Delta\Sigma M$'s noise shaping filter or noise transfer function (NTF). For the $\Delta\Sigma M$ shown in Figure 14.15,

$$NTF(s) = \frac{Y(s)}{E_o(s)} = \frac{1}{1 + G(s)H(s)}$$

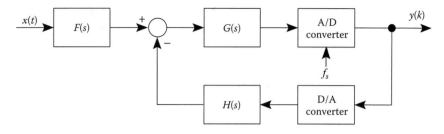

FIGURE 14.14 A block diagram of a typical $\Delta\Sigma$M.

The transfer function that describes the relationship between the $\Delta\Sigma$M's input, $X(s)$, and $\Delta\Sigma$M's output, $Y(s)$, is referred to as the $\Delta\Sigma$M's signal transfer function (STF). For the $\Delta\Sigma$M shown in Figure 14.15,

$$STF(s) = \frac{Y(s)}{X(s)} = \frac{F(s)G(s)}{1 + G(s)H(s)}$$

Because the block diagram in Figure 14.15 is a linear model, the $\Delta\Sigma$M's output, $Y(s)$, can be written as

$$Y(s) = STF(s)X(s) + NTF(s)E_o(s)$$

A $\Delta\Sigma$M's NTF is designed to shape the quantization noise, $E_o(s)$, so that quantization noise is attenuated over the frequencies of interest while $\Delta\Sigma$M's STF is often designed as an antialiasing filter. For example, consider the $\Delta\Sigma$M model in Figure 14.15 where $F(s) = 1$, $G(s) = 1/s$, and $H(s) = 1$ which implies that

$$STF(s) = \frac{Y(s)}{X(s)} = \frac{1}{s+1} \quad \text{and} \quad NTF(s) = \frac{Y(s)}{E_o(s)} = \frac{s}{s+1}$$

Figure 14.16 shows the magnitude response of the $\Delta\Sigma$M's STF and NTF. As shown in the figure, the NTF attenuates the A/D converter's quantization noise at low frequencies, and the STF acts as an antialiasing filter. The $\Delta\Sigma$M's NTF and STF cannot be designed independently, because the NTF and STF share a common denominator. Since the NTF directly affects the $\Delta\Sigma$M's SNR, a $\Delta\Sigma$M's NTF is typically designed before the $\Delta\Sigma$M's STF.

Because a $\Delta\Sigma$M's quantization noise is shaped by the $\Delta\Sigma$M's NTF, a $\Delta\Sigma$M's SNR is a function of its particular NTF. To estimate the SNR of a $\Delta\Sigma$M, consider a simple $\Delta\Sigma$M where

$$F(s) = 1, \quad G(s) = \frac{(2\pi f_c)^N}{s^N}, \quad H(s) = 1$$

where f_c is the 3-dB frequency of the $\Delta\Sigma$M's STF and NTF, and N is an integer value which is referred to as the $\Delta\Sigma$M's order. For this $\Delta\Sigma$M,

$$STF(s) = \frac{Y(s)}{X(s)} = \frac{(2\pi f_c)^N}{s^N + (2\pi f_c)^N} \quad \text{and} \quad NTF(s) = \frac{Y(s)}{E_o(s)} = \frac{s^N}{s^N + (2\pi f_c)^N}$$

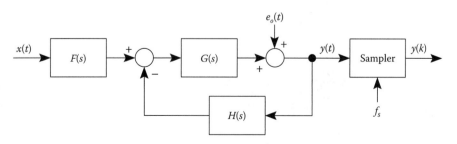

FIGURE 14.15 A linear model of the $\Delta\Sigma$M in Figure 14.14.

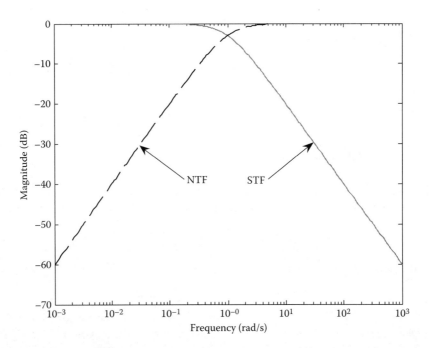

FIGURE 14.16 Magnitude response of the STF and NTF.

Assuming that the $\Delta\Sigma$M's input signal and the quantization noise are zero mean signals,

$$(SNR)_{dB} = 10\log_{10}\frac{P_{x(k)}}{P_e} = 10\log_{10}\frac{\sigma^2_{x(k)}}{\sigma^2_e}$$

where $x(k)$ and $e(k)$ are the output signal and output noise of the sampling rate converter, respectively. If the STF and the digital antialiasing filter pass the input signal with unity gain, the analog input signal power, $P_{x(t)}$ and the digital output signal power, $P_{x(k)}$, are related by

$$P_{x(k)} = f_s^2 P_{x(t)} = f_s^2 \sigma^2_{x(t)} \tag{14.47}$$

The output noise power can be calculated as

$$P_e = \frac{1}{2\pi}\int\limits_{\Omega=-\pi}^{\pi} \Phi_{ee}(e^{j\Omega})\,d\Omega = \frac{1}{2\pi f_s}\int\limits_{\omega=-\pi f_s}^{\pi f_s} \Phi_{ee}(e^{j\omega/f_s})\,d\omega \tag{14.48}$$

where $\Phi_{ee}(e^{j\Omega})$ is the PSD of $e(k)$, Ω represents frequency in radians/sample and ω represents frequency in rad/s. Assuming that the digital antialiasing filter and the STF are linear and time invariant,

$$\begin{aligned}
\Phi_{ee}(e^{j\omega}) &= \left|NTF(e^{j\omega/f_s})\right|^2 \left|H_a(e^{j\omega/f_s})\right|^2 \Phi_{e_oe_o}(e^{j\omega/f_s})\\
&= \left|f_s \cdot NTF(j\omega)\right|^2 \left|H_a(e^{j\omega/f_s})\right|^2 \Phi_{e_oe_o}(e^{j\omega/f_s})
\end{aligned} \tag{14.49}$$

where $H_a(e^{j\omega/f_s})$ is the frequency response of the digital antialiasing filter which is ideally a low-pass filter with a very narrow transition band and a cutoff frequency of $\pi f_s/M$ rad/s. Substituting Equation 14.49 into

Equation 14.48 and assuming that the quantization noise is white which implies that $\phi_{e_o e_o}(n) = \sigma_{e_o}^2 \delta(n)$, or equivalently

$$\Phi_{e_o e_o}(e^{j\omega/f_s}) = \sigma_{e_o}^2$$

$$P_e = \frac{f_s \sigma_{e_o}^2}{2\pi} \int\limits_{\omega = -\pi f_s/M}^{\pi f_s/M} \left| NTF(j\omega) \right|^2 d\omega \tag{14.50}$$

For proper operation of this $\Delta\Sigma M$, $\pi f_s/M \ll 2\pi f_c$, which implies that

$$\left| NTF(j\omega) \right|^2 = \left| \frac{(j\omega)^N}{(j\omega)^N + (2\pi f_c)^N} \right|^2 \approx \left| \frac{(j\omega)^N}{(2\pi f_c)^N} \right|^2 = \frac{\omega^{2N}}{(2\pi f_c)^{2N}} \tag{14.51}$$

Substituting Equation 14.51 into Equation 14.50 and integrating,

$$P_e = \frac{f_s \sigma_{e_o}^2}{2\pi} \frac{2}{(2N+1)(2\pi f_c)^{2N}} \left(\frac{\pi f_s}{M} \right)^{2N+1} = \frac{f_s^{2N+2} \sigma_{e_o}^2}{2N+1} \frac{\pi^{2N}}{(2\pi f_c)^{2N}} \left(\frac{1}{M} \right)^{2N+1} \tag{14.52}$$

Using Equations 14.47 and 14.52,

$$SNR = \frac{P_{x(k)}}{P_e} = \frac{f_s^2 \sigma_{x(t)}^2}{\frac{f_s^{2N+2} \sigma_{e_o}^2}{2N+1} \frac{\pi^{2N}}{(2\pi f_c)^{2N}} \left(\frac{1}{M} \right)^{2N+1}} = \frac{\sigma_{x(t)}^2}{\sigma_{e_o}^2} \frac{(2N+1)M^{2N+1}(2\pi f_c)^{2N}}{(\pi f_s)^{2N}}$$

In dB, the *SNR* becomes

$$(SNR)_{dB} = 10\log_{10}\sigma_{x(t)}^2 - 10\log_{10}\sigma_{e_o}^2 + (20N+10)\log_{10}M + 10\log_{10}(2N+1) + 20N\log_{10}\frac{f_c}{f_s/2} \tag{14.53}$$

Substituting Equation 14.31 into Equation 14.53,

$$(SNR)_{dB} = 10\log_{10}\sigma_{x(t)}^2 + 10.79 + 6.02C + (20N+10)\log_{10}M + 10\log_{10}(2N+1) + 20N\log_{10}\frac{f_c}{f_s/2}$$

or equivalently, in terms of *B*,

$$(SNR)_{dB} = 10\log_{10}\sigma_{x(t)}^2 + 4.77 + 6.02B + (20N+10)\log_{10}M + 10\log_{10}(2N+1)$$
$$+ 20N\log_{10}\frac{f_c}{f_s/2} \tag{14.54}$$

Recall that $B = C + 1$ so that *B* includes all the A/D converter's bits including its sign bit. If, for example, the input signal is a full-scale sine wave, then $\sigma_{x(t)}^2 = 1/2$, and Equation 14.54 can be written as

$$(SNR)_{dB} = 6.02B + 1.76 + +(20N+10)\log_{10}M + 10\log_{10}(2N+1) + 20N\log_{10}\frac{f_c}{f_s/2}$$

Throughout the literature, it is also common to assume that $f_c/(f_s/2) = 1/\pi$, then

$$(SNR)_{dB} = 6.02B + 1.76 + (20N+10)\log_{10}M + 10\log_{10}(2N+1) - 9.94N$$

Unlike the previously described oversampling system which has an SNR that can be limited by the accuracy of its A/D converter, a $\Delta\Sigma M$'s SNR is typically not limited by the accuracy of its A/D converter because the A/D converter's output is filtered by the NTF. Instead, a $\Delta\Sigma M$'s SNR is often limited by the accuracy of its D/A converter, because the D/A converter's output is added directly to the input signal and then filtered by the STF. Typically, $\Delta\Sigma M$s with single bit A/D converters are inherently linear; however, the linearity of a $\Delta\Sigma M$ with multibit A/D converters is typically limited by the DAC's accuracy.

For $\Delta\Sigma$Ms with multibit A/D converters, techniques such as dynamic element matching can be used to improve the linearity of the D/A converter [3].

14.7 Stochastic Analysis of Quantization Errors in Digital Processing

One approach for analyzing roundoff and truncation errors generated in digital processing of controllers is to derive deterministic upper bounds on the maximum errors that can possibly result from roundoff or truncation. In general, however, these bounds are *pessimistic* because the errors usually add up in the worst possible way.

Another popular approach for analyzing roundoff and truncation errors is to develop stochastic noise models of these errors first, and then determine their effects on system performance.

14.7.1 Fixed-Point Arithmetic

It was mentioned earlier that, in fixed-point arithmetic, quantization errors occur in multiplication and not in addition. Figure 14.17a shows a block diagram of a multiplier model in which two C-bit numbers are multiplied and then quantized to produce a C-bit number. An equivalent noise model of the multiplier, useful for analysis, is shown in Figure 14.17b. The quantization error $e(k)$ is modeled as a stationary, additive, white-noise sequence so that

$$E\{e(k)\} = \mu_e$$

$$E\{e(i)e(j)\} = \begin{cases} 0, & i \neq j \\ E\{e^2\}, & i = j \end{cases}$$

where the mean and the variance of the error can be determined from the probability density function shown in Figure 14.4.

Using superposition, the output of a digital controller due to the error $e(k)$ alone is given by the convolution solution

$$y_e(k) = \sum_{m=0}^{k} g(m)e(k - m)$$

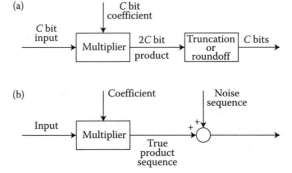

FIGURE 14.17   Model of multiplier. (a) Multiplier model. (b) Statistical model of multiplier.

where $g(m)$ is the unit pulse response of the system whose input is the error source and whose output is the digital controller output. The mean of the output is

$$E\{y_e(k)\} = \sum_{m=0}^{k} g(m)E\{e(k-m)\}$$

$$= \mu_e \sum_{m=0}^{k} g(m)$$
(14.55)

and the variance of $y_e(k)$ is

$$E\{y_e^2(k)\} = E\left\{ \left[\sum_{m=0}^{k} g(m)e(k-m) \right] \cdot \left[\sum_{n=0}^{k} g(n)e(k-n) \right] \right\}$$

$$= \sum_{m=0}^{k} \sum_{n=0}^{k} g(n)g(m)E\{e(k-m)e(k-n)\}$$

Because $e(k)$ is modelled as a white noise sequence

$$E\{e(k-m)e(k-n)\} = E\{e^2\}\delta(m-n)$$

where δ is the unit pulse function. Then as a result,

$$E\{y_e^2(k)\} = E\{e^2\} \sum_{m=0}^{k} g^2(m)$$

Therefore as k approaches infinity, the mean and the variance of the output are

$$E\{y_e(k)\} = \mu_e \sum_{m=0}^{\infty} g(m)$$
(14.56)

and

$$E\{y_e^2(k)\} = E\{e^2\} \sum_{m=0}^{\infty} g^2(m)$$
(14.57)

respectively. Hence, the variance of the output equals the variance of the quantization noise times the noise power gain, *NPG*, where

$$(NPG) = \sum_{m=0}^{\infty} g^2(m)$$
(14.58)

Because digital controller transfer functions are realized using first- and second-order subsystems in parallel or cascade forms, the noise power gains of first- and second-order subsystems are now determined.

The transfer function of a first-order recursive subsystem is

$$G(z) = \frac{A(1+\beta_1 z^{-1})}{1-\alpha_1 z^{-1}} = k_0 + \frac{k_1}{1-\alpha_1 z^{-1}}$$

Hence,

$$g(m) = k_0\delta(m) + k_1\alpha_1^m, \quad m = 0, 1, 2, \ldots$$

and

$$\sum_{m=0}^{\infty} g(m) = k_0 + \frac{k_1}{1-\alpha_1}$$

Therefore, the mean of the output is

$$E\{y_e(k)\} = \mu_e \left[k_0 + \frac{k_1}{1 - \alpha_1} \right] \tag{14.59}$$

Similarly,

$$g^2(m) = \begin{cases} (k_0 + k_1)^2, & m = 0 \\ k_1^2 \alpha_1^{2m}, & m = 1, 2, \ldots \end{cases}$$

and therefore, the noise power gain becomes

$$(NPG) = k_0^2 + 2k_0 k_1 + \frac{k_1^2}{1 - \alpha_1^2} \tag{14.60}$$

On the other hand, the transfer function of a second-order recursive subsystem is

$$G(z) = A \frac{1 + \beta_1 z^{-1} + \beta_2 z^{-2}}{1 - \alpha_1 z^{-1} - \alpha_2 z^{-2}}$$
$$= k_0 + \frac{k_1}{1 - r_1 z^{-1}} + \frac{k_2}{1 - r_2 z^{-1}} \tag{14.61}$$

Therefore,

$$g(m) = k_0 \delta(m) + k_1 r_1^m + k_2 r_2^m, \quad m = 0, 1, 2, \ldots$$

and hence,

$$\sum_{m=0}^{\infty} g(m) = k_0 + \frac{k_1}{1 - r_1} + \frac{k_2}{1 - r_2} \tag{14.62}$$

Similarly,

$$g^2(m) = \begin{cases} (k_0 + k_1 + k_2)^2, & m = 0 \\ (k_1 r_1^m + k_2 r_2^m)^2, & m = 1, 2, \ldots \end{cases}$$

and the noise power gain becomes

$$(NPG) = k_0^2 + 2k_0 k_1 + 2k_0 k_2 + \frac{k_1^2}{1 - r_1^2} + \frac{2k_1 k_2}{1 - r_1 r_2} + \frac{k_2^2}{1 - r_2^2} \tag{14.63}$$

which is real. For the special case

$$G(z) = \frac{1}{1 - \alpha_1 z^{-1} - \alpha_2 z^{-2}}$$

the (NPG), in terms of the polynomial coefficients, is

$$(NPG) = \frac{1 - \alpha_2}{1 + \alpha_2} \frac{1}{(1 - \alpha_2)^2 - \alpha_1^2}$$

Equations 14.59 through 14.63 are used regularly in the stochastic analysis of quantization error.

Another method for calculating the noise power gain is to use contour integration as follows:

$$(NPG) = \sum_{m=0}^{\infty} g^2(m)$$

$$= \sum_{m=0}^{\infty} g(m) \frac{1}{2\pi j} \oint_C G(z) z^{m-1} dz$$

where C is the contour of integration chosen as the unit circle $|z| = 1$. Thus,

$$(NPG) = \frac{1}{2\pi j} \oint_C G(z) z^{-1} \left(\sum_{m=0}^{\infty} g(m) z^m \right) dz$$

but

$$\sum_{m=0}^{\infty} g(m) z^m = G(z^{-1})$$

Therefore,

$$(NPG) = \frac{1}{2\pi j} \oint_C G(z) G(z^{-1}) z^{-1} dz \tag{14.64}$$

which is the sum of residues of $G(z)G(z^{-1})z^{-1}$ within the unit circle. Note that Equation 14.64 can be easily solved numerically [5].

14.7.2 Quantization Noise Model of First-Order Subsystems

Consider the first-order subsystem shown in Figure 14.18. Setting all the noise sources to zero, the transfer function of the ideal subsystem is

$$T(z) = \frac{Y(z)}{X(z)} = \frac{A(1 + \beta_1 z^{-1})}{1 - \alpha_1 z^{-1}}$$

The effect of the A/D converter on the output can be determined by setting all signals but e_0 to zero. The transfer function that relates e_0 to the filter output is

$$Y_0(z) = G_0(z) E_0(z)$$

where

$$G_0(z) = T(z) = k_0 + \frac{k_1}{1 - \alpha_1 z^{-1}}$$

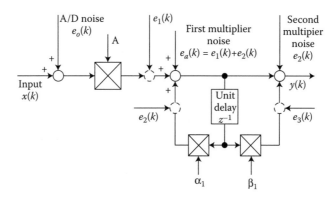

FIGURE 14.18 Fixed-point quantization noise model of first-order subsystem.

Then, the mean of the output is given by Equation 14.56:

$$E\{y_0\} = \mu_0 \left[k_0 + \frac{k_1}{1 - \alpha_1} \right]$$

and using Equation 14.57, the variance of the output is

$$E\{y_0^2\} = \left[k_0^2 + 2k_0k_1 + \frac{k_1^2}{1 - \alpha_1^2} \right] E\{e_0^2\}$$
$$= (NPG)_0 E\{e_0^2\}.$$

The effect of quantization error due to multiplication on system output is determined as follows. Setting all signals but e_a to zero,

$$y_a(z) = G_a(z)E_a(z) = T(z)E_a(z)$$

Therefore,

$$E\{y_a(z)\} = \mu_a \left[k_0 + \frac{k_1}{1 - \alpha_1} \right]$$

and

$$E\{y_a^2\} = \left[k_0^2 + 2k_0k_1 + \frac{k_1^2}{1 - \alpha_1^2} \right] E\{e_a^2\}$$

Assuming the multiplier errors e_1 and e_2 are uncorrelated, then

$$\text{variance}\{e_a\} = \text{variance}\{e_1 + e_2\} = E\{e_1^2\} + E\{e_2^2\}$$

Similarly, the second multiplier noise gain is calculated by setting all sources but e_3 to zero. The transfer function that relates e_3 to the output is

$$Y_3(z) = G_3(z)E_3(z)$$

where

$$G_3(z) = 1$$

Thus,

$$E\{y_3\} = E\{e_3\}$$

and

$$E\{y_3^2\} = E\{e_3^2\}$$

Also,

$$(NPG)_3 = 1$$

Assuming the output noises y_0, y_a, and y_3 are uncorrelated, then

$$\text{variance}\{y_0 + y_a + y_3\} = (NPG)_0 E\{e_0^2\} + (NPG)_a [E\{e_1^2\} + E\{e_2^2\}] + (NPG)_3 E\{e_3^2\}$$

14.7.3 Quantization Noise Model of Second-Order Subsystems

The previous analysis of first-order subsystems can easily be extended to second-order subsystems. The quantization noise model shown in Figure 14.18 for first-order subsystems can easily be modified for second-order subsystems. Setting all the noise sources to zero, the ideal transfer function of a second-order subsystem that relates $X(z)$ to $Y(z)$ is

$$T(z) = \frac{Y(z)}{X(z)} = \frac{A(1 + \beta_1 z^{-1} + \beta_2 z^{-2})}{1 - \alpha_1 z^{-1} - \alpha_2 z^{-2}}$$

Assuming that the output noises y_0, y_a, and y_b are uncorrelated, then

$$\text{variance}\{y_0 + y_a + y_b\} = (NPG)_0 E\{e_0^2\} + (NPG)_a [E\{e_1^2\} + E\{e_2^2\} + E\{e_3^2\}] + (NPG)_b [E\{e_4^2\} + E\{e_5^2\}]$$

where

$$(NPG)_0 = (NPG)_a$$

is determined from $T(z)$ and

$$(NPG)_b = 1$$

14.7.4 Floating-Point Arithmetic

The analysis of quantization errors in floating-point digital controllers is more complicated than in fixed-point digital controllers. It was mentioned earlier that in floating-point arithmetic, errors occur only in the mantissa. It was also mentioned that roundoff and truncation errors are introduced in both addition and multiplication. Let x_1 and x_2 be any two numbers before quantization. Quantizing the sum and the product of these two numbers gives

$$(x_1 + x_2)_q = (x_1 + x_2)(1 + \varepsilon_s) \tag{14.65}$$

and

$$(x_1 \cdot x_2)_q = (x_1 \cdot x_2)(1 + \varepsilon_p) \tag{14.66}$$

respectively, where the relative errors ε_s and ε_p, depending on the number representation, satisfy Equations 14.9 through 14.12. Each arithmetic operation introduces quantization error according to Equations 14.65 and 14.66. Detailed examples of roundoff and truncation errors accumulated in first- and second-order subsystems using floating-point arithmetic are given in [6].

14.8 Limit Cycle and Deadband Effects

When digital controllers are implemented with FWL, *limit cycles*, or sustained oscillations, may appear at the controller output even in the absence of any applied input. Basically, there are two different kinds of limit cycles. One is due to roundoff in multiplication, termed the *deadband* effect, and the other is due to register overflow. Limit cycles exist in fixed-point digital controllers but can be ignored in floating-point controllers.

To illustrate the phenomenon of limit cycle due to roundoff, consider the first-order controller described by the difference equation

$$y(k) = ay(k-1) + x(k) \tag{14.67}$$

Let

$$x(k) = 0.9\delta(k), \quad a = 0.5, \quad y(-1) = 0$$

If the controller equation is implemented with infinite word length registers, then

$$y(k) = 0.9(0.5)^k$$

As k approaches infinity, the steady-state value of the output $y(k)$ approaches zero. However, assuming that the controller equation is implemented with a 3-bit word length, then

$$y_q(k) = Q[0.5y_q(k-1)] + 0.75\delta(k)$$

Using decimal representation, the output can be calculated recursively as follows:

$$y_q(0) = Q[(0.5)(0)] + 0.75 = 0.75$$
$$y_q(1) = Q[(0.5)(0.75)] = 0.375$$
$$y_q(2) = Q[(0.5)(0.375)] = 0.25$$
$$y_q(3) = Q[(0.5)(0.25)] = 0.125$$
$$y_q(4) = Q[(0.5)(0.125)] = 0.125$$

$$\vdots$$

$$y_q(k) = Q[(0.5)(0.125)] = 0.125$$

Hence, as k approaches infinity, the steady-state value of $y_q(k)$ approaches 0.125 and not zero.

As another example, consider the system described by Equation 14.67, where

$$x(k) = 0, \quad a = -0.5, \quad y(-1) = 0.75$$

Again, if the controller equation is implemented with infinite word length registers, then the output,

$$y(k) = 0.75(-0.5)^k$$

which approaches zero as k approaches infinity. Assuming that the controller equation is implemented with a 3-bit word length, then

$$y_q(k) = Q[-0.5y_q(k-1)]$$

The output can be calculated recursively as follows:

$$y_q(0) = Q[(-0.5)(0.75)] = Q[-0.375] = -0.375$$
$$y_q(1) = Q[(-0.5)(-0.375)] = 0.25$$
$$y_q(2) = Q[(-0.5)(0.25)] = -0.125$$
$$y_q(3) = Q[(-0.5)(-0.125)] = 0.125$$
$$y_q(k) = Q[(-0.5)(0.125)] = -0.125$$

and the output oscillates between 0.125 and −0.125 indefinitely.

An interesting example of limit cycle due to register overflow is given in [7]. Limit cycles due to roundoff and overflow are unwanted and their effect on control system performance should be minimized.

Bibliography

1. Koren, I., *Computer Arithmetic Algorithms*, 2nd ed., A K Peters Ltd, Natick, MA, 2001.
2. Widrow, B. and Kollár, I., *Quantization Noise: Roundoff Error in Digital Computation, Signal Processing, Control, and Communications*, Cambridge University Press, New York, 2008.
3. Stubberud, P. and Bruce, J.W., An analysis of dynamic element matching flash digital to analog converters, *IEEE Transactions on Circuits and Systems II: Analog and Digital Signal Processing*, 48(2), 205–213, 2001.
4. Middleton, R. and Goodwin, G., *Digital Control and Estimation: A Unified* Approach, Prentice-Hall, Englewood Cliffs, NJ, 1990.
5. Grove, R. and Hwang, P., *Introduction to Random Signals and Applied Kalman Filtering*, 3rd ed., John Wiley & Sons, New York, 1996.
6. Oppenheim, A.V. and Schafer, R.W., *Digital Signal Processing*, Prentice-Hall, Englewood Cliffs, NJ, 1975.
7. Stubberud, P., *Digital Signal Processing*, Class Notes. University of Nevada, Las Vegas, NV, 2008.
8. Smith, S., *The Scientist & Engineer's Guide to Digital Signal Processing*, California Technical Pub., San Diego, 1997.
9. Franklin, G.F., Powell, J.D., and Workman, M.L., *Digital Control of Dynamic Systems*, 3rd edn, Ellis-Kagle Press, Half Moon Bay, CA, 2006.
10. Mitra, S., *Digital Signal Processing Laboratory Using MATLAB*. McGraw-Hill, New York, 2005.
11. Proakis, J. and Manolakis, D., *Digital Signal Processing*, 4th ed., Prentice-Hall, Englewood Cliffs, NJ, 2006.
12. Shenoi, B.A., *Introduction to Digital Signal Processing and Filter Design*, Wiley-Interscience, New York, 2005.
13. Bauer, P. H., *High-Speed Fixed and Floating Point Implementation of Delta-Operator Formulated Discrete Time Systems*, Electrical Engineering, Notre Dame University, 1994.
14. Santina, M. and Stubberud, A., Basics of sampling and quantization, *Handbook of Networked and Embedded Control Systems*, Birkhäuser, Boston, 2005.

15

Sample-Rate Selection

Michael Santina
The Boeing Company

Allen R. Stubberud
University of California, Irvine

15.1 Introduction

In this chapter, the selection of sampling rate for a digital control system is briefly discussed. As the sampling rate is increased, the performance of the digital controller usually improves. Computer cost also increases because less time is available to process the controller equations. Reducing the sample rate for the sake of reducing cost, however, may degrade system performance or even cause instability. Additionally, for systems with analog-to-digital converters, higher sample rates require faster A/D conversion speed which may also be expensive.

Aside from cost, the selection of sampling rate for digital control systems depends on many factors, including smoothness of the time response, effects of disturbances and sensor noise, parameter variations, and quantization. *The best sampling rate which can be chosen for a digital control system is the slowest rate that meets all performance requirements.* Before we discuss the selection of sampling rate, a statement of the sampling theorem is in order.

15.2 The Sampling Theorem

Sampling is the process of deriving a discrete-time sequence from a continuous-time function. Usually, but not always, the samples are evenly spaced in time. *Reconstruction* is the reverse; it is the formation of a continuous-time function from a sequence of samples. Many different continuous-time functions can have the same set of samples, so a reconstruction is not unique. Figure 15.1 shows two different continuous-time signals with the same samples, illustrating how, except in highly restricted circumstances, a sampled function is not uniquely determined by its samples. One important situation for which samples of a continuous-time function are unique occurs when the function is *bandlimited*. A signal $g(t)$ and its

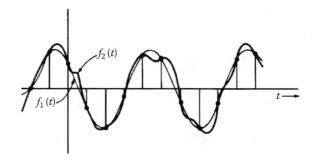

FIGURE 15.1 Two different continuous-time signals with the same samples.

Fourier transform $G(\omega)$ are generally related by

$$G(\omega) = \int_{-\infty}^{\infty} g(t)e^{-j\omega t}\,dt$$

$$g(t) = \frac{1}{2\pi} \int_{-\infty}^{\infty} G(\omega)e^{j\omega t}\,d\omega \tag{15.1}$$

This relationship is similar to Laplace transformation with $s = jw$, but the transform integral of Equation 15.1 extends over all time rather than from $t = 0^-$ on. The Fourier transform $G(\omega)$ is termed the *spectrum* of $g(t)$.

If a signal $g(t)$ is uniformly sampled with sampling period T to form the sequence

$$g(k) = g(t = kT)$$

then the corresponding impulse train that extends both ways in time

$$g^*(t) = \sum_{k=-\infty}^{\infty} g(kT)\delta(t - kT)$$

is a continuous-time (hence it has a Fourier transform) signal that is equivalent to $g(kT)$ and has the Fourier transform

$$G^*(\omega) = \frac{1}{T} \sum_{n=-\infty}^{\infty} G(\omega - n\omega_s) \tag{15.2}$$

where

$$\omega_s = 2\pi f_s = \frac{2\pi}{T}$$

To prove this result, the periodic function

$$s(t) = \sum_{k=0}^{\infty} \delta(t - kT)$$

is represented by an exponential Fourier series of the form

$$s(t) = \sum_{n=-\infty}^{\infty} d_n e^{(jn2\pi/T)t}$$

where

$$d_n = \frac{1}{T} \int_{-T/2}^{T/2} \sum_{k=-\infty}^{\infty} \delta(t - kT)e^{-(jn2\pi/T)t}\,dt = \frac{1}{T}$$

Hence,

$$s(t) = \frac{1}{T} \sum_{n=-\infty}^{\infty} e^{(jn2\pi/T)t}$$

Substituting this result in the impulse train

$$g^*(t) = \sum_{k=-\infty}^{\infty} g(kT)\delta(t - kT)$$

gives

$$g^*(t) = \frac{1}{T}g(t) \sum_{n=-\infty}^{\infty} e^{(jn2\pi/T)t}$$

and taking the Fourier transform yields

$$G^*(\omega) = \frac{1}{T} \sum_{n=-\infty}^{\infty} \int_{-\infty}^{\infty} g(t)e^{(jn2\pi/T)t}e^{-j\omega t}\, dt$$

Therefore,

$$G^*(\omega) = \frac{1}{T} \sum_{n=-\infty}^{\infty} G\left(\omega - n\frac{2\pi}{T}\right)$$

$$= \frac{1}{T} \sum_{n=-\infty}^{\infty} G(\omega - n\omega_s)$$

which completes the proof.

The function $G^*(\omega)$ in Equation 15.2 is periodic in ω, and each individual term in the series has the same form as the original $G(\omega)$, with the exception that the nth term is centered at

$$\omega = n\frac{2\pi}{T} \quad n = \ldots, -2, -1, 0, 1, 2, \ldots$$

In general, then, if $G(\omega)$ is not limited to a finite frequency range, these terms overlap each other along the ω-axis. A signal is bandlimited at (Hertz) frequency f_B if

$$G(\omega) = 0 \quad \text{for } |\omega| > 2\pi f_B = \omega_B$$

as shown in Figure 15.2a. Equation 15.1 becomes

$$g(t) = \frac{1}{2\pi} \int_{-\omega_B}^{\omega_B} G(\omega)e^{j\omega t}\, d\omega$$

If the sampling frequency f_s is more than twice the bandlimit frequency f_B, the individual terms in Equation 15.2 do not overlap as shown in Figure 15.2b, and $G(\omega)$, and thus $g(t)$, can be determined from $G^*(\omega)$, which in turn, is determined from the samples $g(k)$. Furthermore, if the sampling frequency f_s is exactly twice the bandlimit frequency f_B, the individual terms in Equation 15.2 do not overlap as shown in Figure 15.2c.

In terms of the sampling period,

$$\omega_s = 2\omega_B$$

and

$$T = \frac{2\pi}{\omega_s}$$

(a)

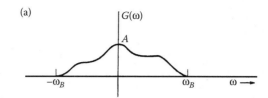

(b)

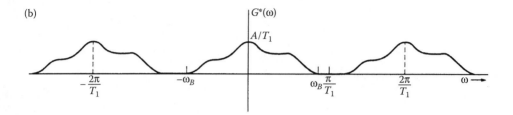

(c)

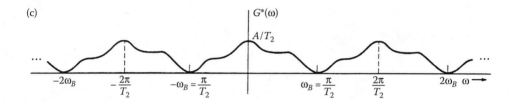

(d)

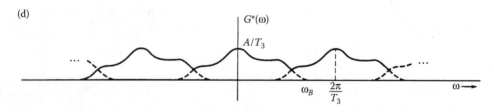

FIGURE 15.2 Frequency spectra of a signal sampled at various frequencies. (a) Frequency spectrum of an analog bandlimited signal $g(t)$. (b) Frequency spectrum of a sampled signal $g^*(t)$ with $f_{s1} > 2f_B(f_{s1} = 1/T_1)$. (c) Frequency spectrum of a sampled signal $g^*(t)$ with $f_{s2} = 2f_B(f_{s2} = 1/T_2)$. (d) Frequency spectrum of a sampled signal $g^*(t) < 2f_B(f_{s3} = 1/T_3)$.

Then

$$T = \frac{\pi}{\omega_B} = \frac{1}{2f_B}$$

which relates the sampling period to the highest frequency f_B in the signal.

A statement of the sampling theorem is the following:

The uniform samples of a signal $g(t)$, that is bandlimited above (Hertz) frequency f_B, are unique if, and only if, the sampling frequency is higher than $2f_B$.

In terms of the sampling period,

$$T < \frac{1}{2f_B} \qquad\qquad (15.3)$$

The frequency $2f_B$ is termed the *Nyquist frequency* for a bandlimited signal. As shown in Figure 15.2d, if the sampling frequency does not exceed the Nyquist frequency, the individual terms in Equation 15.2 overlap, a phenomenon called *aliasing* (or *foldover*).

In digital signal processing applications, selection of the sampling period also depends on the reconstruction method used to recover the bandlimited signal from its samples [1]. Another statement of the sampling theorem related to signal reconstruction states that *when a bandlimited continuous-time signal is sampled at a rate higher than twice the bandlimit frequency, the samples can be used to reconstruct uniquely the original continuous-time signal.*

Åström and Wittenmark [2] suggest, by way of an example, a criterion for the selection of the sample rate that depends on the magnitude of the error between the original signal and the reconstructed signal. The error decreases as the sampling rate is increased considerably higher than the Nyquist rate. Depending on the hold device used for reconstruction, the number of samples required may be several hundreds per sampling period.

Although the sampling theorem is not applicable to most discrete-time control systems, because the signals (e.g., steps and ramps) are not bandlimited and because good reconstruction requires long time delays, it does provide some guidance in selecting the sample rate and in deciding how best to filter sensor signals before sampling them.

15.3 Control System Response and the Sampling Period

The main objective of many digital control system designs is to select a controller so that the system-tracking output, as nearly as possible, tracks or "follows" the tracking command input. Perhaps, the first figure of merit that the designer usually selects is the closed-loop bandwidth, f_c (Hz), of the feedback system because f_c is related to the speed at which the feedback system should track the command input. Also, the bandwidth f_c is related to the amount of attenuation the feedback system must provide in the face of plant disturbances. It is then appropriate to relate the sampling rate to the bandwidth f_c, as suggested by the sampling theorem, because the bandwidth of the closed-loop system is related to the highest frequency of interest in the command input.

Consider the control system shown in Figure 12.9a in the Chapter 12 where the controller,

$$G_c(s) = \frac{0.5(s+5)}{s(s+2)}$$

has been designed so that the resulting feedback system has a 3-dB bandwidth $f_c = 0.228$ Hz. The step response of the digital control system using Euler's and Tustin's approximations for various sampling periods is shown in Figure 12.10. Raising the sample rate tends to decrease the amplitude of each step input change and thus reduces the amplitude of the undesirable between-sample response. As the sampling period is decreased from $T = 0.4$ s to $T = 0.1$ s, or equivalently, the sampling rate is increased from 2.5 Hz (11 times f_c) to 10 Hz (44 times f_c), the step response of the feedback system using either approximation approaches the step response of the continuous-time system. However, as discussed in Chapter 12, Tustin's approximation usually gives better results than Euler's approximation for the same sampling period.

As a general rule of thumb, the sampling period should be chosen in the range

$$\frac{1}{40f_c} < T < \frac{1}{10f_c} \tag{15.4}$$

Of course, other design requirements may require even higher sample rates, but sampling rates less than 10 times f_c are not desirable and should be avoided if possible.

An interesting problem involving the sample rate selection is encountered in the control system design of flexible spacecraft. The spacecraft has a large number of bending modes of which the lowest bending mode frequency may be a fraction of 1 Hz and the highest frequency of interest may be 100 Hz or even higher. Typically, the closed-loop bandwidth of the spacecraft is an order of magnitude less than the lowest mode frequency, and as long as the controller does not excite any of the flexible modes, the sampling period may be selected solely based on the closed-loop bandwidth. Otherwise, these modes need to be attenuated* or controlled, and therefore, their frequencies will impact the sampling rate selection [3].

Another criterion for selecting the sampling period is based on the *rise time* of the feedback system so as to provide smoothness in the time response. It can easily be shown that the rise time (10–90%), T_r, of a first-order system of the form

$$H(s) = \frac{1}{\tau s + 1}$$

is

$$T_r = 2.2\tau$$

The sampling period, in terms of the rise time, can be selected according to

$$0.095 T_r < T < 0.57 T_r$$

which is derived from Equation 15.4. Similarly, the rise time of the second-order system,

$$H(s) = \frac{\omega_n^2}{s^2 + 2\zeta\omega_n s + \omega_n^2}$$

is

$$T_r = \frac{\pi - \beta}{\omega_d}$$

where

$$\omega_d = \omega_n \sqrt{1 - \zeta^2}$$

and

$$\beta = \sin^{-1} \sqrt{1 - \zeta^2}$$

For a damping ratio $\zeta = 0.707$, the rise time is

$$T_r = \frac{3.33}{\omega_n}$$

Based on Equation 15.4, the sampling period is

$$0.05 T_r < T < 0.19 T_r \tag{15.5}$$

Continuing with the previous example shown in Figure 12.10, according to Equation 15.5, a sampling period $T = 0.11$ s is selected which agrees with the previous results.

In digital control systems, a time delay of up to a full sample period may be possible before the digital controller can respond to the next input command. Franklin et al. [3] suggest that the time delay be kept to about 10% of the rise time. Then, the sampling period should satisfy

$$T < \frac{0.05}{f_c}$$

Yet another criterion for selecting the sampling period, which depends on the frequency response of the continuous-time system, is selected so that

$$0.08 < T\omega_0 < 0.3 \tag{15.6}$$

where ω_0 is the gain crossover frequency of the continuous-time system in rad/s.

---

* When the modal parameters are well known, notch filters can be used to attenuate the modes, if necessary. In this case, the notch frequency of the filter attenuating the mode with the highest frequency may dictate the sampling rate of the digital controller.

Continuing with the previous example, the open-loop transfer function is

$$G(s)H(s) = \frac{2.5(s+5)}{s(s^2+4s+8)(s+2)}$$

The frequency response of GH is shown in Figure 15.3 where the gain crossover frequency is $\omega_0 = 0.74$ rad/s. According to Equation 15.6, the sampling period should be chosen in the range

$$0.1 < T < 0.4$$

which also agrees with the previous results.

15.4 Control System Response to External Disturbances

Plant disturbances are undesired, inaccessible plant inputs that the plant should *not* track. An example of disturbance is wind gusts buffeting a positioning system for a microwave antenna. Like initial conditions, the specific disturbance signals are normally unknown, although something is probably known about their character, their statistics, or both. As far as the selection of sampling rate is concerned, the most important plant disturbance to consider is random white noise because of its high-frequency contents.

In general, when the controller is implemented digitally, it will perform less well than the analog controller in the face of white-noise disturbance inputs. As the sampling rate is increased, the response of the digital controller usually approaches the response of the continuous-time controller.

Consider the continuous-time system described by

$$\dot{x} = \mathcal{A}x(t) + \mathcal{B}u(t) + Lw(t) \tag{15.7}$$

where $u(t)$ is the control input and $w(t)$ is a white-noise process with covariance matrix

$$E\{w(t)w^\dagger(t+\tau)\} = Q\delta(\tau)$$

where E denotes expected value. If the control input is given by

$$u(t) = E_c x(t) \tag{15.8}$$

then

$$\dot{x} = (\mathcal{A} + \mathcal{B}E_c)x(t) + Lw(t) = \mathcal{A}_c x(t) + Lw(t) \tag{15.9}$$

Let the state covariance matrix be

$$P_c(t) = E\{x(t)x^\dagger(t)\}$$

It can be shown [4] that the steady-state solution of the state covariance matrix P_c is

$$\mathcal{A}_c P_c + P_c \mathcal{A}_c^\dagger + LQL^\dagger = 0 \tag{15.10}$$

where P_c is a measure of the variation of the state vector about its mean. The solution of Equation 15.10 can be easily obtained using MATLAB® (see LYAP.m) or some other computer-aided design tools.

When the controller is implemented digitally, however, the covariance of the discretized state vector will, in general, be higher than the covariance of the continuous-time state vector for identical disturbance inputs. This is to say that the amplitude of the state will be higher with the discrete controller than its

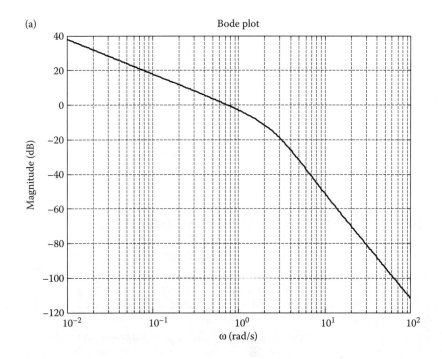

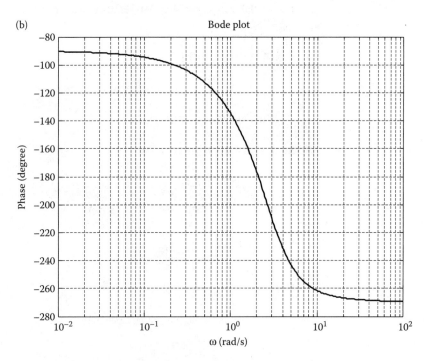

FIGURE 15.3 Frequency response of open-loop transfer function of the example system. (a) Magnitude response and (b) phase response.

continuous-time counterpart. When the continuous-time system described by Equation 15.7 is sampled with sampling period T, its discrete-time equivalent is (see Chapter 12).

$$x(k+1) = \Phi x(k) + Bu(k) + \omega(k) \tag{15.11}$$

where

$$\Phi = e^{AT} = I + AT + \frac{A^2 T^2}{2!} + \frac{A^3 T^3}{3!} + \cdots \tag{15.12}$$

and

$$B = \left(\int_0^T e^{A\eta}\, d\eta \right) B = \left[IT + \frac{AT^2}{2!} + \frac{A^2 T^3}{3!} + \cdots \right] \tag{15.13}$$

The discrete-time noise is given by the integral

$$\omega(k) = \int_{kT}^{kT+T} \Phi(kT + T - \tau) Lw(\tau)d\tau$$

The covariance of the discrete-time noise is

$$
\begin{aligned}
Q_d = E\{\omega(k)\omega^\dagger(k)\} &= \int_{kT}^{kT+T} \int_{kT}^{kT+T} \Phi(kT + T - \tau) \\
&\quad \times LE\{w(\tau)w^\dagger(\lambda)\}L^\dagger \Phi^\dagger(kT + T - \lambda)\, d\lambda\, d\tau \\
&= \int_{kT}^{kT+T} \int_{kT}^{kT+T} \Phi(kT + T - \tau) \\
&\quad \times LQ\delta(\tau - \lambda)L^\dagger \Phi^\dagger(kT + T - \lambda)\, d\lambda\, d\tau \\
&= \int_{kT}^{kT+T} \Phi(kT + T - \tau) \\
&\quad \times LQL^\dagger \Phi^\dagger(kT + T - \tau)\, d\tau
\end{aligned}
$$

Let

$$\gamma = kT + T - \tau$$

Then

$$Q_d = \int_0^T \Phi(\gamma) LQL^\dagger \Phi^\dagger(\gamma)\, d\gamma \tag{15.14}$$

Returning to Equation 15.11, the state feedback

$$u(k) = E_d x(k)$$

gives

$$
\begin{aligned}
x(k+1) &= (\Phi + BE_d)x(k) + \omega(k) \\
x(k+1) &= \Phi_c x(k) + \omega(k)
\end{aligned}
\tag{15.15}
$$

Therefore, using Equation 15.15, the discrete-time state covariance matrix is

$$
\begin{aligned}
P_d(k+1) &= E\{x(k+1)x^\dagger(k+1)\} \\
&= E\{[\Phi_c x(k) + \omega(k)][\Phi_c x(k) + \omega(k)]^\dagger\} \\
&= \Phi_c P_d(k)\Phi_c^\dagger + Q_d
\end{aligned}
$$

where Q_d is given by Equation 15.14. Hence, the steady-state covariance discrete-time matrix is

$$P_d = \Phi_c P_d \Phi_c^\dagger + Q_d \tag{15.16}$$

which can easily be solved using MATLAB (see DLYAP.m).

To illustrate the ideas involved in the selection of the sampling rate of a system driven by a white-noise disturbance, consider the following simplified model for the roll attitude control of a spacecraft:

$$\dot{x}(t) = \begin{bmatrix} 0 & 1 \\ 0 & 0 \end{bmatrix} x(t) + \begin{bmatrix} 0 \\ 1/J \end{bmatrix} u(t) + \begin{bmatrix} 0 \\ 1/J \end{bmatrix} w(t)$$

$$y(t) = [1 \quad 0] x(t)$$

where x_1 is the roll attitude of the spacecraft in radians, x_2 is the roll rate in rad/s, u is the control torque about the vehicle roll axis produced by the spacecraft actuators in foot-pounds, w is the disturbance torque acting on the spacecraft in foot-pounds, and J is the moment of inertia of the vehicle about the roll axis at the vehicle center of mass in slug-feet squared.

For simplicity, we assume that $J = 1$. Suppose that it is desired to have both eigenvalues of the state feedback system at $s_{1,2} = -4.6$. Then the feedback gain vector is

$$e^\dagger = [-21.16 \quad -9.2]$$

If the noise covariance $Q = 1$, then the steady-state solution of the state covariance matrix given by Equation 15.10 is

$$P_c = \begin{bmatrix} 0.002568 & 0 \\ 0 & 0.0543476 \end{bmatrix}$$

and, therefore, the RMS of the spacecraft attitude,* x_1, is 0.0507, and the RMS of the spacecraft attitude rate, x_2, is 0.2231.

If the continuous-time model of the spacecraft is discretized with sampling period T, then, according to Equations 15.12, 15.13, and 15.14,

$$\Phi = \begin{bmatrix} 1 & T \\ 0 & 1 \end{bmatrix}$$

$$b = \begin{bmatrix} \dfrac{T^2}{2} \\ T \end{bmatrix}$$

and

$$Q_d = \begin{bmatrix} \dfrac{T^3}{3} & \dfrac{T^2}{2} \\ \dfrac{T^2}{2} & T \end{bmatrix}$$

If the eigenvalues of the discrete-time system are located at $z_{1,2} = e^{-4.6T}$, then the feedback gain vector for the discrete-time system is

$$e^\dagger = \begin{bmatrix} \dfrac{2e^{-4.6T} - e^{-9.2T} - 1}{T^2} & \dfrac{2e^{-4.6T} + e^{-9.2T} - 3}{2T} \end{bmatrix}$$

Figure 15.4 shows the RMS values of the states x_1 and x_2 in terms of the sampling period generated with Equation 15.16. As the sampling period is increased, the RMS values of the states increase, and, as the sampling period is decreased, the RMS values decrease and eventually approach the RMS values of the continuous-time state variables calculated earlier. Examining the figure, an appropriate value of the sampling period is $T = 0.05$ s. The performance of the digital controller degrades as T is increased. This value of the sampling period also agrees with inequality Equation 15.4.

---

* Root-mean-square (RMS) of a random variable X is the square root of the mean-square value (second moment) of X. If the random variable X has zero mean, then the RMS value and standard deviation of X are equal.

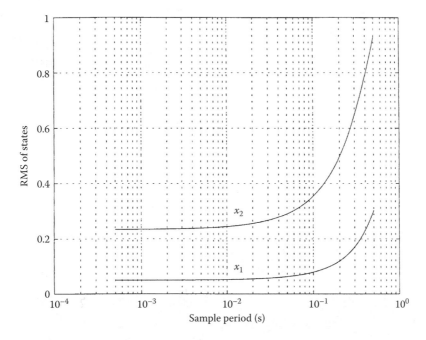

FIGURE 15.4 Response of digital control system as a function of a sampling period. As the sampling period is increased, the response degrades and, as the sampling period is decreased, response approaches continuous-time response.

15.5 Measurement Noise and Prefiltering

The sampling theorem is important to control system design, because when A/D conversion is done on noisy signals with significant frequency components above half the sampling frequency, the high frequencies produce errors in the sampling *indistinguishable* from lower frequency errors. For this reason, low-pass filters, termed *prefilters,* or *antialiasing filters,* are used to reduce the high frequencies in sensor signals before their A/D conversion as in Figure 15.5.

For example, consider the noisy voltage signal, in Figure 15.6a, composed of a 2 Hz sinusoidal signal of amplitude 2 V and an 80 Hz sinusoidal noise of amplitude 0.2 V. If this signal is sampled with $T = 1/45$ s, then according to the sampling theorem, the sampled signal will be aliased as in Figure 15.6b.

The first-order low-pass filter,

$$H(s) = \frac{50}{s + 50}$$

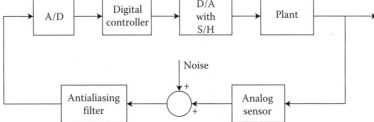

FIGURE 15.5 Antialiasing filters reduce high-frequency noise in the sensor signal before A/D conversion.

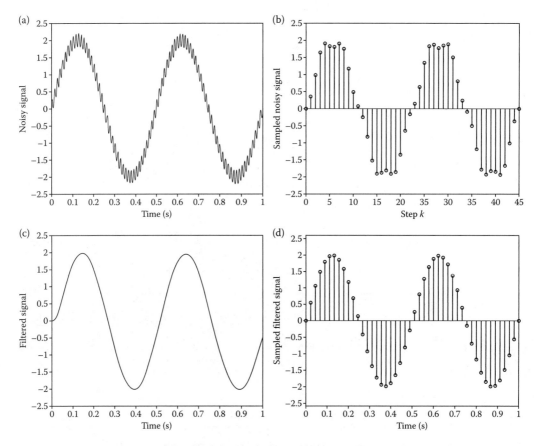

FIGURE 15.6 Filtering of measurement noise using antialiasing filters. (a) Noisy signal; (b) aliased signal; (c) filtered signal; and (d) sampled-filtered signal.

selected with a corner frequency of about one-sixth the sampling frequency, will attenuate the 80 Hz noise, as shown in Figure 15.6c. Samples of the filtered analog signal are shown in Figure 15.6d.

In feedback control systems, the phase lag of antialiasing filters may be significant enough to cause system instability. However, if the corner frequency of the filter is sufficiently higher than the control system bandwidth, the phase lag of the filter may not affect the performance of the system. On the other hand, if noise attenuation requires higher-order filters or filters with corner frequencies close to the control system bandwidth, the filter should be treated as if it is part of the plant controlled by the discrete-time controller.

15.6 Effect of Sampling Rate on Quantization Error

Quantization errors in digital control systems are discussed in detail in the chapter entitled "Quantization Effects." In some applications, quantization errors cannot be ignored and their effect on system output depends on the sampling period.

As an example, consider the analog controller described by

$$H(s) = \frac{10^4}{s+1} \tag{15.17}$$

Its discrete-time equivalent, using the impulse invariant approximation, is

$$H(z) = \frac{10^4 z}{z - e^{-T}} = \frac{10^4}{1 - e^{-T}z^{-1}}$$

$$= k_0 + \frac{k_1}{1 - \alpha z^{-1}}$$

where

$$k_0 = 0, \quad k_1 = 10^4, \quad \alpha = e^{-T}$$

As discussed in the chapter entitled "Quantization Effects," the variance of the controller output equals the variance of the roundoff noise times the noise power gain. For the first-order controller, the variance of the output is determined with Equations 15.20, 15.29, 15.30, and 15.32. Figure 15.7 shows the RMS of the controller output in terms of the sampling period and the word length C. As the sampling period is decreased, the RMS value of the output is increased. Also shown in the figure, increasing the word length C decreases the RMS value of the output. The RMS output does not necessarily increase as the sampling period is decreased. If the analog controller described by Equation 15.17 is discretized using Tustin's approximation, the discrete-time controller becomes

$$H(z) = \frac{10^4(z + 1)}{\left(\dfrac{2}{T} + 1\right)z + 1 - \dfrac{2}{T}}$$

Repeating the previous RMS analysis on this controller gives the results in Figure 15.8. Although the RMS of the controller output increases as T increases, the RMS value is less with Tustin's approximation than with the impulse-invariant approximation. Also, as shown in the figure, the RMS of the output increases as the word length is decreased.

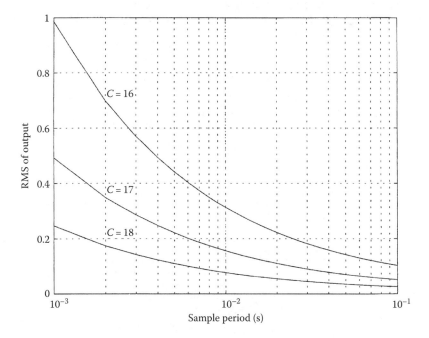

FIGURE 15.7 Response of controller using impulse invariant approximation as a function of sampling period for various word lengths. As the sampling is increased, the error is decreased, and, as the word length is increased, the error is decreased.

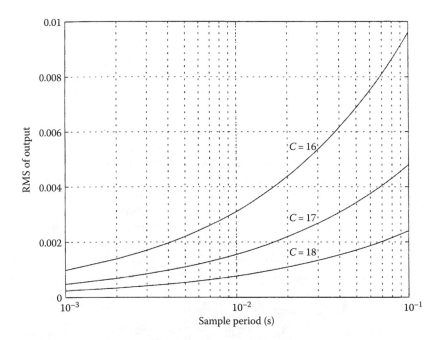

FIGURE 15.8 Response of controller using Tustins's approximation as a function of sampling period for various word lengths.

It is through computer simulation of the plant and the controller that the best sample rate is achieved. It is good practice to investigate carefully the behavior of the controlled system for various sample rates when the arithmetic precision of the controller is reduced, when disturbances and noises are injected into the system at likely points, and when the plant model is changed in ways that might occur in practice.

References

1. Santina, M.S., Stubberud, A.R., and Hostetter, G.H., *Digital Control System Design,* 2nd edn., Saunders College Publishing, Philadelphia, 1994.
2. Åström, K.J. and Wittenmark, B., *Computer Controlled Systems: Theory and Design,* 3rd ed., Prentice-Hall, Englewood Cliffs, NJ, 1996.
3. Franklin, G.F., Powell, J.D., and Workman, M.L., *Digital Control of Dynamic Systems,* 3rd ed., Prentice-Hall, Englewood Cliffs, NJ, 1997.
4. Glasson, D. and Dowd, J., *Research in Multirate Estimation and Control–Optimal Sample Rate Selection,* Analytic Sciences Corporation, MA, 1981.

16

Real-Time Software for Implementation of Feedback Control

David M. Auslander
University of California, Berkeley

John R. Ridgely
California Polytechnic State University

Jason C. Jones
SunPower Corporation

16.1 An Application Context

Digital computers are the primary means of implementing feedback control for physical systems. Real-time software is the medium in which these solutions are expressed. Computers are sequential devices, and, as such, can act only as sampled-data controllers, with time discretized. The fundamental characteristics of the real-time software, as distinct from "regular" software, are that the control algorithms must be run at their scheduled sample intervals (with some specified tolerance) and that associated software components, which interact with the sensors and actuators, can have critical time-window constraints.

The nature and difficulty of producing real-time software depends on the complexity and timing constraints of the problem. The tighter the time constraints with respect to the computer's basic computing speed limitations and the more things that need to be serviced simultaneously (at least "simultaneously" from the viewpoint of the control object), the more difficult it will be to complete the software design and implementation successfully. Truly simultaneous operation is provided by using multiple communicating processors; this does not change any of the basic system designs, but can add complexity and delay due to communication.

16.2 The Software Hierarchy

Implementing software in a layered, or hierarchical, manner makes far more maintainable, more readable, and more reliable software. To the extent that the layers are truly independent, this modularity allows changes to be made at any layer without any software changes needed at other layers. This model follows the extremely successful layered model used for Internet software.

Up to four distinct hierarchical levels are used in this model for software implementing feedback control. The extent to which all of these levels are used and the degree of interaction within and across levels determines the degree of complexity referred to above. The methods described here are appropriate for the types of demands placed on each of these levels by typical feedback control problems. Other problems, which might require more complex relationships within or across hierarchical levels, would require additional and/or different formalities for organizing the software design.

The four potential hierarchical levels are:

1. Instrument/actuation activities: The lowest level of interaction with the instruments and actuators. Software at this level is usually short and might need to run frequently. It can be activated either by time or by an external signal. This level is also shared with peripheral controllers, such as disk, network, or printer controllers.
2. Feedback control algorithms: Use the information from the instruments to compute actuation commands. Normally run on a specified time schedule.
3. Supervisory/sequence control: Provide set point information to the feedback algorithms. Can also be used for corrections and adjustments to the feedback algorithms, such as adaptive control or set point scheduling.
4. Other: Data logging, operator interface, communication with other computers, and so on.

16.3 The Ground Rules

The following items describe the "ground rules" used to formulate the design methodology presented here.

16.3.1 Generic Solution

The methods and software developed here are designed for generic implementation. Although a number of proprietary means exist for implementing feedback control, for example, code translators that produce real-time code from a design, analysis, or simulation package, none of them are used widely enough to represent a de facto standard for feedback control implementation. For that reason, the methods here are intended for a broad variety of environments, without purchasing anything beyond basic computing tools.

16.3.2 Stand-Alone Feedback Control

The designs implemented are intended for stand-alone feedback control systems. When feedback control is a part of a larger software system, a variety of design constraints will be imposed based on the environment in which that system has been designed. The software methods described here will often be amenable to implementation in such environments, but, in some cases, may need substantial modification to be compatible.

16.3.3 Single Processor

The design formality used here is presented in a format that is suitable for running in a single computer. However, the formality itself is applicable to multiprocessor systems and a number of methods exist for the extension of applications to multiprocessor environments.

16.4 Portability

Software portability, the ease with which software written in one environment can be used in another, is a major factor in overall development costs, time to market, maintainability, and upgradability. It has thus been a major focus in formulating the feedback control implementation methodology presented here.

16.4.1 Design Cycle

The design cycle for a control product typically goes through phases starting with simulation, followed by laboratory prototypes, preproduction prototypes, and finally, production systems. Each of these phases can use different computers and different development environments. Unless care is taken to assure software portability, large sections of code may need rewriting for each phase. In addition to the obvious cost consequences, there is a significant probability of introducing new bugs with each rewrite. To compound this problem, it is likely that software responsibility will rest with different people at each phase, placing a premium on consistent design methodology and documentation.

16.4.2 Life Cycle

While current commercial life cycles for computers are two to three years, with overall capability approximately doubling in each new generation, the commercial lifetime of the equipment being controlled can be as long as 20 years. To keep a product up-to-date, it is often necessary to introduce new product models with minor changes in the physical design, but with new computational equipment greatly increasing the system's functionality, diagnostic abilities, communications, and so on.

To accomplish this in a timely, cost-effective manner, it must be possible to build on existing software, as new models are introduced, and rapidly port old software to new platforms and operating environments.

16.4.3 Execution Shell

There are always parts of a control program that depend on the specific computational hardware and software being used. The *execution shell* isolates these environment-specific portions from the rest of the program. This part of the program will normally include interfaces to the:

- Instrumentation/actuation hardware
- Real-time operating system (if any)
- Interrupt system
- Clock
- Operator interface

Concentrating this type of code in the execution shell insulates the control engineer from the need to do system-level programming, a critical part of portable software design.

16.4.4 Language

The general structure described here is independent of language, but *some* language must be chosen for the actual implementation. As of this writing, there are three top choices for languages to write portable control system implementation code: C, C++, and Java.

Of these, C (currently the most widely used) is now an old design, which has been brought up-to-date with C++, a superset of C. Java was originally invented at Sun Microsystems, Inc., as an embedded computing language for use in TV set-top boxes. It was then adapted for use as a client-side computing language for World-Wide-Web Internet applications, then extended for server-side applications and stand-alone use (see http://java.sun.com/features/1998/05/birthday.html for a more detailed history of Java). Because of its origins as an embedded systems language, Java is an excellent candidate for a language to implement feedback control.

C++ has been chosen for the samples in this chapter. The style of program writing that it encourages is much more conducive to writing code that can be maintained and modified more easily than C, and, because C++ is a superset of C, applications can be created by C programmers with only a recent introduction to C++. Java has a basic language structure that is very similar to C++ but is simpler and easier to use. On the other hand, its implementation structure via a virtual machine and internal memory recovery ("garbage collection") make its real-time performance more questionable.

There are a large number of C++ instructional books currently available. The useful tutorial book by Lippman [1] introduces the C++ language without requiring previous background in C. The reference manual to C++ is written by Stroustrup [2], the inventor of the language.

Software for implementing the design model described here is available in C, C++, and Java from one of the authors (Auslander, dma@me.berkeley.edu). This software was written for instructional rather than commercial purposes and is available on an as-is basis.

16.5 Software Structure: Scan Mode

Successful control system design requires programming paradigms beyond the basic structure of algorithmic languages. In particular, the parallelism inherent in the control of physical systems and the notion of *duration,* are not concepts included within the syntax of standard languages.

16.5.1 Parallelism

Except for the simplest of control systems, several activities must be carried out simultaneously. A single computer, however, is a strictly sequential device, so that the parallelism viewed from the outside must be constructed by a rapid succession of sequential activities on the inside of the computer. The paradigm used for control system software must deal with the need for pseudo-simultaneous execution of program components, and it must do that in a way that meets the portability requirements. Multiple computers combined to do a single control job can exhibit true parallelism. In most practical cases, though, even when multiple computers are used, many of them must carry out more than one activity at a time.

A number of mechanisms, both commercial and *ad hoc*, exist for realizing parallelism in computing. The focus in this chapter will be on exploiting those mechanisms without compromising portability.

16.5.2 Nonblocking Code

The first step in constructing the paradigm for control system software design is to recognize the difference between *blocking* and *nonblocking* code. For a segment of code to be *nonblocking*, it must have a predictable computing time. The computing time need not be short, but it must be predictable.

A typical example of blocking code in many control programs is the "wait-for-something" statement based on a *while* loop,

```
while(-check-for-something-) ;
                    //Wait for an event
```

Because the event, in general, is asynchronous, there is no way to know when (or whether) the event will happen. The code is blocking, because its execution time is not predictable. Likewise the *scanf* (or *cin*) statement is blocking, because its execution time depends on when the user completes typing the requested input.

The methodology presented here is based on code that is completely nonblocking. Waiting for events, however, is fundamental to system control, so that a higher level of structure will be provided to accommodate such waits without the blocking code.

16.5.3 Scanned Code

The restriction to nonblocking code permits adopting a *scan* model for all software. In this model, all software elements are designed to be operated through repeated execution. It is the rapid repetition that gives the illusion of parallel execution even within a strictly sequential environment. By using the scan structure, parallelism can be maintained in computational environments that do not normally support real-time multitasking, or across competitive and, therefore, usually incompatible real-time environments. In many respects, this software structure is an extension of the model used by programmable logic controllers (PLCs), which have successfully solved an important class of industrial control problems.

16.6 Control Software Design

Most important in assuring that software design will meet the engineering specifications of a project is a design structure that matches the problem reasonably well and allows for separating the solution into components matching the designers' interests and skills. In feedback control software, the system engineering of the problem and its computational structure are partitioned. This facilitates designing portable software, and matching the skills for control system design to control the engineer's knowledge.

16.6.1 System Engineering Structure: Tasks and States

A two-level structure characterizes the system engineering of the control job. *Tasks,* expressing the parallelism in the job, are a partition of the job into a set of semi-independent, activities, which, in general, are all active simultaneously. Tasks are internally characterized by *states,* indicating the particular action a task is carrying out.

The major creative engineering effort consists of selecting and describing tasks and states. Once done, creating the actual computer code is relatively straightforward, because the formal definition of tasks and states modularizes the code, with code sections specifically connected to these design elements. This is very important for maintenance and upgrading where the code will need to be modified by people uninvolved in its original creation.

An additional benefit of the task/state paradigm is that the system operation is described in terms understandable to any engineers involved in the project. This opens the door to broad design review, sorely lacking in many software control projects.

Characterizing tasks as "semi-independent" recognizes that they are part of a system with a well-defined objective. Tasks will exchange data with other tasks and must synchronize their operations with other tasks. In all other aspects, they operate independently and asynchronously. They are asynchronous, because their operation can be governed by activities outside of the computer, which, in general, are not synchronous. Except for explicit synchronization, there is no *a priori* way to know how tasks will relate to each other computationally. This has important consequences for debugging and system reliability. In conventional software, erroneous situations can be repeatedly recreated until the cause of the problem is found. This is not possible in real-time systems, because with asynchronous operation of tasks, with respect to each other and to external events, it may not be possible to reproduce an erroneous situation except in a statistical sense.

The concept of *state* in control theory means capturing information about the operation of a system in a set of variables. The state variable for a task also serves this purpose. Using the scan model for computing, the state provides the task with information indicating what action is required at each scan. It thus captures the scan history for the task. In this context, states are represented as integer variables, with each task recognizing only a finite number of state variable values (i.e., states). The task is thus a *finite-state machine* and the whole program is a set of finite-state machines.

16.6.2 Computational Structure: Threads

The system engineering part of the design effort is founded on the operational description of the machine, but the computational structure is determined by detailed performance specifications. As long as the program is constructed from nonblocking code and the scan model for software structure, it is theoretically possible to accommodate *any* control job with a single, fast computer without special real-time hardware or software. However, computers are never as fast or inexpensive as desired. Fortunately, in most problems only a few tasks lack computational resources. In these cases, a computational structure is necessary to allow shifting resources from those tasks with excess to those that have need.

This is done by *threads*. Threads represent separate computing entities that can run asynchronously with respect to each other. They are activated by the *interrupt* mechanism of the computer. Resource shifting can be accomplished by putting selected groups of tasks into threads that are activated to receive more computational resources than would otherwise be the case.

In computing terminology, threads are distinguished as *lightweight* and *heavyweight* threads. Lightweight threads are executed asynchronously but share the same address space, whereas heavy-weight threads do not share address spaces (also called "processes" in operating system terminology). In this chapter, the term *thread* will always refer to a lightweight thread.

Real-time environments are characterized by the types of thread structures implemented. By separating the system engineering from the computational structure, exactly the same application source code can be used in a variety of thread structures for the most effective implementation.

16.7 Design Formalism

Design formalisms are used in engineering to organize the design process, allow for effective communication, and to specify the documentation needed for modifying or analyzing the object designed.

Software design has traditionally been a notable exception to this rule! Because real-time software adds the element of asynchronous operation, appropriate design formalism is even more important in designing control system software than conventional engineering software.

A major purpose for formalizing design is to establish a mechanism for modularizing the software. Modular software clearly connects sections of software with specific machine operations, keeps those sections short and readable, and clearly identifies interactions between software elements.

Given the state/scan model for control software, many tasks will have only a single action repeated on a fixed schedule. For those tasks, no further internal structure is needed. The task itself will have only one small or modest-sized code module, and the task's description will relate very closely to that code.

16.7.1 Formality/Complexity

The degree of formality required is related to the complexity of the problem. Overly formalistic procedures can result in a rigid design environment absorbing too much overhead; too little formalism can create a chaotic environment and can fail to meet delivery and/or reliability commitments.

The stand-alone control system software described in this chapter falls into the range of low to moderate complexity. Some formalism is needed, but how it is applied is left to the designer's discretion. In particular, the state-transition notation described below will be used in a manual construction mode to achieve maximum portability and flexibility. Because no specialized design software is required, the method can be used to design software for any target computer in any host environment. On the other hand, if more complex designs are attempted, a more organized approach is needed to control the design process.

The structure for control system software based on tasks and states has already been established. No further formal structure will be applied to tasks, but state transition logic will be used to characterize the relationships of states.

16.7.2 State-Transition Logic

A task can be in only one state at a time, characterized by the integer value of its state variable. *State-transition logic* describes how states change within a task. It is most commonly shown in diagrammatic form, with circles for states and arrows for transitions between the states. These diagrams are widely used for sequential logic design, from which they can be directly converted to logic equations, and also to design software, for operating systems design [3–6]. A distinction in the usage described here, as well as in the more formal real-time usage referred to below, is that the software derived from the diagrams and the portable real-time implementation are tightly connected.

To make computer graphics easier, the transition logic diagram in Figure 16.1, for pulse-width modulation (PWM) generation, is drawn with rectangles for states and ovals for the transition conditions. In addition, the transition conditions are connected by dashed lines to the transition they describe. This avoids ambiguity if the transition description is close to more than one transition line. If none of the indicated transition conditions is true, no change of state takes place. Each state is identified by a name, and information about what the state does, if that is necessary.

Software-generated PWM can be used whenever the required frequency is not too high, for example, for running a heater. This logic diagram shows the task states needed to implement PWM, including the special cases of duty cycles of 0 or 1. This diagram translates directly into highly modular code, is specific enough so that it specifies in detail how the system should operate, and yet can be read by any engineer familiar with the application. Thus, it is a primary document and a template for generating code in a relatively mechanical way.

PWM would normally be a low level task, forming the interface between a feedback control algorithm and an actuator. It would need to run often and would have a tight tolerance on its actual run times, because errors in the run times will change the output power delivered by the actuator (and appear as noise to the control system).

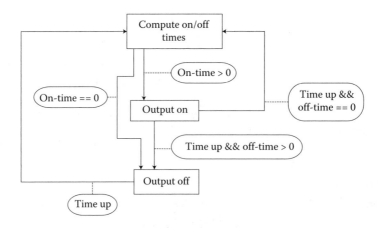

FIGURE 16.1   State transition logic for pulse-width modulation.

A higher level supervisory task, shown in Figure 16.2, would be used to generate the set point for a servo (mechanical position) controller or other types of controllers. The task would run less frequently and have more tolerant requirements for its actual run times.

The profile is based on a trapezoidal velocity, a constant acceleration section followed by a constant velocity section and, finally, a constant deceleration. The associated velocity and position are shown in Figure 16.3. In this case, the "path" is the motion of the motor under control; in the case of an XY motion, for example, the path velocity could be along a line in XY space with the X and Y motions derived from the path velocity and position.

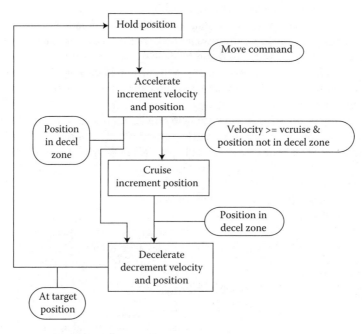

FIGURE 16.2   Motion profile generator.

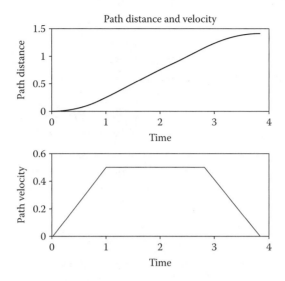

FIGURE 16.3 Trapezoidal velocity profile.

16.7.3 Transition Logic Implementation

Implementation of the transition logic in code is up to the user, as is the maintenance of the transition logic diagrams and other design documentation. The execution shell will tend to the storage of state information and the scanning of the task code, but the user is responsible for all other aspects. This recommended design procedure is thus a semi formal method, and the user is responsible for keeping the design documentation up-to-date. However, by avoiding specialized design tools, the method is much more widely applicable.

Transition logic code can be implemented directly with the *switch* statement of C++ (and C or Java). Each *case* in the switch represents a state. The code associated with the state can be placed in-line if it is short, or as a separate function if it is longer. Within the code for the state, a decision is made as to which state should be executed next. The code is usually set up so that the default is to stay in the same state.

An example of implementing the PWM task is shown in Figure 16.4. In this implementation, it is assumed that the task is invoked at timed intervals, and that the task can control when it will next be invoked. Only the relevant transition logic code is shown here; variable definitions, and so on are not shown.

Each scan of this code executes only one state, and so the scan will be extremely short. When `set_next_time()` is executed, the intention is that no more scans will take place until that time is reached. All code used is nonblocking as specified above. Any operations requiring waiting for some event to take place are coded at the transition logic level, rather than at the C++ code level. These logic elements tend to be the critical parts of control programs, so that placing them in the higher level of specification makes it easier to understand and critique a program.

This is not the only possible implementation. For example, the task could be written for an environment where the task is scanned continuously (see *continuous* tasks below). In that case, the transition conditions would include a test of time as well as velocity or position, and there would be no `set_next_time()` call. For the control system software implementation recommended here, a form roughly equivalent to the above code would be used.

```
...
switch(state)
    {
    case 1: // Compute on/off times
            on_time = duty_cycle * period;
            off_time = period - on_time;

            if(on_time > 0)next_state = 2;  // Turn output on
            else next_state = 3;  // Turn output off directly
            break;  // Completes the scan for this state
    case 2: // Output on
            digital_out(bit_no,1);  // Turn on the relevant bit
            set_next_time(current_time + on_time);
            if(off_time > 0)next_state = 3;  // to output off
            else next_state = 1;  // Duty cycle is 1.0; no off time
            break;
    case 3: // Output off
            digital_out(bit_no,0);  // Turn off the relevant bit
            set_next_time(current_time + off_time);
            next_state = 1;  // Back to compute times
            break;
    }  // End of switch
...
```

FIGURE 16.4 Implementation of the PWM task.

16.7.4 Further State Structure

While the software structure shown in Figure 16.4 is adequate for many feedback control problems, a further level of formalization is used in the instructional software mentioned above (Section 16.4.4). In these cases, the state structure is further defined by specification of *entry, action, transition test,* and *exit* functions. This provides for further modularization of the code.

The *entry* function (or code section) is executed only when a state is activated, that is, on a transition from another state. It modularizes whatever initialization is needed for that state. The *action* function is executed on every scan and thus performs the ongoing activities of the state. The *transition test* functions check to see whether a state transition is called for and to what state, while the *exit* function associated with each transition test function is executed only if the transition is taken.

16.8 Scheduling

The computational technology needed to insure that all of the tasks meet their timing specifications is generally referred to as *scheduling*. The subsections below present various scheduling technologies in order of increasing complexity, and thus increasing cost and difficulty in development and utilization.

Because of the scan mode/nonblocking software paradigm, scheduling takes on a somewhat different flavor than it does in more general discussions. In particular, choosing a specific scheduling method is strictly a matter of performance. Any scheduling method could meet performance specifications if a fast enough computer were available. While this is a general statement of the portability principles given above, it is also an important design element, because commitments to specific hardware and/or software execution environments can be delayed until sufficient software testing has been done to provide accurate performance data.

There are two general criteria for performance specification:

1. General rate of progress
2. Execution within the assigned time slot

All tasks have a specification for (1). The job cannot be accomplished unless all tasks get enough computing resources to keep up with the demand for results. There is a large variation among tasks in the second performance specification. As noted above, some tasks have a very strict tolerance defining the execution time slot; others have none at all. The failure to meet time slot requirements is usually referred to as *latency* error, that is, the delay between the event that triggers a task and when the task is actually executed. The event is most often time, but can also be an external or internal signal.

16.8.1 Cooperative Scheduling

This is by far the simplest of all of the scheduling methods, because it requires no special-purpose hardware or software. It will run on any platform that supports a C++ (or other appropriate language) compiler and has a means of determining time (and even the time determination requirement can be relaxed in some cases). Cooperative scheduling utilizes a single computing thread encompassing the scheduler and all of the tasks. For the scheduler to gain access to the CPU, the tasks have to relinquish their control of the CPU voluntarily at timely intervals, thus the name "cooperative." In conventional programming, the most difficult aspect of using cooperative scheduling is building the "relinquishment points" into the code. For control system code built with the scan/nonblocking rules, however, this is not an issue because control is given up after every scan.

The performance problems with cooperative scheduling are not general progress but latency. The cooperative scheduler is extremely efficient because it has no hardware-related overhead. If some tasks exceed general progress requirements and others are too slow, the relative allocation of computing resources can be adjusted by changing the number of scans each task gets. However, if the overall general progress requirements cannot be met, then either a faster processor must be found, the job must be redesigned for multiple processors, or the job requirements must be redefined.

The most efficient form of cooperative scheduler gives each task in turn some number of scans in round-robin fashion. The latency for any given task is thus the worst-case time it takes for a complete round-robin. This can easily be unacceptable for some low-level tasks. A *minimum-latency* cooperative scheduler can be designed. At the expense of efficiency, it checks high-priority tasks between every scan. In this case, the maximum latency is reduced to the worst-case execution time for a single scan.

Cooperative scheduling is also a mechanism that allows for portability of the control code to a simulation environment. Because it will run on any platform with the appropriate compiler, all that needs to be done is replace the time-keeping mechanism with a computed (simulated) time and add a hook to the simulation of the control object.

When cooperative scheduling is used, all tasks occupy a single computing thread.

16.8.2 Interrupt Scheduling

When general progress requirements can be met, but latency specifications cannot, then some form of scheduling beyond cooperative must be used. All of these start with *interrupts*. In effect, the interrupt mechanism is a hardware-based scheduler. In its simplest usage, it can reduce the latency of the highest-priority task from the worst-case single scan time in cooperative, minimum latency scheduling down to a few microseconds or less. It does this by operating in parallel with the CPU to monitor external electrical signals. When a specified signal changes, the interrupt controller signals the CPU to stop its current activity and change *context* to an associated interrupt task. Changing context requires saving internal CPU information and setting up to execute the interrupt task. When the interrupt task is done, the CPU resumes its former activity. Each interrupt occurrence activates a new computing thread.

Used by itself, the interrupt system can reduce the latency for a group of tasks that have very tight latency tolerances. The design rule for interrupt-based tasks is that they also have short execution times. This is important if there are several interrupt tasks, so that interference among them is minimal. An interrupt task will often prevent any other interrupt tasks from running until it is done. In other cases, an interrupt task will prevent interrupt tasks at equal or lower priority from running until it is done. Any priority system in this environment must be implemented as part of the interrupt hardware. The PWM task illustrated above would often be implemented as an interrupt task.

Many computer systems have fewer actual interrupt inputs than needed, but latency can still be reduced by "sharing" the existing interrupts. To do that, there must be software to determine which task to activate when the interrupt occurs. This scheme cannot match the low latencies achieved when each task is attached to its own interrupt, but there still can be a large improvement over cooperative scheduling.

16.8.3 Preemptive Scheduling

When tasks have latencies comfortably met by cooperative scheduling but too long (in execution time) for interrupt tasks, a preemptive scheduler must be used. The preemptive scheduler is itself activated by interrupts. It checks a set of tasks for priority, then runs the highest priority task that has requested computing resource. It then resets the interrupt system so that further interrupts can occur. If, at the time of a later interrupt, a task of priority higher than the currently running task becomes ready to run, it will *preempt* the existing task, which will be suspended until the higher priority task completes. In the same manner, if an interrupt-based task is activated, it will take precedence over a task scheduled by the preemptive scheduler.

The preemptive scheduler is thus a software version of the interrupt controller. Because of the time needed for the scheduler itself to run, the latencies for the tasks it schedules are longer than for interrupt tasks. The priority structure, in this case, is software based.

In order for preemptive scheduling to work, interrupt-activated code must be allowed to be reentrant or to overlap in time. Each interrupt activates a new computing thread. If several interrupts overlap, there will be several computing threads active simultaneously.

The simplest form of preemptive scheduler, adequate for control problems, requires that, once a task is activated, it will run to completion before any other tasks at the same or lower-priority level run. Other than suspension due to preemptive activity, the task cannot "suspend itself" until it completes (by executing a *return* statement).

16.8.4 Time Slice Scheduling

If several tasks with only weak latency requirements have scan execution times that vary greatly from one state to another, it may not be possible to balance the resource allocation by giving different numbers of scans to each. In the worst circumstances, it might be necessary to use some blocking code, or one task might be technically nonblocking (i.e., predictable), but have such a long execution time per scan that it might as well be blocking. A *time slicing* scheduler can balance the computing resource in these cases by allocating a given amount of computing time to each task in turn, rather than allocating scans as a cooperative scheduler does. This can be quite effective, but the time slicing scheduler itself is quite complex and thus better avoided, if possible. Time slice schedulers are usually part of more general scheduling systems that also include the ability of tasks to suspend themselves, for example, to wait for a resource to become available.

16.9 Task Type Preferences

Tasks that follow the state model (scan/nonblock) need no further differentiation, in theory, to solve all control software problems. Designing and building a code on this basis would, however, be bad practice.

If a scheduling mode other than cooperative was required, changes to the code would be necessary for implementation, thereby destroying the important portability. For this reason, task types are defined for using the code in any scheduling environment needed to meet the performance specifications. Designation of a task type is an indication of *preference.* The actual implementation might use a simpler scheduling method if it can meet performance demands.

Tasks are broken down into two major categories, intermittent and continuous, and into several subcategories within the intermittent tasks.

16.9.1 Intermittent

Tasks that run in response to a specific event, complete their activity, and are then dormant until the next such events are categorized as *intermittent.* Most control system tasks fit into this category. Following the types of schedulers available, they are subcategorized as *unitary* or *preemptable.*

Unitary tasks can be scheduled as direct interrupt tasks. To simplify scheduling and make execution as fast as possible, each time a unitary task is run it gets only a single scan. Control is then returned to the interrupt system. This design is based on the assumption that most tasks of this type will be single-state tasks and will thus only need one scan. The PWM task shown above is an exception to this rule. The "compute on/off times" state is transient. Before returning from the interrupt, either the "output on" or "output off" state must be executed. Therefore, it sometimes uses a single scan when it is invoked, and sometimes uses two scans. To accommodate this, the repetition through the scans must be handled internally, a structural compromise, but one that improves the overall efficiency of these very high-priority tasks.

No specific task priorities are associated with unitary tasks. Because any priority structure must exist in the interrupt controller, from the software perspective they are all considered to be of the same priority, the highest priority of any tasks in the project.

Preemptable tasks generally use more computing time than unitary tasks and are often more complex. For that reason, they are given as many scans as needed to complete their activity. When they are finished, they must signal the scanner that they are ready to become dormant. Software-based preemption is the characteristic of this class of tasks, so that each task must be assigned a priority.

The external control of scans is very important for preemptable tasks, and somewhat in contrast to the policy for unitary tasks. It is the scans that allow for effective cooperative scheduling. The external capture of state-transition information is necessary for an automatically produced audit trail of operation. The audit trail is a record of transitions that have taken place. It is particularly useful in verifying proper operation or in determining what each task has been doing just prior to a system malfunction. The exception for internal transitions in unitary tasks is based on assuming that many of them will not do any state transitions, and, when they do, they will be few and the total execution time will still be very short. State transitions in unitary tasks that occur across invocations are still tracked externally and are available for audit purposes.

16.9.2 Continuous

Continuous tasks have no clean beginning or ending points. They will absorb all the computing resource they can get. The scan/nonblock structure is absolutely essential for continuous tasks; otherwise they would never relinquish the central processing unit (CPU)!

Continuous tasks represent the most primitive form of task. Given the state transition structure, continuous tasks are all that is needed to implement the "universal" solution referred to above. If all of the timing and event detection is done internally using transition logic, the same functionality can be produced as can be realized from unitary and preemptable tasks. However, that information will be entered in an *ad hoc* manner and so cannot be used to implement any scheduler other than cooperative.

16.9.3 Task Type Assignment

Control tasks are dominated by unitary and preemptable tasks. The choice between the two is based on the nature of the task and defines the type of scheduling used with the task to meet performance issues. Over the life of the software, it is likely that it will actually be executed in several different environments.

Unitary tasks are used for activities that are very short and of high priority. Most low-level interactions with instruments and actuators fall into this category. Separate tasks are used when the data requires some processing as in PWM, pulse timing, step-rate generation for a stepping motor and so on, or when the data are needed by several tasks and conflicts could arise from multiple access to the relevant I/O device. Use of analog-to-digital converters could fall into that category. Very little computation should be done in the unitary task, enough to insure the integrity of the data, but no more. Further processing is left to lower-priority tasks.

Preemptable tasks are used for longer computations. This would include the control algorithms, supervisory tasks to coordinate several feedback loops, batch process sequencers, adaptation or identification algorithms, and so on. Some tasks, such as a PID controller, could still be single state, but the amount of computation needed makes them better grouped with preemptable tasks rather than unitary tasks.

Continuous tasks are usually used for functions, such as the operator interface, data logging, communication with a host computer, generation of process statistics, and other such activities. They form the lowest-priority group of activities and only receive CPU resource when preemptable and interrupt tasks are not active.

16.10 Intertask Communication

Tasks cooperate by exchanging information. Because the tasks can execute asynchronously (in separate threads) in some scheduling domains, the data exchange must be done to assure correct and timely information. The methods discussed here are valid for single computer systems; other methods for data exchange must be used for multicomputer systems.

16.10.1 Data Integrity

The primary danger in data exchange is that a mixture of old and new data will be used in calculations. For example, if a task is transferring a value to a variable in an ordinary C or C++ expression, an interrupt may occur during this transfer. If the interrupt occurs when the transfer is partially complete, the quantity represented will have a value consisting of some of the bits from its earlier value and some from the later value. For most C data types, that quantity will not necessarily lie anywhere between those two values (earlier, later) and might not even be a valid number.

At a higher level, if a task is carrying out a multiline calculation, an interrupt could occur somewhere in the middle. As a result of the interrupt, if a value that is used both above and below the point of interrupt is changed, the early part of the calculation will be carried out with the old value and the later part with the new value. In some cases, this could be benign, because no invalid data are used, but in other cases it could lead to erroneous results.

The most insidious factor associated with these errors is that, whether or not the error will occur is statistical—sometimes it happens, sometimes not. Because interrupts arise from sources whose timing is not synchronized with the CPU, the relative timing will never repeat. The "window" for such an error occurrence could be as small as a microsecond. Debugging in the conventional sense then becomes impossible. The usual strategy of repeating the program over and over again, each time watching how the error occurs and extracting different data, does not work here because the error occurrence has a probability of only one chance in several hundred thousand per second. However, while it is almost impossible to catch such an error in lab testing, or even know that it exists, on a system installed for long-term operation the probability that the error will occur sooner or later approaches 100%! Careful program design is the only antidote.

16.10.2 Mutual Exclusion

In single processor systems, the general method of guarding data is to identify critical data transactions and to protect them from interruption by globally disabling the computer's interrupt system. In effect, this temporarily elevates the priority of the task, in which this action is taken, to the highest priority possible, because nothing can interrupt it. This is a drastic action, so that the time spent in this state must be minimized. Other less drastic ways can be used to protect data integrity, but they are more complex to program requiring substantially more overhead.

Assuring data integrity in information interchange has two components:

1. Defining sets of variables in each task used solely for data exchange (exchange variables).
2. Establishing regions of mutual exclusion in each task for protected exchange activities.

The exchange variables can only be accessed under mutual exclusion conditions, which must be kept as short as possible. The best way to keep them short is to allow nothing but simple assignment statements in mutual exclusion zones. Under this method, the exchange variables themselves are never used in computations. This controls both low-level problems (mid-data-value interrupts) and high-level problems (ambiguous values), because the exchange values are changed only when data are being made available to other tasks.

16.10.3 Exchange Mechanisms

There are two direct mechanisms for implementing data exchange in C/C++. One is the use of global variables for the exchange variables; the other is to use local variables for the exchange variables and implement data exchange through function calls. Global variables (*statics* if task functions are in the same file, *externs* otherwise) are more efficient, but functions encapsulate data better and are more easily generalizable (e.g., to multiprocessor systems). Unless computing efficiency is the critical design limitation, the function method is recommended. An example using the function call method follows.

16.10.3.1 Function Exchange Method

Assume that Task1 and Task2 are in the same file. Each of the tasks is defined as a C++ class in Figure 16.5.

The "exchange variables" are designated by the prefix `exch_`. They are defined within the *class* used for each task, and thus have scope only within that class.

A similar definition exists for Task2.

Within the Task1 function, the exchange now can be defined as in Figure 16.6.

The functions `_disable()` and `_enable()`, which act directly on the interrupt system, are compiler dependent. These definitions are from Microsoft Visual C++.

16.11 Prototyping Platform

The software developed to implement this design methodology is used primarily in an instructional environment. The environment used is similar to what might be used for early-stage prototyping in an industrial environment. PC-class computers running a standard operating system are the basis for program development, simulation, and, in many cases, are also used to run real-time control of physical systems. In other cases, similar computing hardware running real-time operating systems are used for the

```
...
   static
   class task1_class
        {
        private:
        float exch_a,exch_b;      // Exchange variables
        ...      // Other stuff
        public:
        float get_a(void)
             {
             float a;            // Local variable to hold copied data

             _disable();
             a = exch_a;
             _enable();
             return a;
             }
        float get_b(void)
             ...
        ...      // Other stuff
        }task1;
```

FIGURE 16.5   Task1 class.

real-time control applications. The move to a real-time operating system was often essential as the time-reproducibility of standard operating systems can be quite bad by control standards—tens of milliseconds in many cases. This is not a criticism of these operating systems as they are not designed or advertised for real-time operation. However, the recent change to multicore processors has changed that equation. In the examples showing control of actual physical objects (Section 16.13), performance using dual core computers built in 2007 running Windows XP Professional^{TM} achieved time reproducibility (or from the negative perspective, time jitter) of a few hundred microseconds.

Although space limitations preclude a detailed discussion here of implementation software, the use of one of these packages is shown in some detail. Five implementation packages have been produced: two in C++ (TranRun4 and GrpCpp), two in Java (TranRunJ and TranRunJLite) and one in C (Tran-RunC). As noted earlier, these software packages were written for instructional rather than commercial use. However, they are available in full source form for interested readers (contact David Auslander, dma@me.berkeley.edu). These packages are written in standard forms of their respective languages and so it should be relatively easy to port them to other environments.

A detailed study of an example control program is given below as a model of software usage and control program construction. Task structures and relevant details are given for several other sample problems.


```
...
// Get information from Task2
a = task2.get_a(); // The disable/enable is inside the 'get' function
b = task2.get_b();
...      // Do some computation
// Copy results to exchange variables
_disable();
exch_a = a;
exch_b = b;
_enable();
...
```

FIGURE 16.6   Task2 class.

16.11.1 Control Hardware

For the PC environment, laboratory I/O boards are used for communicating with physical systems. In general, these boards provide analog-to-digital conversion, digital-to-analog conversion, digital input, and digital output. A variety of such boards have been used over the years.

16.11.2 Compilers

C++ and C programs tend not to be as portable as one would like them to be! As a result, which compiler to use often becomes an issue. The C++ packages used here have been compiled with the Microsoft and Borland compilers and the C package has been compiled with the National Instruments and GCC compilers. Java programs tend to be more portable; the Java packages were primarily compiled with the Sun compiler, but the IBM Java compiler was also used.

16.11.3 Time-Keeping

The time resolution depends entirely on the nature of the physical objects under control and the operation of the actuation and sensor elements. The time resolution of most standard operating systems, 1–10 ms, is adequate for slow systems, such as many thermal systems, but is too crude for most motion or other small-motor applications. Real-time operating systems may give better resolutions, but not always. In the Windows^{TM} environment, the Windows Performance Timer is a convenient way to get relative timing with much better resolution—1 μs or better. Note that the time resolution is not necessarily related to the time reproducibility. The time reproducibility is a measure of how accurate the timing will be in real-time operation, which is a critical measure of control software performance. It can be much worse than the time resolution! Other computing environments will have other ways to get timing information, but, in general, they will depend on the presence of time-keeping hardware.

For systems with very loose timing accuracy requirements, the most portable mode for timing is to run the system as if it were a simulation, but calibrate the "tick time" so that computed time in the program matches actual time. This is completely portable, because no special facilities are required. It will actually solve a surprisingly large number of practical problems. Recalibration is needed whenever the program or the processor is changed, but the recalibration process is very easy.

16.11.4 Operator Interface

The operator interface is an extremely important part of any control system but far beyond the scope of this chapter to discuss adequately. The primary problem in prototype development is that nothing in C or C++ syntax is helpful in developing a suitable, portable operator interface package. Like every other part of the program, the operator interface must be nonblocking, but all of the basic console interaction in C or C++ is blocking (*scanf, cin,* etc.). Java does include graphical user interface (GUI) construction as part of its language definition, and so it is easier to construct a native operator interface when programming in Java.

A simple, character-based interface is provided for use with one of the C++ packages (TranRun4). It is suitable for prototype use and is relatively easy to port. The motor and robot examples (Section 16.13) use LabVIEW^{TM} for operator interfaces.

16.12 Program Structure: The Anatomy of a Control Program

A relatively simple control problem will illustrate the construction of a control program. It is a single-loop position control of an inertial object with a PID controller. To simplify the job, it is assumed that the position is measured directly (from an analog signal or from a wide-enough decoder not requiring frequent scanning) and that the actuation signal is a voltage. Thus, no special measurement or actuation tasks are needed.

Four tasks are required:

1. Feedback (PID) control, sample-time
2. Supervisor, which gives set points to the feedback task, sample time
3. Operator interface, continuous
4. Data logger, continuous

The supervisory task generates a set point profile for the move. The state-transition logic diagram is shown in Figure 16.2. The profile is trapezoidal, with constant acceleration, constant velocity, and constant deceleration sections.

Tasks are defined as C++ classes. The tasks communicate among themselves by making calls to public functions for data exchange, as discussed above. However, any task could call a function in any other task, so the order in which the parts of the class are defined is important. In implementing a small control job, it is convenient to put all of the tasks into a single file. For larger jobs, however, each task should have its own file, as should the "main" section.

To meet these needs, the following program structure is recommended:

1. Define all of the classes for all of the tasks in a header file.
2. Put global references to each task (pointers) in the same header file.
3. The functions associated with each task can go in separate files or can be collected in fewer files.
4. The system setup information and the *UserMain()* function can go in the same or a separate file.
5. Use the *new* operator to instantiate all of the tasks in the same file as *UserMain()*.
6. Make sure the header file is *included* in all files.

The function *UserMain()* plays the role of a program *main* function. The actual main function definition will be in different places, depending on the environment to which the program is linked.

Because this is a relatively small job, only two files are used: *mass1.hpp* for the header and *mass1.cpp* for everything else.

16.12.1 Task Class Definition

Classes for the tasks are all derived from an internal class called *CTask*. This task contains all the information needed to schedule and run tasks as well as a *virtual* function called *Run*. This function must have a counterpart in the derived class for the task, where it becomes the function that is executed each time the task is scanned. Other class-based functions are normally defined for data exchange and other additional computation.

The supervisory task is a good starting point. Its class definition is as in Figure 16.7.

16.12.2 Instantiating Tasks

Tasks represent instantiations of the task classes. In most cases, each class has only one associated task. However, as noted below, several tasks can be defined from the same class. All task references are made by pointers to the tasks. These are declared as global (*extern*) in the header file,

```
#ifdef CX_SIM_PROC
   extern CMassSim *mass_sim;
#endif
extern Mass1Control *Mass1;
extern CDataLogger *DataLogger;
extern CSupervisor *Supervisor;
extern COpInt *OpInt;
```

The pointers are defined in the file (or section of the file) containing *UserMain()*,

```
#ifdef CX_SIM_PROC
   CMassSim *mass_sim;
#endif
```

```
class CSupervisor : public CTask
  {
  private:
    float position_set;
    float vprof,xprof;   // Profile velocity and position
    float xinit,xtarget,dir;
    float dt;
    int newtarget;
    real_time t4,thold;      // Used to time the hold period
    float accel,vcruise;     // Accel (and decel), cruise velocity
    float exch_t,exch_v,exch_x,exch_target; // Exchange variables
  public:
    CSupervisor(real_time dtt,int priority);    // Constructor
    void Run(void);        // The 'run' method -- virtual from CTask
    void SetTarget(float xtrgt);
  };
```

FIGURE 16.7 Supervisory task class.

```
Mass1Control *Mass1;
CDataLogger *DataLogger;
CSupervisor *Supervisor;
COpInt *OpInt;
```

Then, within *UserMain()*, the actual tasks are instantiated with the memory allocation operator, *new,*

```
#ifdef CX_SIM_PROC
   mass_sim = new CMassSim;
#endif
Mass1 = new Mass1Control;
DataLogger = new CDataLogger;
Supervisor = new CSupervisor(0.02,5);
// Send sample time and priority
OpInt = new COpInt(0.2);
// Set the print interval.
```

Each of these variables has now been allocated memory, so that tasks can call functions in any of the other task classes. The simulation, *mass_sim,* is treated differently from the other tasks and will be discussed after the control tasks. The simulation material is only compiled with the program when it is compiled for simulation, as noted by the *#ifdef* sections.

16.12.3 Task Functions

The functions in the *public* section of *CSupervisor* are typical of the way tasks are defined. The constructor function for the *Supervisor* task (*CSupervisor* is the name of the class; *Supervisor* is the name of the task) sets initial values for variables and passes basic configuration information on to the parent *CTask* by a call to its constructor. This is the constructor for *Supervisor*:

```
CSupervisor::CSupervisor(real_time dtt,
  int priority) // Constructor
  : CTask("Supervisor", SAMPLE_TIME,
  priority,dtt)
  {

  dt = dtt;
  xinit = 0.0;
  xtarget = 1.2;
  dir = 1.0;
  accel = 0.5;
```

```
vcruise = 0.6;
thold = 0.1;
// Hold for 100 ms after the
// end of the profile
t4 = 0.0;
vprof = 0.0;
xprof = xinit;
newtarget = 0;
State = 0;
};
```

The *Run* function is where all the work is done. In *CSupervisor* it is based on a transition logic structure moving through the various stages of the profile, accelerate, cruise, decelerate, and hold. It will then wait for a new target position command, or time-out and stop the program if the command does not come. The virtualized *Run* function is called from the parent *CTask* every time the task is activated (in this case, every sample time). See Figure 16.8.

The *Idle()* at the end of the function is extremely important. It is the signal to the scheduler that this task needs no attention until the next time increment elapses. If this is left out of a *sample-time* or *event task,* the scheduler will continue to give it attention, at the expense of any lower-priority tasks and all of the continuous tasks. Leaving out the *Idle()* statement or putting it in the wrong position is a common bug.

References to other tasks are made by calling *public* functions in those tasks. Both *Mass1* and *DataLogger* are referenced here. When functions from other tasks are called, they run at the priority level of the calling task, rather than the priority level of the task where they are defined.

The *SetTarget* function allows other tasks to change the target value, thus defining a new move. This would most likely be done from the operator interface task. Because the state structure is set up so that the check for a new target is made only in state 4, a move will be completed before the next move is started. This also means that, if two new moves are sent while a previous move is still in progress, only the last will be recognized.

```
void CSupervisor::SetTarget(float trgt)
  {
  DisableInterrupts();
  exch_target = trgt;
  newtarget = 1;
  EnableInterrupts();
  }
```

16.12.4 A Generic Controller Class

Where several tasks of similar function will be used, the inheritance property of C++ can be used to great advantage. PID controllers are so common that a class for PID controllers has been defined that can act as a parent class for any number of actual PID controllers. The class definition for this is in Figure 16.9.

The arguments to the *constructor* function have the data needed to customize a task to the scheduler. All the variables are *protected* rather than *private* so that they can be referred to readily from the derived class. The major properties, distinguishing one actual derived PID control task from another, are where it gets its process data from and where it sends its actuation information. These are specified in the *GetProcVal* and *SetActVal* functions, listed here as *virtual*, and must be supplied in the derived class because the versions here are just dummies.

The *Run* function in the *PIDControl* class is not virtual so that the control calculation can be completely defined in the parent class (it is, of course, virtual in the higher level parent class, *Ctask*). Its only connection to the derived class is in getting the process value and sending out the actuation value (Figure 16.10).

```
void CSupervisor::Run(void)
  {
  float d_end,d_decel;
  real_time tt;    // Curent time

  tt = GetTimeNow();
  switch(State)
    {
    case 0:      // Initial call
     Mass1->SetSetpoint(xprof); // Send a new setpoint to the
      controller
     Mass1->SetStart(); // Send a start message to the controller
     if(xtarget > xinit)dir = 1.0; // Determine direction of
      the move
     else dir = -1.0;
     State = 1;
     break;

    case 1:      // Start profile -- Accelerate
     vprof += dir * accel * dt;// Integrate the velocity and
      position
     if(fabs(vprof) >= vcruise)
        {
        // This is the end of the acceleration section
        vprof = dir * vcruise;
        State = 2;        // Go on to next state
        }
     // Check whether
     // cruise should be skipped because decel should be started
     d_end = fabs(xprof - xtarget); // Absolute distance to end
     d_decel = vprof * vprof / (2.0 * accel); // Distance to
      decelerate
              // to stop at current velocity
     if(d_decel >= d_end)
        {
        // Yes, go straight to decel
        vprof -= dir * accel * dt;  // Start decel
        State = 3;
        }
     xprof += vprof * dt;
     break;

    case 2:      // Cruise -- constant velocity
      xprof += vprof * dt;
      d_end = fabs(xprof - xtarget);  // Absolute distance to end
      d_decel = vprof * vprof / (2.0 * accel); // Distance to
       decelerate
          // to stop at current velocity
      if(d_decel >= d_end)
      {
         // Yes, go to decel
         vprof -= dir * accel * dt;  // Start decel
         State = 3;
        }
       break;
      case 3:      // Deceleration
```

FIGURE 16.8 Run function.

```
        d_end = fabs(xprof - xtarget); // Absolute distance
         to end
        vprof = dir * sqrt(2.0 * d_end * accel);// Velocity that
         will get
              // to stop in desired distance
        xprof += vprof *dt;
        if(fabs(xprof - xinit) >= fabs(xtarget - xinit))
          {
              // This is the end of the profile
          xprof = xtarget;
          vprof = 0.0;     // Stop

          t4 = GetTimeNow(); // Start a timer for the hold state
          State = 4;         // Go to HOLD state
          }
      break;

     case 4: // Hold final position until either a command for a
             // new target is sent, or time runs out
          // Check for new target
          DisableInterrupts();
          if(newtarget)
           {

           xinit = xtarget; // Start new profile where this
            one ended
           xtarget = exch_target;  // New target position
           newtarget = 0;
           vprof = 0.0;
           xprof = xinit;
           State = 0;       // Start the profile again
           break;
           }
           if((GetTimeNow() - t4) >= thold) // Check for timeout
            {
            // End the program
            TheMaster->Stop();
            }
           break;
          }
     DisableInterrupts();      // Copy data to exchange variables
     exch_t = tt;
     exch_v = vprof;
     exch_x = xprof;
     EnableInterrupts();
     Mass1->SetSet point(exch_x); // Send a new set point to
      the controller
     DataLogger->LogProfileVal(exch_t,exch_v,exch_x);
     Idle();
     };
```

FIGURE 16.8   *Continued.*

The use of the *waiting* state (state 0) prevents the controller from becoming active before appropriate initialization and setup work has been done by other tasks. No call to *Idle* is in the portion of state 0 that is doing the transition to state 1. This is done to prevent a delay between the start signal and the actual beginning of control.

```
class PIDControl : public CTask
  {
  protected:// This is just the generic part of a control task
    // so all variables are made accessible to the derived class
    float integ;
    float set,val;       // Set point and output (position) value
    float prev_set,prev_val;
    float kp,ki,kd,min,max; // Controller gains, limits
    real_time dt;
    float mc;            // Controller output
    int start_flag;      // Used for transition from initial state

    // Exchange variables, set in this class
    float exch_val,exch_mc;
    // Exchange variables obtained from other tasks
    float exch_set,exch_kp,exch_ki,exch_kd,exch_min,exch_max;
    int exch_newgains;// Use this as a flag to indicate that new gains
        // have been set
    int exch_start_flag;

  public:
    PIDControl(char *name,int priority,float dtt); // Constructor
    void Run (void);        // Run method
    float PIDCalc(void);      // Do the PID calculation
    void SetGains(float kpv,float kiv,float kdv,float minv,float maxv);
    void SetStart(void);     // Set the start flag to 1

    virtual void SetActVal(float val){}// Set the actuation value --
    // The real version of this must be supplied in the derived class
    void SetSetpoint(float sp);
    void GetData(float *pval,float *pmc,float *pset);

    virtual float GetProcVal(void){return 0.0;}
    //Get the process value --
    //  The real version of this must be supplied in the derived class
};
```

FIGURE 16.9 PID controller class.

Defining a class for an actual PID control is very simple, and has very little in it. Here is the definition for the position control task used in this sample problem:

```
class Mass1Control : public PIDControl
  {
  public:
    Mass1Control();        // Constructor
    void SetActVal(float val);
    // Set the actuation value
    float GetProcVal(void);
    // Get the process value --
  };
```

It has no private data at all, and only defines a constructor and the two virtual functions for getting process data and setting the actuation value. Its constructor is

```
Mass1Control::Mass1Control() :
  PIDControl("Mass1",10,0.02)
  // Call base class constructor also
  {
  kp = 1.5;
```

```
     void PIDControl::Run (void) // Task function
      {
     // This task has two states - a ''waiting'' state when it
     // first turns on and
     // a ''runnning'' state when other tasks have properly
      initialized
     // everything
     // The variable 'State' is inherited as a 'protected'
     // variable from the parent class, CTask.

     switch(State)
      {
     case 0:      // Waiting for 'start' signal
      DisableInterrupts(); // copy relevant exchange
                // variables for this state
      start_flag = exch_start_flag;
      EnableInterrupts();
      if(start_flag)
        {
       State = 1;
       return;      // Set new state and return.
                    // Next scan will go to 'run' state
        }
       else
        {
       // Stay in 'wait' state
       Idle(); // Indicate that the task can inactivate
                // until next sample time
       return;
        }
     case 1:      // Run the control algorithm

       {
      DisableInterrupts(); // copy relevant exchange variables
      if(exch_newgains) // Using this flag minimizes interrupt
                        // disabled time
        {
       kp = exch_kp;
       ki = exch_ki;
       kd = exch_kd;
       min = exch_min;
       max = exch_max;
       exch_newgains = 0;   //Turn off the flag
        }
      set = exch_set;
      EnableInterrupts();
      val = GetProcVal(); // Get the process output
      mc = PIDCalc(); // Do the PID calculation
      SetActVal(mc);   // Send out the actuation value
      DisableInterrupts();
      // Set output exchange values
      exch_val = val;
      exch_set = set;
      exch_mc = mc;
      EnableInterrupts();
      Idle();      // Wait for next sample interval
        }
       }
     }
```

FIGURE 16.10   Run function for the PID control class.

```
// Initial controller gains
ki = 0.0;
kd = 2.0;
set = 0.0;
// Initial set point
min = -1.0;
max = 1.0;
}
```

It first calls the constructor for the parent class to set the task name, priority, and sample time, and then sets the controller gains, limits and initial set point. These variables are *protected* in the parent class, and so are freely accessible from this derived class.

The *GetProcVal* and *SetActVal* functions are where much of the hardware-dependent code goes, at least as relates to the control system hardware. The versions shown here use *define* statements for the simulation code so that other sections can be easily added as the environment changes.

```
void Mass1Control::SetActVal(float val)
// Set the actuation value
  {
  #ifdef CX_SIM_PROC
  // The following code is for
  // simulation
  mass_sim->SetSimForce(val);
  // No mutual exclusion is needed for
  // simulation
  #endif
  }

float Mass1Control::GetProcVal(void)
  {
  float x,v;

  #ifdef CX_SIM_PROC
  // The following code is for
  // simulation
  mass_sim->GetSimVal(&x,&v);
  return(x);
  // Position is the controlled
  // variable
  #endif
  }
```

Other such generic definitions can be used to great advantage when multiple elements with similar function are in a control system. For example, the *CSupervisor* class could easily be generalized in a similar manner to allow for several simultaneous profile generating tasks.

Because the remaining tasks are constructed in a similar manner and do not introduce any new elements, they will not be discussed in detail here.

16.12.5 The Simulation Function

Simulation is extremely important to control system development, but because it appears as "overhead," it is often neglected. In addition to whatever simulation has been done on the algorithm side, it is very important that the actual control system software be tested in simulation. The environment of simulation is so much more friendly than the actual control environment that there is a large time saving for every bug or misplaced assumption found while simulating.

The simulation part of the software is treated as a separate category. In principle, it is handled similarly to the tasks, but is kept separate so that the task structure is not distorted by the simulation.

There are two ways to handle simulation. The method here is to define an internal simulation class, *CSim,* from which a class is derived to define the particular simulation. The simulation itself is then written in C++, with the simulation function called whenever time is updated. An alternative method is to provide an interface to a commercial simulation system, most of which have the ability to call C or C++ functions. To do this, the internal computing paradigm must be changed. The simulation system is in charge, so that the call to the simulation must also trigger the operation of the control system. This latter method is not implemented in the current software.

The simulation class defined for the sample position control problem is

```
#ifdef CX_SIM_PROC
class CMassSim : public CSim
  {
  private:
    float dt;
    float x,v;   // position and velocity
    float m;     // mass
    float f;     // force
    FILE *file;
    public:
    CMassSim();
    void RunSimStep(real_time t,real_time dt);
    // The simulation function
    void GetSimVal(float *px,float *pv);
    void SetSimForce(float ff);
  };
#endif   // End of simulation section
```

Similarly to the tasks, the *RunSimStep* function is declared *virtual* in the base class and must be defined in the derived class. This is the function called when time is updated. For the position control problem, it contains a very simple simulation of an inertial object, based on the parameter values set in the constructor:

```
CMassSim::CMassSim() : CSim()
// Call base class constructor also
  {

  x = v = 0.0;
  m = 0.08;   // Mass value
  f = 0.0;    // Initial value of force
  file = fopen("mass_sim.dat","w");
  }

void CMassSim::RunSimStep(real_time t,
                          real_time dtt)
  {
  // Calculate one step
  // (Modified Euler method)

  dt = dtt;
  // Set internal value for step size
  v += f * dt / m;
  x += v * dt;
  fprintf(file,"%lg %g %g %g\n",t,v,x,f);
  }
```

The simulation output is also created directly from this function. Because the simulation is not present in other conditions, it is more appropriate to put this output here than to send it to the data logging task. The other functions in the class exchange data. Because the simulation is not used in a real-time environment, it is not necessary to maintain mutual exclusion on the simulation side of these exchanges.

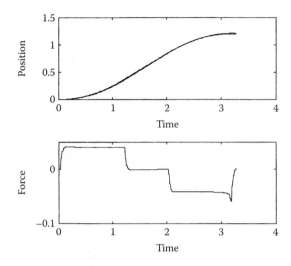

FIGURE 16.11   Profile position change, inertial system.

16.12.6 Results

The graphs in Figure 16.11 show the results for the data sets given in the listings above. These results are for a simulation. The top graph shows the position of the mass (solid line) and the set point generated by the nearly overlaid profile (dashed line). The bottom graph shows the force applied to the mass to achieve these results.

These results were plotted from the output of the data logging task and the output generated directly by the simulation.

In addition to these, there is also a transition audit trail that is generated. The transition audit trail keeps track of the most recent transitions, which, in this case, includes all of the transitions that take place. The file is produced when the control is terminated:

```
CTL_EXEC State Transition Logic Trace File

Time      Task            From     To

0         OpInt           0        1
0.01      Supervisor      0        1
0.03      Mass1           0        1
0.2       OpInt           1        0
0.21      OpInt           0        1
0.41      OpInt           1        0
0.42      OpInt           0        1
0.62      OpInt           1        0
0.63      OpInt           0        1
0.83      OpInt           1        0
0.84      OpInt           0        1
1.04      OpInt           1        0
1.05      OpInt           0        1
1.23      Supervisor      1        2
1.25      OpInt           1        0
1.26      OpInt           0        1
1.46      OpInt           1        0
1.47      OpInt           0        1
1.68      OpInt           1        0
1.69      OpInt           0        1
```

```
1.9      OpInt         1         0
1.91     OpInt         0         1
2.01     Supervisor    2         3
2.11     OpInt         1         0
2.12     OpInt         0         1
2.32     OpInt         1         0
2.33     OpInt         0         1
2.53     OpInt         1         0
2.54     OpInt         0         1
2.74     OpInt         1         0
2.75     OpInt         0         1
2.96     OpInt         1         0
2.97     OpInt         0         1
3.15     Supervisor    3         4
3.18     OpInt         1         0
3.19     OpInt         0         1
```

This shows the *Supervisor* task turning on and then moving through states 1, 2, 3, and 4. Task *Mass1*, the control task, starts in its "waiting" state, 0, and goes to the "running" state, 1, in response to the command from the *Supervisor*. It also shows the transitions in the operator interface task going back and forth between a state that prints the ongoing results (state 0) and a state that waits (0.2 s in this case) for the next time to print its progress report. Because standard C/C++ input/output functions are used for this implementation, a progress check is about all that can be done in the operator interface because the input functions (*scanf* or *cin*) are blocking.

The audit trail is a fundamental debugging tool. It can be used to find out where a program went astray, whether critical timing constraints are being met, and so on.

16.13 Some Real Examples

Two motor-based examples are used to illustrate the low-level operation of control software using the task-state structure to do profiled position control. The two systems are a bare motor (i.e., no load attached to the motor's shaft), Figure 16.12, and one axis of an articulated robot, Figure 16.13.

These examples are controlled by a PC system running Windows connected to a National Instruments 7833R lab IO system. The field-programmable gate array (FPGA) on the 7833R is used to decode the

FIGURE 16.12   Bare motor.

FIGURE 16.13 Robot.

quadrature and also to generate the PWM signals to drive the H-bridge amplifiers. It is programmed in LabVIEW. The software uses the methodology described above, but is written in C rather than in C++.

The program is structured so that the real-time portion can be run either in Windows or in LabVIEW-Real-Time, an operating system that runs on PC hardware. Although performance is better on the

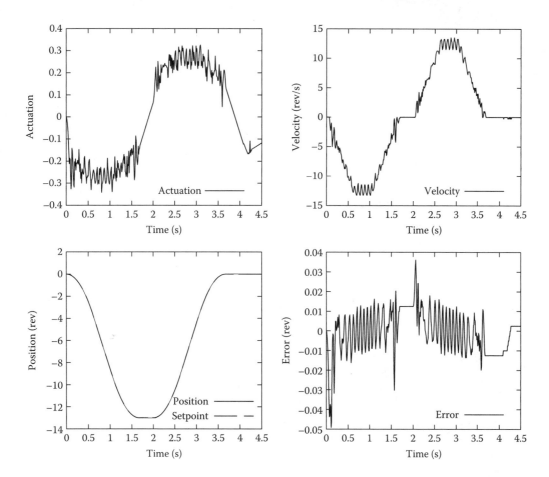

FIGURE 16.14 Bare motor point-to-point moves: Tuned for minimum error.

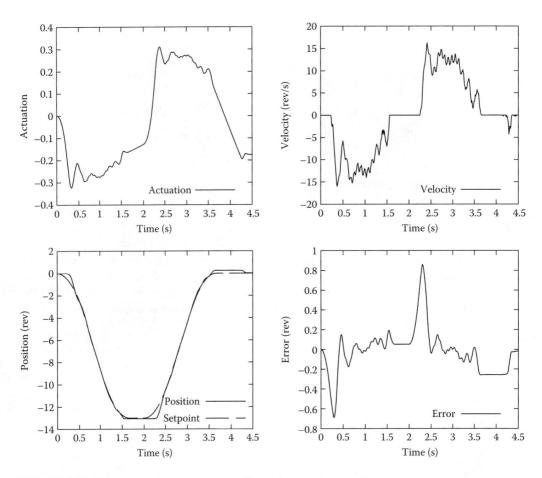

FIGURE 16.15   Bare motor point-to-point moves: Tuned for smoother actuation.

real-time operating system, time jitter of only a few hundred microseconds when running in Windows on a dual-core PC system is low enough for excellent control performance. The data shown below were taken from systems running Windows.

16.13.1 Bare Motor

Position control of this motor is dominated by stick-friction ("stiction") as it stops or when it is moving at very low velocity. As the velocity approaches zero, the motion will suddenly stop. The stopping point is not predictable so, in point-to-point motion, the final error will be different each time the system is run. As shown below, the bounds for this error are affected by how the controller is tuned and what performance trade-offs are selected.

In this example, a pair of point-to-point moves is made using a trapezoidal profile (see Figure 16.3). The advantage to a profiled move is that control is maintained throughout the move, even for relatively large moves such as these. In contrast, if a step change is made to the set point for a large move, the actuation will saturate and the system will run open loop until it gets close enough to the target for the controller output values to unsaturate. Such a move is also nonlinear, necessitating, for example, integrator windup protection if the integrator term is non zero. No such problems exist when the move is profiled, because the controller never saturates and stays within small-signal operating bounds.

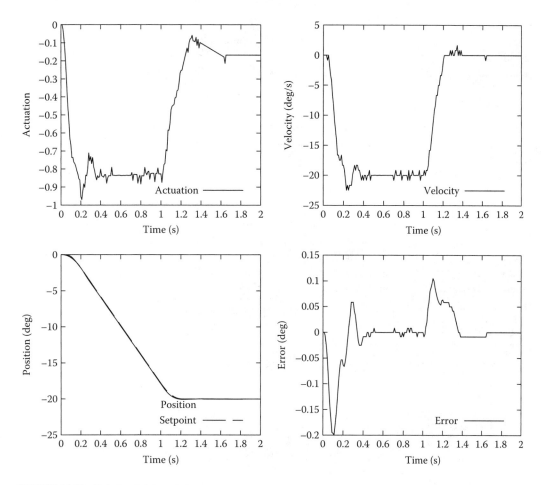

FIGURE 16.16 Robot point-to-point move.

Figure 16.14 shows the performance using an aggressive controller tuning to achieve minimum error. Two profiled moves are shown, one from the initial position to −13 revolutions and a second move back to the initial location.

The residual error at the end of the move is approximately 0.01 rev (about 4 degrees). This is only approximate, because with stiction in the system the error will be different each time. A better-specified range can be established by doing enough experiments to give a statistically significant data set. The sticking behavior can be observed at the end of the second profile by looking at the error plot. The movement initially stops at about $t = 3.7$ s with an error of a bit more that −0.01 rev. It stays there for a while, then jumps to 0.002 rev. Unless the controller is turned off at this point, there can be small jumps of this sort at any time. Just turning off the integral term after the end of the profile will reduce these jumps greatly, although jumps triggered by small disturbances would still be possible.

This tuning produces low error but at the expense of rather rough actuation (see the Actuation plot on Figure 16.14). If the particular application can tolerate larger errors, but would suffer because of the rough actuation (e.g., because it would cause vibration in the system structure) the controller can be retuned for smoother performance. Figure 16.15 shows the performance with such a tuning. The actuation is indeed much smoother, but the error at the end of the profile is much larger (about 0.2 rev compared to 0.01 rev). This is ultimately corrected by the integral action, however, to leave much smaller residual error. This tuning uses only the proportional and integral terms; the tighter tuning uses the derivative term as well.

The smoother tuning shows the stiction action very graphically. Looking at the Position graph in Figure 16.15, the motor does not move at all at the beginning of each of the profiles until the actuation builds up to a high-enough value to unstick it. The same phenomenon happens with the tighter tuning, but, because of the high controller gains, it is too fast to be visible on the graphs.

16.13.2 Robot

The robot of Figure 16.13 uses motors connected through gearboxes to drive the articulated links. The software is set up to control all of the robot's motions, but for this example only the upper arm is actually moved (the link with the label "Intelitek" in the figure).

The major challenge in controlling this axis is the large gravity load. The gearing for this axis is a fairly high ratio, but not so high that the gravity load cannot drive the axis away from its setpoint when the power is turned off (i.e., the axis is back drivable). Figure 16.16 shows the behavior of this system using a controller with proportional, integral, and derivative control terms.

The motion is a single-profiled move of 20 deg (moving up to approximately the position shown in Figure 16.13). The control is very effective, the integral action canceling the gravity effects, with an error of 0.1 degree at the end of the move settling quickly to no measurable error (note that the scale on these graphs is in degrees; it is revolutions on the graphs for the bare motor).

Acknowledgment

This work was supported in part by the Synthesis Coalition, sponsored by the U.S. National Science Foundation.

References

1. Lippman, S. B., *C++ Primer,* 2nd Ed., Addison-Wesley, Reading, MA, 1991.
2. Stroustrup, B., *The C++ Programming Language,* 2nd Ed., Addison-Wesley, Reading, MA, 1991.
3. Auslander, D. M. and Sagues, P., *Microprocessors for Measurement and Control,* Osborne/McGraw-Hill, Berkeley, CA, 1981.
4. Dornfeld, D. A., Auslander, D. M., and Sagues, P., Programming and optimization of multimicroprocessor controlled manufacturing processes, *Mech. Eng.,* 102(13), 34–41, 1980.
5. Simms, M. J., Using state machines as a design and coding tool, *Hewlett-Packard J.,* 45(6), 1994.
6. Ward, P. T. and Mellor, S. J., *Structured Development for Real-Time Systems,* Prentice-Hall, Englewood Cliffs, NJ, 1985.

17

Programmable Controllers

Gustaf Olsson
Lund University

17.1 Introduction

Programmable logical controllers (PLCs) have been in use since the 1960s and are still the basis for the low-level control in many automation systems. Today PLCs can handle not only the lowest levels of control but also advanced control of hybrid systems, where time-driven continuous controllers have to be integrated with event-driven controllers. The state concept is of fundamental importance to understand sequencing control.

Binary combinatorial and sequencing control is the basis of this chapter. Switching theory, which provides the foundation for binary control, is used not only in automation technology but is also of fundamental importance in many other fields. This theory provides the very principle on which the function of digital computers is based. In general, binary combinatorial and sequencing control is simpler than conventional feedback (analog and digital) control, because both the measurement values and the control signals are binary. However, also binary control has its specific properties that have to be considered in more detail.

Logical circuits have traditionally been implemented with different techniques; until the mid-1970s when most circuits were built with electromechanical relays and pneumatic components. During the 1970s PLCs became more and more commonplace, and today sequencing control is almost exclusively implemented in software. Despite the change in technology, the symbols for the description of switching operations, known as *ladder diagrams* (LDs) that derive from earlier relay technology, are still used to describe and document sequencing control operations implemented in software. Another important type of description language that can be used not only for programming but also as documentation tool is *sequential function charts* (SFCs). Programming of a modern PLC can be realized in five different programming languages, text oriented or graphical. An international standard IEC 61131-3 forms the basis for advanced automation system programming.

In Section 17.2, the finite-state concept is introduced as the fundamental importance for discrete event systems. In Section 17.3, hardware is described for binary sensors and actuators. Section 17.4 gives an elementary description of Boolean algebra. LDs are still used to describe logical circuits and are described in Section 17.5. The structure of *PLCs* is outlined in Section 17.6, and their programming is described in Section 17.7. Their role in large automation systems is shown in Section 17.8, where communication is emphasized.

17.2 The Finite-State Concept

The concept of a finite state is fundamental for the understanding of discrete event systems. Such a system can be described as always being in one well-defined state. For example, a machine can be operating or idle, which means that it is always in one of these two states. A buffer storage can be in many states, equal to the number of places (N) plus one. The complexity of a discrete state system is not the sum of all the states, but is closer to the product of all the states, if all the variants have to be described.

Some condition has to be satisfied in order to make a transfer between two states. Such a condition can, for example, be an external event, an operator command, or a timer signal. When a machine operation is finished, its state will change from *operating* to *idle*. In our models we assume that such a state transition takes place immediately. Likewise, a timer can indicate a state transfer, for example, the start of a pump. Consequently, the pump condition will (immediately) change from idle to operating.

In a discrete event control system there are two basic elements, *states* and *transitions*. While the system is in one state there will be some action (operation) taking place. We will illustrate this by a simple example.

A tank is to be filled with a liquid. When the tank is full its content must be heated up to a predefined temperature and the liquid has to be well mixed. After a specified time, the tank is emptied, and the process starts all over again. First we will consider the states. It is important to note that the states describe the transportation or progress of the *product*. This means that actuators (motors and valves) and sensors are not part of the state definition:

- A sensor signal *empty* signals that the tank is empty and can be filled again. This is defined as the *initial* state of the operation.
- A signal *start* will initiate the filling of the tank. The start signal then indicates the transition from the *initial* state to the *filling* state. In this state there are two actions being performed. First the bottom valve of the tank is closed. Then a filling pump is started.
- The next transition signal is a sensor signal, indicating that the tank is full. This transition will bring the tank into the state *heating*. Now there are another two actions being started. First the filling pump is turned off, and then a heater is switched on.

- The tank will remain in the *heating* state and the heater will stay on until the temperature has reached the predefined setpoint. When the preset temperature has been reached, another transition takes place. At this point the state will be transferred to the state *wait*.
- In the *wait* state the heater is first switched off and a timer is initiated. The timer will run a predefined waiting time "time out" to make sure that the temperature in the liquid has become homogeneous. A mixer may also be running in this state.
- The timer initiates the next state transition to the *emptying* state. The action *open the discharge valve* is initiated and will stay on until the tank is empty. Once the tank is empty a sensor signal will indicate *empty*, and there is a state transition to return to the *initial* state.

We have now defined the four states *filling, heating, waiting*, and *emptying* as well as the initial state. We have also defined the *control signals* that initiate the transitions from one state to another. The signals *empty, start, full, temp*, and *wait_time* are usually global variables. We have also seen that there are certain activities at each step. In the initial step, usually nothing happens, since the step is a "resting state." In the *filling* state, the discharge valve is closed and the pump turned on. Similarly, there are other activities related to the other steps. This structure of dividing the sequence in *states, control signals* or *transitions*, and *actions* is important for all sequencing control.

We summarize the typical features:

- The system is only in one state at a time;
- The state transition is initiated by some sensors, timers, or operator signals and it takes place *immediately*;
- While the system is in one state, some action will take place. There may be more than one action at the same time. The actions have to stop at the next state transition.

In order to implement the various states and the state transitions, the software has to guarantee that the three conditions above are satisfied at all times.

17.3 Binary Sensors and Actuators

In sequential control, measurements are of the *on/off* type and depend on binary sensors. In a typical process or manufacturing industry, there are literally thousands of *on/off* conditions that have to be recorded. Binary sensors are used to detect the position of contacts, count discrete components in material flows, detect alarm limits of levels and pressures, and find end positions of manipulators.

17.3.1 Limit Switches

Limit switches have been used for decades to indicate positions. They consist of mechanically actuated electrical contacts. A contact opens or closes when some variable (position, level) has reached a certain value. There are hundreds of types of limit switches. Limit switches represent a crucial part of many control systems, and the system reliability depends, to a great extent, on them. Many process failures are due to limit switches. They are located "where the action is" and are often subject to excessive mechanical forces or too large currents.

A normally opened (n.o.) and a normally closed (n.c.) switch contact are shown in their normal and actuated positions in Figure 17.1. A switch can have two outputs, called change-over and transfer contacts. In a circuit diagram it is common practice to draw each switch contact the way it appears with the system at rest.

The simplest type of sensor consists of a *single-pole, single-throw* (SPST) mechanical switch. Closing a mechanical switch causes a problem because it "bounces" for a few milliseconds. When it is important to detect only the first closure, such as in a limit switch or in a "panic button" the subsequent opening and closing bounces need not be monitored. When the opening of the switch must be detected after a closing,

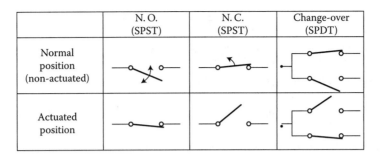

	N. O. (SPST)	N. C. (SPST)	Change-over (SPDT)
Normal position (non-actuated)			
Actuated position			

FIGURE 17.1   Limit switch symbols for different contact configurations.

the switch must not be interrogated until after the switch "settling time" has expired. A programmed delay is one means of overcoming the effects of switch bounce.

A *transfer contact* (sometimes called *single-pole, double-throw*, SPDT) can be classified as either "break-before-make" (BBM) or "make-before-break" (MBB) (Figure 17.2). In a BBM contact, both contacts are open for a short moment during switching. In an MBB contact, there is a current in both contacts briefly during a switch.

Contact debouncing in an SPDT switch can be produced in the hardware. When the grounded moving contact touches either input, the input is pulled low and the circuit is designed to latch the logic state corresponding to the first contact closure and to ignore the subsequent bounces.

17.3.2 Point Sensors

There are many kinds of measurement sensors that switch whenever a variable (level, pressure, temperature, or flow) reaches a certain point. Therefore, they are called *point sensors*. They are used as devices that actuate an alarm signal or shut down the process whenever some dangerous situation occurs. Consequently, they have to be robust and reliable.

17.3.3 Digital Sensors

Digital measuring devices (*digital sensors*) generate discrete output signals, such as pulse trains or encoded data that can be directly read by the processor. The sensor part of a digital measuring device is usually quite similar to that of their analog counterparts. There are digital sensors with microprocessors to perform numerical manipulations and conditioning locally, and to provide output signals in either digital or analog form. When the output of a digital sensor is a pulse signal, a counter is used to count the pulses or to count clock cycles over one pulse duration. The count is first represented as a digital word according to some code and then read by the computer.

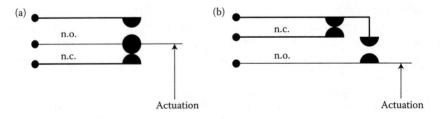

FIGURE 17.2   (a) BBM contact. (b) MBB contact.

17.3.4 Binary Actuators

In many situations, sufficient control of a system can be achieved if the actuator has only two states: one with electrical energy applied (*on*) and the other with no energy applied (*off*). In such a system, no digital-to-analog converter is required and amplification can be performed by a simple switching device rather than by a linear amplifier.

Many types of actuators, such as magnetic valves controlling pneumatic or hydraulic cylinders, electromagnetic relays controlling electrical motors, and lamps, can receive digital signals from a controller. There are two main groups of binary actuators, monostable and bistable units. A *monostable* actuator has only one stable position and gives only one signal. A contactor for an electric motor is monostable. As long as a signal is sent to the contactor, the motor rotates, but, as soon as the signal is broken, the motor will stop.

A *bistable* unit remains in its given position until another signal arrives. In that sense the actuator is said to have a memory. For example, in order to move a cylinder, controlled by a bistable magnet valve, one signal is needed for the positive movement and another one for the negative movement.

17.3.5 Switches

The output lines from a computer output port can supply only small amounts of power. Typically, a high-level output signal has a voltage between 2 and 5 V and a low-level output of less than 1 V. The current capacity depends on the connection of the load but is generally less than 20 mA, so that the output can switch power of less than 100 mW.

For higher power handling capability, it is more important to avoid a direct electrical connection between a computer output port and the switch. The switch may generate electrical noise which would affect the operation of the computer if there were a common electrical circuit between the computer and the switch. Also, if the switch fails, high voltage from the switch could affect the computer output port and damage the computer.

The most common electrically isolated switch in control applications has been the *electromechanical relay*. Low-power reed relays are available for many computer bus systems and can be used for isolated switching of signals. Relays for larger power ratings cannot be mounted on the computer board. A relay is a robust switch that can block both direct and alternating currents. Relays are available for a wide range of power, from reed relays used to switch millivolt signals to contactors for hundreds of kilowatts. Moreover, their function is well understood by maintenance personnel. Some of the disadvantages are that relays are relatively slow, switching in the order of milliseconds instead of microseconds. They suffer from contact bouncing problems that can generate electric noise, and, in turn, may influence the computer.

The switching of high power is easily done in solid-state switches which avoid many deficiencies of relays. A solid-state switch has a control input which is coupled optically or inductively to a solid-state power switching device. The control inputs to solid-state switches, designed to be driven directly from digital logic circuits, are quite easily adaptable to computer control.

17.4 Elementary Switching Theory

In this section, we describe elementary switching theory that is relevant for process control applications.

17.4.1 Notations

An electric switch or relay contact and a valve intended for logic circuits are both binary, designed to operate in the *on/off* mode. A transistor can also be used as a binary element operating only in on/off mode, either conducting or not conducting current.

The state of a binary element is indicated by a binary variable that can consequently only take two values, conventionally indicated as "0" or "1." For a switch contact, relay contact, or a transistor (represented by a Boolean variable x), the statement $x = 0$ means that the element is open (does not conduct current) and $x = 1$ means closed (conducts). For a push button or a limit switch, $x = 0$ means that the switch is not being actuated, and $x = 1$ indicates actuation.

Often a binary variable is represented as a voltage level. In *positive* logic, the higher level corresponds to logical 1 and the lower level to logical 0. In transistor–transistor logic (TTL), logical 0 is typically defined by levels between 0 and 0.8 V and logical 1 by any voltage higher than 2 V. Similarly, in pneumatic systems, $x = 0$ may mean that the line is exhausted to atmospheric pressure while $x = 1$ means a pressurized line.

Standardized symbols are used to represent logic (combinatorial and/or sequencing) circuits independently of the practical implementation (with electric or pneumatic components). This type of representation is called a *function block*. There are international standards for the logic symbols, IEC 113-7 and IEC 617; many other national standards are also defined on their basis.

17.4.2 Basic Logical Gates

Combinatorial circuits consist of several logical connections, in which the output y depends only on the *current* combination of input signals $u = (u1, u2, \ldots)$ or

$$y(t) = f[u(t)].$$

Gates have *no memory,* so the network is a *static* system. Therefore, there are no states defined. Here follows a brief recapitulation of Boolean algebra. The simplest logical operation is the negation (*NOT*), with one input and one output. If the input $u = 0$, then the output $y = 1$, and vice versa. We denote negation of x by $\bar{x}$. The behavior of a switching circuit can be represented by *truth tables*, where the output value is given for all possible combinations of inputs. The symbol and the truth table for *NOT* is shown in Figure 17.3.

Two n.o. switch contacts connected in series constitute an *AND* gate defined by *Boolean multiplication* as

$$y = u_1 \cdot u_2.$$

$y = 1$ only if both u_1 and u_2 are equal to 1, otherwise $y = 0$ (Figure 17.4). The multiplication sign is often omitted, just as in ordinary algebra. An *AND* gate can have more than two inputs, because any number of switches can be connected in series. Adding a third switch results in $y = u_1 \cdot u_2 \cdot u_3$. We use the International Standards Organization (ISO) symbol for the gate.

A common operation is a logical *AND* between two bytes in a process called *masking*. The first byte is the input register reference while the other byte is defined by the user to mask out bits of interest. The

u	y
0	1
1	0

FIGURE 17.3   The ISO symbol and its truth table for *NOT*.

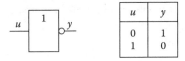

$u1$	$u2$	y
0	0	0
0	1	0
1	0	0
1	1	1

FIGURE 17.4   An *AND* gate, its ISO symbol, and its truth table.

Input register	1 1 0 1 1 0 0 0
mask	0 1 1 0 1 1 0 1
Output	0 1 0 0 1 0 0 0

FIGURE 17.5 Masking two bytes with an *AND* operation.

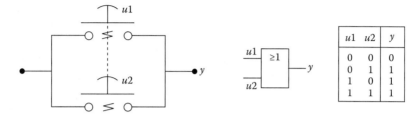

FIGURE 17.6 An *OR* gate, its ISO symbol, and its truth table.

AND operation is made bit by bit of the two bytes (Figure 17.5). In other words, only where the mask byte contains "ones" is the original bit of the reference byte copied to the output.

If two switches u_1 and u_2 are connected in parallel, the operation is a *Boolean addition* and the function is of the *OR* type. Here, $y = 1$ if either u_1 or u_2 is actuated; otherwise $y = 0$. The logic is denoted (Figure 17.6) by

$$y = u_1 + u_2.$$

As for the *AND* gate, more switches can be added (in parallel), giving $y = u_1 + u_2 + u_3 \ldots$. The ">1" designation inside the *OR* symbol means that gate output is "high" if the number of "high" input signals is equal to or greater than 1.

A logical *OR* between two bytes also makes a bit-by-bit logical operation (Figure 17.7). The *OR* operation can be used to set one or several bits unconditionally to 1.

There are some important theorems for one binary variable x, such as

$$x + x = x,$$
$$x \cdot x = x,$$
$$x + \bar{x} = 1,$$
$$x \cdot \bar{x} = 0.$$

Similarly, for two variables we can formulate and easily verify

$$x + y = y + x,$$
$$x \cdot y = y \cdot x,$$
$$x + xy = x,$$

Input register	1 1 0 1 1 0 0 0
mask	0 1 1 0 1 1 0 1
Output	1 1 1 1 1 1 0 1

FIGURE 17.7 Masking two bytes with an *OR* operation.

$$x \cdot (x + y) = x,$$
$$(x + \bar{y}) \cdot y = x \cdot y,$$
$$x \cdot \bar{y} + y = x + y,$$
$$xy + \bar{y} = x + \bar{y}.$$

The *De Morgan* theorems are useful in manipulating Boolean expressions:

$$\overline{(x + y + z + \cdots)} = \bar{x} \cdot \bar{y} \cdot \bar{z} \cdots$$
$$\overline{(x \cdot y \cdot z \cdots)} = \bar{x} + \bar{y} + \bar{z} \cdots$$

The theorems can help in simplifying complex binary expressions, thus saving components for the actual implementation.

17.4.3 Additional Gates

Two n.c. gates in series may define a *NOR* gate, that is, the system conducts if *neither* the first *nor* the second switch is actuated. According to De Morgan's theorem, this can be expressed as

$$y = \overline{u_1} \cdot \overline{u_2} = \overline{(u_1 + u_2)},$$

showing that the *NOR* gate can be constructed from the combination of a *NOT* and an *OR* gate (Figure 17.8). The circle at an input or output line of the symbol represents Boolean inversion.

A *NOR* gate is easily implemented electronically or pneumatically. Moreover, any Boolean function can be obtained only from a *NOR* gate, which makes it a *universal gate*. For example, a *NOT* gate is a *NOR* gate with a single input. An *OR* gate can be obtained by connecting a *NOR* gate and a *NOT* gate in series. An *AND* gate, obtained by using two *NOT* gates and one *NOR* gate (Figure 17.9), is written as

$$y = \overline{\overline{u_1} + \overline{u_2}} = \overline{\overline{u_1}} \cdot \overline{\overline{u_2}} = u_1 \cdot u_2.$$

A *NAND* gate is defined by

$$y = \overline{u_1 \cdot u_2} = \overline{u_1} + \overline{u_2}.$$

The system does *not* conduct if both u_1 *and* u_2 are actuated, that is, it conducts if either switch is not actuated. Like the *NOR* gate, the *NAND* gate is a universal gate (Figure 17.10).

The *NAND* and *NOR* operations are called *complete operations*, because all others can be derived from either of them. No other gate or operation has the same property.

A circuit with two switches, each having double contacts (one n.o. and the other n.c.), is shown in Figure 17.11. This is an *exclusive OR (XOR)* circuit, and the output is defined by

$$y = u_1 \cdot \overline{u_2} + \overline{u_1} \cdot u_2.$$

The circuit conducts *only* if *either* $u_1 = 1$ or $u_2 = 1$, but $y = 0$ if *both* u_1 *and* u_2 have the same sign (compare with the *OR* gate). For example, such a switch can be used to control the room light from

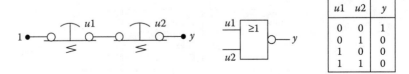

u1	u2	y
0	0	1
0	1	0
1	0	0
1	1	0

FIGURE 17.8 A *NOR* gate, its ISO symbol, and its truth table.

$$y = \overline{\overline{u1} + \overline{u2}} = \overline{\overline{u1}} \bullet \overline{\overline{u2}} = u1 \bullet u2$$

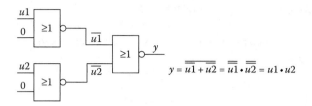

FIGURE 17.9 Three *NOR* gates acting as an *AND* gate (this is not the minimal realization of an *AND* gate).

u1	u2	y
0	0	1
0	1	1
1	0	1
1	1	0

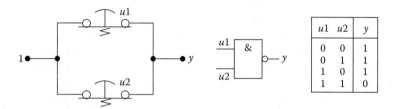

FIGURE 17.10 A *NAND* gate, its ISO symbol, and its truth table.

two different switch locations u_1 and u_2. In digital computers, *XOR* circuits are extremely important for binary addition.

An *exclusive OR (XOR)* between two bytes will copy the 1 in the input register only where the mask contains 0. Where the mask contains 1, the bits of the first operand are inverted. In other words, in the positions where the operands are equal, the result is 0 and, conversely, where the operands are not equal, the result is 1 (Figure 17.12). This is often used to determine if and how a value of an input port has been changed between two readings.

u1	u2	y
0	0	0
0	1	1
1	0	1
1	1	0

$$y = \overline{u1} \bullet u2 + u1 \bullet \overline{u2}$$

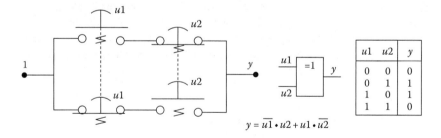

FIGURE 17.11 An *exclusive-OR* gate, its ISO symbol, and its truth table.

Input register mask	1 1 0 1 1 0 0 0 0 1 1 0 1 1 0 1
Output	1 0 1 1 0 1 0 1

FIGURE 17.12 Masking two bytes with an *XOR* operation.

Example 17.1: Simple Combinatorial Network

A simple example of a combinatorial circuit expressed in ISO symbols is shown in Figure 17.13. The logical expressions are

$$y_3 = u_1 \cdot \overline{u_{12}},$$
$$y_4 = u_2 \cdot y_2,$$
$$y_2 = y_4 + \overline{u_1}.$$

The ISO organization that deals specifically with questions concerning electrotechnical and electronic standards is the International Electrotechnical Commission (IEC). Standards other than IEC are often used to symbolize switching elements. The IEC symbols are not universally accepted and in the United States there are at least three different sets of symbols. In Europe, the DIN (the German standardization organization) standard is common. Three common standards are shown in Figure 17.14.

In principle, all switching networks can be tested by truth tables. Unfortunately, the number of Boolean functions grows rapidly with the number of variables n, because the number of combinations becomes 2^n. It is outside the scope of this text to discuss different simplifications of Boolean functions. A method known as the Karnaugh map may be used if the number of variables is small. For systems with many variables (more than about 10), there are numerical methods to handle the switching network. The method by Quine–McCluskey may be the best known, and is described in standard textbooks on switching theory (e.g., [4,5]).

17.4.4 Flip-Flops

Hitherto we have described *combinatorial networks*, that is, the gate output Y depends only on the *present* combination of input signals. In other words, the gates have *no memory*, so the network is a *static* system. In a sequencing network instead it is possible to store signal values and states and to use them later in the course of another operation. The memory function can be realized with *flip-flop* elements or bistable switches. The flip-flop has two stable output states (hence the term "bistable") that depend not only on the present state of the inputs but also on the previous state of the flip-flop output.

The basic type of flip-flop is the Set-Reset (SR) flip-flop. The two inputs S and R can be either 1 or 0. Both are, however, not permitted to be 1 or 0 at the same time. The output is called y and normally $\bar{y}$ also is an output. If $S = 1$, then $y = 1$ ($\bar{y} = 0$) and the flip-flop becomes *set*. If S returns to 0, then the gate remembers that S had been 1 and keeps $y = 1$. If R becomes 1 (assuming that $S = 0$), the flip-flop is *reset*, and $y = 0$ ($\bar{y} = 1$). Again R can return to 0, and y remains 0 until a new S signal appears. Let us call the states at consecutive moments y_n and y_{n+1}. Then the operation can be written as

$$y_{n+1} = \overline{R} \cdot (S + y_n).$$

An *SR* flip-flop can be illustrated by two logical elements (Figure 17.15).

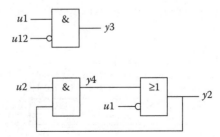

FIGURE 17.13 Simple combinatorial circuit.

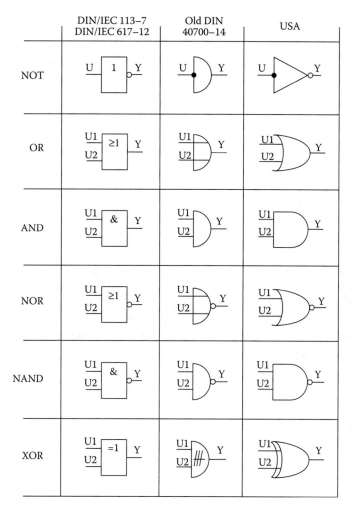

FIGURE 17.14   Commonly used logical gate symbols.

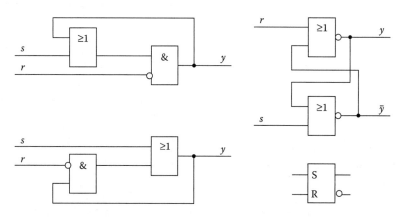

FIGURE 17.15   Three different illustrations of a flip-flop gate and its DIN/IEC symbol.

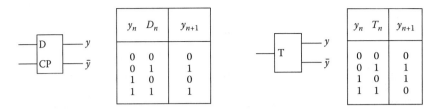

FIGURE 17.16   A delay (*D*) flip-flop and a trigger (*T*) flip-flop.

By adding two *AND* gates and a clock-pulse (CP) input to the flip-flop, we obtain a delay (*D*) flip-flop or a *latch*. The delay flip-flop has two inputs, a data (*D*) and a *CP* (Figure 17.16). Whenever a CP appears, the output *y* accepts the *D* input value that existed before the appearance of the CP. In other words, the *D* input is delayed by one CP. The new state y_{n+1} is always independent of the old state.

By introducing a feedback from the output to the flip-flop input and a time delay in the circuit, we obtain a *trigger* or a *toggle* (T) flip-flop. The T flip-flop (Figure 17.16) is often used in counting and timing circuits as a "frequency divider" or a "scale-of-two" gate. It has only one input, *T*. Whenever an upgoing pulse appears in *T*, the output *y* flips to the other state.

All three types of flip-flops can be realized in a *JK (Master–Slave) flip-flop*, with *J* being the set signal and *K* the reset signal. It frequently comes with a CP input. Depending on the input signals, the *JK* flip-flop can be an *SR* flip-flop, a latch, or a trigger.

17.4.5 Realization of Switching

Electronic logic gates, for example, of *AND* and *OR* types can be implemented in a straightforward way with diodes. The cascade connection of several diode gates in series brings, however, about problems, among others, because the signals are strongly attenuated at each step (diodes are passive elements), so that this solution is not particularly attractive. A common way to implement gate circuits to avoid this problem is by using transistor logic, since the output signals at each step are amplified back to the full logical level.

Today gates realized on integrated circuits (ICs), also known as "chips" are mostly used. There are several types of ICs for the realization of logical operations; each type is characterized by particular power consumption and speed. Conventional, simple TTL circuits have been used for long time but are now being replaced by other product families.

The low-power Schottky TTL (LS-TTL) elements contain so-called Schottky diodes, which in comparison with conventional diodes are faster and in addition use considerably less power than the older TTL types. Largely used are also the complementary metal-oxide semiconductor (CMOS) ICs that are based on field-effect transistors (FETs) rather than on bipolar transistors. The power consumption of a CMOS circuit is about three orders of magnitude less than for a corresponding TTL element. In addition, the CMOS circuits are less sensitive to electrical noise and variations in the supply voltage. On the other hand, CMOS circuits are more sensitive for static electricity and are also slower than corresponding TTL circuits. A solution will be probably represented by a new generation of CMOS circuits, the high-speed CMOS (HC) Logic.

Complex circuits can also be manufactured as medium-scale (MSI) or large-scale (LSI) ICs; this type of production is, however, economically justifiable only for large quantities (i.e., a minimum of some thousands of components). An alternative is the use of so-called programmable logic devices (PLDs) that allow the inexpensive production of semicustomized ICs. PLDs mostly belong to the LS-TTL family. They contain a large array of gates that are interconnected by microscopic fuses. By using special programming equipment, these fuses can be selectively blown, so that the result is an IC with the desired logical properties.

There are several types of ICs in the PLD family: the programmable array logic (PAL), the field-programmable logic arrays (FPLA), and the programmable read-only memory (PROM). A PAL circuit is built with a programmable AND-gate array, where the AND gates are connected to an OR gate. An FPLA circuit has a similar structure, with the difference that both the AND and the OR gates are programmable. Special PAL and FPLA chips with other gates such as NOR, XOR, and D-flip-flops are available, so that a complete sequencing control system can be realized by the user with one or few chips (Figure 17.17).

The programming of PLDs is made easier by using software packages that are also available for personal computers. These programs convert the description of a control sequence in the form of Boolean relations into the data for the programming unit. Also, the testing of the programmed chips is usually carried out by this software.

The function of a PLD circuit can be freely defined by the user. The basic structure of the PLD consists of an AND and of an OR matrix, programming takes place by "burning" fast connections in the AND and in the OR matrices.

In a PAL circuit, the AND matrix is programmable, while the connections between the AND and the OR gates are fixed. In a PROM, the AND-matrix is fixed and the OR matrix is programmable. In this case, for example, each combination of the input bits (the "address") leads to the activation of a single AND gate, and the programmable state of the cell in the OR matrix reflects the stored logical value. In an FPLA, the AND as well as the OR matrix can be freely programmed.

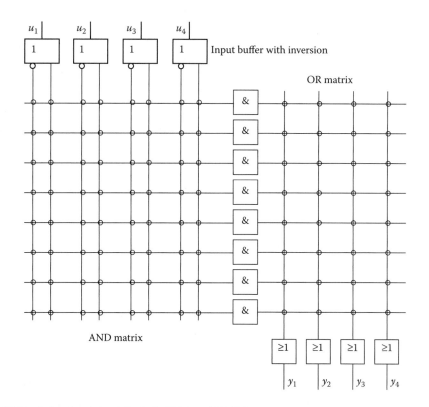

FIGURE 17.17 A functional circuit for PAL, FPLA, and PROM circuits.

17.5 Ladder Diagrams

The implementation of Boolean expressions can be programmed in LDs, which now make up a part of the international standard IEC 61131-3. An LD consists of graphic symbols, representing logic expressions and contacts and coils, representing outputs.

Many switches are produced from solid-state gates, but electromechanical relays are still used in many applications. Statistics show that the share of electromechanical relays versus the total number of gates in use is decreasing. This does not mean that their importance is dwindling; relays remain, in fact, a necessary interface between the control electronics and the powered devices.

Relay circuits are usually drawn in the form of *LDs*. Even if the relays are replaced by solid-state switches or programmable logic, they are still quite popular for describing combinatorial circuits or sequencing networks. They are also a basis for writing programs for programmable controllers.

17.5.1 Basic Description

An LD reflects a conventional wiring diagram (Figure 17.18). A wiring diagram shows the physical arrangement of the various components (switches, relays, motors, etc.) and their interconnections, and is used by electricians to do the actual wiring of a control panel. The LDs are more schematic and show each branch of the control circuit on a separate horizontal row (the rungs of the ladder). They emphasize the function of each branch and the resulting sequence of operations. The base of the diagram shows two vertical lines, one connected to a voltage source and the other to ground.

Relay contacts are either n.o. or n.c., where *normally* refers to the state in which the coil is not energized. Relays can implement elementary circuits such as *AND* and *OR* as well as *SR* flip-flops. The relay symbols are shown in Figure 17.19.

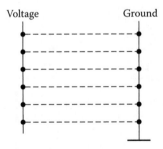

FIGURE 17.18 Framework of an LD.

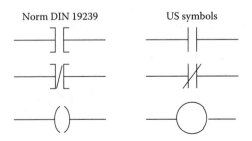

FIGURE 17.19 Relay symbols for n.o. and n.c. contacts, and the relay coil.

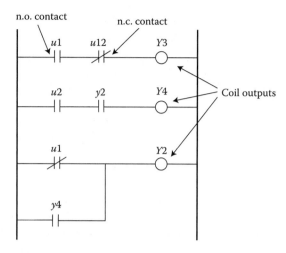

FIGURE 17.20   The combinatorial circuit in Figure 17.13 represented by an LD.

Example 17.2: Combinatorial Circuit

The combinatorial circuit of Figure 17.13 can be represented by an LD (Figure 17.20). All of the conditions have to be satisfied simultaneously. The series connection is a logical *AND* and the parallel connection a logical *OR*. The lower-case characters (u, y) denote the switches and the capital symbols (Y; the ring symbol) denote the coil.

The relay contacts usually have negligible resistance, whether they are limit switches, pressure, or temperature switches. The output element (the ring) could be any resistive load (relay coil) or a lamp, motor, or any other electrical device that can be actuated. Each rung of the LD must contain at least one output element, otherwise a short circuit would occur.

Example 17.3: A Flip-Flop as an LD

A flip-flop (Figure 17.15) can also be described by an LD (Figure 17.21). When a set signal is given, the *S* relay conducts a current that reaches the relay coil *Y*. Note that the *R* is not touched. Energizing the relay coil closes the relay contact *y* in line 2. The *set* push button can now be released and current continues to flow to coil *Y* through the contact *y*, that is, the flip-flop remains *set*. Thus, the *y* contact provides the 'memory' of the flip-flop. In industrial terminology, the relay is a *self-holding* or *latched* relay. At the moment the *reset* push button is pressed, the circuit to *Y* is broken and the flip-flop returns to its former *reset* state.

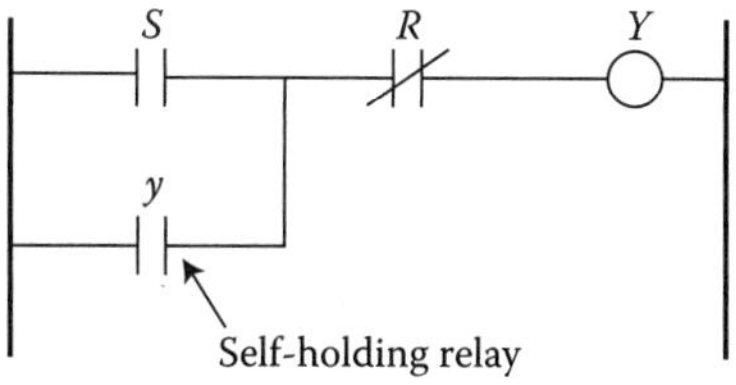

FIGURE 17.21   An SR flip-flop described by an LD.

17.5.2 Sequencing Circuits

In a combinatorial network the outputs depend only on the momentary values of the inputs. In a sequence chart, however, the outputs also depend on earlier inputs and states of the system. The related graphical representation of the operation must therefore contain signals and states at different times. Many sequence operations can be described by LDs and can be defined by a number of states, where each state is associated with a certain control action.

Note that only one state at a time can be active. In the LD this will correspond to the fact that only one rung (step) at a time can be executed. Therefore some kind of execution control signal is necessary in order to transfer from one state to another. This type of control signal can be given when a condition is satisfied (the condition could of course also be a complex combination of control signals). The conditional order acts at the same time as *reset* (R) signal for one step and as *set* signal for the following step (compare with Figure 17.21). The sequencing control execution can be described as a series of SR flip-flops, where each step (= state) corresponds to a rung of the ladder (Figure 17.22). At each execution control signal, the next flip-flop is set. The execution proceeds one step at a time and after the last step it returns to the beginning of the sequence (Step 1).

In practical execution, Step 1 can be initiated with a *start* button. When running in an infinite loop, it can also be started from the *last step*. When the *last step* is active together with a new condition for the startup of *Step 1*, then the *Step 1* coil is activated, and the self-holding relay keeps it set. When the condition for *Step 2* is satisfied, the relay *Step 2* latches circuit 2 and at the same time guarantees that circuit 1 is broken. This is then continued in the same fashion. In order to insure a repetitive sequence, the last step has to be connected to *Step 1* again.

This is an example of an *asynchronous* execution. In switching theory there are also *synchronous* charts, where the state changes are caused by a CP. In industrial automation applications, we mostly talk about asynchronous charts, because the state changes do depend not on CPs but on several conditions in different parts of the sequence. In other words, an asynchronous system is *event based*, while a synchronous system is *time based*. Moreover, we are dealing with design of asynchronous systems with *sustained input signals* rather than pulse inputs.

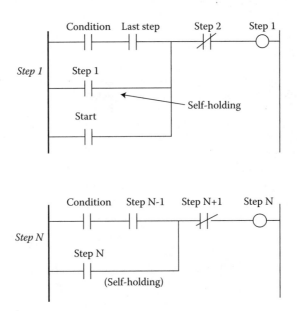

FIGURE 17.22 A sequence described by an LD.

17.6 Programmable Controllers

PLCs are particular microcomputers designed to carry out Boolean switching operations in industrial environments. The name is actually a misnomer, because PLCs can today perform much more than simple logic operations. However, the abbreviation has been retained in order to avoid confusion between the more general term Programmable Controller and Personal Computer (both PC). A PLC generates on/off signal outputs for the control of actuators like electric motors, valves, lights, etc. that can be found in all industrial branches as vital parts of automation equipment.

The *PLC* was initially developed by a group of engineers from General Motors in 1968, where the initial specification was formulated: it had to be easily programmed and reprogrammed, preferably in-plant, easily maintained and repaired, smaller than its relay equivalent, and cost-competitive with the solid-state and relay panels then in use. This provoked great interest from engineers of all disciplines using the PLC for industrial control. A microprocessor-based PLC was introduced in 1977 by Allen-Bradley Corporation in the United States, using an Intel 8080 microprocessor with circuitry to handle bit logic instructions at high speed.

The early PLCs were designed only for logic-based sequencing jobs (on/off signals). Today there are hundreds of different PLC models on the market. They differ in their memory size and I/O capacity. The difference also lies in the features they offer. The smallest PLCs serve just as relay replacers with added timer and counter capabilities. Many modern PLCs also accept proportional signals. They can perform arithmetic calculations and handle analog input and output signals and PID controllers. This is the reason why the letter *L* was dropped from PLC, but the term PC may cause confusion with personal computers so we keep the *L* here.

The logical decisions and calculations may be simple in detail, but the decision chains in large plants are very complex. This naturally raises the demand for structuring the problem and its implementation. Sequencing networks operate asynchronously, that is, the execution is not directly controlled by a clock. The chain of execution may branch for different conditions, and concurrent operations are common. This makes it crucial to structure the programming and the programming languages and an international software standard IEC 61131-3 has been defined. Applications of function charts in industrial control problems are becoming more and more common. Graphical tools for programming and operator communication are also becoming a standard of many commercial systems. Furthermore, in a large automation system, communication between the computers becomes crucial for the whole operation.

17.6.1 Basic Structure

The basic operation of a PLC corresponds to a software-based equivalent of a relay panel. However, a PLC can also execute other operations, such as counting, delays, and timers. Because a PLC can be programmed in easy-to-learn languages, it is naturally more flexible than any hardware relay system. PLCs are more flexible than programmable logical devices but usually slower, so that PLDs and PLCs often coexist in industrial installations offering the best and most economical solutions.

Figure 17.23 shows the basic structure of a PLC. The input signals are read into a buffer, the input memory register. This function is already included in the system software in the PLC. An input–output register could consist of only a bit but is often a full byte. Consequently, one input instruction gives the status of eight different input ports.

The instructions fetch the value from the input register and operate on only this or on several operands. The central processing unit (CPU) works toward a result register or accumulator (A). The result of an instruction is stored either in some intermediate register or directly in the output memory register that is written to the outputs. The output function is usually included in the system programs in a PLC.

A PLC is specifically made to fit an industrial environment where it is exposed to hostile conditions, such as heat, humidity, unreliable power, mechanical shocks, and vibrations. A PLC also comes with input–output modules for different voltages and can be easily interfaced to hundreds of input and output

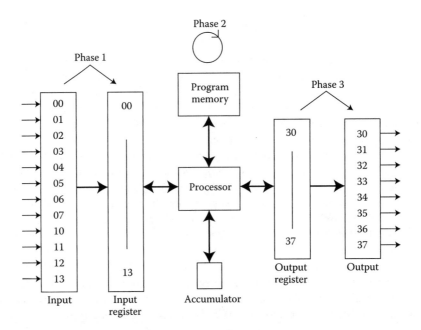

FIGURE 17.23   Basic structure of a PLC.

lines. PLCs have both hardware and software features that make them attractive as controllers of a wide range of industrial equipment. They are specially built computers with three functions, memory, processing, and input/output.

17.6.2 Basic Instructions and Execution

To make a PLC system for industrial automation, it has to work in real time. Consequently, the controller has to act on external events very quickly, with a short response time. There are two principal ways to sense the external signals, by *polling* (repeated requests) the input signals regularly or by using *interrupt* signals. The polling method's drawback is that some external event may be missed if the processor is not sufficiently fast. On the other hand, such a system is simple to program.

A system with interrupts is more difficult to program but the risk of missing some external event is much smaller. The polling method is usually used in simpler automation systems while interrupts are used in more complex control systems.

"Programming" a PLC consists mainly of defining sequences. The input and output functions are already prepared. The instructions from an LD, a logical gate diagram, or Boolean expressions are translated into machine code. At the execution, the program memory is run through cyclically in an infinite loop. In this way, it is simulating the parallelism inherent in the wired relay logic. The *read-execute-write* cycle is called a scan cycle. Every scan may take some 15–30 ms in a small PLC, and the scanning time is approximately proportional to the memory size. In some PLCs the entire memory is always scanned even if the code is shorter. In other systems the execution stops at an *end* statement that concludes the code; thus the loop time can be shortened for short programs.

The response time of the PLC of course depends on the processing time of the code. While the instructions and the output executions are executed, the computer system cannot read any new input signals. Usually this is not a big problem, since most signals in industrial automation are quite slow or last for a relatively long time.

The LD can be considered as if every rung were executed at the same time. Thus it is not possible to visualize that the LD is executed sequentially on a row-by-row basis. The execution has to be very fast compared to the timescale of the process.

A small number of basic machine instructions can solve most sequencing problems. A program that contains these instructions is called *instruction list* (IL). Some of the fundamental machine instructions are listed here; usually they can operate on bits as well as on bytes:

ld, ldi A number from the computer input memory is loaded (LD) or inverted (LDI) before it is read into the accumulator (A).

and, ani An *AND* or *AND Inverse* instruction executes an *AND* logical operation between A and an input channel, and stores the result in A.

or, ori An *OR* or *OR Inverse* instruction executes an *OR* logical operation between A and an input channel, and stores the result in A.

out The instruction outputs A to the output memory register. The value remains in A, so that the same value can be sent to several output relays.

Example 17.4: Translation from an LD to Machine Code

The translation from the LD to machine code is illustrated in Figure 17.24. The gate *y*11 gives a self-holding capability. Note that *y*11 is a memory cell, and *Y*11 is an output.

A logical sequence or LD is often branched. Then there is a need to store intermediate signals for later use. This can be done with special help relays, but in a *PLC* it is better to use two instructions *orb* (*OR* Block) or *anb* (*AND* Block). They use a memory stack area (last in, first out) in the PLC to store the output temporarily.

Example 17.5: Using the Block Instruction and Stack Memory

The LD (Figure 17.25) can be coded with the following machine code:

ld x1 Channel 1 is read into the accumulator (A).
and x2 The result of the *AND* operation is stored in A.
ld x3 The content of A is stored on the stack. Channel 3 is read into A.
and x4 The result of lines 3 and 4 is stored in A.
orb An *OR* operation between A and the stack. The result is stored in A. The last element of the stack is eliminated.
out Y1 Output of A on Channel 1.

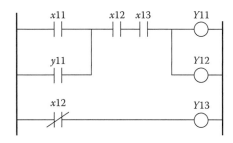

FIGURE 17.24   Translation of an LD into a machine code.

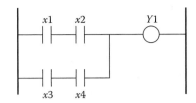

FIGURE 17.25   Example of using a stack memory.

Example 17.6: Using the Block Instructions and the Stack Memory

The logical gates in Figure 17.26 are translated to machine code by using block instructions. The corresponding machine code is as follows:

ld x1 Load Channel 1.
and x2 The result is stored in *A*.
ld x3 The content of *A* is stored on the stack. Status of Channel 3 is loaded into *A*.
and x4 The result of lines 3 and 4 is stored in *A*.
ld x5 The content of *A* is stored on the stack. Status of Channel 5 is loaded into *A*.
and x6 The result of lines 5 and 6 is stored in *A*.
orb Operates on the last element in the stack (the result of lines 3 and 4) and the content of *A*. The result is stored in *A*. The last element of the stack is removed.
anb Operates on the last element in the stack (the result of lines 1 and 2) and the content of *A*. The result is stored in *A*. The last element of the stack is removed.
out Y1.

17.6.3 Additional PLC Instructions

For logical circuits there are also operations such as *XOR, NAND,* and *NOR* as described earlier. Modern PLC systems are supplied with instructions for alphanumerical or text handling and communication as well as composed functions such as timers, counters, memory, and pulses.

A pulse instruction (PLS) gives a short pulse, for example, to reset a counter. A PLC may also contain delay gates or time channels so that a signal in an output register may be delayed for a certain time. Special counting channels can count numbers of pulses.

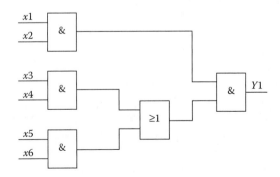

FIGURE 17.26   Example of a logical circuit.

Different signals can be shaped, such as different combinations of delays, oscillators, rectangular pulses, ramp functions, shift registers, or flip-flops. As already mentioned, advanced PLCs also contain floating-point calculations as well as prepared functions for signal filtering and feedback control algorithms.

17.7 PLC Programming and Software Standards

PLCs can be programmed in different ways: with the assembler-like IL or in higher, problem-oriented Structured Text (ST). We have demonstrated that both combinatorial networks and sequences can be described using LDs. The LD has been particularly popular in the United States, while in Europe the use of function block diagrams (FBDs) with the graphical symbols for logical gates is more common. The high-level description of sequencing functions using the Grafcet-like SFC is naturally gaining in popularity. This will be described in the next section.

PLCs are usually programmed via external units. These units as a rule are not needed for the PLC online operation and may be removed when the PLC is in operation. Programming units are typically small hand-held portable units or portable personal computers. A manual PLC programmer looks like a large pocket calculator, with a certain number of keys and a simple display. Each logic element of the LD or program instruction is entered with specific keys or key combinations.

More sophisticated programming can be performed with a PC, offering both graphical and text editors. The editor typically shows several LD lines at a time or the basic structure of an SFC. To make debugging and testing simpler the computer can also indicate the power flow within each line during operation, so that the effect of the input over the output is immediately recognizable. In some cases programming can take place by drawing on the display an FBD with logical gates. The gate symbols are input with key combinations and/or with the mouse, by choosing from a predefined table.

The international standard IEC 61131-3 (earlier called IEC 65A (SEC) 67) is the only global standard for industrial control programming. It harmonizes the way people design and operate industrial controls by standardizing the programming interface. A standard programming interface allows people with different backgrounds and skills to create different elements of a program during different stages of the software lifecycle: specification, design, implementation, testing, installation, and maintenance. Yet all pieces adhere to a common structure and work together harmoniously.

IEC 61131-3 includes the definition of the SFC language, used to structure the internal organization of a program, and four interoperable programming languages: IL, LD, FBD, and ST. ST has a formal syntax similar to that of the programming language Pascal, as shown in this short example:

$$IF \ TEMP1 > 50.0 \ THEN$$
$$Flow\_rate := 65.0 + OFFSET$$
$$ELSE$$
$$Flow\_rate := 75.0; PUMP := ON;$$
$$END \ IF;$$

ST supports a wide range of standard functions and operators. ST and IL represent algorithmic formulations in clear text. The FBD, the LD, and the SFC are instead graphical representations of the function and the structure of logical circuits. The international standard IEC 61131-3 should therefore guarantee a wide application spectrum for PLC programming.

17.7.1 Sequential Function Charts

The need for structuring a sequencing process problem may not be immediately apparent for small applications, but as the complexity of the control action increases, also the need for better functional descriptions becomes more important. We have seen already in the simple transfer line example that structuring becomes necessary. As a tool for an appropriate *top-down* analysis and the representation of control sequences, function charts have been introduced. Today, function charts are offered as programming tools by several PLC producers, such as Grafcet, GRAPH-5 (Siemens), HI-FLOW (Hitachi), and others. The basic ideas at the basis of these languages are similar and the differences are only of secondary importance. On the surface we may say that the function charts are implementations of Petri nets.

17.7.2 Describing States and Transitions Using SFCs

An SFC can be considered as a special purpose language for the description of control sequences in the form of a graphical scheme. Toward the late 1970s the first function chart language, Grafcet (GRAphe de Commande Etape-Transition "Function chart—step transition"), was developed in France and has later provided the basis for the definition of the international standard IEC 848 (Preparation of function charts for control systems). Now there is an international standard for the control of sequences, IEC 61131-3. This standard lists a number of alternative languages. Of these the SFC is the most important. It may be noted that the IEC 61131-3 standard does not really consider an SFC to be a programming language, but rather a program-structuring element used to organize the program written in one or more of the other languages. Here we will consider the SFC a programming language. The SFCs have evolved through Grafcet from safe Petri nets.

Function charts describe control sequences with the help of predefined rules for

- The controls that must be carried out and in the order in which they are carried out.
- The execution details of each instruction.

The function diagram is correspondingly divided into two parts (Figure 17.27). The "sequence" part describes the order between the major control steps (left part of Figure 17.27). It consists of the states (marked by the five boxes to the left), also called *steps* in the SFC.

The vertical lines that connect each box with the following one represent active connections (directed links). Each transition from a step to the following one is connected with a logical condition, the transition condition or receptivity. The Boolean expression for the transition condition is written in proximity of a small horizontal line that intersects the link from one box to the next. When the logical condition is satisfied, that is, the related Boolean expression is true, the transition takes place, and the system proceeds with the following step. The actions taking place in each state (step) are described by the "object" or "control" part of the diagram. This part consists of the boxes to the right of the sequence steps. Every action has to be connected to a step and can be described either by an LD, a logical circuit, a Boolean expression, or even a continuous control action such as a PID controller.

The use of function charts will be illustrated with the example of control of the batch tank in Figure 17.27. The states can now be recognized as boxes 1–5, while the state transitions are the marked signals between the states.

In the function charts syntax, a step (=state) at any given time can be either active or inactive. "Active" means that this step is currently being executed. The initial step is represented in the function chart by a double-framed box. An "action" is a description of the commands that have to be executed at each step. A logical condition can be associated with a step, so that the related commands are executed only when the step is active and the condition is fulfilled. The association with a condition represents, therefore, a security control. Several commands can be associated with a step. These commands can be simple controls but also represent more complex functions such as timer, counters, regulators, filtering procedures, or commands for the external communication. As we have already seen, in the function chart there is also a function for transition, that is, a kind of obstacle between two steps to which only an active step can follow. After a transition, a new step becomes active and the earlier one inactive. The transition is controlled by a logical condition and takes place only when the condition is satisfied.

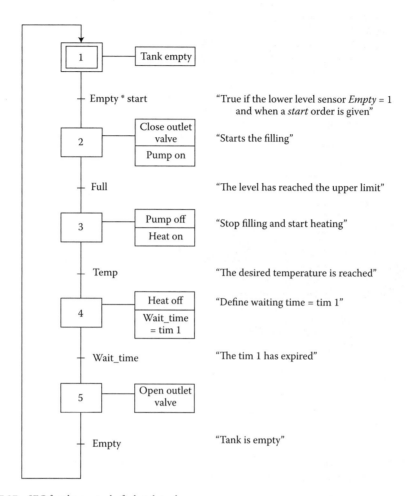

FIGURE 17.27   SFC for the control of a batch tank process.

The function chart syntax allows much more than just the iterative execution of the same control instructions. The three functional blocks initial step, step(s), and transitions can be interconnected in many different ways, thus allowing the description of a large number of complex functions. Three types of combinations are possible—in analogy with Petri nets:

- Simple sequences
- Execution branching (alternative parallel sequence)
- Execution splitting (simultaneous parallel sequence)

In the simple sequence, there is only one transition after a step and only one step after a transition. No branching takes place. In the alternative parallel sequence (Figure 17.28), there are two or more transitions after one step. In this way, the execution flow can take alternative routes depending on external conditions. Often this is an *if-then-else* condition and is useful to describe, for example, alarm situations.

In the alternative parallel sequence, it is very important to verify that the condition for the selection of one of the program execution branches is consistent and unambiguous; in other words, the alternative branches should not be allowed to start simultaneously. Each branch of an alternative parallel sequence must always start and end with logical conditions for a transition.

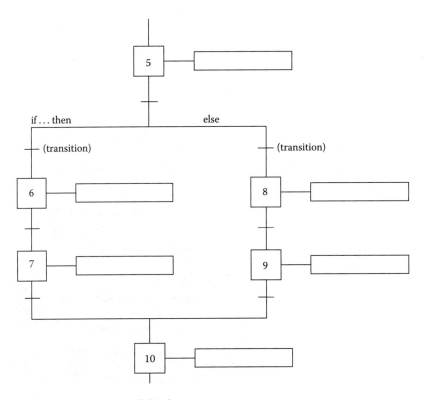

FIGURE 17.28   SFC for alternative parallel paths.

In the simultaneous parallel sequence (Figure 17.29) two or more steps are foreseen after a transition, and these steps can be simultaneously active. The simultaneous parallel sequence represents therefore the concurrent (parallel) execution of several actions.

The double horizontal lines indicate the parallel processing. When the condition for the transition is satisfied, both branches become simultaneously active and are executed separately and concurrently. The transition to the step below the lower double horizontal line can take place only after the execution of all concurrent processes has been terminated. This corresponds to the simultaneous execution of control instructions and is comparable with the notation *cobegin-coend*, used in real-time programming.

The three types of sequence processing can be also used together. However, one should act carefully in order to avoid potential conflicts. For example, if two branches of an alternative execution sequence are terminated with the graphic symbol for the end of parallel execution (the double horizontal bars), then further execution is locked, since the computer waits for both branches to terminate their execution, while only one branch was started because of the alternative condition. Also the opposite error is possible. If parallel branches that have to be executed simultaneously are terminated with an alternative ending (a single horizontal bar), then many different steps may remain active, so that further process execution might no longer take place in a controlled way.

Of course a compiler would recognize such a mismatch of beginning and end clauses and would thus alarm the user before the code is executed. But even with the best compiler around, many errors remain tricky and undetectable. A structured and methodical approach on the part of the programmer is always an important requirement.

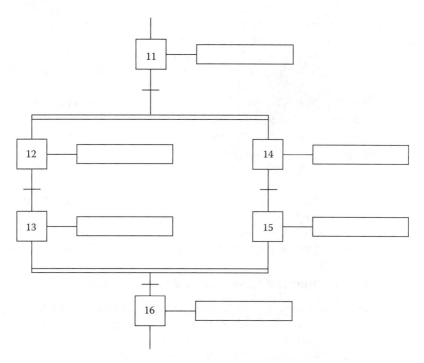

FIGURE 17.29   SFC for simultaneous parallel paths.

17.7.3 Computer Implementation of SFCs

Programs written with the help of functions charts operate under real-time conditions; so each implementation must exhibit real-time capabilities. Usually, the realization of real-time systems requires intensive efforts with considerable investments in time and personnel.

However, in this specific case, the designer of the function chart language compiler carries most of the burden, while the user can describe complex control sequences in a comparatively simple way. The aspects of real-time programming are also valid for the design of PLCs, but concern the final user only indirectly and in a limited way.

Compilers for function charts are available for many different industrial control computers. The programming and program compilation on PCs is commonplace. After compilation, the code in the form of control instructions is transferred to a PLC for execution. The PC is then no longer necessary during the real-time PLC operation. Some compilers can also operate as simulation tools and show the execution flow on the computer screen without needing to be connected to the object PLC. There are also PLCs with the compiler already built into their software.

The obvious advantage of abstract descriptions in the form of function charts is their independence from any specific hardware and their orientation to the task to be performed rather than to the computer. Unfortunately, it must be said that high-level languages like function charts do not yet enjoy the success they deserve. It seems odd that so many programmers always start anew with programming in low-level languages, even for those applications that would be much easier to solve with function chart description languages.

As in any complex system description, the diagram or the code has to be structured suitably. A function chart implementation should allow the division of the code into smaller parts. For example, each machine of a complex line to be controlled may have its own graph, and the graphs for several machines could then be assembled together. Such hierarchical structuring is of fundamental importance when programming the operation of large, complex systems.

Function charts are not only suitable for complex operations, but can also be useful for simpler tasks. A function chart is quite easy for the nonspecialist to understand. An accepted standard for the description of automated operations has also the advantage that more computer codes can be maintained and reutilized and do not need to be written anew each time, as it would be the case with incompatibles devices and languages.

The translation of function charts to computer code depends on the specific PLC and its tools, as not all devices have such compilers. Still, even if the function charts cannot be transformed in programming code, the diagrams are very useful, since they provide the user with a tool to analyze and structure the problem. Some companies use function charts to describe the function and use of their equipment. Of course, it would be much simpler if function charts would be used all the way from functional description to actual programming.

The importance of structuring is obvious. The sequencing operations of an industrial operation could have been written in machine code or as an LD. However, long codes in low-level languages are not meant for people to read, understand, debug, or maintain. In a high-level language like SFC, the code itself is a good documentation. Finally, remember that *comments* are crucial parts of the documentation!

17.7.4 Application of Function Charts in Industrial Control

The use of function charts for sequential programming is demonstrated for a manufacturing cell in a flexible manufacturing system. The cell consists of three NC machines (e.g., a drill, lathe, and mill), a robot for material handling, and a buffer storage (Figure 17.30).

At the cell level we do not deal with the individual control loops of the machines or of the robot. They are handled by separate systems. The cell computer sends on/off commands to the machines and its main tasks are to control the individual sequencing of each machine (and the robot) and to synchronize the operations between the machines and the robot. The control task is a mixture of sequencing control and real-time synchronization. We will demonstrate how an SFC expresses the operations. The implementation of the function chart is then left to the compiler.

The manufactured product has to be handled in the three machines in a predefined order (like a transfer line). The robot delivers new parts to each machine and moves them between the machines.

17.7.4.1 Synchronization of Tasks

The synchronization of the different machines is done by a *scheduler* graph with a structure indicated in Figure 17.31.

The scheduler communicates with each machine and with the robot and determines when they can start or when the robot can be used. It works like a scheduler in a real-time operating system, distributing

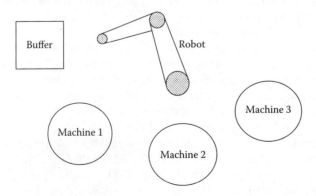

FIGURE 17.30 Layout of the manufacturing cell.

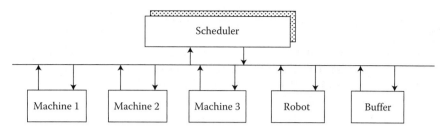

FIGURE 17.31 Logical structure of the machine cell.

the common resource, the robot, as efficiently as possible. The scheduler has to guarantee that the robot does not cause any deadlock. If the robot has picked up a finished part from a machine and has nowhere to place it, then the system will stop. Consequently, the scheduler has to match demands from the machines with the available resources (robot and buffer capacity).

The scheduler graph is described by a number of parallel branches, one for each machine, robot, and buffer. Because all of the devices are operating simultaneously, the scheduler has to handle all of them concurrently by sending and receiving synchronization signals of the type *start* and *ready*. When a machine gets a *start* command from the scheduler, it performs a task defined by its function chart. When the machine has terminated, it sends a *ready* signal to the scheduler. Figure 17.31 shows that no machine communicates directly with the robot. Instead all the communication signals go via the scheduler. The signals are transition conditions in each function chart branch. By structuring the graph in this hierarchical way, new machines can be added to the cell without reprogramming any of the sequences of the other machines. The robot has to add the new operations to serve the new machines.

A good implementation of SFCs supports a hierarchical structuring of the problem. The total operation can first be defined by a few complex operations, each one consisting of many steps. Then it is possible to go on to more and more detailed operations.

17.7.5 Analog Controllers

Many *PLCs* can handle not only binary signals, but also analog-to-digital (A/D) and digital-to-analog (D/A) converters that can be added to the *PLC* rack. The resolution, that is, the number of bits to represent the analog signal, varies between the systems. The converted analog signal is placed in a digital register in the same way as a binary signal and is available for the standard PLC arithmetic and logical instructions.

In the event that plant signals do not correspond to any of the standard analog ranges, most manufacturers provide signal conditioning modules. Such a module provides buffering and scaling of plant signals to standard signals (typically 0–5 V or 0–10 V).

A PLC equipped with analog input channels may perform mathematical operations on the input values and pass the results directly to analog output modules to drive continuous actuators in the process directly. The sophistication of the control algorithms may vary, depending on the complexity of the PLC, but most systems today offer proportional-integral-derivative (PID) controller modules. The user has to tune the regulator parameters. To obtain sufficient capacity, many systems provide add-on PID function modules, containing input and output analog channels together with dedicated processors to carry out the necessary control calculations. This processor operates in parallel with the main CPU. When the main CPU requires status data from the PID module, it reads the relevant locations in the I/O memory where the PID processor places this information each time it completes a control cycle.

Many PLC systems also supply modules for digital filtering, for example, first-order smoothing (exponential filter) of input signals. The user gives the time constant of the filter as a parameter.

A PLC system may provide programming panels with a menu of questions and options relating to the setup of the control modules, such as gain parameters, integral and derivative times, filter time constants, sampling rate, and engineering unit conversions.

17.8 PLCs as Part of Automation Systems

The demands for communication in any process or manufacturing plant are steadily increasing. Any user today demands flexible and open communication, following some standard. Here we will just mention some of the crucial concepts essential for any nontrivial PLC installation.

17.8.1 Communication

A distributed system is more than simply connecting different units in a network. Certainly, the units in such a system can communicate, but the price is too much unnecessary communication, and the capacity of the systems cannot be fully used. Therefore, the architecture of the communication is essential. Reasons for installing network instead of point-to-point links are that

- All devices can access and share data and programs.
- Cabling for point-to-point becomes impractical and prohibitively expensive.
- A network provides a flexible base for contributing communication architectures.

To overcome the difficulties of dealing with a large number of incompatible standards, the ISO has defined the *open systems interconnection* (OSI) scheme. OSI itself is not a standard, but offers a framework to identify and separate the different conceptual parts of the communication process. In practice, OSI does not indicate which voltage levels, transfer speeds, or protocols need to be used to achieve compatibility between systems. It says that there *has* to be compatibility of voltage levels, speed, and protocols as well as for a large number of other factors. The practical goal of OSI is optimal network interconnection, in which data can be transferred between different locations without wasting resources for conversion and creating related delays and errors.

PLC systems are an essential part of most industrial control systems. In the following, we will illustrate how they are connected at different levels of a plant network (Figure 17.32).

17.8.2 Fieldbus: Communication at the Sensor Level

There is a trend to replace conventional cables from sensors with a single digital connection. Thus, a single digital loop can replace a large number of 4–20 mA conductors. This has been implemented not only in manufacturing plants but also in aircraft and automobiles. Each sensor needs an interface to the bus, and standardization is necessary. This structure is known as *Fieldbus*.

The term "fieldbus" is subject to a wide variety of definitions. A European standard exists that defines certain characteristics of a network that make it a fieldbus. To further confuse the issue, there is also a fieldbus called Fieldbus. The various definitions of fieldbus do have some common ground.

- The network is open and supported by multiple vendors.
- The network interconnects components of an automated industrial system for control and monitoring purposes. (Note that this definition does not mention performance, network services, or intended use, which are not factors in a network's designation as a fieldbus. They simply affect the suitability of a particular fieldbus for a specific application.)

There is another network type called a "sensor bus" that is either a subset of fieldbus or a completely independent category, depending on the vendor. A sensor bus is a low-level network whose primary purpose is to replace I/O wiring.

Two fundamental types of fieldbus can be identified: those that define a software architecture in addition to a communication protocol, and those that define only the communication protocol.

When all communicating units located in a close work cell are connected to the same physical bus, there is no need for multiple end-to-end transfer checks as if the data were routed along international networks. To connect computers in the restricted environment of a factory, the data exchange definition

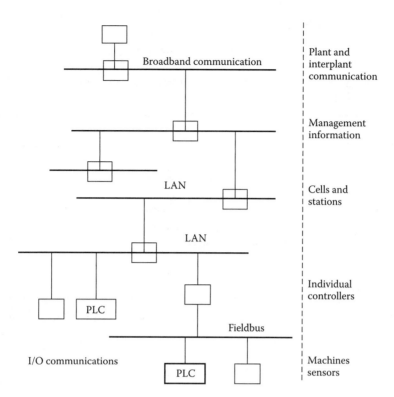

FIGURE 17.32 Structure of a plant network.

of OSI layers 1 (physical layer) and 2 (data link layer) and an application protocol at the OSI level 7 are enough.

Fieldbuses open notable possibilities. A large share of the intelligence required for process control is moved out to the field. The maintenance of sensors becomes much easier because test and calibration operations can be remotely controlled and require less direct intervention by maintenance personnel. And as we have already pointed out, the quality of the collected data influences directly the quality of process control.

There is now an international standard for fieldbuses, called IEC 61158. Under the general title *Digital data communications for measurement and control—Fieldbus for use in industrial control systems,* the following parts are defined:

1. Overview and guidance for the IEC 61158 series
2. Physical Layer specification and service definition
3. Data Link Service definition
4. Data Link Protocol specification
5. Application Layer Service definition
6. Application Layer Protocol specification

The standard should ensure interconnectivity of different devices connected to the same physical medium. Some examples of fieldbuses are Interbus, LonWorks, PROFIBUS, BITBUS. All of them are based on industrial Ethernet. Other examples include the CAN bus, widely used, for example, in the automotive industry.

17.8.3 Local Area Networks (LANs)

To communicate between different PLC systems and computers within a plant there is a clear trend to use Ethernet as the medium. Ethernet is a widely used local area network (LAN) for both industrial and office applications. Jointly developed by the Xerox, Intel, and Digital Equipment, Ethernet was introduced in 1980. Ethernet follows the IEEE 802.3 specifications.

Ethernet has a bus topology with branch connections. Physically, Ethernet consists of a screened coaxial cable to which peripherals are connected with "taps." Ethernet does not have a network controlling unit. All devices decide independently when to access the medium. Consequently, because the line is entirely passive, there is no single-failure point on the network. Ethernet supports communication at different speeds, as the connected units do not need to decode messages not explicitly directed to them.

Ethernet's concept is flexible and open. There is little capital bound up in the medium, and the medium itself does not have active parts like servers or network control computers which could break down or act as bottlenecks to tie up communication capacity. Some companies offer complete Ethernet-based communication packages which may also implement higher-layer services in the OSI hierarchy.

Acknowledgments

The author has enjoyed numerous discussions on this topic with Tekn. Lic. Lars Ericson, Dr. Gunnar Lindstedt, and Thomas Gillblad.

Further Reading

A wealth of literature exists on PLCs and their applications. One good representative is

1. Bolton, W., *Programmable Logic Controllers*, 5th edn, Newnes, Oxford, 2009.

There are several books with good coverage on, not only PLCs and the IEC 61131 standard, but their application in discrete manufacturing:

2. John, K.-H. and Tiegelkamp, M., *IEC 61131-3: Programming Industrial Automation Systems. Concepts and Programming Languages, Requirements for Programming Systems, Decision-Making Aids*, Springer-Verlag, Berlin, 2001, ISBN: 978-3-540-67752-9.

The international standard, IEC 61131-3 is documented in

3. International Electrotechnical Commission (IEC), *Programmable Controllers-Part 3: Programming Languages*, 2nd edn, IEC, 2003.

For updated information on IEC standards, see the webpage http://www.iec.ch
The manuals from different PLC manufacturers provide full details of the facilities and programming methods for a given model.
Switching theory is described in numerous textbooks, such as

4. Lee, S.C., *Modern Switching Theory and Digital Design*, Prentice Hall, Englewood Cliffs, NJ, 1978.

5. Fletcher, D.I., *An Engineering Approach to Digital Design*, Prentice Hall, Englewood Cliffs, NJ, 1980.

A good overview of sensors, actuators, and switching elements for both electric environment and pneumatic environment is contained in

6. Pessen, D.W., *Industrial Automation: Circuit Design and Components*, John Wiley & Sons, New York, NY, 1989.

There is an overwhelming literature on communication, in textbooks, articles, and on the web. Since this field is so rapidly changing, we just give references to the fundamental principles. Information about the current development of networks and fieldbuses is best followed via the standards organizations and the vendor information.

7. Tanenbaum, A.S., *Computer Networks*, 3rd edn, Prentice Hall, Upper Saddle River, NJ, 1996.

Describes almost everything that is to be mentioned about computer communication, at a very high level and yet not boring, while

8. Tanenbaum, A.S., *Distributed Operating Systems*, Prentice Hall, Englewood Cliffs, NJ, 1995.

Deals with computer communication networks as fundamental components in distributed computer systems. Tanenbaum provides a very solid technical foundation, breaking established writing patterns to give new insights.

Several articles on components, PLCs, and market reviews appear regularly in journals such as *Control Engineering, Control Engineering Practice* (Elsevier), *Instruments and Control Systems* (Springer).

V

Analysis and Design Methods for Nonlinear Systems

18

Analysis Methods

Derek P. Atherton
The University of Sussex

18.1 Introduction

18.1.1 The Describing Function Method

The describing function method, abbreviated as DF, was developed in several countries in the 1940s [1], to answer the question: "What are the necessary and sufficient conditions for the nonlinear feedback system of Figure 18.1 to be stable?" The problem still remains unanswered for a system with static nonlinearity, $n(x)$, and linear plant $G(s)$, All of the original investigators found limit cycles in control systems and observed that, in many instances with structures such as Figure 18.1 the waveform of the oscillation at the input to the nonlinearity was almost sinusoidal. If, for example, the nonlinearity in Figure 18.1 is an ideal relay, that is, has an on–off characteristic, so that an odd symmetrical input waveform will produce a square wave at its output, the output of $G(s)$ will be almost sinusoidal when $G(s)$ is a low-pass filter which attenuates the higher harmonics in the square wave much more than the fundamental. It was, therefore, proposed that the nonlinearity should be represented by its gain to a sinusoid and that the conditions for sustaining a sinusoidal limit cycle be evaluated to assess the stability of the feedback loop. This gain of the nonlinearity in response to a sinusoid is a function of the amplitude of the sinusoid and is known as the DF. Because DF methods can be used for other than a single sinusoidal input to distinguish this DF it is also referred to as the single sinusoidal DF or sinusoidal DF.

18.1.2 The Sinusoidal DF

For the reasons explained above, if we assume in Figure 18.1 that $x(t) = a \cos \theta$, where $\theta = \omega t$ and $n(x)$ is a symmetrical odd nonlinearity, then the output $y(t)$ will be given by the Fourier series

$$y(\theta) = \sum_{n=0}^{\infty} a_n \cos n\theta + b_n \sin n\theta \qquad (18.1)$$

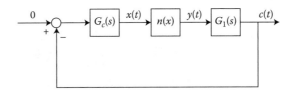

FIGURE 18.1   Block diagram of a nonlinear system.

where

$$a_0 = 0 \tag{18.2}$$

$$a_1 = (1/\pi) \int_0^{2\pi} y(\theta) \cos\theta \, d\theta \tag{18.3}$$

and

$$b_1 = (1/\pi) \int_0^{2\pi} y(\theta) \sin\theta \, d\theta \tag{18.4}$$

The fundamental output from the nonlinearity is $a_1 \cos\theta + b_1 \sin\theta$, so that the DF, defined as the fundamental output divided by the input amplitude, is complex and is given by

$$N(a) = \frac{(a_1 - jb_1)}{a} \tag{18.5}$$

which may be written

$$N(a) = N_p(a) + jN_q(a) \tag{18.6}$$

where

$$N_p(a) = \frac{a_1}{a} \quad \text{and} \quad N_q(a) = -\frac{b_1}{a} \tag{18.7}$$

Alternatively, in polar coordinates,

$$N(a) = M(a)e^{j\Psi(a)} \tag{18.8}$$

where

$$M(a) = \frac{(a_1^2 + b_1^2)^{1/2}}{a}$$

and

$$\Psi(a) = -\tan^{-1}\frac{b_1}{a_1} \tag{18.9}$$

If $n(x)$ is single valued, then $b_1 = 0$ and

$$a_1 = \frac{4}{\pi} \int_0^{\pi/2} y(\theta) \cos\theta \, d\theta \tag{18.10}$$

giving

$$N(a) = \frac{a_1}{a} = \frac{4}{a\pi} \int_0^{\pi/2} y(\theta) \cos\theta \, d\theta \tag{18.11}$$

Although Equations 18.3 and 18.4 are an obvious approach to evaluating the fundamental output of a nonlinearity, they are indirect, because one must first determine the output waveform $y(\theta)$ from the known nonlinear characteristic and sinusoidal input waveform. This is avoided if the substitution $\theta = \cos^{-1}(x/a)$ is made. After some simple manipulations,

$$a_1 = \frac{4}{a} \int_0^a x n_p(x) p(x)\, dx \tag{18.12}$$

and

$$b_1 = \frac{4}{a\pi} \int_0^a n_q(x)\, dx \tag{18.13}$$

The function $p(x)$ is the amplitude probability density function of the input sinusoidal signal given by

$$p(x) = \frac{1}{\pi}(a^2 - x^2)^{-1/2} \tag{18.14}$$

The nonlinear characteristics $n_p(x)$ and $n_q(x)$, called the in-phase and quadrature nonlinearities, are defined by

$$n_p(x) = \frac{n_1(x) + n_2(x)}{2} \tag{18.15}$$

and

$$n_q(x) = \frac{n_2(x) - n_1(x)}{2} \tag{18.16}$$

where $n_1(x)$ and $n_2(x)$ are the portions of a double-valued characteristic traversed by the input for $\dot{x} > 0$ and $\dot{x} < 0$, respectively. When the nonlinear characteristic is single valued, $n_1(x) = n_2(x)$, so $n_p(x) = n(x)$ and $n_q(x) = 0$. Integrating Equation 18.12 by parts yields

$$a_1 = \frac{4}{\pi} n(0^+) + \frac{4}{a\pi} \int_0^a n'(x)(a^2 - x^2)^{1/2} dx \tag{18.17}$$

where $n'(x) = dn(x)/d(x)$ and $n(0^+) = \lim_{\varepsilon \to 0} n(\varepsilon)$ a useful alternative expression for evaluating a_1.

An additional advantage of using Equations 18.12 and 18.13 is that they yield proofs of some properties of the DF for symmetrical odd nonlinearities. These include the following:

1. For a double-valued nonlinearity, the quadrature component $N_q(a)$ is proportional to the area of the nonlinearity loop, that is,

$$N_q(a) = -\frac{1}{a^2 \pi}(\text{area of nonlinearity loop}) \tag{18.18}$$

2. For two single-valued nonlinearities $n_\alpha(x)$ and $n_\beta(x)$, with $n_\alpha(x) < n_\beta(x)$ for all $0 < x < b$, $N_\alpha(a) < N_\beta(a)$ for input amplitudes less than b.
3. For a single-valued nonlinearity with $k_1 x < n(x) < k_2 x$ for all $0 < x < b$, $k_1 < N(a) < k_2$ for input amplitudes less than b. This is the sector property of the DF; a similar result can be obtained for a double-valued nonlinearity [2].

When the nonlinearity is single valued, from the properties of Fourier series, the DF, $N(a)$, may also be defined as

1. The variable gain, K, having the same sinusoidal input as the nonlinearity, which minimizes the mean-squared value of the error between the output from the nonlinearity and that from the variable gain and
2. The covariance of the input sinusoid and the nonlinearity output divided by the variance of the input.

18.1.3 Evaluation of the DF

To illustrate the evaluation of the DF, several simple examples are considered.

18.1.3.1 Cubic Nonlinearity

For this nonlinearity, $n(x) = x^3$ and using Equation 18.3,

$$a_1 = \frac{4}{\pi} \int_0^{\pi/2} (a \cos \theta)^3 \cos \theta \, d\theta$$

$$= \frac{4}{\pi} a^3 \int_0^{\pi/2} \cos^4 \theta \, d\theta$$

$$= \frac{4}{\pi} a^3 \int_0^{\pi/2} \frac{(1 + \cos 2\theta)^2}{4} \, d\theta$$

$$= \frac{4}{\pi} a^3 \int_0^{\pi/2} \left(\frac{3}{8} + \frac{\cos 2\theta}{2} + \frac{\cos 4\theta}{8} \right) d\theta = \frac{3a^3}{4}$$

giving $N(a) = 3a^2/4$.

Alternatively from Equation 18.12,

$$a_1 = \frac{4}{a} \int_0^a x^4 p(x) \, dx$$

The integral $\mu_n = \int_{-\infty}^{\infty} x^n p(x) \, dx$ is known as the nth moment of the probability density function and, for the sinusoidal distribution with $p(x) = (1/\pi)(a^2 - x^2)^{-1/2}$, μ_n has the value

$$\mu_n = \begin{cases} 0, & \text{for } n \text{ odd, and} \\ a^n \dfrac{(n-1)}{n} \dfrac{(n-3)}{(n-2)} \cdots \dfrac{1}{2}, & \text{for } n \text{ even} \end{cases} \tag{18.19}$$

Therefore,

$$a_1 = \left(\frac{4}{a} \right) \frac{1}{2} \cdot \frac{3}{4} \cdot \frac{1}{2} a^4$$

$$= 3a^3/4 \text{ as before.}$$

18.1.3.2 Saturation Nonlinearity

To calculate the DF, the input can alternatively be taken as $a \sin \theta$. For an ideal saturation characteristic, the nonlinearity output waveform $y(\theta)$ is as shown in Figure 18.2. Because of the symmetry of the nonlinearity, the fundamental of the output can be evaluated from the integral over a quarter period so that

$$N(a) = \frac{4}{a\pi} \int_0^{\pi/2} y(\theta) \sin \theta \, d\theta$$

which, for $a > \delta$, gives

$$N(a) = \frac{4}{a\pi} \left[\int_0^\alpha ma \sin^2 \theta \, d\theta + \int_\alpha^{\pi/2} m\delta \sin \theta \, d\theta \right]$$

where $\alpha = \sin^{-1} \delta/a$. Evaluation of the integrals gives

$$N(a) = \frac{4m}{\pi} \left[\frac{\alpha}{2} - \frac{\sin 2\alpha}{4} + \delta \cos \alpha \right]$$

which, on substituting for δ, gives the result

$$N(a) = \frac{m}{\pi}(2\alpha + \sin 2\alpha) \tag{18.20}$$

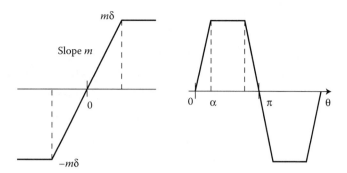

FIGURE 18.2 Saturation nonlinearity.

Because, for $a < \delta$, the characteristic is linear, giving $N(a) = m$, the DF for ideal saturation is $mN_s\,(\delta/a)$ where

$$N_s(\delta/a) = \begin{cases} 1, & \text{for } a < \delta \quad \text{and} \\ \dfrac{1}{\pi}(2\alpha + \sin 2\alpha), & \text{for } a > \delta \end{cases} \tag{18.21}$$

where $\alpha = \sin^{-1} \delta/a$.

Alternatively, one can evaluate $N(a)$ from Equation 18.17, yielding

$$N(a) = \frac{a_1}{a} = \left(\frac{4}{a^2\pi}\right) \int_0^\delta m(a^2 - x^2)^{1/2} dx$$

Using the substitution $x = a \sin \theta$,

$$N(a) = \frac{4m}{\pi} \int_0^\alpha \cos^2 \theta \, d\theta = \frac{m}{\pi}(2\alpha + \sin 2\alpha)$$

as before for $a > \delta$.

18.1.3.3 Relay with Dead Zone and Hysteresis

The characteristic is shown in Figure 18.3 together with the input, assumed to be equal to $a \cos \theta$, and the corresponding output waveform.

Using Equations 18.3 and 18.4 over the interval $-\pi/2$ to $\pi/2$ and assuming that the input amplitude a is greater than $\delta + \Delta$,

$$a_1 = \frac{2}{\pi} \int_{-\alpha}^\beta h \cos \theta \, d\theta$$
$$= \frac{2h}{\pi}(\sin \beta + \sin \alpha)$$

where $\alpha = \cos^{-1}[(\delta - \Delta)/a]$ and $\beta = \cos^{-1}[(\delta + \Delta)/a]$, and

$$b = \frac{2}{\pi} \int_{-\alpha}^\beta h \sin \theta \, d\theta$$
$$= \frac{-2h}{\pi}\left(\frac{(\delta + \Delta)}{a} - \frac{(\delta - \Delta)}{a}\right) = \frac{4h\Delta}{a\pi}$$

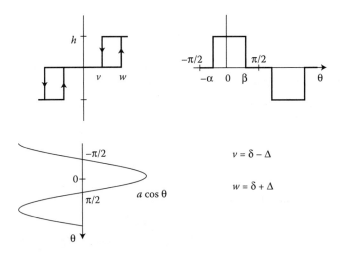

FIGURE 18.3   Relay with dead zone and hysteresis.

Thus

$$N(a) = \frac{2h}{a^2\pi}\left\{\left[a^2 - (\delta+\Delta)^2\right]^{1/2} + \left[a^2 - (\delta-\Delta)^2\right]^{1/2}\right\} - \frac{j4h\Delta}{a^2\pi} \tag{18.22}$$

For the alternative approach, one must first obtain the in-phase and quadrature nonlinearities shown in Figure 18.4. Functions $n_p(x)$ and $n_q(x)$ for the relay of Figure 18.3. Using Equations 18.12 and 18.13,

$$a_1 = \frac{4}{a}\int_{\delta-\Delta}^{\delta+\Delta} x\frac{h}{2}p(x)\,dx + \int_{\delta+\Delta}^{a} xhp(x)\,dx$$

$$= \frac{2h}{a\pi}\left\{\left[a^2 - (\delta+\Delta)^2\right]^{1/2} + \left[a^2 - (\delta-\Delta)^2\right]^{1/2}\right\}$$

and

$$b = \frac{4}{a\pi}\int_{d-\Delta}^{\delta+\Delta}\left(\frac{h}{2}\right)dx = \frac{4h\Delta}{a\pi} = \frac{\text{Area of nonlinearity loop}}{a\pi}$$

as before.

The DF of two nonlinearities in parallel equals the sum of their individual DFs, a result very useful for determining DFs, particularly of linear segmented characteristics with multiple break points. Several

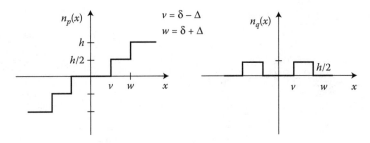

FIGURE 18.4   Functions $n_p^{(x)}$ and $n_q^{(x)}$ for the relay of Figure 18.3.

TABLE 18.1 DFs of Single-Valued Nonlinearities

General quantizer	$a < \delta_1$ $\delta_{M+1} > a > \delta_M$	$N_p = 0$ $N_p = (4/a^2\pi) \sum_{m=1}^{M} h_m(a^2 - \delta_m^2)^{1/2}$
Uniform quantizer $h_1 = h_2 = \cdots h$ $\delta_m = (2m-1)\delta/2$	$a < \delta$ $(2M+1)\,\delta > a > (2M-1)\delta$ $n = (2m-1)/2$	$N_p = 0$ $N_p = (4h/a^2\pi) \sum_{m=1}^{M} (a^2 - n^2\delta^2)^{1/2}$
Relay with dead zone	$a < \delta$ $a > \delta$	$N_p = 0$ $N_p = 4h(a^2 - \delta^2)^{1/2}/a^2\pi$
Ideal relay		$N_p = 4h/a\pi$
Preload		$N_p = (4h/a\pi) + m$
General piecewise linear	$a < \delta_1$ $\delta_{M+1} > a > \delta_M$	$N_p = (4h/a\pi) + m_1$ $N_p = (4h/a\pi) + m_{M+1}$ $\quad + \sum_{j=1}^{M} (m_j - m_{j+1})N_s(\delta_j/a)$
Ideal saturation		$N_p = mN_s(\delta/a)$
Dead zone		$N_p = m[1 - N_s(\delta/a)]$
Gain changing nonlinearity		$N_p = (m_1 - m_2)N_s(\delta/a) + m_2$

continued

TABLE 18.1 (continued) DFs of Single-Valued Nonlinearities

Saturation with dead zone		$N_p = m[N_s(\delta_2/a) - N_s(\delta_1/a)]$

$N_p = -m_1 N_s(\delta_1/a) + (m_1 - m_2)N_s(\delta_2/a) + m_2$

| $a < \delta$ | $N_p = 0$ |
| $a > \delta$ | $N_p = 4h(a^2 - \delta^2)^{1/2}/a_2\pi + m - mN_s(\delta/a)$ |

| $a < \delta$ | $N_p = m_1$ |
| $a > \delta$ | $N_p = (m_1 - m_2)N_s(\delta/a) + m_2 + 4h(a^2 - \delta^2)^{1/2}/a^2\pi$ |

| $a < \delta$ | $N_p = 4h/a\pi$ |
| $a > \delta$ | $N_p = 4h/[a - (a^2 - \delta^2)^{1/2}]/a^2\pi$ |

Limited field of view

$N_p = (m_1 + m_2)N_s(\delta/a) - m_2 N_s[(m_1 + m_2)\delta/m_2 a]$

| $a < \delta$ | $N_p = m_1$ |
| $a > \delta$ | $N_p = (m_1 N_s(\delta/a) - 4m_1\delta(a^2 + \delta^2)^{1/2}/a^2\pi$ |

$y = x^m$

$m > -2$ Γ is the gamma function

$$N_p = \frac{\Gamma(m + 1)a^{m-1}}{2^{m-1}\,\Gamma[3 + m)/2]\Gamma\,[(1 + m)/2]}$$

$$= \frac{2}{\sqrt{\pi}}\frac{\Gamma(m + 2)/2]a^{m-1}}{\Gamma[(m + 3)/2]}$$

procedures [1] are available for approximating the DF of a given nonlinearity either by numerical integration or by evaluating the DF of an approximating nonlinear characteristic defined, for example, by a quantized characteristic, linear segmented characteristic, or Fourier series. Table 18.1 gives a list of DFs for some commonly used approximations of nonlinear elements. Several of the results are in terms of the DF for an ideal saturation characteristic of unit slope, $N_s(\delta/a)$, defined in Equation 18.21.

18.1.4 Limit Cycles and Stability

To investigate the possibility of a limit cycle in the autonomous closed-loop system of Figure 18.1, the input to the nonlinearity $n(x)$ is assumed to be a sinusoid so that it can be replaced by the amplitude-dependent DF gain $N(a)$. The open-loop gain to a sinusoid is thus $N(a)G(j\omega)$ and, therefore, a limit cycle exists if

$$N(a)G(j\omega) = -1 \qquad (18.23)$$

where $G(j\omega) = G_c(j\omega)G_1(j\omega)$. As in general, $G(j\omega)$ is a complex function of ω and $N(a)$ is a complex function of a, solving Equation 18.23 will yield both the frequency ω and the amplitude a of a possible limit cycle.

Various procedures can be used to examine Equation 18.23; the choice is affected to some extent by the problem, for example, whether the nonlinearity is single valued or double valued or whether $G(j\omega)$ is available from a transfer function $G(s)$ or as measured frequency response data. Usually the functions $G(j\omega)$ and $N(a)$ are plotted separately on Bode or Nyquist diagrams, or Nichols charts. Alternatively, stability criteria (e.g., the Hurwitz-Routh) or root locus plots may be used with the characteristic equation

$$1 + N(a)G(s) = 0 \qquad (18.24)$$

although the equation is appropriate only for $s \approx j\omega$.

Figure 18.5 illustrates the procedure on a Nyquist diagram, where the $G(j\omega)$ and $C(a) = -1/N(a)$ loci are plotted intersecting for $a = a_0$ and $\omega = \omega_0$. The DF method indicates therefore that the system has a limit cycle with the input sinusoid to the nonlinearity, x, equal to $a_0 \sin(\omega_0 t + \phi)$, where ϕ depends on the initial conditions. When the $G(j\omega)$ and $C(a)$ loci do not intersect, the DF method predicts that no limit cycle will exist if the Nyquist stability criterion is satisfied for $G(j\omega)$ with respect to any point on the $C(a)$ locus. Obviously, if the nonlinearity has unit gain for small inputs, the point $(-1, j0)$ will lie on $C(a)$ and may be used as the critical point, analogous to a linear system.

For a stable case, it is possible to use the gain and phase margin to judge the relative stability of the system. However, a gain and phase margin can be found for every amplitude a on the $C(a)$ locus, so it is usually appropriate to use the minimum values of the quantities [1]. When the nonlinear block includes dynamics so that its response is both amplitude and frequency dependent, then it may be represented by an amplitude- and frequency-dependent DF $N(a, \omega)$, and a limit cycle will exist if

$$G(j\omega) = -1/N(a, \omega) = C(a, \omega) \qquad (18.25)$$

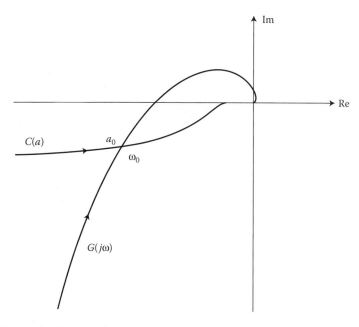

FIGURE 18.5 Nyquist plot showing solution for a limit cycle.

To check for possible solutions of this equation, a family of $C(a,\omega)$ loci, usually as functions of a for fixed values of ω, is drawn on the Nyquist diagram.

When a solution to Equation 18.23 exists, an additional point of interest is whether the predicted limit cycle is stable. This is important if the control system is designed to have a limit cycle operation, as in an on–off temperature control system. It may also be important in other systems, because, if an unstable limit cycle condition is obtained, the signal amplitudes will not become bounded but continue to grow. Provided that only one possible limit cycle solution is predicted by the DF method, the stability of the limit cycle can be assessed by applying the Nyquist stability criterion to points on the $C(a)$ locus on both sides of the solution point. For this perturbation approach, usually known as the Loeb criterion, if the stability criterion indicates instability (stability) for the point on $C(a)$ with $a < a_0$ and stability (instability) for the point on $C(a)$ with $a > a_0$, the limit cycle is stable (unstable).

When multiple solutions exist, the situation is more complicated and the Loeb criterion is a necessary but not sufficient result for the stability of the limit cycle [3].

Normally in these cases, the stability of the limit cycle can be ascertained by examining the roots of the characteristic equation

$$1 + N_{i\gamma}(a)G(s) = 0 \tag{18.26}$$

where $N_{i\gamma}(a)$ is known as the incremental describing function (IDF) and is the gain to an unrelated small signal in the presence of the sinusoid of amplitude a. $N_{i\gamma}(a)$ for a single-valued nonlinearity can be evaluated from

$$N_{i\gamma}(a) = \int_{-a}^{a} n'(x)p(x)\,dx \tag{18.27}$$

where $n'(x)$ and $p(x)$ are as previously defined. $N_{i\gamma}(a)$ is related to $N(a)$ by the equation

$$N_{i\gamma}(a) = N(a) + (a/2)\,dN(a)/da \tag{18.28}$$

Thus, for example, for an ideal relay, taking $\delta = \Delta = 0$ in Equation 18.22 gives $N(a) = 4h/a\pi$, which may also be found directly from Equation 18.17 and, substituting this value in Equation 18.28, yields $N_{i\gamma}(a) = 2h/a\pi$. Some examples of feedback system analysis using the DF follow.

18.1.4.1 Autotuning in Process Control

In 1943, Ziegler and Nichols [4] suggested a technique for tuning the parameters of a PID controller. Their method was based on testing the plant in a closed loop with the PID controller in the proportional mode. The proportional gain was increased until the loop started to oscillate and then the value of gain and the oscillation frequency were measured. Formulae were given for setting the controller parameters based on the gain named the critical gain, K_c, and the frequency called the critical frequency, ω_c.

Assuming that the plant has a linear transfer function $G_1(s)$, then K_c is its gain margin and ω_c the frequency at which its phase shift is 180°. Performing this test in practice may prove difficult. Even if the plant has a linear transfer function, which is very unlikely, and the gain is adjusted too quickly, a large amplitude oscillation may start to build up. In 1984 Astrom and Hagglund [5] suggested replacing the proportional control by a relay element to control the amplitude of the oscillation. Consider therefore the feedback loop of Figure 18.1 with $n(x)$ a relay with hysteresis Δ, $G_c(s) = 1$, and the plant with a transfer function $G_1(s) = 10/(s+1)^3$. The $C(a)$ locus, $-1/N(a)$, in Equation 18.33 is a line parallel to the negative real axis and at a distance $\pi\Delta/4h$ below it, as illustrated in Figure 18.6, which also shows the Nyquist locus of $G(j\omega)$. For the ideal relay, $\Delta = 0$; so $C(a)$ lies on the negative real axis and is given by $C(a) = -a\pi/4h$. The values of a and ω at the intersection can be calculated from

$$-a\pi/4h = \frac{10}{(1+j\omega)^3} \tag{18.29}$$

which can be written as

$$\mathrm{Arg}\left(\frac{10}{(1+j\omega)^3}\right) = 180° \tag{18.30}$$

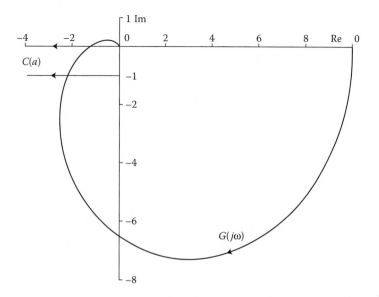

FIGURE 18.6 Nyquist plot for $10/(s+1)^3$ and $C(a)$ locus for relay with hysteresis.

and

$$\frac{a\pi}{4h} = \frac{10}{(1+\omega^2)^{3/2}} \tag{18.31}$$

The solution for ω_c from Equation 18.30 is $\tan^{-1}\omega_c = 60°$, where $\omega_c = \sqrt{3}$. Because the DF solution is approximate, the actual measured frequency of oscillation will differ from this value by an amount which will be smaller the closer the oscillation is to a sinusoid. The exact frequency of oscillation in this case is 1.708 rad/s, which is in error by a relatively small amount. For a square wave input to the plant at this frequency, the plant output signal will be distorted by a small percentage. The distortion, d, is defined by

$$d = \left[\frac{\text{M.S. value of signal} - \text{M.S. value of fundamental harmonic}}{\text{M.S. value of fundamental harmonic}}\right]^{1/2} \tag{18.32}$$

Solving Equation 18.31 gives the amplitude of oscillation a as $5h/\pi$. The gain through the relay is $N(a)$ equal to the critical gain K_c. In the practical situation where a is measured, typically taken as the peak value of the limit cycle although strictly it should be the amplitude of the fundamental component, K_c is then calculated from $4h/a\pi$. Its value should be close to the known value of 0.8 for this transfer function as the distortion in the limit cycle is quite small.

If the relay has an hysteresis of Δ, then from Equation 18.22

$$N(a) = \frac{4h(a^2 - \Delta^2)^{1/2}}{a^2\pi} - j\frac{4h\Delta}{a^2\pi}$$

from which

$$C(a) = \frac{-1}{N(a)} = \frac{-\pi}{4h}\left[(a^2 - \Delta^2)^{1/2} + j\Delta\right] \tag{18.33}$$

Taking $\Delta = 1$ and $h = \pi/4$ gives $C(a) = -(a^2 - 1)^{1/2} - j$. If the same transfer function is used for the plant, then the limit cycle solution is given by

$$-(a^2 - 1)^{1/2} - j = \frac{10}{(1+j\omega)^3}$$

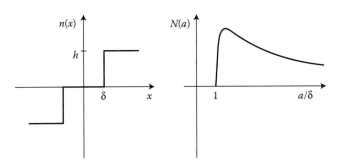

FIGURE 18.7 $N(a)$ for ideal relay with dead zone.

where $\omega = 1.266$, which compares with an exact solution value of 1.254, and $a = 1.91$. For the oscillation with the ideal relay, Equation 18.26 with $N_{i\gamma}(a) = 2h/a\pi$ shows that the limit cycle is stable. This agrees with the perturbation approach which also shows that the limit cycle is stable when the relay has hysteresis.

18.1.4.2 Feedback Loop with a Relay with Dead Zone

For this example, the feedback loop of Figure 18.1 is considered with $n(x)$ a relay with dead zone and $G(s) = 2/s(s+1)^2$. From Equation 18.22, with $\Delta = 0$, the DF for this relay is given by

$$N(a) = \frac{4h(a^2 - \delta^2)^{1/2}}{a^2\pi} \quad \text{for } a > \delta \tag{18.34}$$

and is real, because the nonlinearity is single valued. A graph of $N(a)$ against a is given in Figure 18.7, and shows that $N(a)$ starts at zero, when $a = \delta$, increases to a maximum, with a value of $2h/\pi\delta$ at $a = \delta\sqrt{2}$, and then decreases toward zero for larger inputs. The $C(a)$ locus, shown in Figure 18.8, lies on the negative real axis starting at $-\infty$ and returning there after reaching a maximum value of $-\pi\delta/2h$. The given transfer function $G(j\omega)$ crosses the negative real axis, as shown in Figure 18.8, at a frequency of $\tan^{-1}\omega = 45°$, that is, $\omega = 1$ rad/s and, therefore, cuts the $C(a)$ locus twice. The two possible limit cycle amplitudes at

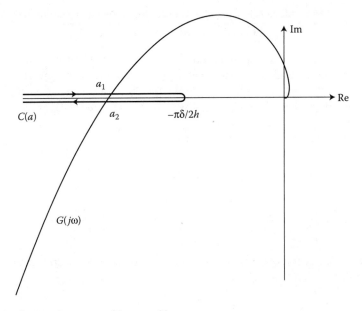

FIGURE 18.8   Two limit cycles: a_1, unstable; a_2, stable.

this frequency can be found by solving

$$\frac{a^2 \pi}{4h(a^2 - \delta^2)^{1/2}} = 1$$

which gives $a = 1.04$ and 3.86 for $\delta = 1$ and $h = \pi$. Using the perturbation method or the IDF criterion, the smallest amplitude limit cycle is unstable and the larger one is stable. If a condition similar to the lower amplitude limit cycle is excited in the system, an oscillation will build up and stabilize at the higher-amplitude limit cycle.

Other techniques, see Section 18.1.8, show that the exact frequencies of the limit cycles for the smaller and larger amplitudes are 0.709 and 0.989, respectively. Although the transfer function is a good low-pass filter, the frequency of the smallest amplitude limit cycle is not predicted accurately, because the output from the relay, a waveform with narrow pulses, is highly distorted.

If the transfer function of $G(s)$ is $K/s(s+1)^2$, then no limit cycle will exist in the feedback loop, and it will be stable if

$$\left.\frac{K}{\omega(1+\omega^2)}\right|_{\omega=1} < \frac{\pi\delta}{2h}$$

that is, $K < \pi\delta/h$. If $\delta = 1$ and $h = \pi$, $K < 1$, which may be compared with the exact result for stability of $K < 0.96$.

18.1.4.3 Feedback Loop with a Polynomial Nonlinearity

In this example, the possibility of a limit cycle in a feedback loop with $n(x) = x - (x^3/6)$ and $G(s) = K(1-s)/s(s+1)$ is investigated. For the nonlinearity $N(a) = 1 - (a^2/8)$, and the $C(a)$ locus starts at -1 on the Nyquist diagram. As a increases, the $C(a)$ locus moves along the negative real axis to $-\infty$ for $a = 2\sqrt{2}$. For a greater than $2\sqrt{2}$, the locus returns along the positive real axis from ∞ to the origin as a becomes large. For small signal levels, $N(a) \approx 1$, and an oscillation will start to build up, assuming that the system is initially at rest with $x(t) = 0$, only if the feedback loop with $G(s)$ alone is unstable. The characteristic equation

$$s^2 + s + K - Ks = 0$$

must have a root with a positive real part, that is, $K > 1$. The phase of $G(j\omega) = 180°$, when $\omega = 1$, and the corresponding gain of $G(j\omega)$ is K. Thus the DF solution for the amplitude of the limit cycle is

$$|G(j\omega)|_{\omega=1} = \frac{1}{1 - (a^2/8)}$$

resulting in

$$K = \frac{8}{(8 - a^2)}$$

and

$$a = 2\sqrt{2}[(K-1)/K]^{1/2} \tag{18.35}$$

As K increases, the limit cycle becomes more distorted because of the shape of the nonlinearity. For example, if $K = 2.4$, the DF solution gives $\omega = 1$ and $a = 2.10$. If four harmonics are balanced [1], the limit cycle frequency is 0.719 and the amplitudes of the fundamental, third, fifth, and seventh harmonics at the input to the nonlinearity are 2.515, 0.467, 0.161, and 0.065, respectively.

Because the DF approach is a method for evaluating limit cycles, it is sometimes argued that it cannot guarantee the stability of a feedback system, when instability is caused by an unbounded, not oscillatory, signal in the system. Fortunately, another result is helpful with this problem [6]. This states that, in the feedback system of Figure 18.1, if the symmetric odd nonlinearity $n(x)$ is such that $k_1 x < n(x) < k_2 x$, for $x > 0$, and $n(x)$ tends to $k_3 x$ for large x, where $k_1 < k_3 < k_2$, then the nonlinear system is either stable or possesses a limit cycle, provided that the linear system with gain K replacing N is stable for $k_1 < K < k_2$. For this situation, often true in practice, the nonexistence of a limit cycle indicates stability.

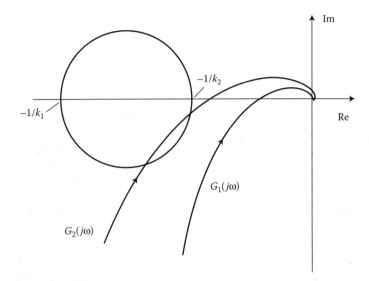

FIGURE 18.9 Circle criterion and stability.

18.1.5 Stability and Accuracy

Because the DF method is an approximate procedure, it is desirable to be able to judge its accuracy. Predicting that a system will be stable, when in practice it is not, may have unfortunate consequences. Many attempts have been made to solve this problem, but those obtained are difficult to apply or produce too conservative results [7].

The problem is illustrated by the system in Figure 18.1 with a symmetrical odd single-valued nonlinearity confined to a sector between lines of slope k_1 and k_2, that is, $k_1 x < n(x) < k_2 x$ for $x > 0$. For absolute stability, the circle criterion essentially requires satisfying the Nyquist criterion for the locus $G(j\omega)$ for all points within a circle having its diameter on the negative real axis of the Nyquist diagram between the points $(-1/k_1, 0)$ and $(-1/k_2, 0)$, as shown in Figure 18.9. On the other hand, because the DF for this nonlinearity lies within the diameter of the circle, the DF method requires satisfying the Nyquist criterion for $G(j\omega)$ for all points on the circle diameter, if the autonomous system is to be stable.

Therefore, for a limit cycle in the system of Figure 18.1, errors in the DF method relate to its inability to predict a phase shift, which the fundamental harmonic may experience in passing through the nonlinearity, rather than an incorrect magnitude of the gain. When the input to a single-valued nonlinearity is a sinusoid together with some of its harmonics, the fundamental output is not necessarily in phase with the fundamental input, that is, the fundamental gain has a phase shift. The actual phase shift varies with the harmonic content of the input signal in a complex manner, because the phase shift depends on the amplitudes and phases of the individual input harmonic components.

From an engineering viewpoint, one can judge the accuracy of DF results by estimating the distortion, d, in the input to the nonlinearity. This is straightforward when a limit cycle solution is given by the DF method; the loop may be considered opened at the nonlinearity input, the sinusoidal signal corresponding to the DF solution can be applied to the nonlinearity, and the harmonic content of the signal fed back to the nonlinearity input can be calculated. Experience indicates that the percentage accuracy of the DF method in predicting the fundamental amplitude and frequency of the limit cycle is less than the percentage distortion in the fed back signal. In the previous section, the frequency of oscillation in the autotuning example, where the distortion was relatively small, was given more accurately than in the third example. Due to the relatively poor filtering of the plant in this example, the distortion in the fed back signal was much higher. As mentioned previously, the DF method may incorrectly predict stability.

To investigate this problem, the procedure above can be used again, by taking, as the nonlinearity input, a sinusoid with amplitude and frequency corresponding to values of those parameters where the phase margin is small. If the distortion calculated for the fed back signal is high, say greater than 2% per degree of phase margin, the DF result should not be relied on.

The limit cycle amplitude predicted by the DF is an approximation to the fundamental harmonic. Use of the peak value of a distorted limit cycle for the fundamental amplitude is clearly therefore not appropriate for assessing the DF accuracy. It is possible to estimate the limit cycle more accurately by balancing more harmonics, as mentioned earlier. Although this is difficult algebraically, other than with loops where the nonlinearity has a simple mathematical description, for example, a cubic, software has been written for this purpose [8]. The procedure involves solving sets of nonlinear algebraic equations but good starting guesses can usually be obtained for the magnitudes and phases of the other harmonic components from the waveform fed back to the nonlinearity, assuming its input is the DF solution. This procedure was used to balance four harmonics to obtain the better solution given for the distorted limit cycle in example 3 of the previous section.

18.1.6 Compensator Design

Although the design specifications for a control system are often in terms of step-response behavior, frequency-domain design methods rely on the premise that the correlation between a frequency response and a step response is such that a less oscillatory step response is obtained if the gain and phase margins are increased. Therefore, the design of a suitable linear compensator for the system of Figure 18.1 using the DF method is usually done by selecting, for example, a lead network to provide adequate gain and phase margins for all amplitudes. This approach may be used in example 2 of the previous section where a phase lead network could be added to stabilize the system, say for a gain of 1.5, for which it is unstable without compensation. Other approaches are the use of additional feedback signals or modification of the nonlinearity $n(x)$ directly or indirectly [1,9].

When the plant is nonlinear, its frequency response also depends on the input sinusoidal amplitude; hence, it can be represented as $G(j\omega, a)$ or $N(a, \omega)$ as mentioned in Section 18.1.4. In recent years, several approaches [10,11] use the DF method to design a nonlinear compensator for the plant, with the objective of producing closed-loop performance approximately independent of the input amplitude.

18.1.7 Closed-Loop Frequency Response

When the closed-loop system shown in Figure 18.1 has a sinusoidal input $r(t) = R\sin(\omega t + \theta)$, it is possible to evaluate the closed-loop frequency response using the DF. If the feedback loop has no limit cycle when $r(t) = 0$ and, in addition, the sinusoidal input $r(t)$ does not induce a limit cycle, then, provided that $G_c(s)G_1(s)$ gives good filtering, $x(t)$, the nonlinearity input, almost equals the sinusoid $a\sin\omega t$. Balancing the components of frequency ω around the loop,

$$g_c R\sin(\omega t + \theta + \phi_c) - ag_1g_cM(a)$$

$$\sin[\omega t + \phi_1 + \phi_c + \Psi(a)] = a\sin\omega t \tag{18.36}$$

where $G_c(j\omega) = g_c e^{j\phi_c}$ and $G_1(j\omega) = g_1 e^{j\phi_1}$. In principle, Equation 18.36, which can be written as two nonlinear algebraic equations, can be solved for the two unknowns a and θ and the fundamental output signal can then be found from

$$c(t) = aM(a)g_1\sin[\omega t + \Psi(a) + \phi_1] \tag{18.37}$$

to obtain the closed-loop frequency response for R and ω.

Various graphical procedures have been proposed for solving the two nonlinear algebraic equations resulting from Equation 18.36 [12–14]. If the system is lightly damped, the nonlinear equations may have more than one solution, indicating that the frequency response of the system has a jump resonance. This phenomenon of a nonlinear system has been studied by many authors, both theoretically and practically [15,16].

18.1.8 Limit Cycles in Relay Systems

The examples presented in Sections 18.1.4.1 and 18.1.4.2 contained relay nonlinearities and accurate values for the limit cycle frequencies were given for comparison with the DF results. Exact methods [17–21] are available for evaluating limit cycles, and also their stability, in systems with a relay element. The reason for this is that the relay is a unique form of nonlinearity in that the output does not depend continuously on the input, as between switches the input has no control of the output.

The approach used to obtain a limit cycle solution is to start with an assumed relay output wave-form. For the simplest case, namely an on–off relay operating symmetrically, the output will be a symmetrical periodic square wave. The waveform that this produces when passed through a transfer function $G(s)$ can be calculated by any one of the three methods mentioned below. One, due to Tsypkin [17], is to use a Fourier series representation for the square wave and then combine the individual harmonics at the output, which can be obtained in a closed-form expression. Alternatively, one can sum a limited number of harmonics and if the first only is used, the DF solution is obtained. The second is to use a Laplace transform approach, as done by Hamel [18], and the third is to use a state-space formulation as introduced by Chung [19]. If this output of $G(s)$ is the signal fed back through -1 to the relay, then for the assumed square wave to exist the signal must pass through zero, if the relay is ideal, with the correct slope, to switch it at the correct time to generate the assumed relay output square wave. This switching condition yields a nonlinear algebraic equation which can be solved for the limit cycle frequency, which is the only unknown. The solution is best obtained using computational methods [22] so that the actual waveform at the relay input can also be shown. This is important, because the solution will only be valid provided it does not pass through the switching level at other time instants, which can happen for some $G(s)$, for example, if $G(s)$ has a resonance.

If the relay has a dead zone and a symmetrical odd limit cycle is considered, then there will be two switching conditions, one for the relay switching on from zero and the other for the relay switching off. The resulting two nonlinear algebraic equations yield the two unknowns, namely the limit cycle frequency and the pulse width of the relay output. With software solving the switching equations and showing the resultant limit cycle waveform, it is possible to extend the approach to cover quite complex limit cycles in relay systems. Studies [22–27] have included limit cycles with multiple pulses per period, limit cycles in multivariable systems, and limit cycles with sliding. Forced oscillations in relay systems, including situations resulting in multiple pulses per period, have also been studied [22].

Further Information

Many control engineering textbooks contain material on nonlinear systems where the DF is discussed. The coverage, however, is usually restricted to the basic sinusoidal DF for determining limit cycles in feedback systems. The basic DF method, which is one of quasilinearization, can be extended to cover other signals, such as random signals, and also to cover multiple input signals to nonlinearities. These descriptions are required to investigate phenomena such as subharmonic resonance, synchronization of a loop limit cycle with an external signal, and other unique phenomena that may occur in a nonlinear feedback loop. The two books with the most comprehensive coverage of this are Gelb and van der Velde [9] and Atherton [28]. More specialized books on nonlinear feedback systems usually cover the phase plane method, the subject of the next article, and the DF together with other topics such as absolute stability, exact linearization, and so on.

References

1. Atherton, D.P., *Non Linear Control Engineering*, Student Ed., Van Nostrand Reinhold, New York, 1982.
2. Cook, P.A., Describing function for a sector nonlinearity, *Proc. Inst. Electr. Eng.*, 120, 143–44, 1973.
3. Choudhury, S.K. and Atherton, D.P., Limit cycles in high order nonlinear systems, *Proc. Inst. Electr. Eng.*, 121, 717–24, 1974.
4. Ziegler, J.G. and Nichols, N.B., Optimal setting for automatic controllers, *Trans. ASME*, 65, 433–44, 1943.
5. Astrom, K.J. and Haggland, T., Automatic tuning of single regulators, in *IFAC Congress*, pp. 267–72, Budapest, Vol 4, 1984.
6. Vogt, W.G. and George, J.H., On Aizerman's conjecture and boundedness, *IEEE Trans. Autom. Control*, 12, 338–39, 1967.
7. Mees, A.I. and Bergen, A.R., Describing function revisited, *IEEE Trans. Autom. Control*, 20, 473–78, 1975.
8. McNamara, O.P. and Atherton, D.P., Limit cycle prediction in free structured nonlinear systems, *IFAC Congress*, Munich, Vol. 8, 23–8, July 1987.
9. Gelb, A. and van der Velde, W.E., *Multiple Input Describing Functions and Nonlinear Systems Design*, McGraw-Hill, New York, 1968.
10. Nanka-Bruce, O. and Atherton, D.P., Design of nonlinear controllers for nonlinear plants, *IFAC Congress*, Tallinn, Vol. 6, 75–80, 1990.
11. Taylor, J.H. and Strobel, K.L., Applications of a nonlinear controller design approach based on the quasilinear system models, *Proc ACC*, San Diego, 817–24, 1984.
12. Levinson, E., Some saturation phenomena in servo-mechanisms with emphasis on the techometer stabilised system, *Trans. Am. Inst. Electr. Eng.*, Part 2, 72, 1–9, 1953.
13. Singh, T.P., Graphical method for finding the closed loop frequency response of nonlinear feedback control systems, *Proc. Inst. Electr. Eng.*, 112, 2167–70, 1965.
14. West, J.C. and Douce, J.L., The frequency response of a certain class of nonlinear feedback systems, *Br. J. Appl. Phys.*, 5, 201–10, 1954.
15. Lamba, S.S. and Kavanagh, R.J., The phenomenon of isolated jump resonance and its applications, *Proc. Inst. Electr. Eng.*, 118, 1047–50, 1971.
16. West, J.C., Jayawant, B.V., and Rea, D.P., Transition characteristics of the jump phenomenon in nonlinear resonant circuits, *Proc. Inst. Electr. Eng.*, 114, 381–92, 1967.
17. Hamel B, Etude Mathematique des Systemes a plusieurs Degrees de Liberte Decrits par des Equations Lineaires avec un Terme de Commande Discontinu, *Proc. Journees d'Etude des Vibrations*, AERA, Paris, 1950.
18. Tsypkin Ya, Z., *Relay Control Systems*, Cambridge University Press, Cambridge, England, 1984. (English version of a much earlier Russian edition.)
19. Chung, J.K.-C. and Atherton, D.P.,The determination of periodic modes in relay systems using the state space approach, *Int. J. Control*, 4, 105–26, 1966.
20. Willems, J.L., *Stability Theory of Dynamical Systems*, 112–13, Nelson, London, 1970.
21. Balasubramanian, R., Stability of limit cycles in feedback systems containing a relay, *IEE Proc. D Control Theory Appl.*, 128, 24–9, 1981.
22. Wadey, M., *Extensions of Tsypkin's Method for Computer Aided Control System Design*, DPhil Thesis, University of Sussex, England, 1984.
23. Atherton, D.P. and Wadey, M.D., Computer aided analysis and design of relay systems, *IFAC Symposium on CAD of MV Tech. Systems*, 355–60, Purdue, USA, September 15–7, 1982.
24. Atherton, D.P, *Conditions for Periodicity in Control Systems Containing Several Relays*, Paper 28E, *3rd IFAC Congress*, London, 16pp., June 1966,
25. Atherton, D.P., Oscillations in relay systems, *Trans. InstMC*, 3(4), 171–84, 1982.
26. Atherton, D.P., Analysis and design of relay control systems. *CAD for Control Systems*, Chapter 15, 367–94, Marcel Dekker, New York, NY, 1993.
27. Rao, U.M. and Atherton, D.P., *Multi-Pulse Oscillations in Relay Systems. Proc. 7th IFAC World Congress*, Helsinki, Vol. 3, Paper No. 42.4, pp. 1747–54, 1978.
28. Atherton, D.P., *Nonlinear Control Engineering, Describing Function Analysis and Design*, Van Nostrand Reinhold, London, 1975.

18.2 The Phase Plane Method

18.2.1 Introduction

The phase plane method was the first method used by control engineers for studying the effects of nonlinearity in feedback systems. The technique which can only be used for systems with second-order models was examined and further developed for control engineering purposes for several major reasons.

1. The phase plane approach had been used for several studies of second-order nonlinear differential equations arising in fields such as planetary motion, nonlinear mechanics, and oscillations in vacuum tube circuits.
2. Many of the control systems of interest, such as servomechanisms, could be approximated by second order nonlinear differential equations.
3. The phase plane was particularly appropriate for dealing with nonlinearities with linear segmented characteristics which were good approximations for the nonlinear phenomena encountered in control systems.

The next section considers the basic aspects of the phase plane approach but later concentration is focused on control engineering applications where the nonlinear effects are approximated by linear segmented nonlinearities.

18.2.2 Background

Early analytical work [1], on second-order models assumed the equations

$$\dot{x}_1 = P(x_1, x_2)$$
$$\dot{x}_2 = Q(x_1, x_2) \tag{18.38}$$

for two first-order nonlinear differential equations. Equilibrium, or singular points, occur when

$$\dot{x}_1 = \dot{x}_2 = 0$$

and the slope of any solution curve, or trajectory, in the $x_1 - x_2$ state plane is

$$\frac{dx_2}{dx_1} = \frac{\dot{x}_2}{\dot{x}_1} = \frac{Q(x_1, x_2)}{P(x_1, x_2)} \tag{18.39}$$

A second-order nonlinear differential equation representing a control system can be written

$$\ddot{x} + f(x, \dot{x}) = 0$$

If this is rearranged as two first-order equations, choosing the phase variables as the state variables, that is $x_1 = x$, $x_2 = \dot{x}$, then Equation 18.2.2 can be written as

$$\dot{x}_1 = x_2 \quad \dot{x}_2 = -f(x_1, x_2) \tag{18.40}$$

which is a special case of Equation 18.39. A variety of procedures has been proposed for sketching state [phase] plane trajectories for Equations 18.39 and 18.40. A complete plot showing trajectory motions throughout the entire state (phase) plane is known as a state (phase) portrait. Knowledge of these methods, despite the improvements in computation since they were originally proposed, can be particularly helpful for obtaining an appreciation of the system behavior. When simulation studies are undertaken, phase plane graphs are easily obtained and they are often more helpful for understanding the system behavior than displays of the variables x_1 and x_2 against time.

The step response of the second-order linear system is discussed in section 8.1.8 and Figure 8.4 shows the unit step response against the normalized time $\omega_n t$ for various values of the damping ratio ζ. For no input the equation with ω_n taken as unity is $\ddot{x} + 2\zeta\dot{x} + x = 0$ and in state space form it can be written as

$$\dot{x}_1 = x_2 \quad \dot{x}_2 = -x_1 - 2\zeta x_2. \tag{18.41}$$

The response to a unit step can be shown in a phase plane, where the final rest, or equilibrium position, will be the origin by using the initial condition of $(-1, 0)$. For no damping, that is $\zeta = 0$, the plot will be a circle about the origin, which is called a center and for $0 < \zeta < 1$ the trajectory spirals into the origin, which is known as a focus. For $\zeta \geq 1$ the trajectory moves to the origin, which is now called a node, without overshoot.

Many investigations using the phase plane technique were concerned with the possibility of limit cycles in the nonlinear differential equations. When a limit cycle exists, this results in a closed trajectory in the phase plane. Typical of such investigations was the work of Van der Pol, who considered the equation

$$\ddot{x} - \mu(1 - x^2)\dot{x} + x = 0 \tag{18.42}$$

where μ is a positive constant. The phase plane form of this equation can be written as

$$\dot{x}_1 = x_2$$
$$\dot{x}_2 = -f(x_1, x_2) = \mu(1 - x_1^2)x_2 - x_1 \tag{18.43}$$

The slope of a trajectory in the phase plane is

$$\frac{dx_2}{dx_1} = \frac{\dot{x}_2}{\dot{x}_1} = \frac{\mu(1 - x_1^2)x_2 - x_1}{x_2} \tag{18.44}$$

and this is only singular (that is, at an equilibrium point), when the right-hand side of Equation 18.44 is $0/0$, that is $x_1 = x_2 = 0$.

The form of this singular point which is obtained from linearization of the equation at the origin depends upon μ, being an unstable focus for $\mu < 2$ and an unstable node for $\mu > 2$. All phase plane trajectories have a slope of r when they intersect the curve

$$rx_2 = \mu(1 - x_1^2)x_2 - x_1 \tag{18.45}$$

One way of sketching phase plane behavior is to draw a set of curves given for various values of r by Equation 18.45 and marking the trajectory slope r on the curves. This procedure is known as the method of isoclines and has been used to obtain the limit cycles shown in Figure 18.10 for the Van der Pol equation with $\mu = 0.2$ and 4.

18.2.3 Piecewise Linear Characteristics

When the nonlinear elements occurring in a second-order model can be approximated by linear segmented characteristics then the phase plane approach is usually easy to use because the nonlinearities divide the phase plane into various regions within which the motion may be described by different linear second-order equations [2]. The procedure is illustrated by the simple relay system in Figure 18.11.

The block diagram represents an "ideal" relay position control system with velocity feedback. The plant is a double integrator, ignoring viscous (linear) friction, hysteresis in the relay, or backlash in the gearing. If the system output is denoted by x_1 and its derivative by x_2, then the relay switches when $-x_1 - x_2 = \pm 1$; the equations of the dotted lines are marked switching lines on Figure 18.12.

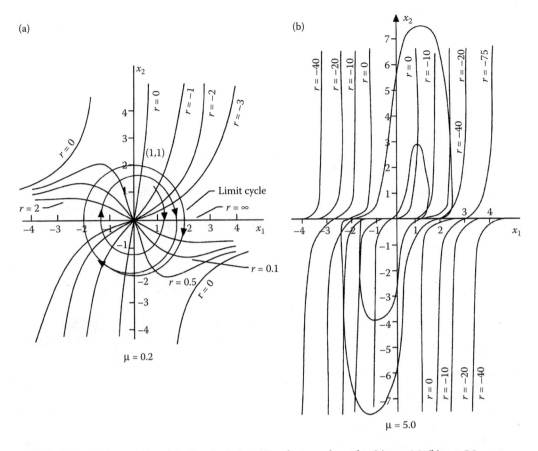

FIGURE 18.10   Phase portraits of the Van der Pol equation for two values of μ. (a) $\mu = 0.2$; (b) $\mu = 5.0$.

Because the relay output provides constant values of ± 2 and 0 to the double integrator plant, if the constant value is denoted by h, then the state equations for the motion are

$$\dot{x}_1 = x_2$$
$$\dot{x}_2 = h \tag{18.46}$$

which can be solved to give the phase plane equation

$$x_2^2 - x_{20}^2 = 2h(x_1 - x_{10}) \tag{18.47}$$

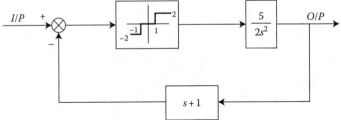

FIGURE 18.11   Relay system.

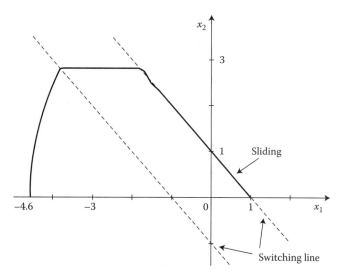

FIGURE 18.12 Phase plane for relay system.

which is a parabola for h finite and the straight line $x_2 = x_{20}$ for $h = 0$, where x_{20} and x_{10} are the initial values of x_2 and x_1. Similarly, more complex equations can be derived for other second-order transfer functions. Using Equation 18.47 with the appropriate values of h for the three regions in the phase plane, the step response for an input of 4.6 units can be obtained as shown in Figure 18.12.

In the step response, when the trajectory meets the switching line $x_1 + x_2 = -1$ for the second time, trajectory motions at both sides of the line are directed toward it, resulting in a sliding motion down the switching line. Completing the phase portrait by drawing responses from other initial conditions shows that the autonomous system is stable and also that all responses will finally slide down a switching line to equilibrium at $x_1 = \pm 1$.

An advantage of the phase plane method is that it can be used for systems with more than one nonlinearity and for those situations where parameters change as functions of the phase variables. For example, Figure 18.13 shows the block diagram of an approximate model of a servomechanism with nonlinear effects due to torque saturation and Coulomb friction.

The differential equation of motion in phase variable form is

$$\dot{x}_2 = f_s(-x_1) - (1/2)\,\mathrm{sgn}\,x_2 \qquad (18.48)$$

where f_s denotes the saturation nonlinearity and sgn the signum function, which is $+1$ for $x_2 > 0$ and -1 for $x_2 < 0$. There are six linear differential equations describing the motion in different regions of the

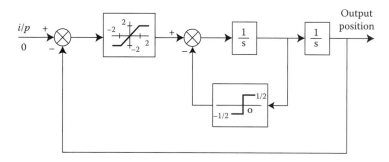

FIGURE 18.13 Block diagram of servomechanism.

phase plane. For x_2 positive, Equation 18.48 can be written

$$\ddot{x}_1 + f_s(x_1) + 1/2 = 0$$

so that for

a. x_2+ve, $x_1 < -2$, we have $\dot{x}_1 = x_2, \dot{x}_2 = 3/2$, a parabola in the phase plane.
b. x_2+ ve $|x_1| < 2$, we have $\dot{x}_1 = x_2, \dot{x}_2 + x_1 + 1/2 = 0$, a circle in the phase plane.
c. x_2+ve, $x_1 > 2$, we have $\dot{x}_1 = x_2, \dot{x}_2 = -5/2$, a parabola in the phase plane.
   Similarly for x_2 negative,
d. x_2−ve, $x_1 - 2$, we have $\dot{x}_1 = x_2, \dot{x}_2 = 5/2$, a parabola in the phase plane.
e. x_2−ve, $|x_2| < 2$, we have $\dot{x}_1 = x_2, \dot{x}_2 + x_1 - 1/2 = 0$, a circle in the phase plane.
f. x_2−ve,$x_1 > 2$, we have $\dot{x}_1 = x_2, \dot{x}_2 = -3/2$, a parabola in the phase plane.

Because all the phase plane trajectories are described by simple mathematical expressions, it is straightforward to calculate specific phase plane trajectories.

18.2.4 Discussion

The phase plane approach is useful for understanding the effects of nonlinearity in second-order systems, particularly if it may be approximated by a linear segmented characteristic. Solutions for the trajectories with other nonlinear characteristics may not be possible analytically so that approximate sketching techniques were used in early work on nonlinear control. These approaches are described in many books, for example, [3–10]. Although the trajectories are now easily obtained with modern simulation techniques, knowledge of the topological aspects of the phase plane are still useful for interpreting the responses in different regions of the phase plane and appreciating the system behavior.

References

1. Andronov, A.A., Vitt, A.A., and Khaikin, S.E., *Theory of Oscillators*, Addison-Wesley, Reading, MA, 1966. (First edition published in Russia in 1937.)
2. Atherton, D.P., *Nonlinear Control Engineering*, Student Ed., Van Nostrand Reinhold, New York, 1982.
3. Blaquiere, A., *Nonlinear Systems Analysis*, Academic Press, New York, 1966.
4. Cosgriff, R., *Nonlinear Control Systems*, McGraw-Hill, New York, 1958.
5. Cunningham, W.J., *Introduction to Nonlinear Analysis*, McGraw-Hill, New York, 1958.
6. Gibson, J.E., *Nonlinear Automatic Control*, McGraw-Hill, New York, 1963.
7. Graham, D. and McRuer, D., *Analysis of Nonlinear Control Systems*, John Wiley & Sons, New York, 1961.
8. Hayashi, C., *Nonlinear Oscillations in Physical Systems*, McGraw-Hill, New York, 1964.
9. Thaler, G.J. and Pastel, M.P., *Analysis and Design of Nonlinear Feedback Control Systems*, McGraw-Hill, New York, 1962.
10. West, J.C., *Analytical Techniques of Nonlinear Control Systems*, E.U.P., London, 1960.

19

Design Methods

R.H. Middleton
National University of Ireland, Maynooth

Stefan F. Graebe
PROFACTOR GmbH

Anders Ahlén
Uppsala University

Jeff S. Shamma
The University of Texas at Austin

19.1 Dealing with Actuator Saturation

R. H. Middleton

19.1.1 Description of Actuator Saturation

Essentially all plants have inputs (or manipulated variables) that are subject to hard limits on the range (or sometimes also rate) of variations that can be achieved. These limitations may be due to restrictions deliberately placed on actuators to avoid damage to a system and/or physical limitations on the actuators themselves. Regardless of the cause, limits that cannot be exceeded invariably exist. When the actuator has reached such a limit, the actuator is said to be "saturated" since no attempt to further increase the control input gives any variation in the actual control input. The simplest case of actuator saturation in a control system is to consider a system that is linear apart from an input saturation as depicted in Figure 19.1.

Mathematically, the action of a saturating actuator can be described as:

$$\bar{u} = \left\{ \begin{array}{l} u_{max} : u \geq u_{max} \\ u : u_{min} < u < u_{max} \\ u_{min} : u_{min} \geq u \end{array} \right\}$$

Heuristically, it can be seen that once in saturation, the incremental (or small signal) gain of the actuator becomes zero. Alternatively, from a describing function viewpoint, a saturation is an example of a sector nonlinearity with a describing function as illustrated in Figure 19.2.

This describing function gives exactly a range of gains starting at 1 and reducing to zero as amplitude increases. From both perspectives, actuator saturation can be seen to be equivalent to a nonlinear reduction in gain.

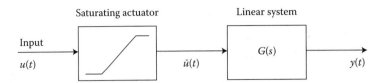

FIGURE 19.1   Linear plant model including actuator saturation.

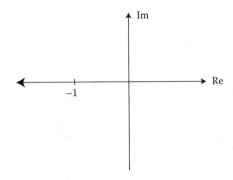

FIGURE 19.2   Describing function for a saturation.

19.1.2 Effects of Actuator Saturation

The main possible effects of actuator saturation on a control system are poor performance and/or instability to large disturbances. These effects are seen as "large" disturbance effects since for "small" disturbances, actuator saturation may be averted, and a well-behaved linear response can occur. The following two examples illustrate the possible effects of actuator saturation.

Example 19.1: Integral Windup

In this example, we consider the control system depicted in Figure 19.3, with $u_{min} = -1$ and $u_{max} = 1$

The simulated response for this system with a step change in the setpoint of 0.4 is illustrated in Figure 19.4. Note that this step change corresponds to a "small" step where saturation is evident, but only to a small extent. In this case, the step response is well behaved, with only a small amount of overshoot occurring.

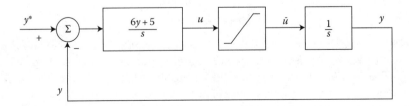

FIGURE 19.3   Example control system for saturating actuators.

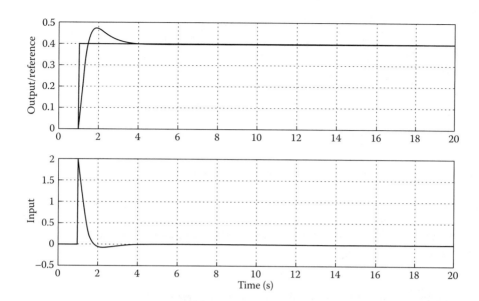

FIGURE 19.4   Response to a small step change for Example 19.1.

In contrast to this, Figure 19.5 shows simulation results for a step change of four units in the set-point. In this case, note that the response is very poor with large overshoot and undershoot in the response. The input response shows why this is occurring, where the unsaturated input reaches very large values due to integral action in the controller. This phenomenon is termed integral (or reset*) windup.

An even more dramatic effect is shown in Figure 19.6 where an open loop system is strictly unstable.[†]

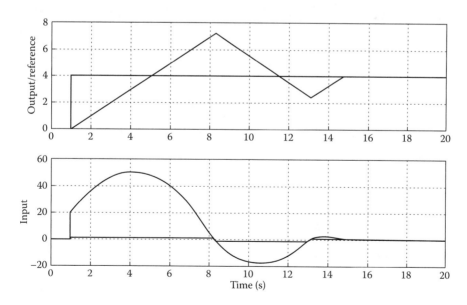

FIGURE 19.5   Response to a large step change.

---

*   The term *reset* is commonly used in the process control industry for integral action.
[†]   The strict sense is that the plant has an open-loop pole with positive real part.

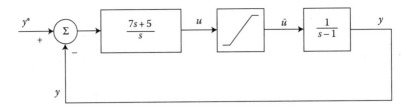

FIGURE 19.6   Open-loop unstable system with actuator saturation.

Example 19.2:   Controller Saturation for an Unstable Plant

In this case (where again we take $u_{min} = -1$ and $u_{max} = 1$), a step change in the reference of 0.8 units causes a dramatic failure* of the control system as illustrated in Figure 19.7. This instability is caused solely by saturation of the actuator since, for small step changes, the control system is well behaved.

19.1.3 Reducing the Effects of Actuator Saturation

The effects of actuator saturation cannot always be completely avoided. However, there are ways of reducing some of the effects of actuator saturation, as indicated below.

1. *Where possible, avoid conditionally stable[†] control systems.* Conditionally stable feedback control systems are undesirable for several reasons. Included in these reasons is the effect of actuator saturations. Simple describing function arguments show that the combination of a conditionally stable control system and a saturating actuator give rise to limit cycle behavior. In most cases, this

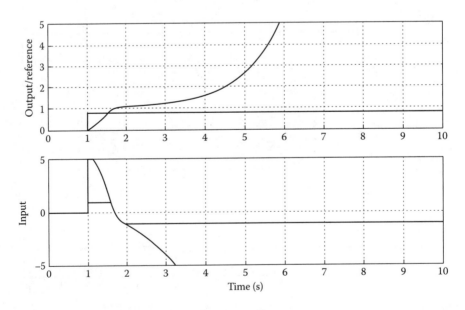

FIGURE 19.7   Actuator saturation causing instability in Example 19.2.

---

*   It has been reported (e.g., Stein, G., [1]) that this type of failure was one of the factors that caused the Chernobyl nuclear disaster (in this case, a limit on the rate of change of the actuator exacerbated an already dangerous situation).

†   A conditionally stable control system is one in which a reduction in the loop gain may cause instability.

limit cycle behavior is unstable. Instability of such a limit cycle generally means that for slightly larger initial conditions or disturbances, compared with the limit cycle, the output diverges; and, conversely, for smaller initial conditions, stable convergence to a steady state occurs. This is clearly undesirable, but cannot always be avoided. Note that a controller for a plant can be designed that gives unconditional stability if and only if the plant has:

a. No poles with positive real part

b. No purely imaginary poles of repetition greater than 2

Therefore, a plant that is open loop strictly unstable can be only conditionally stabilized.

2. *Avoid applying unrealistic reference commands to a control system.* Note that in the examples presented previously, the reference commands were in many cases unrealistic. Take, for example, the situation shown in Figure 19.5. In this case, an instantaneous change of 4 units is being commanded. Clearly, however, because of the actuator saturation, the output, y, can never change by more than 1 unit per second. Therefore, we know that the commanded trajectory can never be achieved. A more sensible reference signal would be one that ramps up (at a rate of 1 unit per second) from 0 to 4 units. If this reference were applied instead of the step, a greatly reduced overshoot, etc. would be obtained. The implementation of this type of idea is often termed a *reference governor* or *reference conditioner* for the system. See, for example, [2] for more details.

3. *Utilize saturation feedback to implement the controller.* To implement saturation feedback, we note that any linear controller of the form

$$U(s) = C(s)E(s) \tag{19.1}$$

can be rewritten as

$$U(s) = \frac{P(s)}{L(s)}E(s) \tag{19.2}$$

where $L(s) = s^n + l_{n-1}s^{k-1} + \cdots + l_o$ is a monic polynomial in s, and $P(s) = p_n s^n + p_{n-1}s^{n-1} + \cdots + p_o$ is a polynomial in s. Let the closed-loop poles be at $s = -\alpha_i; i = 1 \ldots N > n$. Then the controller can be implemented via saturation feedback as shown in Figure 19.8.

In the above implementation $E_1(s)$ can, in principle, be any stable monic polynomial of degree n. The quality of the performance of this anti-integral windup scheme depends on the choice of $E_1(s)$. A simple choice that gives good results in most cases is

$$E_1(s) = (s + \alpha_{m_1})(s + \alpha_{m_2}) \ldots (s + \alpha_{m_n}) \tag{19.3}$$

where $\alpha_{m_1} \ldots \alpha_{m_n}$ are the n fastest closed-loop poles. Note that when the actuator is not saturated we have that $U(s) = \frac{P(s)}{E_1(s)}E(s) - \frac{L(s)-E_1(s)}{E_1(s)}U(s)$ and so $\frac{U(s)}{E(s)} = \frac{P(s)}{L(s)}$, which is precisely the desired linear transfer function. When the actuator does saturate, the fact that E_1 is stable improves the controller behavior. The following examples illustrate the advantages of this approach.

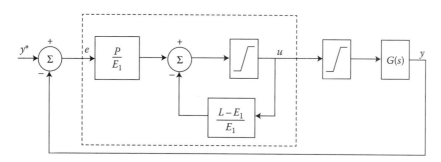

FIGURE 19.8   Controller implementation using saturation feedback.

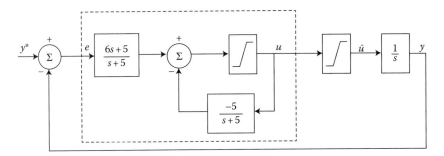

FIGURE 19.9 Control system for Example 19.3.

Example 19.3: Anti-Integral Windup (Example 19.1 Revisited)

In this case $L(s) = s$; $P(s) = 6s + 5$; and the closed-loop poles are at $s = -1$ and $s = -5$. We therefore choose $E_1(s) = (s + 5)$ and obtain the control system structure illustrated in Figure 19.9.

Figure 19.10 compares the performance of this revised arrangement with that of Figure 19.5, showing excellent performance in this case.

Example 19.4: Improved Control of Unstable Systems (Example 19.2 Revisited)

Let us now consider Example 19.2 again. In this case, $P(s) = 7s + 5$; $L(s) = s$ and the closed-loop poles are again at $s = -1$ and $s = -5$. This suggests $E_1(s) = (s + 5)$ giving the following control system structure:

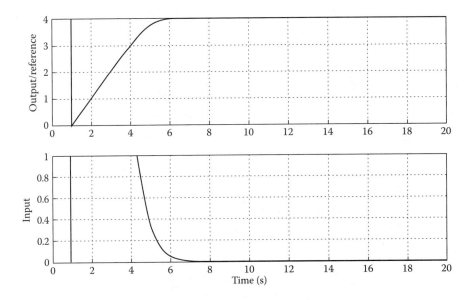

FIGURE 19.10 Response to a large step change for Example 19.3.

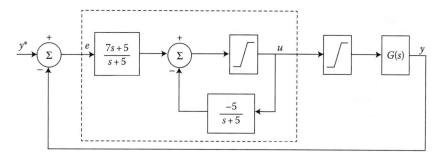

FIGURE 19.11   Controller structure for Example 19.4.

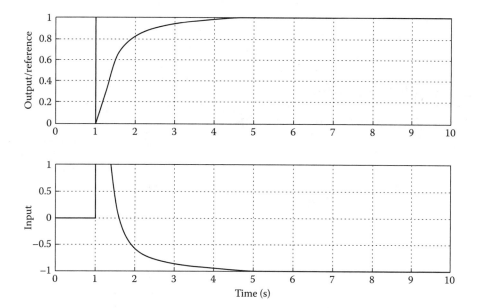

FIGURE 19.12   Response for Example 19.4.

The comparative step response to Figure 19.7 is given for a slightly larger step (in this case, 1.0 units) in Figure 19.12. Note in this case that previously where instability arose, in this case very good response is obtained.

References

1. Stein, G., Bode lecture, 28th IEEE Conference on Decision and Control, Tampa, FL, 1989.
2. Seron, S.G.M. and Goodwin, G., All stabilizing controllers, feedback linearization and anti-windup: A unified review, *Proc. Am. Control Conf.,* Baltimore, 1994.
3. Astrom, K. and Wittenmark, B., *Computer Controlled Systems: Theory and Design,* 2nd ed., Prentice Hall, Englewood Cliffs, NJ, 1990.
4. Gilbert, E., Linear control systems with pointwise in time constraints: What do we do about them?, *Proc. Am. Control Conf.,* Chicago, 1992.
5. Braslavsky, J. and Middleton, R., On the stabilisation of linear unstable systems with control constraints, *Proc. IFAC World Congr.,* Sydney, Australia, 1993.

Further Reading

The idea of using saturation feedback to help prevent integral windup (and related phenomena) has been known for many years now. Astrom and Wittenmark [3] give a description of this and an interpretation in terms of observer design with nonlinear actuators.

More advanced, constrained optimization-based procedures are the subject of current research by many authors. Gilbert [4] gives an overview of this area (together with the problem of maintaining system states within desired constraints).

Another approach that may be useful where actuator saturation is caused by large changes in the reference signal (as opposed to disturbances or other effects) is that of a *reference governor* or *reference conditioner*. Seron and Goodwin [2] explore this and its relationship with the technique of *saturation feedback*. Also, as mentioned previously, it has long been known that strictly unstable systems with actuator constraints can never be globally stabilized; see, for example, [5] for a recent look at this problem.

19.2 Bumpless Transfer

Stefan F. Graebe and Anders Ahlén

19.2.1 Introduction

Traditionally, the problem of bumpless transfer refers to the instantaneous switching between manual and automatic control of a process while retaining a smooth ("bumpless") control signal. As a simple example illustrating this issue, we consider a typical start-up procedure.

Example 19.5:

Consider a system with open-loop dynamics

$$\dot{x}(t) = -0.2x(t) + 0.2u(t), \quad x(t_o) = x_o$$
$$y(t) = x(t) \tag{19.4}$$

where $u(t)$ is the control signal and $y(t)$ is a noise-free measurement of the state $x(t)$. With s denoting the Laplace transform complex variable, we also consider the proportional-integral (PI) controller

$$C(s) = 2.3 \left(1 + \frac{1}{4.2s}\right)$$

digitally approximated as

$$X^I(t + \Delta) = X^I(t) + \Delta 0.547 e(t)$$
$$u_{\text{ctrl}}(t) = X^I(t) + 2.3 e(t) \tag{19.5}$$

In Equation 19.5, $X^I(t)$ denotes the integrator state of the PI controller, Δ is the sampling period, $e(t) \triangleq r(t) - y(t)$ is the control error, and $u_{\text{ctrl}}(t)$ is the control signal generated by the feedback controller. The reference signal is assumed to have a constant value of $r \equiv 4$. Then the following procedure, although simplified for this example, is typical of industrial start-up strategies.

The system is started from rest, $x_o = 0$, and the control is manually held at $u_{\text{man}}(t) = 4$ until, say at time t_s^-, the system output is close to the desired setpoint

$$y(t_s^-) \approx r = 4$$

At that point, control is switched to the PI controller for automatic operation. Figure 19.13 illustrates what happens if the above start-up procedure is applied blindly without bumpless transfer and the controller state is $X^I(t_s) = 0$ at switching time $t_s = 16$. With a manual control of $u_{\text{man}}(t) = 4$, $t \in [1,2)$, the control error at switching time can be computed from Equation 19.4 as

$$e(t_s^-) = y(t_s^-) - r(t_s^-) = 0.2$$

Hence, from Equation 19.5, the automatic control at switching time for $X^I(t_s) = 0$ yields $u_{\text{ctrl}}(t_s^+) = 0.46$. As a result, there is a "bump" in the control signal, u, when switching from manual control, $u = u_{\text{man}}(t_s^-) = 4$, to automatic control, $u = u_{\text{ctrl}}(t_s^+) = 0.46$, and an unacceptable transient follows.

Avoiding transients after switching from manual to automatic control can be viewed as an initial condition problem on the output of the feedback controller. If the manual control just before switching is $u_{\text{man}}(t_s^-) = u_{\text{man}}^o$, then bumpless transfer requires that the automatic controller take that same value as initial condition on its output, u_{ctrl}, so that $u_{\text{ctrl}}(t_s^+) = u_{\text{man}}^o$. By mapping this condition to the controller states, bumpless transfer can be viewed as a problem of choosing the appropriate initial conditions on the controller states.

In the case of the PI controller in Equation 19.5, the initial condition on the controller state X^I that yields an arbitrary value u_{man}^o is trivially computed as

$$X^I(t_s^-) = u_{\text{man}}^o - 2.3e(t_s^-) \tag{19.6}$$

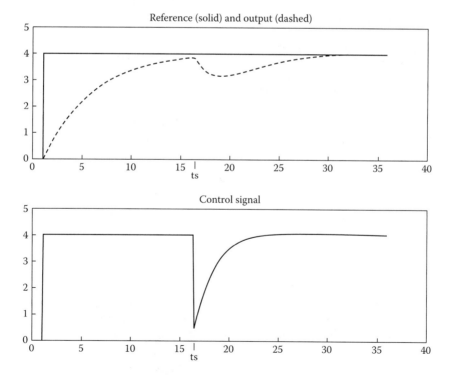

FIGURE 19.13 Without bumpless transfer mechanism, poor performance occurs at switching from manual to automatic control at time t_s. (Top) Reference signal, $r(t)$, (solid); output, $y(t)$, (dashed). (Bottom) Control signal, $u(t)$.

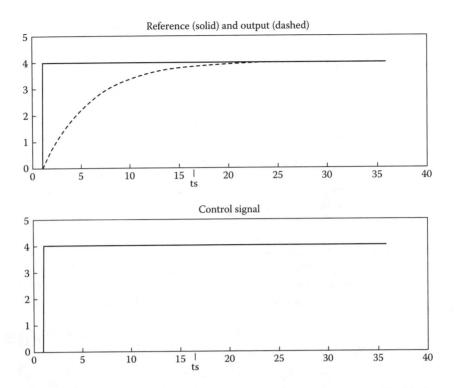

FIGURE 19.14 Bumpless transfer from manual to automatic control at switching time t_s. (Top) Reference signal, $r(t)$, (solid); output, $y(t)$, (dashed). (Bottom) Control signal, $u(t)$.

By including Equation 19.6 as initial condition at the switching time on the PI controller in Equation 19.5, bumpless transfer is achieved as shown in Figure 19.14.

Taking a more general point of view, there are several practical situations that call for strategies that could be classified as bumpless transfer. We list these and their associated constraints in Section 19.2.2. In Section 19.2.3, we review a general framework in which bumpless transfer is considered to be a tracking control problem. Section 19.2.4 presents a number of other techniques and Section 19.2.5 provides a brief summary.

19.2.2 Applications of Bumpless Transfer

In this section, we present several scenarios that may all be interpreted as bumpless transfer problems. Since each of theses scenarios is associated with different constraints, they tend to favor different bumpless transfer techniques.

19.2.2.1 Switching between Manual and Automatic Control

The ability to switch between manual and automatic control while retaining a smooth control signal is the traditional bumpless transfer problem. Its essence is described in Example 19.5, although the general case allows arbitrary controller complexity instead of being restricted to PI controllers. This is probably the simplest bumpless transfer scenario, as it tends to be associated with three favorable factors.

Firstly, the switching scheme is usually designed for a particular feedback loop. Thus, the controller and its structure are known and can be exploited. If it is a PI controller, for example, the simple strategy of Section 19.2.1 suffices. Other techniques take advantage of the particular structures of *observer-based controllers* [1], *internal model controllers* (*IMC*) [3,4], or controllers implemented in incremental form [1].

Secondly, in contrast to the scenario of Section 19.2.1, manual to automatic switching schemes are usually implemented in the same process control computer as the controller itself. In that case, the exact controller state is available to the switching algorithm, which can manipulate it to achieve the state associated with a smooth transfer. The main challenge is therefore to compute the appropriate state for higher-order controllers.

Thirdly, switching between manual and automatic controllers usually occurs under fairly benign conditions specifically aimed at aiding a smooth transfer. Many strategies implemented in practice (see Section 19.2.4) are simple, because they implicitly assume constant signals.

19.2.2.2 Filter and Controller Tuning

It is frequently desired to tune filter or controller parameters on-line and in response to experimental observations.

Example 19.6:

Consider Figure 19.15, which shows the sinusoid

$$s(t) = 5\sin(0.4\pi t) + 2$$

filtered by a filter F_1. Until the switching time, the filter is given by

$$F_1(s) = \frac{1}{0.5s + 1}$$

Assume that, at time $t_s \approx 10$, it is desired to retune the time constant of the filter to obtain

$$F_2(s) = \frac{1}{2s + 1}$$

Then, merely changing the filter time constant without adjusting the filter state results in the transient shown in Figure 19.15.

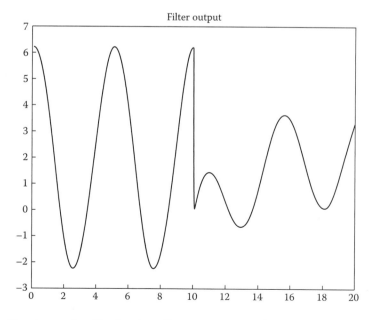

FIGURE 19.15 Transient produced by changing a filter time constant without adjusting the filter state.

The scenario of Example 19.6 can be considered as a bumpless transfer problem between two dynamical systems, F_1 and F_2. Although these systems have the meaning of filters in Example 19.6, the same considerations apply, of course, to the retuning of controller parameters.

Assuming that the bumpless transfer algorithm is implemented in the same computer as the filter or controller to be tuned, this problem amounts to an appropriate adjustment of the state as discussed in the previous section. The main difference is now that one would like to commence with the tuning even during transients. Although the general techniques of Sections 19.2.3 and 19.2.4 could be applied to this case, a simpler scheme, sufficient for low-order filters and controllers, can be derived as follows.

Let the signal produced by the present filter or controller be denoted by u_1 and let the retuned filter or controller be implemented by the state-space model

$$\dot{x}(t) = Ax(t) + Be(t)$$
$$u_2(t) = Cx(t) + De(t) \tag{19.7}$$

Bumpless retuning at time t_s requires that the state of Equation 19.7 be such that

$$u_2(t_s) \approx u_1(t_s) \tag{19.8}$$

and

$$\left.\frac{d^k u_2(t)}{dt^k}\right|_{t=t_s} \approx \left.\frac{d^k u_1(t)}{dt^k}\right|_{t=t_s} \qquad k = 1, \ldots, n \tag{19.9}$$

for n as large as possible. This ensures that the retuning not only avoids discontinuous jumps, but also retains smooth derivatives. Substituting Equation 19.7 into Equation 19.8 yields, at time $t = t_s$,

$$u_1 = Cx + De$$
$$\frac{du_1}{dt} = CAx + CBe + D\frac{de}{dt}$$
$$\frac{d^2 u_1}{dt^2} = CA^2 x + CABe + CB\frac{de}{dt} + D\frac{d^2 e}{dt^2}$$
$$\vdots$$
$$\frac{d^{n-1} u_1}{dt^{n-1}} = CA^{n-1}x + CA^{n-2}Be + \cdots + CB\frac{d^{n-2}e}{dt^{n-2}} + D\frac{d^{n-1}e}{dt^{n-1}}$$

Hence, assuming that the system in Equation 19.7 is observable and x has dimension n, the observatibility matrix

$$\mathcal{O} \triangleq \begin{pmatrix} C \\ CA \\ CA^2 \\ \vdots \\ CA^{n-1} \end{pmatrix}$$

is nonsingular and

$$x = \mathcal{O}^{-1}\left\{ \begin{pmatrix} u_1 \\ \dfrac{du_1}{dt} \\ \vdots \\ \dfrac{d^{n-1}u_1}{dt^{n-1}} \end{pmatrix} - \begin{pmatrix} D \\ CB \\ CAB \\ \vdots \\ CA^{n-2}B \end{pmatrix} e - \begin{pmatrix} 0 \\ D \\ CB \\ \vdots \\ CA^{n-3}B \end{pmatrix}\frac{de}{dt} - \begin{pmatrix} 0 \\ 0 \\ D \\ CB \\ \vdots \\ CA^{n-4}B \end{pmatrix}\frac{d^2 e}{dt^2} - \cdots - \begin{pmatrix} 0 \\ 0 \\ 0 \\ \vdots \\ 0 \\ D \end{pmatrix}\frac{d^{n-1}e}{dt^{n-1}} \right\}$$

$$\tag{19.10}$$

uniquely determines the state, x, that will match the $(n-1)$ first derivatives of the output of the retuned system to the corresponding derivatives of the output from the original system. Of course, the state cannot be computed directly from Equation 19.10, as this would require $(n-1)$ derivatives of u_1 and e, which could be noisy. A standard technique, however, is to approximate the required derivatives with band-pass filters as

$$\frac{d^k e}{dt^k} \approx \mathcal{L}^{-1}\left\{\frac{s^k}{e_m s^m + \cdots + e_1 s + 1}E(s)\right\}$$

where $E(s)$ denotes the Laplace transform of the control error, $\mathcal{L}^{-1}$ is the inverse Laplace transform, and $[e_m s^m + \cdots + e_1 s + 1]$, $m \geq n-1$, is an observer polynomial with roots and degree selected to suit the present noise level. Filters and controllers with on-line tuning interface can easily be augmented with Equation 19.10 to recompute a new state whenever the parameters are changed by a user.

Clearly, as the order n of Equation 19.7 increases, Equation 19.10 becomes increasingly noise sensitive. Therefore, it is not the approach we would most recommend, although its simplicity bears a certain attraction for low-order applications. Our primary reason for including it here is because it captures the essence of bumpless transfer as being the desire to compute the state of a dynamical system so its output will match another signal's value and derivatives. Indeed, Equation 19.10 can be interpreted as a simple observer that reconstructs the state by approximate differentiation of the output. As we will see in Section 19.2.4, this idea can be extended by considering more sophisticated observers with improved noise rejection properties.

Consider the setup of Example 19.6. Retuning the filter constant and adjusting the state according to Equation 19.10 yields the smooth performance shown in Figure 19.16.

19.2.2.3 Scheduled and Adaptive Controllers

Scheduled controllers are controllers with time-varying parameters. These time variations are usually due to measured time-varying process parameters (such as a time delay varying with production speed) or due to local linearization in different operating ranges. If the time variations are occasional and the controller remains observable for all parameter settings, the principle of the previous section could be applied.

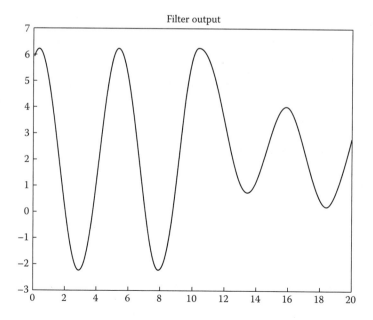

FIGURE 19.16 Smooth retuning of the filter constants by adjustment of the state according to Equation 19.10.

If, however, the controller order becomes large, the noise sensitivity of Equation 19.10 can become prohibitive. Furthermore, due to the inherent bandwidth limitations of evaluating the filtered derivatives, Equation 19.10 is not suitable for bumpless transfer if the parameters change significantly at every sampling interval, such as in certainty equivalence adaptive controllers. In that case, the techniques of Sections 19.2.3 and 19.2.4 are preferable.

19.2.2.4 Tentative Evaluation of New Controllers

This is a challenging, and only recently highlighted (see [5,6]), bumpless transfer scenario. It is motivated by the need to test tentative controller designs safely and economically on critical processes during normal operation.

Consider, for example, a process operating in closed loop with an existing controller. Assume that the performance is mediocre, and that a number of new controller candidates have been designed and simulated. It is then desired to test these controllers, tentatively, on the plant to assess their respective performances. Frequently it is not possible or feasible to shut down the plant intermittently, and the alternative controllers therefore have to be brought on-line with a bumpless transfer mechanism during normal plant operation.

This is not a hypothetical situation; see [7] for a full-scale industrial example. Indeed, this scenario has considerable contemporary relevance, since economic pressures and ecological awareness require numerous existing plants to be retrofitted with high-performance advanced controllers during normal operation.

There are four primary factors that make the tentative evaluation phase particularly challenging for bumpless transfer: safety, economic feasibility, robustness, and generality.

Safety. The actual performance of the new controllers is not reliably known, even if they have been simulated. In a worst case, one of them might drive the process unstable. It is then of overriding concern that bumpless transfer back to the original, stabilizing, controller is still possible. Due to this safety requirement, the technique should not rely on steady-state or constant signals, but be dynamic.

Economic feasibility. Since the achievable improvement due to the new controllers is not accurately known in advance, there tends to be a reluctance for costly modifications in hardware and/or software during the evaluation phase. Therefore, the bumpless transfer algorithm should be external to the existing controller and require only the commonly available signals of process input, output and reference. In particular, it should not require manipulating the states of the existing controller, as they may be analog. The technique should not only provide smooth transfer to the new controller, but also provide the transfer back to the existing controller. Thus, it should be bidirectional.

Robustness. Since the existing controller could very well be analog, it might be only approximately known. Even digital controllers are commonly implemented in programmable logic controllers (PLC) with randomly varying sampling rates that change the effective controller gains. Hence, the technique should be insensitive to inaccurately known controllers.

Generality. To be applicable as widely as possible, the bumpless transfer technique should not require the existing controller to have a particular order or structure, such as the so-called velocity form or such as constant feedback from a dynamical observer.

These objectives can be achieved by considering the tentative (also called the idle or latent) controller itself as a dynamic system and forcing its output to track the *active controller* by means of a tracking loop [5,6]. This recasts bumpless transfer into an associated tracking problem to which systematic analysis and design theory may be applied.

19.2.3 Robust Bidirectional Transfer

In this section, we describe a general framework in which the problem of bumpless transfer is recast into a tracking problem. The solution is specifically aimed at the scenario described above. Beyond providing a practical solution to such cases, the framework also provides useful insights when analyzing other techniques described later in Section 19.2.4.

Consider Figure 19.17, where G denotes the transfer function of a single-input single-output (SISO) plant currently controlled by the active controller C_A. The bold lines in Figure 19.17 show the active closed loop

$$y = \frac{C_A G}{1 + C_A G}\, r \tag{19.11}$$

The regular lines make up an additional feedback configuration governed by

$$u_L = \frac{F_L T_L C_L}{1 + T_L C_L Q_L}\, u_A + \frac{C_L}{1 + T_L C_L Q_L}(r - y) \tag{19.12}$$

which describes the *two-degree-of-freedom* tracking loop of Figure 19.18. Within this configuration, the *latent controller*, C_L, takes the role of a dynamical system whose output, u_L, is forced to track the active control signal, u_A, which is the reference signal to the tracking loop. Tracking is achieved by means of the tracking controller triplet (F_L, T_L, Q_L). Frequently, a *one-degree-of-freedom* tracking controller, in which $F_L = Q_L = 1$, is sufficient; we include the general case mainly for compatibility with other techniques, which we will mention in Section 19.2.4. Note that the plant control error, $r - y$, acts as an input disturbance in the tracking loop. Its effect can be eliminated by an appropriate choice of F_L, or it can be attenuated by designing (F_L, T_L, Q_L) for good input disturbance rejection.

While u_L is tracking u_A, the plant input can be switched bumplessly from the active controller to the latent controller (for graphical clarity, this switch is not shown in Figure 19.17). Simultaneously, the effect

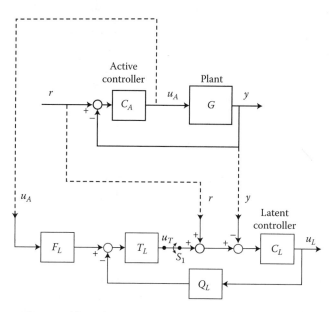

FIGURE 19.17 The unidirectional bumpless transfer diagram, in which the plant, G, is controlled by the active controller, C_A. The output, u_L, of the latent controller, C_L, is forced to track the active control signal, u_A, by means of the tracking controller, (F_L, T_L, Q_L). Any time u_L is tracking u_A, the plant input can be switched from the active controller to the latent controller and bumpless transfer is achieved. Simultaneously, the effect of the tracking loop (F_L, T_L, Q_L) is removed from the plant loop by opening switch S_1. Complementing the diagram with a second tracking circuit allows bidirectional transfer.

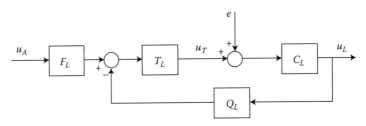

FIGURE 19.18 Tracking loop with latent control signal u_L tracking the active control signal u_A. The plant control error, $e = r - y$, acts as input disturbance to the latent controller C_L, here regarded as a "system" to be controlled.

of the tracking loop is removed by opening the switch S_1 in Figure 19.17. Thus, C_L becomes the now-active controller regulating the plant, while the tracking loop (F_L, T_L, Q_L) is disconnected and never affects the plant control loop. Clearly, a second tracking loop (also not included in Figure 19.17 for clarity) can then be switched in to ensure that the previously active controller now becomes a latent controller in tracking mode.

The control problem associated with bumpless transfer, then, is the design of the triplet (F_L, T_L, Q_L) to guarantee a certain tracking bandwidth in spite of noise and controller uncertainty in C_L, which is the "plant" of the tracking loop.

Note that this strategy achieves the objectives set out in Section 19.2.2. Firstly, assume that a newly designed controller is temporarily activated for performance analysis (becoming C_A) and the existing controller is placed into tracking mode (becoming C_L). Then, if the new controller inadvertently drives the plant unstable, the existing controller still retains stable tracking of the (unbounded) active control, u_L. Therefore, control can be bumplessly transferred back to the existing controller for immediate stabilization of the plant; see [5] for an example. This guarantees safety during the testing of newly designed controllers.

Secondly, the only signals accessed by the tracking loop are the plant reference, the plant output and the active control signals, all of which are commonly available. Furthermore, by adding a second tracking loop to the diagram of Figure 19.17, the scheme becomes bidirectional and does not require the existing controller to feature bumpless transfer facilities. Thirdly, a distinguishing feature compared to alternative techniques is that the tracking controllers are operating in closed loop for $Q_L \neq 0$. Thus, they can be designed to be insensitive to inaccurately known controllers. Fourthly and finally, the technique does not presuppose the plant controllers to have any particular structure. As long as one can conceive of a tracking controller for them, there is no requirement for them to be biproper, minimum phase, digital or linear.

Once bumpless transfer has been associated with a control problem in this way, tracking controllers can be created by considering the plant controllers as being, themselves, "plants" and designing regulators for them by any desired technique. For a discussion of tracking controller design and the associated robustness issues, see [6].

19.2.4 Further Bumpless Transfer Techniques

In this section, we outline a number of further bumpless transfer techniques commonly referred to in the literature. Most of these techniques are presented in the context of antiwindup design. To be consistent with the literature, we adopt this context here.

If a process exhibits actuator saturations that were neglected in a linear control design, the closed loop may suffer from unacceptable transients after saturation. This is due to the controller states becoming inconsistent with the saturated control signal and is known as *windup*.

Techniques designed to combat windup are known as *antiwindup*. Their aim is to ensure that the control signal never attempts to take a value beyond the saturation limits. In the sense that this requires the control output to track a signal (i.e., the saturation curve), the problem of antiwindup is structurally equivalent to the problem of bumpless transfer.

The perhaps most well-known techniques for antiwindup and bumpless transfer design are the *conditioning technique* of Hanus [8] and Hanus et al. [9] and the observer-based technique by Åström and Wittenmark [1]. The fundamental idea of the conditioning technique is to manipulate the reference signal such that the control signal under consideration is in agreement with the desired control signal, that is, the saturated signal in an antiwindup context and the active control signal in the bumpless transfer context.

The observer-based technique is built on the idea of feeding an observer with the desired (active) control signal and thereby obtaining controller states that are matched to the states of the active controller. Both the above-described techniques have recently been found to be special cases of a more general antiwindup and bumpless transfer structure presented by Rönnbäck [10] and Rönnbäck et al. [11]. We present the conditioning and observer-based techniques in this context next.

Consider the system

$$y = \frac{B}{A}v \tag{19.13}$$

and the controller

$$Fu = (F - PR)v + PTr - PSy \tag{19.14}$$

$$v = \text{sat}(u) \overset{\Delta}{=} \begin{cases} u_{\min} & u \leq u_{\min} \\ u & u_{\min} \leq u \leq u_{\max} \\ u_{\max} & u \geq u_{\max} \end{cases} \tag{19.15}$$

where v is a signal caused by actuator saturations, r is the reference signal, and P and F constitute additional design freedom to combating windup. See Figure 19.19.

When $u(t)$ does not saturate, we obtain the nominal two-degree-of-freedom controller

$$Ru = Tr - Sy \tag{19.16}$$

whereas when $u(t)$ does saturate, feeding back from $v(t)$ prevents the controller states from winding up.

The observer-based technique of Åström and Wittenmark [1] is directly obtained by selecting the polynomials P and F above as $P = 1$ and $F = A_{\circ}$, where $A_{\circ}$ is the characteristic polynomial of the observer. For a particular choice of observer polynomial, namely $A_{\circ} = T/t_0$, where t_0 represents the *high-frequency gain* of T/R in the nominal controller in Equation 19.16, we obtain the conditioning technique.

In the context of antiwindup design, bumpless transfer between an active control signal, u_A, and a latent control signal, u_L, is achieved by setting $u = u_L$, and $u_{\min} = u_{\max} = u_A$. Choosing bumpless transfer technique is thus a matter of choosing polynomials P and F in Equation 19.14. For details about the relations between antiwindup design and bumpless transfer tracking controller design, the reader is

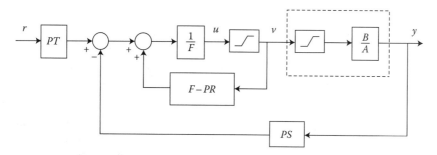

FIGURE 19.19   General antiwindup and bumpless transfer structure.

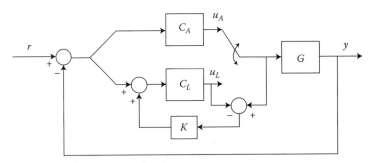

FIGURE 19.20 Conventional high-gain antiwindup and bumpless transfer strategy, with $K = K_0 \gg 1$ or $K = K_0/s$ with $K_0 \gg 1$.

referred to Graebe and Ahlén [6]. More details about antiwindup design can be found in, e.g., [10,11] as well as in another chapter of this book.

Another technique that is popularly used for both antiwindup and bumpless transfer is depicted in the diagram of Figure 19.20, where G is a plant being controlled by the active controller, C_A, and C_L is a latent alternative controller. If

$$K = K_0 \gg 1 \tag{19.17}$$

is a large constant [or a diagonal matrix in the multiple-input multiple-output (MIMO) case], then this technique is also known as *high gain conventional antiwindup and bumpless transfer*. In a slight variation, Uram [12] used the same configuration but proposed K to be designed as the high-gain integrator

$$K = \frac{K_0}{s}, \quad K_0 \gg 1 \tag{19.18}$$

Clearly, the configuration of Figure 19.20 is a special case of the general tracking controller approach of Figure 19.17 with the particular choices $F_L = 1$, $Q_L = 1$ and $T_L = K$, where K is given by either Equation 19.17 or Equation 19.18. One of the advantages of viewing bumpless transfer as a tracking problem is that we can immediately assess some of the implications of the choices in Equation 19.17 or Equation 19.18. The performance of these two schemes is determined by how well the latent controller C_L, viewed as a system, lends itself to wide bandwidth control by a simple proportional or integral controller such as in Equation 19.17 or Equation 19.18.

Campo et al. [2] and Kotare et al. [13] present a general framework that encompasses most of the known bumpless transfer and antiwindup schemes as special cases. This framework lends itself well for the analysis and comparison of given schemes. It is not as obvious, however, how to exploit the framework for synthesis. The fact that different design choices can indeed have a fairly strong impact on the achieved performance is nicely captured by Rönnbäck et al. [11] and Rönnbäck [10]. These authors focus on the problem of controller windup in the presence of actuator saturations. They gain interesting design insights by interpreting the difference between the controller output and the saturated plant input as a fictitious input disturbance. As discussed by Graebe and Ahlén [6], the proposal of Rönnbäck [10] is structurally equivalent to the tracking controller configuration of Section 19.2.3. It is rarely pointed out, however, that the design considerations for bumpless transfer and antiwindup can be rather different.

19.2.5 Summary

This section has discussed the problem of bumpless transfer, which is concerned with smooth switching between alternative dynamical systems. We have highlighted a number of situations in which this problem arises, including switching from manual to automatic control, on-line retuning of filter or controller parameters and tentative evaluation of new controllers during normal plant operation.

If it is possible to manipulate the states of the controller directly, there are several techniques to compute the value of the state vector that will give the desired output. If it is not possible to manipulate the states directly, such as in an analog controller, then the input to the controller can be used instead. Viewed in this way, bumpless transfer becomes a tracking problem, in which the inputs to the controller are manipulated so that its output will track an alternative control signal. Once bumpless transfer is recognized as a tracking problem, systematic design techniques can be applied to design appropriate tracking controllers. We have outlined several advantages with this approach and showed that some other known techniques can be interpreted within this setting.

19.2.6 Defining Terms

Active controller: A regulator controlling a plant at any given time. This term is used to distinguish the active controller from an alternative standby controller in a bumpless transfer context.

Conditioning technique: A technique in which the reference signal is manipulated in order to achieve additional control objectives. Typically, the reference signal is manipulated in order to avoid the control signal's taking a value larger than a known saturation limit.

High-frequency gain: The high-frequency for gain of a strictly proper transfer function is zero; of a biproper transfer function, it is the ratio of the leading coefficients of the numerator and denominator; and for an improper transfer function, it is infinity. Technically, the high-frequency gain of a transfer function H(s) is defined as $\lim_{s \to \infty} H(s)$.

Internal model controller (IMC): A controller parameterization in which the model becomes an explicit component of the controller. Specifically, $C = Q/(1 - Q\hat{G})$, where $\hat{G}$ is a model and Q is a stable and proper transfer function.

Latent controller: A standby controller that is not controlling the process, but that should be ready for a smooth takeover from an active controller in a bumpless transfer context.

Observer-based controller: A controller structure in which the control signal is generated from the states of an observer.

One-degree-of-freedom controller: A control structure in which the reference signal response is uniquely determined from the output disturbance response.

Two-degree-of-freedom controller: A control structure in which the reference signal response can be shaped, to a certain extent, independently of the disturbance response. This is usually achieved with a setpoint filter.

References

1. Åström, K.J. and Wittenmark, B., *Computer Controlled Systems, Theory and Design*, Prentice Hall, Englewood Cliffs, NJ, 2nd ed., 1984.
2. Campo, P.J., Morari, M., and Nett, C. N., Multivariable anti-windup and bumpless transfer: A general theory, *Proc. ACC '89*, 2, 1706–1711, 1989.
3. Morari, M. and Zafiriou, E., *Robust Process Control*, Prentice Hall, Englewood Cliffs, NJ, 1989.
4. Goodwin, G.C., Graebe, S. F., and Levine, W.S., Internal model control of linear systems with saturating actuators, *Proc. ECC '93*, Groningen, The Netherlands, 1993.
5. Graebe, S. F. and Ahlén, A., Dynamic transfer among alternative controllers, *Prepr. 12th IFAC World Congr.*, Vol. 8, Sydney, Australia, 245–248, 1993.
6. Graebe, S. F. and Ahlén, A., Dynamic transfer among alternative controllers and its relation to anti-windup controller design, *IEEE Trans. Control Syst. Technol.* To appear. Jan. 1996.
7. Graebe, S. F., Goodwin, G.C., and Elsley, G., Rapid prototyping and implementation of control in continuous steel casting, Tech. Rep. EE9471, Dept. Electrical Eng., University of Newcastle, NSW 2308, Australia, 1994.
8. Hanus, R., A new technique for preventing control windup, *Journal A*, 21, 15–20, 1980.
9. Hanus, R., Kinnaert, M., and Henrotte, J–L., Conditioning technique, a general anti-windup and bumpless transfer method, *Automatica*, 23(6), 729–739, 1987.

10. Rönnbäck, S., Linear control of systems with actuator constraints, Ph.D. thesis, Luleå University of Technology, Sweden, 1190, 1993.
11. Rönnbäck, S. R., Walgama, K. S., and Sternby, J., An extension to the generalized anti-windup compensator, in *Mathematics of the Analysis and Design of Process Control*, Borne, P., et al., Eds., Elsevier/North-Holland, Amsterdam, 1992.
12. Uram, R., Bumpless transfer under digital control, *Control Eng.*, 18(3), 59–60, 1971.
13. Kotare, M.V., Campo, P.J., Morari, M., and Nett, C.N., A unified framework for the study of anti-windup designs, CDS Tech. Rep. No. CIT/CDS 93-010, California Institute of Technology, Pasadena, 1993.

19.3 Linearization and Gain-Scheduling

Jeff S. Shamma

19.3.1 Introduction

A classical dilemma in modeling physical systems is the trade-off between model accuracy and tractability. While sophisticated models might provide accurate descriptions of system behavior, the resulting analysis can be considerably more complicated. Simpler models, on the other hand, may be more amenable for analysis and derivation of insight, but might neglect important system behaviors. Indeed, the required fidelity of a model depends on the intended utility. For example, one may use a very simplified model for the sake of control design, but then use a sophisticated model to simulate the overall control system.

One instance where this dilemma manifests itself is the use of linear versus nonlinear models. Nonlinearities abound in most physical systems. Simple examples include saturations, rate limiters, deadzones, and backlash. Further examples include the inherently nonlinear behavior of systems such as robotic manipulators, aircraft, and chemical process plants. However, methods for analysis and control design are considerably more available for linear systems than nonlinear systems.

One approach is to directly address the nonlinear behavior of such systems, and nonlinear control design remains an topic of active research. An alternative is to linearize the system dynamics, i.e., to approximate the nonlinear model by a linear one. Some immediate drawbacks are that (1) the linear model can give only a local description of the system behavior and (2) some of the intricacies of the system behavior may be completely neglected by the linear approximation—even locally. In some cases, these consequences are tolerable, and one may then employ methods for linear systems.

One approach to address the local restriction of linearization-based controllers is to perform several linearization-based control designs at many operating conditions and then interpolate the local designs to yield an overall nonlinear controller. This procedure is known as *gain-scheduling*. It is an intuitively appealing but heuristic practice, which is used in a wide variety of control applications. It is especially prevalent in flight control systems.

This chapter presents a review of linearization and gain scheduling, exploring both the benefits and practical limitations of each.

19.3.2 Linearization

19.3.2.1 An Example

To illustrate the method of linearization, consider a single link coupled to a rotational inertia by a flexible shaft (Figure 19.21). The idea is to control the link through a torque on the rotational inertia. This physical system may be viewed as a very simplified model of a robotic manipulator with flexible joints.

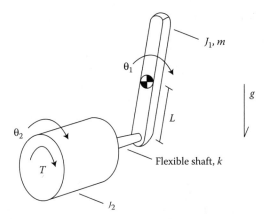

FIGURE 19.21 Rotational link.

The equations of motion are given by

$$\frac{d}{dt}\begin{pmatrix}\theta_1(t)\\ \theta_2(t)\\ \dot{\theta}_1(t)\\ \dot{\theta}_2(t)\end{pmatrix} = \begin{pmatrix}0\\ 0\\ (mgL\sin(\theta_1(t)))/J_1 - c\dot{\theta}_1(t)\,|\dot{\theta}_1(t)|\\ 0\end{pmatrix}$$

$$+ \begin{pmatrix}0 & 0 & 1 & 0\\ 0 & 0 & 0 & 1\\ -k/J_1 & k/J_1 & 0 & 0\\ k/J_2 & -k/J_2 & 0 & 0\end{pmatrix}\begin{pmatrix}\theta_1(t)\\ \theta_2(t)\\ \dot{\theta}_1(t)\\ \dot{\theta}_2(t)\end{pmatrix} + \begin{pmatrix}0\\ 0\\ 0\\ 1/J_2\end{pmatrix}T(t) \qquad (19.19)$$

Here $\theta_1(t), \theta_2(t)$ are angles measured from vertical, $T(t)$ is the torque input, k is a rotational spring constant, c is a nonlinear damping coefficient, J_1, J_2 are rotational inertias, L is the link length, and m is the link mass.

Now suppose the link is to be controlled in the vicinity of the upright stationary position. First-order approximations near this position lead to $\sin\theta_1 \simeq \theta_1$ and $c\dot{\theta}_1\,|\dot{\theta}_1| \simeq 0$. The state equation 19.19 is then approximated by the equations,

$$\frac{d}{dt}\begin{pmatrix}\theta_1(t)\\ \theta_2(t)\\ \dot{\theta}_1(t)\\ \dot{\theta}_2(t)\end{pmatrix} \simeq \begin{pmatrix}0 & 0 & 1 & 0\\ 0 & 0 & 0 & 1\\ (mgL-k)/J_1 & k/J_1 & 0 & 0\\ k/J_2 & -k/J_2 & 0 & 0\end{pmatrix}\begin{pmatrix}\theta_1(t)\\ \theta_2(t)\\ \dot{\theta}_1(t)\\ \dot{\theta}_2(t)\end{pmatrix} + \begin{pmatrix}0\\ 0\\ 0\\ 1/J_2\end{pmatrix}T(t) \qquad (19.20)$$

The simplified dynamics are now in the general *linear* form

$$\dot{x}(t) = Ax(t) + Bu(t) \qquad (19.21)$$

Two consequences of the linearization are

- Global behavior, such as full angular rotations, are poorly approximated.
- The nonlinear damping, $c\dot{\theta}_1\,|\dot{\theta}_1|$, is completely neglected, even locally.

Despite these limitations, an analysis or control design based on the linearization can still be of value for the nonlinear system, provided that the state vector and control inputs are close to the upright equilibrium.

19.3.2.2 Linearization of Functions

This section reviews some basic concepts from multivariable calculus. For a vector $x \in \mathcal{R}^n$, $|x|$ denotes the Euclidean norm, i.e.,

$$|x| = \left(\sum_{i=1}^{n} x_i^2 \right)^{1/2} \tag{19.22}$$

Let $f : \mathcal{R}^n \to \mathcal{R}^p$, i.e., f is a function which maps vectors in $\mathcal{R}^n$ to values in $\mathcal{R}^p$. In terms of individual components,

$$f(x) = \begin{pmatrix} f_1(x_1, \ldots, x_n) \\ \vdots \\ f_p(x_1, \ldots, x_n) \end{pmatrix} \tag{19.23}$$

where the x_i are scalar components of the $\mathcal{R}^n$-vector x, and the f_i are scalar valued functions of $\mathcal{R}^n$.

The *Jacobian matrix* of f is denoted Df and is defined as the $p \times n$ matrix of partial derivatives

$$Df = \begin{pmatrix} \dfrac{\partial f_1}{\partial x_1} & \cdots & \dfrac{\partial f_1}{\partial x_n} \\ \vdots & \ddots & \vdots \\ \dfrac{\partial f_p}{\partial x_1} & \cdots & \dfrac{\partial f_p}{\partial x_n} \end{pmatrix} \tag{19.24}$$

In case f is continuously differentiable at x_o, then the Jacobian matrix can be used to approximate f. A multivariable Taylor series expansion takes the form

$$f(x) = f(x_o) + Df(x_o)(x - x_o) + r(x) \tag{19.25}$$

where the remainder, $r(x)$, represents higher-order terms which satisfy

$$\lim_{h \to 0} \frac{|r(x_o + h)|}{|h|} = 0 \tag{19.26}$$

Now let $f : \mathcal{R}^n \times \mathcal{R}^m \to \mathcal{R}^p$, i.e., f is a function which maps a pair of vectors in $\mathcal{R}^n$ and $\mathcal{R}^m$, respectively, to values in $\mathcal{R}^p$. The notations $D_1 f$ and $D_2 f$ denote the Jacobian matrices with respect to the first variable and second variables, respectively. Thus, if

$$f(x, u) = \begin{pmatrix} f_1(x_1, \ldots, x_n, u_1, \ldots, u_m) \\ \vdots \\ f_p(x_1, \ldots, x_n, u_1, \ldots, u_m) \end{pmatrix} \tag{19.27}$$

then $D_1 f$ denotes the $p \times n$ matrix

$$D_1 f = \begin{pmatrix} \dfrac{\partial f_1}{\partial x_1} & \cdots & \dfrac{\partial f_1}{\partial x_n} \\ \vdots & \ddots & \vdots \\ \dfrac{\partial f_p}{\partial x_1} & \cdots & \dfrac{\partial f_p}{\partial x_n} \end{pmatrix} \tag{19.28}$$

and $D_2 f$ denotes the $p \times m$ matrix

$$D_2 f = \begin{pmatrix} \dfrac{\partial f_1}{\partial u_1} & \cdots & \dfrac{\partial f_1}{\partial u_m} \\ \vdots & \ddots & \vdots \\ \dfrac{\partial f_p}{\partial u_1} & \cdots & \dfrac{\partial f_p}{\partial u_m} \end{pmatrix} \tag{19.29}$$

Example 19.7: Rotational Link Jacobian Matrices

Consider again the rotational link example in Equation 19.19. Let x denote the state vector, u denote the torque input, and $f(x, u)$ denote the right-hand side of Equation 19.19. The Jacobian matrices $D_1 f$ and $D_2 f$ are given by

$$D_1 f(x, u) = \begin{pmatrix} 0 & 0 & 1 & 0 \\ 0 & 0 & 0 & 1 \\ (mgL \cos \theta_1 - k)/J_1 & k/J_1 & 2c \, |\dot{\theta}_1| & 0 \\ k/J_2 & -k/J_2 & 0 & 0 \end{pmatrix}$$

$$D_2 f(x, u) = \begin{pmatrix} 0 \\ 0 \\ 0 \\ 1/J_2 \end{pmatrix} \tag{19.30}$$

19.3.2.3 Linearization about an Equilibrium

Approximation of System Dynamics

A general form for nonlinear differential equations is

$$\dot{x}(t) = f(x(t), u(t)) \tag{19.31}$$

where $x(t)$ is the state vector, $u(t)$ is the input vector, and $f : \mathcal{R}^n \times \mathcal{R}^m \to \mathcal{R}^n$. The existence and uniqueness of solutions will be assumed.

The pair (x_o, u_o) is called an *equilibrium* if

$$0 = f(x_o, u_o) \tag{19.32}$$

The reasoning behind this terminology is that, starting from the initial condition $x(0) = x_o$ with a constant input $u(t) = u_o$, the solution *remains* at $x(t) = x_o$.

Assuming that f is continuously differentiable at (x_o, u_o), a multivariable Taylor series expansion yields

$$\dot{x}(t) = f(x_o, u_o) + D_1 f(x_o, u_o)(x(t) - x_o)$$
$$+ D_2 f(x_o, u_o)(u(t) - u_o) + r(x(t), u(t)) \tag{19.33}$$

where the remainder, $r(x, u)$, satisfies

$$\lim_{(x,u) \to (x_o, u_o)} \frac{r(x, u)}{\sqrt{|x - x_o|^2 + |u - u_o|^2}} = 0 \tag{19.34}$$

Thus, the approximation is accurate up to first order. Define the deviation-from-equilibrium terms

$$\tilde{x}(t) = x(t) - x_o \tag{19.35}$$
$$\tilde{u}(t) = u(t) - u_o \tag{19.36}$$

Assuming that the equilibrium is *fixed*, i.e., $\frac{d}{dt} x_o = 0$, along with the condition $0 = f(x_o, u_o)$, leads to

$$\dot{\tilde{x}}(t) \simeq D_1 f(x_o, u_o) \tilde{x}(t) + D_2 f(x_o, u_o) \tilde{u}(t) \tag{19.37}$$

Equation 19.37 represents the *linearization* of the nonlinear dynamics (Equation 19.31) about the equilibrium point (x_o, u_o).

Example 19.8: Rotational Link Linearization

Consider again the rotational link example of Equation 19.19. In addition to the upright equilibrium, there is a *family* of equilibrium conditions given by

$$x_o = \begin{pmatrix} q \\ (q - mgL \sin q)/k \\ 0 \\ 0 \end{pmatrix} \qquad u_o = -mgL \sin q \tag{19.38}$$

where q denotes the equilibrium angle for θ_1. When $q = 0$, Equation 19.38 yields the upright equilibrium.

For a fixed q, the linearized dynamics about the corresponding equilibrium point may be obtained by substituting the Jacobian matrices from Example 19.7 into Equation 19.37 to yield

$$\dot{\tilde{x}}(t) = \begin{pmatrix} 0 & 0 & 1 & 0 \\ 0 & 0 & 0 & 1 \\ (mgL \cos q - k)/J_1 & k/J_1 & 0 & 0 \\ k/J_2 & -k/J_2 & 0 & 0 \end{pmatrix} \tilde{x}(t) + \begin{pmatrix} 0 \\ 0 \\ 0 \\ 1/J_2 \end{pmatrix} \tilde{u}(t) \tag{19.39}$$

As before, the nonlinear damping is completely neglected in the linearized equations.

Stability

This section discusses how linearization of a nonlinear system may be used to analyze stability. First, some definitions regarding stability are reviewed.

Let x_o be an equilibrium for the unforced state equations:

$$\dot{x}(t) = f(x(t)) \tag{19.40}$$

i.e., $f(x_o) = 0$.

The equilibrium x_o is *stable* if for each $\varepsilon > 0$, there exists a $\delta(\varepsilon) > 0$ such that

$$|x(0) - x_o| < \delta(\varepsilon) \implies |x(t) - x_o| < \varepsilon, \quad \forall t \geq 0 \tag{19.41}$$

It is *asymptotically stable* if it is stable and for some δ^*

$$|x(0) - x_o| < \delta^* \implies |x(t) - x_o| \to 0, \text{ as } t \to \infty \tag{19.42}$$

It is *unstable* if it is not stable. Note that the conditions for stability pertain to a neighborhood of an equilibrium.

The following is a standard analysis result. Let $f : \mathcal{R}^n \to \mathcal{R}^n$ be continuously differentiable. The equilibrium x_o is asymptotically stable if all of the eigenvalues of $Df(x_o)$ have negative real parts. It is unstable if $Df(x_o)$ has an eigenvalue with a positive real part.

Since the eigenvalues of $Df(x_o)$ determine the stability of the linearized system

$$\dot{\tilde{x}}(t) = Df(x_o)\tilde{x}(t) \tag{19.43}$$

this result states that the linearization can provide *sufficient conditions* for stability of the nonlinear system in a neighborhood of an equilibrium. In case $Df(x_o)$ has purely imaginary eigenvalues, then nonlinear methods are *required* to assess stability of the nonlinear system.

Example 19.9: Rotational Link Stability

Consider the *unforced* rotational link equations, i.e., torque $T(t) = 0$. In this case, the two equilibrium conditions are the upright position, $x_o = 0$, or the hanging position, $x_o = (\pi\ 0\ 0\ 0)^T$. Intuitively, the upright equilibrium is unstable, while the hanging equilibrium is stable.

Let $J_1, J_2, k, m, g, L, c = 1$. For the upright equilibrium, the linearized equations are

$$\dot{\tilde{x}}(t) = \begin{pmatrix} 0 & 0 & 1 & 0 \\ 0 & 0 & 0 & 1 \\ 0 & 1 & 0 & 0 \\ 1 & -1 & 0 & 0 \end{pmatrix} \tilde{x}(t) \tag{19.44}$$

which has eigenvalues of $(\pm 0.79,\ \pm 1.27j)$. Since one of the eigenvalues has a positive real part, the upright equilibrium of the original nonlinear system is unstable.

For the hanging equilibrium, the linearized equations are

$$\dot{\tilde{x}}(t) = \begin{pmatrix} 0 & 0 & 1 & 0 \\ 0 & 0 & 0 & 1 \\ -2 & 1 & 0 & 0 \\ 1 & -1 & 0 & 0 \end{pmatrix} \tilde{x}(t) \tag{19.45}$$

Note that $\tilde{x}(t)$ represents *different* quantities in the two linearizations, namely the deviation from the different corresponding equilibrium positions. The linearization now has eigenvalues of $(\pm 1.61j,\ \pm 0.61j)$. In this case, the linearization *does not* provide information regarding the stability of the nonlinear system. This is because the nonlinear damping term $c\dot{\theta}_1 |\dot{\theta}_1|$ is completely neglected. If this term were replaced by linear damping in (Equation 19.19), e.g., $c\dot{\theta}_1$, then the linearized dynamics become

$$\dot{\tilde{x}}(t) = \begin{pmatrix} 0 & 0 & 1 & 0 \\ 0 & 0 & 0 & 1 \\ -2 & 1 & -1 & 0 \\ 1 & -1 & 0 & 0 \end{pmatrix} \tilde{x}(t) \tag{19.46}$$

and the resulting eigenvalues are $(-0.35 \pm 1.5j,\ -0.15 \pm 0.63j)$. In this case, the linearized dynamics are asymptotically stable, which in turn implies that the nonlinear dynamics are asymptotically stable. In the case of the nonlinear damping, one may use alternate methods to show that the hanging equilibrium is indeed asymptotically stable. However, this could not be concluded from the linearization.

Stabilization

Linearization of a nonlinear system also may be used to design stabilizing controllers. Let (x_o, u_o) be an equilibrium for the nonlinear equations 19.31. Let

$$y(t) = g(x(t)) \tag{19.47}$$

denote the outputs available for measurement. In case f and g are continuously differentiable, the linearized equations are

$$\dot{\tilde{x}}(t) = D_1(x_o, u_o)\tilde{x}(t) + D_2(x_o, u_o)\tilde{u}(t) \tag{19.48}$$

$$\tilde{y}(t) = Dg(x_o)\tilde{x}(t) \tag{19.49}$$

where $\tilde{x}(t) = x(t) - x_o$, $\tilde{u}(t) = u(t) - u_o$, and $\tilde{y}(t) = y(t) - g(x_o)$.

Using the results from the previous section, if the controller

$$\dot{z}(t) = Az(t) + B\tilde{y}(t) \tag{19.50}$$

$$\tilde{u}(t) = Cz(t) + D\tilde{y}(t) \tag{19.51}$$

asymptotically stabilizes the linearized system, then the control

$$u(t) = u_o + \tilde{u}(t) \tag{19.52}$$

asymptotically stabilizes the nonlinear system at the equilibrium (x_o, u_o).

Conversely, a linearized analysis under certain conditions can show that no controller with continuously differentiable dynamics is asymptotically stabilizing. More precisely, consider the controller

$$\dot{z}(t) = F(z(t), y(t)) \tag{19.53}$$

$$u(t) = G(z(t)) \tag{19.54}$$

where $(0, y_0)$ is an equilibrium and $u_o = G(0)$. Suppose $[D_1 f(x_o, u_o), D_2 f(x_o, u_o)]$ either is *not* a stabilizable pair or $[Dg(x_o), D_1 f(x_o, u_o)]$ is not a detectable pair. Then *no* continuously differentiable F and G lead to an asymptotically stablilizing controller.

Example 19.10: Rotational Link Stabilization

Consider the rotational link equations with $J_1, J_2, k, m, g, L, c = 1$. The equilibrium position with the link at 45 degrees is given by

$$x_o = \begin{pmatrix} \frac{\pi}{4} \\ \frac{\pi}{4} - \frac{1}{\sqrt{2}} \\ 0 \\ 0 \end{pmatrix}, \quad u_o = \frac{-1}{\sqrt{2}} \tag{19.55}$$

For this equilibrium, the linearized equations are

$$\dot{\tilde{x}}(t) = A\tilde{x}(t) + B\tilde{u}(t) \tag{19.56}$$

where

$$A = \begin{pmatrix} 0 & 0 & 1 & 0 \\ 0 & 0 & 0 & 1 \\ \frac{1}{\sqrt{2}} - 1 & 1 & 0 & 0 \\ 1 & -1 & 0 & 0 \end{pmatrix} \tilde{x}, \quad B = \begin{pmatrix} 0 \\ 0 \\ 0 \\ 1 \end{pmatrix} \tilde{u} \tag{19.57}$$

One can show that the state feedback (which resembles proportional-derivative feedback)

$$\tilde{u}(t) = -K\tilde{x}(t) = -(2\ 4\ 4\ 2)\tilde{x}(t) \tag{19.58}$$

is stabilizing. Therefore the control

$$u(t) = u_o - K(x(t) - x_o) \tag{19.59}$$

stabilizes the nonlinear system at the $45°$ equilibrium.

Now suppose only $\theta_1(t)$ is available for feedback, i.e.,

$$y(t) = Cx(t) = (1\ 0\ 0\ 0)x(t) \tag{19.60}$$

Let $\tilde{y}(t) = y(t) - Cx_o$. A model-based controller which stabilizes the linearization is

$$\dot{z}(t) = Az(t) - BKz(t) + H(\tilde{y}(t) - Cz(t)) \tag{19.61}$$

$$\tilde{u}(t) = -Kz(t) \tag{19.62}$$

where

$$H = \begin{pmatrix} 1 \\ 2 \\ 2 \\ 1 \end{pmatrix} \tag{19.63}$$

Therefore, the control

$$\dot{z}(t) = Az(t) - BKz(t) + H(y(t) - y_o - Cz(t)) \tag{19.64}$$

$$u(t) = u_o - Kz(t) \tag{19.65}$$

stabilizes the nonlinear system.

19.3.2.4 Limitations of Linearization

This section presents several examples that illustrate some limitations in the utility of linearizations.

Example 19.11: Hard Nonlinearities

Consider the system

$$\dot{x}(t) = Ax(t) + BN(u(t)) \tag{19.66}$$

This system represents linear dynamics where the input u first passes through a nonlinearity, N. Some common nonlinearities are saturation:

$$N(u) = \begin{cases} 1 & u \geq 1 \\ u & -1 \leq u \leq 1 \\ -1 & u \leq -1 \end{cases} \tag{19.67}$$

deadzone:

$$N(u) = \begin{cases} u - 1 & u \geq 1 \\ 0 & -1 \leq u \leq 1 \\ u + 1 & u \leq -1 \end{cases} \tag{19.68}$$

and relay:

$$N(u) = \begin{cases} 1 & u > 0 \\ -1 & u < 0 \end{cases} \tag{19.69}$$

Other nonlinearities are backlash and hysteresis. All of these nonlinearities do not lend themselves to linearization-based analysis. Even in the regions where the nonlinearities are differentiable, a linearization *completely removes* the intricacies that the nonlinearities cause.

Example 19.12: Local Nature of Linearization

Consider the scalar equation

$$\dot{x}(t) = -\sin(x(t)) \tag{19.70}$$

The equilibrium $x_o = 0$ is asymptotically stable. However, the equilibrium $x_o = \pi$ is not.

Example 19.13: Linearization not Asymptotically Stable

Consider the scalar equation

$$\dot{x}(t) = -x^3(t) \tag{19.71}$$

The equilibrium $x_o = 0$ is globally asymptotically stable, i.e., $x(t) \to 0$ as $t \to \infty$ for any initial condition. However, the linearization yields

$$\dot{\tilde{x}}(t) = 0 \tag{19.72}$$

which is inconclusive. A similar phenomenon was seen with the rotational link in the hanging equilibrium.

Example 19.14: Linearization not Stabilizable

Consider the scalar equation

$$\dot{x}(t) = x(t) + x(t)u(t) \tag{19.73}$$

At the equilibrium $x_o, u_o = 0$, the linearization is

$$\dot{\tilde{x}}(t) = \tilde{x}(t) \tag{19.74}$$

This is not stabilizable, since no input term appears. However, the constant feedback $u(t) = -2$ in Equation 19.73 yields

$$\dot{x}(t) = -x(t) \tag{19.75}$$

which is asymptotically stable.

Example 19.15: Non-Differentiable Feedback

Consider the second-order nonlinear system

$$\dot{x}_1(t) = u(t) \tag{19.76}$$
$$\dot{x}_2(t) = x_2(t) - x_1^3(t) \tag{19.77}$$

For the equilibrium $x_o, u_o = 0$, any stabilizing feedback law, $u = g(x)$, must satisfy $g(0) = 0$. Suppose that g is continuously differentiable. The linearization of the closed-loop system yields

$$\dot{\tilde{x}}(t) = \begin{pmatrix} \partial g/\partial x_1(0) & \partial g/\partial x_2(0) \\ 0 & 1 \end{pmatrix} \tilde{x}(t) \tag{19.78}$$

which is unstable. Therefore, no continuously differentiable feedback is stabilizing. However, one can show that the *nondifferentiable* feedback

$$u(x) = -x_1 + x_2 + \frac{4}{3}x_2^{1/3} - x_1^3 \tag{19.79}$$

is stabilizing.

19.3.2.5 Linearization about a Trajectory

The previous sections addressed linearization about a single equilibrium point. Another situation in which linearization can be useful is where the nonlinear system is to follow a prescribed trajectory. Possible sources for this trajectory are repeated maneuvers of the nonlinear system or the outcome of some trajectory optimization (e.g., a robot following a specified optimal path or an aerospace vehicle executing an optimal flight path). The objective of the linearization is then to study the behavior of the system near the prescribed trajectory.

Let $x^*(t)$ and $u^*(t)$ satisfy the nonlinear differential equation 19.31. Let f be continuously differentiable. The objective is to examine the behavior of the nonlinear system near the *trajectory* $(x^*(t), u^*(t))$. Toward this end, let

$$\tilde{x}(t) = x(t) - x^*(t), \quad \tilde{u}(t) = u(t) - u^*(t) \tag{19.80}$$

Assuming that f is continuously differentiable, it may be approximated near the trajectory $(x^*(t), u^*(t))$ by

$$f(x(t), u(t)) \simeq f(x^*(t), u^*(t)) + D_1 f(x^*(t), u^*(t))\tilde{x}(t)$$
$$+ D_2 f(x^*(t), u^*(t))\tilde{u}(t) \tag{19.81}$$

Substituting this approximation into Equation 19.31 leads to

$$\dot{x}(t) \simeq f(x^*(t), y^*(t)) + D_1 f(x^*(t), u^*(t))\tilde{x}(t)$$
$$+ D_2 f(x^*(t), u^*(t))\tilde{u}(t) \tag{19.82}$$

Using $\dot{x}^*(t) = f(x^*(t), u^*(t))$ then leads to the linear *time-varying* dynamics:

$$\dot{\tilde{x}}(t) = D_1 f(x^*(t), u^*(t))\tilde{x} + D_2 f(x^*(t), u^*(t))\tilde{u}(t) \tag{19.83}$$

The time-varying nature of the linearization occurs even though the original nonlinear system is time-invariant.

As in the case of linearization about an equilibrium, the linearized dynamics may be used to infer properties of the nonlinear system when the state-trajectory is close to $x^*(t)$ and the input history is close to $u^*(t)$. Linearization along a trajectory does not restrict the nonlinear system to stay close to a single equilibrium point. The cost of this advantage is that the situation is more complicated, in that one must establish stability properties and/or design stabilizing controllers for linear—but time-varying—system dynamics. These issues are discussed in other articles.

Example 19.16: Rotational Link along a Trajectory

Consider the rotational link equations with $J_1, J_2, k, m, g, L, c = 1$. The nominal trajectory of interest is the link at a constant rate of rotation, i.e., $\theta_1^*(t) = \omega_0 t$. Solving for the remaining states and torque leads to

$$\theta_2^*(t) = \omega_0 + \omega_0^2 + \omega_t o - \sin \omega_0 t \tag{19.84}$$

$$\dot{\theta}_1^*(t) = \omega_0 \tag{19.85}$$

$$\dot{\theta}_2^*(t) = \omega_0 - \omega_0 \cos \omega_0 t \tag{19.86}$$

$$T^*(t) = 2\omega_0 + \omega_0^2 - \omega_0 \cos \omega_0 t - \sin \omega_0 t \tag{19.87}$$

Linearizing along this trajectory yields

$$\dot{\tilde{x}}(t) = \begin{pmatrix} 0 & 0 & 1 & 0 \\ 0 & 0 & 0 & 1 \\ \cos \omega_0 t - 1 & 1 & 2\omega_0 & 0 \\ 1 & -1 & 0 & 0 \end{pmatrix} \tilde{x}(t) + \begin{pmatrix} 0 \\ 0 \\ 0 \\ 1 \end{pmatrix} \tilde{u}(t) \tag{19.88}$$

As expected, the linearized dynamics are time-varying (note the $\cos \omega_0 t$ term).

19.3.3 Gain Scheduling

19.3.3.1 Motivation

A major drawback of linearization is that a control design based on the linearized dynamics need not exhibit good performance or even be stabilizing when operating away from the equilibrium. One possibility is to linearize along a trajectory which is not restricted to a local operating region. However, this trajectory must be known in advance in order to perform the control design. Such advance knowledge of the trajectory is often not available.

One approach to address the local restriction in linearization is a design procedure called *gain scheduling*. The main idea is to break the control design process into two steps. First, one designs local linear controllers based on linearizations of the nonlinear system at several different equilibria, usually called in this context operating conditions. Second, a global nonlinear controller for the nonlinear system is obtained by interpolating, or "scheduling," the local operating point designs.

One example is flight control. Here, the linearized systems correspond to the aircraft in a particular flight condition characterized by the atmospheric conditions, aircraft orientation, and aircraft velocity. The local linear controllers are adequate to control the aircraft near a particular operating condition. The global controller, formed by patching together local controllers, is needed to provide transitions between flight conditions.

While intuitively appealing, gain scheduling is an ad hoc practice guided by heuristic rules of thumb. Nevertheless, it does enjoy widespread usage in a variety of applications, such as aircraft control, missile autopilots, jet-engine control, and process control.

This section provides an outline of gain scheduling, its advantages, and limitations.

19.3.3.2 Gain-Scheduled Control Design

Nonlinear Systems

Consider the nonlinear system

$$\dot{x}(t) = f(x(t), u(t)) \tag{19.89}$$

$$y(t) = g(x(t)) \tag{19.90}$$

where $y(t)$ denotes the measured output. Assume that there exists a parameterized family of equilibrium points (x_{eq}, u_{eq}), i.e.,

$$0 = f(x_{eq}(s), u_{eq}(s)) \tag{19.91}$$

where s takes its values in some specified operating region. The variable s, called the *scheduling variable*, will be measured upon operation of the control system and will be used to infer the equilibrium to which the system is near.

Now assuming that s is *fixed* leads to a family of linearizations

$$\dot{\tilde{x}}(t) = A(s)\tilde{x}(t) + B(s)\tilde{u}(t) \tag{19.92}$$

$$\tilde{y}(t) = C(s)\tilde{x}(t) \tag{19.93}$$

where

$$A(s) = D_1 f(x_{eq}(s), u_{eq}(s))$$
$$B(s) = D_2 f(x_{eq}(s), u_{eq}(s))$$
$$C(s) = Dg(x_{eq}(s)) \tag{19.94}$$
$$\tilde{x}(t) = x(t) - x_{eq}(s)$$
$$\tilde{u}(t) = u(t) - u_{eq}(s)$$
$$\tilde{y}(t) = y(t) - g(x_{eq}(s)) \tag{19.95}$$

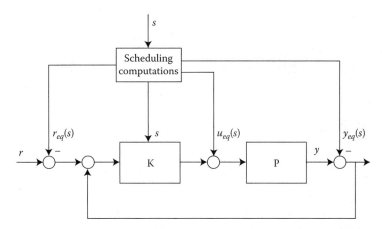

FIGURE 19.22   Gain-scheduled command following.

Under the appropriate stabilizability/detectability assumptions, one can design stabilizing linear controllers (using any of a variety of linear design methods). The result is an indexed collection of controllers,

$$\dot{z}(t) = \overline{A}(s)z(t) + \overline{B}(s)\tilde{y}(t) \tag{19.96}$$

$$\tilde{u}(t) = \overline{C}(s)z(t) + \overline{D}(s)\tilde{y}(t) \tag{19.97}$$

In practice, controllers usually are not designed at every value of s but rather at several operating points indexed by selected values $\{s_1, s_2, \ldots, s_N\}$. In between these points, the controller matrices are interpolated according to the scheduling variable, s.

Although the family of controllers is designed assuming that s is fixed, upon operation of the control system the controller matrices *vary* in time according to the evolution of s. The method of changing the controller matrices can either be smooth or discontinuous switching. In either case, the desired effect is to alleviate the restriction to the local operating region of any individual linearized design. Therefore, depending on the current region of operation (according to s), appropriate controller gains are employed. For example, Figure 19.22 shows a block diagram of gain scheduling for command following. The scheduling variable, s, can either be endogenous to the plant (e.g., a particular state-variable) or an exogenous signal (e.g., a function of some reference command r).

Example 19.17: Gain-Scheduled Design for Simplified Rotational Link

Consider the rotational link example with the simplification that the rotational flexibility is now rigid. In this case, $\theta_1 = \theta_2$ and $J = J_1 + J_2$. Dropping the subscript on the angles, the equations simplify to

$$\frac{d}{dt}\begin{pmatrix} \theta(t) \\ \dot{\theta}(t) \end{pmatrix} = \begin{pmatrix} \theta(t) \\ mgL\sin(\theta(t))/J - c\dot{\theta}(t)\,|\dot{\theta}(t)| + T(t) \end{pmatrix} \tag{19.98}$$

which are in the form $\dot{x}(t) = f(x(t), u(t))$. Let $r(t)$ denote the reference command for $\theta(t)$. A family of equilibrium conditions is parameterized by

$$x_{eq}(s) = \begin{pmatrix} s \\ 0 \end{pmatrix}, \quad u_{eq}(s) = -mgL\sin s/J \tag{19.99}$$

For a fixed s, the linearization is

$$\dot{\tilde{x}}(t) = A(s)\tilde{x}(t) + B(s)\tilde{u}(t) \tag{19.100}$$

where

$$A(s) = \begin{pmatrix} 0 & 1 \\ mgL\cos s/J & 0 \end{pmatrix}, \quad B(s) = \begin{pmatrix} 0 \\ 1 \end{pmatrix}$$ (19.101)

Let $m, g, L, J = 1$. At any *fixed* equilibrium, the proportional–derivative feedback

$$\tilde{u}(t) = -\left(\cos s + 2 \quad 2\right)\tilde{x}(t) + 2\tilde{r}(t)$$ (19.102)

places the closed-loop poles at $-1 \pm j$ and has zero steady-state error to step commands, where $\tilde{r}(t) = r(t) - s$. The family of linearization-based controllers is then

$$u(t) = u_{eq}(s) - \left(\cos s + 2 \quad 2\right)(x(t) - x_{eq}(s)) + 2(r(t) - s)$$ (19.103)

At this point, the scheduling variable s will vary in time according to some scheduling variable. The decision now becomes how to schedule the gains and what to use as a scheduling variable. More precisely, the choices are $s(t) = \theta(t)$ vs. $s(t) = r(t)$ for the scheduling variable and smooth vs. switched scheduling. These options are described as follows:

Switched Scheduling on $\theta(t)$ In this case, the operating range is divided into several regions $\{R_1, \ldots, R_N\}$. Within each region is a representative equilibrium, say $\{\theta_1^*, \ldots, \theta_N^*\}$ and the scheduling variable varies according to

$$s(t) = \theta_i^* \quad \text{whenever } \theta(t) \in R_i$$ (19.104)

Smooth Scheduling on $\theta(t)$ Rather than switch between operating points, let $s(t) = \theta(t)$. A peculiar consequence of such scheduling is that the linearization of the overall closed-loop system *differs* from the original linearized plant and linear controller. This is due to terms that were constant, but now vary, such as $u_{eq}(s)$ and $\cos s$.

Switched Scheduling on $r(t)$ This scheduling is based on the *anticipation* that the angle $\theta(t)$ will follow the reference command. Similarly to switched scheduling on $\theta(t)$, set

$$s(t) = \theta_i^* \quad \text{whenever } r(t) \in R_i$$ (19.105)

Smooth Scheduling on $r(t)$ Again an inherent assumption is that the angle $\theta(t)$ does not lag the reference command. One possibility is $s(t) = r(t - T)$.

Note that the gain-scheduled design allows larger variations in $\theta(t)$ than would a design based on a single equilibrium. However, effects such as the nonlinear damping are still neglected. In case $\dot{\theta}(t)$ is large, the approximation accuracy of the fixed linearizations (upon which the gain-scheduled designs are based) suffers. Fast changes in the scheduling variable also increase the discrepancy between the resulting system dynamics and the design model linearization.

Linear Parameter Varying Systems

A "linear parameter varying" (LPV) system is defined as a linear system whose coefficients depend on an exogenous time-varying parameter, e.g.,

$$\dot{x}(t) = A(\theta(t))x(t) + B(\theta(t))u(t)$$ (19.106)
$$y(t) = C(\theta(t))x(t)$$ (19.107)

The exogenous parameter, $\theta(t)$, is unknown *a priori*; however, it can be measured or estimated upon operation of the system. The reason for the special nomenclature is to distinguish LPV systems from linear time-varying systems for which the time variations are known beforehand (as in periodic systems). Typical *a priori* assumptions on $\theta(t)$ are bounds on its magnitude and rate of change.

LPV systems form a useful paradigm for the study of gain-scheduled control. Gain-scheduled control design traditionally starts with a family of linearizations of a nonlinear system indexed by a scheduling variable. This naturally leads to the LPV structure. An LPV structure also arises from simplifying assumptions on the internal structure of a nonlinear model. Rather than model the dynamic evolution of a particular variable, one can treat it as an exogenous independent parameter. For example, in flight control, the dynamic pressure is a dynamic function of the aircraft maneuvers. However, it is useful to model it as an independent time-varying variable.

Example 19.18: Rotational Link as LPV

Recall that the rotational link model has a family of equilibrium points:

$$x_{eq}(s) = \begin{pmatrix} s \\ (s - mgL \sin s)/k \\ 0 \\ 0 \end{pmatrix} \qquad u_{eq}(s) = -mgL \sin s \qquad (19.108)$$

Define the new state and input variables:

$$x_{new}(t) = \begin{pmatrix} \theta_1(t) \\ \theta_2(t) - \theta_{2,eq}(t) \\ \dot{\theta}_1(t) \\ \dot{\theta}_2(t) \end{pmatrix}$$

$$u_{new}(t) = u(t) - u_{eq}(\theta_1(t)) \qquad (19.109)$$

Then, neglecting the nonlinear damping, the state dynamics can be written as:

$$\dot{x}_{new}(t) = \begin{pmatrix} 0 & 0 & 1 & 0 \\ 0 & 0 & mg\cos\theta_1(t)/K & 0 \\ 0 & k/J_2 & 0 & 0 \\ 0 & -k/J_2 & 0 & 0 \end{pmatrix} x_{new}(t) + \begin{pmatrix} 0 \\ 0 \\ 0 \\ 1 \end{pmatrix} u_{new}(t) \qquad (19.110)$$

Note that the original state equations are *transformed* into a quasi-LPV form, with the "exogenous" parameter actually the angle $\theta_1(t)$. It is interesting to note that this quasi-LPV family *is not* the same family obtained by performing linearizations about equilibrium conditions.

Now suppose that the parameter in the LPV plant (Equation 19.106) satisfies the bounds $|\theta(t)| \leq 1$. A traditional gain-scheduled design approach is to assume that the parameter is *constant* and design a family parameter-dependent controller to achieve desired stability and performance specifications. This results in an LPV controller such as

$$\dot{z}(t) = \overline{A}(\theta(t))z(t) + \overline{B}(\theta(t))y(t) \qquad (19.111)$$

$$u(t) = \overline{C}(\theta(t))z(t) + \overline{D}(\theta(t))y(t) \qquad (19.112)$$

In practice, the LPV controller gains come from an interpolation of several designs throughout the parameter range of values. Upon operation of the control system, the controller gains are updated according to the parameter time variations.

Because the parameter is actually time varying, the gain-scheduled design can experience degradation of performance or even loss of stability. However, one can show that if the parameter variations are sufficiently *slow*, then the desired properties are maintained.

Example 19.19: Time-Varying Oscillator

A classical example of parameter-varying instability from frozen parameter stability is the time-varying oscillator:

$$\dot{x}(t) = \begin{pmatrix} 0 & 1 \\ -(1 + \theta(t)/2) & -.2 \end{pmatrix} x(t) \tag{19.113}$$

These equations can be viewed as a mass-spring-damper system with time-varying spring stiffness. For *fixed* parameter values, $\theta(t) = \theta_o =$ a constant, the equilibrium $x_o = 0$ is asymptotically stable. However, for the parameter trajectory $\theta(t) = \cos 2t$, it becomes unstable. An intuitive explanation is that the stiffness variations are timed to pump energy into the oscillations.

19.3.3.3 Discussion

Conceptually, gain scheduling allows for greater operating regions than a design based on a single equilibrium. However, since the scheduling variable is no longer constant, the gain schedule introduces *time variations* in the overall control system. Such time variations typically are not addressed in the original frozen-equilibrium design. One consequence is possible degradation in performance or even instability of the gain-scheduled system. Another consequence is that the state of the nonlinear system while in transition need not be close to any of the equilibrium points, and hence outside of the design regions of the linearized controllers. However, the effects of these phenomena are reduced in the case of slow transitions among the operating conditions. In the end, the quality of a gain-scheduled design is typically inferred from extensive computer simulations.

Despite its widespread popularity, gain scheduling has received relatively little theoretical attention. Some references are stated in the section for further information. However, the theoretical basis for gain scheduling can be summarized as follows. The overall design is based on a collection of *linearizations* at *fixed* equilibrium conditions. If these design models are reasonable representations of the system dynamics, then one can expect that the stability and performance properties of the linearized designs should carry over to the overall gain-scheduled design. If the nonlinearities dominate or if the transitions between operating conditions are fast, and if these phenomena are not recognized in the design process, then one should not expect that the gain-scheduled design will perform satisfactorily. This reasoning leads to the popular heuristic guideline for successful gain-scheduled designs to "schedule on a slow-variable which captures the nonlinearities."

Further Reading

References discussing linearization as well as other methods for nonlinear systems:

1. Khalil, H.K., *Nonlinear Systems*. Macmillan, New York, 1992.
2. Vidyasagar, M., *Nonlinear Systems Analysis*, 2nd ed. Prentice Hall, Englewood Cliffs, NJ, 1993.

For more specialized results when linearization methods are inconclusive (including a discussion of Example 19.15.

3. Bacciotti, A., *Local Stabilizability of Nonlinear Control Systems,* World Scientific Publishing, Singapore, 1992.

References which give an overview of gain scheduling as well as some theoretical analyses:

4. Rugh, W.J, Analytical framework for gain-scheduling. *IEEE Control Syst. Mag.*, 11(1), 79–84, 1991.
5. Shamma, J.S. and Athans, M., Analysis of nonlinear gain-scheduled control systems. *IEEE Trans. Autom. Control.* 35(8), 898–907, 1990.
6. Shamma, J.S. and Athans, M., Gain scheduling: Potential hazards and possible remedies. *IEEE Control Syst. Mag.*, 12(3), 101–107, 1992.

References discussing methods for linear parameter varying systems:

7. Packard, A., Gain scheduling via linear fractional transformations. *Syst. Control Lett.,* 22, 79–92, 1994.
8. Shahruz, S.M. and Behtash, S., Design of controllers for linear parameter-varying systems by the gain scheduling technique. *J. Math. Anal. Appl.,* 168(1), 125–217, 1992.
9. Shamma, J.S. and Athans, M., Guaranteed properties of gain scheduled control of linear parameter varying plants. *Automatica,* 27(3), 559–564, 1991.

References presenting new approaches to gain scheduling:

10. Kaminer, I., Pascoal, A.M., Khargonekar, P.P, and Coleman, E., A velocity algorithm for the implementation of gain-scheduled controllers, *Proceedings of the 1993 European Control Conference,* pp. 787–792, to appear in *Automatica,* 1993.

References with example applications of gain scheduling:

11. Apkarian, P., Gahinet, P., and Biannic, J.-M., Self-scheduled H-infinity control of a missile via LMIs. *Proceedings of the 33rd IEEE Conference on Decision and Control.* pp. 3312–3317, 1994.
12. Astrom, K.J. and Wittenmark, B., *Adaptive Control,* Addison-Wesley, New York, 1989, chapt. 9.
13. Nichols, R.A., Reichert, R.T., and Rugh, W.J., Gain scheduling for H-infinity controllers: A flight control example, *IEEE Trans. Control Syst. Technol.,* 1(2), 69–79, 1993.
14. Shamma, J.S. and Cloutier, J.R., Gain-scheduled missile autopilot design using linear parameter varying methods. *J. Guidance Control Dynam.,* 16(2), 256–263, 1993.
15. Whatley, M.J. and Pott, D.C., Adaptive gain improves reactor control, *Hydrocarbon Processing,* May, pp. 75–78, 1984.

References discussing other linearization-based methods:

16. Baumann, W.T. and Rugh, W.J., Feedback control of nonlinear systems by extended linearization, *IEEE Trans. Autom. Control,* 31(1), 40–46, 1986.
17. Reboulet, C. and Champetier, C., A new method for linearizing nonlinear systems: The pseudolinearization. *Int. J. Control,* 40(4), 631–638, 1984.
18. Lawrence, D.A. and Rugh, W.J., Input-output pseudolinearization for nonlinear systems, *IEEE Trans. Autom. Control,* 39(11), 2207–2218, 1994.

Index

Note: n = Footnote

A